BUILDING CONTRACTS

AUSTRALIA
The Law Book Company Ltd.
Sydney : Melbourne : Brisbane

CANADA AND U.S.A.
The Carswell Company Ltd.
Agincourt, Ontario

INDIA
N. M. Tripathi Private Ltd.
Bombay

ISRAEL
Steimatzky's Agency Ltd.
Jerusalem : Tel Aviv : Haifa

MALAYSIA : SINGAPORE : BRUNEI
Malayan Law Journal (Pte.) Ltd.
Singapore

NEW ZEALAND
Sweet & Maxwell (N.Z.) Ltd.
Wellington

PAKISTAN
Pakistan Law House
Karachi

P. R. MORRIS.

BUILDING CONTRACTS

including a commentary on the J.C.T. Standard
Form of Building Contract

DONALD KEATING, Q.C., B.A.
of Lincoln's Inn; a Recorder

FOURTH EDITION

With a Commentary on the
I.C.E. Conditions of Contract

by

JOHN UFF
Ph.D., B.Sc.(Eng.), C.Eng., M.I.C.E.
of Gray's Inn, Barrister

LONDON
SWEET & MAXWELL
1978

First Edition 1955
Second Impression.................................... 1956
Second Edition 1963
Third Edition 1969
Fourth Edition....................................... 1978

By the same author

R.I.B.A. FORMS OF CONTRACT

first published in 1959

Published in 1978 by
Sweet & Maxwell Ltd. of
11, New Fetter Lane, London
and printed in Great Britain
by The Eastern Press Ltd.
of London and Reading

ISBN 0 421 19720 X

Semper in hoc libro vivat carissima coniunx

PREFACE

FOR various reasons the preparation of this edition has been delayed. The opportunity has been taken to make certain alterations, while maintaining the essential features of earlier editions, namely to deal in a practical and reasonably concise way with many of the legal problems which can arise where building works are carried out, and to include a commentary on the current standard form of building contract.

The first sentence of this book has always included a definition of the term "building contract" which includes an engineering contract. This reflects the opinion that there is no essential difference between the two traditional branches of construction contracts. The Banwell Committee of 1964 expressed this view and recommended the use of one common form of contract for construction work. I welcomed this recommendation, lectured on the subject and published (in the third edition of this book) my lecture notes. I intended to work them up into a commentary, but never found the time. Now, in 1977, there is no common form of construction contract and there seems less prospect of one than in the past. The differences between the standard forms of building and engineering contracts have in fact widened in recent years. But it still seemed desirable that this book should deal with the two major forms of construction contract in use. In these circumstances when John Uff, Barrister, of these Chambers (and my former pupil), expressed his willingness to prepare a commentary on the 1973 I.C.E. Conditions of Contract, I, knowing his extensive knowledge of engineering practice, readily agreed. Hence the new Chapter 17 which he prepared in draft. We went through it together and, although the wording in general remains his, we are agreed in substance upon the many difficult questions of construction which arise.

There is a new short Chapter 16 with commentaries upon the Form of Agreement between employer and nominated sub-contractor and the Form of Warranty to be given by a nominated supplier, and some suggestions by John Uff as to the preparation of comparable documents for use where the I.C.E. Conditions apply. Other major alterations concern the Appendices. In order to keep the book to a reasonable length they had to be shortened. After much thought I decided to omit the glossary of building terms prepared by Paul Badcock, consoled by the thought that there must be over 11,000 copies in existence. Do not throw away an earlier edition of this book. Other deletions from the appendices are: the definition of prime cost of dayworks; the code of procedure for selective tendering; the form of direct contract and my lecture notes referred to above. The R.I.B.A. Conditions of Engagement are retained because in practice they state, or form the basis for, the terms of contract upon which most architects are engaged in England. The "Green Form" of sub-contract is retained because it is in such extensive use. There have been various statements that it is about to be superseded, but as yet a new form has not arrived and even when it does the problems of the Green Form will be with us for a long time.

Other alterations cannot all be referred to but it may be helpful to mention some. Statute steadily encroaches. As regards dwellings we now have the Defective Premises Act 1972 (p. 44). Possibly more important when it comes into full effect will be section 71 of the Health and Safety at Work Act 1974, and in any event it seems that a cause of action may already lie against builders for breach of statutory duty for failure to comply with the Building Regulations or by-laws (p. 47). Having regard to the advancement of liability in statute, coupled with the attitude of the Court of Appeal to exclusion clauses (p. 142), it may be that the great days of contract are numbered and that we are on the way to a position where our affairs will be governed not by our promises, but by what Parliament in prospect, or the courts in retrospect, consider to be just. But that time has not fully arrived. There are still wide areas where parties of substantially equal bargaining power may regulate their own business affairs by promises which the courts will enforce.

Some of the detailed alterations and additions include: design and build contracts (p. 5); letter of intent (p. 11); implied terms (p. 35); time (p. 60); disqualification of the certifier (p. 85); inaccurate statements—taking into account the recent developments in the law of tort on this subject (p. 89); repudiation, a term which I still find convenient to use generally in a wide sense while acknowledging that there are occasions when it is useful to treat it more narrowly as in contrast with a fundamental breach (p. 111); injunction to enforce forfeiture (p. 123); liability to third parties (p. 134); risk, indemnity and exclusion clauses (p. 138); breach of confidence (p. 142); contractor's " claims "—dealt with at some length in the hope that it may assist in making them shorter and more to the point (pp. 152, 260, 266, 302 and 316); interest (pp. 154, 254); limitation of actions (p. 165); nominated sub-contractors (p. 178); architect's liability for certificates (p. 216); architect's liability in contract and/or tort (pp. 217, 227); refusal of stay under section 4, Arbitration Act 1950 (p. 242); summary judgment (p. 278). The commentary on the Standard (J.C.T.) form of building contract has in many areas been substantially re-written and the format somewhat altered to make it easier, it is hoped, to follow.

Less emphasis has been given to older authorities and some of them have been cut out, but in the law of contract many must still be treated with due respect. The reference to Commonwealth authorities has been increased where they are of assistance in giving examples of, or filling gaps in, English law.

Of the assistance I have received in the preparation of this edition I have already referred to the contribution of John Uff. Jonathan Gaunt, Barrister, gave me great assistance in the initial stages of preparation by reviewing the authorities; John Marrin, Barrister, has prepared the Index ; my son Roland composed the dedication for me. The publishers have kindly arranged for the preparation of Tables of Cases and Statutes. Members of Chambers and many others have given me much help in various discussions I have had with them and, on occasions, by writing to me. I am indebted to the various bodies who have given permission to publish the documents appearing in Chapters 15, 16 and 17 and the Appendices. The publishers and printers have applied their usual care and skill to the difficult and intricate task of the preparation of a new edition.

My wife Betty encouraged me from the earliest days in the 1950s; more than that she read through and discussed everything. Although not a lawyer, her acute mind sharpened mine and compelled me to re-think and clarify much of what I had prepared. I have tried to maintain the standards she set.

*   *   *

The text was generally prepared as at November 1, 1976, but an endeavour has been made to incorporate new law up to August 1, 1977. A few references after August 1, 1977, have been inserted. The J.C.T. Standard Form of Building Contract printed and discussed is the July 1977 revision.

It is intended to issue supplements from time to time.

DONALD KEATING

11, *King's Bench Walk,*
Temple,
London.
September 13, 1977

ADDENDUM

The Unfair Contract Terms Act 1977

This Act, coming into force on February 1, 1978, is of great importance. The relevant excerpts are printed *post*, p. 617 as Appendix C. Its title indicates its principle, namely that relief is required from certain terms in contracts or notices which would otherwise affect liability in tort because such terms or the effect of such notices are unfair. It thus carries a stage further the approach to exclusion clauses developed by the courts and recent statutes designed to protect consumers from others, and business men from each other. Its scope is wide and includes building contracts. The arguments in favour of the Act appear in the Law Commission's report on exemption clauses, number 69 (1975) to which reference may usefully be made with the warning that, as yet, such documents are not admissible to assist in the construction of statutes.

The Act follows the Supply of Goods (Implied Terms) Act 1973 (which it amends) in various ways. Thus it draws a distinction between consumers and business men and gives the former a higher degree of protection than the latter; it does not apply to what may loosely be termed foreign contracts; it requires the court to pronounce upon the reasonableness of certain terms, giving it guidelines as to its approach in certain circumstances (s. 11, Sched. 2).

The most important part of the Act is its avoidance or restriction upon the operation of exemption clauses (ss. 1, 2, 3, 5, 6, 7); but there are other alterations in the law. A consumer is protected against unreasonable indemnities which he gives (s. 4); section 3 of the Misrepresentation Act 1967 is amended (s. 8); an exemption clause which can take effect if reasonable does not lose this quality upon termination by, or acceptance of, breach (s. 9)—thus substantially modifying the *Harbutt's Plasticine* case [1970] 1 Q.B. 447). The Act does not apply to contracts made before February 1, 1978; but subject to this it applies to liability for any loss or damage which is suffered on or after that date (s. 31 (1), (2)).

The passage in the text which will be most affected by the Act will be that dealing with risk, indemnity and exclusion clauses (p. 138). But the subject matter of other passages may require reconsideration to some degree. These include: design and build contracts (p. 5); notice of terms (p. 16); the *contra proferentem* rule (p. 34); warranties of fitness (p. 39); exactly defined contracts (p. 64)—such contracts may not attract the Act whereas widely expressed obligations subject to exceptions may; architect's final certificates (pp. 70, 80)— consider section 13; Misrepresentation Act 1967, s. 3 (p. 96); negligent misstatement where there is no assumption of a duty to be careful (p. 98)—consider section 2 (2); *Harbutt's Plasticine* case (p. 114); defects and maintenance clauses (p. 127); liquidated damages (p. 155)—consider sections 2, 13; nominated sub-contractors and nominated suppliers (p. 178); architect's responsibility for design (p. 212); architect's duties to contractor and others (p. 227).

As to the JCT Standard Form of Building Contract, various interesting questions arise. Thus, because of its method of preparation (see p. 281), is it never to be regarded as "the other's written standard terms of business" (s. 3 (1))? And if it is not, is the position affected by amendments made, for example, by a local authority and contained in their invitation to tender? Does the architect's final certificate where quality and standards are to be to his reasonable satisfaction (see cll. 1, 30) fall within sections 2 and 13? Consider also clauses 4 (1) (e), 5, 15 (2) (and note at p. 327), 18, 19, 23, 26, 35. Analogous problems may arise under the I.C.E. Conditions; consider clauses 20, 22, 24, 26 (2), 59A (3) (6), 59B (4) (6), 60 (2), 66.

CONTENTS

xiii

TABLE OF CASES

TABLE OF STATUTES

Table of References in Chapters 1–14 to
J.C.T. Standard Form of Building Contract 1977 Revision

ABBREVIATIONS

C.A.	= Court of Appeal
D.C.	= Divisional Court
H.L.	= House of Lords
P.C.	= Privy Council
Hals.	= Halsbury's Laws of England
H.B.C.	= Hudson's Building Contracts
R.S.C.	= Rules of the Supreme Court

For other abbreviations, *e.g.* to reports of Commonwealth cases, reference should be made to the table of abbreviations in other works such as the Current Law Year Book.

CROSS-REFERENCES

ante	= earlier
supra	= earlier, but not far away
post	= later
infra	= later, but not far away

CHAPTER I

THE NATURE OF A BUILDING CONTRACT

1. Relation to the general law of contract

The phrase " building contract " in this book is used to include any contract where one person[1] agrees for valuable consideration to carry out building or engineering works for another. It thus covers every contract from, *e.g.* a simple oral agreement to repair a garage roof to elaborate public works contracts. The law of building contracts is a part of the general law of contract[2] and is not governed by any codifying statute. All the elements of a simple contract must be present[3] and the general rules of performance and discharge of contract apply to a building contract. Where the general principles of the law of contract apply to problems of common occurrence in building contracts they have been dealt with in detail in this book; but where they apply to problems which rarely arise in building contracts, *e.g.* mistake avoiding a contract, a mere outline has been given and for further information reference must be made to one of the standard works on the law of contract.[4]

2. The persons concerned in a building contract

(a) *The employer and the contractor*

The employer for whose benefit the work is carried out and the contractor who must carry out the work are the parties to a building contract. The employer has frequently been termed " the building owner," and the contractor the " builder,"[5] or the " building contractor." For the sake of clarity and because the Standard form of building contract[6] uses the terms, the parties are always referred to in this book as " employer " and " contractor," unless a quotation from a decided case involves the use of the other terms. In addition to the parties, there are usually in a large contract several other persons involved. Although they are not parties they may materially affect the legal relationship between the contractor and the employer, and a short note on their position will be given in this chapter although a full account, where appropriate, will be found in other chapters.

---

[1] This includes a corporation. See Chap. 2 (*post*, p. 24).

[2] English law is referred to throughout the book unless the contrary is indicated. A reader who has not previously studied any law may find it useful to read *Learning the Law*, by Professor Glanville Williams.

[3] See Chap. 2 (*post*, p. 9).

[4] *e.g. Chitty on Contracts* (24th ed., 1977).

[5] This term has received judicial consideration under a now repealed statute formerly relating to bankruptcy law—see H.B.C. (7th ed., 1946), p.105. It is also defined for the purpose of the London Building Acts 1930–39. See London Building Acts (Amendment) Act 1939, s. 4.

[6] 1963 edition. See further Chap. 15. This is the document referred to in previous editions of this book, and generally, as the " R.I.B.A. form." Further references will normally be to the " Standard form of building contract." It has now (August 1977) been stated that it should be called the " JCT Form " (see *post*, p. 281).

I

(b) *The architect*

The term is ordinarily used in this book to describe the person who is engaged by the employer to carry out the duties of an architect referred to in Chapter 12.[7] In the broadest sense his duties are to prepare plans and supervise the execution of the works on behalf of the employer so that they may be completed in accordance with the contract. He is therefore the agent of the employer and owes him a duty of care as agent and professional man notwithstanding that the employer and contractor ordinarily contract on the understanding that many matters may arise under the contract where the architect has to make a decision in a fair and unbiased manner.[8]

In an engineering contract the person who carries out the duties and occupies a position similar to that of an architect in a building contract is termed the engineer.[9]

A surveyor or some other person may carry out the duties and occupy the position of an architect in a building contract.[10]

(c) *The surveyor*

The surveyor is employed by or on behalf of the employer to estimate the quantities of the proposed works and set them out in the form of bills of quantities for the purposes of tender. He may also be employed to measure and value variations and to do such other works of measurement and valuation as the architect may require.[11] In some cases the architect takes out his own quantities and does his own measuring of the works.

There is no requirement of the law that there shall be an architect or surveyor employed in a building contract and many contracts are entered into and completed without their employment.

(d) *Sub-contractors* [12]

The contractor frequently sub-contracts, or, as it is sometimes termed, sublets, much of the work to sub-contractors. If the architect or the employer has nominated the sub-contractor he is termed a nominated sub-contractor.

(e) *Sureties*

Persons known as sureties may guarantee the performance of the works by the contractor or payment by the employer or the good faith of the architect or other persons having control of money.[13]

---

[7] See Chap. 12, Section 1. Thus references to " architect " do not refer to a person who, though he may have qualified as an architect, does not practise as such but acts as, say, an expert adviser or salaried assistant to a large contractor.

[8] *Sutcliffe* v. *Thackrah* [1974] A.C. 727 (H.L.). See further *post*, p. 85. When making such a decision he was in the past often termed a " quasi-arbitrator " but the term is now better forgotten.

[9] See generally Chap. 12, Section 2. And note that surveyors, engineers and others may not term themselves architects unless registered under the Architects Registration Acts 1931 to 1969.

[10] *Ibid.*

[11] See generally Chap. 12.

[12] See generally Chap. 10.

[13] See generally Chap. 8.

3. A typical building operation—traditional procedure

The scope of the law of building contracts and some of the problems which may arise are illustrated in the following summary.[14]

The employer obtains a sufficient interest in the land upon which he proposes to build to enable him to give the contractor possession of the site. If the contract is of any substance he usually employs an architect who, if authorised to do so, engages a quantity surveyor on behalf of the employer. The architect translates the employer's wishes into detailed plans and the quantity surveyor measures the amount of work and materials necessary to complete the plans and sets this out in detail in bills of quantities. Any appropriate planning consents are obtained and all statutory requirements taken into account.[15] The contract must then be placed. This is done either by open or limited competition[16] or by individual negotiation. Forms of tender are sent out with bills of quantities, specification, plans and conditions of contract, or such other information as the architect may think desirable. If good practice is followed the documents intended to form part of the contract will be clearly stated. The contractor estimates the cost of the works on the basis of the amount of work and materials shown on the plans and described in the bills of quantities and the specification or other documents supplementing or replacing the bills of quantities. He then submits his estimate, *i.e.* tender, and it is the normal practice that such tender is an offer capable of acceptance. Where there is more than one tender the employer with the advice and assistance of his architect selects one of the tenders, usually, though not necessarily, the lowest, and upon notifying the contractor of his unqualified acceptance a binding contract comes into existence. The contract is frequently expressed in a formal document such as the Standard form of building contract or, for engineering works, the I.C.E. conditions.

The works are carried out by the contractor subject to the supervision of the architect. It is usual for interim payments of the contract sum to be made from time to time as the architect certifies that work of a certain value has been carried out, such payments being subject to the retention of a percentage of the value of the work carried out, the money retained (" retention money ") being held by way of security until completion of the works. After the completion of the works it is usual for the contractor to be under an express obligation to make good defects which appear during a certain period often termed the " defects liability " or " maintenance " period.[17]

Numerous difficulties may arise. For example, the completion of the works may be delayed by the ordering of variations, by late or inadequate instructions, by shortage of materials or delay on the part of sub-contractors. The contractor

---

[14] For further reading see the Government publications: *The Placing and Management of Building Contracts* (H.M.S.O., 1944) (the Simon Report); *The Placing and Management of Contracts for Building and Civil Engineering Work* (H.M.S.O., 1964) (the Banwell Report). Procedure can, of course, vary at the will of the parties.

[15] Town and country planning law, Building Regulations and special statutory controls are outside the scope of this work.

[16] For a suggested procedure for limited competitive tendering, see *A Code of Procedure for Single Stage Selective Tendering* 1977, obtainable from the National Joint Consultative Committee of Architects, Quantity Surveyors and Builders.

[17] See Chap. 8, Section 4.

may find that he is required to execute more work or spend more money to complete than he originally estimated. A third party injured by a falling object or annoyed by dust and noise may make a claim against the contractor or the employer. One or both of the parties may become bankrupt or go into liquidation. A breach of contract by one of the parties may give rise to a claim by the other for damages or even to bring the contract to an end, either by virtue of an express clause or under the general law of contract. After completion, disputes may arise about defects alleged to exist or to have appeared during the defects liability or maintenance period. Any of these events may affect the obligations of a surety.[18]

When the works have been completed the architect usually has the power to decide the amount of money payable to the contractor and to state his decision in the form of a certificate. In arriving at the final sum he may have to take into account many matters, some of which are mentioned above. The result may be that the final certificate is for a very different sum from that stated as the original contract sum.[19] If the parties cannot agree on the amount to be paid under the contract or on some other matter which has arisen during the course of the building operations, there must be legal proceedings, but the parties may have expressly limited their right to litigate in two ways: first, by providing that the architect's decision on certain matters shall be binding and conclusive; secondly, by providing for the arbitration of any disputes arising between them.[20]

4. Contract documents

Contract documents contain the terms of the contract and are to be distinguished from other documents such as invitations to tender [21] or mere representations [22] not intended to form part of the contract. It is a question partly of fact and partly of construction [23] to determine which documents are contractual. The following may form part of the contract [24]:

(a) *Agreement or articles of agreement*

This usually sets out the date, the parties, the intended works and the consideration. It may also name the architect and surveyor.

(b) *Conditions*

Elaborate conditions are often made part of the contract and attempt to provide for the various problems which can arise during and after the execution of the works.

---

[18] Discussion of the various matters appearing in this paragraph appears in subsequent chapters.

[19] Under the Standard form of building contract, the contract sum can be varied by the operation of numerous clauses.

[20] See *supra*, note 18.

[21] See Chap. 2, Section 1.

[22] See Chap. 7, Section 1.

[23] See Chap. 3.

[24] See Chap. 15 for a discussion of the contract documents in the various versions of the Standard form of building contract.

(c) *Architects' plans and drawings*

(d) *Bills of quantities*

The purpose of bills of quantities " is to put into words every obligation or service which will be required in carrying out the building project." [25] They may not form part of the contract although submitted to the contractor for tender,[26] and in general it may be said that it requires express words in the articles of agreement to make the bills of quantities a contract document.[27] Bills of quantities quantify the works in detail and are ordinarily prepared in accordance with an agreed standard method of measurement.

(e) *Specification*

This term is much less exact in meaning than " bills of quantities." It usually means a document which describes the work to be done and the goods to be supplied. There is no standard or customary method for the preparation of a specification and its meaning must be considered in each contract. Thus in the Standard form of building contract where quantities do not form part of the contract,[28] it has the meaning, just stated, whereas it does not have such meaning in the version where quantities form part.[29] In the I.C.E. conditions it partakes of the meaning stated above but does not give quantities of work, this function being formed by a separate Bill of Quantities.

(f) *Other documents*

A variety of other documents such as letters, estimates, memoranda, and in some cases the tender or invitation to tender, may contain terms of the contract, although where parties have apparently finally stated their agreement in a document it may be difficult to show that other documents (or oral statements) not expressly incorporated or referred to contain terms of the contract.[30]

5. **Design and build contracts**

The traditional procedure outlined above has developed since the early days of railway contracts in the nineteenth century, and is probably still applied today in a majority of large building or engineering contracts. It is based upon the principle that, in general, the employer, through his agents, provides the design which the contractor carries out. This principle is often not consistently applied,[31] and in any event there have always been some contracts where the contractor has, to greater or less degree, accepted responsibility for design.

In recent times it has become increasingly common for contractors to offer, in addition to building the works, to perform some or all of the of duties of architect,

---

[25] Simon Report, para. 55, see *supra*, note 14.
[26] See *post*, pp. 64, 101.
[27] See *post*, p. 64.
[28] See Chap. 15.
[29] *Ibid*. In the 1976 and 1977 versions it is not mentioned.
[30] See Chap. 3.
[31] See *post*, p. 39 for fitness for purpose; p. 181 for design by specialist sub-contractors; p. 290 for the contractor's position under the Standard form of building contract.

engineer or even surveyor, as performed in traditional contracts. The commercial argument for such an approach is either that it is necessary because the contractor alone possesses the specialist knowledge and skill to design and carry out specialist works or, in other cases, that there will be savings of costs or time or both compared with the traditional procedure. Such contracts are sometimes termed " package deal " contracts.

The Standard form of building contract and the I.C.E. Conditions (see Chaps. 15 and 17) are each wholly unsuitable for use as design and build contracts. There is no standard form of contract available in the sense of a document produced by a committee representative of the interests involved. There is a form of contract issued by a body representative of builders, the National Federation of Building Trades Employers. Documents proffered for consideration by contractors require scrutiny to see whether they afford reasonable protection for the employer. In particular it should be considered how far, if at all, by express terms they affect the term of suitability for purpose which is ordinarily implied (see *infra*). This implied term is valuable. If the design turns out to be unsuitable it is no defence to the contractor that he had exercised reasonable skill and care in its preparation. It thus affords greater protection to the employer than he obtains under the traditional procedure where, ordinarily, it is a defence for the architect or engineer to show that he used reasonable skill and care in preparing the design. (See *post*, Chap. 12.)

(a) *Construction of contract*

The ordinary rules of construction apply.[32] In so far as the contractor is performing duties of architects, engineers and surveyors, cases dealing with such duties when performed by persons as independent professional men [33] may be of some assistance although care must be taken to have regard to the different subject-matter. Again when considering the performance of duties similar to those carried out by a contractor in a traditional contract some of the decided cases may help but caution is required. Apparently identical terms of contract may, because of the different subject-matter, have a different meaning. An obvious example is that relating to a " defects " clause.[34] In the traditional type of contract it ordinarily excludes defects of design, while in design and build contracts it ordinarily includes them.

(b) *Suitability for purpose*

In a design and build contract there is ordinarily implied a term that the finished work will be reasonably suitable for the purpose for which the contractor knows it is required.[35] But one must always exercise caution when placing a contract into a particular category. Thus it appears that where a contractor is invited to tender to design and build, and submits a design which is adopted by

---

[32] See Chap. 3.
[33] See Chap. 12.
[34] See Chap. 8, Section 4.
[35] *Greaves & Co. Ltd.* v. *Baynham Meikle* [1975] 1 W.L.R. 1095, 1098 (C.A.). See further, *post*, p. 41, " Fitness for purpose of completed works."

the employer's architect acting in the traditional sense and thereafter a formal contract is entered into whereby the contractor's obligations are limited to performing specific works described in the contract, the contractor is not liable in damages for breach of contract if the works do not fulfil the result which, to his knowledge, was sought to be achieved by the employer.[36] But he may be liable if he has expressly guaranteed the result.[37]

(c) " *Turnkey* "

This term is sometimes used in design and build contracts. It may be that it is intended to indicate that upon completion the key can be turned and everything will be ready. It has been said that it is not a term of art,[38] *i.e.* it has no precise legal meaning (see *post*, p. 31).

(d) *Professional men*

Contractors sometimes engage independent professional men to carry out design and other services to be provided to the employer. Arising out of such engagement they owe the contractor a duty to carry out their work properly.[39] The requisite standard will be at least one of reasonable skill and care.[40] Where they know that the contractor is under an absolute duty to provide finished work reasonably suitable for a purpose of which they have knowledge they are, it seems, ordinarily under such a duty themselves to the contractor in carrying out design or other services.[41] Further, although not engaged by the employer, they may in certain circumstances owe him a duty of care in tort.[42]

The employer may himself engage independent professional men to protect his interests. They owe him a duty of care arising out of the engagement.[43] An architect or engineer in such a position usually is, or is supposed to be, an inspector only, whether of the design, the works or both. If such architect or engineer is given authority as agent of the employer the terms of such agency and his actions should be such as not to interfere with the duties of design, administration and otherwise undertaken by the contractor towards the employer. If there is such interference, confusion as to liability may arise and the employer may find himself deprived of, or hindered in the enforcement of, his remedy against the contractor by the operation of doctrines of law such as variation, waiver or estoppel.[44]

---

[36] *Cable (1956) Ltd.* v. *Hutcherson Ltd.* (1969) 43 A.L.J.R. 321, High Court of Australia.
[37] *Steel Co. of Canada Ltd.* v. *Willand Management Ltd.* [1966] S.C.R. 746, Supreme Court of Canada.
[38] *Cable (1956) Ltd.* v. *Hutcherson Ltd., supra*, at p. 324.
[39] See *post*, p. 206.
[40] *Ibid.*
[41] *Greaves & Co. Ltd.* v. *Baynham Meikle, supra.*
[42] *Donoghue* v. *Stevenson* [1932] A.C. 562 (H.L.); *Dutton* v. *Bognor Regis U.D.C.* [1972] 1 Q.B. 373 (C.A.) and other cases referred to *post*, p. 98. " Negligent Misstatements." For the position of salaried architects and comparable employees of the contractor see *post*, p. 200.
[43] See *ante*, note 39.
[44] See *post*, p. 119.

Employers sometimes engage a surveyor to check the contractor's valuations, to price variations and to perform some other duties comparable to those carried out by a quantity surveyor in the traditional procedure. This can usually be accommodated without great difficulty in a design and build contract and some contractors encourage it, presumably, on the basis that it will help to avoid disputes over payment.

CHAPTER 2

FORMATION OF CONTRACT

Section 1: *Elements of Contract*

The essence of a building contract, like any other contract, is agreement. In deciding whether there has been an agreement and what its terms are the court looks for an offer to do or forbear from doing something by one party and an acceptance[1] of that offer by the other party, turning the offer into a promise.[2] The law further requires that a party suing on a promise must show that he has given consideration for the promise, unless the promise was given under seal. There is consideration where " an act or forbearance of the one party or the promise thereof is the price for which the promise of the other is bought." [3] In the ordinary building contract the consideration given by the employer is the price paid or the promise to pay,[4] and by the contractor is the carrying out of the works or promise to carry them out. The parties must have the capacity to make a contract,[5] and any formalities required by law must be complied with.[6] Both the consideration and the objects of the contract must not be illegal.[7] If there is fraud or misrepresentation the contract may be voidable,[8] while if there is a mutual mistake about some fundamental matter of fact this may have the effect of making the contract void.[9]

Section 2: *Offer and Acceptance*

1. Invitation to tender

The employer, normally acting through his architect, sends out an invitation to tender for the proposed works. This document usually includes the proposed

---

[1] See further *infra*, Section 2.

[2] See Chitty, 23rd ed., Vol. 2, Chap. 2.

[3] *Pollock on Contracts* (12th ed.), p. 130, adopted by Lord Dunedin, *Dunlop* v. *Selfridge* [1951] A.C. 847, 855 (H.L.). See also *Currie* v. *Misa* (1875) L.R. 10 Ex. 153, 162.

[4] For the consideration given where the employer was not to pay, see *Charnock* v. *Liverpool Corporation* [1968] 1 W.L.R. 1498, 1505 (C.A.). For performance of an existing contractual duty as consideration, see *New Zealand Shipping Co. Ltd.* v. *A. M. Satterthwaite & Co. Ltd.* [1975] A.C. 154 (P.C.).

[5] See *infra*, Section 4.

[6] See *infra*, Section 3.

[7] See *post*, Chap. 7, Section 7.

[8] See *post*, Chap. 7, Section 1.

[9] See *Bell* v. *Lever Bros.* [1932] A.C. 161 (H.L.); Chitty (24th ed.), 1977, Vol. 1, Chap. 5. Mistake in this sense does not appear to have been argued in a reported building contract case, neither does it seem important in practice. For these reasons it is not discussed further in this book. For rectification where there has been a mistake in expressing the contract, see *post*, Chap. 3, Section 3. For the equitable powers of the court in cases of mistake, see *Solle* v. *Butcher* [1950] 1 K.B. 671 (C.A.); *Grist* v. *Bailey* [1967] 1 Ch. 532; *Curtin* v. *G.L.C.* (1970) 114 S.J. 932 (C.A.). See also *Magee* v. *Pennine Insurance Co. Ltd.* [1969] 2 Q.B. 507 (C.A.). For mistake as to the nature of a document, see *Saunders* v. *Anglia Building Society* [1971] A.C. 1004 (H.L.).

conditions of contract, plans, and a specification and bills of quantities in blank, *i.e.* with the quantities of work set out but the price column in blank. An invitation to tender is not normally an offer binding the employer to accept the lowest or any tender. It is comparable to an advertisement that one has a stock of books to sell or houses to let and such advertisements have been described as " offers to negotiate—offers to receive offers—offers to chaffer." [10] It follows that the clause frequently inserted in tenders to the effect that the employer does not undertake to accept the lowest or any tender is probably unnecessary in law.[11] But an express offer to accept the lowest tender might be binding and have the effect of turning the invitation to tender into an offer.[12]

Statements of fact in the invitation to tender about such matters as the quantities or the site or existing structures may, if a contract is entered into, have no legal effect at all, or they may take effect as representations, or they may form collateral warranties or they may give rise to a claim for negligent misstatement or they may subsequently become incorporated into the contract.[13] It is a question partly of fact, partly of construction to determine the nature of such statements.

2. Tender

The contractor's offer to carry out the works is usually termed a tender.[14] It may well happen that as a result of negotiation [15] it is the employer who eventually makes the offer. In any event a statement, to amount to an offer, must be definite and unambiguous.[16] The person making the offer is for the purposes of this part of the law termed the offeror; the person to whom it is made, the offeree.

Costs of tendering

The cost to the contractor of preparing his tender, including any amended tender necessitated by bona fide alterations in the bills of quantities and plans, may be considerable, but in ordinary circumstances there is no implication that he will be paid for this work [17]; ". . . he undertakes this work as a gamble, and its cost . . . he hopes will be met out of the profits of such contracts as are made

---

[10] *Per* Bowen L.J., *Carlill* v. *Carbolic Smoke Ball Co.* [1893] 1 Q.B. 256, 268 (C.A.); *Grainger & Son* v. *Gough* [1896] A.C. 325, 334 (H.L.). See also the building contract case of *Moore* v. *Shawcross* [1954] C.L.Y. 342; [1954] J.P.L. 431.

[11] *Cf. Pauling* v. *Pontifex* (1852) 2 Saund. & M. 59.

[12] *Cf. South Hetton Coal Co.* v. *Haswell Coal Co.* [1898] 1 Ch. 465 (C.A.); *Warlow* v. *Harrison* (1859) 1 E. & E. 309.

[13] For a full discussion, see *post,* Chap. 7.

[14] The term is here used in a completely different sense from that of the tender of goods or money which may amount to a defence under a contract—see Chitty (24th ed., 1977), Vol. 1, para. 1322.

[15] See *post,* p. 17, " Lengthy negotiations for a contract," *post,* p. 17, " Unconditional acceptance " and *post,* p. 21, " Negotiations after contract."

[16] *Falck* v. *Williams* [1900] A.C. 176 (P.C.); *Harvey* v. *Facey* [1893] A.C. 552 (P.C.); *Bigg* v. *Boyd Gibbins Ltd.* [1971] 1 W.L.R. 913 (C.A.); *Peter Lind & Co. Ltd.* v. *Mersey Docks and Harbour Board* [1972] 2 Lloyd's Rep. 234.

[17] *William Lacey (Hounslow) Ltd.* v. *Davis* [1957] 1 W.L.R. 932. The principle is based on custom; *ibid.* at pp. 934, 935.

as a result of tenders which prove to be successful . . ." [18] But the law implies a promise on the part of the employer to pay a reasonable sum to the contractor for work done at the employer's request which falls outside the normal work which a contractor performs gratuitously.[19] Thus in *William Lacey (Hounslow) Ltd.* v. *Davis* [20] the contractor submitted a tender for the rebuilding of war-damaged premises. The tender · was not accepted, but in the belief that the contract would be placed with him the contractor subsequently prepared various further estimates, schedules and the like which the employer made use of in negotiation with the War Damage Commission, but he never placed any contract with the contractor. It was held that the contractor was entitled to a reasonable sum for the work carried out subsequent to the tender.

If the employer invites a tender without any intention of entering into a contract and the contractor, believing the invitation to tender to be genuine, incurs expense in tendering, the contractor may have a claim for damages in fraud against the employer.[21]

Design costs

As part of the process of tendering, specialist contractors sometimes carry out works of design. It is thought that, in the absence of agreement,[22] the costs of such works are part of the costs of tendering so that by analogy with the *Lacey* case if there is no contract they are irrecoverable unless the employer makes some use of the design or causes the contractor to carry work beyond what is normal in the circumstances.

3. Letter of intent

Documents so described are frequently sent. They ordinarily express an intention to enter into a contract in the future but create no liability in regard to that future contract.[23] In one case a design and build contractor [24] offered to the employer to undertake certain urgent works of design necessary to obtain estimates and planning permission provided he obtained an assumption of liability for such work. He indicated that he would regard receipt of a letter of intent as an acceptance of such offer. The employer sent a letter of the kind referred to above. Held, he was liable for the work carried out. [25]

It is a question upon the facts of each case whether the sending of a letter of intent can give rise to any, and if any, what, liability.

---

[18] *Ibid.* at p. 934.

[19] *Ibid.*

[20] *Supra.*

[21] *Cf. Richardson* v. *Silvester* (1873) L.R. 9 Q.B. 34. For fraud, see *post*, p. 92.

[22] For an example of such an agreement arising out of a letter of intent, see *infra*.

[23] *Turriff Construction Ltd.* v. *Regalia Knitting Mills Ltd.* (1971) 222 E.G. 169, His Honour, Judge Fay Q.C., Official Referee. The text is based upon the transcript of the judgment.

[24] For a short discussion of such contracts see *ante*, p. 5.

[25] See *supra*, note 23.

4. Estimates

There may be an offer although the contractor makes it on a document called an estimate. Thus in *Croshaw* v. *Pritchard* [26] the architect for the employer sent a letter to the defendants asking them if they " would be willing to give us a tender in competition for the work," and wrote later enclosing information required for tender. In a letter headed " Estimate " the defendants wrote to the architect saying " our estimate . . . amounts to the sum of £1,230." This was accepted. The defendants then purported to withdraw their estimate and, when sued for the difference between £1,230 and the cost incurred in having the work executed by another contractor, claimed that they had made no offer capable of acceptance. This defence was rejected and the damages claimed awarded to the plaintiff. It was further held that there was no custom that a letter such as that written by the defendants could not amount to an offer, and it was said that if such a custom existed it would be bad and not enforceable. [27]

5. Revocation of offer

An offer may be revoked at any time before acceptance unless consideration has been given to keep it open [27a] or the offeror is, in special circumstances, estopped from acting inconsistently with the existence of the offer. [28] Revocation of the offer is not effective until it has been communicated to the offeree. [29] In the ordinary case an offeree who has acted on the offer cannot recover damages in tort if the offer is revoked before acceptance. [30]

Lapse of offer. The offer can expressly state the time within which it is open for acceptance. If that time expires without acceptance the offer lapses. Where negotiations were taking place between contractor and employer in the course of which the contractor's offer lapsed but the contractor proceeded to carry out the work without ever concluding an express contract with the employer the contractor recovered a reasonable sum. [31] If no time is stated the offer remains in force for a reasonable time and upon the expiry of that time it lapses. [32]

Death. " It is admitted law that, if a man who makes an offer dies, the offer cannot be accepted after he is dead." [33]

---

[26] (1899) 16 T.L.R. 45; *sub nom. Crowshaw* v. *Pritchard*, H.B.C. (4th ed.), Vol. 2, p. 274.

[27] At p. 276 of the H.B.C. Report.

[27a] See note 29, *infra.*

[28] *Watson* v. *Canada Permanent Trust Co.* (1972) 27 D.L.R. (3d) 735, British Columbia Supreme Court applying the principle sometimes known as that stated in *Central London Property Trust Ltd.* v. *High Trees House Ltd.* [1947] K.B. 130. See *post*, p. 67.

[29] *Byrne* v. *Van Tienhoven* (1880) 5 C.P.D. 344; *cf. Dickinson* v. *Dodds* (1876) 2 Ch.D. 463 (C.A.).

[30] *Holman Construction Ltd.* v. *Delco Timber Co. Ltd.* [1972] N.Z.L.R. 1081, New Zealand Supreme Court, discussed *post*, p. 187.

[31] *Peter Lind & Co. Ltd.* v. *Mersey Docks & Harbour Board* [1972] 2 Lloyd's Rep. 234.

[32] *Ramsgate Victoria Hotel Co.* v. *Montefiore* (1866) L.R. 1 Ex. 109. See also *Manchester Diocesan Council for Education* v. *Commercial & General Investments Ltd.* [1970] 1 W.L.R. 241.

[33] *Per* Mellish L.J., *Dickinson* v. *Dodds, supra*, at p. 475.

6. Standing offers

Tenders are sometimes invited for the periodic carrying out of work.[34] If the contractor tenders and there is an acceptance, the result depends upon the construction of the documents, but can have one of three well-known consequences.[35] First, a contract for the carrying out of a definite amount of work during a certain period. Secondly, a contract in which the employer agrees to order such work as he needs during the period. In such a case the employer is in breach if during the period he places orders for the work elsewhere, and the contractor is in breach if he refuses to carry out the work during the period.[36] Thirdly, there may be a standing offer on the part of the contractor to carry out certain work during the period if and when the employer chooses to give an order.[37] The contractor may revoke his offer in respect of future orders unless there is consideration to keep it open or the documents are under seal,[38] but if before revocation an order in the terms of the agreement is given, a contract comes into existence in respect of that order and the contractor must carry it out.[39] If no order at all is given during the period,[40] or if less work is ordered than the probable amount indicated in the invitation to tender, whether because the work is given to another contractor or otherwise,[41] the contractor has no action for breach of contract.[42]

7. Restrictive tendering agreements

Contractors have, it seems, often entered into agreements, formal or informal, with other contractors restricting their prices or otherwise limiting their freedom to tender in a freely competitive manner.[43] Until 1956 such agreements were solely governed by common law.[44] Now these agreements may be registrable under the Restrictive Trade Practices Act 1976 and important consequences follow. The matter will first be considered at common law and then under the Act.[45]

---

[34] Most of the cases deal with the supply of goods but the principle applies to contracts of work and labour, *R.* v. *Demers* [1900] A.C. 103 (P.C.).

[35] *Percival Ltd.* v. *L.C.C. Asylums Committee* (1918) 87 L.J.K.B. 677, 678.

[36] *Ibid.* at p. 679. *Cf. Att.-Gen.* v. *Stewards & Co. Ltd.* (1901) 18 T.L.R. 131 (H.L.).

[37] *Percival's* case, *supra*, at p. 679.

[38] *Offord* v. *Davies* (1862) 12 C.B.(N.S.) 748; *G.N. Ry.* v. *Witham* (1873) L.R. 9 C.P. 16, 19. Compare contracts of indefinite duration which may be subject to determination upon reasonable notice; see *Re Spenborough U.D.C.'s Agreement*; *Spenborough Corporation* v. *Cooke Sons & Co. Ltd.* [1968] Ch. 139.

[39] *G.N. Ry.* v. *Witham*, *supra*.

[40] See *R.* v. *Demers*, *supra*.

[41] See *Att.-Gen.* v. *Stewards & Co. Ltd.*, *supra*.

[42] *Ibid.*; *Gilmour* v. *McLeod*, 12 N.Z.L.R. S.C. 334; *Pitcaithly & Co.* v. *Mclean & Son* (1911) 31 N.Z.L.R. 648.

[43] See: the account of the operations of the London Builders' Conference given in paper 7 of the *Placing of Management of Building Contracts* (H.M.S.O., 1944); the report of the Monopolies Commission, July 29, 1954, and, as a recent example, *Re Electrical Installations Agreement* [1970] 1 W.L.R. 1391.

[44] The term is used here in its broadest sense in contrast with statute law.

[45] The subject is vast. Only an introduction is given here; for further reference see Wilberforce, Campbell and Elles, *Restrictive Trade Practices and Monopolies* (2nd ed., 1966) and supplement; Chitty (24th ed., 1977), Vol. 2 and supplement.

Contractor's position at common law [46]

Agreements between contractors limiting their rights to tender are in restraint of trade.[47] They are therefore prima facie invalid, but if supported by consideration will be enforced by the courts if they are reasonable both in the interests of the parties and of the public. It is a question for the court in each case to decide whether such a contract is reasonable, but in considering what is reasonable between the parties the law " regards the parties as the best judges of what is reasonable as between themselves." [48] There is no reported case of an agreement between contractors restricting their right to tender being held to be unreasonable, although price fixing and marketing arrangements in other branches of trade and industry have from time to time been held to be unreasonable.[49] There is apparently no case reported in which an agreement reasonable in the interests of the parties has been held to be unenforceable because it is unreasonable in the interests of the public.[50] An agreement between two contractors that one should not tender has been enforced.[51]

Employer's position at common law [52]

An employer whose interests are adversely affected because of a restrictive tendering agreement is unlikely to have any effective remedy,[53] unless he has previously taken from a contractor, and can prove a breach of, a warranty not to enter into such an agreement.[54]

The Restrictive Trade Practices Act 1976

This statute [55] makes registrable a wide range of agreements under which specified restrictions are accepted by two or more parties in respect of the

---

[46] See *supra*, note 44.

[47] For the principle, see *Nordenfelt* v. *Maxim Nordenfelt Co.* [1894] A.C. 535, 565 (H.L.); *Esso Petroleum Co. Ltd.* v. *Harpers Garage (Stourport) Ltd.* [1968] A.C. 269 (H.L.). For a detailed discussion of the whole subject, see Wilberforce, Campbell and Elles, *Restrictive Trade Practices and Monopolies* (2nd ed., 1966) and supplement.

[48] *North Western Salt Co.* v. *Electrolytic Alkali Co.* [1914] A.C. 461, 471 (H.L.).

[49] *e.g. Urmston* v. *Whitelegg* (1891) 7 T.L.R. 295 (C.A.); *Evans* v. *Heathcote* [1918] 1 K.B. 418 (C.A.); *McEllistrim* v. *Ballymacelligott Co-op. Society* [1919] A.C. 548 (H.L.).

[50] *Cf. Att.-Gen. of Australia* v. *Adelaide Steamship Co.* [1913] A.C. 781, 796 (P.C.).

[51] *Metcalf* v. *Bouck* (1871) 25 L.T. 539; *Jones* v. *North* (1875) L.R. 19 Eq. 426; *cf. Rawlings* v. *General Trading Co.* [1921] 1 K.B. 635 (C.A.)—knock-out agreement at auction not illegal. Now subject to the Auctions (Bidding Agreements) Act 1927 and the Auctions (Bidding Agreements) Act 1969.

[52] See *supra*, note 44.

[53] Conspiracy is the form of action which comes to mind but it will fail if the defendants show that their predominant object was to further or protect their interests. See *Mogul Steamship Co.* v. *M'Gregor Gow & Co.* [1892] A.C. 25 (H.L.); *Crofter Hand Woven Harris Tweed Co.* v. *Veitch* [1942] A.C. 435 (H.L.); *Byrne* v. *Kinematograph Renters Society Ltd.* [1958] 1 W.L.R. 762. There is no building contract case reported in which this form of action against contractors has been attempted.

[54] For breach of warranty, misrepresentation and fraud, see Chap. 7.

[55] Only the briefest introduction is given here. Earlier legislation governing restrictive trade practices has now been consolidated in the Restrictive Trade Practices Act 1976 and the Restrictive Practices Court Act 1976. Both Acts are consolidating Acts although there has been substantial re-wording and re-arrangement. References are to the Restrictive Trade Practices Act 1976 unless otherwise stated. For loan finance and credit facilities, see the Restrictive Trade Practices Act 1977.

supply of goods [56] or services [57] and also in respect of the construction or carrying out of buildings, structures and other works by contractors.[58] The term
" agreement " includes an arrangement not intended to be legally enforceable
and the term " restriction " includes any negative obligation, whether express
or implied, and whether absolute or not.[59] The restrictions specified include
those relating to prices, terms and conditions,[60] quantities and descriptions,
processes and the persons or classes of persons to, for or from whom or the areas
or places in which goods are to be supplied or acquired or works carried out.[61]
Certain information agreements are also registrable.[62] The agreements must in
due course be proved to the satisfaction of the Restrictive Practices Court not
to be contrary to the public interest.[63] If they are found to be contrary to the
public interest they become void and the court can prevent their enforcement.[64]
If the agreements are not registered in accordance with the acts they are void
and unenforceable.[65]

Restrictions are deemed to be contrary to the public interest unless the court
is satisfied of certain matters.[66] The result is, to take one example, that an agreement as to the adjusting of tender prices is registrable and it is most unlikely
that the Restrictive Practices Court would be satisfied that it was not contrary to
the public interest. An agreement between a few contractors made ad hoc for
one occasion is, it is submitted, registrable.

Remedies of employer by reason of the Acts

Any person affected by a breach of the duty to register an agreement can bring
an action for breach of that duty.[67] It seems, therefore, that an employer who,
after entering into a contract discovers that the price has been increased, or the
terms of the contract made adverse to his interests, by reason of an unregistered,

---

[56] ss. 1 (1) and 6 (1). For restriction of price maintenance on goods, see the Resale
Prices Act 1964. For the over-riding effect of the Treaty of Rome and regulations made
thereunder and of European Community Law, see the European Communities Act 1972,
in particular, ss. 1, 2 and 10. For the prohibition of agreements which may affect trade
between Member States of the Community or prevent, restrict or distort competition
within the Common Market, see Article 85 of the Treaty discussed in *Esso Petroleum
Co. Ltd.* v. *Kingswood Motors (Addlestone) Ltd.* [1974] Q.B. 142; *Application des Gaz S.A.* v.
Falks Veritas Ltd. [1974] Ch. 381 (C.A.).
[57] ss. 1 (1) and 11; Restrictive Trade Practices (Services) Order 1976. (S.I. 1976 No.
98.)
[58] s. 43 (3).
[59] s. 43 (1).
[60] See *Re Birmingham Association of Building Trades Employers' Agreement* [1963] 1
W.L.R. 484—agreement that members shall press for inclusion of R.I.B.A. form; *Re
Electrical Installations Agreement* [1970] 1 W.L.R. 1391—agreement between seven
electrical contractors not to tender until after discussions registrable.
[61] ss. 6 (1), 11 and 43 (3).
[62] ss. 7 and 12; Restrictive Trade Practices (Information Agreements) Order 1969 (S.I.
1969 No. 1842).
[63] ss. 1 (3), 2 and 35.
[64] *Ibid.*
[65] s. 35 (1).
[66] ss. 10, 19 and 31 (9). These are set out *in extenso* and have been frequently considered
by the court in reported cases.
[67] s. 35 (2).

but registrable, agreement, has a remedy in damages [68] against the persons who should have registered.

An injunction has been granted restraining persons from giving effect to an agreement interfering with the plaintiffs' trade, which the court [69] was satisfied would be held to be void by the Restrictive Practices Court.[70]

Government departments and local authorities

They sometimes obtain undertakings from tenderers against collusive tendering practices.[71]

8. Notice of terms

A party to a contract cannot be bound by a term of which he has no notice.

In particular, if the other party seeks to rely on some clause limiting his common law obligations he must prove strictly that the clause on which he seeks to rely is part of the contract.[72] He can do this by a signed contract which includes the clause, for a person who has signed a contract is ordinarily,[73] in the absence of fraud or misrepresentation,[74] bound by its terms whether he has read it or not.[75] Failing such a document it is a question in each case whether reasonably sufficient steps have been taken to give notice of the terms.[76] Where work was done in pursuance of a typewritten offer which had on the face of it, in very small print, an exemption clause relating to delay caused by strikes, it was held that the employer was bound by the clause although he had never read it.[77] If in such a case the clause had been on the back of the document with no reference to its existence on the face of the document the contractor might not have been able to rely on it.[78]

---

[68] Or, in the appropriate case, by way of set-off, see *post*, p. 276.

[69] Pennycuick V.-C., Chancery Division.

[70] *Brekkes Ltd.* v. *Cattel* [1972] Ch. 105.

[71] See further, the 3rd ed. of this book giving the historical background.

[72] *Olley* v. *Marlborough Court* [1949] 1 K.B. 532, 549 (C.A.); *Mendelssohn* v. *Normand Ltd.* [1970] 1 Q.B. 177 (C.A.); *Thornton* v. *Shoe Lane Parking Ltd.* [1971] 2 Q.B. 163 (C.A.); *Hollier* v. *Rambler Motors (A.M.C.) Ltd.* [1972] 2 Q.B. 71 (C.A.).

[73] For the position where a special relationship between the parties gives rise to a fiduciary duty owed by one party to the other, see *Lloyds Bank Ltd.* v. *Bundy* [1975] Q.B. 326 (C.A.).

[74] See *e.g. Curtis* v. *Chemical Cleaning Co.* [1951] 1 K.B. 805 (C.A.); *Mendelssohn* v. *Normand*, *supra*. See also *Dietz* v. *Lennig Chemicals Ltd.* [1969] 1 A.C. 170 (H.L.).

[75] *Parker* v. *S.E. Ry.* (1877) 2 C.P.D. 416, 421; *L'Estrange* v. *Graucob* [1934] 2 K.B. 394, 403 (C.A.). For mistake as to nature of document, see *Saunders* v. *Anglia Building Society* [1971] A.C. 1004 (H.L.); *United Dominions Trust* v. *Western* [1976] Q.B. 513 (C.A.).

[76] *Parker* v. *S.E. Ry.*, *supra*; *Thornton* v. *Shoe Lane Parking Ltd.*, *supra*. For a person not bound by notice in foreign language, see *Geier* v. *Kujawa Western & Warne Bros. (Transport) Ltd.* [1970] 1 Lloyd's Rep. 364.

[77] *Frederick Sage & Co. Ltd.* v. *Spiers & Ponds Ltd.* (1915) 31 T.L.R. 204; *Cf. L'Estrange* v. *Graucob*, *supra*. See also *British Road Services* v. *Crutchley (Arthur V.) Ltd.* [1968] 1 All E.R. 811 (C.A.).

[78] *Parker* v. *S.E. Ry.*, *supra*; *Olley* v. *Marlborough Court* [1949] 1 K.B. 532 (C.A.) and other authorities referred to in footnote 72 above; see also *Chapelton* v. *Barry U.D.C.* [1940] 1 K.B. 532 (C.A.).

advantage of the absence of the certificate.[42] Thus a contract provided that payment should be made upon the issue by the employer's surveyor (acting in the position of an architect) of a certificate that the work had been satisfactorily carried out. The surveyor contended that his function of certification extended to economy of time, labour and materials, and demanded, with the concurrence of the employer, certain information for this purpose. It was not given and the surveyor therefore refused to certify. The court held that on the construction of the contract his function of certifying was confined to whether the work had been satisfactorily carried out and that an " illegitimate condition precedent to any consideration of the granting of a certificate," [43] had been insisted on by the surveyor and the employer. Consequently the employer could not rely on the absence of the certificate.[44] If the employer had, contrary to the surveyor, taken the correct view of the surveyor's functions it would have been his duty to appoint another surveyor; failure to appoint another surveyor would have absolved the contractor from the necessity of obtaining a certificate.[45]

Where the interference of the employer's agents delayed works so that they were not completed to time, the employer was not allowed to resist payment on the ground that the absence of the architect's certificate extending time made the contractor liable to penalties.[46]

5. Death or incapacity of the certifier

The position upon the death or incapacity of the architect depends upon the construction of the contract. The matter may be dealt with expressly,[47] or it may be an implied term that the employer has a right to appoint a new architect.[48] If there is no term express or implied providing for the appointment of a new architect, it is submitted that the employer cannot rely upon the absence of a certificate caused by the death or incapacity of the architect.[49]

6. Award superseding a certificate

An arbitration clause may be so worded that on a proper reference the arbitrator can decide whether the architect was right in withholding a certificate, and if he finds in the contractor's favour award a sum of money to be paid notwithstanding the absence of a certificate, a condition precedent.[50]

---

[42] *Panamena, etc.* v. *Frederick Leyland & Co. Ltd., supra; Hotham* v. *East India Co.* (1787) 1 T.R. 638; *Oshawa (City of)* v. *Brennan Paving Co. Ltd.* [1955] S.C.R. 76, Supreme Ct. of Canada.

[43] *Panamena, etc.* v. *Frederick Leyland & Co. Ltd., supra*, at p. 435; see also *Kellett* v. *New Mills U.D.C., supra*, at p. 300.

[44] *Ibid.*

[45] *Panamena etc.* case, *supra*, at p. 436.

[46] *Russell* v. *Sa da Bandeira* (1862) 13 C.B.(N.S.) 149; see also *Roberts* v. *Bury Commissioners, supra*. For penalties, used here in the sense of liquidated damages, see *post*, p. 155.

[47] See: articles of agreement, Standard form of building contract; cases cited under " Given by wrong person," *post*, p. 83.

[48] *Cf. Panamena, etc.* v. *Frederick Leyland & Co. Ltd., supra*, at p. 436.

[49] *Cf. Kellett* v. *New Mills, supra*, at p. 300.

[50] *Brodie* v. *Cardiff Corp.* [1919] A.C. 337 (H.L.); *Neale* v. *Richardson, supra; Prestige & Co.* v. *Brettell* [1938] 4 All E.R. 346 (C.A.). *Cf.* clause 35, Standard form of building contract. See further, *post*, p. 87.

The refusal of the arbitrator to deal with the matter may enable the contractor to sue in the courts. Thus where an architect was also the arbitrator he refused both to grant a certificate and to arbitrate. The arbitration clause was wide enough to override the clause making certificates a condition precedent. It was held that, neither party having taken any steps to stay the proceedings or appoint a new arbitrator, the court had jurisdiction to deal with the contractor's claim.[51]

Section 4: *Binding and Conclusive Certificates*

1. Architect's decision may be binding and conclusive

An architect in granting ordinary certificates in building contracts is not an arbitrator.[52] But his decision as stated in his certificate may be as binding and conclusive between the parties as if it were an award.[53] Because he is not an arbitrator he cannot be compelled to state a case for the decision of the court upon a point of law, although it may be that the court would entertain an application for a declaration and other relief in respect of an allegation that a decision on a point of law in an architect's certificate is wrong,[54] particularly if it is a " speaking " certificate, *i.e.,* where the point appears on its face or the architect has given his reasons.[55] It has been said that the object of making the architect's decision in a final certificate binding and conclusive is to have the benefit of his skill and knowledge as an independent man to decide what is finally due between the parties without recourse to the enormous expense and trouble often involved in judicial proceedings.[56]

2. Effect of such certificates

When parties have agreed that the architect shall decide matters finally between them they are not allowed in the absence of fraud or special circumstances [57] to go behind his decision upon such matters merely because they are dissatisfied with it. Thus, if the question of extras is within the architect's jurisdiction to decide finally, the contractor cannot recover more than the

---

[51] *Neale* v. *Richardson, supra.* For arbitration, see *post,* pp. 233 *et seq.*

[52] *Sutcliffe* v. *Thackrah* [1974] A.C. 727 (H.L.).

[53] *Goodyear* v. *Weymouth Corporation* (1865) 35 L.J.C.P. 12, 17. It is difficult to be consistent in the use of terms, for words such as " conclusive," " final and conclusive," " binding and conclusive," " final," have been used to describe an architect's certificate or decision against which there is no appeal to the courts, or to arbitration. Such certificates or decisions are frequently called " final," but this is not completely satisfactory, as a final certificate in the sense defined *ante* (p. 77) may be conclusive on some matters only, or not conclusive at all—*Robins* v. *Goddard* [1905] 1 K.B. 294 (C.A.), discussed in *East Ham Borough Council* v. *Bernard Sunley & Sons Ltd.* [1966] A.C. 406, 434, 441, 447 (H.L.).

[54] Consider *Re Davstone Estates Ltd.'s Leases* [1969] 2 Ch. 378 appplying *Lee* v. *Showmen's Guild of Great Britain* [1952] 2 Q.B. 329, 342, 354 (C.A.). See also *Kaye (P. & M.) Ltd.* v. *Hosier & Dickinson* [1972] 1 W.L.R. 146, 157 (H.L.). " The court retains ultimate control in seeing that the architect acts properly and honestly and in accordance with the contract "—*per* Lord Wilberforce.

[55] See *Campbell* v. *Edwards* [1976] 1 W.L.R. 403, 407, 408 (C.A.).

[56] *Sharpe* v. *San Paulo Ry.* (1873) L.R. 8 Ch.App. 597, 609, 611; *Tullis* v. *Jacson* [1892] 3 Ch. 441, 444.

[57] See *post,* p. 82.

amount allowed by the certificate.[58] Conversely, the employer cannot go behind the certificate and say that a lesser sum is due [59] If a certificate is given which the contract shows was intended to be a final expression of the architect's satisfaction, the employer is not allowed subsequently to allege that there are any defects.[60]

3. When is a certificate binding and conclusive?

It is submitted that it is impossible to formulate a comprehensive test to determine whether a certificate is binding and conclusive,[61] and that the most useful approach is to ask whether any particular certificate, when given, can be successfully attacked and set aside. This involves a consideration not only of the parties' intention to be derived from the contract, but of their conduct and that of the architect, and is considered in the following section.

Section 5: *Attacking a Certificate*

1. Not intended to be conclusive

It is a question of construction in each case to determine whether it was intended that a particular certificate should be conclusive upon the matter with which it purports to deal. Express words are frequently used such as, for example that " the certificate of the engineer . . . shall be binding and conclusive on both parties." [62] It seems that prima facie a final certificate which is a condition precedent to payment is conclusive.[63] Progress certificates are usually not conclusive.[64] An arbitration clause may prevent a certificate from being conclusive.[65]

" *And* " clauses

A contractor must frequently complete work according to a specification or to a certain standard *and* to the satisfaction of the architect. It is a question in each case whether this imposes a double obligation on the contractor. In some cases it has been held that it does not, in the sense that when the architect had given a certificate of satisfaction the employer was not permitted to give evidence showing that the work was not according to the specification or the required standard.[66] Where work was to be completed in accordance with plans and a specification, the work to be carried out to the satisfaction of the surveyor and local sanitary inspector, and they expressed their satisfaction upon completion, and defects

---

[58] *Sharpe* v. *San Paulo Ry.*, *supra*; *Brunsden* v. *Staines Local Board* (1884) 1 Cab. & El. 272.

[59] See *ante*, p. 70.

[60] *Harvey* v. *Lawrence* (1867) 15 L.T. 571; *Bateman* (*Lord*) v. *Thompson* (1875) H.B.C. (4th ed.), Vol. 2, p. 36 (C.A.); *Dunaberg Ry.* v. *Hopkins* (1877) 36 L.T. 733; *Kaye* (*P. & M.*) *Ltd.* v. *Hosier & Dickinson* [1972] 1 W.L.R. 146 (H.L.), discussed further, *post*, p. 395; *cf. Billyack* v. *Leyland Construction Co. Ltd.* [1968] 1 W.L.R. 471.

[61] *Cf.* Halsbury (4th ed.), Vol. 4, p. 616, para. 1209.

[62] *Kennedy* v. *Barrow-in-Furness Corp.* (1909) H.B.C. (4th ed.), Vol. 2, pp. 411, 413 (C.A.). See *post*, p. 393, for a discussion of the effect of clause 30 (7) of the Standard form of building contract.

[63] *Sharpe* v. *San Paulo Ry.*, *supra*.

[64] See *ante*, p. 76.

[65] See *post*, p. 87.

[66] *Harvey* v. *Lawrence*, *supra*; *Bateman* (*Lord*) v. *Thompson*, *supra*; *Dunaberg Ry.* v. *Hopkins*, *supra*.

then appeared, it was held that their approval was merely an added protection to the employer. It was not conclusive, so that the employer was entitled to recover damages.[67]

Expiration of time

The contract may provide that a certificate is not to become conclusive until the expiration of a certain period from its issue.[68]

2. Not within the architect's jurisdiction

The architect's decision is not conclusive upon matters which are not within his jurisdiction to decide.[69]

Employer's breach of contract

It seems that clear language is required before an architect's powers of certification extend to cover breaches of contract by the employer,[70] so that in the absence of such words if an architect gives a certificate for payment and the contractor has been put to loss and expense by the employer's breach of contract, such as delay or interference, the certificate is not conclusive upon the contractor's claim for damages unless it is an award upon a proper reference of the dispute.[71] However, the contractor may waive his rights to object to the certificate by expressly seeking the architect's decision on the matter.[72]

New contract

The architect's power of certifying under a contract cannot extend to matters which have arisen under a new and independent contract.[73]

Conclusive on some matters only

The architect's decision may be conclusive on some matters but not on others.[74]

---

[67] *Newton Abbot Development Co. Ltd.* v. *Stockman Bros* (1931) 47 T.L.R. 616, followed in *Billyack* v. *Leyland Construction Co. Ltd.* [1968] 1 W.L.R. 471; see also *Bird* v. *Smith* (1848) 12 Q.B. 786; *Petrofina S.A. of Brussels* v. *Compagnia Italiana, etc.* (1937) 53 T.L.R. 650, 653 (C.A.)—dealing with the construction of a charterparty; *Minster Trust Ltd.* v. *Traps Tractors* [1954] 1 W.L.R. 963.

[68] See, *e.g. Bateman (Lord)* v. *Thompson, supra.*

[69] See *Northampton Gaslight Co.* v. *Parnell* (1855) 24 L.J.C.P. 60; *Russell* v. *Sa da Bandeira* (1862) 13 C.B.(N.S.) 149; *Roberts* v. *Bury Commissioners* (1870) L.R. 5 C.P. 310; *Ashwell and Nesbit* v. *Allen* (1912) H.B.C., 4th ed., ii, p. 462 (C.A.); *Panamena, etc.* v. *Frederick Leyland & Co. Ltd.* [1947] A.C. 428 (H.L.).

[70] *Russell* v. *Sa da Bandeira, supra; Roberts* v. *Bury Commissioners, supra; Lawson* v. *Wallasey Local Board* (1883) 48 L.T. 507 (C.A.).

[71] *Lawson* v. *Wallasey Local Board, supra,* upholding the Divisional Court (1883) 11 Q.B.D. 229, but on different grounds; *cf. Sattin* v. *Poole* (1901) H.B.C. (4th ed.), Vol. 2, p. 306 (D.C.). For extension of time, see Chap. 9 (*post,* p. 162). Note that a valid extension of times does not necessarily extinguish the contractor's claim for damages—*Roberts* v. *Bury Commissioners, supra,* at p. 327.

[72] *Sattin* v. *Poole* (1901) H.B.C. (4th ed.), Vol. 2, p. 306 (D.C.).

[73] *Trade Indemnity Co.* v. *Workington Harbour and Dock Board* [1937] A.C. 1 (H.L.).

[74] See *Re Meadows and Kenworthy* (1896) H.B.C. (4th ed.), Vol. 2, p. 265 (C.A. affd. H.L.); *East Ham Borough Council* v. *Bernard Sunley & Sons Ltd.* [1966] A.C. 406 (H.L.), explaining the final certificate under the 1939 ed., R.I.B.A. form (pre-1957 revision). See also *Kaye (P. & M.) Ltd.* v. *Hosier & Dickinson* [1972] 1 W.L.R. 146 (H.L.) for an important question raised but not decided on the effect of the final certificate in the then Standard form of building contract. See, further, *post,* p. 395.

Course of dealing

" If two parties have made a series of similar contracts each containing certain conditions, and then they make another without expressly referring to those conditions it may be that those conditions ought to be implied. " [79] It is thought that this principle is unlikely to be of frequent application in building contracts because of the complexity of both subject-matter and of the terms usually contained in express written contracts.

9. Unconditional acceptance

There is an acceptance of the offer bringing a binding contract into existence when the offeree makes an unconditional acceptance.[80] If he suggests any new terms there cannot be an acceptance and this may amount to a fresh offer,[81] although a mere request for information about the terms of an offer does not amount to a counter-offer.[82]

Lengthy negotiations for a contract

It is sometimes difficult to determine whether a concluded contract has come into existence when there have been lengthy negotiations between the parties but no formal contract has ever been signed. It is suggested that a useful approach is to ask whether the following can be answered in the affirmative[83]: (i) in the relevant period of negotiation did the parties intend to contract?[84] (ii) at the time when they are alleged to have contracted had they agreed with sufficient certainty upon the terms which they then regarded as being required in order that a contract should come into existence? (iii) did those terms include all the terms which, even though the parties did not realise it, were in fact essential[85] to be agreed if the contract was to be commercially workable?

---

[79] *Per* Lord Reid, *McCutcheon* v. *David MacBrayne Ltd.* [1964] 1 W.L.R. 125, 127 (H.L.); see also *Hardwick Game Farm* v. *Suffolk, etc., Association* [1969] 2 A.C. 31, 90, 104, 105, 130 (H.L.); *Mendelssohn's* and *Hollier's* cases *supra*; *Roberts* v. *Elwells Engineers Ltd.* [1972] 2 Q.B. 586 (C.A.); *British Crane Hire Corporation Ltd.* v. *Ipswich Plant Hire Ltd.* [1975] Q.B. 303 (C.A.).

[80] See *Nicolene* v. *Simmonds* [1953] 1 Q.B. 543 (C.A.). Problems as to the existence of a contract and as to its meaning when it is decided that there is a contract are closely related and it may be useful to refer to the next chapter on construction of contracts.

[81] *Hyde* v. *Wrench* (1840) 3 Beav. 334; *cf. Leslie & Co.* v. *Commissioners of Works* (1914) 78 J.P. 462; *Peter Lind & Co. Ltd.* v. *Mersey Docks & Harbour Board* [1972] 2 Lloyd's Rep. 235.

[82] *Stevenson* v. *McLean* (1880) 5 Q.B.D. 346.

[83] See propositions applied by Megaw J., *Trollope & Colls Ltd.* v. *Atomic Power Constructions Ltd.* [1963] 1 W.L.R. 333, 336. See also *British Guiana Credit Corp.* v. *Da Silva* [1965] 1 W.L.R. 248 (P.C.); *Bushwall Properties Ltd.* v. *Vortex Properties Ltd.* [1976] 1 W.L.R. 591, 603 (C.A.), and *post*, pp. 21 and 32.

[84] For the requirement that the parties intended to create a legal relationship, see *Edwards* v. *Skyways Ltd.* [1964] 1 W.L.R. 349 and authorities referred to therein (offer to make *ex gratia* payment does not carry necessary or even probable implication that the agreement is to be without legal effect). See also *Horrocks* v. *Forray* [1976] 1 W.L.R. 230 (C.A.).

[85] " If some particulars essential to the agreement still remain to be settled afterwards there is no contract "; Lord Blackburn, *Rossiter* v. *Miller* (1878) 3 App.Cas. 1124, 1151 (H.L.).

(iv) was there a sufficient indication of acceptance by the offeree of the offer as then made complying with any stipulation in the offer itself as to the manner of acceptance? Some of these matters are discussed further below.

All negotiations should be considered. This is important, especially where there have been meetings at which oral statements were made showing that essential terms, not referred to in certain correspondence, were still awaiting agreement at the time of such correspondence.[86]

Essential terms

It is thought that agreement as to parties, price, time and description of works is the minimum necessary to make the contract commercially workable. Lack of agreement as to parties can arise when companies have common directors and there is confusion as to which company is intended to be a contracting party. It can also arise when there is an issue as to whether a contract is with an agent or his principal.[87] Price is discussed *post*, p. 20, and time *post*, p. 59. It is submitted that silence by the parties as to time does not prevent a contract coming into existence, for if the other essential terms are agreed a reasonable time for completion will be implied. The description of the works may be in wide terms (*post*, p. 62) and it may be subject to the retrospective operation of a variation clause. Thus in *Trollope & Colls Ltd.* v. *Atomic Power Constructions Ltd.*[88] an offer was made in February 1959 to carry out certain works for £X. In June 1959, while the parties were still negotiating the terms of the contract, work commenced and was still continuing in April 1960 when the parties agreed upon all the essential terms including a clause providing for the variation of the contract work. By April 1960, as a result of variations, the work to be carried out and the price to be paid if there was a contract differed from the work and price referred to in the February 1959 offer, but it was held that a contract came into existence upon the terms finally agreed in April 1960.[89]

"Subject to contract"

If a purported acceptance is expressed to be "subject to contract," or some other words[90] are used which show that further negotiations are contemplated, there is no concluded contract.[91] But there may be a concluded contract where

---

[86] *Hussey* v. *Horne-Payne* (1879) 4 App.Cas. 311, 316 (H.L.); *Panorama Developments (Guildford) Ltd.* v. *Fidelis Furnishing Fabrics Ltd.* [1971] 2 Q.B. 711 (C.A.).

[87] See the *Panorama Development* case, *supra*. See also *post*, p. 202.

[88] [1963] 1 W.L.R. 333.

[89] It followed that the contractors were bound to complete and accept payment in accordance with the contract and were not entitled to stop work and claim payment on a *quantum meruit*.

[90] *e.g.* "subject to strike and lock-out clauses": *Love & Stewart Ltd.* v. *S. Instone* (1917) 33 T.L.R. 475; "subject to surveyor's report": *Marks* v. *Board* (1930) 46 T.L.R. 424; "subject to satisfactory survey": *Astra Trust Ltd.* v. *Adams & Williams* [1969] 1 Lloyd's Rep. p. 81; "subject to satisfactory running trials": *John Howard & Co.* v. *J. P.Knight Ltd.* [1969] 1 Lloyd's Rep. 364; "subject to satisfactory mortgage": *Lee Parker* v. *Izzet* (No. 2) [1972] 1 W.L.R. 775. *Cf. Mackay* v. *Dick* (1881) 6 App.Cas. 251 (H.L.). For a contract subject to ratification by the principal of the person purporting to make it, see *Warehousing, etc., Ltd.* v. *Jafferali & Sons Ltd.* [1964] A.C. 1 (P.C.) and *post*, p. 202.

[91] *Rossiter* v. *Miller* (1878) 3 App.Cas. 1124 (H.L.); *Bozson* v. *Altrincham U.D.C.* (1903) 67 J.P. 397 (C.A.); *Chillingworth* v. *Esche* [1924] 1 Ch. 97 (C.A.); *cf. Branca* v.

the parties have agreed upon all the terms and merely agree that these shall later be embodied in a formal document.[92]

Certainty of terms

" In order to constitute a valid contract the parties must so express themselves that their meaning can be determined with a reasonable degree of certainty." [93] Thus agreements for the sale of goods " on hire-purchase terms over two years," [94] and " subject to war clause," [95] have been held to be too vague.[96] But " a distinction must be drawn between a clause which is meaningless and a clause which is yet to be agreed. A clause which is meaningless can often be ignored while still leaving the contract good; whereas a clause which has yet to be agreed may mean that there is no contract at all because the parties have not agreed on all the essential terms." [97]

Reference to reasonable requirements or standards is not too vague; the court (or arbitrator when there is an arbitration clause) can, in default of agreement, determine what is reasonable.[98]

Meaningless words

Where an otherwise clear acceptance was expressed to be subject to " the usual conditions of acceptance " and there were no usual conditions of acceptance so that the words were meaningless, the court ignored the words quoted and held that the writer of the words was bound by the clear terms of the rest

---

Cobarro [1947] K.B. 854 (C.A.); *Smallman* v. *Snallman* [1972] Fam. 25, 32 (C.A.); *Brown* v. *Gould* [1972] Ch. 53; *Tiverton Estates Ltd.* v. *Wearwell Ltd.* [1975] Ch. 146 (C.A.); *Munton* v. *G.L.C.* [1976] 1 W.L.R. 649 (C.A.). For a discussion of the principles relating to a contract subject to a suspensive condition, see *Cranleigh Precision Engineering Ltd.* v. *Bryant* [1965] 1 W.L.R. 1293.

[92] *Rossiter* v. *Miller, supra*; *Lewis* v. *Brass* (1877) 3 Q.B.D. 667 (C.A.); *Love & Stewart Ltd.* v. *S. Instone, supra*. See also the very special case of *Richards (Michael) Properties Ltd.* v. *St. Saviours Parish, Southwark* [1975] 3 All E.R. 416 where the words " subject to contract " were added by mistake to a concluded contract and therefore were rejected as surplusage. In *Munton* v. *G.L.C.* (*supra*) the Court of Appeal emphasised that this was a very special case which in no way altered the " sanctity of the words subject to contract " (p. 656).

[93] *Per* Lord Maugham, *Scammell* v. *Ouston* [1941] A.C. 251, 255 (H.L.); *Bushwall Properties Ltd.* v. *Vortex Properties Ltd.* [1976] 1 W.L.R. 591 (C.A.); *cf.* Lord Wright, *Hillas & Co. Ltd.* v. *Arcos Ltd.* (1932) 147 L.T. 503, 504 (H.L.) and see *post*, p. 32, " Valid meaning."

[94] *Scammell* v. *Ouston, supra*.

[95] *Bishop & Baxter* v. *Anglo-Eastern Co. Ltd.* [1944] 1 K.B. 12 (C.A.); see also *British Electrical, etc., Ltd.* v. *Patley Pressings Ltd. and Others* [1953] 1 W.L.R. 280—" Subject to *force majeure* conditions " too vague; but see Denning L.J., *Nicolene* v. *Simmonds, supra*, at p. 552, and *cf. Hong Guan & Co. Ltd.* v. *R. Jumabhoy & Sons Ltd.* [1960] A.C. 684, 700 (P.C.), a sale, " Subject to *force majeure* and shipment," valid; *Three Rivers Trading Co.* v. *Gwinear and District Farmers* (1967) 111 S.J. 831 (C.A.)—sale of " 400 tons (approx.) " of barley not void for uncertainty; and see *Edwards* v. *Skyways Ltd.* [1964] 1 W.L.R. 349.

[96] *Supra*, notes 94 and 95.

[97] *Per* Denning L.J., *Nicolene* v. *Simmonds, supra*, at p. 551; *Lovelock Ltd.* v. *Exportles* [1968] 1 Lloyd's Rep. 163 (C.A.). See also " subject to contract," *supra*; *cf. Shamrock SS. Co. Ltd.* v. *Storey* (1899) 81 L.T. 413 (C.A.); *Hobbs Padgett and Co. (Reinsurance)* v. *J. C. Kirkland Ltd.* [1969] 2 Lloyd's Rep. 547 (C.A.).

[98] See *Sweet & Maxwell Ltd.* v. *Universal News Services Ltd.* [1964] 2 Q.B. 699 (C.A.).

of the document and there was a contract in existence.[99] But a contract made upon " usual terms " as to one of its incidents is valid, and the term will be enforced if the contents of the term and its general use by the class of persons who make such contracts is proved.[1]

No price fixed

An agreement to do specified work which is silent as to price is binding, the law implying a promise on the part of the employer to pay a reasonable price.[2] If there is an agreement to do work " at a price to be agreed " by the parties the position differs according to whether the agreement is executory (*i.e.* nothing done by either party in performance of the agreement) or executed, wholly or in part. In the former case there is no binding contract[3]; in the latter case the employer must pay a reasonable price for work done, and it seems, the terms of the contract are binding.[4] An agreement to do work at a price to be agreed or, in default of agreement, to be determined by arbitration is valid,[5] it being a question of construction whether a document containing an arbitration clause has this effect.[6] Where there is such an agreement there is an obligation on the parties to try to agree the price and refusal to discuss is a repudiation.[7]

The parties may agree upon the price after work has commenced but so as to show that they intended the agreement to operate retrospectively to cover the work already carried out.[8]

Contract to negotiate

The law does not recognise such a contract because it would be too uncertain to have any binding force.[9]

Price payable by another

There can be a contract with the employer to do work for the employer,

---

[99] *Nicolene* v. *Simmonds, supra.*

[1] *Shamrock SS. Co. Ltd.* v. *Storey* (1899) 81 L.T. 413 (C.A.).

[2] See *post*, p. 58.

[3] *May & Butcher Ltd.* v. *The King* [1934] 2 K.B. 17n. (H.L.); *British Bank for Foreign Trade Ltd.* v. *Novinex Ltd.* [1949] 1 K.B. 623; *King's Motors (Oxford) Ltd.* v. *Lax* [1970] 1 W.L.R. 426; *Smith* v. *Morgan* [1971] 1 W.L.R. 803; *Brown* v. *Gould* [1972] Ch. 53; *Courtney & Fairbairn Ltd.* v. *Tolaini Brothers (Hotels) Ltd.* [1975] 1 W.L.R. 297 (C.A.).

[4] *Foley* v. *Classique Coaches* [1934] 2 K.B. 1 (C.A.); *National Coal Board* v. *Galley* [1958] 1 W.L.R. 16 (C.A.); see also *Hillas* v. *Arcos* (1932) 147 L.T. 503 (H.L.).

[5] *May & Butcher Ltd.* v. *The King, supra*; *Hillas* v. *Arcos, supra*; *Foley* v. *Classique Coaches, supra*; *F. & G. Sykes (Wessex) Ltd.* v. *Fine Fare Ltd.* [1966] 2 Lloyd's Rep. 205, affd. [1967] 1 Lloyd's Rep. 53 (C.A.), although reversed in part on different grounds. See also *Brown* v. *Gould, supra*; *Courtney & Fairbairn Ltd.* v. *Tolaini Brothers (Hotels) Ltd., supra.*

[6] *Ibid.*

[7] *F. & G. Sykes (Wessex) Ltd.* v. *Fine Fare Ltd.* [1966] 2 Lloyd's Rep. 205, 213; *Foley* v. *Classique Coaches, supra*, at p. 12. For meaning and effect of repudiation, see *post*, p. 111. See also *Smallman* v. *Smallman* [1972] Fam. 25, 32 (C.A.).

[8] See *Trollope & Colls Ltd.* v. *Atomic Power Ltd.* [1963] 1 W.L.R. 333.

[9] *Courtney & Fairbairn Ltd.* v. *Tolaini Brothers (Hotels) Ltd., supra*; unanimous decision of the court disapproving the dictum of Lord Wright in *Hillas & Co. Ltd.* v. *Arcos Ltd.* (1932) 147 L.T. 503, 515 (H.L.). But in principle why should there not be such a contract where the parties contemplate negotiations by skilled and experienced persons? As to damages, consider *Chaplin* v. *Hicks* [1911] 2 K.B. 786 (C.A.).

subject to implied duties owed to the employer to exercise reasonable skill and complete within a reasonable time, but upon terms that payment is not to be by the employer but by another person,[10] and such a contract can exist at the same time as a separate contract with that other person.[11]

Negotiations after contract

" Where negotiations are in progress between parties intending to enter into a contract the whole of those negotiations must be looked at to determine when if at all, the contract comes into being. . . . Once the contract comes into being, however, subsequent negotiations by either party seeking, for example, to obtain better terms will not affect the existence of the previously concluded contract." [12]

10. Acceptance by conduct

The offeror cannot bind the offeree by a stipulation that silence will amount to acceptance,[13] but acceptance can be by conduct showing an intention to accept the terms of the other.[14] It is a question in each case whether conduct, known to the offeror, shows such an intention.[15] Thus if an offer is made to a contractor for the performance of certain work upon stated terms, and without making any express acceptance, or counter-offer, the contractor carries out the work, he is bound by the terms of the offer.[16] The same principles apply to an employer who, without any express acceptance, or counter-offer, permits a contractor to carry out work the subject matter of an offer subject to certain terms. He is bound by those terms.[17]

11. Acceptance by post

The general rule is that acceptance must be communicated to the offeror, and does not become effective until it reaches him [18] but if the acceptance is by post special rules may apply. Where the circumstances are such that the parties must have contemplated that the post might be used as a means of communication the acceptance is complete as soon as it is posted [19] although the ordinary

---

[10] *Charnock* v. *Liverpool Corporation* [1968] 1 W.L.R. 1498 (C.A.), a car repair case. For the consideration given, see *ibid.* p. 1505. See also *Brown and Davis* v. *Galbraith* [1972] 1 W.L.R. 997 (C.A.). For the implied duty of skill, see *post*, p. 38, and the implied duty as to time, *post*, p. 59.

[11] *Ibid.*

[12] *British Guiana Credit Corp.* v. *Da Silva* [1965] 1 W.L.R. 248, 255 (P.C.). See further, *post*, p. 30, " Subsequent conduct of parties."

[13] *Felthouse* v. *Bindley* (1862) 11 C.B.(N.S.) 869.

[14] *Charnock* v. *Liverpool Corporation* [1968] 1 W.L.R. 1498, 1507 (C.A.).

[15] See for recent examples considering the matter, *Shields Furniture Ltd.* v. *Golt* [1973] 2 All E.R. 655; *Fairline Shipping Corporation* v. *Adamson* [1975] Q.B. 180.

[16] *Brogden* v. *Metropolitan Railway* (1877) 2 App.Cas. 666 (H.L.); *Dartford Union* v. *Trickett* (1888) 59 L.T. 754; *Greenwood* v. *Hawkings* (1907) 23 T.L.R. 72; *Robophone Facilities Ltd.* v. *Blank* [1966] 1 W.L.R. 1428; *Peter Lind & Co. Ltd.* v. *Mersey Docks & Harbour Board* [1972] 2 Lloyd's Rep. 235; *New Zealand Shipping Co. Ltd.* v. *Satterthwaite & Co. Ltd.* [1975] A.C. 154 (P.C.).

[17] *Ibid.* For the position where no price is fixed see *post*, p. 58.

[18] *Entores Ltd.* v. *Miles Far East Corporation* [1955] 2 Q.B. 327 (C.A.).

[19] *Henthorn* v. *Fraser* [1892] 2 Ch. 27, 33; *Holwell Securities Ltd.* v. *Hughes* [1974] 1 W.L.R. 155 (C.A.).

rule still applies that revocation of the offer is not effective until it reaches the offeree. Therefore if in such circumstances an acceptance is posted before the offeree knows of the revocation of the offer a contract comes into existence.[20]

This artificial concept [21] of acceptance by posting yields to the express terms of the offer. Thus where an option to purchase a freehold was exercisable by " notice in writing to . . ." the defendant it was held that actual communication of the acceptance was necessary notwithstanding that the parties contemplated that the post might be used.[21] Accordingly there was no valid exercise of the option where the notice was lost in the post.[22]

Section 3 : *Formalities of Contract*

The general rule of law is that no formalities are required for the formation of a contract. To this rule there are, for the purposes of the law relating to building contracts, two exceptions: first, contracts which are not enforceable unless evidenced in writing; secondly, contracts with a corporation made before July 29, 1960, which required a seal.

1.   **Contracts to be evidenced in writing**

(a) *Contracts for the sale of land*

A contract for the sale or other disposition of land or any interest in land cannot be enforced unless it is in, or evidenced by, writing, signed by the person to be charged or his agent.[23] Such a contract does not include the normal building contract because the contractor is merely given a licence to enter the land.[24] A contract to grant a building lease, *i.e.* a lease, part of the consideration for the grant of which consists of a promise by the lessee to build on the demised land must be evidenced in writing.[25]

(b) *Contracts of suretyship and guarantee*

A contract to answer for the debt, default or miscarriage of another must be contained in, or evidenced by, writing signed by the surety or his agent,[26] although the writing need not show the consideration for which the promise was given.[27]

---

[20] *Byrne* v. *Van Tienhoven* (1880) 5 C.P.D. 344; see also *Household Insurance Co.* v. *Grant* (1879) 4 Ex.D. 216.

[21] *Holwell Securities, supra*, at p. 157. Lord Justice Lawton in his judgment at p. 161 set out other qualifications to the concept. See also *Yates Building* v. *Pulleyn, The Times,* Feb. 26, 1975 (C.A.).

[22] *Ibid.*

[23] Law of Property Act 1925, s. 40 (1). *Tiverton Estates Ltd.* v. *Wearwell Ltd.* [1975] Ch. 146 (C.A.). For the exception of part performance of the contract see *Steadman* v. *Steadman* [1976] A.C. 536.

[24] *Camden* v. *Batterbury* (1860) 7 C.B.(N.S.) 864; *cf. Lavery* v. *Pursell* (1888) 39 Ch.D. 508. See also *Hounslow, London Borough Council* v. *Twickenham Gardens Development Ltd.* [1971] 1 Ch. 233 discussed *post,* p. 123.

[25] See Law of Property Act 1925, s. 205 (1).

[26] Statute of Frauds 1677, s. 4.

[27] Mercantile Law Amendment Act 1856, s. 3. There were formerly certain other contracts required to be evidenced in writing, and a large body of law, some of which is

2. Contracts under seal

Contracts under seal are sometimes called specialty contracts and the document by which they are effected is termed a deed.[28]

Section 4: *Capacity of Parties*

There are certain categories of persons whose capacity to make contracts is less than that of the ordinary person, or in respect of whose contracts special rules apply.

1. Infants (or minors)

An infant (or minor [29]) is a person under 18 years of age.[30] The general position is that contracts made by an infant are not binding on him save that he must pay a reasonable sum for necessaries which he has received.[31] So far as goods are concerned necessaries are defined as " goods suitable to the condition in life of such infant and to his actual requirements at the time of the sale and delivery." [32] Necessaries can include services, such as lodging and even attendance if the infant's position in society warrants it; and an infant was held liable on a contract for the burial of his wife and children.[33] There seems therefore to be no reason in principle why a contract for building work could not be a contract for necessaries,[34] but a contractor is ill-advised to enter into a contract with an infant. Apart from the perils of the law relating to necessaries, an infant cannot hold any legal estate in land,[35] and might not therefore be able to grant possession of the site for completion of the works. The contractor's better course where he has been negotiating with an infant is to make the contract with the trustees of the land in which the infant has a beneficial interest.

2. Aliens

An alien is liable upon, and can enforce, contracts within the jurisdiction of the English courts unless he is an alien enemy. For the purpose of the enforceability of contracts an alien enemy is a person, whatever his nationality, resident in or carrying on business in the enemy's country, or in a country within the effective control of the enemy.[36]

---

now obsolete, developed in relation to contracts required to be evidenced in writing. As the subject is now of such limited application to building contracts, it is not further dealt with here.

[28] For contracts made by corporations before July 29, 1960, see *post*, p. 25; for estoppel by deed, see *post*, p. 43; for the 12-year limitation period, see *post*, p. 165.

[29] Family Law Reform Act 1969, s. 12.

[30] Family Law Reform Act 1969, s. 1.

[31] Sale of Goods Act 1893, s. 2 (goods sold and delivered)—it is not clear whether the same rule applies to necessaries which are not goods, *i.e.* that he pays a reasonable sum, not necessarily the contract sum; probably it does—*Nash* v. *Inman* [1908] 2 K.B. 1, 8 (C.A.).

[32] Sale of Goods Act 1893, s. 2.

[33] *Chapple* v. *Cooper* (1844) 13 M. & W. 252.

[34] There are no such cases reported.

[35] Law of Property Act 1925, s. 1 (6).

[36] *Porter* v. *Freudenberg* [1915] 1 K.B. 857 (C.A.); see further, Chitty (24th ed., 1977), Vol. 1, para. 658.

3. Bankrupts

An undischarged bankrupt may enter into any contract provided that he does not commit the criminal offences of:

(i) either alone or jointly with any other person obtaining credit to the extent of £50 or upwards from any person without informing that person that he is an undischarged bankrupt; or

(ii) engaging in any trade or business under a name other than that under which he was adjudicated bankrupt without disclosing to all persons with whom he enters into any business transaction the name under which he was adjudicated bankrupt.[37]

4. Members of local authorities and river authorities

Where they have a pecuniary interest in any contract or proposed contract they must disclose the fact, and must not take part in the consideration or discussion of, or vote on any question with respect to, the contract.[38] A person failing to comply with these requirements commits a criminal offence.[39]

5. Corporations

A corporation is a legal person, that is to say it is regarded by the law as a legal entity quite distinct from the person or persons who may for the time being be the member or members of the corporation. A corporation sole is composed of one office holder such as a bishop or the Public Trustee, a corporation aggregate of more than one person such as the members of a company.[40]

The ultra vires doctrine

Corporations may be broadly classified into those incorporated by charter, either express or implied and those incorporated as the result of an Act of Parliament. A chartered corporation has at common law the power " to bind itself to such contracts as an ordinary person can bind himself to." [41] But a statutory corporation being a statutory creature can only enter into such contracts as are necessary to attain the objects of its creation. Any purported contract for purposes beyond these objects is *ultra vires* (*i.e.* beyond its powers) and therefore void. For a person dealing with a company registered under the Companies Act 1948, which includes the ordinary commercial company, the position has recently been much simplified. Since January 1, 1973, any transaction decided on by the directors shall, in favour of a person dealing with the company in good faith, be deemed to be one which it is in the capacity of the company to enter into, and the power of the directors to bind the company is deemed to be free of any limitation under the memorandum or articles of association; and a

---

[37] Bankruptcy Act 1914, s. 155; Insolvency Act 1976, s. 1. See *Fisher* v. *Raven* [1964] A.C. 210 (H.L.); *R* v. *Hartley* [1972] 2 Q.B. 1 (C.A. Crim. Div.).

[38] See Local Government Act 1972, ss. 94–98, 105, 272; Water Resources Act 1963, s. 6 (6).

[39] *Ibid.* s. 94 (2). See *Brown* v. *D.P.P.* [1956] 2 Q.B. 369; *Rands* v. *Oldroyd* [1959] 1 Q.B. 204.

[40] See Chitty (24th ed., 1977), Vol. 1, para. 562.

[41] *Wenlock (Baroness) Ltd.* v. *River Dee Co.* (1887) 36 Ch.D. 674, 685 (C.A.).

party to a transaction so decided on is not bound to inquire as to the capacity of the company to enter into it or as to any such limitation on the powers of the directors, and is presumed to have acted in good faith unless the contrary is proved.[42]

Corporation acts through its agents

A corporation acts through its agents and even though the contract purported to be entered into is within the powers of the corporation to be binding it must be within the express or ostensible authority of the agent. Directors of a company registered under the Companies Act 1948 have authority to bind the company and the position of a person dealing with them has recently been clarified, see *supra*. At one time the secretary of such a company had very little authority on its behalf. It has now been held that he has ostensible authority to make contracts on its behalf in connection with the administrative side of its affairs such as employing staff and ordering cars and so forth.[43] It is thought that such authority extends to works of repair and maintenance of a company's premises, but it must be a matter of doubt how far it extends to building contracts generally.

Where a person represents that he has authority which he does not possess, and thereby induces another to enter into a contract which is void for want of authority, that other person, if he has suffered loss, can sue the first for breach of warranty of authority.[44]

Requirement of a seal—contracts before July 29, 1960

The general rule was that a corporation could only contract under seal but the rule was subject to many exceptions.[45]

Requirement of a seal abolished—contracts after July 28, 1960

The Corporate Bodies' Contracts Act 1960 abolished the former rule as regards contracts made after July 28, 1960, and equated the formalities required in respect of such contracts made by a corporation with those required in respect of contracts made by private persons [46] (see Section 3, *ante*, p. 22).

6. Contracts by local authorities

A local authority must make standing orders with regard to the making of

---

[42] European Communities Act 1972, s. 9 (1). Formerly a person was at risk of it being held that a contract was not within the powers of the company as set out within its memorandum of association; or of the directors as limited by the articles; see *Royal British Bank* v. *Turquand* (1856) 6 E.B. 327; *Ashbury Railway* v. *Riche* (1875) L.R. 7 H.L. 653 (H.L.) ;*Bell Houses Ltd.* v. *City Wall Properties Ltd.* [1966] 2 Q.B. 656 (C.A.). See also s. 9 (2) of the 1972 Act which makes the promoter of a company personally liable where the company has not been formed.

[43] *Panorama Development (Guildford) Ltd.* v. *Fidelis Furnishing Fabrics Ltd.* [1971] 2 Q.B. 711 (C.A.).

[44] See cases cited at p. 203, *post*. For the authority of the architect, see *post*, p. 202. For the authority of senior officers of local authorities, see, *infra*.

[45] The rule and the exceptions are discussed and the relevant authorities referred to at pp. 22–24 of the 2nd ed. of this book.

[46] The relevant provisions for companies are in s. 32 of the Companies Act 1948.

contracts by them or on their behalf for the execution of works,[47] but a person entering into a contract with a local authority is not bound to inquire whether the standing orders have been complied with and non-compliance does not invalidate any contract entered into by or on its behalf.[48]

Local authorities are corporations. Contracts made with them after July 28, 1960, do not have to be under seal.[49] In practice local authorities still frequently enter into substantial contracts under seal. This gives them the benefit of a 12-year limitation period[50] and avoids questions as to whether a person had authority to contract on their behalf. Senior officers are likely to be held to have such authority.[51]

7. Government contracts

Contracts with government departments can be enforced by or against the Crown.[52]

8. Partnerships

Partnership is " the relationship which subsists between persons carrying on business in common with a view to profit."[53] A partnership is not a corporation and in general a partner acting for the purposes of the partnership can enter into a contract on behalf of the other partners so as to make them liable.[54]

9. Unincorporated associations

An unincorporated association is a general term given to a group of persons neither a corporation nor a partnership who act together for certain purposes.[55] A typical example is a club. As it has no corporate existence contracts cannot be made on its behalf and a member is not liable unless he made the contract or authorised or ratified it.[56] Members of the committee who enter into the contract are liable, and in general a person dealing with a club is normally entitled to look to the members of the committee of management.[57]

---

[47] Local Government Act 1972, s. 135 (2). The London Government Act 1963 applies to London.
[48] Local Government Act 1972, s. 135 (4).
[49] The Corporate Bodies' Contracts Act 1960.
[50] See *post*, p. 165.
[51] See *e.g. Carlton Contractors* v. *Bexley Corporation* (1962) L.G.R. 331—borough surveyor; *Roberts & Co. Ltd.* v. *Leicestershire County Council* [1961] Ch. 555—officers of county architect's department. See also *post*, p. 200.
[52] For the right to sue the Crown and the procedure, see the Crown Proceedings Act 1947.
[53] Partnership Act 1890, s. 1; *Keith Spicer Ltd.* v. *Mansell* [1970] 1 W.L.R. 333 (C.A.).
[54] See generally *Lindley on Partnership* (13th ed., 1971). For a recent case, see *Mann* v. *D'Arcy* [1968] 1 W.L.R. 893.
[55] See further Chitty (24th ed., 1977), Vol. 1, para. 600.
[56] *Bradley Egg Farm Ltd.* v. *Clifford* [1943] 2 All E.R. 378 (C.A.); *Wise* v. *Perpetual Trustee Co.* [1903] A.C. 139 (P.C.).
[57] *Steele* v. *Gourley* (1887) 3 T.L.R. 118, 119; *Bradley Egg Farm Ltd.* v. *Clifford, supra*; *cf. Draper* v. *Manvers (Earl)* (1892) 9 T.L.R. 73.

CHAPTER 3

CONSTRUCTION OF CONTRACTS AND RECTIFICATION

Section 1: *Construction of Contracts*

MANY of the problems likely to arise under a building contract are concerned with the meaning to be given to words in a written contract. The process by which the courts arrive at this meaning is termed construing a contract, and the meaning, as determined by the court, the construction of the contract.[1]

1. Expressed intention

In construing a contract the court applies the rule of law that, " while it seeks to give effect to the intention of the parties, [it] must give effect to that intention as expressed, that is, it must ascertain the meaning of the words actually used." [2] For the purposes of construction " intention " does not mean motive, purpose, desire or a state of mind but intention as expressed, and the common law adopts an objective standard of construction excluding general evidence of actual intention of the parties.[2] Thus while, as will be shown below, it permits evidence of the circumstances in which the contractual document was made, of the special meaning of words, of customs and certain other matters to assist the court in arriving at the expressed intention of the parties, nevertheless the fundamental rule is that the words must speak for themselves.[3] The parties cannot come into court to give evidence of what they intended to say.[4] If the rule were otherwise " all certainty would be taken from the words in which the parties have recorded their agreement." [5]

(a) *Extrinsic evidence—not normally admissible*

It follows from the principle just stated that no evidence outside the document itself, *i.e.* extrinsic evidence, may normally be adduced to add to, vary, modify or contradict the written terms.[6]

BLANKS. Where a complete blank is left in a material part of the contract evidence is not admissible to fill it.[7] Thus where the date for completion was omitted, and to insert it would result in the imposition of an onerous obligation

---

[1] For a full exposition, see *Norton on Deeds*, and for a shorter account, see *Odgers on the Construction of Deeds and Statutes*. Problems of construction and of formation of contract are often closely related and reference should also be made to Chap. 2, *ante*, especially the sections dealing with offer and acceptance.

[2] *Inland Revenue Commissioners* v. *Raphael* [1935] A.C. 96, 142 (H.L.); *Schuler A.G.* v. *Wickman Machine Tool Sales Ltd.* [1974] A.C. 235 (H.L.).

[3] *Ibid.*

[4] But note that in certain circumstances equity will rectify a contract mistakenly expressed. See *post*, p. 47.

[5] *Inland Revenue Commissioners* v. *Raphael, supra*, at p. 142; see also *London County Council* v. *Henry Boot & Sons Ltd.* [1959] 1 W.L.R. 1069, 1077 (H.L.).

[6] *O'Connor* v. *Hume* [1954] 1 W.L.R. 824, 830 (C.A.).

[7] *In the goods of De Rosaz* (1877) 2 P.D. 66, 69.

under a liquidated damages clause, the court refused to admit evidence that each party had been told of the date.[8]

PRELIMINARY NEGOTIATIONS. When the parties have entered into a final concluded contract in writing the preliminary negotiations such as letters cannot be referred to for the purpose of explaining their intention.[9] In one case [10] a covering letter sent with the tender stated that the tender was " subject to " a certain term. After negotiations a formal contract was signed which defined the contract documents without including the letter, although in the contract documents, as defined, there was a reference to part of the letter for a different purpose. It was held that that term was not part of the contract.

DELETIONS FROM PRINTED DOCUMENTS. " When the parties use a printed form and delete parts of it one can, in my opinion, pay regard to what has been deleted as part of the surrounding circumstances in the light of which one must construe what they have chosen to leave in "—*per* Lord Cross, stating the majority view, *Mottram Consultants Ltd.* v. *Bernard Sunley & Sons Ltd.*[11] Applying this principle, the House of Lords referred to the deletion in an existing contract of a clause giving the employer an express right of set-off as part of the surrounding circumstances showing that the parties directed their minds to the question of set-off and decided that it should not be allowed. The decision must be taken as stating the current law, although some earlier cases in the House of Lords seem to be to contrary effect.[12] In any event the principle must be carefully reconciled with that which prevents reference to preliminary negotiations, see *supra*.

(b) *Extrinsic evidence—when admissible*

To the general rule that extrinsic evidence is not admissible to interpret a written contract there are many qualifications and exceptions relevant to building contracts of which the following are the most important:

(i) *To show meaning of words*

SURROUNDING CIRCUMSTANCES. The court does not ascertain the intention of the parties by an interpretation based purely on internal linguistic considerations.[13] It must ". . . inquire beyond the language and see what the circumstances were with reference to which the words were used, and the object, appearing

---

[8] *Kemp* v. *Rose* (1858) 1 Giff. 258.

[9] *Inglis* v. *Buttery* (1878) 3 App.Cas. 552 (H.L.); *Leggott* v. *Barrett* (1880) 15 Ch.D. 306, 311 (C.A.); *Kinlen* v. *Ennis U.D.C.* [1916] 2 I.R. 299 (H.L.); *Davis Contractors Ltd.* v. *Fareham Urban District Council* [1956] A.C. 696 (H.L.); *L.C.C.* v. *Boot, supra*; *Prenn* v. *Simmonds* [1971] 1 W.L.R. 1381 (H.L.). But see " Parties' own dictionary," *infra*.

[10] *Davis Contractors'* case, *supra*.

[11] [1975] 2 Lloyd's Rep. 197, 209 (H.L.), Lords Hodson and Wilberforce agreeing; Lords Morris and Salmon dissenting, but on another point and not dealing with this issue. See further, *post*, pp. 279, 348.

[12] *Inglis* v. *Buttery, supra*; *Ambatielos* v. *Jurgens* [1923] A.C. 175, 185 (H.L.); and see also, *City & Westminster Properties* (1934) *Ltd.* v. *Mudd* [1959] Ch. 129; *Prenn* v. *Simmonds supra*. For earlier cases consistent with the *Mottram* case, see *Chitty on Contracts* (24th ed., 1977), Vol. 1, para. 715, note 85. For the Standard Form see *post*, p. 348.

[13] *Prenn* v. *Simmonds, supra, per* Lord Wilberforce at p. 1384, and see *supra*, note 9.

from those circumstances, which the person using them had in view." [14] Thus evidence can be given to identify persons and things referred to in the document and to explain the circumstances existing at the time of its making. [15] " In construing all instruments you must know what the facts were when the agreement was entered into." [16] " Such facts give very little help in the construction if the words of the deed [17] are clear, but they will help very much if the words are ambiguous." [18] They may assist in deciding what terms, if any, are to be implied. [19] But this principle does not permit evidence to be made available of negotiations before contract or of the parties' subjective intentions. [20]

FOREIGN AND TECHNICAL WORDS. The meaning of foreign and technical words can be proved. [21] If there is no dispute about the meaning of technical words the court can inform itself about them by any means that is reliable and ready to hand, such as an explanation by counsel or by reference to dictionaries or in cases of difficulty by calling in aid an assessor. [22]

CUSTOM AND USAGE. Evidence is admissible to show that words were used according to a special custom or usage attaching to the trade or locality applicable to the contract. [23] The custom must be strictly proved and must be notorious so that everybody in the trade or locality concerned enters into a contract with that custom as an implied term. [24] It must not be inconsistent with the express or necessarily implied terms of the contract, [25] must be reasonable and not against the law. [26]

PARTIES' OWN DICTIONARY. Where words in a contract are capable of more than one meaning but the parties have negotiated on the agreed basis that the words have one meaning only, it is permissible to look at the negotiations to find out " their own dictionary meaning." [26a]

---

[14] *Ibid.*

[15] *Shore* v. *Wilson* (1842) 9 Cl. & F. 355 (H.L.).

[16] *Per* Jessel M.R., *Cannon* v. *Villars* (1878) 8 Ch.D. 415, 419; *Hart* v. *Hart* (1881) 18 Ch.D. 670, 693. See also, *Bank of New Zealand* v. *Simpson* [1900] A.C. 182, 187 (P.C.); *A/s Tankexpress* v. *Compagnie Financiere Belges des Petroles S.A.* [1949] A.C. 76 (H.L.).

[17] The principle applies to all written instruments, *Shore* v. *Wilson, supra.*

[18] *Per* Lord Esher M.R., *Roe* v. *Siddons* (1888) 22 Q.B.D. 224, 233 (C.A.).

[19] See *British Movietonews Ltd.* v. *London and District Cinemas Ltd.* [1952] A.C. 166 (H.L.).

[20] See *supra,* note 13.

[21] *Shore* v. *Wilson, supra*; *R.* v. *Patents Appeal Tribunal, ex p. Baldwin & Francis Ltd.* [1959] 1 Q.B. 105 (C.A.); affd. *sub nom. Baldwin & Francis Ltd.* v. *Patents Appeal Tribunal* [1959] A.C. 663 (H.L.).

[22] *Baldwin & Francis Ltd.* v. *Patents Appeal Tribunal, supra,* at pp. 679, 691.

[23] See *Symonds* v. *Lloyd* (1859) 6 C.B.(N.S.) 691 (" Reduced brickwork " means brickwork 9 inches thick); *cf. Crowshaw* v. *Pritchard* (1899) H.B.C. (4th ed., 1914), Vol. 2, p. 274; also 16 T.L.R. 45. See *ante,* p. 12.

[24] *Nelson* v. *Dahl* (1879) 12 Ch.D. 568, 575 (C.A.); *Brown* v. *Inland Revenue Commissioners* [1965] A.C. 244, 258, 262, 266 (H.L.).

[25] *Ibid.*; *London Export Corporation Ltd.* v. *Jubilee Coffee Roasting Co. Ltd.* [1958] 1 W.L.R. 661, 675, 677 (C.A.).

[26] *Ibid.*; *Crowshaw* v. *Pritchard, supra*; *Three Rivers Trading Co.* v. *Gwinear and District Farmers* (1967) 111 S.J. 831 (C.A.), sale of " 400 tons (approx.) " of barley; custom that seller could deliver 10 per cent. less than figure stated held to be unreasonable.

[26a] *The Karen Oltmann* [1976] 1 Lloyd's Rep. 708, 712.

SUBSEQUENT CONDUCT OF PARTIES INADMISSIBLE. ". . . it is not legitimate to use as an aid in the construction of the contract anything which the parties said or did after it was made," [27] and this rule applies even where there is an ambiguity in the contract. [28] The rule applies particularly where a contract is arrived at in the course of correspondence. All the letters must be looked at in order to determine whether, and at what stage, the parties intended to make themselves contractually liable, but when that stage has been established subsequent correspondence cannot be taken into account in construing the contract. [29]

(ii) *To attack the contract*

Evidence is admissible to show an unfulfilled condition precedent, *e.g.* to show that an obligation was not to arise until an event, such as a third person's approval, had taken place and the event has not occurred either at all or within the time limited by the agreement [30]; or that a contract is void or voidable because of misrepresentation, fraud, mistake, illegality, infancy or made by a mentally disordered person [31]; or that the contract has been varied, rescinded, or is subject to an estoppel. [32] Evidence of rescission can be given even though the original contract was by deed and the rescission was oral, [33] or if the original contract was required to be evidenced in writing and has been orally rescinded. [34]

(iii) *To claim rectification* [35]

(iv) *To show collateral warranties*

Where it is alleged that the written contract does not contain the whole of the agreement entered into and that there is a collateral warranty (sometimes termed collateral contract) [36] evidence of such a warranty is admissible. [37] But so far as there is a collateral contract the sole effect of which is to vary or add to the terms of the principal contract, it is viewed with suspicion by the law and must be strictly proved, for any laxity on this point would enable parties to escape from the full performance of their obligations and lessen the authority of written contracts. [38]

---

[27] *Whitworth Street Estates (Manchester) Ltd.* v. *James Miller & Partners Ltd.* [1970] A.C. 583, *per* Lord Reid at p. 603 (H.L.).

[28] *Schuler A.G.* v. *Wickman Machine Tool Sales Ltd.* [1974] A.C. 235, 252, 265 (H.L.). In the *Schuler* case doubt was expressed as to the authority of *Watcham* v. *East African Protectorate* [1919] A.C. 533 (P.C.) which dealt with an ambiguous title to land.

[29] *Bushwall Properties Ltd.* v. *Vortex Properties Ltd.* [1976] 1 W.L.R. 591, 603 (C.A.).

[30] *Pym* v. *Campbell* (1856) 6 E. & B. 370, 374.

[31] *Norton on Deeds* (2nd ed.), p. 151; Chitty (24th ed., 1977), Vol. 1, para. 535.

[32] *Schuler* v. *Wickman*, *supra*, at p. 261. For variation and estoppel, see *post*, p. 119.

[33] *Berry* v. *Berry* [1929] 2 K.B. 316.

[34] *Morris* v. *Baron* [1918] A.C. 1 (H.L.).

[35] See *post*, p. 47.

[36] For meaning, see *post*, p. 100.

[37] *De Lassalle* v. *Guildford* [1901] 2 K.B. 215 (C.A.).

[38] *Heilbut Symons & Co.* v. *Buckleton* [1913] A.C. 30, 47 (H.L.); *Routledge* v. *McKay and Others* [1954] 1 W.L.R. 615 (C.A.); *Dick Bentley Ltd.* v. *Harold Smith (Motors) Ltd.* [1965] 1 W.L.R. 623 (C.A.); *Quickmaid Rental Services Ltd.* v. *Reece* (1970) 114 S.J. 372 (C.A.).

CONTRACT ONLY PARTLY IN WRITING. The reluctance of the courts to admit extrinsic evidence of collateral warranties or collateral contracts as just stated has been said to be out of date.[39] Whether this is so or not the position must be distinguished from that which arises where it cannot be said that the main contract is in writing but that there is a contract partly in writing and partly oral and/or by conduct.[40] In such a case the special rules about collateral warranties do not apply and the court construes the contract according to all its terms gathered from the documents, words and/or conduct comprising the contract.[41]

2. Rules of construction [42]

A rule of law takes effect although the parties may have expressed a contrary intention, but a rule of construction merely " points out what a court shall do in the absence of express or implied intention to the contrary." [43] It is therefore only applied to assist the court where there is some ambiguity or inconsistency, for if the words are plain the court gives effect to them.[44]

Ordinary meaning

The court in construing a document gives words any special technical, trade, or customary meaning which the parties must have intended the words to bear [45]; subject to this, " the grammatical and ordinary sense of the words is to be adhered to unless that would lead to some absurdity, or some repugnance or inconsistency with the rest of the instrument, in which case the grammatical and ordinary sense of the words may be modified so as to avoid that absurdity or inconsistency, but no further." [46] The " absurdity " in the last sentence is, apparently, limited to that which would arise if reading the document as a whole it is clear that the parties intended a special meaning to be given to a word and that to give it its ordinary meaning would therefore create an absurdity.[47]

Term of art

This is used to describe words or phrases which have acquired a precise legal meaning ordinarily applied by the courts, but " where a word or phrase which is

---

[39] *Evans & Sons, J. (Portsmouth) Ltd.* v. *Andrea Merzario Ltd.* [1976] 1 W.L.R. 1078, 1081 (C.A.), *per* Lord Denning.

[40] *Evans* v. *Merzario, supra.* The majority of the Court of Appeal did not seem to agree wholly with the approach of Lord Denning and dealt with the case as a contract partly in writing, partly oral and partly by conduct.

[41] *Ibid.*

[42] Sometimes termed " canons of construction."

[43] *Re Coward* (1887) 57 L.T. 285, 291 (C.A.).

[44] *Leader* v. *Duffey* (1888) 13 App.Cas. 294, 301 (H.L.); *James Archdale & Co.* v. *Comservices* [1954] 1 W.L.R. 459, 463 (C.A.).

[45] *Produce Brokers Co. Ltd.* v. *Olympia Oil & Cake Co. Ltd.* [1916] 1 A.C. 314, 324 (H.L.). See also " Parties' own dictionary," *supra.* See further, *ante,* p. 29, as to custom and usage.

[46] *Per* Lord Wensleydale, *Grey* v. *Pearson* (1857) 6 H.L.Cas. 61, 106 (H.L.); *cf. per* Lord Blackburn, *Caledonian Ry.* v. *North British Ry.* (1881) 6 App.Cas. 114, 131 (H.L.).

[47] *Rhodes* v. *Rhodes* (1882) 7 App.Cas. 192, 205 (H.L.). See also *Sir Lindsay Parkinson & Co. Ltd.* v. *Commissioners of Works* [1949] 2 K.B. 632, 662 (C.A.), but *cf. British Movietonews Ltd.* v. *London and District Cinemas Ltd.* [1952] A.C. 166, 183 (H.L.).

a ' term of art ' is used by an author who is not a lawyer, particularly in a document which he does not anticipate may have to be construed by a lawyer, he may have meant by it something different from its meaning when used by a lawyer as a term of art." [48]

Reasonable meaning

" When the terms of a contract are ambiguous and one construction would lead to an unreasonable result, the court will be unwilling to adopt that construction." [49] If there is no ambiguity the court will enforce the terms however unreasonable they may be,[50] and there are numerous examples of contractors being held to the harsh terms of plainly worded contracts.[51]

Valid meaning

" Where a clause is ambiguous a construction which will make it valid is to be preferred to one which will make it void." [52] This principle is sometimes expressed by the maxim *verba ita sunt intelligenda ut res magis valeat quam pereat*, which has particular reference to documents prepared by businessmen who " often record the most important agreements in crude and summary fashion." [53]

Clerical errors corrected

Where there is a manifest error in a document, the court will put a sensible meaning on it by correcting or reading the error as corrected.[54]

Contract read as a whole

" The contract must be construed as a whole, effect being given, so far as practicable, to each of its provisions." [55] Construing a contract may involve two stages; first, the court may have to determine which documents are contractual [56]; secondly, having decided which documents form part of the contract it must give effect to all the terms and endeavour to reconcile inconsistencies by the rules of construction. In doing this it will be assisted if the parties have

---

[48] *Sydall* v. *Castings Ltd.* [1967] 1 Q.B. 302, 314 (C.A.), *per* Diplock L.J. In *L.C.C.* v *Boot* (*Henry*) & *Sons Ltd.* [1959] 1 W.L.R. 1069, 1075 (H.L.) it was said that, in construing a contract, no help could be got from the then Working rule agreement, which was "not an artistically drawn document."

[49] *Per* Lord Esher M.R., *Dodd* v. *Churton* [1897] 1 Q.B. 562, 566 (C.A.); *Schuler A.G.* v. *Wickman Machine Tool Sales Ltd.* [1974] A.C. 235, 251, 256, 265, 272 (H.L.).

[50] *Jones* v. *St. John's College, Oxford* (1870) L.R. 6 Q.B. 115, distinguished in *Dodd* v. *Churton, supra.*

[51] *e.g. Jackson* v. *Eastbourne Local Board* (1886) H.B.C. (4th ed., 1914), Vol. 2, p. 88 (H.L.), and see further Chaps. 4 and 5 *post.*

[52] *Per* Kay L.J., *Mills* v. *Dunham* [1891] 1 Ch. 576, 590 (C.A.).

[53] *Per* Lord Wright, *Hillas* v. *Arcos* (1932) 147 L.T. 503, 514 (H.L.); *Adamastos Shipping Co. Ltd.* v. *Anglo-Saxon Petroleum Co. Ltd.* [1959] A.C. 133, 161 (H.L.); *Nea Agrex S.A.* v. *Baltic Shipping Co. Ltd.* [1976] 2 W.L.R. 925, 934 (C.A.); for a building contract case, see *Gold* v. *Patman* & *Fotheringham Ltd.* [1958] 1 W.L.R. 697, 702 (C.A.); *cf. ante*, p. 19, " Certainty of terms."

[54] *Burchell* v. *Clark* (1876) 11 C.P.D. 88, 97. For an example, see *Townsends* (*Builders*) *Ltd.* v. *Cinema News, etc., Ltd.* [1959] 1 W.L.R. 119, 122 (C.A.).

[55] *Per* Lord Atkinson, *Brodie* v. *Cardiff Corp.* [1919] A.C. 337, 355 (H.L.). See also *supra*, " Ordinary meaning."

[56] *e.g.* whether quantities are part of the contract. See Chap. 5. See also, as to offer and acceptance, Chap. 2.

expressly [57] or impliedly [58] indicated that certain clauses or documents are to prevail in the event of an inconsistency.

Written words prevail

Where there is a contract contained in a printed form with clauses inserted or filled in inconsistent with printed words, " the written words are entitled to have a greater effect attributed to them than the printed words, inasmuch as the written words were the immediate language and terms selected by the parties themselves for the expression of their meaning." [59] If printed words by an express clause state that the printed form is to prevail over written words a difficult question of construction arises,[60] but it seems that if the clause is sufficiently clear the printed form will prevail.[61]

Ejusdem generis rule

This rule is that where there are words of a particular class followed by general words, the general words are treated as referring to matters of the same class. Thus a ship was exempted from liability for non-delivery of a cargo if the port was unsafe " in consequence of war, disturbance or any other cause." It was held that danger from ice was not within the meaning of " any other cause " which must be limited to causes similar to war or disturbance.[62] In a clause permitting an extension of time to be granted to the contractor, if the works were " delayed by reason of any alteration or addition, . . . or in case of combination of workmen, or strikes, or by default of the sub-contractors . . . or other causes beyond the contractor's control," the " other causes " were limited to those *ejusdem generis* with the causes particularised, and did not therefore include the employer's own default in failing to give possession of the site.[63] Where the words " et cetera " were inserted between words describing a particular class and general words it was held that their meaning was too vague to prevent the operation of the rule.[64]

" The tendency of the more modern authorities is to attenuate the application of the rule of *ejusdem generis*," [65] and it does not apply if the parties show that

---

[57] *e.g. Brodie* v. *Cardiff Corp., supra,* at p. 344; and see Standard form of building contract, clauses 1 and 12.

[58] As where there are written and printed parts of the contract—see *infra.*

[59] Lord Ellenborough in *Robertson* v. *French* (1803) 4 East 130; *Glynn* v. *Margetson* [1893] A.C. 351 (H.L.); *Adamastos Shipping Co. Ltd.* v. *Anglo-Saxon Petroleum Co. Ltd.* [1959] A.C. 133 (H.L.); *Renton (G. H.) & Co. Ltd.* v. *Palmyra Trading Corp.* [1957] A.C. 149 (H.L.); *The Brabant* [1967] 1 Q.B. 588; *A. Davies & Co. (Shopfitters)* v. *William Old* (1969) 113 S.J. 262 (C.A.). See also *Neuchatel Asphalte Co. Ltd.* v. *Barnett* [1957] 1 W.L.R. 356 (C.A.).

[60] *English Industrial Estates Corporation* v. *George Wimpey & Co. Ltd.* [1973] 1 Lloyd's Rep. 118 (C.A.); see, further, *post,* p. 320.

[61] *Gold* v. *Patman & Fotheringham Ltd.* [1958] 1 W.L.R. 697, 701 (C.A.); *N.W. Metropolitan Regional Hospital Board* v. *T. A. Bickerton & Son Ltd.* [1970] 1 W.L.R. 607, 617 (H.L.); *English Industrial Estates* v. *Wimpey, supra.*

[62] *S. S. Knutsford Ltd.* v. *Tillmans & Co.* [1908] A.C. 406.

[63] *Wells* v. *Army & Navy Co-op. Society* (1902) H.B.C. (4th ed., 1914), Vol. 2, pp. 353 357 (C.A.). See, further, *post,* p. 160.

[64] *Herman* v. *Morris* (1919) 35 T.L.R. 574 (C.A.).

[65] *Allen* v. *Emmerson* [1944] K.B. 362, 367 (D.C.). Note that this decision dealt with the construction of a statute.

they intended its exclusion.[66] " Where you find the word ' whatsoever ' follow-ing . . . upon certain substantives, it is often intended to repel . . . the implication of the so-called doctrine of *ejusdem generis*." [67] Thus where there was a clause somewhat similar to those cited above save that the general words " of what kind soever " were used the rule did not apply, and these words were given their ordinary unrestricted meaning.[68]

Recitals

A recital is the part of a document, usually beginning " whereas," which indicates what the parties are desirous of effecting by their contract. " If there is any doubt about the construction of that document the recital may be looked at in order to determine what is the true construction; but if there is no doubt about the construction the rights of the parties are governed entirely by the operative part of the writing or deed." [69]

The contra proferentem rule

If there is an ambiguity in a document which all the other methods of con-struction have failed to resolve so that there are two alternative meanings to certain words the court may construe the words against the party seeking to rely on them and give effect to the meaning more favourable to the other party.[70] Where there was a clause enabling the architect to extend the time for completion and the employer sought to rely on the clause to enable him to claim liquidated damages,[71] it was held for various reasons that the employer could not rely on it. It was said that in case those reasons were wrong, then in any event the employer could not rely on the clause for it was ambiguous and would therefore be given the construction favourable to the contractor,[72] and more recently it has been said " The liquidated damages and extension of time clauses in printed forms of contract must be construed strictly *contra proferentem*." [73] But in each of the cases just referred to the form of contract was, apparently, one put forward by the employer. It seems that the rule has little, if any, application where there is a common form document prepared and revised jointly by bodies which include representatives of employers and contractors.[74]

---

[66] *Jersey (Earl of)* v. *Guardians of Neath* (1889) 22 Q.B.D. 555 (C.A.).
[67] Per Fry L.J., *Jersey (Earl of)* v. *Guardians of Neath, supra*, at p. 566.
[68] *Larsen* v. *Sylvester* [1908] A.C. 295 (H.L.).
[69] Per Brett L.J., *Leggott* v. *Barrett* (1880) 15 Ch.D. 306, 311 (C.A.). For misstatement of facts in recitals in deeds, see *post*, p. 43.
[70] *Lee & Son* v. *Railway Executive* (1949) 65 T.L.R. 604, 605–606 (C.A.); *Billyack* v. *Leyland Construction Co. Ltd.* [1968] 1 W.L.R. 471, 477. See also *Adams* v. *Richardson & Starling Ltd.* [1969] 1 W.L.R. 1645 (C.A.), discussed, *post*, p. 129.
[71] See Chap. 9 *post*.
[72] *Miller* v. *L.C.C.* (1934) 50 T.L.R. 479, 482, discussed further, *post*, p. 161.
[73] *Peak Construction (Liverpool) Ltd.* v. *McKinney Foundations Ltd.* (1971) 69 L.G.R. 1, 11 (C.A.), *per* Salmon L.J.
[74] *Tersons Ltd.* v. *Stevenage Dev. Corporation* [1963] 2 Lloyd's Rep. 333, 368 (C.A.) and consider the approach to demurrage clauses—*Suisse Atlantique* v. *N.V. Rotterdamsche, etc.* [1967] A.C. 361 (H.L.). The Standard form of building contract and the I.C.E. conditions are documents of the kind last referred to in the text.

Irreconcilable clauses

The court always endeavours to resolve an apparent inconsistency,[75] but if two clauses cannot be reconciled it will give effect to that which states the intention of the parties. Finally, if it is unable otherwise to ascertain which clause should prevail, it will give effect to the earlier clause and reject the later.[76]

3. Alterations

Alterations made before a contract is signed are binding. For the effect of deletions in a printed form, see *ante*, p. 28. It is presumed that alterations apparent on a document were made before execution.[77] Alterations or erasures made after execution by one party without the consent of the other are of no effect.[78] If both parties alter a document after execution by agreement this is a variation of the contract.

4. Implied terms

It has been said that there are three different senses in which the expression " implied term " is used.[79] The first is a term which does not depend on the actual intention of the parties but on a rule of law such as the implied terms in a contract for the sale of goods.[80] The second is where the law in some circumstances holds that a contract is dissolved if there is a vital change of conditions.[81] The third is where a term is sought to be implied based on an intention imputed to the parties from their actual circumstances. The third sense is discussed here.

There are varieties of implication which the courts make in seeking to establish the imputed or presumed intention of the parties,[82] but two broad areas or categories may be discerned.[83] The first appears where the parties have drawn up a detailed contract but it is necessary to insert a term to make it work. This

---

[75] *Bush* v. *Watkins* (1851) 14 Beav. 425, 432.

[76] *Forbes* v. *Git* [1922] 1 A.C. 256, 259 (P.C.). For meaningless clauses, see *ante*, p. 19, " Certainty of terms."

[77] *Doe* d. *Tatum* v. *Catomore* (1851) 16 Q.B. 745.

[78] See *Pattinson* v. *Luckley* (1875) L.R. 10 Ex. 330.

[79] *Per* Lord Wright, *Luxor (Eastbourne) Ltd.* v. *Cooper* [1941] A.C. 108, 137 (H.L.); *cf. Sterling Engineering Co. Ltd.* v. *Patchett* [1955] A.C. 534, 547 (H.L.). See also *Tsakiroglou & Co. Ltd.* v. *Noblee Thorl G.m.b.H.* [1962] A.C. 93, 122 (H.L.); *Young and Marten* v. *McManus Childs* [1969] 1 A.C. 454, 465, H.L.; *Gloucestershire County Council* v. *Richardson* [1969] 1 A.C. 480, 503, H.L. For another use of the term see *ante*, p. 29, " Custom and Usage." For an interesting discussion of terminology, see Halsbury (4th ed.), Vol. 9, paras. 351 *et seq.* For a recent classification by Lord Denning, see *Shell U.K. Ltd.* v. *Lostock Garage Ltd.* [1976] 1 W.L.R. 1187, 1195 (C.A.).

[80] See Sale of Goods Act 1893, ss. 12–14; Supply of Goods (Implied Terms) Act 1973, ss. 1–3. A building contract is not a contract for the sale of goods, but see *infra*, " Implication of ' usual ' terms."

[81] This approach to the judicial basis of frustration is, it seems, now in disfavour, see *post*, p. 120, " Frustration generally," but it is convenient to retain this classification of meaning given to the expression " implied term " as there are numerous reported cases, including decisions of the House of Lords, where the words are used in this second sense.

[82] *Liverpool City Council* v. *Irwin* [1977] A.C. 239, 253 (H.L.). Lord Wilberforce refers to a " continuous spectrum."

[83] *Ibid.* p. 255, 257. See also *Greaves & Co. Ltd.* v. *Baynham Meikle* [1975] 1 W.L.R. 1095, 1099, 1103 (C.A.).

is sometimes termed " *The Moorcock*[84] approach." The second is where in all contracts of a certain type, such as building contracts, the law implies certain usual terms unless the parties have shown an intention to exclude or modify them.[85] In each category the implication must be necessary and the term must be reasonable, but terms which are not necessary will not be implied merely because the court regards them as reasonable.[86] It seems that in the first category it is for the party putting forward the term to establish its existence, and in the second for the party denying the term to establish its absence.

Implication to make contract work

The court does not make or improve contracts. Its " function is to interpret and apply the contract which the parties have made for themselves. If the express terms are perfectly clear and free from ambiguity, there is no choice to be made between different possible meanings; the clear terms must be applied even if the court thinks some other terms would have been more suitable. An unexpressed term can be implied if and only if the court finds that the parties must have intended that term to form part of their contract; it is not enough for the court to find that such a term would have been adopted by the parties as reasonable even if it had been suggested to them; it must have been a term that went without saying, a term *necessary* to give business efficacy to the contract, a term which, though tacit, formed part of the contract which the parties made for themselves."[87] The implication may be necessary because " language is imperfect and there may be, as it were, obvious interstices in what is expressed which have to be filled up."[88]

A term sought to be implied under this heading cannot be what the parties must have intended and therefore is not implied if it provides only one of several possible solutions to the matter in question.[89]

In *Trollope & Colls Ltd.* v. *North West Metropolitan Regional Hospital Board*,[90] the contract provided for the work to be carried out in phases.[91] Phase III was to commence six months after the issue of the certificate of practical completion of phase I and was to be completed on April 30, 1972. There was delay in completing phase I, so that the period for completion of phase III which would have been 30 months if there had been no such delay became 16 months. There

---

[84] (1889) 14 P.D. 64 (C.A.).
[85] See *supra*, note 83.
[86] *Liverpool City Council* v. *Irwin*, *supra*, at p. 262.
[87] *Trollope & Colls Ltd.* v. *North West Metropolitan Regional Hospital Board* [1973] 1 W.L.R. 601 (H.L.), *per* Lord Pearson, p. 609. See also *The Moorcock* (1889) *supra*; *Reigate* v. *Union Manufacturing Co. Ltd.* [1918] 1 K.B. 592, 605 (C.A.). See *Shirlaw* v. *Southern Foundries* (1926) *Ltd.* [1939] 2 K.B. 206, 227 (C.A.) for the test of what answer the parties would give to the " officious bystander."
[88] *Luxor (Eastbourne) Ltd.* v. *Cooper* [1941] A.C. 108, 137 (H.L.).
[89] *Trollope & Colls Ltd.* v. *North West Metropolitan Regional Hospital Board*, *supra*, at pp. 610, 614. See also *Brown & Davis Ltd.* v. *Galbraith* [1972] 1 W.L.R. 997 (C.A.).
[90] *Supra*.
[91] Such contracts seem to give rise to problems of construction; for the position where the Standard form of building contract is used, see *post*, p. 330. In the *Trollope & Colls* case there was one contract but three sets of conditions based on the Standard form of building contract.

was express provision for extension of time for the completion of phase I but
not for the extension of the completion date for phase III if phase I was delayed.
The House of Lords refused to imply a term extending the time for completion
of phase III by a period equal to the extension of time for completion of phase I
properly allowable in respect of phase I. The express terms were clear so there
was no room for implication and the parties must be taken to have accepted the
risks involved; in any event there were four or five different ways in which the
question of extension of time in the event of delay in completion of phase I might
have been dealt with.

Implication of " usual " terms

Where there is a comprehensive written contract such as the Standard form
of building contract there may be very little room for the implication of any
terms, for if the parties have dealt expressly with a matter in the contract, no
term dealing with the same matter can be implied.[92] But where there is no express
contract, or its terms do not deal with the matters about to be mentioned, certain
terms are usually implied. Such terms are usually referred to as " warranties."

(a) *By employer*

Co-operation. " Generally speaking, where B is employed by A to do a piece
of work which requires A's co-operation . . . it is implied that the necessary
co-operation will be forthcoming." [93] The employer impliedly agrees to do all
that is necessary on his part to bring about completion of the contract.[94] For
example, he must give possession of the site within a reasonable time [95]; if
instructions, nominations, information, plans or details are required, supply them
at reasonable times [96]; if an architect is to supervise the work he must appoint an
architect,[97] and, if to his knowledge, the architect persists in applying the contract
wrongly in regard to those matters where the architect must act fairly between
the parties, he must dismiss him and appoint another.[98]

Not to prevent completion. " In general . . . a term is necessarily implied in any
contract, the other terms of which do not repel the implication, that neither

---

[92] See, *e.g. Lynch* v. *Thorne* [1956] 1 W.L.R. 303 (C.A.); *Jones* v. *St. John's College,
Oxford* (1870) L.R. 6 Q.B. 115, 126. The principle is sometimes expressed *expressio unius
est exclusio alterius*. The principle cannot be relied on to imply warranties as to matters
where warranties are not expressly excluded, *Thorn* v. *London Corp.* (1876) 1 App.Cas.
120, 131 (H.L.). See generally Hals. (4th ed.), Vol. 12, para. 1476.

[93] Per Lord Simon, *Luxor* v. *Cooper, supra,* at p. 118. See article by L. F. Burrows (1968)
31 M.L.R. 390.

[94] *Mackay* v. *Dick* (1881) 6 App.Cas. 251, 263 (H.L.); *cf. Hamlyn* v. *Wood* [1891] 2
Q.B. 488 (C.A.).

[95] *Freeman* v. *Hensler* (1900) H.B.C. (4th ed.), Vol. 2, p. 292 (C.A.).

[96] *Roberts* v. *Bury Commissioners* (1870) L.R. 5 C.P. 310, 325; *McAlpine* v. *Lanarkshire,
etc., Ry.* (1889) 17 R. 113; *Wells* v. *Army & Navy Co-op. Society* (1902) H.B.C. (4th ed.),
Vol. 2, pp. 346, 352 (C.A.); *Neodox Ltd.* v. *Swinton and Pendlebury B.C.* (1958) unrep. save
in H.B.C. (10th ed.), p. 137. For claims for damages by the contractor arising out of breach
of these duties, see *post,* p. 152 and *post,* p. 260.

[97] *Hunt* v. *Bishop* (1853) 8 Ex. 675.

[98] *Panamena Europa Navegacion* v. *Frederick Leyland & Co. Ltd.* (*J. Russell & Co.*)
[1947] A.C. 428, 436 (H.L.), discussed *post,* p. 78.

party shall prevent the other from performing it." [99] A term of this nature cannot be implied if it is illegal,[1] or contrary to an express term of the contract.[2] The particular implied term relied on should be expressly pleaded, and " except possibly in the rare cases where the wrongful act alleged is independent of the contract " it is circumlocution to add a general allegation of prevention.[3] The employer does not impliedly warrant the fitness of the site,[4] nor, it seems, that there will be no wrongful interference by third parties.[5] The employer is not, in the absence of fraud or collusion, responsible for delay caused by a sub-contractor, even though under the terms of the contract the sub-contractor was nominated by the employer or his architect without consultation with the contractor.[6] The employer is liable for delay caused by his servants and agents.[7] This liability extends, it is submitted, to delay on the part of the architect or the engineer under a typical building or engineering contract.[8] There is, it is submitted, an implied term that the employer will not so act as to disqualify the architect.[9]

Unjustified interference by the employer in the supply of goods necessary for the contract is a breach of the implied term, notwithstanding that the supplier has no contract direct with the contractor.[10]

(b) *By contractor*

Workmanship. The contractor must do the work with all proper skill and care.[11] In deciding what degree of skill is required the court will, it is submitted, consider all the circumstances of the contract including the degree of skill expressly or impliedly professed by the contractor.[12] Breach of the duty includes the use of materials containing patent defects, even though the source of such materials has been chosen by the employer.[13]

---

[99] *Per* Lord Asquith, *Cory Ltd.* v. *City of London Corp.* [1951] 2 K.B. 476, 484; see *Lawson* v. *Wallasey Local Board* (1883) 48 L.T. 507 (C.A.); *Bywaters* v. *Curnick* (1906) H.B.C. (4th ed.), Vol. 2, p. 393 (C.A.).

[1] *Cory Ltd.* v. *City of London Corp., supra.*

[2] *Farr* v. *The Admiralty* [1953] 1 W.L.R. 965.

[3] *Per* Devlin J., *Mona, etc., Ltd.* v. *Rhodesia Rys. Ltd.* [1949] 2 All E.R. 1014, 1016; *Luxor* v. *Cooper, supra,* at p. 149; *cf.* Lord Atkin, *Southern Foundries* v. *Shirlaw* [1940] A.C. 701, 717 (H.L.).

[4] *Appleby* v. *Myers* (1867) L.R. 2 C.P. 651, and see *post,* pp. 63, 101.

[5] *Porter* v. *Tottenham U.D.C.* [1915] 1 K.B. 776 (C.A.).

[6] *Leslie & Co. Ltd.* v. *The Managers of the Metropolitan Asylums District* (1901) 68 J.P. 86 (C.A.); *Kirk & Kirk* v. *Croydon Corporation* [1956] J.P.L. 585.

[7] *Russell* v. *Sa da Bandeira* (1862) 13 C.B.(N.S.) 149.

[8] See *Sutcliffe* v. *Thackrah* [1974] 2 A.C. 727 (H.L.). *Re de Morgan Snell & Co.* (1891) H.B.C. (4th ed.), Vol. 2, p. 185 (C.A.) so far as it is to contrary effect should no longer be considered as of general application.

[9] See *Sutcliffe* v. *Thackrah* (*supra*). For the meaning of disqualification of a certifier see *post,* p. 85. For whether acts by an employer in breach of the implied term can be treated as repudiation see *post,* p. 118.

[10] *Acrow* (*Automation*) *Ltd.* v. *Rex Chainbelt Inc.* [1971] 1 W.L.R. 1676, 1680 (C.A.).

[11] *Young and Marten Ltd.* v. *McManus Childs Ltd.* [1969] 1 A.C. 454, 465 (H.L.); *Charnock* v. *Liverpool Corporation* [1968] 1 W.L.R. 1498 (C.A.); *Duncan* v. *Blundell* (1820) 3 Stark. 6; *Pearce* v. *Tucker* (1862) 3 F. & F. 136. The duty is sometimes expressed as one to do the work in a good and workmanlike manner. For compliance with by-laws (and Building Regulations), see *post,* pp. 43, 72, 108.

[12] *Duncan* v. *Blundell, supra,* at p. 7; *Harmer* v. *Cornelius* (1858) 5 C.B.(N.S.) 236; *Young and Marten Ltd.* v. *McManus Childs Ltd., supra,* at pp. 468, 472.

[13] *Young and Marten's* case, *supra,* pp. 469, 470.

Materials. The law as to implied warranties [14] was considered fully by the House of Lords in the cases of *Young and Marten Ltd.* v. *McManus Childs Ltd.*[15] and *Gloucestershire County Council* v. *Richardson.*[16]

(i) *Warranties of fitness for purpose and good quality.* If the contractor is to supply materials he warrants that the materials he will use are: (1) reasonably fit for the purpose for which they will be used; (2) of good quality, unless the express terms of the contract and any admissible surrounding circumstances show that the parties intended to exclude either or both warranties.[17] The two warranties correspond substantially with the warranties formerly [18] implied by section 14 of the Sale of Goods Act 1893 upon a contract for the sale of goods,[19] so that where there is no limitation upon the contractor's right of recourse against his supplier he will in the case of defective materials [20] or goods have his remedy over against his supplier who in turn will have his remedy over and so on down the chain of contracts until the manufacturer or other author of the defect is reached.[21]

The effect of the second warranty is, in the ordinary case, to make the contractor liable for latent defects even though the employer may have chosen the materials or nominated the supplier and there has been no lack of care and skill on the part of the contractor.[22] Thus in *Young and Marten Ltd.* v. *McManus Childs Ltd.*, *supra,* the agent of the employer,[23] a highly skilled and experienced person, relied on his own skill and judgment in the choice of " Somerset 13 " tiles to be fixed by the contractor. After fixing and exposure to weather, defects appeared which required the replacement of the tiles. It was held that although the warranty of fitness for purpose was excluded the warranty of quality was not, and the contractor was liable. Lord Pearce stated by way of illustration of the general principle: " It is frequent for builders to fit baths, sanitary equipment, central heating and the like, encouraging their clients to choose from the wholesaler's display rooms which they prefer. It would, I think, surprise the average householder if it were suggested that simply by exercising a choice he had lost

---

[14] Used here in the sense of " implied terms."

[15] *Supra.*

[16] [1969] 1 A.C. 480 (H.L.).

[17] *Young and Marten's case, supra; Gloucestershire C.C.* case, *supra. Myers* v. *Brent Cross Service Co.* [1934] 1 K.B. 46 (D.C.), explnd. in *Young and Marten's* case at pp. 467, 471, 474, 478; *Samuels* v. *Davies* [1943] 1 K.B. 526 (C.A.); *Stewart* v. *Reavell's Garage* [1952] 2 Q.B. 545; *Ingham* v. *Emes* [1955] 2 Q.B. 366 (C.A.).

[18] The warranties have been amended by the Supply of Goods (Implied Terms) Act 1973 which came into operation on May 18, 1973.

[19] *Ibid.* noting that Lord Upjohn's view, at p. 475, was that the obligation upon the contractor in respect of materials was higher than that upon the seller of goods.

[20] Materials are " goods " for the purposes of the Sale of Goods Act 1893 and the Supply of Goods (Implied Terms) Act 1973.

[21] *Young and Marten's* case, *supra,* at p. 466; *Gloucestershire C.C.* case, *supra.*

[22] It makes him ". . . an insurer in respect of the defects which he could not prevent and of which he did not know ": *per* Lord Wilberforce, *Young and Marten Ltd.* v. *McManus Childs Ltd.*, *supra,* at p. 479.

[23] In fact the main contractor contracting with his sub-contractor, but the principle is the same as in the case of employer and contractor, and for convenience the usual terminology is used.

all right of recourse in respect of the quality of the fittings against the builder who normally has a better knowledge of these matters." [24]

(ii) *Exclusion of warranties.* This depends upon the contract and surrounding circumstances,[25] and, so far as it is a question of surrounding circumstances, it is a question of fact and degree,[26] but some principles can be stated.

The warranty of fitness for purpose is excluded if in the selection of the materials in question the employer placed no reliance on the contractor's skill and judgment.[27] The facts of *Young and Marten Ltd.* v. *McManus Childs, supra,* illustrate the point. The warranty is excluded where the employer fails to disclose some relevant abnormal circumstance relating to the use of the material and thus does not make known to the contractor the particular purpose for which the materials are required.[28] The warranty is excluded if it is inconsistent with express terms [29] or if the contractor contracts on the basis of a disclaimer of responsibility.[30]

It seems that the court will readily infer an intention to exclude the warranty of quality if the employer chooses materials which he and the contractor know can only be obtained from a supplier upon terms which remove or substantially limit the contractor's right of recourse against his supplier in respect of defects of quality in such materials.[31] In *Gloucestershire County Council* v. *Richardson, supra,* the contractor was directed to enter into a contract for the supply of concrete columns at a price and upon terms which had been fixed by the employer. One of the terms limited liability for defective goods to free replacement and excluded liability for consequential loss. It was held [32] that this showed an intention that the contractor should not be liable for latent defects due to bad manufacture. Other circumstances which influenced the House were that the design, materials, specification, quality and price of the columns were fixed without reference to the contractor,[33] and that under the form of contract used [34] there was no right to object to the nomination or to insist on an indemnity.[35]

As to repairing contracts it has been said: ". . . less cogent circumstances may

---

[24] p. 470.

[25] See *supra,* note 17.

[26] *Young and Marten's* case, *supra,* at p. 471.

[27] See cases cited in note 17, *supra; Cammell Laird & Co. Ltd.* v. *Manganese Bronze and Brass Co. Ltd.* [1934] A.C. 402 (H.L.), a case on the Sale of Goods Act 1893, s. 14, but the principle is the same; *Young and Marten's* case, *supra,* at pp. 472, 479. Some assistance may be got from *Bower* v. *Chapel-en-le-Frith R.D.C.* (1910) 75 J.P. 122; (1911) 9 L.G.R. 661—leaking bricks; *Adcock's Trustee* v. *Bridge R.D.C.* (1911) 75 J.P. 241—unsatisfactory windmill.

[28] *Ingham* v. *Emes* [1955] 2 Q.B. 366 (C.A.).

[29] *Young and Marten's* case, *supra,* at p. 471; *Gloucestershire County Council* case, *supra,* at p. 495. See also *Lynch* v. *Thorne* [1956] 1 W.L.R. 303 (C.A.).

[30] *Young and Marten's* case, *supra,* at p. 471.

[31] *Young and Marten's* case, *supra,* at pp. 466, 471; *Gloucestershire C.C.* case, *supra* at pp. 497, 503, 504, 507.

[32] By a majority, Lord Pearson dissenting.

[33] See the speeches of Lord Pearce and Lord Wilberforce.

[34] R.I.B.A. Form, 1939 ed., printed in the 1st and 2nd editions of this book.

[35] As there was in the case of nominated sub-contractors under cl. 21 of the contract—see Lord Pearce, p. 495, and Lord Wilberforce, p. 507. Lord Upjohn at pp. 502, 503 expressed a different view.

be sufficient to exclude an implied warranty of quality where the use of spare parts is only incidental to what is in essence a repairing operation where the customer's main reliance is on the skill of the tradesman, than in a case where the main element is the supply of an article, the installation being merely incidental." [36]

Fitness for purpose of completed works. Where the employer makes known to the contractor the particular purpose for which the work is to be done and the work is of a kind which the contractor holds himself out as performing, and the circumstances show that the employer relied on the contractor's skill and judgment in the matter, there is an implied warranty that the work as completed will be reasonably fit for the particular purpose. [37] The contractor's duty includes, it is submitted, the performance of any works of design necessary to complete the work to the requisite standard. [38] The warranty may apply only to a part or parts of the works. [39] The warranty is excluded in the same circumstances as those set out above relating to materials alone. [40] It follows that in a building contract where the contractor is bound to complete according to detailed plans and specification or bills of quantities the room for the operation of such an implied warranty is small. [41] But where a contractor had expressly warranted the fitness of the works, it was held that such warranty overrode the duty to comply with the detailed specification [42]; and where a contractor undertook to build a house in accordance with plans supplied by the employer's architect, but there was no supervision by the architect and the employer relied upon the experience, judgment and skill of the contractor to supervise the construction, it was held that the contractor was liable to the employer for failing to warn him of obvious defects in the plans which resulted in defects. [43]

Specialist contractors. The implication of fitness more readily arises where a contractor is employed because of a specialist skill he professes. [44]

Sale of buildings. Where there is a contract for the sale of a house to be erected, or in the course of erection, there is, subject to the express terms of the contract,

---

[36] *Per* Lord Reid, *Young and Marten Ltd.* v. *McManus Childs Ltd.*, *supra*, at p. 468, commenting on dictum of du Parcq J., *Myers* v. *Brent Cross* [1934] 1 K.B. 46, 55 (D.C.). See also Lord Wilberforce, *Young and Marten's* case, *supra*, at p. 476.

[37] *Greaves & Co. Ltd.* v. *Baynham Meikle* [1975] 1 W.L.R. 1095, 1098 (C.A.); see also cases cited in note 27, *supra*; *Smith & Snipes Hall Farm Ltd.* v. *River Douglas Catchment Board* [1949] 2 K.B. 500 (C.A.) and *Cable (1956) Ltd.* v. *Hutcherson Ltd.* (1969) 43 A.L.J.R. 321, discussed *ante*, pp. 6, 7.

[38] See note 37, *supra*.

[39] *Cammell Laird & Co. Ltd.* v. *Manganese Bronze & Brass Co. Ltd.* [1934] A.C. 402 (H.L.).

[40] *Supra*, see note 37.

[41] See *Lynch* v. *Thorne* [1956] 1 W.L.R. 303, 311 (C.A.). Note comment on the application of this principle to this case, *infra*, p. 42.

[42] *Steel Co. of Canada Ltd.* v. *Willand Management Ltd.* [1966] S.C.R. 746, Supreme Court of Canada.

[43] *Brunswick Construction* v. *Nowlan* (1974) 49 D.L.R. (3d), Supreme Court of Canada. Upon the particular facts, the dissenting judgment of Dickson J. is attractive.

[44] See cases cited in note 27, *supra*. See also *Adams* v. *Richardson and Starling Ltd.* [1969] 1 W.L.R. 1645 (C.A.)—specialists in eradication of dry rot. See also: *ante*, p. 5, for design and build contracts, *post*, p. 181, for nominated sub-contractors, *post*, p. 212, for the analogous position of the professional man.

". . . a threefold implication: that the builder will do his work in a good and work-manlike manner; that he will supply good and proper materials; and that it will be reasonably fit for human habitation." [45] The implication extends to the whole house including the parts, if any, built before the contract of sale was made, [46] but it does not arise if the contractor merely sells a house which he has already completed. [47]

Since January 1, 1974, the contractor may become liable for breach of statutory duty in respect of dwellings provided to the order of a person, to that person and to others who acquire an interest in the dwelling. [48] He has a statutory duty to visitors under the Occupiers' Liability Act 1957. [49] He has a liability in tort to persons who suffer injury or loss by his negligence. [50]

Effect of express terms. ". . . if a builder has done his work badly, and defects afterwards appear, he is not to be excused from liability except by clear words." [51] So where a house was sold and defects appeared it was held that a clause providing for the making good of defects discovered within six months did not deprive the purchaser of damages in respect of defects discovered after six months. [51] But in *Lynch* v. *Thorne* [52] a builder agreed to sell a house to be completed according to a detailed specification which provided for nine-inch solid brick walls with no rendering. One of the walls when completed permitted the entry of driving rain which made a room uninhabitable. It was held that there was no implied warranty of fitness with regard to the wall because it had been constructed exactly in accordance with the specification and to have built it with cavities or with rendering so as to make it waterproof would not have been a compliance with the contract. Accordingly the purchaser failed in his claim against the builder in respect of the wall. [53] It is possible that the case is distinguishable where no sale of land is involved and the employer does not rely on his own skill and judgment in the choice of the specified item. [54]

---

[45] *Hancock* v. *B. W. Brazier (Anerley) Ltd.* [1966] 1 W.L.R. 1317, 1332 (C.A.). See also *Miller* v. *Cannon Hill Estates* [1931] 2 K.B. 113; *Perry* v. *Sharon Development Co.* [1937] 4 All E.R. 390 (C.A.); *Jennings* v. *Taverner* [1955] 1 W.L.R. 932; *Lynch* v. *Thorne, supra.*

[46] *Ibid.*

[47] See cases cited in note 45, *supra*; *Hoskins* v. *Woodham* [1938] 1 All E.R. 692; *Minster Trust Ltd.* v. *Traps Tractors Ltd.* [1954] 1 W.L.R. 963, 975.

[48] Defective Premises Act 1972; see *post*, p. 44.

[49] As amended by the Defective Premises Act 1972.

[50] See *A. C. Billings Ltd.* v. *Riden* [1958] A.C. 240 (H.L.); *Gallagher* v. *McDowell Ltd.* [1961] N.I. 26 (Northern Ireland); *Sharpe* v. *E. T. Sweeting & Son Ltd.* [1963] 1 W.L.R. 665; *Dutton* v. *Bognor Regis U.D.C.* [1972] 1 Q.B. 373 (C.A.); *Anns* v. *Merton London Borough Council* [1977] 2 W.L.R. 1024, 1038, 1047 (H.L.). For the letting of houses at a low rent, see the Housing Act 1957, s. 6.

[51] *Hancock* v. *Brazier, supra*, at p. 1334; see also *Billyack* v. *Leyland Construction Co. Ltd.* [1968] 1 W.L.R. 471.

[52] [1956] 1 W.L.R. 303 (C.A.).

[53] Note that in respect of a claim relating to a window it was conceded that the purchaser could succeed for breach of implied warranty because the specification did not so precisely identify the nature of the work, p. 309. See also Lord Parker C.J. at p. 311.

[54] Contracts for the sale of land and for work and labour differ in that in the former but not the latter the rule of *caveat emptor* still applies fundamentally (*Young & Marten Ltd.* v. *McManus Childs Ltd.* [1969] 1 A.C. 454 (H.L.), *per* Lord Upjohn at p. 472). Further, *quaere*, whether on the facts failure to comply with by-laws could not have been alleged as a breach of an implied term?

Breach of by-laws.[55] An action can be brought against the builder, " in effect, for breach of statutory duty by any person for whose benefit or protection the by-law was made." [56]

Effect of price. The standard of work to be carried out depends upon the terms of the contract express or implied. The question whether the price is high or low in relation to the work is, it is submitted, irrelevant in considering what standard of work is required,[57] unless the parties have expressly or impliedly agreed that the amount of the price is to affect the standard of the work.

5. Construction of deeds

Deeds are construed in the same way as other documents save that where one party wishes to deny the truth of a statement in the deed he may be estopped (*i.e.* prevented) from doing so by the application of a further rule known as estoppel by deed. This is a rule of evidence founded on the principle that a solemn and unambiguous statement or engagement in a deed must be taken as binding between parties and privies and therefore as not admitting any contradictory proof.[58] Statements of fact in recitals are subject to the rule.[59] Statements in the deed may bind all or only one or some of the parties according to the construction of the deed.[60] The estoppel does not operate where the deed was fraudulent or, in general, where it was illegal,[61] nor where there was a common mistake of fact giving rise to a right to rectification,[62] nor where the party seeking to set up the estoppel caused the misstatement of fact to appear in the deed.[63]

6. The value of previous decisions

(a) *Common form contracts*
The court is very reluctant to depart from a previous long-standing decision upon the meaning of a common form of contract in constant use because the parties are taken thereafter to have contracted upon the basis of that meaning.[64]

(b) *Similar contracts*
" To some extent decisions on one contract may help by way of analogy or illustration in the decision of another contract. But however similar the contracts

---

[55] Now replaced by the Building Regulations 1965, 1972, 1976 outside London. See further, *post*, pp. 72, 108.

[56] *Anns* v. *Merton London Borough Council, supra,* p. 1039, *per* Lord Wilberforce.

[57] *Cf. Cotton* v. *Wallis* [1955] 1 W.L.R. 1168 (C.A.), but note that this case, discussed *post*, p. 296, was a dispute between architect and employer and is not, it is submitted, an authority on disputes between employer and contractor.

[58] *Per* Lord Maugham, *Greer* v. *Kettle* [1938] A.C. 156, 171 (H.L.).

[59] *Ibid.*

[60] *Ibid.*; see also generally Halsbury (4th ed.), Vol. 12, paras. 1353 *et seq.*

[61] *Ibid.*

[62] *Wilson* v. *Wilson* [1969] 1 W.L.R. 1470 and for rectification, see *infra*, Section 3.

[63] *Supra*, see note 61.

[64] *Dunlop & Sons* v. *Balfour Williamson & Co.* [1892] 1 Q.B. 507, 518 (C.A.); *The Annefield* [1971] P. 169, 183, 185 (C.A.).

may appear the decision as to each must depend on the consideration of the language of the particular contract read in the light of the material circumstances of the parties in view of which the contract is made." [65] Nevertheless there are many instances in which one building contract has been used to assist in the construction of another.[66]

Section 2: *Civil Liability for Breach of Statutory Duty*

1. The Defective Premises Act 1972

The Defective Premises Act 1972 came into operation on January 1, 1974. By section 1 it imposes duties upon persons taking on work for or in connection with the provision of a dwelling. The duties are additional to any duty otherwise owed [67] and cannot be excluded or restricted.[68] Thus it is still necessary to consider the position in contract [69] and in tort [70] and, when it comes into effect, section 71 of the Health and Safety at Work Act 1974 (see *post*, p. 47). Breach of the duty imposed by the statute gives rise, it is submitted, to a claim for damages at the suit of a person to whom the duty is owed and who has suffered damage by reason of the breach.[71] The Act does not apply to the repair or enlargement of a dwelling,[72] and cannot, in broad terms, be relied on by the purchaser who has the benefit of the National Housebuilders' Registration Council Scheme.[73] In practice this effects a major restriction upon the operation of the Act.

The duty to build dwellings properly

Section 1.—(1) A person taking on work for or in connection with the provision of a dwelling (whether the dwelling is provided by the erection or by the conversion or enlargement of a building) owes a duty—

(*a*) if the dwelling is provided to the order of any person, to that person; and

(*b*) without prejudice to paragraph (*a*) above, to every person who acquires an interest (whether legal or equitable) in the dwelling;

to see that the work which he takes on is done in a workmanlike or, as the case may be, professional manner, with proper materials and so that as regards that work the dwelling will be fit for habitation when completed.

---

[65] *Per* Lord Wright, *Luxor* v. *Cooper* [1941] A.C. 108, 130 (H.L.).

[66] *e.g. Re Ford & Bemrose* (1902) H.B.C. (4th ed.), Vol. 2, p. 324 (C.A.); also (1902) 18 T.L.R. 443. In H.B.C. (7th ed.), Chap. IV, there is a collection of words and phrases judicially interpreted. See also index of this book.

[67] s. 6 (2). For a highly critical article by J. R. Spencer see [1974] C.L.J. 307.

[68] s. 6 (3).

[69] See *ante*, p. 41.

[70] See *ante*, p. 42 and in particular *Anns* v. *Merton London Borough Council* [1977] 2 W.L.R. 1024 (H.L.)—Local Authority may be liable for negligence of its building inspector in passing house with defective foundations or otherwise in relation to its duties under the Public Health Act 1936.

[71] This is not expressed in the statute but arises from the general principles relating to actions for breach of statutory duty. See Salmond, *Torts* (17th ed., 1977), pp. 242 *et seq.*

[72] See s. 1 (1).

[73] See s. 2 (1) and the House-building Standards (Approved Scheme, etc.) Order 1973 (S.I. 1973 No. 1843) and 1977 (S.I. 1977 No. 642). To get the exact position reference must be made to the Act, the Statutory Instrument and the Scheme. Other schemes can be approved under the Act. For the position where there is compulsory acquisition, see s. 2 (7).

The standard to be achieved corresponds substantially with that required by the usual implied terms upon sale of a house in the course of erection.[74] It is a question in each case whether a person has taken on work within the meaning of the Act. Thus ordinarily it will include the main contractor and any professional person, such as an architect, engineer or quantity surveyor and any subcontractor specifically employed on or in connection with the provision of the dwelling. But, it seems, other persons may be liable. Thus a person, not a qualified architect, who provides plans for the dwelling is, it is submitted, within the section, and so also is a supplier who makes up some component specifically for the dwelling in question. How far other suppliers who provide such goods as boilers suitable for use in dwellings to a builder can be liable must depend upon the application of the section to the facts of a particular case. It is not clear whether a local authority performing its duties under the building by-laws or the Building regulations can ever come within the section.[75]

A person within the section doing professional work has to do it in a professional manner. Does this mean that, as in the tort of negligence, it is a defence to show that he exercised proper care ? It is thought not and that all persons coming within the section are under a strict duty to fulfil its requirements.

Defence of " Instructions "

Section 1.—(2) A person who takes on any such work for another on terms that he is to do it in accordance with instructions given by or on behalf of that other shall, to the extent to which he does it properly in accordance with those instructions, be treated for the purposes of this section as discharging the duty imposed on him by subsection (1) above except were he owes a duty to that other to warn him of any defects in the instructions and fails to discharge that duty.

(3) A person shall not be treated for the purposes of subsection (2) above as having given instructions for the doing of work merely because he has agreed to the work being done in a specified manner, with specified materials or to a specified design.

There is no definition of the word " instructions " and it is difficult to understand its ambit. Subsection (3) is of assistance in that it contrasts a mere agreement that work shall be done in a specified way with the giving of instructions. At least it seems fairly clear that a person in the position of the purchaser in *Lynch* v. *Thorne* [76] would ordinarily have a remedy under the Act. At the other end of the spectrum a contractor performing work under the Standard form of building contract according to architect's instructions or in accordance with the contract documents is, it is submitted, and subject to the exception as to warning, (see *infra*) entitled to the defence provided by subsection (2). Between these extremes it must be a question in each case whether the communication relied on is an instruction or a mere agreement.

The Act does not itself impose a duty of warning and, it is to be presumed,

---

[74] See *ante*, p. 41.
[75] *Sparham-Souter* v. *Town & Country Development (Essex) Ltd.* [1976] 1 Q.B. 858, 870 (C.A.).
[76] [1956] 1 W.L.R. 303 (C.A.), discussed *ante*, p. 42.

does not refer to a mere moral duty. The result is, it is thought, that the exception as to a duty to warn must refer to a duty arising in tort or contract. The extent of an express or implied duty to warn arising under the Standard form of building contract is discussed *post*, p. 290. It may be that in some cases a comparable duty in tort arises, although if the plaintiff is in contract with the defendant the problem arises, discussed *post*, p. 217 in relation to architects, whether liability can exist in contract and in tort at the same time in respect of the same subject-matter.

Persons owing the duty

Section 1.—(4) A person who—

(*a*) in the course of a business which consists of or includes providing or arranging for the provision of dwellings or installations in dwellings;

or (*b*) in the exercise of a power of making such provision or arrangements conferred by or by virtue of any enactment;

arranges for another to take on work for or in connection with the provision of a dwelling shall be treated for the purposes of this section as included among the persons who have taken on the work.

This includes the person often termed "the developer" within the ambit of those who take on work.

Limitation of actions

Section 1.—(5) Any cause of action in respect of a breach of the duty imposed by this section shall be deemed, for the purposes of the Limitation Act 1939, the Law Reform (Limitation of Actions, etc.) Act 1954 and the Limitation Act 1963, to have accrued at the time when the dwelling was completed, but if after that time a person who has done work for or in connection with the provision of the dwelling does further work to rectify the work he has already done, any such cause of action in respect of that further work shall be deemed for those purposes to have accrued at the time when the further work was finished. (See *post*, p. 165.)

Continuing duty upon disposal of premises

Section 3 of the Act imposes a continuing duty of care upon persons who have carried out works of construction, repair, maintenance or demolition upon premises after they have disposed of them. There are exceptions to the duty.[77]

Measure of damages

This is not dealt with by the Act. It is thought unlikely that the intention is to exclude economic loss from damages recoverable under the Act.[78]

2. Civil liability for breach of building regulations

The Health and Safety at Work, etc., Act 1974, s. 71 provides that, subject to the provisions of the section, breach of a duty imposed by building regulations,

---

[77] See also *ante*, p. 42. The matter is dealt with summarily as it is more suitable to a work on torts, see *e.g.* Salmond, *Torts* (17th ed., 1977), pp. 291 *et seq.*

[78] See for damages in tort for negligent acts *S.C.M. (United Kingdom) Ltd.* v. *W. J. Whittall & Sons Ltd.* [1971] 1 Q.B. 337 (C.A.); *Spartan Steel Ltd.* v. *Martin & Co. Ltd.* [1973] 1 Q.B. 27 (C.A.).

shall, so far as it causes damage, be actionable except in so far as the regulations provide otherwise. There is provision for the regulations to provide prescribed defences. Liability does not arise until a regulation is made bringing the section into force and this had not yet been done.[78a] It will be necessary to consider the regulations when they appear, in particular the prescribed defences but the effect could be to provide for a very wide range of matters in respect of which civil proceedings can be brought.[78b] Actions under the Act may well effectively supersede those under the Defective Premises Act 1972 and must be considered in relation to actions for breach of contract.

Section 3: *Rectification*

After a written contract has been entered into one of the parties may discover that it does not correctly set out his intention. He may find, for example, that a price or a period of time is wrong. He can approach the other party and ask him to agree that the contract shall be altered so as to correct the mistake. Alternatively he may be able to rely on an express term for the rectification of certain errors.[79] If he is unable to reach such an agreement or rely on a contractual right he may have grounds for seeking the discretionary [80] remedy of rectification.

" Rectification is a remedy which is available where parties to a contract, intending to reproduce in a more formal document the terms of an agreement upon which they are already *ad idem*, use, in that document, words which are inapt to record the true agreement reached between them. The formal document may then be rectified so as to conform with the true agreement which it was intended to reproduce and enforced in its rectified form." [81]

The party seeking rectification must produce " convincing proof " [82] that there was a common intention, of which there was some outward expression of accord, in regard to a particular provision or aspect of the agreement continuing up to the moment of execution of the document which intention, by a mistake, the document failed to express.[83] He does not have to show that there was before the execution of the document a binding and conclusive contract.[84] Where two

---

[78a] But it seems that it (" in effect ") exists already—*Anns* v. *Merton London Borough Council* [1977] 2 W.L.R. 1024, 1039 (H.L.).

[78b] The section came into operation for the purpose only of enabling regulations to be made on March 17, 1977, S.I. 1977 No. 294.

[79] *e.g.* clause 12, Standard form of building contract.

[80] See *Whiteside* v. *Whiteside* [1950] Ch. 65, 71 (C.A.). The discretion is not exercised arbitrarily. If the plaintiff proves his case in accordance with the principles set out in the text rectification will ordinarily be granted unless he is guilty of conduct such as delay or acquiescence which is a ground for refusing relief or there are some special circumstances such as arose in the *Whiteside* case in connection with tax considerations in a family matter. On the discretion of the court in equitable matters generally see the standard textbooks such as *Snell* or *Hanbury*. For fundamental mistake avoiding the contract at common law, for other equitable powers of the court not amounting to rectification and for mistake as to the nature of a document, see *ante*, p. 9.

[81] *American Airlines Inc.* v. *Hope* [1974] 2 Lloyd's Rep. 301, 307 (H.L.), *per* Lord Diplock. See also *Re Butlin's Settlement Trusts* [1976] 1 Ch. 251, 260.

[82] *Joscelyne* v. *Nissen* [1970] 2 Q.B. 86, 98 (C.A.); *Ernest Scragg & Sons Ltd.* v. *Perseverance Banking Ltd.* [1973] 2 Lloyd's Rep. 101, 104 (C.A.), *cf. ibid.* p. 103 " satisfied beyond reasonable doubt."

[83] *Joscelyne* v. *Nissen, supra.*

[84] *Ibid.*

persons have agreed expressly upon the meaning of a particular phrase but do not record the definition in the contract itself, if one of them seeks to enforce the agreement on the basis of some other meaning he can be prevented by an action for rectification.[85]

Unilateral mistake

If one party makes a mistake in expressing the contract and the other party has no knowledge of that mistake rectification is not granted.[86] Where the other party at the time of the contract knows of and takes advantage of the mistake rectification is granted.[87]

In *Roberts & Co. Ltd.* v. *Leicestershire County Council*[88] the plaintiffs submitted a tender which specified a period for completion of 18 months. The defendants wrote stating that the tender was accepted and that a formal contract would shortly be submitted for execution. The contract was sent to the plaintiffs who sealed it believing the completion period to be 18 months. In fact they had failed to notice that the period inserted by the defendants was 30 months. Before the defendants sealed the contract there were two meetings by the end of the second of which the defendants must have known that the plaintiffs believed the period to be 18 months. However, the defendants sealed the contract without ever telling the plaintiffs of their mistake. It was held that, although the plaintiffs could not have rectification on the ground that there was a mistake in expressing a common intention, they were entitled to rectification on the ground that the defendants were prevented or estopped by their conduct from saying that there was no such mistake. It appears that the basis of the court's approach is that the plaintiff in such circumstances must prove a degree of sharp practice on the part of the defendant.[89]

A typical case where rectification will not be granted is where a contractor makes arithmetical errors, operating against his interest in the tender, which are incorporated in the contract price without either party being aware of the errors before the contract is signed.[90]

Checking the bills of quantities

It is common practice for the employer or his agent, *e.g.* the surveyor, to check the calculations in the priced bills of quantities of the successful tenderer before a contract is finally concluded. If the employer discovers arithmetical errors operating against the contractor's interest but these errors remain and are incorporated into the contract price, is the contractor entitled to rectification ? The answer requires the application of the principles set out above to the facts of each case, but as a guide the following approach is suggested:

---

[85] *Per* Megaw J., *London Weekend Television Ltd.* v. *Harris and Griffith* (1969) 113 S.J. 222, referred to in *Joscelyne* v. *Nissen* at p. 98.

[86] *Riverlate Properties Ltd.* v. *Paul* [1975] Ch. 133 (C.A.).

[87] *Ibid.*

[88] [1961] Ch. 555.

[89] *Riverlate* case, *supra*, at p. 140.

[90] *Riverlate* case, *supra*; *Ewing & Lawson* v. *Hanbury & Co.* (1900) 16 T.L.R. 140; *Kinlen* v. *Ennis U.D.C.* [1916] 2 I.R. 299, 309 (H.L.); *cf. Webster* v. *Cecil* (1861) 30 Beav. 62.

(i) The errors are clearly brought to the attention of the contractor and he decides to keep to his tender price and conditions. There cannot be rectification.

(ii) As in (i) but the parties agree upon some alteration in the tender price or upon some other alteration in the suggested terms of contract, *e.g.* as to calculation of prices for variations. If by mistake the contract does not express these agreed alterations then upon the agreement being clearly proved there can be rectification.[91]

(iii) The errors are not brought to the attention of the contractor and he, ignorant of them, signs the contract. Proof of these facts alone does not, it is submitted, give the contractor grounds for rectification,[92] but if the contractor proves clearly [93] that at the time of entering into the contract he believed the contract price to consist of the true totals of the properly calculated prices in the bills and that the employer at that time knew of the contractor's belief there can be rectification.[94]

---

[91] See *Carlton Contractors* v. *Bexley Corp.* (1962) 60 L.G.R. 331.

[92] Neither do they give him a right to damages in fraud, see *Dutton* v. *Louth Corp.* (1955) unrep., except in (1955) 116 E.G. 128 (C.A.), discussed *post*, p. 93. For the importance of the distinction between lump sum contracts and measure and value contracts, see *post*, p. 66. For the effect of pricing errors on valuation of variations, see *post*, p. 71. For misrepresentation, see *post*, p. 89.

[93] For the standard of proof, see *ante*, p. 47.

[94] See *Riverlate Properties Ltd.* v. *Paul* [1975] Ch. 133 (C.A.).

CHAPTER 4

THE RIGHT TO PAYMENT AND THE TIME FOR COMPLETION

I. THE RIGHT TO PAYMENT

THE contractor's right to payment depends upon the wording of the contract. Within the limits of legality parties can make what arrangements they please, but there are three broad heads under which the right can arise: a lump sum contract; an express contract other than for a lump sum; a claim for a reasonable sum frequently called a *quantum meruit*.

Section 1: *Lump Sum Contracts*

1. The amount payable

A lump sum contract is a contract to complete a whole [1] work for a lump sum, *e.g.* to build a house for £30,000. If the house is completed in every detail required by the contract [2] the contractor is [3] entitled to £30,000, and if extra work was carried out he may be able to recover further payment.[4] If he does not complete the house detailed clauses may provide what amount, if any, he is to receive [5]; but parties entering into a contract do not always contemplate its breach, and in the absence of such clauses, and even to some extent when they are present, a difficult problem may arise. This has two aspects. The first is what payment, if any, the contractor can recover; the second is what claim, if any, the employer has for damages. This chapter is concerned with the first aspect, Chapter 9 with the second.[6]

2. When completion is a condition precedent

If a contractor agrees to do a whole work according to a specification which consists of 40 items for a lump sum of £5,000 and fails to carry out 20 of the items, it is obvious that he is not entitled to recover the whole of the £5,000 and that the employer may have an action against him for damages. But is the contractor entitled to recover any of the £5,000? Can the employer say to him, " You agreed to complete the whole and to be paid when the whole was completed.[7] The work is incomplete, therefore you are entitled to nothing " ? And can the employer rely on the same argument where only two out of the 40 items

---

[1] Sometimes termed an " entire " or a " specific " work.

[2] See Chap. 5 for discussion of work which may be impliedly included in the contract.

[3] Subject to some unfulfilled condition precedent such as an architect's certificate. See Chap. 6.

[4] See Chap. 5. The contract may provide many other ways in which a different amount to the original contract sum may eventually become payable—see, *e.g.* Notes to the articles of agreement of the Standard form of building contract, *post*, Chap. 15.

[5] See, *e.g.* Standard form of building contract, clause 25 (4) (*d*).

[6] See also Chap. 14, " Litigation."

[7] See *Appleby* v. *Myers* (1867) L.R. 2 C.P. 651, 661.

are omitted? In other words, is completion a condition precedent to the contractor's right to be paid and what is meant by completion? These problems have greatly exercised the courts.[8] Although each case turns on the construction of the contract,[9] there are certain principles which seem to be well established. Before discussing them in detail it may assist to state the position in summary form.

Summary of position

SUBSTANTIAL COMPLETION. In the ordinary lump sum contract[10] the employer cannot refuse to pay the contractor merely because there are a few defects and omissions. If there is substantial completion he must pay the contract price[11] subject to a deduction by way of set-off or counterclaim for the defects.[12]

ENTIRE COMPLETION. The parties may if they choose by clear language show that they intend that the contractor should be entitled to nothing until he has completed the contract in every detail, or that he should not be entitled to the retention money until he has so completed the contract.[13]

NON-COMPLETION.[14] If the contractor fails to complete, either substantially in the ordinary case, or in every detail in the special case referred to in the last paragraph, he is not entitled to anything unless he can show either:

  (i) A contractual right to unpaid instalments, *or*
  (ii) Prevention of completion by employer, *or*
  (iii) Implied promise to pay for the work done by way of waiver or acceptance, *or*
  (iv) Impossibility or frustration.

(a) *Substantial completion*

  (i) *What it means.* In *Hoenig* v. *Isaacs*[15] the plaintiff was employed to decorate a one-room flat and provide it with certain furniture for a sum of £750, the terms of payment being " net cash, as the work proceeds, and balance on completion." The plaintiff claimed that he had carried out the work under the contract and asked for payment of the unpaid balance of £350 of the contract price. The defendant entered into occupation of the flat and used the furniture but refused to pay, complaining of faulty design and bad workmanship. The Official Referee found that the door of a wardrobe required replacing, and that a bookshelf, which was too short, would have to be remade, which would require alterations being made to a bookcase. He held that there had been substantial compliance with the

---

[8] See cases cited *infra* and in Notes to *Cutter* v. *Powell* (1795) 2 Sm.L.C. 1.
[9] *Hoenig* v. *Isaacs* [1952] 2 All E.R. 176, 178, 180 (C.A.).
[10] Denning L.J. said of the contract in *Hoenig* v. *Isaacs* (for the facts see *infra*) " I think this contract should be regarded as an ordinary lump sum contract."
[11] Subject to some unfulfilled condition precedent imposed by the contract, *e.g.* an architect's certificate.
[12] *Hoenig* v. *Isaacs*, *supra*; see further, *infra*, " Substantial completion," and for set-off and counterclaim, see *post*, p. 275.
[13] *Appleby* v. *Myers*, *supra*; *Hoenig* v. *Issacs*, *supra*, and see *post*, pp. 53 *et seq.*
[14] See *post*, p. 54.
[15] *Supra.*

contract and awarded £294, being the £350 claimed less £56, the cost of remedying the defects. This finding was upheld by the Court of Appeal which rejected the defendant's contention that there had been non-performance of an entire contract, and that therefore the plaintiff was not entitled to any further payment under the contract, but could only claim on a *quantum meruit*.[16]

(ii) *The principle.* The principle applied by the court in *Hoenig* v. *Isaacs* is usually traced back to Lord Mansfield C.J., who said: " Where mutual covenants go to the whole of the consideration on both sides, they are mutual conditions, the one precedent to the other. But where they go only to a part, where a breach may be paid for in damages, there the defendant has a remedy on his covenant and shall not plead it as a condition precedent." [17] The principle is illustrated by the law relating to sale of goods which makes a distinction between conditions, breach of which gives a right to reject the goods, and warranties, breach of which gives no right to reject the goods but leaves the buyer to his claim for damages.[18] In applying the principle to building contracts, although the question is one of construction, " when a contract provides for a specific sum to be paid on completion of specified work, the courts lean against a construction which would deprive the contractor of any payment at all simply because there are some defects or omissions." [19]

(iii) *Application of the principle.* One test to be applied is whether the work was " finished " or " done " in the ordinary sense even though part of it is defective.[20] And ". . . it is relevant to take into account both the nature of the defects and the proportion between the cost of rectifying them and the contract price." [21] Thus it is not sufficient to consider the cost of rectification alone. In one case where the contract price was £520 and the cost of remedying the defects was £200 the Court of Appeal upheld a finding that there had been substantial completion.[22] But in *Bolton* v. *Mahadeva* [23] where the cost of remedying defects was £174 against a contract price of £560 the Court of Appeal allowed an appeal against

---

[16] It was not necessary for the defendant to argue that the plaintiff was entitled to nothing because apparently the defendant was satisfied that if the work were measured and valued on a *quantum meruit* basis the amount found due to the plaintiff would not exceed the £400 already paid. (See pp. 181, 183 of the report.) Presumably it was conceded by the defendant that he had accepted the work and impliedly promised to pay a reasonable sum.

[17] *Boone* v. *Eyre* (1777) 1 Hy.Bl. 273, cited by Somervell L.J., *Hoenig* v. *Isaacs* [1952] 2 All E.R. 176, 178 (C.A.).

[18] *Hoenig* v. *Isaacs, supra,* at p. 178.

[19] *Per* Denning L.J., *Hoenig* v. *Isaacs, supra,* at p. 181. For the analogy with the grant of specific performance in contracts for the sale of land when a vendor fails to make title to some insignificant part, see the judgment of Romer L.J. in *Hoenig* v. *Isaacs, supra.*

[20] See *Hoenig* v. *Isaacs, supra,* at p. 179, referring to *Dakin* v. *Lee, supra,* and *Appleby* v. *Myers* (1867) L.R. 2 C.P. 651, 661. See also *Foreman & Co. Proprietary* v. *The Ship* " *Liddesdale* " [1900] A.C. 190, 200 (P.C.).

[21] *Bolton* v. *Mahadeva* [1972] 1 W.L.R. 1009, 1013 (C.A.).

[22] *Kiely & Sons* v. *Medcraft* (1965) 109 S.J. 829 (C.A.). See also *Ibmac Ltd.* v. *Marshall* (*Homes*) *Ltd.* 208 E.G. 851, November 23, 1968 (C.A.)—one-third value *not* substantial completion.

[23] *Supra.*

a finding that there had been substantial completion. The contract was to provide central heating and the defects were such that the system did not heat the house adequately and fumes were given out so as to make living rooms uncomfortable. The work was ineffective for its primary purpose.[24]

(b) *Entire completion*

(i) *Condition precedent to any payment.* This type of contract is sometimes termed an entire contract and has been defined as follows. " An entire contract is an indivisible contract, one where the entire fulfilment of the promise by either party is a condition precedent to the right to call for the fulfilment of any part of the promise by the other." [25]

CLEAR WORDS are needed to bring an entire contract into existence.[26] In the absence of such words the ordinary lump sum contract is not an entire contract.[27] The type of contract which may be entire is that where the contractor undertakes some simple clear obligation such as to put some broken article or part of a house in working order and completely fails to do so. In such a case he may be entitled to nothing although he has expended much work and labour, for the main purpose of the contract is that the article or part of the house shall work and there is no scope in the contract for terms collateral to the main purpose.[28]

INSTALMENTS. A contract which gives the contractor an enforceable right to instalments cannot be an entire contract because the contractor has the right to call for fulfilment of part of the employer's promise before he has entirely completed his own promise.[29]

The right to instalments may arise expressly or by implication. The nature of the right under an express provision is a matter of construction of the words. The inference of an implied right is governed by the usual rules,[30] but it has been said that, " a man who contracts to do a long costly piece of work does not contract, unless he expressly says so that he will do all the work, standing out of pocket until he is paid at the end. He is entitled to say, '. . . there is an understanding all along that you are to give me from time to time at the reasonable times payments for work done.' " [31] It was held that a shipwright who had undertaken to put a ship into thorough repair (apparently no price was agreed) was entitled to demand payment for part of the work he had carried out.[32] It

---

[24] See also *Technistudy Ltd.* v. *Kelland* [1976] 1 W.L.R. 1042, 1045 (C.A.).

[25] H.B.C. (7th ed.), p. 165, citing *Cutter* v. *Powell* (1795) 2 Sm.L.C. 1; note Somervell L.J.'s interpretation of *Cutter* v. *Powell* in *Hoenig* v. *Isaacs, supra,* at p. 178.

[26] *Appleby* v. *Myers, supra,* at p. 661; *Hoenig* v. *Isaacs, supra,* at p. 180; Smith's L.C. Notes to *Cutter* v. *Powell* (13th ed.), Vol. 2, p. 26.

[27] *Hoenig* v. *Isaacs, supra.*

[28] *Duncan* v. *Blundell* (1820) 3 Stark. 6; *Sinclair* v. *Bowles* (1829) 9 B. & C. 92; *Portman* v. *Middleton* (1858) 4 C.B.(N.S.) 322; see *Hoenig* v. *Isaacs, supra,* at p. 178. See also *Vigers* v. *Cook* [1919] 2 K.B. 475 (C.A.)—an undertaker's contract.

[29] *Terry* v. *Duntze* (1878) 2 H.Bl. 389.

[30] See Chap. 3.

[31] Per Phillimore J., *The Tergeste* [1903] P. 26, 34. *Cf. Rees* v. *Lines* (1837) 8 C. & P. 126.

[32] *Roberts* v. *Havelock* (1832) 3 B. & Ad. 404.

seems to be implied in the absence of clear words that a person carrying out repairing work to another's property is entitled to payment from time to time before completion.[33]

RECOVERY OF MONEY PAID. If the employer pays money under a contract which the contractor fails to complete the employer can recover that money in an action for money had and received if there has been a total failure of consideration.[34] If the employer has received any value from performance by the contractor there has not been a total failure of consideration,[35] and any payment is not recoverable even though it was made in advance of performance.[36] In such circumstances the employer can claim damages for breach of contract.[37] In his claim he has to give credit for the value of work performed by the contractor.[38] The combined effect of the rules relating to recovery of money paid and damages for breach of contract can operate somewhat capriciously to favour employer or contractor according to whether the contract price was high or low, and the value of the partial performance by the contractor in relation to the amount of money, if any, paid to the contractor by way of instalments or advance upon the contract price.[39]

(ii) *Condition precedent to payment of retention money.* The contract may provide " for progress payments to be made as the work proceeds, but for retention money to be held until completion. Then entire performance is usually a condition precedent to payment of the retention money but not, of course, to the progress payments. The contractor is entitled to payment *pro rata* as the work proceeds, less a deduction for rention money. But he is not entitled to the retention money until the work is entirely finished, without defects or omissions." [40]

(c) *Non-completion*

(i) *The general rule*

WHEN ENTIRE COMPLETION CONDITION PRECEDENT. The contractor cannot recover anything either under the contract or on a *quantum meruit* if he has failed to complete in every detail.[41]

---

[33] *Menetone* v. *Athawes* (1764) 3 Burr. 1592; *Roberts* v. *Havelock, supra; Appleby* v. *Myers, supra,* at p. 660.

[34] *Fibrosa Spolka Akcyjna* v. *Fairbairn Lawson Ltd.* [1943] A.C. 32 (H.L.).

[35] *Ibid.; The Julia* [1949] A.C. 293 (H.L.).

[36] *Ibid.; Whincup* v. *Hughes* (1871) L.R. 6 C.P. 78.

[37] See *post,* Chap. 9.

[38] *Ibid.* and see especially *Mertens* v. *Home Freeholds Co.* [1921] 2 K.B. 526 (C.A.) the facts of which are set out *post,* p. 148.

[39] See the Law Commission's Working Paper No. 65; Goff and Jones, *The Law of Restitution;* and consider the effect of applying various different figures to the facts of *Mertens* v. *Home Freeholds Co., supra.*

[40] *Per* Denning L.J., *Hoenig* v. *Isaacs, supra,* at p. 181, where it was also held that the £400 not paid by the employer could not be treated as retention money because it formed so large a proportion of the contract sum. If the parties so desired they could expressly make such a large sum retention money. For the Standard form of building contract, see *post,* Chap. 15.

[41] *Cutter* v. *Powell* (1795) 2 Sm.L.C. 1; *Sinclair* v. *Bowles, supra;* see also *Ellis* v. *Hamlen* (1810) 3 Taunt. 52, but see Ridley J., *Dakin* v. *Lee, supra,* at p. 572; *Stegmann* v. *O'Connor* (1899) 81 L.T. 627 (C.A.); *Vigers* v. *Cook, supra.*

ORDINARY LUMP SUM CONTRACT. The contractor cannot recover anything either under the contract or on a *quantum meruit* unless he shows substantial completion (see *supra*).

(ii) *Payment where non-completion.* There are certain cases where the contractor can recover payment despite non-completion, as follows:

UNPAID INSTALMENTS. Unless there is some provision to the contrary, not amounting to a penalty,[42] the employer cannot, it is submitted, refuse to pay unpaid instalments which have, on the terms of the contract, become payable to the contractor.[43]

PREVENTION. If the employer's own wrongful acts have prevented completion,[44] the contractor can recover a reasonable sum on a *quantum meruit* for the work he has carried out.[45] It has been said that the rule that a contractor who has not substantially completed cannot recover payment does not work hardly upon him if only he is prepared to remedy the defects before seeking to resort to litigation to recover the lump sum.[46] It seems to follow that ordinarily there is an implied duty upon an employer to give a willing contractor an opportunity to remedy defects, breach of which duty amounts to prevention. Such duty does not, it is submitted, arise if the defects are so grave as to show that the contractor is unable to perform the contract.[47]

IMPLIED PROMISE TO PAY. An implied promise to pay a reasonable sum for the work done can arise from acceptance or waiver.

Acceptance. If the contractor can prove a fresh contract to pay for the work done he can recover on that contract.[48] Such a contract may be inferred from the acceptance by the employer of the work done with the full knowledge of the failure to complete,[49] but it is difficult to prove acceptance from mere occupation and use of the building works.[50] Thus in *Sumpter* v. *Hedges* [51] S contracted to build a house for H for £565. S did work to the value of £333, received part payment and then abandoned the contract. H completed the house, incorporating the work carried out by S. S sued for the difference between the money he had received and the value of his work, but did not recover.[52] It was said that

---

[42] See Chap. 9, " Liquidated Damages."
[43] See further Chap. 8, " Forfeiture Clauses."   [44] See *post*, pp. 117, 160.
[45] *Appleby* v. *Myers, supra; Hoenig* v. *Isaacs, supra.* For damages, see *post*, Chap. 9.
[46] *Bolton* v. *Mahadeva* [1972] 1 W.L.R. 1009, 1015 (C.A.).
[47] See *post*, p. 111. " Repudiation—generally."
[48] *Hoenig* v. *Isaacs, supra*, at p. 181.
[49] *Munro* v. *Butt* (1858) 8 E. & B. 738; *Appleby* v. *Myers, supra.* Acceptance does not prevent the employer counterclaiming for damages for defects—see *post*, p. 119.
[50] " In the case of goods sold and delivered, it is easy to show a contract from the retention of goods; but this is not so where work is done on real property "; *per* Bramwell B., *Pattinson* v. *Luckley* (1875) 10 Ex. 330, 334.   [51] [1898] 1 Q.B. 673 (C.A.).
[52] In earlier editions of this book it was suggested that it was just possible that this case might be decided differently today having regard to developments in the doctrine of restitution and unjust enrichment. But the approach of the Court of Appeal in *Bolton* v. *Mahadeva* [1972] 1 W.L.R. 1009 suggests that this is unlikely. It seems that any change in the law to deal with a principle which can sometimes cause injustice must come from Parliament—see The Law Commission's Working Paper No. 65. For statutory recognition of the doctrine of unjust enrichment in relation to goods, see the Torts (Interference with Goods) Act 1977, s. 7 (4) (not yet in force).

" there are cases in which though the plaintiff has abandoned the performance of a contract, it is possible for him to raise the inference of a new contract to pay but in order that that may be done the circumstances must be such as to give an option to the defendant to take or not to take the benefit of the work done. . . . The mere fact that a defendant is in possession of what he cannot help keeping or even has done work upon it affords no ground for such an inference. He is not bound to keep unfinished a building which in an incomplete state would be a nuisance on his land." [53]

Waiver. " It is always open to a party to waive a condition which is inserted for his benefit." [54] On the facts in *Hoenig* v. *Isaacs*,[55] the court held that even if entire performance was a condition precedent the employer by entering into occupation and using the furniture had waived the condition and could no longer rely on it. This is not inconsistent with *Sumpter* v. *Hedges* [56] because the contract included a number of chattels which the employer could have avoided using, but did in fact use and therefore put himself " in the same position as a buyer of goods who by accepting them elects to treat a breach of condition as a breach of warranty." [57]

IMPOSSIBILITY OR FRUSTRATION. If the failure to complete is due to impossibility of performance or frustration and the employer has obtained a valuable benefit from the work done, the contractor can recover from the employer such sum as the court considers just.[58]

Section 2: Contracts Other than for a Lump Sum

1. Generally

The manner of payment can be arranged in a variety of ways and it is impossible to attempt any exhaustive classification. A contract to do a whole work in consideration of the payment of different sums for different parts of the work is prima facie subject to the same rules about completion as an ordinary lump sum contract.[59] A contract to do a whole work with a provision for payment of each completed part of the whole may be a divisible contract in the sense that if the whole is not completed through the default of the contractor, he may be entitled to payment under the contract for those parts he has completed subject to the employer's right to counterclaim for non-completion of the whole.[60]

---

[53] *Sumpter* v. *Hedges, supra,* at p. 676, *per* Collins L.J. See also *Whitaker* v. *Dunn* (1887) 3 T.L.R. 602 (D.C.), a very strong case, where substantial performance might perhaps have been argued, and *Wheeler* v. *Stratton* (1911) 105 L.T. 786.
[54] *Per* Denning L.J., *Hoenig* v. *Isaacs, supra,* at p. 181.
[55] *Supra.*
[56] *Supra.*
[57] *Per* Somervell L.J., *Hoenig* v. *Isaacs* [1952] 2 All E.R. 176, 180 (C.A.). In *Sumpter* v. *Hedges (supra)* the employer was held liable to account for the value of unfixed materials which he had used in the works. For goods see also, when it comes into effect, the Torts (Interference with Goods) Act 1977.
[58] Law Reform (Frustrated Contracts) Act 1943: and see *post,* p. 106.
[59] *Appleby* v. *Myers* (1867) L.R. 2 C.P. 651.
[60] *Newfoundland Government* v. *Newfoundland Ry.* (1888) 13 App.Cas. 199 (P.C.).

2. Measurement and value contracts

A contract where the amount of work when completed is to be measured and valued according to a schedule, or formula, or at cost plus a fixed fee,[61] or a percentage of the cost, or at a reasonable price, is usually described as a measurement and value contract and is contrasted with a lump sum contract.[62] If the agreement is to do a whole work to be measured and valued and paid for on completion, entire completion may be a condition precedent to payment.[63] But it is submitted that normally the rule of substantial completion will apply, so that if the work is substantially performed, the contractor will be entitled to have it measured and valued and to be paid at the contract rate for the work done, subject to the employer's counterclaim for damages for defects and omissions.[64] It seems that in a repairing or jobbing contract the contractor is prima facie entitled to be paid for the work he has carried out,[65] though he may if he chooses contract not to be paid until completion.[66]

3. Cost plus percentage contracts

Such contracts sometimes contain an elaborate description of the method of calculating the cost. Where they do not and there is a simple agreement to pay a percentage upon the cost of labour and materials, " cost " means, it is submitted, the actual cost honestly and properly expended in carrying out the works. The contractor is, it is submitted, not disentitled to such cost merely because through some lack of efficiency on his part it exceeds a reasonable sum, but is disentitled to the cost of labour or materials expended in so wasteful a manner that it cannot truly be said to be part of the cost of the contract works, the question being one of fact and degree in each case.[67] A formally drafted cost plus contract will usually have a clause intended to protect an employer against waste or extravagance on the part of the contractor.[68]

Section 3: *Quantum Meruit*

This term is used in various senses. The proper meaning is where there is no contract between the parties but in the circumstances of the case the law imposes a duty, said to arise in quasi contract, to pay a reasonable sum for goods received or for the benefit of services rendered.[69] It is also sometimes, and perhaps rather loosely, used where there is a contract but the claim is for a reasonable sum.

---

[61] For an example, see " The Fixed Fee Form of Prime Cost Contract " issued under the sanction of the Royal Institute of British Architects, the National Federation of Building Trades Employers and other bodies.

[62] See *Re Ford & Bemrose* (1902) H.B.C. (4th ed.), Vol. 2, pp. 324, 333 (C.A.) (18 T.L.R. 443). See also *post*, p. 65.

[63] *Whitaker* v. *Dunn* (1887) 3 T.L.R. 602 (D.C.).

[64] Cf. *Newfoundland Government* v. *Newfoundland Ry.*, *supra*.

[65] *Appleby* v. *Myers*, *supra*.

[66] *Ibid.* at p. 661.

[67] This paragraph is largely based on remarks by His Honour Sir Brett Cloutman, V.C., senior official referee in an unreported case.

[68] For an example see the Fixed Fee form of contract referred to, *supra*.

[69] See *William Lacey (Hounslow) Ltd.* v. *Davis* [1957] 1 W.L.R. 932. See also Goff & Jones, *The Law of Restitution* and an article by Professor Jones, (1977) 93 L.Q.R. 273.

Assessment of a reasonable sum

The courts have laid down no rules limiting the way in which a reasonable sum is to be assessed. Useful evidence [70] in any particular case may include: abortive negotiations as to price [71]; a calculation based on the net cost of labour and materials used plus a sum for overheads and profit; measurements of work done and materials supplied; the opinion of quantity surveyors, experienced builders or other experts as to a reasonable sum. Although expert evidence is often desirable there is no rule of law that it must be given and in its absence the court does the best it can on the materials before it to assess a reasonable sum.

A claim on a *quantum meruit* cannot arise if there is an existing contract between the parties to pay an agreed sum.[72]

Examples of where a claim for a reasonable sum may arise are as follows:

1.   Express agreement to pay a reasonable sum

2.   Where no price fixed

If the contractor does work under a contract express or implied and no price is fixed by the contract, he is entitled to be paid a reasonable sum for his labour and the materials supplied.[73] If such a contract, or an express agreement to be paid a reasonable sum, is for a whole work, completion may be made a condition precedent to payment, but it seems that in the absence of clear words the contractor is entitled from time to time to demand payment on account of the value of the work he has done.[74]

Failed negotiations. If work is carried out while negotiations as to the terms of the contract are proceeding but agreement is never reached upon essential terms the contractor is entitled to be paid a reasonable sum for the work carried out.[75]

3.   Prevention of completion by employer—see *ante*, p. 55.

4.   Acceptance or waiver—see *ante*, p. 55.

5.   Work outside the contract

Where there is a contract for specified work but the contractor does work outside the contract at the employer's request the contractor is entitled to be paid a reasonable sum for the work outside the contract.[76] In *Parkinson* v. *Commissioners*

---

[70] For preparation of evidence, see *post*, p. 250.

[71] *Way* v. *Latilla* [1937] 3 All E.R. 759, 764, 766 (H.L.).

[72] *Gilbert & Partners* v. *Knight* [1968] 2 All E.R. 248 (C.A.), the facts of which appear *post*, p. 221.

[73] *Moffatt* v. *Laurie* (1855) 15 C.B. 583; *Turriff Construction Ltd.* v. *Regalia Knitting Mills Ltd.* (1971) 222 E.G. 196, His Honour Judge Edgar Fay Q.C., discussed *ante*, p. 11.

[74] *Roberts* v. *Havelock* (1832) 3 B. & Ad. 404; *Appleby* v. *Myers*, *supra*.

[75] *Trollope & Colls Ltd.* v. *Atomic Power Constructions Ltd.* [1963] 1 W.L.R. 333; *Peter Lind & Co. Ltd.* v. *Mersey Docks and Harbour Board* [1972] 2 Lloyd's Rep. 234.

[76] *Thorn* v. *London Corp.* (1876) 1 App.Cas. 120, 127 (H.L.); *Sir Lindsay Parkinson & Co. Ltd.* v. *Commissioners of Works* [1949] 2 K.B. 632 (C.A.); see also *Cana Construction Co.* v. *The Queen* (1973) 37 D.L.R. (3d) 418, Supreme Ct. of Canada.

of Works [77] the contractor agreed under a varied contract to carry out certain work to be ordered by the Commissioners on a cost plus profit basis subject to a limitation as to the total amount of profit. The Commissioners ordered work to a total value of £6,600,000 but it was held that on its true construction the varied contract only gave the Commissioners authority to order work to the value of £5,000,000.[78] The additional work therefore had to be paid for by a *quantum meruit* so that the contractor recovered more than the total fixed profit.[79]

II. TIME FOR COMPLETION

1. Generally

Formal contracts usually make elaborate provisions as to the date of completion and its extension, progress and failure to proceed diligently. If no time is specified for completion of the contract a reasonable time for completion will be implied.[80] What is a reasonable time is a question of fact,[81] but it has been said of a person who has such a duty to complete that he " fulfils his obligation, notwithstanding protracted delay, so long as such delay is attributable to causes beyond his control, and he has acted neither negligently nor unreasonably." [82] Thus a strike occurring after the contract was entered into was taken into account.[83] But car repairers who were under-staffed, were aware of the holiday period approaching and that for commercial reasons they had to give priority to certain other work were not able to rely on delay caused by these factors in assessing a reasonable time.[84]

Requirements to complete which have been considered by the courts include, " as soon as possible," [85] " within a reasonable time," [86] " as speedily as possible," [87] " directly," [88] " forthwith," [89] ". . . such delivery date cannot be guaranteed." [90]

It has been suggested that where there is no express provision as to progress, business efficacy requires the implication of a term that the contractor will

---

[77] [1949] 2 K.B. 632 (C.A.).
[78] As explained by Lord Simon, *British Movietonews Ltd.* v. *London & District Cinemas Ltd.* [1952] A.C. 166, 184 (H.L.). It should be noted that *Bush* v. *Whitehaven Trustees* (1888) H.B.C. (4th ed.), Vol. 2, p. 130 (C.A.), which was referred to in the *Parkinson* case should now be regarded as expressing no principle of law—see *Davis Contractors Ltd.* v. *Fareham U.D.C.* [1956] A.C. 696 (H.L.).
[79] See also *post*, p. 71, " Rate of Payment."
[80] *Charnock* v. *Liverpool Corp.* [1968] 1 W.L.R. 1498 (C.A.); *Startup* v. *Macdonald* (1843) 6 M. & G. 593.
[81] *Startup* v. *Macdonald, supra*; *Hick* v. *Raymond & Reid* [1893] A.C. 22, 32, 33(H.L.); *McDougall* v. *Aeromarine Ltd.* [1958] 1 W.L.R. 1126—a shipbuilding case. The test is objective—*Charnock* v. *Liverpool Corp., supra.*
[82] *Hick* v. *Raymond & Reid, supra*, at p. 32.
[83] *Ibid.*
[84] *Charnock* v. *Liverpool Corp., supra.*
[85] *Attwood* v. *Emery* (1856) 26 L.J.C.P. 73.
[86] *Hydraulic Engineering Co.* v. *McHaffie* (1878) 4 Q.B.D. 670 (C.A.).
[87] *Penarth Dock Engineering Co.* v. *Pounds* [1963] 1 Lloyd's Report 359.
[88] *Duncan* v. *Topham* (1849) 8 C.B. 225.
[89] *Roberts* v. *Brett* (1865) 34 L.J.C.P. 241.
[90] *McDougall* v. *Aeromarine Ltd., supra.*

proceed with reasonable diligence and maintain reasonable progress.[91] It is thought that while such a term may have to be implied in some cases each contract and its surrounding circumstances must be considered, and that there is no such rule of general application. It may well be that in some cases the contractor's only duty is to complete by the due date.

2. Time of the Essence

If the contractor fails to comply with the terms of the contract as to time he is in breach of contract and liable in damages and the employer may have express remedies under the contract. Whether or not the breach enables the employer to treat the contract as at an end requires a consideration of the principles of repudiation of contracts and of time being of the essence of the contract. It may well be that failure to complete to time where it is of the essence is an aspect of repudiation,[92] but having regard to the historical development of the law relating to time being of the essence it is convenient to treat it separately. If time is of the essence and the contractor is in breach the employer can treat the contract as at an end.[93]

" Time will not be considered to be of the essence unless: (1) the parties expressly stipulate that conditions as to time must be strictly complied with; or (2) the nature of the subject-matter of the contract or the surrounding circumstances show that time should be considered to be of the essence; . . ." [94]

The normal rule is that time is not of the essence in building contracts.[95] It seems that ordinarily it is not of the essence where the contract includes provisions for extension of time and the payment of liquidated damages for delay.[96] But in one case where there were such provisions, but also the words " time shall be considered as of the essence of the contract on the part of the contractor . . ." it was said " no doubt this gave the Corporation the right to determine the contract at the end of the 24 months period as extended by the architect." [97]

Notice making time of the essence. Where a reasonable time for performance has elapsed and either time was not originally of the essence or has ceased to be of the essence by waiver or agreement, the employer can serve a notice requiring completion by a certain date.[98] He is really telling the contractor that unless he completes by such date he will treat his failure as a repudiation of the contract.[99]

---

[91] H.B.C. (10th ed.), p. 314.

[92] *United Scientific Holdings Ltd.* v. *Burnley Council* [1977] 2 W.L.R. 806, 830-832. See *post*, p. 111 for repudiation generally.

[93] *Charles Rickards Limited* v. *Oppenheim* [1950] 1 K.B. 616 (C.A.); *Peak Construction (Liverpool) Limited* v. *McKinney Foundations Limited* (1971) 69 L.G.R. 1, 10 (C.A.).

[94] Hals. (4th ed.), Vol. 9, para. 481, approved in *United Scientific* case at pp. 823, 829, 842 (H.L.).

[95] *Lucas* v. *Godwin* (1837) 3 Bing. N.C. 737, 744.

[96] *Lamprell* v. *Billericay Union* (1849) 3 Ex. 283, 308; *Felton* v. *Wharrie* (1906) H.B.C. (4th ed.), Vol. 2, pp. 398, 400 (C.A.).

[97] *Peak Construction (Liverpool) Limited* v. *McKinney Foundations Limited, supra,* at p. 10, *per* Salmon L. J. The other members of the court did not deal with the point.

[98] *United Scientific,* case at p. 832, *Rickards* v. *Oppenheim, supra, Felton* v. *Wharrie, supra,* the facts of which appear *post,* p. 115.

[99] *United Scientific* case, at p. 832.

If the notice was not given prematurely, and the date for completion was not unreasonably soon in all the circumstances, judged at the time when the notice was given,[1] and the contractor fails to complete by such date, the employer can treat the contract as at an end and dismiss the contractor from the site.[2]

---

[1] *Rickards* v. *Oppenheim, supra,* at p. 624. Thus a subsequent strike may not excuse the contractor, *ibid.*

[2] *Felton* v. *Wharrie, supra, Rickards* v. *Oppenheim, supra, United Scientific* case, *supra.*

CHAPTER 5

VARIED WORK

Section 1: Payment for Extras—Introduction

A CONTRACTOR frequently carries out, or is asked to carry out, work for which he considers he is entitled to payment in excess of the original contract sum. To recover such payment he must be prepared to prove: (1) that it is extra work not included in the work for which the contract sum is payable; (2) that there is a promise express or implied to pay for the work; (3) that any agent who ordered the work was authorised to do so, and (4) that any condition precedent to payment imposed by the contract has been fulfilled. If he cannot prove these requirements, or those of them which are in dispute and are relevant, he may be able to recover payment if he can rely upon an architect's final and conclusive certificate or an arbitrator's award in his favour.

The subject-matter of the above paragraph is dealt with in detail below.

Section 2: What is Extra Work?

1. Meaning

There is no generally accepted definition of extra work, but in a lump sum contract it may be defined as work not expressly or impliedly included in the work for which the lump sum is payable.[1] If work is included in the original contract sum the contractor must carry it out and cannot recover extra payment for it, although he may not have thought at the time of entering into the contract that it would be necessary for the completion of the contract.[2] The question is one of construction in each case, but lump sum contracts may be broadly classified into those in which the contractor's obligation is defined in wide terms, such as " to build a house," and those in which it is defined in exact terms, such as " to execute so many cubic metres of digging."

2. Lump sum contract for whole work—widely defined

Indispensably necessary works

Where the contractor must complete a whole work,[3] such as a house, or a railway from A to B, for a lump sum, the courts readily infer a promise on his part to provide everything indispensably necessary to complete the whole work.[4] Such necessary works are not extras for they are impliedly included in the lump sum.[5] Examples of the applications of this principle are as follows:

---

[1] *Cf. Kemp* v. *Rose* (1858) 1 Giff. 258, 268.
[2] *Sharpe* v. *San Paulo Ry.* (1873) L.R. 8 Ch.App. 597.
[3] Sometimes referred to as a " specific " or an " entire " work.
[4] *Williams* v. *Fitzmaurice* (1858) 3 H. & N. 844; *Coker* v. *Young* (1860) 2 F. & F. 98; *Sharpe* v. *San Paulo Ry., supra.*
[5] *Ibid.*

(i) Work not expressly specified. In *Williams* v. *Fitzmaurice* [6] there was a contract to build a house " to be completed and dry and fit for Major Fitzmaurice's occupation by August 1, 1858." In the specification the contractor undertook to provide " the whole of the materials mentioned or otherwise in the foregoing particulars necessary for the completion of the work," and " to perform all the works of every kind mentioned and contained in the foregoing specification for the sum of £1,100." Flooring was omitted from the specification, and the contractor, on this ground, refused to put it in unless it was paid for as an extra. The employer thereupon turned the contractor off the site, seized flooring boards brought upon the site by the contractor and used them to complete. It was held that the contractor could recover neither the amount outstanding under the contract nor the cost of the floorboards, for, although they were omitted from the specification, " it was clearly to be inferred from the language of the specification that the plaintiff was to do the flooring." [7]

(ii) Work not taken out on the quantities supplied to the contractor for tender,[8] or wrongly stated on the drawings. Thus in *Sharpe* v. *San Paulo Ry.*[9] the contractor had undertaken to make a railway line " from terminus to terminus complete." [10] In carrying out the work it was found that the engineer's original plan was quite inadequate and had to be replaced by another. As a result the contractor, upon the engineer's orders, carried out nearly two million cubic yards of excavation in excess of the quantities of work set out in a schedule to the contract, and thus nearly doubled the excavation originally contemplated. It was held that these works were not " in any sense of the words extra works." [11]

(iii) Unexpected labour caused by difficulties of the terrain,[12] or by the proposed method of carrying out the works.[13]

(iv) Work caused by the lawful and not unreasonable exercise by the employer of statutory powers existing at the time of entering into the contract.[14]

No implied warranties by employer

A contractor who has been put to unexpected expense because of inaccurate quantities or drawings or impracticable plans cannot usually recover the expense by bringing an action for breach of an implied warranty that the plans, drawings or bills of quantities are accurate or practicable. No such warranties are implied merely from the fact that these documents are submitted to the contractor for

---

[6] *Supra.*

[7] *Ibid.* at p. 851.

[8] *Scrivener* v. *Pask* (1866) L.R. 1 C.P. 715; *Re Ford and Bemrose* (1902) H.B.C. (4th ed.), Vol. 2, p. 324 (C.A.).

[9] *Supra.*

[10] *Ibid.* at p. 608, and note cl. 25 of the contract set out at p. 599.

[11] *Ibid.* See also *Thorn* v. *London Corp.* (1876) 1 App.Cas. 120 (H.L.), explained in *Re Ford and Bemrose, supra,* at p. 332.

[12] *Bottoms* v. *York Corp.* (1892) H.B.C. (4th ed.), Vol. 2, p. 208 (C.A.); *McDonald* v. *Workington Corp.* (1892) H.B.C. (4th ed.), Vol. 2, p. 228 (C.A.); *Re Nuttall & Lynton & Barnstaple Ry.* (1899) H.B.C. (4th ed.), Vol. 2, p. 279 (C.A.). See *Jackson* v. *Eastbourne Local Board* (1885) H.B.C. (4th ed.), Vol. 2, p. 81 at p. 90 (H.L.), for statement of principle.

[13] *Thorn* v. *London Corp., supra; Tharsis Sulphur & Copper Co.* v. *M'Elroy & Sons and others* (1878) 3 App.Cas. 1040 (H.L.). *Cf. Bacal Construction (Midland) Ltd.* v. *Northampton Dev. Corp.* (1976) 237 E.G. 955 (C.A.) discussed *post,* p. 101.

[14] *Rigby* v. *Bristol Corp.* (1860) 29 L.J.Ex. 359.

tender,[15] nor even from their attachment to the contract as a schedule [16] but the express words of the invitation to tender may show an intention to warrant the accuracy of statements which it contains.[16a]

3. Lump sum contract for whole work—exactly defined

Bills of quantities contract

The term bills of quantities contract is used here to describe a contract where the bills of quantities form part of the contract and describe the work to be carried out for which a lump sum is payable.[17] The contractor may be, and usually is, bound by the terms of the contract to carry out work in excess of that stated in the bills of quantities if it is necessary to complete the contract, but in a bills of quantities contract such excess work is extra work.[18] This type of contract has been said to be " obviously unsafe " for an employer because it can hardly ever be known beforehand what exact quantities of work may be necessary to complete [19]; conversely it may save the contractor much trouble and loss.[20]

It is sometimes a difficult question of construction to determine whether the quantities form part of the contract. The mere fact that quantities are submitted to the contractor for the purposes of tender does not make them form part [21]; and they have been held not to form part: where there was an express power reserved under the contract to the architect to rectify any mistakes in the quantities upon which the contractor's tender was based [22]; where they were a schedule to a contract to do work according to a specification [23]; where there was a contract to erect certain buildings in accordance with plans and specification, and the specification consisted of a bill of quantities with rather fuller description than usual.[24] But they were held to form part where the contract was to do work " according to the plans and the quantities there given by the architect." [25] In *Patman & Fotheringham* v. *Pilditch* [26] the contract was to erect and complete, fit for occupation, a block of flats for a lump sum " according to the plans, invitation to tender, specification and bills of quantities signed by the contractors." Channell J. decided, though not " by any means without doubt," [27]

---

[15] *Thorn* v. *London Corp., supra*; *Re Ford and Bemrose, supra.* See further *post*, p. 89, for a full discussion of inaccurate statements and when they give rise to a remedy.

[16] *Ibid.*

[16a] See *Bacal Construction* case, *supra.*

[17] Bills of quantities may be an important contract document without describing the work for which a lump sum is payable; see *post*, p. 66. In such a case the contract is not a bills of quantities contract in the sense used here.

[18] *Kemp* v. *Rose, supra*; *Patman and Fotheringham Ltd.* v. *Pilditch* (1904) H.B.C. (4th ed.), Vol. 2, p. 368.

[19] *Kemp* v. *Rose, supra*, at p. 268.

[20] See the advice of Stephen J. to contractors in *Priestley* v. *Stone* (1888) H.B.C. (4th ed.), Vol. 2, p. 134 at p. 140.

[21] *Re Ford and Bemrose, supra.*

[22] *Young* v. *Blake* (1887) H.B.C. (4th ed.), Vol. 2, p. 110.

[23] *Sharpe* v. *San Paulo Ry.* (1873) L.R. 8 Ch.App. 597.

[24] In *Re Ford and Bemrose* (1902) H.B.C. (4th ed.), Vol. 2, p. 324 (C.A.).

[25] *Kemp* v. *Rose* (1858) 1 Giff. 258.

[26] (1904) H.B.C. (4th ed.), Vol. 2, p. 368.

[27] At p. 372. It was necessary to distinguish the very strong case of *Ford and Bemrose, supra*, and the dictum therein of Lord Esher at p. 330.

that the quantities signed by the contractors formed part of the contract. They were therefore entitled to recover for all work done by them in completing the contract which had been omitted from or understated in the bills of quantities. The learned judge expressly reserved the case of " things that everybody must understand are to be done, but which happen to be omitted from the quantities," and thought that it would be covered by *Williams* v. *Fitzmaurice.*[28]

Other exactly defined contracts

A bills of quantities contract is the most exact way of describing the contractor's obligation, but other forms of contract may describe his obligation with considerable precision. It was said of the contract in *Williams* v. *Fitzmaurice* [29] that, " the contract was that the house should be complete and fit for occupation by August 1, 1858, and not that the works therein before mentioned should be completed by that day." [30] It follows that the contractor may, by clear words, limit his obligation to works expressly described in the specification. Work not so described will therefore be extra work.[31]

Contractor's implied obligations

Even in an exactly defined contract there is usually room for the implication of some obligations by the contractor so that their performance is not extra work. But the greater the detail used in the bills of quantities or other contract documents to describe the obligation for which the lump sum is payable the less scope there is for such implication.[32]

4. Measurement and value contracts

In a measurement and value contract [33] it is usually immaterial whether any particular item of work that a contractor has to do is in the contract or not, because the contractor is entitled to be paid for it at the contract rate if it is applicable, or at a reasonable price if it is not.[34] But where such a contract provides for the payment of a specified sum of money for a specified item of work, it is a question of construction to determine what work is impliedly included in that item of work and is not therefore extra work. It is submitted that the principles of construction applicable to lump sum contracts apply to each item.[35] And it may be important to determine whether work is of the type contemplated by the contract and therefore governed by the conditions of the contract including price, or is work outside the contract [36] and not therefore subject to the contract conditions or price.

---

[28] *Ante*, p. 63, at p. 373.
[29] *Ante*, p. 63.
[30] (1858) 3 H. & N. 844, 851.
[31] See notes to Standard form of building contract, dealing with the position where quantities do not form part, *post*, p. 292.
[32] See cases cited, *supra* in this chapter, and for implication of terms generally, see Chap. 2. For design and build contracts, see Chap. 1.
[33] For meaning, see *ante*, p. 57.
[34] *Re Walton-on-the-Naze U.D.C.* v. *Moran* (1905) H.B.C. (4th ed.), Vol. 2, pp. 376, 380.
[35] *Cf. Appleby* v. *Myers* (1867) L.R. 2 C.P. 651.
[36] For meaning, see *post*, p. 70.

Comparison with bills of quantities contract

A bills of quantities contract is not a measurement and value contract in the sense in which the term is used in this book. Where the contract was for a " lump sum," the bills of quantities forming part of the contract, and " all variations to be . . . added to, or deducted from the lump sum . . ." it was held that the final account must be taken by adjusting the lump sum for variations, and not by remeasuring the works as carried out and applying the bill rates to the quantities so found.[37] But bills of quantities can form part of a measurement and value contract,[38] see *infra*.

Pricing errors

The difference between a contract to do the work described in the bills of quantities for a lump sum and a measurement and value contract is important when considering pricing errors.[39] Assume two contracts; one for a lump sum; one measurement and value; bills of quantities are in each case a contract document. Say the contractor in pricing the bills of quantities has correctly stated his price per unit of measurement but has incorrectly extended the price and this error has been incorporated into the lump sum in the one case and into his estimated total of prices in the other case. Neither party noticed the error before the contract was made and there are no grounds for rectification.[40] If the contract is for a lump sum there is, it is submitted, no implied right to have the contract price adjusted to take account of the error.[41] If the contract is one of measure and value where the contractor is to be paid £X per unit of measurement, the error will disappear on valuation of the measured work at completion. Bills of quantities in a measurement and valuation contract do not define the work for which a lump sum is payable but are a schedule defining units of measurement and rates for pricing those units.

Section 3: *Promise to Pay for Extra Work*

The mere fact that a contractor has carried out extra work does not of itself entitle him to demand payment. He must show a contract to pay.[42] This he may do either by proving that the work was properly ordered under a provision entitling him to payment in the original contract,[43] or that there is a fresh contract to pay for the work.[44]

1. Work done without request

If the contractor has undertaken to do specified work with certain materials for an agreed price, and without request uses better materials or does more work,

---

[37] *London Steam Stone Sawmills Co.* v. *Lorden* (1900) H.B.C. (4th ed.), Vol. 2, p. 301 (D.C.).

[38] See I.C.E. conditions (1955 ed.), cl. 56; (1973 ed.), cl. 56.

[39] See *post*, p. 71, for effect of pricing errors on pricing variations.

[40] See *ante*, p. 47.

[41] The unreported case of *M. V. Gleeson Ltd.* v. *Sleaford U.D.C.* (1953) noted in H.B.C. (10th ed.), p. 521, may be of assistance.

[42] *Tharsis Sulphur and Copper Co.* v. *M'Elroy & Sons and others* (1878) 3 App.Cas. 1040, 1053 (H.L.).

[43] See *infra*.

[44] See *post*, p. 69, " Implied promise to pay "; p. 70, " Work outside the contract."

this does not entitle him to demand extra payment [45]; and if the materials or work are not in accordance with the contract he may not be able to recover the contract price because he has not completed the contract. [46] Mere permission by the employer to do work different from that contracted for must be distinguished from a request. Thus contractors had undertaken as part of a lump sum contract to make certain girders. They found that it was impracticable or very expensive to make them in the specified way and applied for, and were granted, permission to make them thicker. Lord Blackburn said: " I think there is nothing in that to imply that there was to be a payment for that additional thickness." [47]

Emergency

It is possible that a contractor who in an emergency, when it is impossible to obtain instructions from the employer, expends money in preserving the employer's property can recover payment. [48]

2. No knowledge of increased expense

If the contractor has undertaken to carry out certain work at an agreed price and the employer consents to the execution of different work, the employer is not liable for any increased cost unless he knows, or must be taken to know, that the different work will cost more. [49]

3. Absence of consideration [50]

The contractor may refuse to carry out certain work unless the employer promises to pay for it as extra work. If it appears that the work is not in fact extra work because it is included in the work for which the original contract sum is payable, the employer's promise is not binding for lack of consideration.

---

[45] *Wilmot* v. *Smith* (1828) 3 C. & P. 453; *Bottoms* v. *York Corp.* (1892) H.B.C. (4th ed.), Vol. 2, p. 208 (C.A.).

[46] *Forman* v. *The Liddesdale* [1900] A.C. 190 (P.C.); *Ashwell and Nesbit Ltd.* v. *Allen & Co.* (1912) H.B.C. (4th ed.), Vol. 2, p. 462 (C.A.).

[47] *Tharsis Sulphur and Copper Co.* v. *M'Elroy & Sons and others* (1878) 3 App.Cas. 1040, 1053 (H.L.). The majority of the House of Lords seem to have decided the case on the grounds that: (a) the work was not an extra; (b) there was no order in writing as required by the contract. See also *Kirk & Kirk Ltd.* v. *Croydon Corp.* [1956] J.P.L. 585; *Simplex Concrete Piles Ltd.* v. *St. Pancras Borough Council* (1958) unrep. save in H.B.C. (10th ed.), p. 526.

[48] *Falcke* v. *Scottish Imperial Insurance Co.* (1886) 34 Ch.D. 234, 248-249 (C.A.) appears to be an authority to the contrary, but the court might imply a request; consider the authorities collected and the comment in *The Law of Restitution*, Chap. 14, by Goff and Jones. And see the statements of principle in *Owen* v. *Tate* [1976] Q.B. 402 (C.A.). See also (1977) 93 L.Q.R. 273.

[49] *Lovelock* v. *King* (1831) 1 Moo. & Rob. 60; *Johnson* v. *Weston* (1859) 1 F. & F. 693; *Thames Iron Works, etc., Co.* v. *Royal Mail, etc., Co.* (1861) 13 C.B.(N.S.) 358, 378.

[50] *Sharpe* v. *San Paulo Ry.* (1873) L.R. 8 Ch.App. 597, 608, *per* James L.J., " It is perfectly *nudum pactum*." For the right to set up by way of an equitable defence, despite the absence of consideration, a promise acted upon by the promisee so as to affect his legal position, see *Central London Property Trust Ltd.* v. *High Trees House Ltd.* [1947] K.B. 130; *Combe* v. *Combe* [1951] 2 K.B. 215 (C.A.); *Tool Metal Manufacturing Co. Ltd.* v. *Tungsten Electric Co. Ltd.* [1955] 1 W.L.R. 761 (H.L.); *Ajayi* v. *R. T. Briscoe (Nigeria) Ltd.* [1964] 1 W.L.R. 1326 (P.C.); *Woodhouse Ltd.* v. *Nigerian Produce Ltd.* [1972] A.C. 741 (H.L.). See also *City and Westminster Properties (1934) Ltd.* v. *Mudd* [1959] Ch. 129. For a promise to perform a duty owed to another, see *New Zealand Shipping Co. Ltd.* v. *A. M. Satterthwaite & Co. Ltd.* [1975] A.C. 154 (P.C.).

Section 4: *Agent's Authority*

An architect or other agent of the employer in the position of the architect has
no implied power to vary the terms of the contract,[51] or to vary the contract works
such as by ordering extras,[52] or to order as extras works impliedly included in
the work for which the contract sum is payable.[53] If therefore the contractor has
carried out extra work under the authority of the architect he must show: (a)
that the architect had an authority to order extra work, and (b) that the particular work for which he is claiming was properly ordered within the scope of
that authority.[54] If there is a written contract the question is one of construction,
otherwise it is a question of fact.[55]

A contract to do a whole work usually gives the architect an express power to
order extras. Such a power will not, in the absence of express words, be construed as extending beyond the scope of the original whole work [56]; but if work
outside the contract is ordered to the knowledge of the employer the contractor
will normally be able to recover a reasonable price for such work on a fresh
contract.[57]

Section 5: *Compliance with Condition Precedent*

1. Written orders

Contracts frequently provide that extras must be ordered in a certain manner.
The purpose of these provisions is usually to prevent unauthorised or extravagant
claims for extras. A frequent requirement is that there must be a written order
signed by the architect and that no extras will be paid for unless so ordered.
In such a contract a proper written order is a condition precedent to payment for
extras.[58]

The form of the order will depend upon the wording of the contract, but in
general the writing relied on must be a clear definite order and not a mere
passing reference to some extra work. Thus progress certificates referring to
work claimed to be an extra,[59] and unsigned sketches and drawings prepared in
the architect's office [60] have been held not to be written orders. The contract
may require that such orders must be given before the extra work is carried

---

[51] *Sharpe* v. *San Paulo Ry., supra.*
[52] *R.* v. *Peto* (1826) 1 Y. & J.Ex. 37; *Cooper* v. *Langdon* (1842) 10 M. & W. 785; *Ranger* v.
G.W. Ry. (1854) 5 H.L.C. 72 (H.L.); *Carlton Contractors* v. *Bexley Corp.* (1962) 60 L.G.R.
331.
[53] See *supra*, note 51.
[54] See *supra*, note 52.
[55] *Wallis* v. *Robinson* (1862) 3 F. & F. 307, N.P.; *R.* v. *Peto, supra*, noting the remarks of
Moulton L.J. in the unreported case of *Stevens* v. *Mewes & Davis* (1910) June 8 (C.A.),
cited H.B.C. (10th ed.), p. 532.
[56] *R.* v. *Peto, supra*; *Russell* v. *Sa da Bandeira* (1862) 13 C.B.(N.S.) 149; *cf. Sir Lindsay
Parkinson & Co. Ltd.* v. *Commissioners of Works* [1949] 2 K.B. 632 (C.A.), discussed *ante*,
p. 58.
[57] *Russell* v. *Sa da Bandeira, supra.*
[58] *Russell* v. *Sa da Bandeira, supra; Taverner & Co. Ltd.* v. *Glamorgan County Council*
(1941) 57 T.L.R. 243.
[59] *Tharsis Sulphur and Copper Co.* v. *M'Elroy & Sons and others, supra.*
[60] *Myers* v. *Sarl* (1860) 3 E. & E. 306.

out,[61] but it may be that if an otherwise valid written order is given retrospectively the courts will treat this as a waiver of the requirement.

2. Recovery without written orders

The general rule is that in the absence of written orders, or other formalities which are conditions precedent to payment, the contractor cannot recover either under the contract, or on a fresh contract to pay a reasonable sum, even though the employer has had the benefit of the extra work.[62] To this rule there are certain exceptions:

(a) *Implied promise to pay*

" When there is a condition in the contract that extras shall not be paid for unless ordered in writing by the architect . . . and the employer orders work which he knows, or is told, will cause extra cost, a jury [63] or an arbitrator may find that there was an implied promise by the employer that the work should be paid for as an extra and especially so in cases where any other inference from the facts would be to attribute dishonesty to the employer." [64] Such a promise may be implied where there has been a waiver of the condition. " In order to constitute a waiver there must be conduct which leads the other party reasonably to believe that the strict legal rights will not be insisted on." [65] Thus in principle a written waiver by the employer would be effective,[66] and even an oral waiver would be sufficient if it were a clear undertaking not to rely on the condition.[67] In *Molloy* v. *Liebe* the contractor maintained that certain work was extra work. The employer said that it was not and insisted on the work being done. Upon arbitration the arbitrator held that the work was extra work, and inferred a promise on the part of the employer to pay for it if it should be found to be extra work, although it was not ordered in writing as an extra in the manner required by the contract. The arbitrator's award was upheld, and it was said that it was difficult to see how he could have drawn any other inference without attributing dishonesty to the employer.[68]

Acceptance of work orally ordered by the architect does not show an implied promise to pay [69]; neither has the architect any implied authority to waive a

---

[61] *Lamprell* v. *Billericay Union* (1849) 18 L.J.Ex. 282.

[62] *Kirk* v. *Bromley Union* (1848) 12 Jur. 85; *Ranger* v. *G.W. Ry.*, *supra*; *Taverner* v. *Glamorgan County Council*, *supra*.

[63] Note that juries do not decide such matters today. The judge decides questions of fact, or remits them to a referee—see Chap. 14.

[64] Extract from H.B.C. (6th ed.), p. 313, referred to with approval by Humphreys J., *Taverner* v. *Glamorgan County Council*, *supra*, at p. 245. See also *Tool Metal Manufacturing Co. Ltd.* v. *Tungsten Electric Co. Ltd.* [1955] 1 W.L.R. 761 (H.L.) and other authorities referred to at footnote 50 *supra* and consider restitution referred to in footnote 48 *supra*.

[65] *Per* Denning L.J., *Rickards* v. *Oppenheim* [1950] 1 K.B. 616, 626 (C.A.).

[66] See *Taverner* v. *Glamorgan County Council*, *supra*, at p. 245.

[67] See *Molloy* v. *Liebe* (1910) 102 L.T. 616; *cf. Franklin* v. *Darke* (1862) 6 L.T. 291. In certain circumstances the contractor may be able to say that the employer is estopped from setting up the absence of a written order as a defence. For estoppel there must be a clear unequivocal representation by the employer, see *Woodhouse Ltd.* v. *Nigerian Produce Ltd.* [1972] A.C. 741 (H.L.). See also *post*, p. 119.

[68] *Molloy* v. *Liebe*, *supra*, at p. 617.

[69] *Taverner* v. *Glamorgan County Council* (1941) 57 T.L.R. 243; *Kirk* v. *Bromley Union* (1848) 12 Jur. 85.

term of the contract requiring extras to be ordered in writing.[70] But the court may find that where the employer has desired the execution of extra works, and has stood by and seen the expenditure on them, and taken the benefit of that expenditure, that it would be a fraud on the part of the employer to refuse to pay on the ground that the work was not properly ordered; in such a case the employer will be ordered to account for the value of the extra work.[71] The mere fact that work was done on oral orders by an agent and payment is then refused is by itself no indication of fraud.[72]

(b) *Work outside the contract*

Extra work may be of the kind contemplated by clauses of the contract which provide for the ordering of extras or it may be so peculiar and so different that it is outside the contract.[73] It may be work outside the contract if it is carried out after completion of the original contract work.[74] Extra work outside the contract is not governed by the terms of the contract, and need not therefore be ordered in writing. The employer is liable to pay a reasonable price for such work carried out at his request.[75]

(c) *Arbitration clause*

If there is an arbitration clause in the contract it may be so worded as to give the arbitrator, on a proper reference being made, " power to dispense with the conditions precedent and to order that, notwithstanding the non-performance of those conditions precedent a liability may be established on which money may be ordered to be paid." [76] In *Brodie* v. *Cardiff Corporation* the architect refused to issue a written order for extras on the ground that work required to be carried out was included in the contract price. On a reference the arbitrator awarded sums of money to be paid in respect of the extras despite the absence of an order in writing and his decision was upheld.

(d) *Architect's final certificate—See infra.*

Section 6: *Effect of Architect's Final Certificate*

The parties may by the contract give the architect power to decide various matters finally between them, and to state his decision in the form of a certificate. Having given the architect this power they are not allowed by the courts, in the absence of fraud or other special circumstances, to attack his decision upon such matters; his certificate is binding and conclusive upon them.[77] Thus where

---

[70] *Sharpe* v. *San Paulo Ry.* (1873) L.R. 8 Ch.App. 597.
[71] *Hill* v. *South Staffs Ry.* (1865) 12 L.T.(N.S.) 63, 65; *cf. Kirk* v. *Bromley Union* (1848) 12 Jur. 85.
[72] *Taverner* v. *Glamorgan County Council, supra*, at p. 246.
[73] *Thorn* v. *London Corp.* (1876) 1 App.Cas. 120 (H.L.); *Goodyear* v. *Weymouth Corp.* (1865) 35 L.J.C.P. 12; *cf. Sir Lindsay Parkinson & Co. Ltd.* v. *Commissioners of Works* [1949] 2 K.B. 632 (C.A.), discussed *ante*, p. 58.
[74] *Russell* v. *Sa da Bandeira* (1862) 13 C.B.(N.S.) 149.
[75] *Reid* v. *Batte* (1829) Moo. & M. 413; *Russell* v. *Sa da Bandeira, supra.*
[76] Greer L.J., *Prestige & Co. Ltd.* v. *Brettell* [1938] 4 All E.R. 346, 354 (C.A.), referring to *Brodie* v. *Cardiff Corp.* [1919] A.C. 337 (H.L.), and *Neale* v. *Richardson* [1938] 1 All E.R. 753 (C.A.). For arbitration generally, see Chap. 13.
[77] See generally Chap. 6.

architects have certified by final certificates that certain money is due to the contractor, employers have not been allowed to go behind the certificate and say that a smaller sum was due because the certificate included payment for extras not ordered in writing [78] or not otherwise ordered properly,[79] or for work which was not extra work,[80] or for work which had not been done at all.[81] It is a matter of construction in each case to determine whether the architect's decision is intended to be binding and conclusive on matters relating to extras.[82] For mistakes of law by architects and grounds for attacking their certificates see *post*, p. 80.

Section 7: *Rate of Payment*

1. **Work within the contract**

Payment for extra work of the kind contemplated by the contract will be at the contract rates if any; if there are none it will be a reasonable sum.[83] Where the contract provided that " Additional work shall be paid for at rates pro rata to the estimate " and there were no rates in the estimate which was for a lump sum, it was held that these words had no application to the pricing of extra work for which a reasonable sum was therefore payable.[84]

Effect of pricing errors

When the contractor has made an error in his pricing of the tender for a lump sum contract and there are no grounds for rectification,[85] and the contract provides for payment of variations at rates shown in the tender, a difficult question can arise when pricing variations and the error is apparent. Should any, and if any, what, adjustment be made in the rates shown in the tender to arrive at the new rate for pricing variations ? [86] Many surveyors in practice claim to make an adjustment.[87] It is thought that there is no generally accepted custom and that the question must always be one of construction. The matter can conveniently be dealt with by an express term.[88]

2. **Work outside the contract**

For work outside the contract the contractor is entitled to a reasonable

---

[78] *Goodyear* v. *Weymouth Corp., supra; Laidlaw* v. *The Hastings Pier Co.* (1874) H.B.C. (4th ed.), Vol. 2, p. 13.

[79] *Lapthorne* v. *St. Aubyn* (1885) 1 Cab. & El. 486.

[80] *Laidlaw* v. *The Hastings Pier Co., supra; Richards* v. *May* (1883) 10 Q.B.D. 400 (D.C.).

[81] *Laidlaw* v. *The Hastings Pier Co., supra.* For a sale of goods case where a mistake subsequently admitted by an official certifier did not invalidate the certificate, see *Toepfer, Alfred C.* v. *Continental Grain Co.* [1974] 1 Lloyd's Rep. 11 (C.A.), but note that this was a certificate *in rem, i.e.* upon which numerous persons in a chain of transactions would rely.

[82] See *Pashby* v. *Birmingham Corp.* (1856) 18 C.B. 2; *Re Meadows & Kenworthy* (1896) H.B.C. (4th ed.), Vol. 2, p. 265 (C.A., aff. H.L.). See further, Chap. 6.

[83] *Thorn* v. *London Corp., supra,* at p. 126.

[84] *Reliance Shopfitters Ltd.* v. *Hyams* (1960) unrep., H.H. Percy Lamb Q.C., Official Referee.   [85] See *ante*, p. 47.

[86] The lump sum is not altered, see *ante*, p. 66.

[87] *e.g.* assume a lump sum of £1,000 is £45 less than it would have been but for an error in extending a unit price; it is sometimes said that a reduction of 45/1045 should be made in pricing variations.

[88] See, *e.g.* The Code of Procedure for Selective Tendering, issued by the National Joint Consultative Committee of Architects, Quantity Surveyors and Builders—a new code was issued in 1977.

sum.[89] The normal rule is that however great the amount of work outside the contract, the work within the contract is paid for at the contract rates and only work outside the contract is paid for at a reasonable rate[90]; but " if a man contracts to work by a certain plan, and that plan is so entirely abandoned that it is impossible to trace the contract, and to what part of it the work shall be applied, in such a case the workman shall be permitted to charge for the whole work done by measure and value, as if no contract at all had ever been made." [91]

Section 8: *Production of Agreement*

Where the agreement is in writing and the contractor is claiming payment for extra work he must produce the written agreement to prove that the work claimed for is an extra,[92] unless the extra work is work outside the contract and was therefore carried out under a separate contract express or implied.[93]

Section 9: *Appropriation of Payments*

An employer may from time to time pay money generally on account without appropriating it to any particular items or part of the work. If the contractor has carried out extra work for which he has no claim for payment from the employer because of the failure of a condition precedent or for some other reason, the contractor cannot appropriate the money to payment of such extra work.[94] The general rule is that in the absence of a specific appropriation by the debtor the creditor may appropriate payments on account to whatever debts he pleases, but ". . . before such a question can arise, it must be plain that there must be *two* debts. The doctrine never has been held to authorise a creditor receiving money on account to apply it towards the satisfaction of what does not, nor ever did, constitute any legal or equitable demand against the party making the payments." [95]

Section 10: *Omissions*

The contract usually gives the employer or the architect power to order part of the work to be omitted with a consequent adjustment of the contract price. There is very little authority dealing with the exercise of such a power. On the construction of the contract it may not extend to the ordering of variations,[96] and it may not give the employer the right to omit part of the work from the contract with the object of giving it to another contractor.[97]

Section 11: *Effect of By-Laws*

Compliance with by-laws as the works are carried out may require unanticipated

---

[89] See *supra*, note 83.
[90] *Sir Lindsay Parkinson & Co. Ltd.* v. *Commissioners of Works* [1949] 2 K.B. 632 (C.A.).
[91] *Per* Lord Kenyon, *Pepper* v. *Burland* (1792) 1 Peake N.P. 139.
[92] *Vincent* v. *Cole* (1828) 1 M. & M. 257.
[93] *Vincent* v. *Cole, supra*; *Reid* v. *Batte, supra*; *Parton* v. *Cole* (1841) 11 L.J.Q.B. 70.
[94] *Lamprell* v. *Billericay Union* (1849) 18 L.J.Ex. 282.
[95] *Ibid.* A similar rule was applied to work illegal under the Defence Regulations. See *post*, p. 110.
[96] *R.* v. *Peto* (1826) 1 Y. & J.Ex. 37.
[97] See the American case of *Gallagher* v. *Hirsch* (1899) N.Y. 45 App.Div. 467 and the Australian case of *Carr* v. *J. A. Berriman Pty. Ltd.* (1953) 27 A.L.J. 273. For the effect of unauthorised omissions on the contractor's right to payment, see Chap. 4.

work to be done.[98] Subject to express terms dealing with the matter [99] there is, it is submitted, normally an implied term by the contractor not to complete the work in a manner which contravenes relevant by-laws or other statutory requirements as to methods of construction. In any event he may be liable in tort or for breach of statutory duty if the employer suffers loss caused by the contractor's negligent failure to comply with by-laws.[99a] Can the contractor recover extra payment for unanticipated work required to comply with the by-laws ? The answer depends, it is submitted, upon the application of the principles set out earlier in this chapter and in particular on: (a) whether the work for which the contract sum is payable is defined in terms wide enough to include work unspecified but necessary to comply with by-laws,[1] and (b) where the contract work was not defined in such wide terms, whether the contractor sought the employer's instructions before carrying out the unanticipated work necessary to comply with the by-laws or can otherwise show a promise to pay.[2]

Section 12: *Prime Cost and Provisional Sums*

1. P.C. or prime cost sums

The contractor sometimes undertakes to provide an article at the " P.C." or at the " prime cost sum " of £X. These terms are usually defined in the contract; but in the absence of such definition the ordinary meaning [3] of prime cost is, it is submitted, net cost, and the intention is therefore that the contractor should only charge the employer the actual cost to himself. If this cost is less or more than that stated in the contract the contract sum is adjusted accordingly, subject always to the principles relating to extra work set out in this chapter.

2. Provisional sums

The contract sum may include a provisional sum or provisional amount to cover some expenditure the amount of which is not known at the time of entering into the contract. Thus it may be for extras, or for some item which cannot be exactly estimated, or for a sub-contract which has to be placed after the main contract is entered into.[4] The term is usually defined, but the general rule is that the original contract sum is adjusted according to whether the actual expenditure ordered is greater or less than the provisional sum. It is usual to provide for an adjustment of the contractor's profit in accordance with the alteration of the contract sum.

---

[98] For non-compliance with by-laws, see *post*, p. 108, and for implied terms, see *ante*, p. 35. Outside London, the Building Regulations 1965, 1972, 1976 now apply. The term " by-law " is retained for convenience.

[99] *e.g.* clause 4, Standard form of building contract; and see *Townsends (Builder) Ltd.* v. *Cinema News, etc., Ltd.* [1959] 1 W.L.R. 119 (C.A.), discussed *post*, pp. 109 and 227.

[99a] *Anns* v. *Merton London Borough Council* [1977] 2 W.L.R. 1024 (H.L.).

[1] See *ante*, p. 62.

[2] See *ante*, p. 66.

[3] See *ante*, p. 31. For its meaning in the Standard form of building contract see, *post*, p. 314.

[4] See further, *post*, p. 177. For the meaning of the term in the Standard form of building contract, see *post*, p. 314.

CHAPTER 6

EMPLOYER'S APPROVAL; ARCHITECT'S CERTIFICATES

I. EMPLOYER'S APPROVAL

1. Construction against employer

A contract may provide that work must be completed to the approval of the employer. Such a provision if construed in the employer's favour would be very onerous. If goods do not meet with the approval of a buyer and he rejects them the seller can sell the goods for what they are worth. But in the case of a building contract, when the work is fixed to the land it becomes the property of the owner of the land.[1] If the employer is entitled to say, " I do not approve of the work; you have not carried out your contract, therefore you cannot recover on the contract," the contractor is in an unfortunate position. He cannot sell the work to a third party because it is not his property; he cannot recover on the contract because he has not fulfilled it; and he will have great difficulty even in showing an implied promise to pay a reasonable sum, because mere use and occupation of building work is not evidence of such a promise.[2] For these reasons, and perhaps also because of the maxim " no man shall be a judge in his own cause," [3] clauses of this nature are given a reasonable construction and are construed against the employer.[4] Thus, in the absence of the expression of a contrary intention, the following rules of construction are applied:

(a) *Not a condition precedent*

The court leans against a construction making the approval of the employer a condition precedent to payment, and prefers a construction making the promise to complete according to the employer's approval, and the promise to pay independent of one another.[5] In such a case if the work does not meet with the employer's approval he cannot refuse to pay under the contract, but can only seek a reduction in the contract price by way of set-off, or counterclaim for damages.[6]

(b) *Reasonable, honest and not capricious*

The employer's approval must not be unreasonably, or dishonestly, or capriciously withheld.[7] What is reasonable is a question of fact.[8] The right to

---

[1] See *post*, p. 124.
[2] See *ante*, p. 55.
[3] See *Dimes* v. *Grand Junction Canal (Proprietors)* (1852) 3 H.L.Cas. 794 (H.L.).
[4] *Dallman* v. *King* (1837) 7 L.J.C.P. 6; *Stadhard* v. *Lee* (1863) 32 L.J.Q.B. 75.
[5] *Dallman* v. *King, supra*, at pp. 9, 11. Another way of stating the effect of this proposition is to say that the court treats the promise to complete to the employer's approval as a mere term and not as a condition.
[6] *Ibid*. See Chap. 4 (*ante*, p. 51), " Substantial completion."
[7] *Dallman* v. *King, supra*; *Parsons* v. *Sexton* (1847) 4 C.B. 899; *Andrews* v. *Belfield* (1857) 2 C.B.(N.S.) 779. See also *Cammell Laird & Co.* v. *The Manganese Bronze and Brass Co.* [1934] A.C. 402 (H.L.)—sale of goods to be completed to " entire satisfaction " of a third party.
[8] *Ripley* v. *Lordan* (1860) 2 L.T. 154.

withhold approval may by the terms of the contract be limited to certain parts or qualities of the work.[9] If approval is subject to the completion of certain tests which the employer through his own default fails to carry out, he cannot withhold his approval because the work has not satisfied other tests not agreed upon in the contract.[10]

2. Express words

" Where from the whole tenor of the agreement it appears that however unreasonable and oppressive a stipulation or condition may be, the one party intended to insist upon and the other to submit to it, a court of justice cannot do otherwise than give full effect to the terms which have been agreed upon between the parties . . . without stopping to consider how far they may be reasonable or not." [11] Thus a sub-contract provided that if the sub-contract works did not proceed as rapidly and satisfactorily as required by the main contractors or their agent, the main contractors could put on extra men themselves and deduct the additional cost from the money due under the sub-contract. The court was satisfied that the intention was that this power could be exercised if the main contractors " were dissatisfied, whether with or without sufficient reason," with the progress of the work. It was added that, in the circumstances, the clause was not unreasonable as the main contractors were probably under stringent terms themselves to complete to time.[12]

Where a contractor had undertaken to complete a carriage to B's " convenience and taste," it was held (despite a jury's verdict in favour of the contractor) that B was entitled to reject it if it did not accord with his convenience and taste, assuming that his rejection was bona fide and not capricious.[13] But note that the rejection of goods which can be resold differs in substance from that of work to land which cannot.

3. Approval by alter ego

If work is to be done to the satisfaction of an agent it may, on the construction of the contract, " be plain that he is to function only as the *alter ego* of his master," [14] and not to act in the independent manner of an architect as the term " architect " is used in this chapter. Approval by such an agent is, it is submitted, subject to the same principles as apply to the employer's approval.

4. Approval by employer and architect

Where work was to be carried out to the approval of both the employer and the architect and they had expressed their approval, and the architect had given a final certificate of satisfaction which the contract made binding upon the

---

[9] *Ibid.*
[10] *Mackay* v. *Dick* (1881) 6 App.Cas. 251 (H.L.).
[11] *Per* Cockburn C.J., *Stadhard* v. *Lee, supra,* at p. 78 ; *Diggle* v. *Ogston Motor Co.* (1915) 112 L.T. 1029. See also *Minster Trust Ltd.* v. *Traps Tractors* [1954] 1 W.L.R. 963, 973.
[12] *Ibid.*
[13] *Andrews* v. *Belfield, supra.* See also *Docker* v. *Hyams* [1969] 1 W.L.R. 1060, 1065 (C.A.) where the effect of cases dealing with agreements to provide ships or goods to the purchaser's approval is reviewed.
[14] *Minster Trust Ltd.* v. *Traps Tractors Ltd., supra,* at p. 973.

parties, it was held that the employer, in the absence of fraud or collusion, could not claim damages for defects.[15]

II. ARCHITECT'S CERTIFICATES

Section 1: Types of Certificate

The nature and effect of an architect's certificate depends upon the construction of the particular contract, but in general such certificates may be divided into three classes:

1. Progress or interim certificates

These certificates are issued from time to time during the course of the work certifying that, in the opinion of the architect, work has been carried out, and, in some cases, materials supplied, to the value of £X. They are approximate estimates, made in some instances for the purpose of determining whether the employer is safe in making a payment in advance of the contract sum,[16] in others whether he is under a duty to pay an instalment and if so, how much he is to pay.[17] Such certificates are not normally binding upon the parties as to quality or amount and are subject to adjustment on completion.[18]

Create debt due

Subject to the effect of words showing that it is merely intended to make advances on money not legally due until completion,[19] a progress certificate properly given creates a debt due.[20] The contractor upon the issue of a certificate for payment of an instalment, can seek summary judgment in respect of the debt.[21]

Retention money

It is usual to provide for the retention of a percentage of the value, sometimes subject to a limit as to amount, to provide a fund for the payment of amending defects.[22] In one case the contractor was " entitled . . . under certificates . . . to payment by the employer from time to time by instalments, when in the opinion of the architect actual work to the value of £1,000 [had] been executed in accordance with the contract, at the rate of 90 per cent. of the value of the work so executed. . . ." It was held that an arbitration award was wrong on the face of it in finding that the contractor was only entitled to 90 per cent. of each completed £1,000; he was entitled to payment of 90 per cent. of the value of work actually executed at the time of granting a certificate.[23]

---

[15] *Bateman (Lord)* v. *Thompson* (1875) H.B.C. (4th ed.), Vol. 2, p. 36 (C.A.).
[16] See *Tharsis Sulphur and Copper Co.* v. *M'Elroy* (1878) 3 App.Cas. 1040 (H.L.).
[17] See Standard form of building contract, clause 30 (1).
[18] *Lamprell* v. *Billericay Union* (1849) 18 L.J.Ex. 282.
[19] See *ante*, p. 53.
[20] *Pickering* v. *Ilfracombe Ry.* (1868) L.R. 3 C.P. 235.
[21] *Workman, Clark & Co.* v. *Lloyd Brazileno* [1908] 1 K.B. 968 (C.A.). For a summary judgment and for set-off and counterclaim, see *post*, p. 278.
[22] See, *e.g.* clause 30 (3), Standard form of building contract.
[23] *F. R. Absalom Ltd.* v. *Great Western (London) Garden Village Society* [1933] A.C. 592, 611 (H.L.).

2. Final certificates

A final certificate may certify the amount finally payable to the contractor under the contract, or the satisfaction of the architect that work conforms with the contract, or both.[24] Upon these matters the architect's decision embodied in his certificate is often binding and conclusive on the parties. This is discussed below.

3. Other certificates

The contract may empower the architect to certify various matters, such as the happening of an event which entitles the employer to exercise a right of forfeiture,[25] or to record an extension of time given by the architect to the contractor.[26] The architect's decision may be binding and conclusive upon these matters.[27]

Section 2: Certificates as Condition Precedent

The contract may show by express words, or upon reading it as a whole, that the architect's [28] certificate is a condition precedent to payment whether interim or final. It is a condition precedent if upon the true construction of the contract, the employer only agrees to pay what is certified by the architect,[29] or only to pay upon a certificate of satisfactory completion by the architect.[30] In such a case if the contractor fails to obtain the certificate required for payment he has no claim at law or in equity,[31] unless he can show one of the special circumstances discussed in paragraphs 2–6 of the next section.

The contractor may be able to show that on the wording of the contract a certificate is not a condition precedent to payment.[32]

If the contract only requires a certificate showing the architect's satisfaction, the contractor can sue on the contract when he has obtained such a certificate notwithstanding that it does not certify that a sum of money is payable.[33]

Third parties

Assignees of the contractor are in no better position than the contractor,[34]

---

[24] See, *e.g.* clause 30 (7), Standard form of building contract.

[25] For forfeiture, see *post*, p.120.

[26] See *Sattin* v. *Poole* (1901) H.B.C. (4th ed.), Vol. 2, p. 306 (D.C.). For liquidated damages and extension of time, see *post*, pp. 155 *et seq.*

[27] See *post*, p. 80.

[28] " Architect " is used to mean the architect as defined by the contract. See *post*, p. 83, " Given by wrong person."

[29] *Sharpe* v. *San Paulo Ry.* (1873) L.R. 8 Ch.App. 597, 612.

[30] *Morgan* v. *Birnie* (1833) 9 Bing. 672; *Grafton* v. *Eastern Counties Ry.* (1853) 8 Ex. 699; *Scott* v. *Liverpool Corp.* (1858) 28 L.J.Ch. 230; *Westwood* v. *The Secretary of State for India* (1863) 7 L.T. 736; *Wallace* v. *Brandon & Byshottles U.D.C.* (1903) H.B.C. (4th ed.), Vol. 2, p. 362 (C.A.).

[31] *Scott* v. *Liverpool Corp.*, supra, at p. 239; *Glenn* v. *Leith* (1853) 1 Comm. Law Rep. 569; *Stevenson* v. *Watson* (1879) 4 C.P.D. 148; *Morgan* v. *Lariviere* (1875) L.R. 7 H.L. 423 (H.L.).

[32] See the Scottish case of *Howden* v. *Powell Duffryn Steam Coal Co.*, 1912 S.C. 920; *Re Hohenzollern, etc., and City of London Contract Corp.* (1886) 54 L.T. 596, H.B.C. (4th ed.), Vol. 2, p. 100 (C.A.); *cf. London Gas Light Co.* v. *Chelsea Vestry* (1860) 2 L.T. 217.

[33] *Pashby* v. *Birmingham Corp.* (1856) 18 C.B. 2, 32.

[34] *Lewis* v. *Hoare* (1881) 44 L.T. 66 (H.L.).

but third parties whose rights arise on completion may be able to enforce such rights although no certificate of completion has been given. Thus where H lent money to T, a contractor, on the employer's guarantee to repay upon completion of the contract works in accordance with the building contract, it was held that H could recover from the employer when the works were completed in fact, although no certificate of completion had been given as required by the building contract.[35] The contract of guarantee between H and the employer did not require certificated completion.[36]

Section 3: *Recovery of Payment without Certificate*

1.　Certificate not a condition precedent

See *supra*.[37]

2.　Waiver of condition precedent

The requirement of a certificate as a condition precedent to payment is for the benefit of the employer. He may therefore waive his right to insist upon a certificate. It is submitted that the same principles apply as in the case of waiver of a condition requiring written orders.[38]

3.　Disqualification of the certifier

If as a result of fraud, collusion, or otherwise the architect is disqualified as certifier and fails to grant a certificate the condition precedent goes and the contractor can sue.[39]

4.　Prevention by the employer

The mere failure of the architect to certify where there has been no fraud or collusion or wrongful interference by the employer does not of itself enable the contractor to recover.[40] But if the employer or his agent prevents the architect giving a certificate the employer cannot rely on its absence, for " no person can take advantage of the non-fulfilment of a condition the performance of which has been hindered by himself." [41] While if the architect wrongly neglects, or deliberately, as a result of a mistaken view of his powers, refuses to issue a certificate, and the employer concurs in his action and the contractor has done everything necessary for the issue of the certificate, the employer cannot take

---

[35] *Ibid.*
[36] *Ibid.*
[37] See " Certificates as Condition Precedent," *supra*.
[38] See *ante*, p. 69.
[39] *Panamena Europea Navigacion (Compania Limitada)* v. *Frederick Leyland & Co. Ltd. (J. Russell & Co.)* [1947] A.C. 428 (H.L.); *Hickman* v. *Roberts* [1913] A.C. 229 (H.L.); *Brunsden* v. *Beresford* (1883) 1 Cab. & El. 125. For full discussion of disqualification, see *post*, p. 85.
[40] *Neale* v. *Richardson* [1938] 1 All E.R. 753 (C.A.); *Botterill* v. *Ware Guardians* (1886) 2 T.L.R. 621; *Cooper* v. *Uttoxeter Burial Board* (1864) 11 L.T. 565; *cf. Kellett* v. *New Mills U.D.C.* (1900) H.B.C. (4th ed.), Vol. 2, p. 298; *Scott* v. *Liverpool Corp., supra*, at p. 236.
[41] *Per* Blackburn J., *Roberts* v. *Bury Commissioners* (1870) L.R. 5 C.P. 310, 326; *Panamena, etc.* v. *Frederick Leyland & Co. Ltd., supra*.

3. Not properly made

A certificate is invalid, and therefore not conclusive if not properly made in accordance with the contract.[75]

Form of certificate

The certificate must be in the form, if any, required by the contract. Thus a mere checking of the contractor's account for extras was held not to be a certificate of satisfactory completion.[76] Where, under clause 66 of the I.C.E. conditions (1955 ed.), a " decision " was required as a condition precedent to arbitration, an informal rejection of the contractor's claims was held not to amount to such decision.[77] If the architect is merely required to certify his satisfaction with no express requirement for a written certificate an oral statement of satisfaction is sufficient.[78] In all cases it must be clear that there was an intention to issue the certificate in question and that it was in substance what the contract required.[79] The ordinary rules of construction (see Chap. 3) apply, it is submitted, so that the test of intention is objective. It is thought that evidence of surrounding circumstances such as relevant letters or conversations would frequently be admissible in deciding whether a document (or, in the unusual case, an oral statement) was the certificate.

Delegation of duties

In giving his certificate, the architect is entitled to make use of the assistance of others, such as, for example, a quantity surveyor, for detailed matters of measurement and valuation, but the certificate must be his; he cannot delegate his whole function of certifying.[80]

Given by wrong person

A certificate must be given by the person or persons authorised by the contract.[81] Thus, if a particular person is named as the certifier, in the absence of a term indicating the contrary, that person and no other can give the certificate.[82] Where the contract identifies the certifier by description, as, for example, " anyone whom from time to time the (employer) might choose to select as chief engineer," [83] or " AB or other the engineer " of the employer,[84] then the person

---

[75] *Lamprell* v. *Billericay Union* (1849) 18 L.J.Ex. 282; *Coleman* v. *Gittins* (1884) 1 T.L.R. 8, where an employer successfully resisting payment on a technicality was not awarded his costs; *Kaye* (*P. & M.*) *Ltd.* v. *Hosier & Dickinson* [1972] 1 W.L.R. 146, 157 (H.L.).

[76] *Morgan* v. *Birnie* (1833) 9 Bing. 672; *Elmes* v. *Burgh Market Co.* (1891) H.B.C. (4th ed.), Vol. 2, pp. 170, 171.

[77] *Monmouthshire County Council* v. *Costelloe and Kemple* (1965) 63 L.G.R. 429 (C.A.).

[78] *Roberts* v. *Watkins* (1863) 14 C.B.(N.S.) 592; *Elmes* v. *Burgh Market Co., supra.*

[79] See *Token Construction Company Ltd.* v. *Charlton Estates Ltd.* (1973) unrep. (C.A.), not overruled on this point by *Modern Engineering* (*Bristol*) *Ltd.* v. *Gilbert Ash* (*Northern*) *Ltd.* [1974] A.C. 689 (H.L.), applying a passage from *Minster Trust Ltd.* v. *Traps Tractors Ltd.* [1954] 1 W.L.R. 963, 982, *per* Devlin J.

[80] *Clemence* v. *Clarke* (1880) H.C.B. (4th ed.), Vol. 2, pp. 54, 59 (C.A.); *cf. Burden Ltd.* v. *Swansea Corp.* [1957] 1 W.L.R. 1167, 1173 (H.L.).

[81] *Lamprell* v. *Billericay Union, supra.*

[82] *Ess* v. *Truscott* (1837) 2 M. & W. 385; *Att.-Gen.* v. *Briggs* (1855) 1 Jur.(N.S.) 1084.

[83] *Ranger* v. *G.W. Ry.* (1854) 5 H.L.C. 72, 91 (H.L.).

[84] *Kellett* v. *Mayor of Stockport* (1906) 70 J.P. 154. See also *Cunliffe* v. *Hampton Wick Local Board* (1893) H.B.C. (4th ed.), Vol. 2, p. 250 (C.A.); also 9 T.L.R. 378.

who is the properly appointed engineer (or architect as the case may be) at the time when the certificate is required to be given has power to certify.[85]

Parties not heard

The architect is not normally bound to give anything in the nature of a formal hearing. Where the contract provided that the certifier's decision on a wide range of matters was to " be final and without appeal " it was said that he could if he wished hear arguments from the parties, and state his own view to obtain their opinions, provided that he gave equal opportunity to each party and did not allow his judgment to be influenced by directions given to him in his capacity solely as agent for the employer.[86] On the special wording of another contract it was said that he must hear the parties before giving his certificate in relation to any dispute.[87]

Mistake by architect

A certificate of the architect intended to be binding and conclusive " cannot be impeached for mere negligence, or mere mistake or mere idleness on the part of the architect." [88] The position was formerly said to be different where there is a valuation, and in *Collier* v. *Mason* [89] it was stated thus: " This court upon the principle laid down by Lord Eldon must act on that valuation, unless there be proof of some mistake, or some improper motive, I do not say fraudulent one: as if the valuer had valued something not included, or had valued it on a wholly erroneous principle, or had desired to injure one of the parties to the contract; or even, in the absence of any proof of one of these things, if the price were so excessive or so small as only to be explainable by reference to some such cause; in any one of these cases the court would refuse to act on the valuation." [90] It has never been held that the principle of *Collier* v. *Mason* applies to an architect's certificate, and it now seems unlikely that it would be applied.[91]

---

[85] *Ranger* v. *G.W. Ry.*, *supra*.

[86] *Per* Channell J., *Page* v. *Llandaff, etc., District Council* (1901) H.B.C. (4th ed.), Vol. 2, p. 316, at p. 320. For employer's influence, see *infra*, p. 86.

[87] *Eaglesham* v. *McMaster* [1920] 2 K.B. 169, 174—a contract in which the architect was both certifier and arbitrator. Such a contract is unlikely today because of the effect of the Arbitration Act 1950, s. 24 (1), see *post*, p. 240. See also *Armstrong* v. *South London Tramways Co.* (1891) 7 T.L.R. 123 (C.A.).

[88] *Per* Lindley L.J., *Clemence* v. *Clarke* (1880) H.B.C. (4th ed.), Vol. 2, p. 54, at p. 65 (C.A.); *Goodyear* v. *Weymouth Corp.* (1865) 35 L.J.C.P. 12; *Sharpe* v. *San Paulo Ry.*, *supra*. See also *Campbell* v. *Edwards* [1976] 1 W.L.R. 403 (C.A.)—valuation which gave no reasons not impeachable for mistake; *Toepfer* v. *Continental Grain Co.* [1974] 1 Lloyd's Rep. 11 (C.A.)—grain purchase : . . " Official . . . certificates of inspection to be final as to quality." Binding, despite admitted mistake—but it was a certificate *in rem* (see *ante* p. 70).

[89] (1858) 25 Beav. 200; approved in *Dean* v. *Prince* [1954] Ch. 409 (C.A.), distinguished in *Toepfer's* case, *supra*, and said by Lord Denning M.R. in *Campbell* v. *Edwards*, *supra*, to require reconsideration (at p. 407).

[90] *Supra*, *per* Sir John Romilly M.R. at p. 204.

[91] See *Campbell* v. *Edwards*, *supra*, at p. 407 where Lord Denning M.R. says that the law has been transformed because the parties now have a claim for negligence against architects and valuers since the decisions of the House of Lords in *Sutcliffe* v. *Thackrah* [1974] A.C. 747 and *Arenson* v. *Arenson* [1977] A.C. 405. But see *ante*, p. 80, for error in law.

4. Disqualification of the certifier

An architect is not an arbitrator but he has " two different types of function
to perform. In many matters he is bound to act on his client's instructions,
whether he agrees with them or not; but in many other matters requiring professional skill he must form and act on his own opinion "—*per* Lord Reid, *Sutcliffe*
v. *Thackrah*.[92] Lord Reid was there referring to the architect's position under
the standard form of building contract but it may be taken as applicable to most
forms of building contract where the architect performs the traditional duties of
agent for the employer of supervision and of certification.

The employer and the contractor make their contract on the understanding that
in all matters where the architect has to apply his professional skill he will act in a
fair and unbiased [93] manner in applying the terms of the contract.[94] Such matters
include not only certificates for payment and for satisfaction with the works but
many instances where the architect has to form a professional opinion upon matters
which will affect the amount paid to (or to be deducted from) the contractor.

If the architect fails to act fairly where the contract requires him to do so he is
said to be disqualified as a certifier and his certificates are of no effect,[95] and he
remains disqualified unless the contractor with full knowledge of the facts waives
the breach.[96]

Fraud, collusion, dishonesty

Fraud or collusion with one of the parties, or dishonesty, disqualifies the
certifier.[97] It has been held that a clause that the architect's certificate should
not be set aside or attempted to be set aside on the ground of fraud or collusion
was not void as against public policy, and consequently an allegation of fraud
on the part of the architect in giving a certificate was not entertained by the
court.[98] The decision has been strongly criticised by Lord Justice Scrutton,[99]
and is, it is submitted, open to review by the Court of Appeal.[1]

---

[92] [1974] A.C. 727, 737 (H.L.). There are many old cases (excerpts from which are set
out in earlier editions of this book) referring to the dual function or dual capacity as it was
sometimes called of the architect. Their basis is not, it is submitted, essentially different
from the exposition by the House of Lords in *Sutcliffe* v. *Thackrah* but they frequently use
words which speak in terms of the architect as an adjudicator or quasi-arbitrator and
these terms should no longer be used. It is suggested that it is still, however, correct and
useful when referring to those functions where the architect must act fairly to say that he
must act " independently."

[93] See further on this the heading " Known interests do not disqualify," *infra*.

[94] *Sutcliffe* v. *Thackrah*, p. 737. Lord Morris, p. 751 and Lord Salmon, p. 759 also
use the term " impartially " as part of the description of the duty of the architect. But in
the context they do not use it as suggesting that the architect should or could have the
impartiality of an arbitrator. As to this see Lord Dilhorne at pp. 756–757.

[95] *Hickman* v. *Roberts* [1913] A.C. 229 (H.L.).

[96] *Ibid.* pp. 233, 235, 238. See also *Thornton Hall & Partners* v. *Wembley Electrical
Appliances Ltd.* [1947] 2 All E.R. 630, 634 (C.A.).

[97] *South Eastern Ry.* v. *Warton* (1861) 2 F. & F. 457; *Sharpe* v. *San Paulo Ry.* (1873)
L.R. 8 Ch.App. 597.

[98] *Tullis* v. *Jacson* [1892] 3 Ch. 441.

[99] *Czarnikow* v. *Roth, Schmidt & Co.* [1922] 2 K.B. 478, 488 (C.A.); *cf. Scott* v. *Liverpool
Corp.* (1856) 25 L.J.Ch. 230, 232.

[1] See Denning L.J., *Lazarus Estates Ltd.* v. *Beasley* [1956] 1 Q.B. 702, 712 (C.A.)—
" Fraud unravels everything . . . it vitiates judgments, contracts and all transactions
whatsoever." See also *Re Davstone Estates Ltd.'s Leases* [1969] 2 Ch. 378.

Failure to certify independently

Without any fraud or turpitude on the part of the architect he may so mistake his position, or lack the firmness to repel unworthy communications by one of the parties, that he loses the independence required of him and therefore becomes disqualified.[2] Thus in *Hickman* v. *Roberts*[3] the architect failed to issue certificates at the proper time, and wrote to the contractor saying, "had you better not call and see my clients, because in the face of their instructions to me I cannot issue a certificate whatever my own private opinion in the matter." It was held that he had become so much under the influence of the employer that he had lost his independence, and had not recovered it when he gave a final certificate. It was therefore set aside.

Unknown interests

A certifier is disqualified if he conceals any unusual interest which might influence his mind as certifier, such as a promise to the employer (as opposed to a mere estimate) that the final cost should not exceed a certain figure.[4]

Known interests do not disqualify

The known interests of the architect at the time of entering into the contract do not disqualify him.[5] The interests of the architect which the contractor knows of, or, it is submitted, must be taken to know of, at the time of entering into the contract include: that the architect will have given an estimate of the cost to the employer; that he is paid by and is liable to dismissal by or in damages for negligence to the employer; that he " has in divers ways to look after the interest "[6] of the employer, including supervising the works to see that they comply with the contract. There is " neither unfairness nor partisanship in ensuring that the work is properly carried out."[7] A contractor cannot therefore claim that the architect " must be in the position of an independent arbitrator, who has no other duty which involves acting in the interests of one of the parties."[8]

The architect may consult with the employer, and report to him on the quality of the work, and upon the contractor's expenditure and give him an estimate of the proper sum to be paid to the contractor. Such actions, in the absence of fraud or collusion, do not disqualify him, provided he does not submit to the

---

[2] *Hickman* v. *Roberts* [1913] A.C. 229 (H.L.); *Panamena, etc.* v. *Frederick Leyland & Co. Ltd.* [1947] A.C. 428, 437 (H.L.). For a discussion of the position of certifiers who are not, or are not in the position of, architects, see the case, not on a building contract, of *Minster Trust Ltd.* v. *Traps Tractors Ltd.* [1954] 1 W.L.R. 963.

[3] *Supra.*

[4] *Kimberley* v. *Dick* (1871) L.R. 13 Eq. 1; *Kemp* v. *Rose* (1858) 1 Giff. 258. The giving of an estimate is part of an architect's duties under the R.I.B.A. conditions of engagement (printed *post*, p. 560) and this must be known to contractors.

[5] *Bristol Corp.* v. *Aird* [1913] A.C. 241, 258 (H.L.). *Cf.* the position of an arbitrator; Arbitration Act 1950, s. 24 (1). See *post*, p. 240.

[6] *Sutcliffe* v. *Thackrah*, at p. 741.

[7] *Ibid.* See also *Scott* v. *Carluke Local Authority* (1879) 6 R. 616, 617; *Cross* v. *Leeds Corporation* (1902) H.B.C. (4th ed.), Vol. 2, p. 339 (C.A.); but noting that the terminology now should be that used in *Sutcliffe* v. *Thackrah*. See also *post*, p. 206 for a full discussion of the architect's duties to the employer of which duties the contractor must, it is submitted, at least in outline, be treated as having knowledge.

[8] *Panamena, etc.* v. *Frederick Leyland & Co. Ltd.*, *supra*, at p. 437.

employer's control and influence [9] so as to be incapable of forming an independent view when he comes to give his certificate.[10] On a legal matter such as
the construction of the contract the architect is entitled to consult with the
employer's solicitor and counsel,[11] provided, it is submitted, he treats what they
say not as a direction but as advice only which he is in no way bound to follow.[12]

An architect is not disqualified merely because he is a shareholder in,[13] or
even the president of,[14] the employer company, or is the employee of the
employer,[15] provided that the contractor knows, or it seems, must be taken to
know, of the architect's interest.[16]

5. Effect of an arbitration clause

The effect of an arbitration clause in a contract upon the conclusiveness of a
certificate must depend upon the words of each contract.

(a) *No effect*

The certificate may remain conclusive in a court of law and not be within the
scope of the arbitration agreement.[17]

(b) *Certificate not conclusive at all*

Where a contract provides that either the architect's certificate or the arbitrator's award is a condition precedent to payment the courts of law do not have
jurisdiction to open up the architect's certificate merely because there was a right
of appeal against it to an arbitrator.[18] But the wording of the arbitration clause
may show that the certificate was not intended to be binding at all.[19]

(c) *Certificate conclusive except upon arbitration*

The certificate may be conclusive in a court of law, but subject to review by
arbitration under an arbitration clause in the contract,[20] but it seems that clear

---

[9] See *e.g.* the facts of *Hickman* v. *Roberts, supra.*

[10] *Panamena* v. *Leyland, supra,* distinguishing *Hickman* v. *Roberts, supra,* and approving
the Scottish cases of *Scott* v. *Carluke L.A., supra,* and *Halliday* v. *Hamilton's Trustees*
(1903) 5 F. 800. For what amounts to interference or obstruction in the issue of certificates
where the Standard form of building contract is used, see excerpt from *Burden* v. *Swansea
Corporation* [1957] 1 W.L.R. 1167, 1180 (H.L.) *post,* p. 362. Such conduct would, it is
submitted, if acquiesced in by the architect disqualify him.

[11] See *Panamena* v. *Leyland, supra,* at p. 444.

[12] Even so the architect may in some instances think it more appropriate to take independent advice.

[13] *Ranger* v. *G.W. Ry.* (1854) 5 H.L.C. 72 (H.L.); *Hill* v. *South Staffs Ry.* (1865) 12
L.T. 63, 65.

[14] *Panamena* v. *Frederick Leyland & Co. Ltd., supra.*

[15] *Cross* v. *Leeds Corp., supra.*

[16] *Ranger* v. *G.W. Ry., supra.*

[17] *Re Meadows and Kenworthy* (1896) H.B.C. (4th ed.), Vol. 2, p. 265 (C.A., affd. H.L.);
Brunsden v. *Staines Local Board* (1884) 1 Cab. & El. 272; *East Ham Borough Council* v.
Bernard Sunley & Sons Ltd. [1966] A.C. 406 (H.L.).

[18] *Sharpe* v. *San Paulo Ry.* (1873) L.R. 8 Ch.App. 597 at p. 613; *Eaglesham* v. *McMaster*
[1920] 2 K.B. 169.

[19] *Robins* v. *Goddard* [1905] 1 K.B. 294 (C.A.), a decision correct on its facts but not
apparently authority for any general principle—*East Ham Borough Council* v. *Bernard
Sunley & Sons Ltd., supra,* at pp. 434, 441, 447. See also *Ranger* v. *G.W.R., supra.*

[20] *Sharpe* v. *San Paulo Ry., supra; Clemence* v. *Clarke* (1880) H.B.C., 4th ed., ii, pp. 54,
65; *Eaglesham* v. *McMaster, supra.*

words are necessary to achieve this effect.[21] If the parties ignore the arbitration clause and go to court they may have impliedly agreed to give the court the same powers as the arbitrator, although the position is not clear.[22]

(d) *Certificate or award conclusive*

It seems that if a contract provides that either a certificate of the architect or an award of the arbitrator is to be conclusive, the certificate is conclusive if given before a dispute had arisen but not if given thereafter.[23]

6. Point of law

For the possibility of attacking a certificate on the grounds that it embodies a wrong decision on a point of law, see *ante*, p. 88.

---

[21] See *East Ham Borough Council* v. *Bernard Sunley & Sons Ltd.*, *supra*.

[22] See discussion of the position under the Standard form of building contract and the authorities referred to, *post*, p. 414.

[23] *Lloyd Bros.* v. *Milward* (1895) H.B.C. (4th ed.), Vol. 2, p. 262 (C.A.); *Milestone* v. *Yates* [1938] 2 All E.R. 439.

CHAPTER 7

DEFAULT OF THE PARTIES—EXCUSES FOR NON-PERFORMANCE

THIS and the next two chapters must be considered as a whole. This chapter
deals with excuses for non-performance; Chapter 8 discusses various matters
which may arise upon the default of one of the parties; Chapter 9 is concerned
with the remedies available for a breach of contract.

Section 1: Inaccurate Statements Generally
Many disputes arise as to the effect of inaccurate statements. A plaintiff who has
suffered loss in reliance upon such a statement must consider its nature in order
to determine whether he has a remedy. If the statement has been expressly made
part of a contract, *e.g.* where a report about soil conditions is incorporated into
the contract and expressly warranted to be true, the plaintiff's remedy is for
breach of the contract. If this is not the case, it is suggested that it may be useful
to ask whether the statement falls into one or more of the following categories.
  (a) A mere " puff."
  (b) A statement of honest opinion not containing within itself a representation
      of fact.
  (c) A representation of fact.
  (d) A negligent misstatement.
  (e) A collateral warranty (sometimes termed a " collateral contract ").
Of these categories the first two are dealt with briefly in this section and the others
in separate sections of this chapter.
  A " puff " is a statement which by its nature, and in the context in which it is
made, is not intended to have legal effect.[1] Thus an announcement by a builder
that he is the best builder in town is, it is submitted, a mere " puff." [2]
  A mere statement of honest opinion not impliedly involving a statement of
fact is not actionable.[3] Circumstances which can impliedly give rise to a statement
of opinion involving a statement of fact include those where facts are not equally
known to both sides, where a statement of opinion by one who knows the facts
best very often involves a statement by him of material fact.[4] A misstatement of
material fact is one of the elements in a claim for, or based on, misrepresentation.[5]
The category of a mere statement of honest opinion is to be distinguished from
that of the negligent misstatement made by a person in circumstances in which
he owes a duty of care to the plaintiff, discussed, *post,* p. 98.

Section 2: Misrepresentation—before April 22, 1967
In the invitation to tender and other negotiations leading up to the conclusion of

---

[1] *Carlill* v. *Carbolic Smokeball Company* [1893] 1 Q.B. 256, 261, 266 (C.A.).
[2] See *De Beers Limited* v. *International Company Limited* [1975] 1 W.L.R. 972, 978.
[3] *Bisset* v. *Wilkinson* [1927] A.C. 177 (P.C.).
[4] *Ibid.* p. 182; *Esso Petroleum Company Limited* v. *Mardon* [1976] Q.B. 801 (C.A.)
discussed, *post,* p. 100.
[5] See *infra.*

a contract statements of fact may be made by the employer or his agent about such matters as the quantities of work involved, the nature of the site or the methods by which the work can be carried out. Such a statement if intended to be acted upon by the contractor is termed a representation.[6] If it is incorporated into the contract and is inaccurate causing the contractor loss he has a remedy for breach of contract.[7] If such a statement does not form part of the contract, and did not act as an inducement to the contractor to enter into the contract, it is of no legal effect.[8] But if the contractor shows that it operated as an inducement, that it was untrue, and that he has thereby suffered loss, or will suffer loss, if he completes the contract, his remedy, if any, varies according to the nature of the statement. It may take effect either as an innocent misrepresentation or as a fraudulent misrepresentation.[9] Further, the position varies according to the date of the statement. If made before April 22, 1967, it is as set out in this section. If made on or after such date, see Section 3, *post*, p. 93. The statement of the law in this manner is made because there may yet be litigation involving pre-April 22, 1967 statements, many cases still of importance cannot be understood without knowing the old law and the Misrepresentation Act 1967, which is effective in respect of statements made after April 21, 1967, is not a code of law but a statutory amendment to existing law.

1. Innocent misrepresentation

The general principle is that a person induced to enter into a contract by an innocent misrepresentation of a material fact is entitled, upon discovering the falsity of the representation, to rescission of the contract, but is not entitled to recover damages unless the misrepresentation was also a breach of contract.[10] His remedy is to take proceedings for the rescission of the contract, or he can elect to rescind and refuse to carry out the contract, and if sued for breach set up the misrepresentation by way of defence.[11] The rescission is ineffective until communicated to the other party unless that other party prevented communication fraudulently.[12]

Rescission is an equitable remedy and subject to certain bars. Thus there must be *restitutio in integrum*, *i.e.* the parties must be restored to their original positions. If this cannot be done rescission will not be granted. Contractors after completing the contract works, a branch railway, claimed rescission of the contract and a reasonable sum on a *quantum meruit* on the ground of the innocent misrepresentation of the railway company's engineer as to the nature of the strata through which the railway passed. The remedy was refused because the contractors had

---

[6] *Priestley* v. *Stone* (1888) 4 T.L.R. 730; H.B.C. (4th ed.), Vol. 2, p. 134 (C.A.). For a case where a statement of belief as to a fact involved a representation of having reasonable grounds for that belief, see *Brown* v. *Raphael* [1958] Ch. 636 (C.A.). The general principles set out in this section and Section 3 apply equally to the employer seeking a remedy although the difference in circumstances reduces the opportunities for their application.

[7] See Chap. 9, *post*, p. 144.

[8] See *Edginton* v. *Fitzmaurice* (1885) 29 Ch.D. 459 (C.A.).

[9] See *post*, p. 92.

[10] *Redgrave* v. *Hurd* (1881) 20 Ch.D. 1 (C.A.); see also *Dietz* v. *Lennig Chemicals Ltd.* [1969] 1 A.C. 170 (H.L.).

[11] *Ibid.*

[12] *Car and Universal Finance Co. Ltd.* v. *Caldwell* [1965] 1 Q.B. 525 (C.A.).

completed the railway after full knowledge of the facts, and *restitutio* had become impossible.[13] It seems likely that where any building works have been carried out pursuant to the contract materially affecting the employer's land [14] that rescission will be refused even where the work was carried out without knowledge of the untruth of the representation. Affirmation of the contract by taking some substantial [15] step towards its performance after discovery of the untruth bars rescission.[16] Delay in seeking rescission, or the acquisition of an interest in the subject-matter of the contract by a purchaser innocent of the untruth of the representation bars rescission.[17] It has been said that ". . . if the contract has been executed by the completion of a conveyance or lease or the formal assignment of a chattel, then rescission cannot be obtained on the ground of innocent mis-representation by the vendor or lessor." [18] This principle, known as the rule in *Seddon* v. *North Eastern Salt Co.*,[19] probably has little application to building contracts.

If the misrepresentation complained of has become a term of the contract it seems that rescission for innocent misrepresentation will not be granted.[20]

The right of rescission for a misrepresentation not forming part of the contract must be carefully distinguished from the right of the contractor to treat the contract as at an end (also called rescission) if the employer repudiates the contract.[21] Where the contractor accepts such repudiation [22] he is not only released from further performance but is entitled to damages for breach of contract. Damages are never awarded for an innocent misrepresentation which does not form part of the contract.[23]

Protection against misrepresentation

An agreement can be so worded as to afford " . . . complete protection against innocent misrepresentation." [24]

---

[13] *Glasgow and South Western Ry.* v. *Boyd & Forrest* [1915] A.C. 526 (H.L.), reported fully *sub nom. Boyd & Forrest* v. *Glasgow and South Western Ry.*, 1915 S.C.(H.L.) 20; *cf. Moss* v. *Swansea Corp.* (1910) 74 J.P. 351; *Thorn* v. *London Corp.* (1876) 1 App.Cas. 120, 127 (H.L.); *Abram Steamship Co. Ltd.* v. *Westville Shipping Co. Ltd.* [1923] A.C. 773 (H.L.)—a shipbuilding case.

[14] No case deals with partial completion of building works as a bar to rescission on the ground of the impossibility of *restitutio.*

[15] See the *Abram Steamship* case, *supra*; *cf. Senanayake* v. *Cheng* [1966] A.C. 63 (P.C.)—a partnership case.

[16] *Boyd & Forrest* v. *Glasgow and South Western Ry.*, *supra*; *Leaf* v. *International Galleries* [1950] 2 K.B. 86 (C.A.); *Long* v. *Lloyd* [1958] 1 W.L.R. 753 (C.A.).

[17] *Clough* v. *London & N.W. Ry.* (1871) L.R. 7 Exch. 26. See also *Allen* v. *Robles* [1969] 1 W.L.R. 1193 (C.A.)—a repudiation case.

[18] *Per* McCardie J., *Armstrong* v. *Jackson* [1917] 2 K.B. 822, 825.

[19] (1905) 1 Ch. 326. The existence of the rule has been judicially doubted, see *e.g. Leaf's* case, *supra.*

[20] *Pennsylvania Shipping Co.* v. *Cie. Nat. de Navigacion* [1936] 2 All E.R. 1167, 1171; *Leaf's* case, *supra*, pp. 93, 95.

[21] See *post*, p. 113.

[22] *Ibid.*

[23] *Redgrave* v. *Hurd*, *supra*; *Heilbut, Symons & Co.* v. *Buckleton* [1913] A.C. 30, 51 (H.L.); *Hedley Byrne & Co. Ltd.* v. *Heller* [1964] A.C. 465, 539 (H.L.).

[24] *Per* Lord Shaw, *Boyd & Forrest's* case, *supra*, at p. 36 referring to a typical clause warning the contractor that no allowance would be made if the soil should turn out to be different from what it was stated to be.

2. Fraudulent misrepresentation

In the leading case of *Derry* v. *Peek* Lord Herschell said, in his speech, " fraud is proved when it is shown that a false representation has been made (1) knowingly, (2) without belief in its truth, or (3) recklessly, careless whether it is true or false. Although I have treated the second and third as distinct cases, I think the third is but an instance of the second, for one who makes a statement under such circumstances can have no real belief in the truth of what he states. To prevent a false statement being fraudulent there must always be an honest belief in its truth. . . . If fraud be proved, the motive of the person guilty of it is immaterial; it matters not that there was no intention to cheat or injure the person to whom the statement is made." [25] Fraud in this sense always involves dishonesty even if the motive is not personal gain.[26] A statement made carelessly or negligently, as where, for example, a quantity surveyor carelessly misstates the quantities,[27] cannot in itself be fraudulent,[28] neither is it fraud merely because a person honestly but erroneously believes his statement to be true although according to its objective meaning it is not true [29]; but recklessness may be of such a degree as to be dishonesty [30]; and the fact that an alleged belief in the truth of a statement is destitute of all reasonable foundation may suffice in itself to convince the court that it was not really entertained and that the representation was fraudulent.[31]

In *Pearson* v. *Dublin Corporation* [32] the contractor undertook to carry out a high-level outfall sewer and outfall works for a lump sum. It was an elaborate scheme, an essential feature of which was the use of a wall shown on the maps and drawings supplied by the employer's engineers as extending for a depth of nine feet below the ordnance datum line. In fact it rarely extended as much as three feet. It was therefore useless for the purpose proposed and the contractor was put to extra expense. The defendant's engineers responsible for the drawings and description had carried out no accurate survey and had doubts whether the wall existed or was of the description stated, and " rashly and without inquiry represented nine feet of wall where no wall existed." [33] This was held to be evidence of fraud. The case is also of interest in that the contract provided that " the contractor is to satisfy himself as to the dimensions . . . of all existing works," and that the corporation did not hold itself responsible for the accuracy of statements about existing works. It was held that while such clauses might have furnished a complete answer to any claim for breach of

---

[25] (1889) 14 App.Cas. 337, 374 (H.L.).

[26] *Ibid.*; *Armstrong* v. *Strain* [1952] 1 K.B. 232 (C.A.); *Dutton* v. *Louth Corp.* (1955) 116 E.G. 128 (C.A.). For the standard of proof, see *Hornal* v. *Neuberger Products* [1957] 1 Q.B. 247 (C.A.).

[27] *Priestley* v. *Stone* (1888) 4 T.L.R. 730; H.B.C. (4th ed.), Vol. 2, p. 134 (C.A.).

[28] See *supra*, note 25.

[29] *Akerhielm* v. *De Maré* [1959] A.C. 789 (P.C.). See also *Gross* v. *Lewis Hillman Ltd.* [1970] Ch. 445 (C.A.).

[30] See *supra*, note 25.

[31] *Per* Lord Herschell, *Derry* v. *Peek*, *supra*, at p. 375; *Akerhielm* v. *De Maré*, *supra*.

[32] [1907] A.C. 351 (H.L.).

[33] *Pearson* v. *Dublin Corp.* (1906) H.B.C. (3rd ed.), Vol. 2, p. 453. Note: a fuller account of the facts can be found in H.B.C. (3rd ed.), Vol. 2, than in the reports of the House of Lords decision.

warranty[34] they afforded no defence to an action for fraud, for they might well be part of the fraud being inserted in the hope that no tests would be made.[35] Lord Loreburn said, " no one can escape liability for his own fraudulent statements by inserting in a contract a clause that the other party shall not rely upon them. I will not say that a man himself innocent may not under any circumstances, however peculiar, guard himself by apt and express clauses from liability for the fraud of his own agents. It suffices to say that in my opinion the clauses before us do not admit of such a construction." [36]

The plaintiff must always prove that the fraudulent misrepresentation was an inducement, but the defendant cannot succeed in his defence by showing that there were other more weighty causes which contributed to the plaintiff's decision, " for in this field the court does not allow an examination into the relative importance of contributory causes." [37]

Pricing errors

Where before the contract is signed an error is discovered by the employer in the pricing of the contractor's tender operating in favour of the employer, mere failure clearly to draw the contractor's attention to the error before the contract is signed is not in itself evidence of fraud,[38] although the employer is under a *moral* duty to the contractor to draw his attention to the errors.[39]

Remedies for fraud

Where the contractor has been induced to enter into the contract by a fraudulent misrepresentation he can on discovering the fraud avoid the contract [40] and treat it as at an end, or he can affirm the contract and complete. In either event he can recover damages in an action for the tort of deceit.[41]

Section 3: Misrepresentation—from and including April 22, 1967

When a misrepresentation was made on or after April 22, 1967, the law applies

---

[34] [1907] A.C. 351, 366.

[35] *Per* Lord Ashbourne at p. 360.

[36] p. 353. *Cf. Tullis* v. *Jacson* [1892] 3 Ch. 441, discussed *ante*, p. 85. See also *John Carter (Fine Worsteds) Ltd.* v. *Hanson Haulage (Leeds) Ltd.* [1965] 2 Q.B. 495 (C.A.). For a detailed discussion of the liability of the employer where there is an agent such as the architect, see *post*, p. 204.

[37] *Barton* v. *Armstrong* [1976] A.C. 104, 118 (P.C.). The " field " referred to includes duress. This means " the compulsion under which a person acts through fear of personal suffering "—Hals., *Laws of England* (4th ed.), Vol. 9, para. 297. Duress is not further referred to in this book as it is not thought likely to be of much importance in building contracts.

[38] *Dutton* v. *Louth Corp.* (1955) 116 E.G. 128 (C.A.). *Cf. Bottoms* v. *York Corp.* (1892) H.B.C. (4th ed.), Vol. 2, p. 208 (C.A.).

[39] *Dutton's* case, *supra*. See also *ante*, p. 49. The contractor might have had a remedy in rectification had he sought it.

[40] Unless, it seems, *restitutio* is impossible or the rights of innocent purchasers have arisen—see cases cited, *ante*, p. 91.

[41] See generally *Chitty on Contracts* (24th ed.), Vol. 1. For a summary of the elements of an action for deceit, see *Bradford Building Soc.* v. *Borders* [1941] 2 All E.R. 205, 211 (H.L.). In certain circumstances the measure of damages in tort differ from contract; see *McGregor on Damages* (13th ed.), para. 1357; *Doyle* v. *Olby (Ironmongers) Ltd.* [1969] 2 Q.B. 158 (C.A.).

as set out in Section 2, *infra*, but as modified by the provisions of the Misrepresentation Act 1967. The Act is short but, in some aspects, difficult to understand,[42] and it makes the law more complex. It enlarges the right of rescission for innocent misrepresentation (s. 1). It introduces two statutory rights to damages for misrepresentation. One is linked to the right to rescind (s. 2 (2)); the other provides for what may broadly be described as a statutory liability by one contracting party towards the other in damages for misrepresentation unless the misrepresentor proves the absence of negligence (s. 2 (1)). The Act gives the court discretion to avoid, or reduce the effect of, exclusion clauses affecting misrepresentation (s. 3). The Sale of Goods Act 1893 is amended.

The Misrepresentation Act 1967 is printed below with some notes. The paragraph headings are the side headings of the Act.

Removal of certain bars to rescission for innocent misrepresentation

1. Where a person has entered into a contract after a misrepresentation has been made to him, and—

(*a*) the misrepresentation has become a term of the contract; or

(*b*) the contract has been performed;

or both, then, if otherwise he would be entitled to rescind the contract without alleging fraud, he shall be so entitled, subject to the provisions of this Act, notwithstanding the matters mentioned in paragraphs (*a*) and (*b*) of this section.

Subsection (*b*) removes " the rule in *Seddon* v. *North Eastern Salt Co. Ltd.*" For this and the effect upon the right to rescission of a misrepresentation becoming a term of the contract, see *ante*, p. 91. Other bars to rescission, *e.g.* affirmation, impossibility of *restitutio*, delay, rights of the innocent purchaser, remain; see further, notes to Section 2 (2), *infra*.

Damages for misrepresentation

2.—(1) Where a person has entered into a contract after a misrepresentation has been made to him by another party thereto and as a result thereof he has suffered loss, then, if the person making the misrepresentation would be liable to damages in respect thereof had the misrepresentation been made fraudulently, that person shall be so liable notwithstanding that the misrepresentation was not made fraudulently, unless he proves that he had reasonable ground to believe and did believe up to the time the contract was made that the facts represented were true.

(2) Where a person has entered into a contract after a misrepresentation has been made to him otherwise than fraudulently, and he would be entitled, by reason of the misrepresentation, to rescind the contract, then, if it is claimed, in any proceedings arising out of the contract, that the contract ought to be or has been rescinded, the court or arbitrator may declare the contract subsisting and award damages in lieu of rescission, if of opinion that it would be equitable to do so, having regard to the nature of the misrepresentation and the loss that would be caused by it if the contract were upheld, as well as to the loss that rescission would cause to the other party.

---

[42] Many articles have been written about the Act—see *Current Law Statute Citator* 1947–71, p. 432; in particular see Fairest [1967] C.L.J. 239; Atiyah and Treitel (1967) 30 M.L.R. 396—a highly critical analysis. See also Chitty (24th ed.), Vol. 1, para. 351; Cheshire & Fifoot, *Contract* (9th ed.), pp. 260 *et seq.*

(3) Damages may be awarded against a person under subsection (2) of this section whether or not he is liable to damages under subsection (1) thereof, but where he is so liable any award under the said subsection (2) shall be taken into account in assessing his liability under the said subsection (1).

Section 2 (1). The right to damages is not linked to the right of rescission, so that damages under the subsection may be claimed despite the existence of bars to rescission such as affirmation of the contract. Dishonesty is not an essential element as it is in fraud.[43] Facts such as are necessary to establish a claim in negligence, must ordinarily exist but it is not for the plaintiff to prove negligence, but for the defendant to prove the statutory defence.[44]

Knowledge of servants or agents. Problems arise as to the knowledge, or lack of it, of agents. The employer is, it is submitted, liable where his servants lacked reasonable grounds for their belief. Probably also he is liable for his agents, such as independent architects or engineers, who are employed to act for the project in question, so that a claim under the section would lie on the facts of *Pearson* v. *Dublin Corporation,* set out *supra,* p. 92. But it is thought unlikely that the employer is liable for the lack of reasonable grounds of belief of the independent expert whose opinion he sought on some special matter, such as, *e.g.* water levels or strata, provided he, the employer, had reasonable grounds for believing and did believe the expert's report to be true. It is not clear how far the knowledge or absence of it, of the employer and his servants or agents, or of various servants and agents can be added together so as to prevent proof of reasonable grounds of belief.[45]

Section 2 (2). This gives a right to apply for a discretionary remedy of damages where the misrepresentation is wholly innocent, *i.e.* neither fraudulent nor made with that element of fault which prevents a defence being established under section 2 (1). But it is, it is thought, of limited value to a contractor because, it is submitted, he must prove that he would be entitled to rescission if damages were not awarded,[46] and in consequence his claim for damages fails if one of the bars to rescission exists, such as affirmation, delay or impossibility of *restitutio.* If the contractor discovers the misrepresentation before work commences he can, as before the Act, in the ordinary case, rescind, and the section now gives him a right to apply for damages. But if, as seems likely, the right to rescission must exist at the date of hearing,[47] and meanwhile the employer has had the works carried out by another contractor, the court cannot, or alternatively, would not, it is thought, declare the contract subsisting so as to be able to award damages.

---

[43] See *ante,* p. 92.
[44] Similar words appear in the Companies Act 1948, s. 43. There are cases decided upon s. 43 and the statutes it replaced but it is thought that because of the different context they are of little assistance in the interpretation of this Act.
[45] *Cf. Armstrong* v. *Strain* [1952] 1 K.B. 232 (C.A.) discussed *post,* p. 205, and in Atiyah and Treitel, *op. cit.* at p. 374. For a case of a principal not liable for an agent acting outside the limits of his authority which limits had been made clear to the plaintiff, see *Overbrooke Estates Ltd.* v. *Glencombe Properties Ltd.* [1974] 1 W.L.R. 1335.
[46] See discussion, Atiyah and Treitel, *op. cit.* at p. 377.
[47] *Ibid.*

Repudiation. The terms " rescind," " rescinded " and " rescission " are not, it is submitted, intended to have the meaning of an acceptance of repudiation [48] so as to bring the contract to an end. (This is also called " rescission," see *post*, p. 113.)

Damages

Section 2 (3) seems to contemplate that the measure of damages under sub-section (1) is greater than under subsection (2). No other guidance is given, but it seems that the court assimilates the award of damages under subsection (1) to that for the tort of deceit.[49] This means that the plaintiff is not merely to be put in the same position as if a contract had been performed but can recover actual loss directly flowing from the misrepresentation.[50] The result is that in some cases he recovers more damages than if his claim is for breach of contract,[51] which, for this purpose, includes breach of collateral warranty.

Misrepresentation incorporated into the contract

A statement of fact by the employer as to the nature of the soil, or the site or the like is often made in the tender documents and is at that stage a representation. When the contract is signed such statements are often expressly incorpor-- ated into the contract. It is thought that the remedies under the Act apply notwithstanding such incorporation, so that the contractor may seek damages for breach of contract or under this section or both in the alternative. Matters to be considered include :

(a) In certain circumstances he may recover more damages under the Act than for breach of contract, *supra*.

(b) If there is no clause in the contract excluding or restricting the employer's liability or the contractor's remedy and he has lost the right to rescission, he may prefer the claim for breach of contract so as not to have to meet the statutory defence of reasonable ground of belief under subsection (1) of section 2.

(c) If there is such a clause he may prefer his remedy under the Act so as to be able to apply to have the clause set aside or reduced in its effect under section 3, *infra*.

Avoidance of certain provisions excluding liability for misrepresentation

3. If any agreement (whether made before or after the commencement of this Act) contains a provision which would exclude or restrict—
(*a*)   any liability to which a party to a contract may be subject by reason

---

[48] See *post*, p. 113.
[49] *Gosling* v. *Anderson* [1972] E.G.D. 709, 713 (C.A.); *Jarvis* v. *Swan's Tours Ltd.* [1973] 1 Q.B. 233, 237 (C.A.); *Davis & Co. Ltd.* v. *Afa-Minerva Ltd.* [1974] 2 Lloyd's Rep. 27; *Watts* v. *Spence* [1976] Ch. 165.
[50] *Jarvis* v. *Swan's Tours Ltd., supra; Davis & Co. Ltd.* v. *Afa-Minerva Ltd., supra; Watts* v. *Spence, supra*—criticised at 91 L.Q.R. p. 308 for awarding damages for loss of bargain, but this is the measure of damages for fraud in contracts for the sale of land, see the review of the authorities at *Watts* v. *Spence*, p. 174.
[51] See *Doyle* v. *Olby (Ironmongers) Ltd.* [1969] 2 Q.B. 158 (C.A.).

of any misrepresentation made by him before the contract was made; or

(b) any remedy available to another party to the contract by reason of such a misrepresentation;

that provision shall be of no effect except to the extent (if any) that, in any proceedings arising out of the contract, the court or arbitrator may allow reliance on it as being fair and reasonable in the circumstances of the case.

The " agreement." It must, it is submitted, be enforceable [52] and can be made before the contract in question and even before the commencement of the Act. It can also, it is submitted, be contained within the " contract " referred to in the section and includes, it is submitted, a provision relating to misrepresentations which have become incorporated into the contract. A difficult question of construction may sometimes arise to determine whether certain words operate as an exclusion clause or so as to prevent the misrepresentation complained of from arising at all.

Agent's authority. The section only applies to a provision which would exclude or restrict liability for a misrepresentation made by a party or his duly authorised agent; it has no application to a notice limiting the authority of an agent.[53]

The court's discretion. The presumption is apparently against allowing reliance on the provision. Subject to this, no principle applies save as stated in the section, that effect is only allowed to the provision to the extent (if any) that may be " fair and reasonable in the circumstances of the case." This may introduce a substantial element of uncertainty as to the ultimate apportionment of risks in some contracts and make the estimator's task very difficult.

Amendments of Sale of Goods Act 1893

4.—(1) In paragraph (c) of section 11 (1) of the Sale of Goods Act 1893 (condition to be treated as warranty where the buyer has accepted the goods or where the property in specific goods has passed) the words " or where the contract is for specific goods, the property in which has passed to the buyer " shall be omitted.

(2) In section 35 of that Act (acceptance) before the words " when the goods have been delivered to him, and he does any act in relation to them which is inconsistent with the ownership of the seller " there shall be inserted the words " (except where section 34 of this Act otherwise provides)."

Section 4 amends the Sale of Goods Act 1893 so that the right to reject specific goods depends upon acceptance and not upon the passing of property, and further ensures that the buyer by doing an act inconsistent with the seller's ownership is not deemed to have accepted the goods until he has had an opportunity of examining them.[54]

---

[52] *Cf.* Atiyah and Treitel, *op. cit.,* p. 379, who canvass the idea of a " subject to contract " agreement (*ante,* p. 18), but this is, it is submitted, in law a nullity. *Cf.* definition of " agreement " in the Restrictive Trade Practices Act 1976, *ante,* p. 13.

[53] *Overbrooke Estates Ltd.* v. *Glencombe Properties Ltd.* [1974] 1 W.L.R. 1335.

[54] These important, if technical, alterations to the Sale of Goods Act 1893 must be considered by reference to the Act and to textbooks such as Chalmers, Benjamin, Atiyah.

Saving for past transactions

5. Nothing in this Act shall apply in relation to any misrepresentation or contract of sale which is made before the commencement of this Act.

Short title, commencement and extent

6.—(1) This Act may be cited as the Misrepresentation Act 1967.

(2) This Act shall come into operation at the expiration of the period of one month beginning with the date on which it is passed.

(3) This Act, except section 4 (2), does not extend to Scotland.

(4) This Act does not extend to Northern Ireland.

Section 4: Negligent Misstatement—Liability in Tort

Before 1964 it was thought that no claim could arise out of a negligent mis-statement which was not a breach of contract or a breach of a fiduciary duty.[55] In *Hedley Byrne & Co. Ltd.* v. *Heller Ltd.* the House of Lords held that such a claim could arise in a wider category; that which exists where there is a special relationship between the parties.[56] The liability is in tort. The majority of the Privy Council in a subsequent decision were of the view that the duty of care was limited to persons who carry on or hold themselves out as carrying on the business or profession of giving advice.[57] This view is not binding upon the English courts and has been rejected by the Court of Appeal.[58] Instances where the duty can arise include the following: when an inquirer consults a businessman who carries on or holds himself out as carrying on the business or profession of giving advice [59]; where inquiry is made of a public servant who "knows, or ought to know, that, others, being his neighbours in this regard, would act on the faith of the statement being accurate " [60]; in the course of negotiation for a contract where the person making the statement has, or professes to have, special knowledge and skill not possessed, or not possessed to the same degree, by the other party.[61] And, it is submitted, a duty may be owed to others, such as the contractor, engaged in a building project where a person, such as the architect or the engineer, makes statements before or during that project which he knows, or ought to know, may be acted upon by those others as being accurate.[62]

No duty arises if the person making the statement shows that he is not assuming or accepting a duty to be careful.[63] Thus a banker in a situation where a duty

---

[55] See the exposition of the law in *Hedley Byrne & Co. Ltd.* v. *Heller Ltd.* [1964] A.C. 465. This approach of the law must be borne in mind when reading pre-1964 cases about inaccurate statements.

[56] *Ibid.*; *Mutual Life Ltd.* v. *Evatt* [1971] A.C. 793, 801 (P.C.); *Esso Petroleum Co. Ltd.* v. *Mardon* [1976] Q.B. 801 (C.A.).

[57] *Mutual Life Ltd.* v. *Evatt, supra.*

[58] *Esso Petroleum case, supra.*

[59] *Hedley Byrne case, supra; Mutual Life case, supra.*

[60] *Ministry of Housing* v. *Sharp* [1970] 2 Q.B. 223, 268 (C.A.). For negligent acts by public bodies see *Dorset Yacht Co. Ltd.* v. *Home Office* [1970] A.C. 1004, *Anns* v. *Merton London Borough Council* [1977] 2 W.L.R. 1024 (H.L.).

[61] *McInerny* v. *Lloyds Bank Ltd.* [1974] 1 Lloyd's Rep. 246 (C.A.); *Esso Petroleum* case, *supra.*

[62] See cases cited *supra*; *Arenson* v. *Arenson* [1977] A.C. 405, 438 (H.L.).

[63] *Hedley Byrne* case, at pp. 492, 504, 511, 533, 540; *McInerny* case, *supra.*

would otherwise have arisen, was not liable to an inquirer where he gave a favourable reference, but said it was given " without responsibility." [64] A trader who, when asked for a reference by a fellow trader about X said, " They are all right " was held to owe a duty of care.[65]

The New South Wales case of *Dillingham Ltd.* v. *Downs* [66] provides an interesting discussion of the principles as applied to a building contract. The plaintiffs, engineering contractors, undertook to carry out certain works in a harbour, involving, *inter alia*, the breaking up and removal of rock on the sea floor. The employer was the Government and had for many years had the management and control of dredging and rock removal from the harbour. For a period of two years prior to calling for tender it had extensive surveys, soundings and borings carried out to provide basic material for incorporation in the specification and attached drawings. Much of this material was incorporated into the tender documents. The question was whether the government department was under an obligation to exercise reasonable care in the assembling and presentation to the plaintiff of material relative to the existence of worked-out mines beneath the surface and other faults in the soil which, in the event, disrupted the contract and caused great expense to the contractor.

It was said that the matters to be considered to decide whether a duty arose and its nature and extent were " the material provisions of the contract documents, the position, conduct, knowledge and intention of each of the parties and the communications passing between them." [67]

There were terms of the specification requiring that the tenderer " should fully inform himself on these aspects " (the site conditions) and that though relevant information was given " in good faith and is believed to be accurate " the tenderer " must satisfy himself regarding the adequacy and accuracy of his information on site conditions." [68] It was apparent from the documents that the information as to site conditions supplied was not intended to be complete and exhaustive. The tenderers made a site visit but never sought advice from a geologist or an engineer and were obviously anxious to obtain the contract. Upon the particular facts it was held that there had been no assumption of liability by the employer so as to make it liable in negligence and in any event there had been no reliance by the contractor.[69]

It seems that an employer who has, or professes himself as having, special knowledge of such matters as soil conditions and represents to the contractor that his statements are accurate or have been made after reasonable care and inquiry and does not use words showing a refusal to assume liability, can be liable to the contractor in damages if he was negligent.[70] Upon such facts the

---

[64] *Hedley Byrne* case, *supra.*

[65] *Anderson (W. B.) & Sons Ltd.* v. *Rhodes Ltd.* [1967] 2 All E.R. 850.

[66] [1972] 2 N.S.W.L.R. 49.

[67] *Ibid. per* Hardie J. at p. 56. This statement of principle appears to be consistent with the approach of the Court of Appeal in the *Esso Petroleum* case, *supra.*

[68] Compare the clause relied on by the employer in *Pearson* v. *Dublin Corp.* [1907] A.C. 351 (H.L.), the facts of which are set out *ante*, p. 92, but note that in the *Dillingham* case fraud was not, and, on the facts appearing in the report, could not have been, alleged.

[69] See also *post*, p. 187, for a New Zealand case concerning sub-contractors.

[70] See the *Esso Petroleum* case, *supra.*

contractor may also have a claim for damages for breach of collateral warranty (see *infra*), and, if the representation was made after April 21, 1967, for damages for misrepresentation.

Damages for breach of the duty of care in the kind of circumstances referred to in this section include economic loss.[71]

Section 5: *Collateral Warranty*

If the contractor can show that, although a statement is not a term of the building contract, nevertheless it was, in effect, a promise by the employer that, in consideration of the contractor entering into the contract, the employer would warrant or guarantee the accuracy of the statement then if the statement is incorrect the contractor can recover damages from the employer for breach of warranty.[72] There is, in effect, a collateral contract. " Such collateral contracts, the sole effect of which is to vary or add to the terms of the principal contract, are viewed with suspicion by the law. They must be proved strictly." [73] The test is one of intention. If the court is satisfied that on the totality of the evidence the parties intended or must be taken to have intended the statement to form part of the basis of the contractual relations between them then it will award damages for its inaccuracy.[74] The court will more readily find such an intention if the party by whom it was made had, or professed to have special knowledge and skill in relation to the subject-matter of the statement and made it with the intention of inducing the other party, who had less knowledge, to enter into the contract.[75]

There is no rule that a statement of future fact or a forecast cannot be a warranty.[76] Thus where the agent of an intending lessor stated to the prospective lessee of a petrol station that there was an anticipated throughput of 200,000 gallons per year this, upon the facts, was found to be a warranty that the

---

[71] See the *Hedley Byrne* case, *supra*; *Dutton* v. *Bognor Regis U.D.C.* [1972] 1 Q.B. 373 (C.A.). Where a negligent act (as opposed to a statement) causes immediate physical injury economic loss (as opposed to the immediate loss caused by the injury) is not recoverable in tort, *S.C.M. (United Kingdom) Ltd.* v. *W. J. Whittall & Son Ltd.* [1971] 1 Q.B. 337 (C.A.); *Spartan Steel Ltd.* v. *Martin & Co. Ltd.* [1973] 1 Q.B. 27 (C.A.). See also 92 L.Q.R. 213.

[72] *Heilbut, Symons & Co.* v. *Buckleton*, [1913] A.C. 30; *Routledge* v. *McKay and others* [1954] 1 W.L.R. 615 (C.A.). *Esso Petroleum Co. Ltd.* v. *Mardon* [1976] Q.B. 801 (C.A.). For employer relying on collateral warranty of fitness, see *Miller* v. *Cannon Hill Estates Ltd.* [1931] 2 K.B. 113 (D.C.); *Birch* v. *Paramount Estates Ltd.* (1956) 16 E.G. 396 (C.A.) —" as good as show house." For employer recovering damages for breach of collateral warranty from manufacturer who supplied contractor, see *Shanklin Pier Ltd.* v. *Detel Products Ltd.* [1951] 2 K.B. 854, discussed *post*, p. 175.

[73] *Heilbut, Symons & Co.* v. *Buckleton, supra*, at p. 47; *Oscar Chess Ltd.* v. *Williams* [1957] 1 W.L.R. 370 (C.A.). See also *ante*, p. 31 and discussion based on *Evans (J.) & Sons (Portsmouth) Ltd.* v. *Andrea Merzario Ltd.* [1976] 1 W.L.R. 1078 (C.A.).

[74] *Esso Petroleum* case, *supra*, at p. 826.

[75] *Esso Petroleum* case, *supra*; *Dick Bentley, etc., Ltd.* v. *Harold Smith (Motors) Ltd.* [1965] 1 W.L.R. 623 (C.A.), where Lord Denning M.R. said that it was unnecessary to speak of the warranty as collateral; but the term is frequently used in the cases and, in any event, is convenient to distinguish the claim for warranty from a claim upon the formal building contract.

[76] *Esso Petroleum* case, *supra*.

statement was arrived at after careful consideration and the lessor was liable in damages for loss resulting from his lack of such care.[77]

It has been frequently held that the mere inclusion of plans, bills of quantities and specification in the invitation to tender does not show that the employer is warranting their accuracy.[78] The same rule has been applied even where such statements have been incorporated in schedules to the contract.[79] The principle upon which these decisions are based is that a contractor who has undertaken to complete a whole work such as a house, a railway line or a bridge, should satisfy himself that any statements about the amount of work involved, or the condition of the site, or the methods of carrying out the work are accurate and practicable; he is not therefore entitled to assume that such statements are warranties.[80] He can accept them as honest representations made by skilled persons, but beyond that they do not go.[81]

Where the employer, an architect, was in breach of his promise either to obtain licences or stop the work if he could not, the contractor recovered damages for breach of warranty.[82]

In *Bacal (Midland) Limited* v. *Northampton Dev. Corp.*[82a] the contract was for the contractor to design and build.[82b] He was instructed by the employer to design the foundations upon certain hypotheses as to the nature of the ground conditions. It was held that there was an implied warranty on the part of the employer that such conditions would accord with the hypotheses.

Section 6: *Frustration and Impossibility*

1. Difficulty or expense no excuse

Unexpected difficulty or expense is in general no excuse for non-performance.[83] The contractor in such a case cannot rely on his ignorance of such matters as

---

[77] *Ibid.*

[78] *Scrivener* v. *Pask* (1866) L.R. 1 C.P. 715; *Thorn* v. *London Corp.*, *supra*; *Jackson* v. *Eastbourne Local Board* (1886) H.B.C. (4th ed.), Vol. 2, p. 81 (H.L.); *Young* v. *Blake* (1887) H.B.C. (4th ed.), Vol. 2, p. 110; *Bottoms* v. *York Corp.* (1892) H.B.C. (4th ed.), Vol. 2, p. 208 (C.A.); *McDonald* v. *Workington Corp.* (1892) 9 T.L.R. 230; H.B.C. (4th ed.), Vol. 2, p. 228 (C.A.); *Re Nuttall and Lynton and Barnstaple Ry.* (1899) H.B.C. (4th ed.), Vol. 2, p. 279 (C.A.); *Re Ford and Bemrose* (1902) H.B.C. (4th ed.), Vol. 2, p. 324 (C.A.).

[79] *Sharpe* v. *San Paulo Ry.* (1873) L.R. 8 Ch.App. 597; *Re Ford*, *supra*.

[80] *McDonald* v. *Workington Corp.*, *supra*, at p. 231; *Re Ford*, *supra*.

[81] *Re Ford and Bemrose*, *supra*, at p. 330; *Dillingham Construction (Pty.) Ltd.* v. *Downs* [1972] 2 N.S.W.L.R. 49, 56, New South Wales High Court. It is thought that the proposition in the text is still the law but: (1) the greater willingness of the courts today to find a warranty is exemplified by the *Esso Petroleum* case, *supra*; (2) in the days before the Misrepresentation Act 1967, *supra*, it was often not necessary for the courts to draw a sharp distinction between a representation as that term is now used in the Act, and a mere honest expression of opinion as that term is explained *ante*, p. 89, because no remedy was available in either case.

[82] *Strongman (1945) Ltd.* v. *Sincock* [1955] 2 Q.B. 525 (C.A.). For the architect's authority as agent, see *post*, p. 200.

[82a] (1976) 237 E.G. 955 (C.A.).

[82b] For such contracts generally, see Chap. 1.

[83] *Bottoms* v. *York Corp.* (1892) H.B.C. (4th ed.), Vol. 2, p. 208 (C.A.); *McDonald* v. *Workington Corp.* (1893) 9 T.L.R. 230; H.B.C. (4th ed.), Vol. 2, p. 228 (C.A.); *Brauer Ltd.* v. *Clark Ltd.* [1952] 2 All E.R. 497 (C.A.); *Davis Contractors Ltd.* v. *Fareham U.D.C.* [1955] 1 Q.B. 302 (C.A.); *The " Captain George K "* [1970] 2 Lloyd's Rep. 21.

defects in the soil [84] nor on any implied warranty by the employer that the bills of quantities, plans and specification are accurate [85] or that the work is capable of performance in the manner set out in the invitation to tender.[86] He should make his own investigations [87] or limit his liability to exactly stated quantities of work.[88]

2. Impossibility at time of contract

Actual physical impossibility of performing the contract, whatever means are employed [89] which exists at the time of entering into the contract is, subject to express terms or warranties, an excuse for non-performance.[90] But the contractor is liable in damages if he has warranted the possibility of the work [91] or if he has positively and absolutely contracted to do the work.[92]

3. Frustration and impossibility of performance

Frustration generally

Very rarely [93] after the contract has been lawfully entered into and is in course of operation there may arise some intervening event or change of circumstances of so catastrophic [94] or fundamental a nature as to prematurely determine the contract [95] by the operation of the doctrine of frustration.[96] This " occurs wherever the law recognises that without default of either party a contractual obligation has become incapable of being performed because the circumstances in which performance is called for would render it a thing radically different from that which was undertaken by the contract. *Non haec in foedera veni.* It was

---

[84]　*Bottoms* v. *York, supra; McDonald* v. *Workington Corp., supra.*

[85]　See Section 5 of this chapter.

[86]　*Thorn* v. *London Corp.* (1876) 1 App.Cas. 120 (H.L.).

[87]　*McDonald* v. *Workington Corp., supra.*

[88]　*e.g. Bryant* v. *Birmingham Hospital Saturday Fund* [1938] 1 All E.R. 503 (defects in soil); and see generally Chap. 5 (*ante*, p. 64), " Bills of quantities contract."

[89]　Note that in the cases of *Thorn, Bottoms* and *McDonald, supra,* and *Jackson* v. *Eastbourne Local Board* (1885) H.B.C. (4th ed.), Vol. 2, p. 81 (H.L.), it was in each case possible to complete the work at great expense. Physical impossibility is an express excuse under the I.C.E. conditions, clause 13, 1955 and 1973 editions.

[90]　*Taylor* v. *Caldwell* (1863) 3 B. & S. 826; *Clifford* v. *Watts* (1870) L.R. 5 C.P. 577; *cf. Tharsis, etc.* v. *M'Elroy and Sons* (1878) 3 App.Cas. 1040, 1052–1053 (H.L.). See also *Bell* v. *Lever Bros.* [1932] A.C. 161 (H.L.) for mistake avoiding the contract.

[91]　*Clifford* v. *Watts, supra,* at p. 588.

[92]　*Jones* v. *St. John's College, Oxford* (1870) L.R. 6 Q.B. 115; *cf. Dodd* v. *Churton* [1897] 1 Q.B. 562 (C.A.), and see further *post,* p. 161.

[93]　" Frustration is a doctrine only too often invoked by a party to a contract who finds performance difficult or unprofitable, but it is very rarely relied on with success. It is in fact a kind of last ditch," *per* Harman L.J., *Tsakiroglou & Co. Ltd.* v. *Noblee Thorl GmbH* [1960] 2 Q.B. 318, 370 (C.A.); affd. [1962] A.C. 93 (H.L.).

[94]　See Asquith L.J., *Sir Lindsay Parkinson & Co. Ltd.* v. *Commissioners of Works* [1949] 2 K.B. 632, 665 (C.A.).

[95]　" It kills the contract itself," *Constantine Ltd.* v. *Imperial Smelting Ltd.* [1942] A.C. 154, 163 (H.L.), *per* Lord Simon L.C.

[96]　*Cricklewood Property & Investment Trust Ltd.* v. *Leighton's Investment Trust Ltd.* [1945] A.C. 221, 228 (H.L.); *British Movietonews Ltd.* v. *London & District Cinemas Ltd.* [1952] A.C. 166 (H.L.); *Davis Contractors Ltd.* v. *Fareham U.D.C.* [1956] A.C. 696 (H.L.). It will be observed that the Law Reform (Frustrated Contracts) Act 1943, treats impossibility of performance as an aspect of frustration.

not this that I promised to do." [97] When frustration occurs both parties are automatically discharged from any further obligation unless the party alleging that there is a breach can prove that the circumstances leading to the frustration were brought about by the default of the other party, in which case that other party will be liable in damages.[98] Thus where a contractor deliberately delayed work hoping that by so doing completion would be prevented by a government prohibition against construction of houses, and it was in fact so prevented, he was held to be liable in damages.[99]

In deciding whether a contract has been frustrated " ' The data for decision are, on the one hand, the terms and construction of the contract, read in the light of the then existing circumstances, and on the other hand the events which have occurred.' [1] In the nature of things there is often no room for any elaborate inquiry. The court must act upon a general impression of what its rule requires. It is for this reason that special importance is necessarily attached to the occurrence of any unexpected event that, as it were, changes the face of things. But, even so, it is not hardship or inconvenience or material loss itself which calls the principle of frustration into play. There must be as well such a change in the significance of the obligation that the thing undertaken would if performed, be a different thing from that contracted for." [2] This was applied in *Davis Contractors Ltd.* v. *Fareham U.D.C.*[3] The contractors entered into a contract to build 78 houses for a fixed price within a contract period of eight months. Attached to their tender had been a letter stating that their tender was subject to adequate supplies of labour being available as and when required, but it was held that this letter was not a contract document.[4] There were unanticipated shortages of labour and materials and, although work never actually stopped, the shortages caused such delay that the contract took 22 months to complete and as a result cost the contractors about £17,000 more than the contract price. These facts were held not to frustrate the contract and the contractors had to bear the loss themselves, even though in one sense it might be said that the " basis " or " footing " of the contract, *i.e.* adequate supplies of labour and materials, had

---

[97] *Davis Contractors'* case, *supra*, at p. 729, *per* Lord Radcliffe, echoing the language of Lord Cairns, *Thorn* v. *London Corp.* (1876) 1 App.Cas. 120, 127 (H.L.); *The Eugenia* [1964] 2 Q.B. 226, 239 (C.A.); *Amalgamated Investment Ltd.* v. *John Walker Ltd.* [1977] 1 W.L.R. 164 (C.A.). Another approach is to ask whether the nature of the contract has been " fundamentally " altered, see *British Movietonews'* case, *supra*, at p. 185 (H.L.). This means the same as " radically different "—*Tsakiroglou's* case, *supra*, at [1962] A.C. 93, 115 (H.L.), *per* Lord Simonds. Yet another explanation of frustration is to base it on an implied term that a particular thing or state of things would continue to exist—see *Tamplin Steamship Co. Ltd.* v. *Anglo-Mexican Petroleum Products Co. Ltd.* [1916] 2 A.C. 397, 403 (H.L.), criticised by Lord Radcliffe in *Davis'* case, *supra*, at p. 728, but *cf.* Lord Guest in *Tsakiroglou's* case, *supra*, at p. 131. Frustration is a " thorny topic," *per* Salmon L.J., *Denmark, etc., Ltd.* v. *Boscobel, etc., Ltd.* [1969] 1 Q.B. 699, 725 (C.A.).
[98] *Constantine Ltd.* v. *Imperial Smelting Ltd., supra.*
[99] *Mertens* v. *Home Freeholds Co.* [1921] 2 K.B. 526 (C.A.). See further, *post*, p. 148, for the measure of damages.
[1] *Denny, Mott & Dickinson Ltd.* v. *James B. Fraser & Co. Ltd.* [1944] A.C. 265, 274–275 (H.L.), *per* Lord Wright.
[2] *Davis Contractors'* case, *supra*, at p. 729, *per* Lord Radcliffe.
[3] *Supra.*
[4] See " Preliminary negotiations," *ante*, p. 28.

gone, for ". . . it by no means follows that disappointed expectations lead to frustrated contracts."[5] and the risk of loss to the contractors caused through the delay which occurred was on the contractors.[6]

Express terms

If the terms of the contract show that the parties contemplated and provided for a particular event that event cannot frustrate the contract.[7] But in considering a clause relied on as expressly providing for the event alleged to frustrate the contract the court determines whether that event is in fact so abnormal or of such a nature as to fall outside what the parties could possibly have contemplated in the clause.[8] An example is given below[9] of a case where the court held that an interruption in the work was of such a kind as to frustrate the contract despite an express provision for interruption.

There follows a discussion of some of the events which might give rise to questions of frustration.

Prohibition by law

If after entering into the contract the completion of the works becomes permanently prohibited by law this normally frustrates the contract.[10] If the prohibition is not permanent the test would seem to be " whether the interruption (caused by the prohibition) was so long as to destroy the identity of the work and service when resumed, with the work and service when interrupted."[11] Thus in *Metropolitan Water Board* v. *Dick Kerr & Co. Ltd.*,[12] the contractors in July 1914, before the outbreak of war, entered into a contract to construct reservoirs at an agreed price within a period of six years. In February 1916 the government prohibited the continuance of the work and seized and sold the contractor's plant and materials, and the work was still prohibited when the case came before the House of Lords in 1918. It was held that the contract was at an end, and this notwithstanding that the engineer had power to extend the time for completion " prospectively or retrospectively " in the event of any " difficulties, impediments, obstructions, oppositions . . . whatsoever and howsoever occasioned."[13]

---

[5] At p. 715, *per* Lord Simonds.

[6] *Bush* v. *Whitehaven Trustees* (1888) H.B.C. (4th ed.), Vol. 2, p. 130 (C.A.) strongly relied on by the contractors (*cf.* pp. 42–44 of the 1st edition of this book) was held not to be an authority for any proposition of law.

[7] *Cricklewood Property & Investment Trust Ltd.* v. *Leighton's Investment Trust Ltd.* [1945] A.C. 221, 228 (H.L.).

[8] *Sir Lindsay Parkinson Ltd.* v. *Commissioners of Works* [1949] 2 K.B. 632, 665 (C.A.); see also *Bank Line Ltd.* v. *Capel* [1919] A.C. 435 (H.L.).

[9] See discussion in the text of *Metropolitan Water Board* v. *Dick Kerr & Co. Ltd.* [1918] A.C. 119 (H.L.).

[10] *Baily* v. *de Crespigny* (1869) L.R. 4 Q.B. 180. *Cf. Walton Harvey Ltd.* v. *Walker and Homfrays Ltd.* [1931] 1 Ch. 274 (C.A.); *Amalgamated Investment Ltd.* v. *John Walker Ltd.* [1977] 1 W.L.R. 164 (C.A.).

[11] *Per* Lord Wright, *Cricklewood's* case, *supra*, at p. 236; *Woodfield Co. Ltd.* v. *Thompson & Sons Ltd.* (1919) 36 T.L.R. 43 (C.A.).

[12] [1918] A.C. 119 (H.L.).

[13] See also *Fibrosa, etc.* v. *Fairbairn, etc., Ltd.* [1943] A.C. 32 (H.L.); *New Zealand Shipping Co.* v. *Ateliers, etc., de France* [1919] A.C. 1 (H.L.).

This case illustrates that war in itself does not bring contracts to an end though circumstances arising out of war may do so.

Delay

Increased cost to the contractor caused by unforeseen delay due to shortage of labour and materials or other causes not the fault of the employer is a risk which, in the absence of express terms to the contrary, is undertaken by the contractor,[14] and it seems unlikely [15] that delay which does not result in a stoppage of work [16] can be sufficient to frustrate the contract.

Change in prices

The contract is not affected by an unanticipated and wholly abnormal rise or fall in prices or a sudden depreciation in the currency unless these events, upon the principles set out above, are of such gravity as to frustrate the contract.[17] It is thought that little short of a catastrophe [18] to the currency would suffice.

Destruction by fire, etc.

Where a contractor undertakes to complete a whole work for a specified price and the contract works are destroyed by fire, flood or the like, he is not, subject to the express terms of the contract, released from his obligation to complete, and normally the contract is not frustrated.[19] If he undertakes to do work on another's building and that building is destroyed the contract may or may not be frustrated according to the application of the principles set out above.[20] If the soil upon which he has undertaken to build is destroyed by flood and the like, then, it is submitted, normally the contract is frustrated.[21]

Weather and acts of God

Bad weather and storms do not generally excuse a contractor for he must be taken to have contemplated their possibility.[22] But if there is an exceptional and

---

[14] See *Davis Contractors Ltd.* v. *Fareham U.D.C.* [1956] A.C. 696 (H.L.). For shipping cases, see *The "Captain George K"* [1970] 2 Lloyd's Rep. 21; *The Angelia* [1973] 1 W.L.R. 210.

[15] In view of the *Davis Contractors'* case, *supra.*

[16] *Cf.* the *Metropolitan Water Board* case, *supra*, where work stopped.

[17] *British Movietonews Ltd.* v. *London & District Cinemas Ltd.* [1952] A.C. 166, 185 (H.L.).

[18] See Asquith L.J., *Sir Lindsay Parkinson & Co. Ltd.* v. *Commissioners of Works* [1949] 2 K.B. 632, 665 (C.A.). There has so far been little enthusiasm to argue that rates of inflation current in 1975 (about 25%) have this effect.

[19] *Appleby* v. *Myers* (1867) L.R. 2 C.P. 651; *Gold* v. *Patman & Fotheringham Ltd.* [1958] 1 W.L.R. 697, 704 (C.A.).

[20] See *Appleby* v. *Myers* where on the facts of the contract it was held to be frustrated (although the modern term "frustration" was not used). But many contracts today expressly provide for fire to the employer's building, see clause 20 of the Standard form of building contract, *post*, p. 338.

[21] See *Taylor* v. *Caldwell* (1863) 3 B. & S. 826.

[22] *Maryon* v. *Carter* (1830) 4 C. & P. 295; *Jackson* v. *Eastbourne Local Board* (1885) H.B.C. (4th ed.), Vol. 2, p. 81 (H.L.); *Electric Power Equipment* v. *R.C.A. Victor Co.* (1964) 46 D.L.R. (2d) 722 (British Columbia C.A.).

extraordinary rainfall [23] or snowfall,[24] or flooding [25] or earthquake,[26] or other weather " such as could not reasonably be anticipated," it may be an act of God.[27] This in itself does not excuse a breach of contract,[28] but the parties may, on the construction of the contract, have intended that an act of God should bring the contract to an end.[29]

Strikes

Strikes do not ordinarily frustrate contracts,[30] although the principles set out above must be applied to the circumstances of any particular case to see whether, exceptionally, there is such certainty of a long duration of the strike or other effect as to frustrate the contract.[31] It is usual to provide expressly for delay caused by strikes.

Compliance with by-laws [32]

This may require heavy expenditure as, for example, when the site is opened up and it is found that elaborate foundations are necessary for a simple building. It is thought that in some cases the expenditure required in comparison with that contemplated may be so heavy that the contract is frustrated.[33]

4. Effect of discharge by impossibility or frustration

Frustration of the contract discharges the parties from further obligations,[34] and at common law the contractor under the ordinary lump sum contract for a whole work cannot recover the cost of work he has carried out.[35]

Law Reform (Frustrated Contracts) Act 1943

The Act provides that where a contract has become impossible of performance or has been otherwise frustrated and the employer has obtained a valuable benefit from the contract works, the contractor can recover from the employer

---

[23] *Dixon* v. *Metropolitan Board of Works* (1881) 7 Q.B.D. 418.

[24] *Briddon* v. *G.N. Ry.* (1858) 28 L.J.Ex. 51.

[25] *Nichols* v. *Marsland* (1876) 2 Ex.D. 1 (C.A.).

[26] *Ibid.*

[27] Per Fry L.J., *Nitrophosphate and Odams' Chemical Manure Co.* v. *London & St. Katherine Docks Co.* (1878) 9 Ch.D. 503, 516 (C.A.). For "*force majeure*" clauses, see *post*, p. 348.

[28] *Baily* v. *de Crespigny* (1869) L.R. 4 Q.B. 180, 185.

[29] *Ibid.*

[30] *Metropolitan Water Board* v. *Dick, Kerr & Co.* [1917] 2 K.B. 1, 35; *Reardon Smith Line* v. *Min. of Agriculture* [1962] 1 Q.B. 42 81 (C.A.). For meaning of "strike" in a charterparty, see *Tramp Shipping Corp.* v. *Greenwich Marine Inc.* [1975] 1 W.L.R. 1042, 1046-1047 (C.A.). For delay caused by a strike where the obligation is to complete within a reasonable time, see *ante*, p. 59.

[31] *Cf. Matsoukis* v. *Priestman & Co.* [1915] 1 K.B. 681.

[32] The Building Regulations 1965, 1972, 1976 now apply outside London.

[33] There is no authority directly in point. The argument would be difficult; see *ante*, p. 101, " Difficulty or expense no excuse," and consider *Walton Harvey Ltd.* v. *Walker and Homfrays Ltd.* [1931] 1 Ch. 274 (C.A.); *Amalgamated Investment Ltd.* v. *John Walker Ltd.* [1977] 1 W.L.R. 164 (C.A.).

[34] See *ante*, p. 103. The ordinary arbitration clause remains effective (see *post*, p. 236).

[35] *Appleby* v. *Myers* (1867) L.R. 2 C.P. 651. For the position of the employer, see *Fibrosa, etc.* v. *Fairbairn, etc. Ltd.* [1943] A.C. 32 (H.L.).

such sum as the court having regard to all the circumstances of the case considers just.[36] Conversely money paid by the employer is recoverable, or if payable, ceases to be payable provided that if the contractor has incurred expenses in performance of the contract the court may, if it considers it just to do so, having regard to all the circumstances of the case, allow the contractor to retain or, as the case may be, recover the whole or any part of the money paid or payable before frustration up to the amount of the expenses incurred.[37]

The Act takes effect subject to the provisions of the contract.[38]

A contractor, under an entire contract,[39] who relies solely on the Act may be in no better position than at common law. He must show that the employer has received a valuable benefit, and the amount he can recover cannot exceed the value of such benefit.[40] So where the contract works are totally destroyed before use or occupation by the employer he may recover nothing. It is probably more in the interest of both parties to deal expressly with matters most likely to frustrate the contract, and in particular to provide for insurance [41] against the more common risks.[42]

Section 7: Illegality

1. No assistance from the court

If at the time of entering into a contract its performance or its object or the consideration for performance is illegal or contrary to sound morals the contract is void.[43] If its completion becomes illegal after performance has commenced the contract is frustrated and both parties are discharged from further performance.[44] No enforceable rights can, in general, accrue under an illegal or immoral contract for " no court will lend its aid to a man who founds his cause of action upon an immoral or illegal act." [45] If the contract is illegal in the sense that its performance is absolutely prohibited, or only permitted under a licence which is not obtained, the general rule is that the contractor cannot recover payment for illegal work even if he did not know of the illegality,[46] while the employer cannot

---

[36] s. 1 (1) (3).

[37] s. 1 (2).

[38] s. 2 (3). *Cf.* Standard form of building contract, clauses 4, 20, 23, 24, 26, 31, 32 and 33.

[39] For meaning, see *ante*, p. 50 and cases there cited, in particular *Appleby* v. *Myers* (1867) L.R. 2 C.P. 651.

[40] s. 1 (3). For difficulties in the interpretation and application of the Act, see *Chitty on Contracts* (24th ed., 1977), Vol. 1, paras. 1454 *et seq.*

[41] Note the effect of s. 1 (5).

[42] For reinstatement by insurers after destruction by fire, see the Fires Prevention (Metropolis) Act 1774, ss. 83, 84. See also *Gold* v. *Patman & Fotheringham Ltd.* [1958] 1 W.L.R. 697 (C.A.), discussed *post*, p. 338.

[43] See generally *Chitty on Contracts* (24th ed., 1977), Vol. 1, Chap. 16.

[44] See *Metropolitan Water Board* v. *Dick, Kerr & Co.* [1918] A.C. 119 (H.L.). See also *ante*, p. 104.

[45] *Per* Lord Mansfield C.J., *Holman* v. *Johnson* (1775) 1 Cowp. 341, 343; *Brightman* v. *Tate* [1919] 1 K.B. 463; *Brown Jenkinson & Co. Ltd.* v. *Percy Dalton (London) Ltd.* [1957] 2 Q.B. 621, 635, 639 (C.A.); *Palaniappa Chettiar* v. *Arunasalam Chettiar* [1962] A.C. 294 (P.C.).

[46] *Re Mahmoud and Ispahani* [1921] 2 K.B. 716 (C.A.); *Bostel Bros. Ltd.* v. *Hurlock* [1949] 1 K.B. 74 (C.A.); *cf. Yin* v. *Sam* [1962] A.C. 304 (P.C.).

recover instalments which he has paid under the contract.[47] If the contract is one where only the object of the building's use is immoral or illegal, it would seem that a contractor who was ignorant of such object would be able to recover for the work done and materials supplied, provided that he ceased work as soon as he became aware of the immoral or illegal purpose.[48] The position might arise in a contract to do work on premises to be used by a prostitute,[49] or to be used as a dwelling by a man's mistress.[50]

2. Return of Goods

The owner of goods the possession of which has passed to another under an illegal contract is entitled to the return of the goods or their value provided that the owner does not have to rely on the illegal contract or plead its illegality to support his claim,[51] but the court will look at the contract when it is just and proper to do so, as when it is necessary for the assessment of damages in an action by the owner for conversion of the goods.[52]

3. Contravention of statute

The general principle is that " what is done in contravention of the provisions of an Act of Parliament cannot be made the subject-matter of action." [53] but the application of this principle involves careful consideration of the statutory provisions in question and the particular breach complained of.

By-laws

Where contravention of building by-laws or similar provisions [54] are alleged a distinction is drawn between contracts illegal as formed and contracts illegal only as performed.[55] The contractor cannot recover payment for carrying out work which on the face of the contract must contravene the statutory provision.[56] In *Stevens* v. *Gourley* [57] the contractor undertook to erect and in fact completed

---

[47] *Smith & Son Ltd.* v. *Walker* [1952] 2 Q.B. 319, 328 (C.A.); *Kearley* v. *Thomson* (1889) 24 Q.B.D. 742, 745–746 (C.A.).

[48] See *Clay* v. *Yates* (1856) 1 H. & N. 73; *Pearce* v. *Brooks* (1866) L.R. 1 Ex. 213; *Upfill* v. *Wright* [1911] 1 K.B. 506 (D.C.); *Kiriri Cotton Co. Ltd.* v. *Dewani* [1960] A.C. 192 (P.C.).

[49] *Cf. Pearce* v. *Brooks, supra.*

[50] *Upfill* v. *Wright, supra.* It may be that a plea of illegality would not succeed today, at any rate, if the man owed a moral duty to provide for the woman; consider *Tanner* v. *Tanner* [1975] 1 W.L.R. 1346 (C.A.). See also (1977) L.Q.R. 386.

[51] *Bowmakers Ltd.* v. *Barnet Instruments Ltd.* [1945] 1 K.B. 65 (C.A.); *Singh* v. *Ali* [1960] A.C. 167 (P.C.), distinguished in *Palaniappa Chettiar* v. *Arunasalam Chettiar, supra.*

[52] *Belvoir Finance Co. Ltd.* v. *Stapleton* [1971] 1 Q.B. 210, 218 (C.A.).

[53] *Per* Lord Ellenborough C.J., *Langton* v. *Hughes*, 1 M. & S. 593, 596; *cf. Howell* v. *Falmouth Boat Construction Co. Ltd.* [1951] A.C. 837 (H.L.); see also *Barton* v. *Piggott* (1874) L.R. 10 Q.B. 86.

[54] Outside London the general statutory control is provided by the Building Regulations 1965, 1972, 1976.

[55] For a full discussion of this principle, see Cheshire and Fifoot, *Contract*, (9th ed.), pp. 323 *et seq.* See also *Archbolds (Freightage) Ltd.* v. *S. Spanglett Ltd.* [1961] 1 Q.B. 374 (C.A.); *Ashmore, Benson Ltd.* v. *Dawson Ltd.* [1973] 1 W.L.R. 828 (C.A.).

[56] Cases cited in note 53, *supra. Stevens* v. *Gourley* (1859) 7 C.B.(N.S.) 99; *Townsend (Builders) Ltd.* v. *Cinema News, etc., Ltd.* [1959] 1 W.L.R. 119 (C.A.).

[57] *Supra.*

a shop made of wood and resting on wooden foundations with the deliberate intention of evading statutory provisions which required that buildings should be made of incombustible material. It was held that he could not recover payment because of the contravention of the statute.[58] But where a contract is on the face of it capable of performance in accordance with the relevant statutory provisions and the contravention arises only in the mode of carrying it out, then the contractor may or may not, according to the particular circumstances, be able to recover payment for work which contravened the statute. Thus in *Townsend (Builders) Ltd.* v. *Cinema News, etc., Ltd.*,[59] the work as specified did not, but as completed did, involve a contravention of a by-law. It was held that there was no " fundamental illegality pervading the whole work and the whole contract " [60] and that the contractor was entitled to recover payment [61] having regard to the following: the by-law in question; the contractor's ignorance until the work was far advanced that there would be a contravention; a temporary waiver of the contravention by the local authority and to the ease with which compliance could be secured by the insertion of a partition. It seems that had it not been for the special circumstances just set out the contractor would not have recovered payment.[62]

In the *Townsend* case (*supra*) the employer recovered as damages the cost of making the works comply with the by-laws on a counterclaim for breach of an express term to comply with by-laws.[63] It is submitted that an employer who knows of, or becomes aware of, the contractor's intention to contravene by-laws cannot recover as damages any loss resulting from the breach.[64] Whether or not he would be debarred from recovering damages for other breaches would seem to depend upon whether the contravention brought about a " fundamental illegality " (see *supra*).

Town and Country Planning Acts

It seems that the same general principles apply to contravention of town and country planning legislation as apply to contravention of by-laws.[65]

Obtaining necessary consents

Where consents, *e.g.* of the local authority, must be obtained to make a contract lawful it depends on the particular contract whether the contractor or the employer must obtain them and whether, if they are not obtained, the parties are

---

[58] The building was, on the magistrate's order, removed.

[59] *Supra.*

[60] [1959] 1 W.L.R. 119, 125.

[61] For the architect's liability, see *post*, pp. 210 and 227.

[62] See in particular the judgment of Lord Cohen at [1959] 1 W.L.R. 119, 126. See also *One Hundred Simcoe Street* v. *Frank Burger Contractors* (1968) 66 D.L.R. (2d) 602, Ontario C.A.; *Fielding & Platt Ltd.* v. *Najjar* [1969] 1 W.L.R. 357 (C.A.).

[63] R.I.B.A. form (1939 ed.), cl. 3, printed on p. 274 of the 2nd edition of this book. For the current clause, see cl. 4, Standard form of building contract.

[64] *Ashmore Benson Ltd.* v. *Dawson Ltd.* [1973] 1 W.L.R. 828 (C.A.).

[65] *Cf. Best* v. *Glenville* [1960] 1 W.L.R. 1198 (C.A.)—a landlord and tenant case; see also *Shaw* v. *Groom* [1970] 1 Q.B. 504 (C.A.).

discharged from further obligations under the contract. It seems that in the absence of express words the party who has to obtain the consent or licence does not give an absolute warranty that he will obtain it, but a warranty to use all due diligence.[66]

Building Licences

Defence Regulations. From 1939 until November 10, 1954 (when the regulations were revoked), most private building work was unlawful unless it was licensed by the appropriate authority,[67] or its cost [68] did not exceed the so-called " free allowance," *i.e.* the amount of work which from time to time the Minister authorised to be carried out without a licence.[69]

Building Control Act 1966. This made unlawful building projects begun after July 27, 1965, or not contracted for before that date unless they were licensed or exempt from control because they were below the cost limit, in 1968, £100,000, or otherwise came within specified exemptions which included housing, industry and public works. Subject to a section of the Act giving a special defence for ignorance of its application,[70] a contractor cannot, it is submitted, recover payment for works which contravene the Act or the terms of a licence.[71] The Building Control (Suspension of Control) Order 1968 (No. 1827), operative on November 20, 1968, suspended until further order the control of building and constructional work imposed by the Act.

The Trade Description Act 1968

This Act imposes criminal sanctions for its contravention. Its purpose as regards goods is to prevent people when selling them from attaching false descriptions to them, and when providing services " to make it an offence if the person providing the services recklessly makes a false statement as to what he has done." [72] In *Beckett* v. *Cohen* [73] a builder who promised to build a garage in about 10 days " as the existing " neighbouring garage did not commit an offence when he did not complete it in 10 days and it was not exactly like the other garage. ". . . Parliament never intended . . . to make a criminal offence out of what is really a breach of warranty." [74]

---

[66] See *Smith* v. *Mayor of Harwich* (1857) 26 L.J.C.P. 257; *Baily* v. *de Crespigny* (1869) L.R. 4 Q.B. 180; *Brauer & Co. Ltd.* v. *Clark, etc., Ltd.* [1952] 2 All E.R. 497 (C.A.); *cf.* cl. 4, Standard form; *Re Northumberland Avenue Hotel Co.* (1887) 56 L.T. 833 (C.A.)— approval of plans by third party not obtained; *A. V. Pound & Co. Ltd.* v. *M. W. Hardy & Co. Inc.* [1956] A.C. 588 (H.L.); *Compagnie Algerienne, etc.* v. *Katana Societa, etc.* [1960] 2 Q.B. 115 (C.A.); *Smallman* v. *Smallman* [1972] Fam. 25, 31 (C.A.); *Agroexport* v. *Cie. Européene* [1974] 1 Lloyd's Rep. 499.

[67] reg. 56A, Defence Regulations 1939.

[68] See para. 4 of reg. 56A; *Young* v. *Buckles* [1952] 1 K.B. 220 (C.A.).

[69] reg. 56A (2), proviso. Cases on the regulations are discussed in the 2nd edition of this book. The reference to the subject is maintained in the text as the cases may be of assistance on questions of illegality.

[70] s. 1 (11).

[71] See *ante*, p. 107, " No assistance from the court." The Act has since been amended but is dealt with here summarily as it is not in operation.

[72] *Beckett* v. *Cohen* [1972] 1 W.L.R. 1593, 1596 (D.C.); see also *British Airways Board* v. *Taylor* [1976] 1 W.L.R. 13 (H.L.).

[73] *Supra.*

[74] See note 72, *supra.*

Section 8 : *Default of Other Party*

1. Repudiation—generally

The word " repudiation " is ambiguous and has several meanings,[75] but it is the most convenient term to describe the position where " one party so acts or so expresses himself as to show that he does not mean to accept the obligations of a contract any further." [76] Such a repudiation if accepted by the innocent party releases the innocent party from further performance.[77]

" Repudiation of a contract is a serious matter, not to be lightly found or inferred." [78] It may consist of a renunciation, an absolute refusal to perform the contract,[79] or it may arise as the result of a breach, or breaches of contract by a party which is or are such as to go to the root of the contract [80] so that " the acts and conduct of the party evince an intention no longer to be bound by the contract." [81] Repudiation before performance is due is termed an anticipatory breach.[82]

Breach of contract as repudiation. A breach of contract always gives rise to a claim for damages but is not necessarily a repudiation.[83] When it is alleged that there is a repudiatory breach difficult questions of law and fact can arise. The court has to consider first the term of which there is a breach and then, depending upon the view which it forms as to such term, the extent of the breach.[84] Thus the court asks whether the term is such that " the parties have agreed either expressly or by necessary implication or . . . the general law regards as a condition going to the root of the contract so that any breach of that term may at once and without further reference to the facts and circumstances be regarded by the innocent party . . ." as a repudiation.[85] Such a term is often called a fundamental term [86] or a condition and is to be contrasted with other terms of the contract variously called mere terms,[87] or collateral terms,[88] or, in contracts for the sale of goods, warranties,[89] breach of which does not in itself amount to a repudiation.[90]

---

[75] See *Heyman* v. *Darwins* [1942] A.C. 356, 378, 398 (H.L.).

[76] *Ibid. per* Lord Simon at p. 361.

[77] *Heyman* v. *Darwins, supra.*

[78] *Ross T. Smyth & Co. Ltd.* v. *T. D. Bailey, Son & Co.* [1940] 3 All E.R. 60, 71 (H.L.), *per* Lord Wright; *James Shaffer Ltd.* v. *Findlay, Durham & Brodie* [1953] 1 W.L.R. 106 (C.A.). See also *Sweet & Maxwell Ltd.* v. *Universal News Services Ltd.* [1964] 2 Q.B. 699 (C.A.).

[79] *Suisse Atlantique, etc.* v. *N.V. Rotterdamsche Kolen Centrale* [1967] 1 A.C. 361, 412, 421 (H.L.); *Mersey Steel & Iron Co. Ltd.* v. *Naylor* (1884) 9 App.Cas. 434, 439, 443 (H.L.).

[80] See note 79.

[81] *General Billposting Co. Ltd.* v. *Atkinson* [1909] A.C. 118, 122 (H.L.).

[82] *The Mihalis Angelos* [1971] 1 Q.B. 164 (C.A.).

[83] *Decro-Wall International S.A.* v. *Practitioners in Marketing* [1971] 1 W.L.R. 361 (C.A.).

[84] *Cehave M.V.* v. *Bremer mbH* [1976] 1 Q.B. 44 (C.A.). *United Scientific Holdings Ltd.* v. *Burnley Council* [1977] 2 W.L.R. 806, 830–831 (H.L.).

[85] *Suisse Atlantique, etc.* v. *N.V. Rotterdamsche Kolen Centrale* [1967] 1 A.C. 361, 422, *per* Lord Upjohn (H.L.).

[86] *Ibid.*

[87] See *Hoenig* v. *Isaacs* [1952] 2 All E.R. 176, 181 (C.A.).

[88] See *Feather & Co. Ltd.* v. *Keighley Corp.* (1953) 52 L.G.R. 30, 32.

[89] *Wallis* v. *Pratt* [1911] A.C. 394 (H.L.).

[90] *Hoenig* v. *Isaacs, supra.* The problem is sometimes approached by asking whether performance of a promise by one party is dependent upon the performance of a promise by

The division of terms of a contract into the two categories set out above is not exhaustive.[91] There is frequently a class of term intermediate between conditions and warranties where the right to treat its breach as a repudiation depends upon the effect or result of the breach, and where " if the breach goes to the root of the contract, the other party is entitled to treat himself as discharged; but, otherwise, not." [92] Other tests are to ask whether the breach is total [93] or fundamental [94] or whether the effect of the breach is such that it would be unfair to leave the injured party to a remedy in damages.[95] In commercial contracts, in particular those relating to shipping, the prime test seems to be whether the commercial purpose of the enterprise is frustrated.[96] It is submitted that, in relation to building contracts, while in any particular case this test may be useful [97] the older test of asking whether the breach goes to the root of the contract is often more helpful. The deliberate character of a breach makes it easier for, but does not compel, the court to find that it was fundamental.[98]

The court is not over ready, unless required by statute or authority so to do, to construe a term in a contract as a condition,[99] and is unlikely to do so where the effect of some breaches of the term is trivial.[1] Thus even where a term is described as a " condition," although this is a strong indication that the parties intended any breach, however small, to be repudiatory, it is not conclusive and yields to the discovery of the parties' intention as disclosed by the contract read as a whole.[2] Thus, where a term was so described but it was apparent that breaches of it would be of small effect the court refused to hold that mere breach, without consideration of whether it was a total or fundamental breach (see *supra*) was a repudiation.[3] Conversely the use of the word " warranty " to describe a term is not conclusive that that term is not a condition. In insurance law breach of warranty is treated as breach of condition [4] and it may well be that in a building contract the parties intend an express " warranty " of performance or as to the result or use of the works to have the effect of a condition.

Erroneous expression of view

There can be repudiation where a party intends to fulfil a contract, but " is

---

the other. Where it is dependent, breach of the promise by the other party releases the first party from his obligation to perform. See *Kingston* v. *Preston* (1773) 2 Doug. 689; *Mac-intosh* v. *Midland Counties Ry.* (1845) 14 L.J.Ex. 338; *cf. Boone* v. *Eyre* (1777) 1 H.Bl. 273 (cited *ante,* p. 52) and notes to *Cutter* v. *Powell* (1795) Smith's L.C. (13th ed.), Vol. 2, p. 1; *Hongkong Fir Shipping Co. Ltd.* v. *Kawasaki, etc., Ltd.* [1962] 2 Q.B. 26 (C.A.).

[91] *Cehave, supra.*
[92] *Ibid.* at p. 60, *per* Lord Denning.
[93] *Heyman* v. *Darwins* [1942] A.C. 356, 397 (H.L.).
[94] *Suisse Atlantique, supra,* at pp. 397, 410 (H.L.).
[95] *Decro-Wall, supra,* at p. 380, *per* Buckley L.J.
[96] *Hongkong Fir Shipping, supra.*
[97] See its application in *Harbutts " Plasticine " v. Wayne Tank & Pump Co.* [1970] 1 Q.B. 447 (C.A.); for frustration generally, see *ante,* p. 102.
[98] *Suisse Atlantique, supra,* at pp. 394, 398, 415, 429, 435.
[99] *Cehave, supra,* p. 70, *per* Roskill L.J.
' *Cehave, supra; Hongkong Fir Shipping Co., supra; Schuler (L.) A.G.* v. *Wickman ·ine Tool Sales* [1974] A.C. 235 (H.L.).
'·uler (L.) A.G.* v. *Wickman, supra.*

determined to do so only in a manner substantially inconsistent with his obligations and not in any other way." [5] But it is not a repudiation merely to put forward
in good faith an interpretation of the contract which is wrong,[6] the more expeci-
ally if it is put forward in such a way as to show that it is open to correction.[7]

Subsequent excuses for alleged repudiation

A party who is alleged to have repudiated a contract can subsequently rely on
any defence notwithstanding that at the time of the alleged repudiation he gave
other or no reasons by way of excuse,[8] unless he is estopped by his conduct and
the other party's reliance thereon from relying upon a reason different from that
which he gave at the time of the alleged repudiation.[9]

Acceptance of repudiation

" Repudiation by one party standing alone does not terminate the contract.
It takes two to end it, by repudiation on the one side, and acceptance of the
repudiation on the other." [10] The innocent party must make it plain that " in
view of the wrongful act of the party who has repudiated he claims to treat the
contract as at an end." [11] By doing this he is usually said to rescind [12] the
contract. He is released from further performance and can sue at once for
damages even though the time for performance has not yet arisen.[13] The terms
of the contract, save only any arbitration clause, cease to apply, but they are
referred to for the purpose of assessing damages.[14] If the innocent party does not
elect to accept the repudiation but affirms the contract the defaulting party may
continue to rely on the terms of the contract, including any agreed damages
clause, unless on its true construction any term excluding or limiting his liability
was not intended to cover the kind of breach which has been committed [15];

---

[5] *Ross T. Smyth & Co. Ltd.* v. *T. D. Bailey, Son & Co.* [1940] 3 All E.R. 60, 72 (H.L.),
per Lord Wright.
[6] *Ibid.*; *Mersey Steel & Iron Co. Ltd.* v. *Naylor* (1884) 9 App.Cas. 434 (H.L.); *James
Shaffer Ltd.* v. *Findlay, Durham & Brodie* [1953] 1 W.L.R. 106 (C.A.); *Sweet & Maxwell
Ltd.* v. *Universal News Services Ltd.* [1964] 2 Q.B. 699 (C.A.).
[7] *Ibid.* especially *Ross T. Smyth, supra,* at p. 72; *Sweet & Maxwell Ltd., supra,* at p. 737.
[8] *Scammell* v. *Ouston* [1941] A.C. 251, 268 (H.L.); *The Mihalis Angelos* [1971] 1 Q.B.
164, 195 (C.A.).
[9] *Panchaud Frères S.A.* v. *Etablissements General Grain Co.* [1970] 1 Lloyd's Rep. 53
(C.A.). For determination under the Standard form of building contract see clauses 25, 26.
[10] *Per* Lord Simon, *Heyman* v. *Darwins* [1942] A.C. 356, 361 (H.L.); *White & Carter
(Councils) Ltd.* v. *McGregor* [1962] A.C. 413, 432 (H.L.); *Lakshmijt* v. *Sherani* [1974] A.C.
605, 616 (P.C.); *Aquis Estates Ltd.* v. *Minton* [1975] 1 W.L.R. 1452 (C.A.). Where there is
no acceptance the injured party cannot obtain a bare declaration that there was conduct
constituting a repudiation—*Howard* v. *Pickford Tool Co. Ltd.* [1951] 1 K.B. 417 (C.A.).
For the position where a repudiatory breach results in a frustrating event, see discussion of
Harbutt's " Plasticine " Ltd. v. *Wayne Tank Ltd., infra.*
[11] *Ibid.*
[12] See *ante,* p. 89, for misrepresentation and *ante,* p. 96, for the view that references in
the Misrepresentation Act 1967 to rescission are not to acceptance of repudiation.
[13] *Ibid.*; *Hochster* v. *de la Tour* (1853) 2 E. & B. 678. This type of repudiation is termed
" anticipatory breach."
[14] *Bloeman (F. J.) Pty. Ltd.* v. *Gold Coast City* [1973] A.C. 115 (P.C.); *Lep Air Services
Ltd.* v. *Rolloswin Ltd.* [1973] A.C. 331 (H.L.). For arbitration, see *post,* p. 236.
[15] *Suisse Atlantique, etc.* v. *N.V. Rotterdamsche Kolen Centrale* [1967] 1 A.C. 361 (H.L.).
See also *Harbutt's " Plasticine " case,* discussed *infra* and *post,* p. 141.

further, where the time for performance has not yet arisen, the defaulting party " has the opportunity of withdrawing from his false position, and even if he does not, may escape ultimate liability because of some supervening event " [16] which frustrates the contract.[17] Where the contract remains in existence the innocent party's obligations continue, and he cannot thereafter terminate the contract for the breach he has waived,[18] although if the breach is not remedied or excused before performance is due, he can then enforce his claim under or for breach of the contract.[19]

The innocent party is not, it seems, in the ordinary case, bound to accept a repudiation before performance is completed in order to reduce the amount ultimately payable to him by the defaulting party,[20] but it is doubtful whether this principle can have any application to repudiation by the employer in the ordinary building contract where the employer's assent is required to go on his land for its performance.[21]

Harbutt's " Plasticine " Ltd. v. Wayne Tank Ltd.[22]

The plaintiffs were owners of a factory where they made Plasticine. The defendants undertook to design, supply and erect equipment for storing and dispensing some of the materials including wax which had to be heated. They used plastic piping for this purpose which they surrounded with an electric heating element and this was, as the Judge found, wholly unsuitable for its purpose. Further, their workmen switched on the heating elements and left the plant unattended overnight whereupon, due to their negligence, it over-heated, burst into flames and destroyed the factory. The cost of the works was £2,330. The damages were £146,000. One of the clauses of the written contract limited liability for damage to the property to the contract price. The issue was whether the contractors could rely on this clause. The Court of Appeal held that they could not. It said that there was a fundamental breach of contract by the contractors of a kind which left no option to the innocent party but to treat the contract as at an end. In those circumstances the clause did not apply.[23]

It is thought that the decision was not intended to be inconsistent with the rule that there must be acceptance of repudiation to bring a contract to an end,

---

[16] *Heyman's* case, *supra*, at p. 361.

[17] *Ibid.*; *Avery* v. *Bowden* (1855) 6 E. & B. 953, where after refusal but before the time for performance the contract became illegal because of the outbreak of war.

[18] *Frost* v. *Knight* (1872) L.R. 7 Ex. 111, 112; *Hughes* v. *Metropolitan Ry.* (1877) 2 App. Cas. 439 (H.L.); *Suisse Atlantique* case, *supra*; *Allen* v. *Robles* [1969] 1 W.L.R. 1193 (C.A.).

[19] *Frost* v. *Knight*, *supra*; *Heyman* v. *Darwins*, *supra*; *White & Carter (Councils) Ltd.* v. *McGregor*, *supra*.

[20] *White & Carter (Councils) Ltd.* v. *McGregor*, *supra*—a Scottish appeal where the majority (3:2) of the House of Lords held that an advertising contractor who did not accept a repudiation was entitled to perform his contract and recover the full contract price. See also *post*, p. 147.

[21] See *Finelli* v. *Dee* (1968) 67 D.L.R. (2d) 393, Ontario C.A.; *Hounslow (London Borough)* v. *Twickenham Garden Developments* [1971] Ch. 233, 252–254, but note comments *post*, p. 123; *Mayfield Holdings* v. *Moana Reef* [1973] 1 N.Z.L.R. 309, New Zealand Supreme Court. See also *Barker, George (Transport) Ltd.* v. *Eynon* [1974] 1 W.L.R. 462 (C.A.); *Attica Sea Carriers* v. *Ferrostaal-Poseidon GmbH* [1976] 1 Lloyd's Rep. 250 (C.A.).

[22] [1970] 1 Q.B. 447 (C.A.).

[23] See further, *post*, p. 141 in relation to exclusion clauses.

and that it can be explained upon the basis that where the breach results in an event of the kind which usually frustrates a contract, such as destruction of the subject-matter of the contract rendering further performance impossible, the law treats the innocent party as having accepted the repudiation. Nevertheless, the decision can give rise to problems of construction particularly where, as in the Standard forms of contract, there are elaborate provisions dealing with what is to happen upon the destruction of or damage to the works.[24]

2. Repudiation by contractor [25]

Refusal or abandonment

An absolute refusal to carry out the work or an abandonment of the work before it is substantially completed, without any lawful excuse, is a repudiation.[26]

Defects

A breach consisting of mere negligent omissions or bad workmanship where the work is substantially completed does not go to the root of the contract in the ordinary lump sum contract,[27] and is not therefore a repudiation.

Can omissions or bad work as they occur during the course of the work be treated as repudiation? It is submitted that in the ordinary case they cannot if they are not such as to prevent substantial completion, but that there is a repudiation where, having regard to the construction of the contract and all the facts and circumstances, the gravity of the breaches is such as to show that the contractor does not intend to or cannot substantially perform his obligations under the contract.[28]

Delay

Delay on the part of the contractor where time is not of the essence of the contract does not amount to a repudiation unless it is such as to show that he will not, or cannot, carry out the contract [29]; and in most cases [30] it is desirable to give notice that continuance of the delay will be treated as repudiation before purporting to accept the repudiation by dismissing the contractor. In *Felton* v. *Wharrie* the contractor had not finished the work by the completion date, and when asked by the employer's agent whether it would take one, two, three or four months, replied that he could not say. He proceeded with the work and two weeks later the employer without any express right under the contract, and without any warning forcibly ejected the contractor from the site. It was held that the

---

[24] See the Standard form, cll. 18 to 20.

[25] The specific matters discussed must be read with " Repudiation—generally," *supra*.

[26] *Mersey Steel & Iron Co. Ltd.* v. *Naylor* (1884) 9 App.Cas. 434 (H.L.); *Marshall* v. *Mackintosh* (1898) 78 L.T. 750; *Hoenig* v. *Isaacs* [1952] 2 All E.R. 176 (C.A.).

[27] *Hoenig* v. *Isaacs, supra*; see generally Chap. 4.

[28] *Suisse Atlantique* case, *supra*, at p. 422. See also the facts in *Harbutt's " Plasticine "* case, *supra*. In *Sutcliffe* v. *Thackrah* [1974] A.C. 727 (H.L.) the judge at first instance, on a matter not appealed against, held that a combination of bad work and delay was a repudiation. See also *Wayne Tank Co. Ltd.* v. *Employers Liability Ltd.* [1974] 1 Q.B. 57, 73 (C.A.).

[29] *Felton* v. *Wharrie* (1906) H.B.C. (4th ed.), Vol. 2, p. 398 (C.A.); *Chandler Bros.* v. *Boswell* [1936] 3 All E.R. 179, 185 (C.A.); *Suisse Atlantique* case, *supra*.

[30] *Cf. Etablissements Chainbaux* v. *Harbormaster Ltd.* [1955] 1 Lloyd's Rep. 303.

employer had no right to determine the contract. " If he were going to act upon the plaintiff's conduct as being evidence of his not going on, he ought to have told him of it, and to have said, ' I treat that as a refusal,' and the man would know of it; but the fact of allowing him to go on cannot be any evidence of justification of re-entry." [31]

Where time is of the essence either by the terms of the contract, or as a result of a notice making it of the essence,[32] and the contractor fails to complete to time the employer is entitled to treat the contract as at an end and to dismiss the contractor from the site.[33]

Other breaches of contract

It is a question in each case whether other breaches of contract go to the root of the contract. For example, it has been held that sub-letting part of the contract works, contrary to an express provision was not a repudiation.[34]

Repudiation and forfeiture

A distinction must be made between the acceptance of repudiation bringing a contract to an end, and the determination of the contract under an express power.[35] If the event giving rise to the exercise of an express right to determine was a repudiation by the contractor, the employer is, subject to the express terms of the contract, entitled to all the damages that flow from non-completion, including the enhanced cost of completing by another contractor.[36] But such damages may not be payable if the event relied on was not a repudiation. Thus in *Feather* v. *Keighley Corporation* [37] the contractor undertook not to sub-let work without consent, the contract providing that for breach of this term the employer could either absolutely determine the contract, or could claim £100 by way of liquidated damages. The contractor sub-let without consent, and the employer thereupon determined the contract and claimed as damages the increased cost of completing by another contractor. It was held that there had only been a breach of a collateral term, not a repudiation, and that the employer having chosen his remedy of determining the contract was not entitled to claim damages.

3. Repudiation by employer [38]

Refusal

An absolute refusal by the employer to carry out his part of the contract, whether made before the works commenced or while they are being carried out is a repudiation of the contract.[39]

---

[31] *Per* Lord Alverstone L.C.J. at p. 400. For express rights, see Chap. 8, " Forfeiture Clauses."

[32] See *ante*, p. 60.

[33] *Rickards* v. *Oppenheim* [1950] 1 K.B. 616, 628 (C.A.).

[34] *Feather* v. *Keighley Corp.* (1953) 52 L.G.R. 30.

[35] See Chap. 8, " Forfeiture Clauses."

[36] See *Marshall* v. *Mackintosh* (1898) 78 L.T. 750. For the measure of damages see *post*, p. 148.

[37] *Supra.*

[38] The specific matters discussed must be read with " Repudiation—generally," *supra.*

[39] *Hochster* v. *de la Tour, supra*; *Mersey Steel & Iron Co. Ltd.* v. *Naylor, supra.*

Rendering completion impossible

It is, in general, a repudiation if the employer wrongfully by his own acts, and without lawful excuse,[40] renders completion impossible.[41]

Possession of site

The employer repudiates the contract if he fails to give possession of the site at all, or without lawful excuse ejects the contractor from the site before completion.[42]

Order not to complete

A clear unjustified order not to complete the works is a repudiation.[43]

Failure to pay instalments

This cannot be a repudiation if there is no contractual duty to pay them.[44] Where there is such a duty it is a question in each case whether failure to pay is a repudiation. Failure to pay one instalment out of many due under the terms of the contract is not ordinarily sufficient to amount to a repudiation.[45] It was held to be a repudiation where a company had only paid £10,000 out of £24,000 then due.[46] A failure to pay is less likely to be a repudiation if it occurs towards the end of a contract [47]

Under-certification

Can an employer who pays certificates issued by the architect be guilty of repudiation if those certificates are substantially too low ? [48] There are difficulties in saying that he can because prima facie he is doing what the contract requires of him. But it has now been settled that the architect is the employer's agent when giving his certificate [49]; and the employer cannot stand by and take advantage of his architect applying a wrong principle in certifying.[50] It may be, therefore, that circumstances can arise where an employer is guilty of repudiation for under-payment notwithstanding that he has paid certificates properly.[51]

---

[40] See *Cory Ltd.* v. *City of London Corp.* [1951] 2 K.B. 476 (C.A.).
[41] *Stirling* v. *Maitland* (1864) 5 B. & S. 840, 852; *Roberts* v. *Bury Commissioners* (1870) L.R. 4 C.P. 755; *Southern Foundries* v. *Shirlaw* [1940] A.C. 701, 717, 741 (H.L.). For the employer's duty to co-operate and not to prevent completion, see *ante*, p. 37.
[42] *Ibid.*; *Felton* v. *Wharrie, supra. Cf. Earth & General Contractors Ltd.* v. *Manchester Corp.* (1958) 108 L.J. 665—interference with possession not repudiation.
[43] *Cort* v. *Ambergate Ry.* (1851) 17 Q.B. 127.
[44] *Rees* v. *Lines* (1837) 8 C. & P. 126; see *ante*, p. 53.
[45] *Mersey Steel & Iron Co. Ltd.* v. *Naylor* (1884) 9 App.Cas. 434 (H.L.); *Decro-Wall International S.A.* v. *Practitioners in Marketing* [1971] 1 W.L.R. 361 (C.A.); *Lakshmijit* v. *Faiz Sherani* [1974] A.C. 605 at p. 616 (P.C.).
[46] *Lep Air Services Ltd.* v. *Rolloswin Ltd.* [1973] A.C. 331 at pp. 344, 346, 353 (H.L.).
[47] *Cornwall* v. *Henson* [1900] 2 Ch. 298 (C.A.).
[48] For certificates generally, see Chap. 6.
[49] *Sutcliffe* v. *Thackrah* [1974] A.C. 727 (H.L.).
[50] *Panamena, etc.* v. *Frederick Leyland & Co. Ltd.* [1947] A.C. 428 (H.L.), discussed *ante*, p. 78.
[51] For the principles which the architect must apply when certifying under the Standard form of building contract, see cl. 30.

Other breaches

It depends upon the construction of the contract and the circumstances whether the acts and conduct of the employer show that he no longer intends to be bound by the contract. Thus, assuming there is a breach, it may or may not be a repudiation if the employer fails to appoint an architect,[52] or to supply plans [53] or materials,[54] or to make a fresh nomination of a sub-contractor.[55]

Where a contractor was carrying out work for an employer under two separate contracts, he was held not to be entitled to stop work on one contract because the employer had not paid on the other contract.[56]

If the employer interferes with the architect in the performance of those functions where he has to act fairly between the employer and the contractor [57] it is, it is submitted, a question in each case depending both upon the nature of the employer's acts and their effect whether such interference amounts to a repudiation.

4. Party cannot rely on own wrong

Where one party has failed to perform a condition of the contract, the other party cannot rely on its non-performance if it was caused by his own wrongful acts.[58] This principle has been considered elsewhere in respect of the contractor's failure to complete the contract work,[59] to obtain written orders for extras,[60] to obtain the architect's certificate,[61] and to complete to time.[62]

A similar principle is applied to the construction of contracts which provide that, upon the happening of certain events, the contract shall become void. When such an event occurs either party may declare the contract void provided he has not himself been the means of bringing about the event [63]; for example, an insolvent contractor cannot rely on his own insolvency to escape from the contract.[64]

---

[52] *Coombe* v. *Greene* (1843) 11 M. & W. 480; *Hunt* v. *Bishop* (1853) 8 Ex. 675; *Ctr. Jones* v. *Cannock* (1852) 3 H.L.C. 700.

[53] *Wells* v. *Army & Navy Co-op. Society* (1902) 86 L.T. 764; H.B.C. (4th ed.), Vol. 2, p. 356 (C.A.); *Trollope & Colls* v. *Singer* (1913) H.B.C. (4th ed.), Vol. 1, p. 849; *cf. Stevens* v. *Taylor* (1860) 2 F. & F. 419.

[54] *Macintosh* v. *Midland Counties Ry.* (1845) 14 L.J.Ex. 338; *cf. Gaze Ltd.* v. *Port Talbot Corp.* (1929) 93 J.P. 89.

[55] See *Bickerton (T. A.) & Son Ltd.* v. *N.W. Regional Hospital Board* [1970] 1 W.L.R. 607 (H.L.), discussed *post*, p. 182.

[56] *Small & Sons Ltd.* v. *Middlesex Real Estates Ltd.* [1921] W.N. 245.

[57] See *Sutcliffe* v. *Thackrah* [1974] A.C. 727, 737 (H.L.) and submission at p. 38, *ante*, that there is an implied term that the employer will not so act as to disqualify the architect. For disqualification, see *ante*, p. 85.

[58] *Roberts* v. *Bury Commissioners* (1870) L.R. 4 C.P. 755. See article by J. F. Burrows (1965) 31 M.L.R. 390. For clauses purporting to exempt a party from the consequences of his own negligence, see *post*, p. 139.

[59] See *ante*, p. 55.

[60] See *ante*, p. 69.

[61] See *ante*, p. 78.

[62] See *post*, p. 160.

[63] *New Zealand Shipping Co.* v. *Ateliers, etc., de France* [1919] A.C. 1 (H.L.).

[64] *Ibid.* at p. 13.

Section 9: *Variation, Waiver, Rescission and Estoppel*

It is a good excuse to an allegation of non-performance to prove that the obligation has been performed according to a varied agreement; and in such a case it is not necessary to show fresh consideration for a promise not to enforce the contract in its original form [65]; a party can always waive a right inserted for his benefit.[66]

The variation of the contract works is usually governed by express terms and has been considered in Chapter 5.

Waiver of claim for defects

The employer does not waive his claim for damages for defective work merely by occupying and using the contract works,[67] nor by paying money on account,[68] nor in full,[69] nor by suffering judgment for the whole contract sum to be entered against him.[70]

Waiver of claim by contractor

If the employer delays the works or otherwise commits a breach of contract while the work is being carried out the contractor does not waive a claim for damages merely by continuing with the work.[71]

Rescission by agreement

The parties may have mutually agreed to rescind the contract and treat it as at an end.

Estoppel

" If a man has led another to believe that a particular state of affairs is settled and correct, he will not be allowed to depart therefrom and assert they were erroneous and incorrect when it would be unjust and inequitable for him to do so." [72]

---

[65] *Per* Denning L.J., *Rickards* v. *Oppenheim* [1950] 1 K.B. 616, 623 (C.A.). But for release of money due, see *Foakes* v. *Beer* (1884) 9 App.Cas. 605 (H.L.); *D. & C. Builders Ltd.* v. *Rees* [1966] 2 Q.B. 617 (C.A.); and see generally *Chitty on Contracts* (24th ed. 1977), Vol. 1, para. 1364. For resumption of suspended legal rights, see *Tool Metal Manufacturing Co.* v. *Tungsten Electrical Co.* [1955] 1 W.L.R. 761 (H.L.); *Woodhouse A.C.* v. *Nigerian Produce Co.* [1972] A.C. 741 (H.L.); *Alan & Co.* v. *El Nasr Co.* [1972] 2 Q.B. 189 (C.A.).

[66] *Hoenig* v. *Isaacs* [1952] 2 All E.R. 176, 181 (C.A.). For waiver of requirements as to notices, see *Lickiss* v. *Milestone Motor Policies* [1966] 2 All E.R. 972, 975 (C.A.).

[67] *Dakin* v. *Lee* [1916] 1 K.B. 566 (C.A.).

[68] *Cooper* v. *Uttoxeter Burial Board* (1864) 11 L.T. 565.

[69] *Davis* v. *Hedges* (1871) L.R. 6 Q.B. 687, unless paid under a binding and conclusive certificate. See *ante*, p. 80.

[70] *Mondel* v. *Steel* (1841) 8 M. & W. 858; *Davis* v. *Hedges, supra*; *Chell Engineering* v. *Unit Tool Co.* [1950] 1 All E.R. 378 (C.A.), and see Chap. 14 (*post*, p. 277). Payment or judgment may be evidence of satisfaction, although not a waiver. It may be advisable to state that they are without prejudice to a claim for defects.

[71] *Lawson* v. *Wallasey Local Board* (1883) 48 L.T. 507 (C.A.). For other applications of waiver reference must be made to the index.

[72] *Per* Lord Denning M.R., *Ismail* v. *Polish Ocean Liners* [1976] 2 W.L.R. 477, 484 (C.A.). For estoppel generally, see Hals. (4th ed.), Vol. 16. See also *Moorgate Mercantile Co. Ltd.* v. *Twitchings* [1976] 3 W.L.R. 66 (H.L.). Estoppel is a doctrine ordinarily relied on by way of defence. In certain circumstances, in relation to land, it may lead to the creation of a proprietary interest, *Crabb* v. *Arun District Council* [1976] Ch. 179 (C.A.). See also *ante*, p. 69, n. 67.

CHAPTER 8

DEFAULT OF THE PARTIES—VARIOUS MATTERS

Section 1: *Forfeiture Clauses*

THE contractor has a licence to occupy the site for the purposes of the contract. [1] If the employer revokes the licence before completion, thus preventing the contractor carrying out the work, there is prima facie a repudiation by the employer. [2] To justify his act he must show either that there was a repudiation by the contractor which he has accepted, [3] or that he has acted under an express power of the contract contained in what is sometimes called a forfeiture clause.

1. The nature of forfeiture clauses

" Forfeiture clause " is a loose term used to describe a clause in a written building contract giving the employer the right upon the happening of an event to determine the contract, or eject the contractor from the site, or otherwise to take the work substantially out of his hands. [4] The right may be stated to arise upon the happening of any event, [5] provided there is no illegality, and the clause is not void because it is contrary to the policy of the law, *e.g.* that materials should be forfeited on bankruptcy. [6] So far as the right arises on bankruptcy, it is discussed in Chapter 12; examples of other events upon which the right can arise are [7]: not proceeding with the works to the satisfaction of the architect [8]; not proceeding with due diligence [9]; not completing to time [10]; not complying with the architect's orders. [11]

2. The mode of forfeiture

Compliance with the contract

The requirements of the contract must be properly complied with, for the courts construe forfeiture clauses strictly, [12] and a wrongful forfeiture by the employer or his agent normally amounts to a repudiation on the part of the

---

[1] *Joshua Henshaw & Sons* v. *Rochdale Corp.* [1944] K.B. 381 (C.A.); *Hounslow, London Borough of* v. *Twickenham Garden Developments Ltd.* [1971] Ch. 233, where Megarry J., in a judgment discussed *post,* p. 123, considered the nature of the contractor's licence under the Standard form of building contract.

[2] See *ante,* p. 117.

[3] See *ante,* p. 113.

[4] In this sense cl. 25 of the Standard form of building contract is a forfeiture clause although the word " forfeiture " is never used. Note that cl. 26 gives the contractor the right of bringing his employment to an end upon the happening of certain events.

[5] *Davies* v. *Swansea Corp.* (1853) 22 L.J.Ex. 297.

[6] See *post,* p. 188.

[7] For a longer list, see H.B.C. (7th ed.), p. 404.

[8] *Davies* v. *Swansea Corp.,* supra.

[9] *Roberts* v. *Bury Commissioners* (1870) L.R. 5 C.P. 310.

[10] *Marsden* v. *Sambell* (1880) 43 L.T. 120.

[11] *Hunt* v. *S.E. Ry.* (1875) 45 L.J.Q.B. 87.

[12] *Roberts* v. *Bury Commissioners,* supra, at p. 326.

employer.[13] There must be some definite unqualified act showing that the power has been exercised, although writing or other formality is not necessary unless expressly required.[14] The contract may require a certain notice to be given, and that such notice must set out the default complained of.[15]

Ascertainment of the event

Where the architect has the power to ascertain whether the event giving rise to the right to forfeit has arisen, his decision can only be attacked in those circumstances, described in Chapter 6,[16] where his certificates can be attacked.[17] If the contract gives the employer the right to ascertain whether the event has arisen, his determination must, in the absence of words to the contrary, be reasonable.[18] But the contract may make it clear that he is entitled to exercise a right of forfeiture when, in his opinion, whether reasonable or not, the event, such as delay, has arisen.[19] It may be reasonable for a main contractor, himself under onerous obligations to complete, to forfeit a sub-contract, where the exercise of a similar power by the employer would be unreasonable.[20]

Time of forfeiture

When an event occurs which gives rise to the right to forfeit, the power of forfeiture must be exercised within a reasonable time or the employer will be deemed to have waived his right unless the event is a continuing breach of contract.[21] And the employer cannot exercise his right if the rights of third parties have intervened, or if he has treated the contract as subsisting.[22] But he can forfeit if a fresh event arises.[23]

Where the contract provides for completion by a certain date and also provides for forfeiture for delay, and the completion date has passed, it is a question of construction whether the forfeiture clause for delay can still be enforced. Thus where the object of the clause was to enable the architect to " have the means of requiring the works to be proceeded with in such a manner and at such a rate

---

[13] *Lodder* v. *Slowey* [1904] A.C. 442 (P.C.), and see *ante*, p. 111, for repudiation, and *post*, p. 151, for the contractor's claim for damages or a reasonable sum.

[14] *Drew* v. *Josolyne* (1887) 14 Q.B.D. 590, 597 (C.A.); see also *Marsden* v. *Sambell*, *supra*, where it was held to be no exercise of a right of forfeiture by the employer merely sending an agent " to keep an eye " on houses being built. For a sale of land case, see *Aquis Estates Ltd.* v. *Minton* [1975] 1 W.L.R. 1452 (C.A.).

[15] See *Pauling* v. *Dover Corp.* (1855) 24 L.J.Ex. 128; *cf. Boot & Sons Ltd.* v. *Uttoxeter U.D.C.* (1924) 88 J.P. 118 (C.A.).

[16] See *ante*, p. 81.

[17] See *Northampton Gas Light Co.* v. *Parnell* (1855) 15 C.B. 630, where the architects' decision was not conclusive.

[18] *Stadhard* v. *Lee* (1863) 3 B. & S. 364; see *ante*, p. 74.

[19] *Stadhard* v. *Lee*, *supra*; *cf. Diggle* v. *Ogston Motor Co.* (1915) 112 L.T. 1029.

[20] See *supra*, note 18.

[21] *Marsden* v. *Sambell*, *supra*; *Aquis Estates Ltd.* v. *Minton*, *supra*.

[22] *Ibid.* at p. 122; *Re Garrud, ex p. Newitt* (1881) 16 Ch.D. 522, 533 (C.A.); *Platt* v. *Parker* (1886) 2 T.L.R. 786, 787 (C.A.); *cf. Hughes* v. *Metropolitan Ry.* (1877) 2 App.Cas. 439 (H.L.); *Tool Metal Manufacturing Co. Ltd.* v. *Tungsten Electric Co. Ltd.* [1955] 1 W.L.R. 761 (H.L.); *Australian Blue Metal Ltd.* v. *Hughes* [1963] A.C. 74 (P.C.); *Alan (W.J.) & Co.* v. *El Nasr Co.* [1972] 2 Q.B. 189 (C.A.); *Woodhouse A.C.* v. *Nigerian Produce Co.* [1972] A.C. 741 (H.L.).

[23] *Stevens* v. *Taylor* (1860) 2 F. & F. 419; *Re Garrud, supra.*

of progress as to ensure their completion at the time stipulated;" [24] it was held that the clause did not apply after the completion date.[25] But in another contract where the clause provided " for the execution of the work with due diligence and as much expedition as the surveyor shall require," it was held that the clause was as much applicable to the fulfilment of the contract within a reasonable time as to its completion by the contract date. Therefore a forfeiture under the terms of the clause was valid although the original date of completion had been ignored by the parties.[26] It should be noted that many contracts provide for the extension of the original completion date.[27]

3. Employer's default

Forfeiture by the employer is wrongful if he or his agents have been the means of bringing about the event which gave rise to the right to forfeit,[28] *e.g.* where delay by the contractor was caused by the wrongful failure of the architect to give proper plans,[29] or by the wrongful and improper withholding of a certificate by the architect.[30]

4. Effect of forfeiture

The parties may agree that any consequences may follow the exercise of a right of forfeiture,[31] provided there is no illegality, nor fraud on the bankruptcy law,[32] and the clause is not so onerous that it will not be enforced on the grounds that it is a penalty.[33] The employer is usually given the right to take possession of the site and complete the works.[34] In addition there is frequently a clause vesting the property in unfixed materials, and perhaps plant, in the employer,[35] or there may be merely a right to seize the material.[36] or hold them by way of lien [37] until they are built into the works, or there may be clauses giving rights of user of the contractor's plant and materials to the employer.[38] Some of these matters are considered further in the next section.

Bills of sale

The ordinary forfeiture clause is not within the Bills of Sale Acts 1878 and

---

[24] *Walker v. L. & N.W. Ry.* (1876) 1 C.P.D. 518, 530.
[25] *Ibid.*
[26] *Joshua Henshaw & Sons v. Rochdale Corp.* [1944] K.B. 381 (C.A.).
[27] See *post*, p. 162.
[28] *Roberts v. Bury Commissioners* (1870) L.R. 5 C.P. 310; *New Zealand Shipping Co. Ltd.* v. *Ateliers de France* [1919] A.C. 1 (H.L.).
[29] *Roberts v. Bury Commissioners, supra*; *cf. Stevens v. Taylor, supra.*
[30] *Smith v. Howden Union* (1890) H.B.C. (4th ed.), Vol. 2, p. 156. For certificates generally, see Chap. 6.
[31] *Davies v. Swansea Corp.* (1853) 22 L.J.Ex. 297.
[32] See *post*, pp. 188 *et seq.*
[33] See *post*, p. 158.
[34] For a comprehensive list of clauses which had then come before the courts, see H.B.C. (7th ed.), p. 432.
[35] See further *post*, p. 125.
[36] See *Baker v. Gray* (1856) 25 L.J.C.P. 161.
[37] See *Re Waugh, ex p. Dickin* (1876) 4 Ch.D. 524.
[38] See *Hawthorn v. Newcastle, etc., Ry.* (1840) 3 Q.B. 734n.; *Re Winters, ex p. Bolland* (1878) 8 Ch.D. 225.

1882.[39] The contract in which it is contained does not therefore require to be registered under the Acts.[40]

Employer's duty to account

When an employer in exercise of his rights under a forfeiture clause enters and completes the work and uses the contractor's materials or plant, or holds retention money due to the contractor, he must, subject to the provisions of the contract, account to the contractor, *i.e.* show that the materials and plant and money were expended reasonably [41]; but the court would, it seems, in principle, make full allowance for extra cost caused by the disruption and delay occasioned by the contractor's default.[42]

Employer's claim for damages

Where the employer determines the contract under a forfeiture clause because of some breach of contract by the contractor, the employer's right to damages depends upon the wording of the contract. He may not be entitled to the enhanced cost of completing by another contractor if the breach for which he determined the contract did not amount to a repudiation.[43]

Injunction to enforce forfeiture

In *Hounslow, London Borough of* v. *Twickenham Garden Development Ltd.*[44] Megarry J. refused an injunction. Notice of determination had been served under clause 25 (1) of the Standard form of building contract based on the alleged failure of the contractor " to proceed regularly and diligently with the works." The allegation was hotly disputed by the contractor and there were affidavits before the court upon the issues of fact. It was held that the case fell considerably short of any standard upon which it would be safe to grant an interlocutory injunction for " what is involved is the application of an uncertain concept to disputed facts." [45] Despite the importance to the borough on social grounds of securing due completion of the works there was a contract in existence and the contractors were not to be stripped of their rights under it however desirable that might be for the borough. An earlier Irish case [46] was distinguished upon the ground that there the engineer's certificate of default had been conclusive whereas in the instant case the architect's opinion was not.

The *Hounslow* case was decided at a time when it was generally accepted that before a plaintiff could obtain an interim injunction he had to show a prima facie case in his favour.[47] The House of Lords in *American Cyanamid Co.* v. *Ethicon*

---

[39] *Brown* v. *Bateman* (1867) L.R. 2 C.P. 272; *Re Garrud, ex p. Newitt* (1881) 16 Ch.D. 522 (C.A.).

[40] *Ibid.*

[41] *Ranger* v. *G.W. Ry.* (1854) 5 H.L.C. 72 (H.L.).

[42] *Cf. Dunkirk Colliery Co.* v. *Lever* (1878) 9 Ch.D. 20, 25 (C.A.); see also *Fulton* v. *Dornwell* (1885) 4 N.Z.L.R. 207; *Dillon* v. *Jack* (1903) 23 N.Z.L.R. 547.

[43] *Feather* v. *Keighley Corp.* (1953) 52 L.G.R. 30, see further *ante*, p. 116. The express words of the clause often give him the right to recover such costs.

[44] [1971] 1 Ch. 233, not followed in *Mayfield Holdings* v. *Moana Reef* [1973] 1 N.Z.L.R. 309, Supreme Court, Auckland.

[45] *Hounslow* case, *ibid.* p. 269.

[46] *Cork Corp.* v. *Rooney* (1881) 7 L.R.Ir. 191.

[47] See *Fellowes* v. *Fisher* [1976] Q.B. 122 (C.A.).

Ltd.[48] in a decision which has caused some comment,[49] has said that there is no such rule and that it is sufficient in the first instance for the plaintiff to show that there is a serious issue to be tried. The court then considers damages. If they are an adequate remedy for the plaintiff an injunction is not granted. If they are inadequate and the defendant would be protected against the effect of an injunction by the plaintiff's undertaking as to damages [50] then it becomes a question of the balance of convenience whether or not an injunction should be granted. The court takes into account the parties' ability to pay damages.[51]

Having regard to the *Cyanamid* case, it is submitted that, on the facts in the *Hounslow* case, and assuming that all contractual requirements as to notices and the like have been complied with, an injunction would now be granted.[52] The contractor would be protected by the undertaking in damages. The injunction would enable the employer to complete his project. The position might be different if it appears that the employer is, or might be, unable to fulfil his undertaking in damages. But even here it does seem a strong step to maintain the contractor in possession for an indefinite period. This achieves a kind of contractor's lien on land such as exists by statute in some jurisdictions,[53] but not in England and Wales.

No injunction restraining forfeiture

It is thought that the *American Cyanamid* case does not affect the principle that in the ordinary case the contractor cannot obtain an injunction restraining forfeiture by the employer, because this would be equivalent to ordering specific performance of the contract and the court does not normally grant this remedy in the case of a building contract.[54] The contractor can be adequately compensated in damages for any wrongful forfeiture.[55]

Section 2: Materials and Plant

1. Ownership of materials and plant

Materials brought onto the site by the contractor remain his property, in the

---

[48] [1975] A.C. 396.

[49] See *New Law Journal*, March 27, 1975, p. 302; (1975) 91 L.Q.R. 168, and judgment of Lord Denning in the *Fellowes* case.

[50] The plaintiff as a condition of obtaining an interim injunction has to give an undertaking to pay to the defendant his damages suffered by the grant of the injunction if the defendant ultimately succeeds. See *Supreme Court Practice,* para. 29/1/20, and for the position of the Crown see *Hoffman-La Roche (F.) & Co. A.G.* v. *Secretary of State for Trade and Industry* [1975] A.C. 295 (H.L.).

[51] *American Cyanamid* case, *supra; Fellowes* v. *Fisher, supra; Hubbard* v. *Pitt* [1976] Q.B. 142 (C.A.).

[52] In the *American Cyanamid* case, *supra,* the *status quo* was protected by the grant of an interim injunction and that would not be the instant case. Nevertheless, it is thought that the principle stated in the text would apply.

[53] *e.g.* Ontario: Mechanics Lien legislation.

[54] *Garrett* v. *Banstead and Epsom Downs Ry.* (1864) 12 L.T. 654; *Munro* v. *Wivenhoe, etc., Ry.* (1865) 12 L.T. 655; *cf. Foster & Dicksee* v. *Hastings Corp.* (1903) 87 L.T. 736—interim injunction granted pending arbitration on whether forfeiture justified; *Garrett* v. *Salisbury, etc., Ry.* (1866) L.R. 2 Eq. 358.

[55] *Munro* v. *Wivenhoe, etc., supra,* at p. 657, and see the cases cited in note 51, *supra.* For the measure of damages, see *post,* p. 151.

absence of a provision to the contrary, until they become affixed to the land, *i.e.* are built into the works,[56] whereupon they become the property of the owner of the freehold.[57] If the employer has an estate or interest less than the freehold, he enjoys the property during such estate or interest.[58] In the case of ships the property passes when the materials are fixed, " or, in a reasonable sense, made part of the corpus." [59]

When the property has passed, the contractor cannot, without an express right, remove the materials even though the employer may have severed them and they are no longer fixed,[60] but if he is under an obligation to keep the property in repair for a certain period then it seems that he will have an implied right to remove and replace defective materials during the period.[61]

Where there is a clause providing for the certification of the amount payable to the contractor in respect of materials upon their delivery, it appears that the property passes upon the making of the relevant certificate even though the materials are not fixed.[62]

In the case of plant erected for the purposes of the work and attached to the land so that technically it is fixed, there is sometimes an express provision for its removal by the contractor. But in any event it is submitted that, by analogy with so-called trade fixtures in the law of landlord and tenant [63] it would not pass to the freeholder. If the property in such plant has, by express agreement, passed to the employer, it is implied that upon completion it re-vests in the contractor and can be removed.[64] In the absence of express agreement to the contrary it seems that hoardings erected by the contractor remain his property so that he can let them for advertising.[65]

2. Vesting clauses

It is common to have a clause which purports to vest materials and sometimes plant in the employer before they are fixed.[66] The principal objects of such a clause are to provide a security to the employer for money advanced and to enable the employer to obtain the speedy completion of the works by another contractor in the event of the original contractor's default, by providing materials and plant on the site ready to use free from the claims of the original contractor,

---

[56] *Tripp* v. *Armitage* (1839) 4 M. & W. 687.
[57] *Elwes* v. *Maw* (1802) 3 East 38.
[58] *Ibid.*
[59] *Seath* v. *Moore* (1886) 11 App.Cas. 350, 381 (H.L.); for other cases on ships, see *Re Salmon and Woods, ex p. Gould* (1885) 2 Morr.Bkptcy.Cas. 137; *Reid* v. *Macbeth & Gray* [1904] A.C. 223 (H.L.); *Re Blyth Shipbuilding, etc., Co. Ltd.* [1926] Ch. 494 (C.A.). *Cf. McDougall* v. *Aeromarine Ltd.* [1958] 1 W.L.R. 1126.
[60] *Lyde* v. *Russell* (1830) 1 B. & Ad. 394.
[61] *Appleby* v. *Myers* (1867) L.R. 2 C.P. 651, 659.
[62] *Banbury, etc., Ry.* v. *Daniel* (1884) 54 L.J.Ch. 265. For the passing of property in contracts for the sale of goods, see s. 17 of the Sale of Goods Act 1893. For an example between contractor and supplier, see *Pritchett, etc., Co. Ltd.* v. *Currie* [1916] 2 Ch. 515 (C.A.), see also Chap. 10.
[63] See Woodfall's *Landlord and Tenant* (27th ed.), para. 1576.
[64] *Hart* v. *Porthgain Harbour Co. Ltd.* [1903] 1 Ch. 690.
[65] *Partington Advertising Co.* v. *Willing & Co.* (1896) 12 T.L.R. 176.
[66] See, *e.g.* cl. 14, Standard form of building contract.

and his creditors, or his trustee in bankruptcy or liquidator. Whether or not the clause achieves its purpose depends upon the words used.[67] If the formula used is " the materials shall become and be," [68] or " be and become " [69] the property of the employer, then normally " the clause means what it says, operates according to its tenor, and effectively transfers the title." [70] If, on the other hand, words like " considered to be," [71] or " deemed to be " (the property of the employer) are used, the clause may be ineffective to achieve its purpose and the property may remain in the contractor,[72] but even so the contract must, it is submitted, be read as a whole to ascertain the intention of the parties.

In one case although the materials had become vested in the employer they were subject to such rights on the part of the contractor that they could not be seized by the sheriff under an execution upon a judgment against the employer.[73]

Where the materials and plant have vested in the employer there are normally implied rights on the part of the contractor to the use of the materials and plant for the purposes of the contract,[74] and to the re-vesting and removal of unused materials and plant upon completion.[75] Further the contractor may have a right and perhaps an obligation to replace defective materials [76]; and in the event of the contractor's bankruptcy, his trustee in bankruptcy may be able to claim the materials under the reputed ownership provisions of the Bankruptcy Act.[77]

Bills of sale
The ordinary vesting clause does not operate as a bill of sale.[78]

Vesting clause operating on bankruptcy
Such a clause is prima facie void.[79]

Section 3 : Lien

A lien in the broad sense of a right of one party to retain the property of the other frequently arises under a building contract. It may be express [80] or implied.[81]

---

[67] *Bennett, etc., Ltd.* v. *Sugar City* [1951] A.C. 786 (P.C.).
[68] *Ibid.*
[69] See *Reeves* v. *Barlow* (1884) 12 Q.B.D. 436 (C.A.).
[70] *Per* Lord Reid, *Bennett, etc., Ltd.* v. *Sugar City, supra,* at p. 814. The clause is not effective if it contravenes the principles of the bankruptcy law, see *post,* p. 188.
[71] See *Re Keen, ex p. Collins* [1902] 1 K.B. 555; *cf. Brown* v. *Bateman* (1867) L.R. 2 C.P. 272; *Hart* v. *Porthgain Harbour Co. Ltd., supra.*
[72] *Bennett, etc., Ltd.* v. *Sugar City, supra,* at p. 814; *cf. Re Winter, ex p. Bolland* (1878) 8 Ch.D. 225.
[73] *Beeston* v. *Marriott* (1863) 8 L.T. 690.
[74] *Bennett, etc., Ltd.* v. *Sugar City* [1951] A.C. 786 (P.C.). For an express right, see *Beeston* v. *Marriott* (1863) 8 L.T. 690.
[75] *Hart* v. *Porthgain Harbour Co. Ltd.* [1903] 1 Ch. 690.
[76] *Appleby* v. *Myers* (1867) L.R. 2 C.P. 651.
[77] See, *e.g. Re Fox, ex p. Oundle and Thrapston R.D.C.* [1948] Ch. 407 (D.C.), and see further, *post,* p. 191.
[78] *Brown* v. *Bateman* (1867) L.R. 2 C.P. 272; *Re Garrud, ex p. Newitt* (1881) 16 Ch.D. 522 (C.A.). See *ante,* p. 122.
[79] See *post,* p. 188.
[80] See *Banbury, etc., Ry.* v. *Daniel* (1884) 54 L.J.Ch. 265. *Cf. Poulton* v. *Wilson* (1858) 1 F. & F. 403. For example, see cl. 26 of the Standard form of building contract, proviso.
[81] *Tripp* v. *Armitage* (1839) 4 M. & W. 687; *Re Waugh, ex p. Dickin* (1876) 4 Ch.D. 524.

An implied lien over unfixed materials may arise in favour of the employer where he has advanced money to the contractor on the security of such materials.[82] Where there is an express provision for a lien on unfixed materials to arise in favour of the employer upon giving notice to the contractor, the right to the lien is lost if before the notice is given the materials have been seized by the sheriff under a *fi. fa.*[83] A lien is ordinarily effective against a receiver of a company,[84] but so far as a right of lien may arise upon the bankruptcy of the contractor it is on general principles, void as against his trustee in bankruptcy.[85]

Section 4: Defects and Maintenance Clauses

There is frequently a clause in building contracts which provides that the contractor shall make good defects, or repair or maintain the works for a certain period after completion. This period is sometimes referred to as the defects liability period or maintenance period. As a security for the contractor's observance of his obligations, part of the contract sum, termed retention money, is usually retained by the employer until the end of the period, and is not released until the architect gives a certificate evidencing his satisfaction that the works accord with the contract.[86]

1. Meaning of terms

The nature and extent of the obligations of the contractor, and the rights of the employer vary according to the terms of each contract, but in general an obligation to maintain the works imposes a wider duty than one merely to make good defects, and extends to matters of wear and tear, whereas a defects clause does not.[87] An obligation to repair the works may include the re-building of parts destroyed by flood or fire,[88] but, subject to express terms, if substantially the whole of the works is destroyed the contract is frustrated and the liability ceases.[89] Under a repairing clause the contractor has ordinarily an implied right, if he thinks it necessary for its fulfilment, to replace parts of the works,[90] though the extent of any duty of replacement must depend, it is submitted, on the nature of the works, the degree to which he had the right of selecting materials or design, and whether there is also an express clause (*e.g.* in the case if an engineering contract) to keep in working order.[91]

There is no statutory lien on land in favour of contractors as in some jurisdictions, *e.g.* Ontario: Mechanics Lien legislation.

[82] *Ibid.*
[83] *Byford* v. *Russell* [1907] 2 K.B. 522 (D.C.).
[84] *George Barker (Transport) Ltd.* v. *Eynon* [1974] 1 W.L.R. 462 (C.A.).
[85] See *post*, p. 188.
[86] *e.g.* see cl. 15, Standard form of building contract, and for certificates, see Chap. 6.
[87] *Sevenoaks, etc., Ry.* v. *London, Chatham, etc., Ry.* (1879) 11 Ch.D. 625.
[88] *Brecknock and Abergavenny Canal Navigation* v. *Pritchard* (1796) 6 Term Rep. 750. For some assistance see *Smith & Snipes Hall Farm Ltd.* v. *River Douglas Catchment Board* [1949] 2 K.B. 500 (C.A.).
[89] *Appleby* v. *Myers* (1867) L.R. 2 C.P. 651, see further *ante*, p. 105.
[90] *Appleby* v. *Myers, supra.*
[91] As in *Appleby* v. *Myers, supra.* Some assistance may be gained from the construction of repairing covenants in leases, see *Proudfoot* v. *Hart* (1890) 25 Q.B.D. 42 (C.A.), *Lurcott* v. *Wakely* [1911] 1 K.B. 905 (C.A.). See further, Woodfall's *Landlord and Tenant*, 27th ed., Chap. 13, Sect. 1.

2. Investigation for defects

The nature and extent of the investigations which the employer should carry out if he is to obtain the full benefit of a defects clause depends upon the circumstances of each case and on the wording of the clause. In a contract to construct sewers the contractor undertook to maintain and keep them in proper working order for three months after completion, and to amend and make good all defects which should appear within that period. After two months a stoppage occurred in a sewer, but the cause, the extensive failure to concrete bends properly, was not discovered until after the end of the three months. It was held that the employer was entitled under the clause to recover the cost of making good the defective work.[92] In a contract to construct a road the contractor was liable to make good bad work discovered within five years. Just before the expiry of the period the employer discovered by means of trial borings that the concrete was defective and not in accordance with the contract. Further investigations after the expiry of the period showed similar defects in many other parts of the road, but it was held that the employer was limited to a claim for the defects actually discovered within the five years.[93]

3. Notice is required

If the contractor is not in possession of the site then in the absence of express terms to the contrary, notice of the defects is it seems necessary before he can become liable and if the employer without giving notice to the contractor amends defects himself he cannot rely on the defects clause as against the contractor[94]; " the reason of the thing is this, that, where there is knowledge in the one party and not in the other there notice is necessary." [95] It has not been decided what the position would be if the contractor is not given notice but has actual knowledge. His liability in such a case would depend on the wording of the contract, *e.g.* the contract may make the giving of notice a condition precedent to the contractor's liability. If the contractor knows of the defect and states that he will not carry on his obligation or disables himself from carrying it out, he may, it seems, be liable despite the absence of notice.[96]

4. Alternative claim in damages

The contractor's liability in damages is not removed by the existence of a defects clause except by clear words,[97] so that in the absence of such words the clause confers an additional right and does not operate to exclude the contractor's liability for breach of contract.[98] Clear words in this context usually require the

---

[92] *Cunliffe* v. *The Hampton Wick Local Board* (1893) 9 T.L.R. 378, H.B.C. (4th ed.), Vol. 2, p. 250 (C.A.).

[93] *Marsden Urban District Council* v. *Sharp* (1931) 47 T.L.R. 549; affd. 48 T.L.R. 23 (C.A.) and explained by Diplock L.J., *Hancock* v. *Brazier, B. W. (Anerley) Ltd.* [1966] 1 All E.R. 1. For the alternative claim in damage, see *infra.*

[94] *London and S.W. Ry.* v. *Flower* (1875) 1 C.P.D. 77.

[95] *Ibid. per* Brett J. at p. 85. This principle has been often applied in the law of landlord and tenant, see, *e.g. McCarrick* v. *Liverpool Corp.* [1947] A.C. 219.

[96] *Johnstone* v. *Milling* (1886) 16 Q.B.D. 460 (C.A.) (landlord and tenant).

[97] *Hancock* v. *Brazier (Anerley) Ltd.* [1966] 1 W.L.R. 1317 (C.A.); see also *Billyack* v. *Leyland Construction Co. Ltd.* [1968] 1 W.L.R. 471.

[98] *Ibid.*

kind of architect's binding and conclusive certificate referred to under the next heading. It is a matter of construction in each case whether a term relating to defects and headed " Guarantee " is an exclusion clause to be construed against the contractor or confers an additional right and is therefore to be construed like any other term.[99]

If the employer fails to give notice, or otherwise to avail himself of a defects clause and brings a claim for damages he may, on the principle of mitigation of loss,[1] be liable to some reduction in the damages which would ordinarily be awarded.

5. Liability after expiry of period

In the absence of words to the contrary effect the contractor's liability for not completing the works in accordance with the contract continues until barred by the Limitation Act 1939 and thus extends for the period of six years in the case of a simple contract, and 12 years in the case of a specialty contract (*i.e.* by deed and under seal) from the date when the cause of action against him arose.[2] But where there is a defects clause and at the end of the defects liability period a binding and conclusive final certificate of satisfaction is given by the architect then, in the absence of fraud or other special circumstances,[3] the contractor's liability in contract for any defects which may appear thereafter comes to an end,[4] and this notwithstanding that the certificate may have been granted after the commencement of legal proceedings in respect of the defects in question.[5]

Liability from end of defects liability period. So far as the contractor's liability does not come to an end at the expiry of such period the limitation periods in respect of defects coming within the clause run, it is submitted, from the date when he should have complied, but did not, with the clause.

Defects concealed by contractor. This may result in an extension of the limitation period, see *post*, p. 166.

Guarantee beyond limitation period. A contractor may be liable under the express terms of a guarantee, or warranty, for many years.[6]

---

[99] *Adams* v. *Richardson & Starling Ltd.* [1969] 1 W.L.R. 1645, 1653 (C.A.).

[1] See *post*, p. 147 and for the position under the Standard form of building contract, see *post*, p. 327.

[2] See Chap. 9, " Limitation Act 1939." Note that in *Cunliffe* v. *Hampton Wick Local Board* (1893) 9 T.L.R. 378, H.B.C. (4th ed.), Vol. 2, p. 250 (C.A.), it was held that completion of " the several works " meant " the whole works " and not each section thereof. The periods in the text do not relate to liability for death or personal injuries.

[3] See Chap. 6.

[4] *Bateman (Lord)* v. *Thompson* (1875) H.B.C. (4th ed.), Vol. 2, p. 36 (C.A.), and see *ante*, p. 80.

[5] *Kaye (P. & M.) Ltd.* v. *Hosier & Dickinson* [1972] 1 W.L.R. 147 (H.L.) discussed *post*, pp. 327 and 395 in relation to the Standard form of building contract, noting that the question whether the contractor can be liable for consequential losses arising from defects appearing before the certificate may arise under other contracts.

[6] See, *e.g. Adams* v. *Richardson & Starling Ltd.*, *supra*, where the period was 10 years. For liability under an indemnity beyond the limitation period, see *post*, p. 140.

Section 5: Guarantees [7]

As a form of security against the default of one of the parties to the contract or of a person concerned with it, an ancillary contract of guarantee or suretyship is frequently entered into. The person giving the guarantee is termed a surety.[8] The surety may guarantee performance by the contractor, payment by the employer, or the fidelity of the architect or any other person connected with the contract who may have responsibility for money.

The ordinary rule is that a surety who discharges the debt is entitled to be indemnified by the debtor but a person who discharges a debt without request from the debtor and under no necessity so to do ordinarily has no right to such indemnity.[9]

1. Bonds

It is common, though not legally necessary, for a surety to enter into a bond. This is a promise by deed whereby the person giving the promise (the obligor) promises to pay another person (the obligee) a sum of money.[10] When entered into by way of a guarantee the obligation is conditioned to determine upon the happening of the event guaranteed, *e.g.* if performance is guaranteed it is subject to a condition that it will come to an end upon performance of the contract. The sum promised to be paid may be greater or less than the estimated loss likely to arise from non-performance, but in an action on the bond " it is well established that the plaintiff has to establish damages occasioned by the breach or breaches of conditions, and, if he succeeds, he recovers judgment on the whole amount of the bond but can only issue execution for the amount of the damages proved. In such a case the judgment remains as security for the recovery of damages for other future breaches not sustained at the date of the commencement of the first action." [11]

Release by obligee. Subject to the express words of the bond, it is submitted that the typical bond guaranteeing performance of the contract according to its conditions extends to any latent defects for which the contractor may be liable in accordance with those conditions. The obligor's liability therefore does not determine upon completion of the works save in so far as the contractor's obligations end at that stage. Further, it is not thought that there is any implied right to require the obligee to release the obligor upon completion of the works.

2. Surety as condition precedent

The wording of the contract may be such that the obtaining of a surety, or sureties, may be a condition precedent to the right to call for payment.[12]

---

[7] The term is not used here in the sense of a guarantee or warranty by a person of his own performance, as to which see *supra*, but as defined *infra*, " Requirement of writing."

[8] For a general account of the law of guarantees and suretyship, see *Chitty on Contracts* (24th ed., 1977), Vol. 2, Chap. 12; *Rowlatt on Principal and Surety.*

[9] *Owen* v. *Tate* [1976] 1 Q.B. 402 (C.A.), discussed 92 L.Q.R. 188.

[10] See Halsbury (4th ed.), Vol. 12, para. 1385.

[11] *Per* Lord Atkin, *Workington Harbour & Dock Board* v. *Trade Indemnity Co. Ltd.* (*No.* 2) [1938] 2 All E.R. 101, 105 (H.L.).

[12] See *Roberts* v. *Brett* (1865) 11 H.L.Cas. 337 (H.L.).

3. Requirement of writing

A contract of guarantee is a promise to answer for the " debt, default or miscarriage of another " and must be evidenced by a note or memorandum in writing signed by the party to be charged or his agent.[13] The consideration need not appear in writing [14] but the promise must be set out.

It may sometimes be difficult to distinguish between a guarantee which is required to be evidenced in writing and an indemnity which is not. A leading case is *Lakeman* v. *Mountstephen* [15] where the contractor, M, was carrying out drainage work for a local board. Certain private persons disregarded notices to make connections with the drain. The chairman of the board, L, said to M: " What objection have you to making the connections ? " M replied: " None, if you or the board will order the work, or become responsible for the payment." L replied: " Go on and do the work, and I will see you paid." It was held that L had given an indemnity and was liable personally despite the absence of writing. He had assumed a primary liability to pay M and had not entered into a mere contract of guarantee, for this requires that there be a principal debtor who is primarily liable, and a surety who becomes liable only on his default.[16]

4. Liability of the surety

The liability of the surety depends upon the wording of the contract in each case.[17] In the absence of some special provision a surety for performance by the contractor is not liable for the repayment of loans made by the employer to enable the contractor to complete, of which loans the surety had no knowledge.[18] A surety for performance by the contractor is normally liable for loss due to the contractor's fraud.[19]

5. Discharge of the surety

Completion

Completion or release [20] of the promise guaranteed discharges the surety from further obligation.[21] In a guarantee of completion by the contractor it seems that completion in fact discharges the surety unless the contract of guarantee expressly requires certificated completion.[22]

---

[13] Statute of Frauds 1677, s. 4.

[14] Mercantile Law Amendment Act 1856, s. 3.

[15] (1874) L.R. 7 H.L. 17 (H.L.). See also *Guild & Co.* v. *Conrad* [1894] 2 Q.B. 885, 895 (C.A.).

[16] For a full discussion of the distinction between guarantee and indemnity, see Chitty (24th ed., 1977), Vol. 2, paras. 4805 *et seq.* See also *Goulston Discount Co. Ltd.* v. *Clark* [1967] 2 Q.B. 493 (C.A.). For a case of novation, *i.e.* substitution of one debtor for another, see *Commercial Bank of Tasmania* v. *Jones* [1893] A.C. 313 (P.C.).

[17] *Lewis* v. *Hoare* (1881) 44 L.T. 66 (H.L.). Note what is said *infra* about the specially favoured position of a surety.

[18] *Trade Indemnity Co.* v. *Workington Harbour and Dock Board* [1937] A.C. 1 (H.L.).

[19] *Kingston-upon-Hull Corporation* v. *Harding* [1892] 2 Q.B. 494 (C.A.).

[20] See *Commercial Bank of Tasmania* v. *Jones, supra.*

[21] *Lewis* v. *Hoare, supra.*

[22] *Ibid.*

Repudiation

An innocent party who accepts the repudiation [23] of his contract by the party whose obligation is guaranteed does not thereby release the surety.[24] Thus a surety guaranteed payment of £40,000 payable by instalments. The debtor defaulted in payment and the creditor accepted his default as a repudiation of the contract at a time when there was £14,000 which had not yet fallen due for payment. It was held that the surety was liable for the total sum due under the contract including the sums not yet payable at the time of acceptance of repudiation.[25]

Fraud

A certificate of completion obtained by the fraud of the contractor does not discharge a surety for completion, because fraud is one of the acts against which the surety has guaranteed, at any rate, it if was guaranteed that the work should be " well and truly " done.[26] Neither in such a case does payment of retention money to the contractor, nor failure by the employer to exercise an option of superintendence, discharge the surety,[27] but it seems that if there is a duty to superintend, and the fraud is permitted because of failure to perform that duty, the surety is discharged.[28] The surety must be immediately informed of the discovery of any fraud or dishonesty.[29]

Invalid payment

A surety for payment is not discharged if the payment made can be set aside by process of law, as, *e.g.* a payment in fraud of creditors under the bankruptcy law.[30]

Non-disclosure

In some cases a surety may be discharged because of the failure of the person for whose benefit the guarantee is made to disclose some fact affecting the obligation. There is not, as in a contract of insurance, a general duty to disclose all material facts which might affect the mind of the surety, *i.e.* contracts of suretyship are not contracts *uberrimae fidei*,[31] but there may be special circumstances which require some disclosure to be made.[32] It has been said that disclosure is necessary if " there is anything that might not naturally be expected to take place between the parties " [33] (referring to the contract of which an obligation

---

[23] See *ante*, p. 111.
[24] *Lep Air Services Ltd.* v. *Rolloswin Ltd.* [1973] A.C. 331 (H.L.).
[25] *Ibid.*
[26] *Kingston-upon-Hull Corp.* v. *Harding, supra*—defective work deliberately covered up on approach of clerk of works; discovered after completion and full payment.
[27] *Ibid.*
[28] *Ibid.*
[29] *Phillips* v. *Foxall* (1872) L.R. 7 Q.B. 666.
[30] *Petty* v. *Cooke* (1871) L.R. 6 Q.B. 790.
[31] *Seaton* v. *Heath* [1899] 1 Q.B. 782, 792 (C.A.); revsd. on another point [1900] A.C. 135 (H.L.); *Provident Accident & White Cross Insurance Co. Ltd.* v. *Dahne and White* [1937] 2 All E.R. 255.
[32] *Ibid.*
[33] *Per* Lord Campbell, *Hamilton* v. *Watson* (1845) 12 Cl. & F. 109, 119 (H.L.).

is guaranteed). The duty depends upon the particular circumstances of each transaction.[34] Thus a surety for performance was released where the employer had not disclosed that the works were to be executed under the joint supervision of his own surveyor and the surveyor of an undisclosed third party.[35] Where there was a fidelity guarantee an employer did not disclose to the surety that a servant had previously been dishonest, the surety having no knowledge of this dishonesty. It was held that the employer could not enforce the guarantee upon the subsequent dishonesty of the servant.[36] But in a guarantee of performance by a contractor who was to carry out certain harbour works, a surety could not escape liability because of undisclosed difficulties of terrain, where the building contract had expressly warned the contractor to make all proper inspections of the site himself.[37]

Conduct to prejudice of surety (laches)

" A surety is undoubtedly and not unjustly the object of some favour both at law and in equity, and . . . is not to be prejudiced by any dealings without his consent between the secured creditor and the principal debtor." [38] Conduct which prejudices the surety's position may discharge the surety's obligation.[39] But " mere omission on the part of the employer, mere passive acquiescence in acts which are improper on the part of the employer, will not release the surety. If there be an omission to do some act which the employer has contracted with the surety to do, or to preserve some security to the benefit of which the surety is entitled, the case is different." [40] Thus a mere failure to exercise an option to superintend on the part of the employer did not release the surety,[41] although it seems that if there had been a duty to superintend and the loss had resulted through failure to exercise that duty the surety would have been released.[42] Similarly a surety will be released from loss resulting from fire if the employer has not carried out a duty to insure the works against fire.[43] In the case of a fidelity guarantee the surety must be immediately informed in the event of any dishonesty on the part of the person whose fidelity is guaranteed.[44]

The release of a co-surety discharges the other sureties to the extent of the contribution which would have been paid by the released surety.[45]

---

[34] *Per* Lord Atkin, *Trade Indemnity Co.* v. *Workington Harbour & Dock Board* [1937] A.C. 1, 17 (H.L.).

[35] *Stiff* v. *Eastbourne Local Board* (1869) 20 L.T. 339; *Ctr. Russell* v. *Trickett* (1865) 13 L.T. 280.

[36] *London General Omnibus Co. Ltd.* v. *Holloway* [1912] 2 K.B. 72 (C.A.).

[37] *Trade Indemnity Co.* v. *Workington Harbour & Dock Board, supra.*

[38] *Per* Lord Selborne in *Re Sherry* (1884) 25 Ch.D. 692, 703 (C.A.); *National Bank of Nigeria Ltd.* v. *Oba M. S. Awolesi* [1964] 1 W.L.R. 1311 (P.C.).

[39] *Kingston-upon-Hull Corp.* v. *Harding* [1892] 2 Q.B. 494 (C.A.).

[40] *Ibid. per* Bowen L.J. at p. 508. For failure to give notice to the surety of the contractors' default as expressly required by a bond, see *Clyde Bank, etc., Trustees* v. *Fidelity & Deposit Co. of Maryland,* 1916 S.C.(H.L.) 69 (H.L.).

[41] *Ibid.* (For facts, see *ante,* p. 132.)

[42] *Ibid.*

[43] *Watts* v. *Shuttleworth* (1861) 7 H. & N. 353.

[44] *Phillips* v. *Foxall* (1872) L.R. 7 Q.B. 666.

[45] *Re Wolmershausen* (1890) 62 L.T. 541.

Material alteration in contract

Any material alteration of the obligation guaranteed releases the surety.[46] Thus extending the time for performance releases a surety for completion,[47] unless there is an express provision for the extension of time [48] (as *e.g.* in the Standard form of building contract, cl. 23). It is submitted that on general principles a material variation of the contract works, not within the scope of a clause permitting the ordering of variations, discharges the surety. The same result follows if there is an overpayment, as when retention money is prematurely advanced to the contractor, for the surety loses the strong inducement which otherwise would have operated on the contractor's mind to induce him to finish on time.[49] But the surety is not released if the retention money was obtained by fraud.[50] Where a building contract provided for the final determination of all questions by the employer's architect, but the parties chose to go to arbitration it was held that the surety was not liable to pay the costs of the arbitration.[51]

Section 6: *Liability to Third Parties*

The carrying out of building operations may involve one of the parties in liability in tort to a third party for injury to his person or property. The possible heads of liability may include trespass, nuisance, negligence, liability under the rule in *Rylands* v. *Fletcher*, and liability to persons coming upon dangerous premises. For the nature of these torts reference must be made to one of the standard works.[52] In this section it is proposed to deal only with the question of who is liable for the torts committed, and to give a short account of certain aspects of the law of nuisance and trespass as they particularly apply to demolition and construction operations.

1. Who is liable

(a) *Contractor*

It is a general principle that a person is always liable for any tort he commits [53]; therefore the contractor can always be sued for torts which he has committed even though the employer may also be liable.[54] The contractor is also liable for the torts of his servants committed in the course of their employment,[55] but so far as torts are committed by sub-contractors or their servants the contractor is only liable in the same manner and to the same extent as the employer is liable for the contractor's torts.[56]

---

[46] See cases cited *infra* and *National Bank of Nigeria Ltd.* v. *Oba M. S. Awolesi* [1964] 1 W.L.R. 1311 (P.C.).

[47] *Rees* v. *Berrington* (1795) 2 Ves.Jun. 540; *Harrison* v. *Seymour* (1866) L.R. 1 C.P. 518.

[48] See *Greenwood* v. *Francis* [1899] 1 Q.B. 312 (C.A.).

[49] *General Steam Navigation Co.* v. *Rolt* (1858) 6 C.B.(N.S.) 550, 595.

[50] *Kingston-upon-Hull Corp.* v. *Harding, supra.*

[51] *Hoole Urban District Council* v. *Fidelity & Deposit Co. of Maryland* [1916] 2 K.B. 568 (C.A.).   [52] *e.g.* Salmond; Winfield; Clerk and Lindsell.

[53] See *Dalton* v. *Angus* (1881) 6 App.Cas. 740, 831 (H.L.).

[54] *Ibid.* The contractor may be liable in negligence to persons to whom he owes no duty in contract; *Anns* v. *Merton London Borough* [1977] 2 W.L.R. 1024, 1038 (H.L.).

[55] See Salmond (17th ed., 1977), para. 170.

[56] *Padbury* v. *Holliday & Greenwood Ltd.* (1912) 28 T.L.R. 494 (C.A.), and see *post,* p. 184. The contractor may expressly agree to indemnify the employer in respect of any

(b) *Employer*

The general rule is that a person is not responsible for the torts of an independent contractor.[57] Thus an employer, who engaged a contractor whom he reasonably believed to be competent, to fell a tree on his land, was held not liable when the negligent felling of the trees caused telephone wires to fall across a highway so that the plaintiff was injured avoiding the anticipated consequences of an approaching car striking the wires.[58] But there are certain important qualifications and exceptions to this rule so that in practice the employer can frequently be sued for a tort committed by the contractor. Thus he is liable:

(i) Where the result of building operations is to cause damage or loss to a third party and the employer does not impose on the contractor the duty of avoiding such damage or loss.[59]

(ii) Where the carrying out of the work gives rise to some duty which the employer himself owes to the plaintiff,[60] for " a person causing something to be done, the doing of which casts on him a duty, cannot escape from the responsibility attaching on him of seeing that duty performed by delegating it to a contractor. He may bargain with the contractor that he shall perform the duty and stipulate for an indemnity from him if it is not performed, but he cannot thereby relieve himself from liability to those injured by the failure to perform it." [61] Whenever the work " of its very nature involves a risk of damage to a third party " the employer is liable.[62] Examples of the application of this principle making the employer liable are: where the contractor interfered with the right of support of an adjoining building [63]; where the works caused much dust and noise [64]; where extra hazardous techniques were used [65]; where electrical installations were so negligently carried out that a house caught fire damaging adjoining property [66]; where persons on the highway were injured by the

claims arising out of the execution of the works, see *post*, p. 140, and *cf.* cl. 18 of the Standard form of building contract.

[57] *Dalton* v. *Angus* (1881) 6 App.Cas. 740, 829 (H.L.); *Honeywill & Stein* v. *Larkin Bros.* [1934] 1 K.B. 191 (C.A.); *Salsbury* v. *Woodland* [1970] 1 Q.B. 324 (C.A.). The term " independent contractor " is used in contrast with " servant " for whose torts he is liable. The architect in the ordinary professional position is an independent contractor: *A.M.F. International Ltd.* v. *Magnet Bowling Ltd.* [1968] 1 W.L.R. 1028.

[58] *Salsbury* v. *Woodland, supra.*

[59] *Robinson* v. *Beaconsfield R.D.C.* [1911] 2 Ch. 188 (C.A.). (Contract for removal of sewage but no provision for its disposal; employer liable for contractor's trespass with sewage.)

[60] *Salsbury* v. *Woodland, supra*, at p. 347.

[61] Per Lord Blackburn, *Dalton* v. *Angus* (1881), *supra*, at p. 829. *Hughes* v. *Percival* (1883) 8 App. Cas. 443, 446 (H.L.). For the statutory right of contribution or indemnity, see the Law Reform (Married Women and Tortfeasors) Act 1935.

[62] Per Romer L.J., *Matania* v. *National Provincial Bank Ltd. and Elevenist Syndicate* [1936] 2 All E.R. 633, 648 (C.A.).

[63] *Bower* v. *Peate* (1876) 1 Q.B.D. 321; *Dalton* v. *Angus, supra.*

[64] *Matania* v. *National Provincial Bank Ltd. and Elevenist Syndicate, supra*; *Andreae* v. *Selfridge & Co. Ltd.* [1938] Ch. 1 (C.A.).

[65] *Honeywill & Stein* v. *Larkin Bros.* [1934] 1 K.B. 191 (C.A.) (open magnesium flash near curtains); *The Pass of Ballater* [1942] P. 112 (oxy-acetylene burner in confined space where danger of petrol fumes).

[66] *Spicer* v. *Smee* [1946] 1 All E.R. 489 where Atkinson J. stated that the employer was always liable for a nuisance created by his independent contractor applying dictum of Scrutton L.J. in *Job Edwards Ltd.* v. *Birmingham Navigations* [1924] 1 K.B. 341, 355 (C.A.).

negligence of independent contractors.[67] The employer's liability for injuries to visitors to the site caused by its dangerous condition resulting from the works carried out by the contractor is governed by the Occupiers' Liability Act 1957.[68]

(iii) Where fire is negligently caused on the site by the contractor in performance of the contract and spreads, causing damage.[69]

The employer is not liable for the casual or collateral negligence of the contractor or his servants.[70] The application of this principle is not always easy but an example occurred where the servant of sub-contractors, employed to insert metallic casements, left a tool on a sill and the wind blew the casement knocking off the tool which injured the plaintiff. It was held that the main contractor was not liable.[71]

2. Nuisance from building operations

Building operations often substantially interfere with adjoining owners' enjoyment of their property because of noise, dust and perhaps vibration. Such matters in some circumstances might be held to be a nuisance and form grounds for an injunction prohibiting their continuance, or an action for damages, or both.[72] If this were the result of ordinary building operations " the business of life could not be carried on," [73] for old buildings could not be pulled down and new erected in their place. But the law takes a common-sense view of the matter and if " operations . . . such as demolition and building . . . are reasonably carried on and all proper and reasonable steps are taken to ensure that no undue inconvenience is caused to neighbours whether from noise, dust, or other reasons, the neighbours must put up with it." [74]

The duty to minimise inconvenience

The duty of those carrying on building operations towards their neighbours has been stated by Sir Wilfrid Greene M.R.[75]: " Those who say that their interference with the comfort of their neighbours is justified . . . are under a specific duty . . . to use reasonable and proper care and skill. It is not a correct attitude to say: ' We will go on and do what we like until somebody complains.' That is not their duty to their neighbours. Their duty is to take proper precautions, and to see that the nuisance is reduced to a minimum. It is no answer for them to say: ' But this would mean that we should have to do the work more slowly than we would like to do it, or it would involve putting us to some extra

---

[67] *Tarry* v. *Ashton* (1876) 1 Q.B.D. 314; *Penny* v. *Wimbledon U.D.C.* [1899] 2 Q.B. 72 (C.A.); *Holliday* v. *National Telephone Co.* [1899] 2 Q.B. 392 (C.A.).
[68] See esp. s. 2 (4) (*b*).
[69] *Balfour* v. *Barty-King* [1957] 1 Q.B. 496 (C.A.); *Emanuel* (*H. & N.*) *Ltd.* v. *G.L.C.* [1971] 2 All E.R. 835 (C.A.); see also *Goldman* v. *Hargrave* [1967] 1 A.C. 645 (P.C.); *Mason* v. *Levy Auto, etc., Ltd.* [1967] 2 Q.B. 530.
[70] *Penny* v. *Wimbledon U.D.C.* [1899] 2 Q.B. 72, 78.
[71] *Padbury* v. *Holliday* (1912) 28 T.L.R. 494 (C.A.). *Cf. Holliday* v. *National Telephone Co.* [1899] 2 Q.B. 392 and see Salmond (17th ed., 1977), p. 492.
[72] See generally Salmond, Chap. 4.
[73] *Per* Vaughan Williams J., *Harrison* v. *Southwark & Vauxhall Water Co.* [1891] 2 Ch. 409, 413.
[74] *Per* Sir Wilfrid Greene M.R., *Andreae* v. *Selfridge & Co. Ltd.* [1938] Ch. 1, 5 (C.A.).
[75] At pp. 9 and 10 of *Andreae's* case, *supra.*

expense.' All those questions are matters of common sense and degree and quite clearly it would be unreasonable to expect people to conduct their work so slowly or expensively, for the purpose of preventing a transient inconvenience that the cost and trouble would be prohibitive. It is all a question of fact and degree. . . . The use of reasonable care and skill in connection with matters of this kind may take various forms. It may take the form of restricting the hours during which work is to be done; it may take the form of limiting the amount of a particular type of work which is being done simultaneously within a particular area; it may take the form of using proper scientific means of avoiding inconvenience."

Where no steps at all were taken to avoid inconvenience the employer [76] was held liable in damages for nuisance.[77] The measure of damages was held to be [78] not all the loss suffered, £4,000, as a result of the building operations, but that part of the loss and inconvenience attributable to the failure to take proper precautions as defined above, £1,000. Damages were awarded against an employer where part of a building was being altered and no steps were taken to minimise the nuisance, although sheets would have reduced the dust and other working arrangements might have reduced the loss caused to the adjoining occupier (a singing instructor) by the noise.[79]

3. Tower cranes—Trespass or Nuisance?

The jib of a modern tower crane travels through a wide area of air space. In a recent case [80] a person through whose air space such a jib travelled was treated [81] as having a remedy for trespass and accordingly awarded an injunction [82] without proof of special damage. However, the point whether such a transient and indeterminate invasion of air space is a trespass or nuisance is, it is submitted open.[83] If it is a nuisance and not a trespass the plaintiff must prove actual loss before being entitled to a remedy.[84]

Section 7: Contractor's Duty of Care towards Employer

Such a duty may arise in tort or by statute or as an implied term of the contract. Two instances are discussed below. In every case the effect of any express clause in the contract dealing with the matter must be considered.

---

[76] This is one of the cases where the employer is liable for the actions of the contractor, see *supra.*

[77] *Andreae* v. *Selfridge & Co. Ltd., supra.* See also *Hoare* v. *McAlpine* [1923] 1 Ch. 167.

[78] By the Court of Appeal reducing the damages originally awarded.

[79] *Matania* v. *National Provincial Bank Ltd. and Elevenist Syndicate* (1936) 155 L.T. 74 (C.A.).

[80] *Woollerton & Wilson Ltd.* v. *Richard Costain Ltd.* [1970] 1 All E.R. 483.

[81] The question whether or not it was trespass or nuisance was not argued and it was presumed that there was a trespass.

[82] Suspended for the period of operation—one year. In *Charrington* v. *Simons & Co. Ltd.* [1971] 1 W.L.R. 598 (C.A.) it was said that such a long suspension should not be granted.

[83] See *Kelsen* v. *The Imperial Tobacco Co. Ltd.* [1957] 2 Q.B. 334 where there was a permanent incursion. See also *Bernstein* v. *Skyviews and General Ltd.* [1977] 3 W.L.R. 136.

[84] *Ibid.*

1. Employer's visits to site

If the contractor knows that the employer is going to walk about on the site it is his duty to make the site reasonably safe.[85]

The employer has no implied contractual right to enter the site at any time without the contractor's knowledge, and to expect to find the site ready and safe for his visit.[86] If the employer does enter the site without warning and without the knowledge of the contractor it seems that he does so at his own risk,[87] so that where an employer in such circumstances trod on an unsafe plank and fell and was injured, he was unable to recover damages from the contractor.[88]

2. Theft of employer's property

Arising out of the contractual relationship between the parties is a duty on the part of the contractor to take reasonable care with regard to the state of the employer's house if he leaves it empty during the performance of his work.[89]

Section 8: *Risk, Indemnity and Exclusion Clauses*

These three types of clauses are dealt with together because their subject-matter overlaps and common problems of construction arise.

1. Risk clauses

In the absence of express provisions dealing with the risk of damage to the works, the contractor's liability depends upon what he has undertaken to perform, but it seems that where he has to complete the works he must, as an incident of the duty to complete, make good any damage to the works occurring before completion,[90] unless the damage is so great and the circumstances such that the contract is frustrated[91] or the damage was caused by the employer's default.[92] In practice written contracts frequently contain a clause stating upon whom rests the risk of damage to the works.[93] It may be coupled with a clause providing for insurance against the risk.[94] It seems that if the contractor is required to effect insurance his duty is to insure himself and does not extend to insuring the employer unless express words are used.[95]

---

[85] *Nabarro* v. *Cope & Co.* [1938] 4 All E.R. 565, 569; Occupiers' Liability Act 1957, s. 2.

[86] *Nabarro* v. *Cope & Co., supra*, at p. 568.

[87] See *supra*, note 85.

[88] *Nabarro* v. *Cope & Co., supra*. It is thought that the result would be the same today under the Occupiers' Liability Act 1957.

[89] *Stansbie* v. *Troman* [1948] 2 K.B. 48 (C.A.)—decorator left alone in house goes out leaving it with catch of Yale lock fastened back; liable for theft while absent.

[90] *Gold* v. *Patman & Fotheringham Ltd.* [1958] 1 W.L.R. 697, 703 (C.A.); *Charon (Finchley)* v. *Singer Sewing Machine Co.* (1968) 112 S.J. 536.

[91] For frustration, see *ante*, p. 102.

[92] See cases cited *infra* to " Loss caused by negligence."

[93] *e.g. Kellett* v. *York Corp.* (1894) 10 T.L.R. 662; cl. 18, Standard form of building contract.

[94] *e.g. James Archdale & Co. Ltd.* v. *Comservices Ltd.* [1954] 1 W.L.R. 459 (C.A.); cll. 19, 20, Standard form of building contract.

[95] *Gold* v. *Patman & Fotheringham Ltd., supra*.

Loss caused by negligence

Express words are required to make a party liable for damage to the works caused by the negligence of the other party,[96] but if one of the parties has clearly undertaken the risk of a certain loss, he is liable for that loss although it was caused by the negligence of the other party.[97] In *Farr* v. *The Admiralty* [98] clause 26 (2) of the contract [99] provided that, " the works . . . shall stand at the risk and be in the sole charge of the Contractor and the Contractor shall be responsible for, and with all possible speed make good, any loss or damage thereto arising from any cause whatsoever. . . ." Part of a jetty being constructed by the contractor was damaged by a ship in the control of the employer. Under an express power in the contract (clause 7) the employer ordered the contractor to repair the damage. This was done and the contractor claimed the cost from the employer, but was not successful as he had undertaken the risk of such loss arising.[1] In *Archdale* v. *Comservices* [2] an employer who had expressly undertaken the risk of loss from fire and a duty to insure the risk was unable to recover the cost of repairs made necessary by a fire caused by the negligence of the contractor's servants. The court referred to principles deriving from the judgment of Scrutton L.J. in *Rutter* v. *Palmer*.[3] These are that a defendant is not exempted from liability for his own negligence unless adequate words are used; that " the liability of the defendant apart from the exempting words must be ascertained; then the particular clause in question must be considered; and if the only liability of the party pleading the exemption is a liability for negligence, the clause will more readily operate to exempt him." [4] For some years the later exposition of these principles by the Court of Appeal in *Alderslade* v. *Hendon Laundry Ltd.* [5] was such that they came to be referred to as the " *Alderslade principle* " but it seems better now not to use this expression.[6] The principles are rules of construction and not rules of law, and have no application where the parties have expressed their intention clearly.[7]

---

[96] *Farr* v. *The Admiralty* [1953] 1 W.L.R. 965, 967.

[97] *Archdale* v. *Comservices, supra. Cf. Bucks County Council* v. *Lovell* [1956] J.P.L. 196. *Gillespie Bros. Ltd.* v. *Roy Bowles Ltd.* [1973] 1 Q.B. 400 (C.A.).

[98] *Supra.*

[99] The General Conditions of Government Contracts—Form CCC/Wks./1, ed. 5.

[1] Later editions of Form CCC/Wks./1 enable the contractor to recover extra payment for loss caused by the neglect or default of a Crown servant. An obligation to indemnify against all " claims whatsoever " extends to those caused by the negligence of the party indemnified—*Gillespie Bros. Ltd.* v. *Roy Bowles Ltd., supra.*

[2] *Supra.* Because of the acceptance of the duty to insure the risk it is thought that this decision is not affected by *Harbutt's " Plasticine " Ltd.* v. *Wayne Tank & Pump Co. Ltd.* [1970] 1 Q.B. 447 (C.A.), discussed *ante*, p. 114 and *post*, p. 141.

[3] [1922] 2 K.B. 87, 92 (C.A.) referred to and applied in *Hollier* v. *Rambler Motors (AMC) Ltd.* [1972] 2 Q.B. 71, 78 (C.A.). See also *Alderslade* v. *Hendon Laundry Ltd.* [1945] K.B. 189 (C.A.); *Canada Steamship Lines Ltd.* v. *The King* [1952] A.C. 192 (P.C.); *Gillespie Bros. & Co. Ltd.* v. *Roy Bowles Ltd., supra; Levisen* v. *Patent Carpet Cleaning Co.* [1977] 3 W.L.R. 90 (C.A.).

[4] *Ibid.*

[5] *Supra.*

[6] See *supra*, note 3.

[7] *Archdale* v. *Comservices Ltd., supra*, at p. 463; *Arthur White (Contractors) Ltd.* v. *Tarmac Civil Engineering Ltd.* [1967] 1 W.L.R. 1508 (H.L.); cases referred to in note 11 *nfra*. For the differences between a rule of construction and a rule of law, see *ante*, p. 31.

2. Indemnity clauses

One of the parties may indemnify the other against certain losses, *i.e.* he may promise to make good any loss suffered by the other party in respect of damage or claims arising out of various matters such as injury to persons or property.[8] It is no defence to an employer sued by a third person that he is indemnified by the contractor,[9] although he can join the contractor as third party to the proceedings between himself and the third person.[10]

Where the contractor gives the employer an indemnity against claims by third parties and there is an exception to the indemnity the courts tend to construe the exception so that it does not operate to deprive the indemnity of effect.[11]

The principles deriving from *Rutter* v. *Palmer*[12] apply to a party seeking to rely on an indemnity clause.[13]

Liability under indemnity clause beyond limitation period. Where there is an obligation to indemnify against loss, the cause of action does not arise until the loss has been established.[14] This may be after the expiry of the ordinary limitation period. Such clauses are therefore potentially very onerous.

Insurance policy—exceptions. Where a loss is caused by two causes, one within the general words describing the risk insured and one within the exception to those words, the insurer is not liable.[15]

3. Exclusion clauses

This term is intended to refer to clauses variously termed exclusion, exception, exemption or limitation clauses.[16] It is a convenient description of clauses relied on by a party who would otherwise be under a certain liability in contract[17] to

---

[8] See, *e.g.* cl. 18, Standard form of building contract. See also *Kirby* v. *Chessum & Sons Ltd.* (1914) 79 J.P. 81 (C.A.), where despite a clause in the *employer's* favour the contractor was entitled to an indemnity from the employer where the adjoining owner recovered damages from the contractor for trespass committed by him while he was obeying the architect's orders. For cases on the indemnity clause in the I.C.E. conditions (1955 ed.), see *C. J. Pearce & Co.* v. *Hereford Corp.* (1968) 66 L.G.R. 647; *Richardson* v. *Buckinghamshire C.C.* (1970) 68 L.G.R. 662.

[9] *Dalton* v. *Angus* (1881) 6 App.Cas. 740, 829 (H.L.).

[10] See R.S.C., Ord. 16.

[11] See *Hosking* v. *De Havilland Ltd.* [1949] 1 All E.R. 540; *Murfin* v. *United Steel Companies Ltd.* [1957] 1 W.L.R. 104 (C.A.), and see further *post*, p. 335.

[12] See *supra.*

[13] *Gillespie Bros. & Co. Ltd.* v. *Roy Bowles Ltd., supra.* See also *A.M.F. International Ltd.* v. *Magnet Bowling Ltd.* [1968] 1 W.L.R. 1028, applying *Walters* v. *Whessoe Ltd.* (1960) unrep. save at [1968] 2 All E.R. 816n. (C.A.). See also *post*, p. 336.

[14] *Collinge* v. *Hayward* (1839) 9 Ad. & El. 633 and other cases cited in Halsbury's Laws (3rd ed.), Vol. 24, para. 393; *County & District Properties* v. *C. Jenner & Son Ltd.* [1976] 1 Lloyd's Rep. 728.

[15] *Wayne Tank & Pump Co. Ltd.* v. *Employers Liability Ltd.* [1974] 1 Q.B. 57 (C.A.).

[16] For a classification see *Kenyon, Son & Craven* v. *Baxter Hoare & Co.* [1971] 1 W.L.R. 519.

[17] In tort the issue is whether the defendant by his words has shown that he does not intend to undertake any liability towards the plaintiff—see *Hedley Byrne & Co.* v. *Heller & Partners* [1964] A.C. 465 (H.L.) and cases which followed.

exclude or limit that liability. Such clauses must be distinguished from those which define the parties' rights and duties, such as agreed damages clauses [18] or defects clauses which confer additional rights,[19] and which are construed like any other clauses in the contract whereas exclusion clauses are construed against the party seeking to rely on them.[20]

Where there is a fundamental breach, or a breach of a fundamental term which breach is accepted by the innocent party so as to bring the contract to an end [21] an exclusion clause cannot be relied on in respect of the breach occasioning the acceptance of the repudiation or in respect of breaches occurring thereafter.[22] If the innocent party does not accept the breach so as to bring the contract to an end, but affirms the contract, it is a question of construction whether the clause can be relied on.[23] Normally the clause cannot be relied on if performance is totally different from that which the contract contemplates.[24]

Harbutt's " Plasticine " v. Wayne Tank Co.[25] The facts of this case are set out *ante*, p. 114. The decision has caused a certain amount of comment [26] in so far as it would appear to say that where a fundamental breach is of such a nature that the innocent party has no alternative but to accept it so as to bring the contract to an end, then, as a matter of law, the defaulting party cannot rely upon an exclusion clause in respect of the breach.[27] Even if there is such a principle of law it is thought that circumstances for its application will not often arise when events such as fire caused by the negligence of one of the parties occurs and the contract is one of the standard forms in use in the construction industry.[28] This is because, the parties having made elaborate provision for such events,[29] it is unlikely that the court would hold that they were in the nature of frustrating

---

[18] *Suisse Atlantique, etc.* v. *N.V. Rotterdamsche Kolen Centrale* [1967] A.C. 361 (H.L.); *Arthur White (Contractors) Ltd.* v. *Tarmac Civil Engineering Ltd.* [1967] 1 W.L.R. 1508, 1520 (H.L.).

[19] See *ante*, p. 128.

[20] *Adams* v. *Richardson & Starling Ltd.* [1969] 1 W.L.R. 1645, 1653 (C.A.).

[21] For the meaning of the various terms used in this sentence, see *ante*, p. 111.

[22] *Suisse Atlantique* case, *supra*. It is not clear how far, after acceptance of repudiation, the exclusion clause governs antecedent breaches. It may well be that rights and liabilities which were accrued before acceptance of repudiation are not affected—see *Suisse Atlantique*, *supra*; *Hirji Mulji* v. *Cheong Yue S.S. Co.* [1926] A.C. 497 (P.C.). In any event it seems that they can be looked at on issues as to damages, *Bloemen (F. J.) Pty. Ltd.* v. *City of Gold Coast Council* [1973] A.C. 115, and *facts* occurring before acceptance of repudiation can be looked at on the issue of damages, see *The Mihalis Angelos* [1971] 1 Q.B. 164 (C.A.).

[23] *Suisse Atlantique*, *supra*, but see the *Wathes* case, *infra*.

[24] *Suisse Atlantique*, *supra*, at pp. 425, 431; *Hardwick Game Farm* v. *S.A.P.P.A.* [1969] 2 A.C. 31 (H.L.); *Harbutt's " Plasticine "* v. *Wayne Tank & Pump Co.* [1970] 1 Q.B. 447 (C.A.); *Farnworth Finance Facilities* v. *Attryde* [1970] 1 W.L.R. 1053 (C.A.); *Kenyon, Son & Craven* v. *Baxter Hoare & Co.*, *supra*; *Wayne Tank & Pump Co. Ltd.* v. *Employers Liability Ltd.* [1974] 1 Q.B. 57, 73 (C.A.); *Wathes (Western) Ltd.* v. *Austin (Menswear) Ltd.* [1976] 1 Lloyd's Rep. 14 (C.A.).

[25] [1970] 1 Q.B. 447 (C.A.).

[26] See *e.g.* 86 L.Q.R. 512; 33 M.L.R. 441; 91 L.Q.R. 380; various comments in the section on the law of contract, *Halsbury's Laws* (4th ed.), Vol. 9.

[27] *Suisse Atlantique*, *supra*, was distinguished, but it is thought that the statements of the House in that case were wide enough to cover the facts in the *Harbutt's " Plasticine "* case so that the matter could have been approached as one of construction.

[28] *e.g.* the Standard form of building contract; the I.C.E. Conditions (5th ed.).

[29] See *Archdale (James)* v. *Comservices Ltd.* [1954] 1 W.L.R. 459 (C.A.) referred to *ante*, p. 139.

events, which is the test for the application of the *Harbutt's "Plasticine"* principle. But difficult questions of construction may arise.

Wathes (Western) Ltd. v. *Austin (Menswear) Ltd.*[30] shows the reluctance of the courts to give effect to exclusion clauses. The plaintiffs undertook to supply and fix an air conditioning plant for the defendants for £1,338. When fitted it was, in breach of contract, so noisy that a neighbour commenced proceedings claiming an injunction restraining the defendants from using it. The proceedings were compromised in circumstances where the defendants were involved in a loss of £1,331 resulting from the breach. Clause 14 of the contract said " Consequential damages. The Company shall be under no liability for any consequential loss, damage, claim or liability of any kind arising from any cause whatsoever."

It was held by the Court of Appeal that although the defendants affirmed the contract after the results of the breach were such that the breach had become fundamental, the plaintiffs could not rely upon the clause. Sir John Pennycuick said, " The current of authority has now set, as I read the cases, in favour of the view that where a contract is affirmed after fundamental breach an exemption clause is treated as inapplicable to liability resulting from that breach, not upon a substantial principle of law, but upon construction, the clause being construed, in the absence of some plain indication of a different intention, as by implication inapplicable to such liability." [31]

Suppliers and installers of goods and equipment still, it is thought, do business relying on clauses which purport to limit their liability in damages. Belief that these terms are effective may be reflected in the price to the employer's advantage.[32] It seems that they would be more prudent to abandon their clauses and put up their prices.[33]

Third Parties. Third parties to a contract cannot rely upon an exclusion clause in the contract,[34] unless it was made as agent for the third party and certain other conditions are fulfilled.[35]

Section 9: *Claim for Breach of Confidence*

Persons involved in building contracts may have a remedy in respect of the unauthorised use or disclosure of confidential matter. It may arise in contract for " if two parties make a contract, under which one of them obtains for the purpose of the contract or in connection with it some confidential matter, then, even though the contract is silent on the matter of confidence, the law will imply an

---

[30] *Supra*, discussed, 92 L.Q.R. 172.

[31] *Ibid.* at p. 25. See also *Levisen* v. *Patent Steam Carpet Cleaning Co.* [1977] 3 W.L.R. 90 (C.A.).

[32] See the remarks of Lord Wilberforce, *Gloucestershire County Council* v. *Richardson* [1969] 1 A.C. 480, 508 (H.L.).

[33] The wider economic considerations involved do not appear to have been considered by the Court of Appeal in the various cases which have resulted in the present position.

[34] *Scruttons Limited* v. *Midland Silicones Ltd.* [1962] A.C. 446 (H.L.).

[35] *New Zealand Shipping Co. Ltd.* v. *Satterthwaite Ltd.* [1975] A.C. 154 (P.C.— majority, 3 : 2).

obligation to treat that confidential matter in a confidential way, as one of the implied terms of the contract." [36] But remedies under this head are not limited to those for breach of contract. Thus where the plaintiff, in the course of negotiations with the defendant where no contract resulted disclosed certain methods of manufacture which the defendant subsequently used, it was held that the plaintiff had a claim in damages,[37] the measure being analogous to that in tort.[38] In a suitable case an injunction can be obtained.[39] The remedies are not confined to commercial or domestic secrets but extend also to public secrets.[40]

The claim is additional to any which may exist for breach of copyright or analogous statutory rights.[41]

---

[36] *Per* Lord Greene M.R., *Saltman Engineering Co. Ltd.* v. *Campbell Engineering Co. Ltd.* (*1948*) [1963] 3 All E.R. 413, 414 (C.A.). For an example of an express clause, see cl. 3 (7) of the Standard form of building contract.

[37] *Seager* v. *Copydex Ltd.* [1967] 1 W.L.R. 923 (C.A.).

[38] *Seager* v. *Copydex Ltd.* (*No. 2*) [1969] 1 W.L.R. 809 (C.A.).

[39] *Peter Pan Manufacturing Corporation* v. *Corsets Silhouette Ltd.* [1963] 3 All E.R. 402; *Attorney-General* v. *Jonathan Cape Ltd.* [1976] Q.B. 752.

[40] *Attorney-General* v. *Jonathan Cape Ltd., supra.*

[41] See *post*, p. 224.

CHAPTER 9

DEFAULT OF THE PARTIES—
DAMAGES AND EQUITABLE REMEDIES

Section 1: Damages

A BREACH of contract which has not been excused gives the injured party the right to bring an action for damages. If he merely proves the breach, but no loss, he is awarded nominal damages, *e.g.* 40s. or less.[1] If he proves actual loss, he is awarded substantial damages as compensation.[2] This section is concerned with the amount of damages recoverable where there is actual loss. A short account of the general principles upon which the courts award damages for breach of contract will be given, followed by a discussion of the application of those principles to breaches by the contractor and the employer.

1. Principles upon which damages are awarded

" The governing purpose of damages is to put the party whose rights have been violated in the same position, so far as money can do, as if his rights had been observed." [3] But if this purpose were relentlessly pursued it would lead to the party in default having to pay " for all loss *de facto* resulting from a particular breach however improbable, however unpredictable." [4] The courts therefore set a limit to the loss for which damages are recoverable, and loss beyond such limit is said to be too remote.

The principle as stated in the case of *Hadley* v. *Baxendale* [5] is as follows: " Where two parties have made a contract which one of them has broken the damages which the other party ought to receive in respect of such breach of contract should be such as may fairly and reasonably be considered [1] either arising naturally, *i.e.* according to the usual course of things from such breach of contract itself, [2] or such as may reasonably be supposed to have been in the contemplation of both parties at the time they made the contract, as the probable result of the breach of it." [6] The figures [1] and [2] have been inserted by the

---

[1] See *Mayne and McGregor on Damages* (13th ed., 1972), Chap. 10. For aggravated and exemplary damages, neither of which normally apply in building contracts, see *Rookes* v. *Barnard* [1964] A.C. 1129 (H.L.).

[2] *Ibid.*

[3] *Per* Asquith L.J., *Victoria Laundry Ltd.* v. *Newman Ltd.* [1949] 2 K.B. 528, 539 (C.A.). See also *Robinson* v. *Harman* (1848) 1 Ex. 850, 855, " where a party sustains a loss by reason of a breach of contract, he is, so far as money can do it, to be placed in the same situation with regard to damages, as if the contract had been performed," said by Lord Pearce to be " the underlying rule of the common law," *Koufos* v. *Czarnikow Ltd.* [1969] 1 A.C. 350, 414 (H.L.), and *The Albazero* [1977] A.C. 774, 841 (H.L.). In tort the measure is the loss caused by the wrong and a greater sum is sometimes recoverable—see, in a claim for deceit, *Doyle* v. *Olby (Ironmongers) Ltd.* [1969] Q.B. 158 (C.A.). See also *Broome* v. *Cassell* [1972] A.C. 1027, 1076, 1080, 1130 (H.L.)—exemplary damages not recoverable for deceit.

[4] *Ibid.*

[5] (1854) 9 Ex. 341.

[6] *Ibid. per* Alderson B. at p. 354.

author to indicate respectively that part of the sentence embodying the so-called
" first " rule, and that part embodying the so-called " second " rule.[7]

In *Victoria Laundry Ltd.* v. *Newman Ltd.*[8] Asquith L.J. in a " classic judgment "[9] explained the principle as stated in *Hadley* v. *Baxendale*. In a recent
shipping case, *Koufos* v. *Czarnikow Ltd.*,[10] the House of Lords expressed differing
views as to this judgment, but two years earlier they applied it without criticism
in a building contract case,[11] and as the judgment is such a " valuable analysis "[12]
expressed in such " felicitous language "[13] there is set out below an excerpt.

" (2)[14] In cases of breach of contract the aggrieved party is only entitled to
recover such part of the loss actually resulting as was at the time of the contract
reasonably foreseeable[15] as liable to result from the breach.[16]

" (3) What was at that time reasonably so foreseeable depends on the knowledge
then possessed by the parties or, at all events, by the party who later commits
the breach.

" (4) For this purpose, knowledge ' possessed ' is of two kinds; one imputed,
the other actual. Everyone, as a reasonable person, is taken to know the ' ordinary
course of things ' and consequently what loss is liable to result from a breach of
contract in that ordinary course. This is the subject-matter of the ' first rule ' in
Hadley v. *Baxendale*. But to this knowledge, which a contract-breaker is assumed
to possess whether he actually possesses it or not, there may have to be added
in a particular case knowledge which he actually possesses, of special circumstances outside the ' ordinary course of things,' of such a kind that a breach in
those circumstances would be liable to cause more[17] loss. Such a case attracts
the operation of the ' second rule ' so as to make additional loss also recoverable.

" (5) In order to make the contract-breaker liable under either rule it is not
necessary that he should actually have asked himself what loss is liable to result[18]
from a breach. As has often been pointed out, parties at the time of contracting
contemplate not the breach of the contract but its performance. It suffices that,
if he had considered the question, he would as a reasonable man have concluded
that the loss in question was liable to result (see certain observations of Lord du

[7] See *Victoria Laundry* case, *supra*, at p. 537; *Koufos* v. *Czarnikow Ltd.*, *supra* at p. 421;
Roth v. *Tyler* [1974] 1 Ch. 30; *Parsons* v. *Uttley*, *The Times*, May 19, 1977.
[8] *Supra*.
[9] *Aruna Mills Ltd.* v. *Gobindram* [1968] 1 Q.B. 655, 668; see also *Allan Peters (Jewellers)
Ltd.* v. *Brocks Alarms Ltd.* [1968] 1 Lloyd's Rep. 387, 392.   [10] *Supra*.
[11] *East Ham Borough Council* v. *Bernard Sunley & Sons Ltd.* [1966] A.C. 406, 440, 445,
450 (H.L.); see also *Overseas Tankship (U.K.) Ltd.* v. *Morts Dock & Engineering Co. Ltd.*
[1961] A.C. 388, 420 (P.C.). Neither case is referred to in the speeches in the *Koufos* case.
[12] *Koufos* case, p. 399.
[13] *Ibid.* at p. 414.
[14] Proposition (1) of Asquith L.J. is included in the paragraph above beginning, "The
governing purpose of damages . . ."
[15] For the test of reasonable foreseeability in tort, see *Overseas Tankship (U K.) Ltd.* v.
Morts Dock & Engineering Co. Ltd., *supra*, not following *Re Polemis* [1921] 3 K.B. 560
(C.A.); and see discussion in *Salmond on Torts* (17th ed., 1977), Chap. 22. Liability is
wider in tort (*Koufos* case, *supra*, at p. 385) and comparison between damages in tort and
contract is not helpful (*ibid.* at p. 413).
[16] Criticised by Lord Reid, *Koufos* case, at p. 389. Lord Upjohn prefers to state the
principle differently (*ibid.* at p. 424): see further, *infra*.
[17] Or less, see Lord Pearce, *Koufos* case, p. 416.
[18] See *supra*, note 16.

Parcq in the recent case of *A/B Karlshamms Oljefabriker* v. *Monarch Steamship Company Ltd.*[19]).

" (6) Nor, finally, to make a particular loss recoverable need it be proved that upon a given state of knowledge the defendant could, as a reasonable man, foresee that a breach must necessarily result in that loss. It is enough if he could foresee it was likely so to result.[20] It is indeed enough, to borrow from the language of Lord du Parcq in the same case at p. 233, if the loss (or some factor without which it would not have occurred) is a ' serious possibility ' or a ' real danger.' For short, we have used the word ' liable ' to result. . . ." [21]

In the *Victoria Laundry* case a laundry company, intending to enlarge its business, ordered a boiler from the defendants, delivery to be on a certain date. Owing to the defendant's default delivery was several months late and as a result the laundry lost certain specially lucrative contracts. It was held that though the defendants were not liable for the loss of profits on these particular contracts of which they had no knowledge, nevertheless they knew, or must be taken to have known from the circumstances and their position as engineers and businessmen, that there was bound to be business loss of some sort, and were liable for such loss on the basis of their knowledge actual or imputed.

Reasonable foreseeability

It is thought that the question whether loss is reasonably foreseeable remains a valuable approach to questions of remoteness provided that one has in mind that some interpretations of the concept may lead to the inclusion of damage which is too remote [22] and considers the alternatives of whether the loss was " not unlikely " [23] or " what was in the assumed contemplation of both parties acting as reasonable men in the light of the general or special facts (as the case may be) known to both parties in regard to damages as the result of a breach of contract." [24]

Impecuniosity

Damage resulting from the impecuniosity of the innocent party known to the defaulting party at the time of making the contract is not too remote.[25]

---

[19] [1949] A.C. 196 (H.L.).

[20] See *supra*, note 16.

[21] There is another sentence which continues, " Possibly the colloquialism ' on the cards ' indicates the shade of meaning with some approach to accuracy." This was applied by Lord Upjohn and Lord Pearson in *East Ham Borough Council* v. *Bernard Sunley & Sons Ltd.*, *supra*, at pp. 445, 451, but is so heavily criticised in the *Koufos* case that it is probably better now not to rely on it. See also *G.K.N. Centrax Gears Ltd.* v. *Marbro Ltd.* [1976] 2 Lloyd's Rep. 555, 579 (C.A.).

[22] See *e.g. Koufos* case at p. 390.

[23] *Ibid.* at p. 383, *per* Lord Reid.

[24] *Ibid.* at p. 424, *per* Lord Upjohn. See also *G.K.N. Centrax Gears Ltd.* v. *Marbro Ltd.*, *supra*, at p. 579. In *Vacwell Engineering Co. Ltd.* v. *B.H. Chemicals Ltd.* [1971] 1 Q.B. 88, Rees J. held that damages were recoverable which were the " direct " consequence of a breach although "much greater than could have been reasonably foreseen." An appeal was compromised—see [1971] 1 Q.B. 111 (C.A.).

[25] *Trans Trust S.P.R.L.* v. *Danubian Trading Co. Ltd.* [1952] 2 Q.B. 297 (C.A.).

Loss of goodwill

This may not be too remote, and if the breach is calculated to cause a loss of goodwill a reasonable sum can be awarded without positive proof,[26] but it is suggested that if it is sought to recover substantial sums under this head particulars should be given and evidence of actual loss adduced.[27] Damages for loss of repeat orders may be recoverable.[27a]

Mitigation of loss

The award of damages as compensation is qualified by a principle, " which imposes on a plaintiff the duty of taking all reasonable steps to mitigate the loss consequent on the breach, and debars him from claiming any part of the damage which is due to his neglect to take such steps." [28] But this " does not impose on the plaintiff an obligation to take any step which a reasonable and prudent man would not ordinarily take in the course of his business." [29] Any gain resulting from the plaintiff's reasonable steps in mitigation must be balanced against the loss caused by the breach.[30] The onus of proof is on the defendant to prove any failure to mitigate.[31] The principle of mitigation is illustrated by the rules governing the award of damages to the employer where the contractor fails to complete.

Tax

In the ordinary case liability to tax is not taken into account in computing damages for breach of contract.[32]

Contingency

A loss dependent upon a contingency is not necessarily too remote, however difficult the assessment of damages may be,[33] but in order to recover more than nominal damages, there must be proof of some loss.[34]

---

[26] *Anglo-Continental Holidays* v. *Typaldos (London)* [1967] 2 Lloyd's Rep. 61 (C.A.).

[27] *Perestrello Ltda.* v. *United Paint Co. Ltd.* [1969] 1 W.L.R. 570 (C.A.).

[27a] *G.K.N. Centrax Gears Ltd.* v. *Marbro Ltd., supra.*

[28] Per Lord Haldane, *British Westinghouse, etc., Co.* v. *Underground Railways Co.* [1912] A.C. 673, 689 (H.L.); *The World Beauty* [1970] P. 144 (C.A.).

[29] *British Westinghouse case, supra,* at p. 689, referring to James L.J., *Dunkirk Colliery Co.* v. *Lever* (1878) 9 Ch.D. 20, 25 (C.A.). For the position where there has been non-acceptance of repudiation, see *White and Carter (Councils) Ltd.* v. *McGregor* [1962] A.C. 413 (H.L.), referred to *ante,* p. 114, but *quaere* whether this case applies to the usual building contract.

[30] *British Westinghouse case, supra,* at p. 691; *The World Beauty, supra; Pagnan (R.) Fratelli* v. *Corbisa Industrial* [1970] 1 W.L.R. 1306 (C.A.).

[31] *British Westinghouse case, supra; Garnac Grain Co. Inc.* v. *Faure & Fairclough Ltd.* [1968] A.C. 1130, 1140 (H.L.).

[32] *Parsons* v. *B.N.M. Laboratories Ltd.* [1964] 1 Q.B. 95 (C.A.). It is taken into account in claims for personal injury or wrongful dismissal (*ibid.*).

[33] *Chaplin* v. *Hicks* [1911] 2 K.B. 786 (C.A.); *Hall* v. *Meyrick* [1957] 2 W.L.R. 458; this case was subsequently reversed but on different grounds [1957] 2 Q.B. 455 (C.A.).

[34] *Sykes* v. *Midland Bank Executor Co.* [1971] 1 Q.B. 113 (C.A.). For the court assessing some substantial damages on unsatisfactory evidence, see *Ashcroft* v. *Curtin* [1971] 1 W.L.R. 1731 (C.A.).

Pre-contract losses

Such losses are recoverable if reasonably in the contemplation of the parties, as likely to be wasted when the contract is entered into and the innocent party elects to claim wasted expenditure instead of loss of profits.[35]

Loss partly plaintiff's fault

It has been held that there can be an apportionment in contract under the Law Reform (Contributory Negligence) Act 1945.[36] Where the plaintiff admits some negligence or expects it to be found against him, he will seek an apportionment lest such negligence prevents him recovering it all.[37] The Act, of course, applies if he can bring his action in tort.[38]

Loss due to more than one cause

If a breach of contract is one of two causes of a loss, both causes co-operating and both of equal efficacy, the breach is, it seems, sufficient to carry judgment for the loss.[39]

One man company

A person may recover his company's losses as damages where that company is personal to him so that its losses are his losses.[40]

Presumption of continuance of existing state of things

This presumption is relied on by the court in estimating future losses where evidence is unsatisfactory.[41]

2. Contractor's breach of contract

Cost of completion

Where the contractor fails to complete,[42] the measure of damages in the first instance is the difference between the contract price, and the amount it would actually cost the employer to complete the contract work substantially as it was originally intended, and in a reasonable manner, and at the earliest reasonable opportunity.[43] Thus in *Mertens* v. *Home Freeholds Co.*[44] the contractor agreed

---

[35] *Anglia Television Ltd.* v. *Reed* [1972] 1 Q.B. 60 (C.A.), discussed 35 M.L.R. 423.

[36] *De Meza* v. *Apple* [1974] 1 Lloyd's Rep. 508, Brabin J.; on appeal [1975] 1 Lloyd's Rep. 498, the Court of Appeal did not have to deal with the point and emphasised that it expressed no view upon it. See also *Government of Ceylon* v. *Chandris* [1965] 3 All E.R. 48.

[37] See *Sole* v. *W. J. Hallt Ltd.* [1973] 1 Q.B. 574, 582; *De Meza* v. *Apple, supra.*

[38] See text of Act.

[39] *Heskell* v. *Continental Express Ltd.* [1950] 1 All E.R. 1033, 1048, Devlin J. If the principle is, as it appears to be, of general application, it can pose problems for a defendant where the other party who caused the loss is not made a defendant by the plaintiff and the defendant is neither jointly liable with him in contract, see R.S.C., Ord. 15, r. 4, and the defendant has no claim against him: see Ord. 16.

[40] *Esso Petroleum Co. Ltd.* v. *Mardon* [1976] Q.B. 801 (C.A.).

[41] *Evans Marshall & Co. Ltd.* v. *Bertola Ltd.* [1976] 2 Lloyd's Rep. 17, 26 (H.L.).

[42] For repudiation by the contractor, see *ante*, p. 115; for the position where the employer determines the contract or ejects the contractor from the site under an express clause, see *ante*, p. 120; and for liquidated damages for non-completion, see *post*, p. 155.

[43] *Mertens* v. *Home Freehold Co.* [1921] 2 K.B. 526 (C.A.).

[44] *Supra.*

to build a house in 1916 for £1,900. It was an unprofitable contract and he therefore deliberately delayed the work so that as a result the work was stopped by
government decree. The earliest moment at which the employer could build was
1919 when it would have cost him £4,153 to complete.[45] It was held that the
employer could recover the difference between £4,153 and £1,900, plus £825
paid to the contractor, less £495 being the value of the work done by the contractor before he ceased work, making a total of £2,583.

In one case the plaintiff recovered the cost of rebuilding the premises where
the defendant had supplied mortar so inferior that the local authority condemned
the building.[46]

Offer to complete

Where a contractor who has repudiated [47] his contract offers to complete
under a new contract it is, it is submitted, a question in each case whether an
employer who does not accept such offer is acting reasonably in mitigation of
his loss.[48]

Loss of value

Where there has been substantial completion the measure of damages is the
amount which the work is worth less by reason of the defects and omissions, and
is normally calculated by the cost of making them good,[49] *i.e.* the cost of reinstatement.[50] In the ordinary case the cost of re-instatement is assessed as at the
date when the breach occurred [51]; but where there are hidden defects the cost is
usually assessed as at the date when the defects are discovered.[52] To the appropriate date there must, it is submitted, in any event, be added a reasonable time
for the employer to survey the defects, obtain any necessary reports and arrange
for the repairs. Sometimes the proper measure of damages is not the cost of reinstatement but the difference in value between the work as it is and as it ought
to have been; see *post*, p. 263.[53]

An employer was awarded as damages the difference between the value of
houses as they ought to have been completed by the contractor and their value
as they were in fact completed, although the only loss the employer had suffered

---

[45] *i.e. all* the contract works, not merely the unfinished works.
[46] *Smith* v. *Johnson* (1899) 15 T.L.R. 179.
[47] For meaning, see *ante*, p. 111.
[48] *Strutt* v. *Whitnell* [1975] 1 W.L.R. 870 (C.A.), and see " Mitigation of loss," *ante*,
p. 147.
[49] *Per* Denning L.J., *Hoenig* v. *Isaacs* [1952] 2 All E.R. 176, 181 (C.A.); for facts, see
ante, p. 51.
[50] *East Ham Borough Council* v. *Bernard Sunley & Sons Ltd.* [1966] A.C. 406.
[51] *Philips* v. *Ward* [1956] 1 W.L.R. 471 (C.A.); *King* v. *Victor Parsons & Co.* [1973] 1
W.L.R. 29 (C.A.). The date of assessment is important because of continually rising
building costs.
[52] *East Ham* case, *supra*; see also *Clark* v. *Woor*, *post*, p. 166.
[53] At H.B.C. (9th ed.), p. 444, there is a passage referred to by Lord Cohen in the *East
Ham* case (at p. 434, not dissenting on this point) where it is said that there is a third
possible basis of assessing damages, namely, " the difference in cost to the builder of the
actual work done and work specified." But can this be appropriate having regard to the
basic principle of damages as compensation ?

was the cost of voluntarily putting right defects at the request of subsequent purchasers.[54]

Inconvenience and discomfort

Damages are recoverable under these heads.[55] It may be that where the employer is the head of a family such a claim could include the loss suffered by its members where the contractor, at the time of entering into the contract, knew that the works were for their benefit.[56] This would be one of the exceptions to the ordinary rule that a plaintiff can only recover for the loss he has himself sustained.[57]

Destruction of premises

Where a breach results in the destruction of the premises or part of them and the innocent party has no option but to re-build, the measure of damages is the cost of replacement.[58] He is not entitled to recover the cost of improvements made voluntarily,[59] but he is not subject to a deduction merely because he gets new for old, nor for improvements which are no more than the result of complying with statutory requirements current at this time of re-building.[60]

Other losses

The liability of the contractor to pay for losses other than those referred to above depends upon his knowledge actual or imputed.[61] Thus it seems that where the works are for industry or commerce the contractor will almost inevitably be liable for some of the further loss; for example, " in a contract to build a mill the builder knows that a delay on his part will result in the loss of business." [62] If he is building houses or blocks of flats for letting, then on analogy with shipping cases he is probably liable for loss of rent,[63] or it may be in some cases where the works, such as public buildings, would not produce rents or profits, that he is liable for loss of interest on capital lying idle.[64] If to the contractor's knowledge the contract works consist of an expansion of a factory

---

[54] *Newton Abbot Development Co. Ltd.* v. *Stockman Bros.* (1931) 47 T.L.R. 616.

[55] *Bolton* v. *Mahadeva* [1972] 1 W.L.R. 1009 (C.A.); *King* v. *Victor Parsons & Co.,* supra.

[56] See *Jackson* v. *Horizon Holidays Ltd.* [1975] 1 W.L.R. 1468 (C.A.); *McCall* v. *Abelesz* [1976] Q.B. 585, 594 (C.A.).

[57] See *The Albazero* [1976] 3 W.L.R. 419, 427 (H.L.).

[58] *Harbutt's " Plasticine " Ltd.* v. *Wayne Tank and Pump Co. Ltd.* [1970] 1 Q.B. 447 (C.A.)—for facts, see *ante*, p. 114.

[59] *Ibid.* p. 476.

[60] See *supra*, note 58.

[61] See the *Victoria Laundry* case, *ante*, p. 145.

[62] *Per* Parke B., *Hadley* v. *Baxendale* (1854) 23 L.J.Ex. 179, 181; and see the *Victoria Laundry* case, *ante*, p. 146. For loss of a machine, see *Sunley & Co. Ltd.* v. *Cunard White Star Ltd.* [1940] 1 K.B. 740 (C.A.).

[63] See *Fletcher* v. *Tayleur* (1855) 25 L.J.C.P. 65; *Wilson* v. *General Iron Screw Colliery Co. Ltd.* (1877) 47 L.J.Q.B. 239; *Steam Herring Fleet Ltd.* v. *Richards & Co. Ltd.* (1901) 17 T.L.R. 731. *Cf. Marshall* v. *Mackintosh* (1898) 78 L.T. 750, but note that this was a building agreement.

[64] See *The Hebridean Coast* [1961] A.C. 545 (H.L.); *Birmingham Corporation* v. *Sowsbery* (1969) 113 S.J. 877.

or other profit-earning structure, he is liable for loss of business resulting from his breach.[65] The extent of his liability varies according to his knowledge, and only includes loss of profit from specially lucrative contracts if he knew of them at the time of entering into the contract.[66] Loss of profit should be expressly pleaded and is inconsistent with a claim for capital expenditure incurred to make that profit.[67]

If as a result of defective work by the contractor not in accordance with the contract, a third party is injured, or suffers loss and sues the employer, the latter may be able to recover from the contractor the amount of the damages or the sum paid by way of settlement of the claim.[68] Where the employer's liability to the third party arising out of the contractor's breach has been established or is admitted by the contractor the amount of any settlement is admissible prima facie evidence of the loss caused by the contractor's breach, although further evidence may be adduced to determine the actual loss.[69] If the employer himself suffers personal injury or loss as a result of the contractor's defective workmanship or materials he may, subject to the terms of the contract, be able to recover damages.[70]

Lack of inspection by architect

The contractor is not, subject to the express terms of the contract, entitled to a reduction in damages because the architect, clerk of works or other agent of the employer failed to discover defects as the work was carried out.[71]

3. Employer's breach of contract

No work carried out

If there is a repudiation of the contract by the employer before any work is carried out the damages recoverable are, it seems, prima facie the amount of profit which the parties knew or must be taken to have assumed, the contractor would have made if he had been permitted to complete in the ordinary way.[72] Further damages resulting from unusual circumstances are recoverable according to the employer's knowledge of those circumstances.[73]

---

[65] By analogy with the *Victoria Laundry* case, *supra*.

[66] *Ibid.*

[67] *Perestrello Ltd.* v. *United Paint Co. Ltd.* [1969] 1 W.L.R. 570 (C.A.); see also *Cullinane* v. *British " Rema " Manufacturing Co.* [1954] 1 Q.B. 292 (C.A.); *Anglia Television* v. *Reed* [1972] 1 Q.B. 60 (C.A.).

[68] *Fisher* v. *Val de Travers* (1876) 45 L.J.Q.B. 479; *Kiddle* v. *Lovett* (1885) 16 Q.B.D. 605; *Mowbray* v. *Merryweather* [1895] 2 Q.B. 640 (C.A.); *Sims* v. *Foster Wheeler Ltd.* [1966] 1 W.L.R. 769 (C.A.); *cf. Biggin & Co. Ltd.* v. *Permanite Ltd.* [1951] 2 K.B. 314 (C.A.). Note that there is usually a clause in the contract providing for an indemnity by the contractor against such claim—see cl. 18, Standard form of building contract, and see *ante*, p. 140, and for sub-contracts, *post*, p. 184.

[69] *Biggin & Co. Ltd.* v. *Permanite Ltd.* [1951] 2 K.B. 314 (C.A.).

[70] *Kimber* v. *William Willett Ltd.* [1947] K.B. 570 (C.A.). See also *ante*, p. 137. For the contractor's liability in tort for failure to comply with statutory requirements, see *Anns* v. *Merton London Borough* [1977] 2 W.L.R. 1024, 1038, 1047 (H.L.).

[71] *East Ham Corp.* v. *Bernard Sunley & Sons Ltd.* [1966] A.C. 406 (H.L.); *A.M.F. International Ltd.* v. *Magnet Bowling Ltd.* [1968] 1 W.L.R. 1028, 1053.

[72] See *Ranger* v. *G.W. Ry.* (1854) 5 H.L.C. 72 (H.L.); *Victoria Laundry Ltd.* v. *Newman Ltd.* [1949] 2 K.B. 528 (C.A.); *Koufos* v. *Czarnikow Ltd.* [1969] 1 A.C. 350, 424 (H.L.).

[73] See *ante*, p. 145.

Election to claim wasted expenditure

The contractor may, it seems, elect to claim wasted expenditure instead of loss of profit, and can include pre-contract expenditure made in preparation for performance provided it was such as would have been reasonably in the contemplation of the employer at the time of entering into the contract.[74]

Work partly carried out

Where the contract work has been partly carried out and the contract is brought to an end by the employer's repudiation, the contractor has the option of either suing for damages, when the measure of damages is normally the loss of profit on the unfinished balance, plus the value of the work done at contract prices,[75] or of ignoring the contract and claiming a reasonable price for work and labour done on a *quantum meruit*.[76] Thus a contractor who is losing money on an unprofitable contract may find himself able to recoup his losses by electing to claim a reasonable price.[77]

Where the employer's breach does not prevent completion the damages recoverable, if any, will vary according to the circumstances. Thus delay caused by the employer may give rise to a claim for damages [78]; it may, for example, turn a summer contract into a winter contract thus causing increased cost of working,[79] or it may keep plant or machinery or even, in some cases, men, idle; productivity of labour may be lowered. All these heads of loss require consideration of the principles set out *supra*, especially foreseeability and mitigation. Ordinarily where completion is delayed the contractor suffers a loss arising from the diminution of his income from the job and hence the turnover of his business, but he continues to incur expenditure, usually called " overheads," [80] which he cannot materially reduce or, in respect of the site, can only reduce, if at all, to a limited extent. Further, he usually suffers a loss of profit arising out of the diminution in turnover, but it seems that to establish this claim he must show that at the time of the delay he could have used the lost turnover profitably.[81] Thus the claim for loss of profit does not, it is submitted, fail merely because the contract in question was unprofitable; the question is what he would have done with the money if he had received it at the proper time. Even if, at that time, the contractor's business was making a loss a sum analogous to loss of profit is, it is submitted, recoverable if the loss of turnover increased the loss of the business.

---

[74] See *Anglia Television Ltd.* v. *Reed* [1972] 1 Q.B. 60 (C.A.)—actor who repudiated contract for leading role in film which was abandoned liable for expenses incurred before contract signed.

[75] *Felton* v. *Wharrie* (1906) H.B.C., 4th ed., ii, p. 398 (C.A.); see further *ante*, p. 115.

[76] *Planché* v. *Colburn* (1831) 8 Bing. 14; *Lodder* v. *Slowey* [1904] A.C. 442, 453 (P.C.); *Chandler Bros. Ltd.* v. *Boswell* [1936] 3 All E.R. 179, 186 (C.A.).

[77] For this reason the scope of the doctrine would benefit from review by the House of Lords. A passage in *Ranger* v. *G.W. Ry.*, *supra*, at p. 96, is difficult to reconcile with the subsequent authorities.

[78] *Lawson* v. *Wallasey Local Board* (1883) 48 L.T. 507 (C.A.).

[79] *Freeman* v. *Hensler* (1900) H.B.C. (4th ed.), Vol. 2, p. 292 (C.A.).

[80] For some assistance see *Shore & Horwitz Construction Co. Ltd.* v. *Franki of Canada Ltd.* [1964] S.C.R. 589 (Supreme Court of Canada).

[81] *Ibid.* and see *Sunley (B.) & Co. Ltd.* v. *Cunard White Star Ltd.* [1940] 1 K.B. 740 (C.A.).

As to idle machinery there are various ways in which loss may arise. One is hiring charges. This is appropriate where the contractor was hiring the machinery.[82] If he owned the machinery it seems that he can only recover hiring charges if he proves that he has lost the opportunity of hiring it, and even then it must not be assumed that the machinery could necessarily have been hired for the whole of the equivalent of the period when it was idle. If he does not prove loss of opportunity of hiring, his loss is assessed by such factors as loss of interest on the capital lying idle, depreciation and maintenance.[83]

Loss of productivity or uneconomic working is a head of claim sometimes made where there has been delay in completion or disturbance of the contractor's regular and economic progress even though, on occasion, the ultimate delay in completion is small or does not occur. As regards machinery and plant it is ordinarily comparatively easy to compare the contemplated periods of use with the actual periods and then to apply the measure of damages as discussed above. Labour is more difficult. Some contractors add an arbitrary percentage to the contemplated labour costs. It is difficult to see how this can be sustained. There can be no custom or general rule because the loss will vary in each case. A better starting point is to compare actual labour costs with those contemplated. Thus a particular activity or part of the works is taken and, where the contract price can be ascertained, as by reference to the priced bills, the labour element is extracted. This is a matter for experienced surveyors and is done by taking the unit price and applying constants which are generally accepted in the trade. From the contractor's records the actual labour content for the activity or part is extracted. From the difference must be deducted any expenditure upon labour which was not caused by the breach, *e.g.* delay or disturbance caused by bad weather, strikes, nominated sub-contractors or the contractor's own inefficiency. If the original contract price was arrived at in a properly organised competition or as the result of negotiation with a skilled surveyor acting on behalf of the employer, the adjusted figure for the difference is some evidence of loss of productivity.

Extension of preliminaries. For the meaning of this somewhat loose term see *post*, p. 317.

Claims under or for breach of the contract. The distinction should always be sharply observed. A claim under the contract may have the advantage of the right to payment as the work proceeds under the architect's certificate but be subject to the fulfilment of certain conditions precedent. Even where the facts cover the same ground there is this difference to be observed: that while in each instance the contractor must prove his case, where his claim arises for breach of contract the court is the more likely to resolve matters of doubt in the contractor's favour and, even where the evidence of loss is very unsatisfactory, if satisfied that there was substantial loss, will make some award.[84]

---

[82] *Ibid.*
[83] *Ibid.*; and see shipping cases applied in *The Hebridean Coast* [1961] A.C. 545 (H.L.).
[84] See, *e.g. Ashcroft* v. *Curtin* [1971] 1 W.L.R. 1731 (C.A.).

Bonus

Where an employer's delay prevented the contractor earning a bonus for completion to time, the whole of the bonus was awarded as damages to the contractor,[85] but it does not seem that for contracts providing for such a bonus the measure of damages is automatically its full amount.[86]

Unusual circumstances

In all cases the employer's liability for damages resulting from such circumstances depends upon his knowledge at the time of entering into the contract: see *ante*, p. 145.

Interest for non-payment of money

If the employer fails to pay money due under the contract the contractor may consider the recovery of interest under one of three heads. The first is under an express term of the contract.[87] The second is where he recovers judgment and applies for interest under the Law Reform (Miscellaneous Provisions) Act 1934 (see, *post*, p. 254). The third is as damages.

The general principle is that interest is not recoverable as damages for non-payment of money.[88] But it has been said that the ground was that interest was generally presumed not to be within the contemplation of the parties, and that if a special loss is foreseeable at the time of the contract as the consequence of the non-payment of money due then such loss may well be recoverable.[89] It is submitted that nowadays the court would easily find, in most instances, that the parties contemplated the contractor losing money if he was not paid at the proper time. It might therefore, be able to distinguish the earlier authorities which appear to state a clear rule that interest is not recoverable as damages.

Local Authorities. When settling a claim for damages, or for money due, by a contractor, local authorities sometimes take the point that they cannot include a sum in respect of interest because such payment is contrary to law rendering them liable to the possibility of surcharge. It is submitted that, provided the local authority acts bona fide in the interests of the ratepayers and with reasonable business acumen the inclusion of such a sum is not contrary to law[90] and, further, if its exclusion results in litigation they would ordinarily be acting against the interest of the ratepayers. This is because, quite apart from the possibility of a claim for interest as damages (*supra*), interest is ordinarily awarded under the Law Reform (Miscellaneous Provisions) Act 1934, so that the contractor would be justified in refusing the sum offered and proceeding to

---

[85] *Bywaters* v. *Curnick* (1906) H.B.C. (4th ed.), Vol. 2, p. 393 (C.A.).

[86] *Ibid.* at p. 397.

[87] *e.g.* cl. 60 (6), I.C.E. conditions (1973 ed.). There is nothing comparable in the Standard form of building contract.

[88] *London, Chatham & Dover Railway Co.* v. *S.E. Railway Co.* [1893] A.C. 429, 437 *et seq.*; *Jefford* v. *Gee* [1970] 2 Q.B. 131, 143 (C.A.).

[89] *Trans Trust S.P.R.L.* v. *Danubian Trading Co. Ltd.* [1952] 2 Q.B. 297, 306, 307 (C.A.); Hals. (4th ed.), Vol. 12, para. 1179.

[90] Decision of *Re Hurle-Hobbs* [1944] 1 All E.R. 249 (D.C.); [1944] 2 All E.R. 261 (C.A.).

trial in order to obtain such interest (see *post*, p. 255), and would normally recover costs.

Section 2: *Liquidated Damages*

The parties often agree that a liquidated (*i.e.* fixed and agreed) sum shall be paid as damages for some breach of a contract. A typical clause provides that if the contractor shall fail to complete by a date stipulated in the contract, or any extended date he shall pay or allow the employer to deduct liquidated damages at the rate of £X per day or week for the period during which the works are uncompleted.[91] Such a clause may result in a considerable saving of costs, because it often arises that " although undoubtedly there is damage the nature of the damage is such that proof of it is extremely complex, difficult and expensive." [92] Where there is such a clause and the contractor has failed to complete to time prima facie he is liable to a claim for the liquidated damages either by way of action or by deduction (if there is an express power) or by a set-off and there can be no inquiry into the actual loss suffered. But he may have a defence to the claim [93] either by proving that the agreed sum is a penalty, or by showing the existence of one of the other matters set out below. If he establishes such a defence it may still leave the employer the right to pursue a claim for unliquidated damages (*i.e.* an ordinary claim for damages) for delay.[94] No general rule can be stated as to the date from which such unliquidated damages are recoverable. This depends upon the contract, the particular defence established and the circumstances. In the case of the employer causing delay (see *post*, p. 160) damages are not, it is submitted, in any event recoverable from a date earlier than that when the works could have been completed but for such delay, but it is not clear whether the delay sets time at large so that the works do not have to be completed before a reasonable time has elapsed.[95]

Defences to claim for liquidated damages

1. Agreed sum is a penalty

If the agreed sum, whatever it is called in the contract, is a penalty it will not be enforced by the courts.[96] In such a case the employer has an option; he may either rely on his claim for the penalty, in which case he cannot recover more

---

[91] *Cf.* cl. 22, Standard form of building contract.
[92] *Clydebank Engineering Co.* v. *Don José Yzquierdo y Castaneda* [1905] A.C. 6, 11 (H.L.).
[93] Where the employer has set up a claim to deduct liquidated damages as a defence the contractor will plead the following matters by way of reply to defence or possibly by an amendment of his claim. For the approach to the construction of liquidated damages clauses see *ante*, p. 34, " The *contra proferentem* rule."
[94] See *Peak Construction (Liverpool) Ltd.* v. *McKinney Foundations Ltd.* (1970) 69 L.G.R. 1, 11, 16 (C.A.).
[95] See *Trollope & Colls Ltd.* v. *N.W. Metropolitan Regional Hospital Board* [1973] 1 W.L.R. 601, 607 (H.L.) where the majority of the House of Lords said that Lord Denning had been wrong when, in the Court of Appeal, he had said that *Dodd* v. *Churton* [1897] 1 Q.B. 562 (C.A.) established that time became at large. The House did not comment on *Wells* v. *Army & Navy Co-operative Society Ltd.* (1902) H.B.C. 346 (C.A.); *Peak Construction (Liverpool) Ltd.* v. *McKinney Foundations Ltd.*, *supra* which were cited to the House in argument in support of the same proposition.
[96] *Watts, Watts & Co. Ltd.* v. *Mitsui & Co. Ltd.* [1917] A.C. 227 (H.L.).

than the actual loss which he proves up to the amount of the penalty, or he can ignore the penalty and sue for unliquidated damages.[97] If he follows the latter course he is entitled to the damages he proves whether they are less or more than the penalty as if the penalty clause did not exist.[98] In theory therefore it may be to the employer's advantage to show that the agreed sum is a penalty. In practice it is likely to be very difficult for an employer to prove both (a) that the agreed sum was so great that it was not a genuine pre-estimate of the loss,[99] and (b) that he has as a result of the breach suffered loss greater than the penalty.[1]

It is sometimes a matter of some difficulty to determine whether agreed damages are, in the particular circumstances of the case, penalties or liquidated damages, but the principles applied by the courts are well established. They were summarised in Lord Dunedin's speech to the House of Lords in *Dunlop Ltd. v. New Garage Co. Ltd.*[2] The relevant extract will be set out in the text followed by notes on the application of the principles to building contracts.[3]

Lord Dunedin's propositions

" 1. Though the parties to a contract who use the words ' penalty ' or ' liquidated damages ' may prima facie be supposed to mean what they say, yet the (A) expression used is not conclusive. The court must find out whether the payment stipulated is in truth a penalty or liquidated damages . . .

" 2. The essence of a penalty is a payment of money stipulated as in terrorem of the offending party; the essence of liquidated damages is a (B) genuine covenanted pre-estimate of damage.[4]

" 3. The question whether a sum stipulated is penalty or liquidated damages is a question of construction to be decided upon the terms and inherent circumstances of each particular contract, judged of as at the time of the making of the contract, not as at the time of the breach.[5]

" 4. To assist this task of construction various tests have been suggested, which if applicable to the case under consideration may prove helpful, or even conclusive. Such are:

" (a) It will be held to be a penalty if the sum stipulated for is (C) extravagant and unconscionable in amount in comparison with the greatest loss that could conceivably be proved to have followed from the breach.[6]

" (b) It will be held to be a penalty if the breach consists only in not paying a sum of money, and the sum stipulated is (D) a sum greater than the sum which

---

[97] *Ibid.*

[98] *Ibid.*; *Feather & Co. v. Keighley Corp.* (1953) 52 L.G.R. 30, 33; but see *Cellulose Acetate Ltd. v. Widnes Foundry (1925) Ltd.* [1933] A.C. 20, 25 (H.L.).

[99] See *infra.*

[1] *Cf.* the *Cellulose Acetate* case, *supra.* See further an article 91 L.Q.R. 25.

[2] [1915] A.C. 79, 86 (H.L.).

[3] The authorities cited by Lord Dunedin will be referred to in the footnotes. Capital letters have been inserted in the extract to show those parts of the speech which are annotated.

[4] Citing *Clydebank, etc., Co. v. Yzquierdo, etc.* [1905] A.C. 6 (H.L.).

[5] Citing *Public Works Commissioner v. Hills* [1906] A.C. 368 (P.C.) and *Webster v. Bosanquet* [1912] A.C. 394 (P.C.).

[6] Citing the *Clydebank* case, *supra.*

ought to have been paid.[7] This though one of the most ancient instances is truly a corollary to the last test . . .

" (c) There is a presumption (but no more) that it is penalty when (E) ' a single lump sum is made payable by way of compensation, on the occurrence of one or more or all of several events, some of which may occasion serious and others but trifling damage.' [8]

" On the other hand:

" (d) It is no obstacle to the sum stipulated being a genuine pre-estimate of damage, that the consequences of the breach are such as to make precise pre-estimation almost an impossibility. On the contrary, that is just the situation when it is probable that pre-estimated damage was the true bargain between the parties." [9]

NOTES

(A) " expression used is not conclusive." There have been several building contract cases in which an agreed sum has been held to be liquidated damages although termed a " penalty," [10] and vice versa.[11]

(B) " genuine . . . pre-estimate." An agreed sum may not be a genuine pre-estimate yet not be a penalty. This arises where a party is unwilling to run the risk of paying the heavy damages which might result from his breach, and therefore deliberately limits his liability to a smaller sum than that which might be awarded as unliquidated damages. Thus where actual loss from delay in installing a machine was £5,850, the defendant's liability was limited by the contract to £600. It was held that the latter sum was the maximum recoverable by the plaintiff although it was not a genuine pre-estimate of the actual damage.[12]

(C) " extravagant and unconscionable." In practice this is probably the most important test to be applied. Lord Halsbury in the *Clydebank* case [13] indicates the approach: " if you agreed to build a house in a year, and agreed that if you did not build the house for £50, you were to pay a million of money as a penalty, the extravagance of that would become at once apparent. Between such an extreme case as I have supposed and other cases, a great deal must depend on the nature of the transaction—the thing to be done, the loss likely to accrue to the person who is endeavouring to enforce the performance of the contract, and so forth."

In considering the question the unequal financial position of the parties is irrelevant; " the word unconscionable . . . does not bring in at all the idea of an unconscionable bargain. It is merely a synonym for something which is extravagant and exorbitant." [14] Also irrelevant is any question of disproportion between

---

[7] Citing *Kemble* v. *Farren* (1829) 6 Bing. 141.

[8] Citing Lord Watson in *Lord Elphinstone* v. *Monkland Iron & Coal Co.* (1886) 11 App.Cas. 332 (H.L.).

[9] Citing the *Clydebank* case, *per* Lord Halsbury; *Webster* v. *Bosanquet*, *per* Lord Mersey.

[10] *e.g. Ranger* v. *G.W. Ry.* (1854) 5 H.L.C. 72 (H.L.); *Crux* v. *Aldred* (1866) 14 W.R. 656; *Re White* (1901) 17 T.L.R. 461 (D.C.); the *Clydebank* case, *supra*.

[11] *e.g. Re Newman* (1876) 4 Ch.D. 724 (C.A.); *Public Works Commissioner* v. *Hills, supra.*

[12] *Cellulose Acetate Silk Co.* v. *Widnes Foundry* (1925) *Ltd.* [1933] A.C. 20 (H.L.); *cf. Suisse Atlantique, etc.* v. *N.V. Rotterdamsche Kolen Centrale* [1967] 1 A.C. 361, 421 (H.L.).

[13] [1905] A.C. 6, 10 (H.L.).

[14] *Per* Lord Wright M.R., *Imperial Tobacco Co.* v. *Parslay* [1936] 2 All E.R. 515, 521. See also *Robert Stewart & Sons Ltd.* v. *Carapanayoti* [1962] 1 W.L.R. 34, 40; *Bridge* v.

the amount of the contract sum and the agreed sum payable on breach.[15] This principle might apply, for example, where building works are required for an important exhibition. In such a case a proper sum for liquidated damages might be so high that after the running of quite a short period of delay it might exceed the contract sum.

(D) " a sum greater." The test stated in this sentence is based on the principle that because the exact amount of the loss is known a greater sum payable cannot be a genuine pre-estimate of the loss.[16]

Forfeiture clauses as liquidated damages

In *Ranger* v. *G.W. Ry.*[17] there was a clause providing that upon forfeiture the contractor was to receive no further payment, that all moneys then due or to become due to him and all tools and materials on the works were to become the property of the employer, and the contractor was to make up any deficiency in the cost of completion.[18] This clause was held to impose a penalty, and Lord Cranworth, referring to the cost to the employers of completion, said [19] that " the amount of their damage was capable of exact admeasurement "; the employers were entitled to recoup themselves out of the property seized to the extent of their loss and no further.

A clause purporting to forfeit retention money upon default is open to the objection that it may be a penalty, for the amount of the retention money increases with no relation to the cost of completion, and in the normal course increases as the cost of completion decreases.[20] A clause forfeiting only the unfixed materials on the site is not open to the same objections, for the amount of materials at any one time is likely to be reasonably constant except that towards completion it will decrease.[21]

In one case it was provided that upon forfeiture for the contractor's default the amount of money already paid to the contractor [22] should be considered the full value of the works which had been executed up to that time and that materials should be forfeited without any further payment. The provision was held to be enforceable although the contractor had not become entitled to any payment for work done at the time of forfeiture.[23]

Campbell Discount Co. Ltd. [1962] A.C. 600, 626 (H.L.), *per* Lord Radcliffe: ". . . the courts of equity never undertook to serve as a general adjuster of men's bargains."

[15] *Ibid.*

[16] *Kemble* v. *Farren, supra.*

[17] (1854) 5 H.L.C. 72 (H.L.).

[18] There was also a clause providing for specified sums increasing each week to be paid as " penalties in case of delay." This was held to provide for liquidated damages.

[19] At p. 108.

[20] See *Public Works Commissioner* v. *Hills* [1906] A.C. 368 (P.C.); *Walker* v. *L. & N.W. Ry.* (1876) 1 C.P.D. 518, 532; *cf. Roach* v. *G.W. Ry.* (1841) 10 L.J.Q.B. 89; *Tooth* v. *Hallett* (1869) L.R. 4 Ch.App. 242; *Bridge* v. *Campbell Discount Co. Ltd., supra.*

[21] See *Re Garrud, ex p. Newitt* (1881) 16 Ch.D. 522 (C.A.). For vesting clauses operating upon forfeiture, see *ante*, p. 125, noting that there is an implied right of revesting in respect of unused materials.

[22] There was an express provision for payment by instalments.

[23] *Davies* v. *Swansea Corp.* (1853) 8 Exch. 808.

Set-off clauses. Clauses providing expressly for a right of set-off are common. They are enforceable provided they do not amount to a penalty. Thus a contractor relied on a clause in a sub-contract to withhold money from a sub-contractor.[24] Part of the clause said: "... if the sub-contractor fails to comply with any of the conditions of this sub-contract the contractor reserves the right to suspend or withhold payment ..." This was unenforceable as a penalty because large sums of money could be withheld for trivial breaches.[25]

(E) " a single lump sum." An example of this principle arose where, in addition to the usual payment of £X per week for delay, there was also a provision that in case the contract should not be in all things duly performed by the contractors they should pay to the employers £1,000 as liquidated damages; it was held that the latter sum was a penalty.[26] But where it was provided that in default of completion by the specified date the contractor should forfeit and pay the sum of £100 and £5 for every seven days of delay it was held that as these sums were payable on a single event only they were liquidated damages.[27]

If a sum of money is payable, not in respect of any breach of contract, but by one of the parties when he exercises a contractual right under the contract, then no question of penalty or liquidated damages arises [28]; but it is not clear whether equity can intervene to relieve against payment of that sum of money where it is such that it would he held to be a penalty if it were payable upon breach of the contract.[29]

2. Omission of date in contract

There must be a date from which liquidated damages can run and if the date is omitted from a written contract the court will not fix a date from conflicting oral evidence.[30]

3. Waiver

There may be a waiver express or implied by the employer. An express undertaking not to enforce the clause at all, or to extend the time from which liquidated damages are to run, is, it is submitted, binding notwithstanding the absence of consideration.[31] Waiver may be implied from circumstances; for

---

[24] *Modern Engineering (Bristol) Ltd.* v. *Gilbert Ash (Northern) Ltd.* [1974] A.C. 689 (H.L.). For set-off generally, see *post*, p. 276.

[25] *Ibid.* pp. 698, 703, 711, 723.

[26] *Re Newman ex p. Capper* (1876) 4 Ch.D. 724 (C.A.); see also *Cooden Engineering Co. Ltd.* v. *Stanford* [1953] 1 Q.B. 86, 98 (C.A.).

[27] *Law* v. *Redditch Local Board* [1892] 1 Q.B. 127 (C.A.); *cf. Pye* v. *British Automobile Commercial Syndicate* [1906] 1 K.B. 425. For a hire case, see *Robophone Facilities Ltd.* v. *Blank* [1966] 1 W.L.R. 1428 (C.A.).

[28] *Associated Distributors Ltd.* v. *Hall* [1938] 2 K.B. 83; considered in *Bridge* v. *Campbell Discount Co. Ltd.* [1962] A.C. 600 (H.L.), where two of their lordships accepted this principle, two did not and the fifth left the matter open. In any event their lordships' views were *obiter*.

[29] See *Bridge* v. *Campbell Discount Co. Ltd.*, *supra*, where their lordships considered the matter *obiter* and differed in their views in the same manner as set out in the last footnote; see also *Stockloser* v. *Johnson* [1954] 1 Q.B. 476 (C.A.); *Phonographic Equipment (1958) Ltd.* v. *Muslu* [1961] 1 W.L.R. 1379 (C.A.).

[30] *Kemp* v. *Rose* (1858) 1 Giff. 258, 266; see *ante*, p. 27, and *post*, p. 344.

[31] See *ante*, p. 119.

example, if the contract provides that the liquidated damages are to be deducted from time to time as progress payments are made and the employer fails to deduct them, he may have no right to deduct them from payments subsequently becoming due. But in the absence of words indicating the contrary, where, after there has been delay, payment of the contract price is made in full, this in itself does not operate as a bar to the recovery of liquidated damages.[32]

4. Final certificate

Where the architect's decision is final and he has power to extend the time and to take into account in his final certificate any liquidated damages due, then he will be presumed to have extended the time for completion unless it is proved or admitted [33] that the matter has not been determined by him, or was not expressly or impliedly within his jurisdiction.[34] But the architect's final certificate certifying the amount due to the contractor in settlement of the contract does not estop the employer from bringing an action for liquidated damages if the contract did not give the architect power to deal with liquidated damages.[35]

5. Employer causing delay

If the employer delays the completion of the works in any way, as, for example, by failing to give possession of the site [36] or of plans [37] at the proper time, or by interfering improperly through his agent in the carrying out of the works,[38] or by ordering extras which necessarily delay the works,[39] or by failing to deliver components he is bound to provide,[40] or by delay in giving essential instructions,[41] the general rule is that he loses the right to claim liquidated damages for non-completion to time, for he " cannot insist on a condition if it is his own fault that the condition has not been fulfilled." [42] There are two exceptions to the above rule, the first to be mentioned being of limited, and the second of general, application:

---

[32] *Clydebank Engineering Co.* v. *Yzquierdo y Castaneda* [1905] A.C. 6 (H.L.).

[33] See *Jones* v. *St. John's College, Oxford* (1870) L.R. 6 Q.B. 115.

[34] *Per* Phillimore J., *British Thomson-Houston Co.* v. *West* (1903) 19 T.L.R. 493, 494, approving passage from H.B.C. (2nd ed.), Vol. 1, p. 332.

[35] *British Thomson-Houston Co.* v. *West, supra.* See also *ante,* p. 81.

[36] *Holme* v. *Guppy* (1838) 3 M. & W. 387; *Felton* v. *Wharrie* (1906) H.B.C. (4th ed.), Vol. 2, p. 398 (C.A.).

[37] *e.g. Roberts* v *Bury Commissioners* (1870) L.R. 5 C.P. 310.

[38] *e.g. Russell* v. *Sa da Bandeira* (1862) 13 C.B.(N.S.) 149.

[39] *e.g.* Dodd v. *Churton* [1897] 1 Q.B. 562 (C.A.).

[40] *Perini Pacific* v. *Greater Vancouver Sewerage and Drainage District* (1966) 57 D.L.R. (2d) 307 (British Columbia C.A.).

[41] *Peak Construction (Liverpool) Ltd.* v. *McKinney Foundations Ltd.* (1971) 69 L.G.R. 1 (C.A.).

[42] *Amalgamated Building Contractors Ltd.* v. *Waltham Holy Cross U.D.C.* [1952] 2 All E.R. 452, 455 (C.A.); *Wells* v. *Army & Navy Co-operative Society* (1902) H.B.C. (4th ed.), Vol. 2, p. 346 at p. 354 (C.A.); *Trollope & Colls Ltd.* v. *N.W. Metropolitan Regional Hospital Board* [1973] 1 W.L.R. 601, 607 (H.L.). For a discussion of the date from which any claim for unliquidated damages runs, see *ante,* p. 155.

(a) *Extras ordered under absolute contract*

The wording of the contract may be such that the contractor binds himself absolutely to complete the contract work with extras within the stipulated time, subject to payment of liquidated damages in default, even though extras may be ordered and no extension of time is granted. Such a contract, though it is very onerous and the contractor may have committed himself to an impossibility, will be enforced provided the extras were such as were contemplated by the contract.[43] But if there is any ambiguity in the contract the courts will not construe it so as to impose such an unreasonable obligation on the contractor and will give effect to the general rule stated above.[44] A provision that the ordering of extras is not to " vitiate " the contract or " is not to vitiate the contract or the claim for penalties " is not sufficient to impose this absolute obligation on the contractor.[45]

(b) *Extension of time for employer's delay*

If the employer makes a valid extension of time in respect of the delay which he has caused, a new date is set for completion and, subject to any further extension or special defences raised by the contractor, liquidated damages will be payable from that extended date in the event of non-completion.

Express power necessary. The employer cannot for his own benefit purport to extend the time because of his own delay if there is no express power in the contract.[46]

Compliance with contract. Any purported extension must comply strictly with the provisions of the contract, and where the extension is for the employer's benefit and the contract was imposed by him on the contractor it will, if it is ambiguous, be construed against the employer.[47]

Retrospective extension for employer's delay. Where delay is solely caused by the employer, then unless clear words are used, a power to extend time because of the employer's delay cannot, it seems, be exercised retrospectively.[48]

---

[43] *Jones* v. *St. John's College, Oxford* (1870) L.R. 6 Q.B. 115; *Tew* v. *Newbold-on-Avon United District School Board* (1884) 1 Cab. & El. 260; *Sattin* v. *Poole* (1901) H.B.C. (4th ed.), Vol. 2, p. 306 (D.C.).
[44] *Dodd* v. *Churton, supra*; *Wells* v. *Army & Navy Co-operative Society, supra.*
[45] *Ibid.*   [46] *Dodd* v. *Churton, supra.*
[47] *Miller* v. *L.C.C.* (1934) 50 T.L.R. 479, 482; *Peak Construction (Liverpool) Ltd.* v. *McKinney Foundations Ltd.* (1971) 69 L.G.R. 1 (C.A.). Cf. *Tersons Ltd.* v. *Stevenage Development Corporation* [1963] 2 Lloyd's Rep. 333, 368 (C.A.). See *ante*, p. 34.
[48] *Amalgamated Building Contractors Ltd.* v. *Waltham Holy Cross U.D.C.* [1952] 2 All E.R. 452, 455 (C.A.); *Miller* v. *L.C.C., supra.* The point is not free from doubt for in *Sattin* v. *Poole* (1901) H.B.C. (4th ed.), Vol. 2, p. 306, referred to with approval in the *Amalgamated Building Contractors* case, an extension was granted after completion to a date in the past and the contractor was not permitted to adduce evidence to show that the delay was caused by the employer. But the contractor had specifically asked the architect to extend the time after he had completed the works, and it was not argued that the architect lacked the power to grant the extension. In *Miller's* case an extension made after completion because of extras ordered by the employer was held to be invalid, but Denning L.J. in the *Amalgamated Building Contractors* case (at p. 455) said that he regarded *Miller's* case as turning on the very special wording of the contract. For an example of clear words, see cl. 23 of the Standard form of building contract.

6. Breach of condition precedent by employer

The contract may provide that some condition precedent must be fulfilled before the employer can claim liquidated damages. Thus in the Standard form of building contract, the architect's certificate under clause 22 is such a condition precedent.[49] Similarly, where a contract imposes a duty on the architect to extend the time and he fails to perform that duty in accordance with the contract the employer is unable to claim liquidated damages.[50]

7. Extension of time

It is common to provide an express power for the extension of the time for completion and if an extension has been granted it operates wholly or partially as a defence to a claim for liquidated damages from the original completion date. The type of events for which provision is made falls under two heads: (*a*) delay caused by the employer, *e.g.* by the ordering of extras. This has been considered above; (*b*) delay not caused by the employer, *e.g.* strikes, *force majeure*,[51] shortage of materials and labour and other events which though they may cause unavoidable delay do not, or might not, excuse the contractor.[52]

Compliance with contract

Extension of time for delay whether caused by the employer[53] or not[54] must comply with the terms of the contract unless one of the parties waives a condition inserted for his benefit.

Retrospective extension—not employer's delay

It is clearly desirable that when possible an extension of time should be granted to a date in the future so that the contractor can plan his work accordingly, and where the delay is caused by the employer an extension can frequently only be granted to a future date[55]; but where the delay is not caused by the employer and an extension is for the benefit of the contractor the position is different. This is illustrated by the facts of *Amalgamated Building Contractors Co. Ltd.* v. *Waltham Holy Cross U.D.C.* The contractors were liable under a modified R.I.B.A. form, 1939 edition, to complete the works by February 7, 1949. On January 19, 1949, the contractors applied for a 12-month extension because of labour and materials difficulties.[56] The architect did not reply beyond a formal

---

[49] *Amalgamated Building Contractors* case at p. 454.

[50] *Ibid.*; *cf.* cl. 23, Standard form of building contract.

[51] For meaning, see *post*, p. 348.

[52] For excuses for non-completion, see Chap. 7. In the Standard form of building contract, (*e*), (*f*), (*h*), (*i*) of cl. 23 correspond roughly to (*a*), and the other sub-clauses correspond to (*b*).

[53] See *ante*, p. 160.

[54] See *Sattin* v. *Poole* (1901) H.B.C. (4th ed.), Vol. 2, pp. 306, 314, construing a clause indistinguishable (*Amalgamated Building Contractors Ltd.* v. *Waltham Holy Cross U.D.C.* [1952] 2 All E.R. 452, 455 (C.A.)) from cl. 18 of the R.I.B.A. form (1939 ed.).

[55] See *ante*, p. 161.

[56] Invoking cl. 18 (ix), the equivalent of cl. 23 (*j*) in the Standard form of building contract.

acknowledgment.[57] The work was completed on August 28, 1950, and on December 20, 1950, the architect wrote to the contractors stating that he had decided that " an addition of 15 weeks bringing the completion date to May 23, 1949, would be a fair and reasonable extension." [58] He also wrote to the employers certifying in accordance with the contract. If the architect's extension had been invalid the employers could not have claimed liquidated damages at all, but it was held that it was valid. Denning L.J. said that this was a case where the cause of delay operated partially but not wholly, every day, until the works were completed. It must follow that an architect could not decide the length of the extension until after completion; therefore the parties must have intended that he could in such circumstances grant an extension retrospectively. And the same principle would apply " where the contractors, near the end of the work, have overrun the contract time for six months without legitimate excuse. . . . Now suppose . . . a strike occurs and lasts a month. The contractors can get an extension of time for that month. The architect can clearly issue a certificate which will operate retrospectively. He extends the time by one month from the original completion date, and the extended time will obviously be a date which is past." [59]

The wording of the contract may indicate that the extension of time must be granted before completion whether the delay is caused by the employer or not. Thus in *Miller* v. *L.C.C.*[60] the architect had the power to grant an extension, and " to assign such other time or times for completion as to him may seem reasonable." It was held that these words were not apt to refer to the fixing of a date *ex post facto*, and that a purported extension made after completion was invalid, and this though the power of extending could be exercised " prospectively or retrospectively," for " retrospectively " referred to after the causes of delay and not to after completion.[61]

Extension and contractor's claim for damages

Where the contractor has a claim for damages against the employer for the latter's delay an extension of time does not, in the absence of express words, release the claim.[62]

8. Rescission or determination

If the contract is brought to an end by determination or otherwise, then

---

[57] It was suggested by Denning L.J. at p. 455 that if the contractors were dissatisfied they should have asked for arbitration. In the unmodified Standard form of building contract this could not have been opened until after the completion of the works unless the employers consented—cl. 35. In any event without co-operation arbitration is rarely quick.

[58] The court rejected a submission that this letter was too informal.

[59] *Per* Denning L.J., *Amalgamated Building Contractors Ltd.* v. *Waltham Holy Cross U.D.C.*, *supra*, at p. 454.

[60] (1934) 50 T.L.R. 479.

[61] The delay was caused by the employer but the case was distinguished by Denning L.J. in the *Amalgamated Building Contractors* case, *supra*, at p. 454, primarily because of the wording of the clause referred to in the text.

[62] *Trollope & Colls* v. *Singer* (1913) H.B.C. (4th ed.), Vol. I, p. 849, and see *Roberts* v. *Bury Commissioners* (1870) L.R. 5 C.P. 310, 327.

prima facie all future obligations cease and no claim can be made for liquidated damages accruing after determination.[63] But there may be some special clause which has the effect of keeping the provision for payment of liquidated damages alive although the work has been taken out of the hands of the contractor.[64]

Section 3: Specific Performance

Specific performance is a decree issued by the court ordering the defendant to perform his promise. It is an equitable remedy granted by the court in its discretion, such discretion being exercised according to well-established principles. Thus it will not grant a decree where the common law remedy of damages will adequately compensate the plaintiff, nor where the court cannot properly supervise performance. For these reasons " it is settled that, as a general rule, the court will not compel the building of houses." [65]

Building agreement. Where there is a purchase or lease of land and an agreement to carry out building works forms part of the consideration on one side or the other the court will order specific performance of the agreement to build if the following conditions are satisfied:

(i) " That the building work . . . is defined by contract; that is to say that the particulars of the work are so far definitely ascertained that the court can sufficiently see what is the exact nature of the work of which it is asked to order the performance." [66]

(ii) " That the plaintiff has a substantial interest in having the contract performed which is of such a nature that he cannot adequately be compensated for breach of the contract by damages." [67]

(iii) " That the defendant is in possession of the land on which the work is contracted to be done." [68]

Building contract—special circumstances. It is suggested that there may be special circumstances, even in the case of the ordinary building contract where there is no element of a transaction in land, where specific performance would be granted. The works would have to be exactly defined,[69] the defendant capable of

---

[63] See *Ex p. Sir W. Hart Dyke, re Morrish* (1882) 22 Ch.D. 410 (C.A.) (forfeiture of a lease); *British Glanzstoff Manufacturing Co. Ltd.* v. *General Accident, etc., Ltd.* [1913] A.C. 143 (H.L.); *Suisse Atlantique, etc.* v. *N.V. Rotterdamsche Kolen Centrale* [1967] 1 A.C. 361, 398 (H.L.), and see generally *ante*, p. 120. *Cf. Wood* v. *Tendring Rural Sanitary Authority* (1886) 3 T.L.R. 272 (D.C.).

[64] See *Re Yeadon Waterworks Co. & Wright* (1895) 72 L.T. 832 (C.A.). In such a case there is no determination of the contract but only a forfeiture of certain rights; see *ante*, p. 120.

[65] *Per* Sir G. Mellish L.J., *Wilkinson* v. *Clements* (1872) L.R. 8 Ch. 96, 112.

[66] *Per* Romer L.J., *Wolverhampton Corp.* v. *Emmons* [1901] 1 K.B. 515, 525 (C.A.). See also *Molyneux* v. *Richard* [1906] 1 Ch. 34.

[67] *Ibid.*

[68] *Per* Farwell J., *Carpenters Estates Ltd.* v. *Davies* [1940] Ch. 160, 164, extending the words of Romer L.J. in the *Wolverhampton* case, who said " that the defendant has *by the contract* obtained possession of land."

[69] See cases cited *supra*; *Morris* v. *Redland Brick Ltd.* [1970] A.C. 652 (H.L.); *Jeune* v. *Queens Cross Properties Ltd.* [1974] 1 Ch. 97—order requiring landlord to reinstate stone balcony.

carrying them out and damages an inadequate remedy. The position might arise in relation to specialised works when no other contractor is available to perform them.[70]

Section 4: *Injunction*

The equitable remedy of an injunction ordering a person to do something or restraining him from continuing a wrong is not normally granted in building contracts for the same reasons that a decree of specific performance is not granted.[71] For injunctions in relation to turning the contractor off the site, see a full discussion, *ante*, p. 123.

Section 5: *Limitation of Actions*

The Limitation Act 1939 imposes limits of time within which actions must be brought or they become barred.[72] The Act must be referred to for its full effect.[73]

Contract

Actions must be brought within six years of the date of the cause of action accruing, or 12 years if the contract is under seal,[74] subject to the provisions of section 26 discussed *infra*. Time runs from the date of breach of duty and not from its discovery.[75] In the case of a contractor who is liable under an entire contract to complete works the limitation period in respect of defect runs, it is submitted, from the date of completion or purported completion, and not from the date, if it is earlier, when that part of the works, the subject-matter of the defects, was carried out. In the case of breaches by the employer time runs from the breach, so that, for example, in the case of failure to supply drawings and instructions at the proper time, it runs, it is submitted, from the date when they should have been supplied. A set-off which is a defence to the claim cannot be defeated by a period of limitation not applicable to the claim.[76]

Tort

Subject to section 26 of the Act, *infra*, actions must be brought within six years of the cause of action accruing.[77] In the case of torts, such as negligence,

---

[70] *Cf. Sky Petroleum Ltd.* v. *VIP Petroleum Ltd.* [1974] 1 W.L.R. 576—contract for supply of petrol enforced at time when other supplies might not be available.

[71] See *supra*. See also *ante*, p. 124. For the principles upon which mandatory injunctions are granted, see *Morris* v. *Redland Bricks Ltd.*, *supra*.

[72] For personal injuries reference should be made to the various subsequent amending statutes.

[73] For architects and professional men see *post*, p. 218.

[74] s. 2.

[75] *Cartledge* v. *Jopling (E. & Sons) Ltd.* [1963] A.C. 758, 782 (H.L.); *Bagot* v. *Stevens Scanlan & Co. Ltd.* [1966] 1 Q.B. 197; *Sparham-Souter* v. *Town & Country Developments (Essex) Ltd.* [1976] Q.B. 858 (C.A.).

[76] *Henriksens Rederi A/S* v. *Rolimpex* [1974] 1 Q.B. 233 (C.A.); see also *Tersons* v. *Alec Colman Investments* (1973) E.G. 2300 (C.A.).

[77] See *supra*, note 74.

where damage is an ingredient of the tort, the cause of action does not accrue until the plaintiff has suffered the damage.[78]

Breach of by-laws. Where the claim is in tort, by persons occupying a house, against a local authority for negligence in performance of its duties under the relevant legislation or, it seems, against the contractor for loss caused by his failure to comply with by-laws, time does not begin to run until there is present or imminent danger to the health or safety of such persons.[78a]

Fraudulent concealment

By clause 26 of the Limitation Act 1939 where the action is based upon the fraud of the defendant or his agent or of any person through whom he claims or his agent or the right of action is concealed by the fraud of any such person the period of limitation does not begin to run until the plaintiff has discovered the fraud or could with reasonable diligence have discovered it. This was applied in *Clark* v. *Woor*.[79] The plaintiffs, who knew nothing of building and had no architect or other person to supervise the works, relied on the defendant, a builder, to treat them in a decent, honest way in building them a house with best Dorking bricks. The defendant, an experienced bricklayer, could not get Dorking bricks and substituted without the plaintiffs' knowledge Ockley bricks containing a substantial portion of seconds which failed after eight years. It was held that there was that special relationship between the parties which brought the defendant's behaviour within the meaning of fraud as it is used in the Act so that the plaintiffs were not barred by expiry of time.

These principles have subsequently been developed in the Court of Appeal. Thus, it has been said ". . . if a builder does his work badly, so that it is likely to give rise to trouble thereafter, and then covers up his bad work so that it is not discovered for some years, then he cannot rely on the statute as a bar to the claim. The right of action is concealed by ' fraud ' in the sense in which 'fraud' is used in this section." [80] The concealment must be deliberate or reckless " like the man who turns a blind eye," [81] but it is not sufficient, in the absence of very special circumstances, that the defendant ought to have known the fact or facts which constituted the cause of action against him.[82] Knowledge can be inferred from the evidence.[83]

Indemnities. See discussion *ante*, p. 140.

---

[78] See *supra*, notes 75 and *infra*, 78a. The result of the recent cases requires, it is suggested, a review by Parliament.

[78a] *Anns* v. *Merton London Borough* [1977] 2 W.L.R. 1024 (H.L.).

[79] [1965] 1 W.L.R. 650.

[80] *Applegate* v. *Moss* [1971] 1 Q.B. 406, 413 (C.A.). A developer is liable for the fraud of his contractor, *ibid.*

[81] *King* v. *Victor Parsons & Co.* [1973] 1 W.L.R. 29, 34 (C.A.).

[82] *Ibid.* at p. 36.

[83] *Ibid.*

CHAPTER 10

ASSIGNMENTS, SUBSTITUTED CONTRACTS AND SUB-CONTRACTS

Section 1: *Assignments*

IN considering assignments it is essential to distinguish between the benefit and the burden of a contract. In the normal building contract the burden on the contractor is the duty to complete the works, and his benefit is the right to receive the contract money when it falls due; the burden on the employer is the duty to pay such money, and the benefit is the right to have the works completed.

1. Assignment by contractor of burden

Vicarious performance

It is a general principle that the burden of a contract cannot be assigned without the consent of the other party.[1] Therefore the contractor cannot assign his liability to complete.[2] But in some cases he is entitled to secure the vicarious performance of his obligations while remaining liable for non-performance himself.[3] The leading case is *British Waggon Co.* v. *Lea.*[4] The Parkgate Co., which was under a contract to Lea to keep certain railway wagons in repair for a number of years, went into liquidation [5] and assigned its contract (*i.e.* both benefit and burden) to the British Waggon Co. It was held that Lea could not refuse to accept performance of the contract by the British Waggon Co. This case has been explained by Lord Simon in the House of Lords where he said,[6] " I may add that a possible confusion may arise from the use of the word ' assignability ' in discussing some of the cases usually cited on this subject. Thus in *British Waggon Co.* v. *Lea* the real point of the decision was that the contract which the Parkgate company had made with Lea for the repair of certain wagons did not call for the repairs being necessarily effected by the Parkgate company itself, but could be adequately performed by the Parkgate company arranging with the British Waggon company that the latter should execute the repairs. Such a result does not depend on assignment at all. It depends on the view that the contract of repair was duly discharged by the Parkgate company by getting the repairs satisfactorily effected by a third party. In other words, the contract bound the Parkgate company to produce a result, not necessarily by its own efforts, but, if it preferred, by vicarious performance through a sub-contract or otherwise."

---

[1] *Nokes* v. *Doncaster Amalgamated Collieries Ltd.* [1940] A.C. 1014 (H.L.).
[2] *Ibid.*
[3] *Ibid.*; *Davies* v. *Collins* [1945] 1 All E.R. 247, 249 (C.A.). For a shipbuilding case see *The " Diana Prosperity "* [1976] 2 Lloyd's Rep. 60 (C.A.).
[4] (1880) 5 Q.B.D. 149 (D.C.).
[5] It was not *dissolved* at the time of trial. See *per* Cockburn C.J. at p. 151.
[6] *Nokes'* case, *supra*, at p. 1019.

It is a matter of construction of the contract whether or not the contractor has the right of securing vicarious performance.[7] He has the right where it is a matter of indifference whose hand should do the work,[8] as in the *British Waggon* case, where the repairs were " a rough description of work which ordinary workmen conversant with the business would be perfectly able to execute." [9] If the employer acquiesces in vicarious performance he may lose the right to any objection he might otherwise possess.[10]

The contractor does not have the right of performing by another if, either there is a prohibition in the contract,[11] or there is some personal element in his obligation. The contract can be personal in this sense if it was made with the contractor because of his skill or special knowledge,[12] or because of his personality or character.[13] A contract to build a lighthouse was held to be personal,[14] and clearly most contracts of a specialist nature are personal. It is submitted that in the case of large contracts involving great experience of site management and the appointment of men to positions of authority and other managerial duties, the administrative duties of the main contractor are personal.[15]

Novation

If by agreement with the employer the contractor ceases to be liable and a fresh contractor takes his place there is a novation and not an assignment.[16] Thus there is a novation where the contractor is a firm but a limited company is formed which takes over the contractor's business and the employer accepts the company in place of the original firm.[17]

2. Assignment by contractor of benefit

It is common for a contractor to assign the retention money or other sums due under the contract in order to secure advances from merchants, bankers and others. The assent of the employer to such assignments is not necessary and if after notice he ignores the assignment and pays the money to the contractor he will have to pay the money again to the assignee.[18]

---

[7] *Davies* v. *Collins* [1945] 1 All E.R. 247, 249 (C.A.); and see *Tolhurst* v. *Associated Portland Cement Manufacturers Ltd.* [1903] A.C. 414 (H.L.), explained by Lord Simon at p. 1020 of *Nokes'* case.

[8] *Davies* v. *Collins, supra.*

[9] *British Waggon* case at p. 153.

[10] *Falle* v. *Le Sueur & Le Huguel* (1859) 12 Moore P.C.C. 501 (P.C.).

[11] See, *e.g.* cl. 17 of the Standard form of building contract.

[12] *Robson* v. *Drummond* (1831) 2 B. & Ad. 303 (an extreme case; see *British Waggon Co.* v. *Lea* (1880) 5 Q.B.D. 149, 153); *Wentworth* v. *Cock* (1839) 10 A. & E. 42, 45; *Knight* v. *Burgess* (1864) 33 L.J.Ch. 727; *Johnson* v. *Raylton* (1881) 7 Q.B.D. 438 (C.A.); *Davies* v. *Collins, supra*; *Edwards* v. *Newland & Co.* [1950] 2 K.B. 534 (C.A.).

[13] See *Kemp* v. *Baerselman* [1906] 2 K.B. 604 (C.A.); *Cooper* v. *Micklefield Coal & Lime Co.* (1912) 107 L.T. 457.

[14] *Anon.*, cited by Patteson J. in *Wentworth* v. *Cock* (1839) 10 A. & E. 42, 45.

[15] *Cf. Edwards* v. *Newland & Co.* [1950] 2 K.B. 534, 542 (C.A.).

[16] See *Commercial Bank of Tasmania* v. *Jones* [1893] A.C. 313 (P.C.); see also *Chatsworth Investments Ltd.* v. *Cussins (Contractors) Ltd.* [1969] 1 W.L.R. 1, 4, 6 (C.A.).

[17] As in *Chandler Bros.* v. *Boswell* [1936] 3 All E.R. 179, 183 (C.A.).

[18] *Brice* v. *Bannister* (1878) 3 Q.B.D. 569 (C.A.). This applies whether the assignment is legal or equitable. Some contracts have clauses purporting to forbid the assignment of the

Such assignments may be legal or equitable. The main difference is that under a legal assignment the assignee can sue in his own name for the money assigned, whereas under an equitable assignment the assignee must sue in the name of the assignor or, if he refuses to be joined as plaintiff, must add him as defendant. The difference is therefore in the main a matter of procedure.[19] The court can overlook non-joinder of the assignor, if no objection is taken,[20] but if any objection is raised it will dismiss the claim of an equitable assignee who does not bring the assignor before the court.[21]

If the assignor of part of a debt sues his debtor for the debt, he must join the assignee in the proceedings before he can recover the debt or even that part of it which was not assigned.[22]

Legal assignment

A legal assignment derives its authority from section 136 of the Law of Property Act 1925. It must be absolute, in writing, and express notice in writing of the assignment must be given to the debtor.[23] In such a case the assignee can sue in his own name for the debt or thing in action assigned and has all legal and other remedies for the same and can give a good discharge without the concurrence of the assignor.[24]

An assignment of retention money or instalments earned, but not payable at the date of assignment, is absolute and within the section.[25] An assignment of unearned instalments is good in equity, and may, if it is of a definite sum, be good at law.[26]

A mortgage of all moneys due or to become due by way of security for advances is absolute and within the section.[27] In such a case there will either be an express right of redemption by the mortgagor [28] (*i.e.* the contractor), or such a right will be implied by equity.[29] But where contractors charged the sum of £1,080 due to them under a contract as security for advances until such advances were

---

benefit. To be effective unauthorised assignment should be a ground for determination. For a discussion of cl. 17 of the Standard form of building contract, see *post*; p. 331.

[19] *Tolhurst* v. *Associated Cement, etc., Ltd., supra.* For procedure when seeking summary judgment on an assignment, see *Les Fils Dreyfus, etc.* v. *Clarke* [1958] 1 W.L.R. 300 (C.A.).

[20] *Brandt's Sons & Co.* v. *Dunlop Rubber Co.* [1905] A.C. 454 (H.L.).

[21] *Williams* v. *Atlantic Assurance Co. Ltd.* [1933] 1 K.B. 81 (C.A.).

[22] *Walter & Sullivan Ltd.* v. *J. Murphy & Sons Ltd.* [1955] 2 Q.B. 584 (C.A.).

[23] s. 136 (1)—" Any absolute assignment by writing under the hand of the assignor (not purporting to be by way of charge only) of any debt or any other legal thing in action, of which express notice in writing has been given to the debtor, trustee or other person from whom the assignor would have been entitled to receive or claim such debt or thing in action, is effectual in law. . . ."

[24] *Ibid.*

[25] *G. & T. Earle (1925) Ltd.* v. *Hemsworth R.D.C.* (1928) 140 L.T. 69 (C.A.).

[26] *G. & T. Earle (1925) Ltd.* v. *Hemsworth, supra,* at p. 71; *Walker* v. *Bradford Old Bank* (1884) 12 Q.B.D. 511; *Jones* v. *Humphreys* [1902] 1 K.B. 10, 13; *Re Tout and Finch* [1954] 1 W.L.R. 178; *cf. Brice* v. *Bannister* (1878) 3 Q.B.D. 569 (C.A.); *Durham Bros.* v. *Robertson* [1898] 1 Q.B. 765 (C.A.); *Drew* v. *Josolyne* (1887) 18 Q.B.D. 590 (C.A.).

[27] *Tancred* v. *Delagoa Bay, etc., Ry.* (1889) 23 Q.B.D. 239.

[28] *Ibid.*

[29] As in *Hughes* v. *Pump House Hotel Co.* [1902] 2 K.B. 190 (C.A.).

repaid, this was held to be a conditional assignment and not within the section,[30] though it was a good equitable assignment.[31]

The assignment of part of an entire debt is not within the section.[32] Therefore if no provision is made in the contract for the payment of instalments or for retention money the contractor cannot make a legal assignment of part of the contract sum.[33]

Equitable assignment

An equitable assignment usually arises where there is a clear intention to assign a debt or other thing in action but one of the requirements of section 136 of the Law of Property Act 1925 has not been satisfied. No particular form is required provided the meaning is clear,[34] and a clearly defined fund is specified.[35] Thus it has been held that clause 11 (*h*) of the standard form of sub-contract[36] operates as a valid equitable assignment to the sub-contractor of that part of the retention money held by the employer under the main contract which is due to the sub-contractor under the sub-contract.[37] And a letter by the main contractor (a limited company) to the employer directing him to pay direct to the sub-contractor a sum of money out of the next certificate to be issued was held to be a valid equitable assignment, and not void as against the receiver of the debenture-holders of the main contractor as being an unregistered charge within the meaning of section 79 [38] of the Companies Act 1929.[39]

An equitable assignment can be of future property,[40] and it need not be absolute but may be by way of a charge.[41]

Exercise of option. An equitable assignee of a contractual option who has not given notice to the grantor of the option cannot exercise the option in his own name so as to bind the grantor.[42]

---

30 *Durham Bros.* v. *Robertson* [1898] 1 Q.B. 765 (C.A.).
31 *Ibid.*
32 *Williams* v. *Atlantic Assurance Co.* [1933] 1 K.B. 81 (C.A.).
33 *Ibid.* though it would be good as an equitable assignment.
34 *Brandt's Sons & Co.* v. *Dunlop Rubber Co.* [1905] A.C. 454, 462 (H.L.); *cf. Re McArdle* [1951] Ch. 669 (C.A.).
35 *Percival* v. *Dunn* (1885) 29 Ch.D. 128; *cf. Re Warren, ex p. Wheeler* [1938] Ch. 725 (D.C.).
36 See *post*, p. 591.
37 *Re Tout and Finch, supra.*
38 Now s. 95 of the Companies Act 1948.
39 *Ashby Warner & Co. Ltd.* v. *Simmons* (1936) 155 L.T. 371 (C.A.); *cf. Re Kent & Sussex Sawmills Ltd.* [1947] Ch. 177; *Walter & Sullivan Ltd.* v. *Murphy & Sons Ltd.* [1955] 2 Q.B. 584 (C.A.). These cases and *Re Tout & Finch, supra*, must be considered in the light of the majority speeches in the House of Lords in *British Eagle Ltd.* v. *Air France* [1975] 1 W.L.R. 758 (H.L.) discussed *post*, p. 189. Their authority on equitable assignments is not affected, but in so far as they are authorities on issues relating to the distribution of insolvent's estates they are open to review.
40 *Tailby* v. *The Official Receiver* (1888) 13 App.Cas. 523, 546 (H.L.); *Re Tout and Finch, supra.*
41 As in *Durham Bros.* v. *Robertson* [1898] 1 Q.B. 765 (C.A.). For what was intended in a charge on " retention money," see *West Yorkshire Bank Ltd.* v. *Isherwood Bros. Ltd.* (1912) 28 T.L.R. 593.
42 *Warner Bros. Records Inc.* v. *Rollgreen Ltd.* [1976] Q.B. 430 (C.A.).

Notice to the employer

Written notice to the debtor of the assignment is necessary to perfect a legal assignment.[43] Such assignment is complete from the date on which the notice is received by the debtor.[44] It should bring to the notice of the debtor with reasonable certainty the fact of the assignment so as to prevent him paying the debt to the original creditor.[45]

Notice to the debtor is not essential to an equitable assignment but is advisable for several reasons. The position as regards exercise of an option is referred to, *supra.* Further, the assignee is bound by any payments the debtor makes to the assignor without notice of the assignment[46]; and, by the rule in *Dearle* v. *Hall*[47] the priority of an assignee against other assignees depends not upon the dates of the assignment but upon the order of giving notice to the debtor.

Consideration

Consideration is not necessary to support a legal assignment,[48] nor an equitable assignment which is absolute and of an existing debt.[49] Where an equitable assignment is conditional or by way of charge or of future property, consideration is, it seems, necessary.[50]

Subject to equities

An assignee, whether legal or equitable, takes subject to equities. This means " subject to all rights of set-off and other defences which were available against the assignor. . . ."[51] Thus the assignee of moneys due under the contract is liable to any claim by the employer arising out of the contract such as for defective work or for damages for delay.[52] Other applications of the principle are given below.

The only exception to the rule that the assignee takes subject to equities is " that after notice of an assignment of a chose in action,[53] the debtor cannot by payment or otherwise do anything to take away or diminish the rights of the assignee as they stood at the time of the notice."[54] Where, after notice, the employer advanced money to the contractor to enable him to complete, it was held that such advance could not be set off against the moneys assigned and he must pay the assignee.[55] It has been held that an employer could not set off a claim under a new contract with the contractor entered into after the notice of

---

[43] See *supra.*
[44] *Holt* v. *Heatherfield Trust Ltd.* [1942] 2 K.B. 1.
[45] *Van Lynn Developments Ltd.* v. *Pelias Construction Co. Ltd.* [1969] 1 Q.B. 607, 613, 615 (C.A.); see also *W. F. Harrison & Co. Ltd.* v. *Burke* [1956] 1 W.L.R. 419 (C.A.).
[46] *Stocks* v. *Dobson* (1853) 4 De G.M. & G. 11.
[47] (1828) 3 Russ. 1.
[48] *Re Westerton* [1919] 2 Ch. 104.
[49] *Holt* v. *Heatherfield Trust Ltd., supra; Re McArdle* [1951] Ch. 669, 674 (C.A.).
[50] *Holt* v. *Heatherfield Trust Ltd., supra; Re McArdle, supra.*
[51] *Per* James L.J., *Roxburghe* v. *Cox* (1881) 17 Ch.D. 520, 526 (C.A.).
[52] *Young* v. *Kitchin* (1878) 3 Ex.D. 127; *Newfoundland Govt.* v. *Newfoundland Ry.* (1888) 13 App.Cas. 199 (P.C.).
[53] *i.e.* a debt or thing in action.
[54] See *supra,* note 51.
[55] *Brice* v. *Bannister* (1878) 3 Q.B.D. 569 (C.A.).

assignment, even though the new contract came into existence as the result of the exercise of an option in the original contract.[56] It was further held that although the employer, by an express clause in the original contract, was empowered to deduct or set off damages due from the contractor to the employer against any money due from the employer to the contractor, this did not enable him to set off claims arising out of concurrent and separate contracts.[57]

Certificated completion. If the building contract requires a certificate of completion before the final balance is payable an assignee is in no better position than the contractor, and cannot therefore recover payment until the certificate has been issued.[58]

Forfeiture. An assignee takes subject to any right of the employer to forfeit or determine the contract, and if upon such forfeiture or determination the money assigned ceased to be payable by the employer the assignee will have no claim against the employer,[59] although he may, according to the terms of the assignment, have a claim for damages against the contractor.[60]

Fraud. Where a certificate for payment was obtained by fraud between the contractor and the architect, it was held that the assignees of retention moneys due on the certificate, although innocent of the fraud, were in no better position than the contractor and the fraud was a good defence against them.[61] If the contract was induced by a fraudulent misrepresentation the employer can rescind the contract or set up the fraud as a defence to a claim by an assignee, but he cannot set off against an innocent assignee a claim for damages for the tort of deceit, for this is a personal claim against the wrongdoer.[62]

Employer cannot counterclaim

The employer's right against the assignee is limited to a set-off up to the amount of the claim. He cannot go beyond this and counterclaim against the assignee.[63]

Arbitration clause

Unless the subject-matter of the contract is such that it cannot be assigned,[64]

---

[56] *Re Asphaltic Wood Pavement Co., ex p. Lee and Chapman* (1885) 30 Ch.D. 216 (C.A.). This is of particular significance in insolvency: see Bankruptcy Act 1914, s. 31; Companies Act 1948, s. 317; *National Westminster Bank Ltd.* v. *Halesowen, etc., Ltd.* [1972] A.C. 785, esp. at p. 821 (H.L.).

[57] *Ibid.*

[58] Per Lord Blackburn, *Lewis* v. *Hoare* (1881) 44 L.T. 66, 67 (H.L.).

[59] *Tooth* v. *Hallett* (1869) L.R. 4 Ch.App. 242; *Drew* v. *Josolyne* (1887) 18 Q.B.D. 590 (C.A.).

[60] See *Humphreys* v. *Jones* (1850) 20 L.J.Ex. 88.

[61] *Wakefield and Barnsley Banking Co.* v. *Normanton Local Board* (1881) 44 L.T. 697 (C.A.).

[62] *Stoddart* v. *Union Trust Ltd.* [1912] 1 K.B. 181, 194 (C.A.).

[63] *Young* v. *Kitchin* (1878) 3 Ex.D. 127; *Newfoundland Govt.* v. *Newfoundland Ry.* (1888) 13 App.Cas. 199 (P.C.).

[64] See *ante*, p. 167. See also *Glegg* v. *Bromley* [1912] 3 K.B. 474 (C.A.); *Compania Colombiana de Seguras* v. *Pacific Steam Navigation Co.* [1965] 1 Q.B. 101.

it seems that the assignee of moneys due can enforce, and will be subject to, an arbitration clause in the contract.[65]

Rival claims

If there are rival claims to moneys due under the contract, either between assignor and assignee, or as a result of claims such as that of a trustee in bankruptcy, the employer should call upon the claimants to interplead, *i.e.* to commence a form of proceeding to determine to whom the money should be paid.[66] If the employer does this he will pay the right claimant and can normally recover his costs out of the moneys assigned; if he does not take this course and chooses to pay the wrong person he will have to pay twice over and may have to pay the costs of the successful claimant.[67]

3. Assignment by employer of burden

The employer cannot without the contractor's assent assign the burden of the contract, *i.e.* the liability to pay the contract moneys.[68] If the contractor's consent is obtained there is a novation, and the original employer is released and a new one takes his place.[69]

4. Assignment by employer of benefit

An employer can assign his rights under the contract provided that there is nothing of a personal nature about the rights.[70]

Section 2: *Substituted Contracts*

If by agreement or because of forfeiture a new contractor is substituted for the original it is a question of construction how far the new contractor is subject to the liabilities of the original contractor. If the contract between the new contractor and the employer expressly or impliedly incorporates or refers to the original contract both contracts must be read together.[71] Thus a new contractor has been held liable in liquidated damages for delay caused by the original contractor.[72] In another case, where the events were similar, but the contracts were

---

[65] *Shayler* v. *Woolf* [1946] Ch. 320, 323 (C.A.); confining statement of Wright J. in *Cottage Club Estates Ltd.* v. *Woodside Estates Co. (Amersham) Ltd.* [1928] 2 K.B. 463 to its particular facts. See also *Aspell* v. *Seymour* [1929] W.N. 152 (C.A.).

[66] s. 136 (1) of the Law of Property Act 1925; this, it seems, applies only to legal assignments, but apart from the section the employer has a right to call for an interpleader issue where the assignment is equitable: R.S.C., Ord. 17; and see, *e.g. Drew* v. *Josolyne, supra.*

[67] *G. & T. Earle (1925) Ltd.* v. *Hemsworth R.D.C.* (1928) 140 L.T. 69, 71 (C.A.). For an example in a company liquidation of a summons to determine the issue in the Chancery Division, see *Re Tout and Finch* [1954] 1 W.L.R. 178.

[68] *Tolhurst* v. *Associated Cement Ltd.* [1903] A.C. 414 (H.L.).

[69] *Commercial Bank of Tasmania* v. *Jones* [1893] A.C. 313 (P.C.).

[70] *Kemp* v. *Baerselman* [1906] 2 K.B. 604 (C.A.); *Cooper* v. *Micklefield Coal & Lime Co.* (1912) 107 L.T. 457; *Tolhurst* v. *Associated Cement Ltd., supra.*

[71] *Cf. Vigers, Sons & Co.* v. *Swindell* [1939] 3 All E.R. 590, 594.

[72] *Re Yeadon Waterworks Co.* v. *Wright* (1895) 72 L.T. 832 (C.A.). The wording of both contracts should be carefully studied.

different, it was held that the liquidated damages clause applied only where the original contractor completed, and that the new contractor was not liable.[73]

When a new contractor is to take over uncompleted building works it is advisable to draw up a schedule showing exactly the work already carried out, including any authorised alterations, and specifying any defects; preferably such schedule should be agreed and signed by the new contractor.

Section 3: Sub-contractors

1. Liability for sub-contractors

The contractor, in the ordinary case and subject to any term to the contrary,[74] is liable to the employer for any default by the sub-contractor in carrying out the terms of the main contract, for the contractor is merely securing the vicarious performance of his own obligations.[75]

" *Or other approved.*" It has been held that a contractor who was required to obtain stone from " the X Co. or other approved firm " was not entitled to extra payment when the employer refused permission to obtain cheaper stone of equal quality from a supplier other than the X Co.[76]

2. Relationship between sub-contractors and employer

A sub-contractor who has entered into a contract with the main contractor to which the employer is not a party has no cause of action against the employer for the price of work done or goods supplied under his contract,[77] unless he sues under a valid assignment.[78] Neither has he any lien on goods supplied, the property in which has passed to the main contractor.[79] Conversely the employer has no claim in contract against a sub-contractor unless he can rely on a collateral warranty.

Collateral warranty

If a sub-contractor or supplier warrants the quality of his work or goods, as the case may be, in consideration of the employer causing the contractor to

---

[73] *British Glanzstoff Manufacturing Co. Ltd.* v. *General Accident, etc., Corporation Ltd.*, 1912 S.C. 591; affd. [1913] A.C. 143 (H.L.); *Wood* v. *Tendring Rural Sanitary Authority* (1886) 3 T.L.R. 272.

[74] *e.g.* cl. 23 (g) of the Standard form of building contract—extension of time for delay on the part of nominated sub-contractors and nominated suppliers.

[75] See *ante*, p. 167. For the default of nominated sub-contractors and nominated suppliers, see *post*, p. 178.

[76] *Leedsford Ltd.* v. *Bradford Corp.*, unrep., March 1960, *Chartered Surveyor* (C.A.), following *Tredegar (Viscount)* v. *Harwood* [1929] A.C. 72 (H.L.).

[77] *Hampton* v. *Glamorgan County Council* [1917] A.C. 13 (H.L.); *Vigers, Sons & Co. Ltd.* v. *Swindell* [1939] 3 All E.R. 590.

[78] See *supra*.

[79] *Pritchett, etc., Co. Ltd.* v. *Currie* [1916] 2 Ch. 515 (C.A.), distinguishing and doubting *Bellamy* v. *Davey* [1891] 3 Ch. 540. There is nothing in English law which corresponds to the mechanics lien legislation in relation to building works which exists in some other common law jurisdictions. It is a matter, at least for discussion, whether there should be.

enter into a contract with the sub-contractor or supplier, the employer can sue the sub-contractor or supplier for loss caused by breach of that warranty.[80]

In *Shanklin Pier Ltd.* v. *Detel Products* the plaintiffs were owners of a pier which they wished to have repaired and repainted, and for this purpose had engaged a contractor and had specified the use of bituminous paint. The defendants then warranted to the plaintiffs that their paint would be suitable for repainting the pier, would give a surface impervious to rust and would have a life of seven to ten years. In reliance on this warranty the plaintiffs instructed the contractor to place an order for, and to use, the defendants' paint in lieu of the bituminous paint. The defendants' paint was a complete failure, and the plaintiffs recovered damages from the defendants[81] for breach of the warranty referred to, although the contract for the sale of the paint was between the defendants and the contractor.

A warranty can be given in respect of time, design or other matters and can be expressed in a formal document. Forms are available and are discussed *post*, p. 427.

Claim in negligence by employer ?

A person who suffers loss resulting from physical damage to property caused by the negligent manufacture of goods or, it is submitted, the negligent carrying out of works, may, in the circumstances of the case, be able to recover damages from the manufacturer or, as the case may be, contractor or sub-contractor, in a claim in tort for negligence.[82] Further, it seems, he does not, upon discovery of a latent defect, have to await the collapse or other physical damage; he can recover the cost of repair.[83]

In an appropriate case the employer may have a remedy against a sub-contractor although consideration must be given to the dictum of Lord Pearce in *Young & Marten Ltd.* v. *McManus Childs Ltd.* who said[84]: " I see great

---

[80] *Shanklin Pier Ltd.* v. *Detel Products* [1951] 2 K.B. 854; *Andrews* v. *Hopkinson* [1957] 1 Q.B. 229; *Yeoman Credit Ltd.* v. *Odgers* [1962] 1 W.L.R. 215, 222 (C.A.); *Wells (Merstham) Ltd.* v. *Buckland Sand & Silica Ltd.* [1965] 2 Q.B. 170; *Bickerton Ltd.* v. *N.W. Hospital Board* [1969] 1 All E.R. 977, 982, 995 (C.A.); *Esso Petroleum Co. Ltd.* v. *Mardon* [1976] Q.B. 801 (C.A.). See also Prof. Wedderburn [1959] C.L.J. 58, and *ante*, p. 100.

[81] The plaintiffs may not have been able to recover damages from the contractors; see *ante*, p. 40, and *post*, p. 181.

[82] *S.C.M. (United Kingdom) Ltd.* v. *Whittall & Son Ltd.* [1971] 1 Q.B. 337 (C.A.); *Dutton* v. *Bognor Regis U.D.C.* [1972] 1 Q.B. 373, 396; *Spartan Steel Ltd.* v. *Martin & Co. Ltd.* [1973] 1 Q.B. 27 (C.A.). For recent consideration by the House of Lords of the ambit of the tort of negligence, see *Dorset Yacht Co. Ltd.* v. *The Home Office* [1970] A.C. 1004 (H.L.); *Anns* v. *Merton London Borough Council* [1977] 2 W.L.R. 1024 (H.L.). The plaintiff cannot recover " economic loss "—see *S.C.M.* case; *Spartan Steel* case.

[83] *Dutton* v. *Bognor Regis, supra*, at p. 396.

[84] [1969] 1 A.C. 454, 469 (H.L.); for facts, see *ante*, p. 39. The point was not argued. It was, however, taken shortly and the two cases mentioned *infra* in the text were read to the House in the course of argument in *Gloucestershire County Council* v. *Richardson* [1969] 1 A.C. 480 (H.L.), which must be read with the *Young & Marten* case (see *Young & Marten*, at p. 468). Lord Upjohn, at p. 475 referred to " a so far scarcely chartered sea of the law of torts in this area." See also the interesting case of *Heaven* v. *Mortimore* (1967) 117 New L.J. 326; *The Times*, March 15, 1967; affd. (1968) 205 E.G. 767 (C.A.); the Canadian case of *Western Processing & Cold Storage* v. *Hamilton Construction Co. & Dow Chemical of Canada* (1965) 51 D.L.R (2d) 245 (Manitoba C.A.); *Weller & Co. Ltd.* v. *Foot & Mouth Disease Research Institute* [1966] 1 Q.B. 569; *A.M.F. International Ltd.* v. *Magnet Bowling Ltd.* [1968] 1 W.L.R. 1028, 1058.

difficulty in extending to an ultimate consumer a right to sue the manufacturer in tort in respect of goods which create no peril or accident but simply result in sub-standard work under a contract which is unknown to the original manufacturer. And if originally, as a term of his contract, the manufacturer limited his liability for defects, there seems no reason (where there is no peril or accident) why a third party should have better rights than the original purchaser. And if his rights are the same there is no need to introduce a rule which would cause various confusions and difficulties since the same result can be achieved in the normal case by third party procedure."

In two shipbuilding cases at first instance a duty of care has been held to exist as between employer and sub-contractor.[85] It is thought that where a sub-contractor or supplier proffers a design upon which the employer relies, the courts may find the existence of a duty of care owed to the employer in respect of the preparation of that design.[86]

Direct payment clause

The contract may provide for direct payment by the employer to the sub-contractor in the event of the main contractor's default.[87] In the absence of an express provision the employer has no such right,[88] and to be effective the words of the contract must be clear.[89] Such a clause is for the benefit of the employer as it encourages specialists to tender at reasonable prices knowing that they have a good chance of being paid.[90] It has never been held that such a clause creates a contractual relationship between employer and a specialist contractor to whom payment can be made under the clause.[91] Neither has it been held to raise the implication of a trust.[92] The right of the employer to pay direct is good as against the main contractor's trustee in bankruptcy or liquidator [93] where the employer continues the contract.

Orders by employer

If the employer gives a direct order to a sub-contractor to carry out work or deliver goods he may make himself liable to the sub-contractor on a promise express or implied to pay the sub-contractor.[94]

---

[85] *Hindustan Steam Shipping Co. Ltd.* v. *Siemans Bros. & Co. Ltd.* [1955] 1 Lloyd's Rep. 167; *The Diamantis Pateras* [1966] 1 Lloyd's Rep. 179. In each case, the claim failed on the facts.

[86] Because the function of designing for a particular purpose for a known person may bring about the " special relationship " referred to in *Hedley Byrne & Co. Ltd.* v. *Heller & Partners Ltd.* [1964] A.C. 465 (H.L.). For such a claim he can recover his " economic loss " —see *ante*, p. 100.

[87] See, *e.g.* cl. 27 (*c*), Standard form of building contract.

[88] *Re Holt, ex p. Gray* (1888) 58 L.J.Q.B. 5.

[89] *Milestone* v. *Yates* [1938] 2 All E.R. 439.

[90] *Re Wilkinson, ex p. Fowler* [1905] 2 K.B. 713.

[91] *Cf. Hobbs* v. *Turner* (1902) 18 T.L.R. 235 (C.A.); *Milestone* v. *Yates, supra*; *Vigers* v. *Swindell, supra*.

[92] Cl. 27 (*f*) of the Standard form of building contract expressly negatives the implication of such a trust.

[93] *Re Tout and Finch Ltd.* [1954] 1 W.L.R. 178, but see *post*, p. 193.

[94] *Dixon* v. *Hatfield* (1825) 2 Bing. 439; *Smith* v. *Rudhall* (1862) 3 F. & F. 143; *cf. Lakeman* v. *Mountstephen* (1874) L.R. 7 H.L. 17 (H.L.). See further *ante*, p. 131.

Orders by architect

An architect has no implied authority to contract on behalf of the employer, and in the absence of express authority any attempt he makes to bind the employer is not effective unless the employer with full knowledge of all the facts ratifies the architect's act.[95] In *Vigers* v. *Swindell* the main contract was in the then current R.I.B.A. form which included a provision for direct payment to sub-contractors,[96] but which expressly provided that the exercise of this right did not create privity of contract between the employer and the sub-contractor. The main contractor went into liquidation and T became the new contractor as a result of a letter from the architect stating, " you yourself will be taking on this contract." The employer, S, knew of this and at the architect's request signed a mandate to her bank authorising it to pay T, " or such nominated sub-contractors as appear on the certificates signed by the architect." Certain sub-contractors, V & Co., then carried out work and the architect purported to pledge S's credit to V & Co. for the cost of the work. S did not know before the work was finished that the architect had pledged her credit. It was held that V & Co. could not recover from S, for the new contract with T was on the same terms as the original contract; therefore the architect had no authority to pledge S's credit, and had not been given authority by the mandate to the bank to pay sub-contractors for this was referable to the provision for direct payment in the original contract; neither had S, with knowledge of the facts, ratified the architect's unauthorised act.[97]

Production of main contract

Where the sub-contractor claims that he is entitled to payment from the employer for work extra to the main contract he may in addition to proving a contract with the employer have to produce the main contract, if it was in writing, to show that it did not include the work for which the claim is made.[98]

3. Prime cost and provisional sums

It is common to describe the sums that the contractor is to pay in respect of the work of sub-contractors as prime cost sums or provisional sums.[99] The contractor is usually only entitled to charge the employer the net sums expended plus a very small percentage [1] and therefore a favourable contract with the sub-contractor is for the employer's benefit. For this reason it was once thought that there was a presumption in such cases that the contractor was acting as agent on behalf of the employer in making such contracts.[2] But it has now been clearly stated by the House of Lords that no inference is to be drawn from the inclusion of a provisional sum for certain goods in a lump sum contract, that the contractor

---

[95] *Vigers, Sons & Co. Ltd.* v. *Swindell* [1939] 3 All E.R. 590; *Ashwell & Nesbitt Ltd.* v. *Allen & Co.* (1912) H.B.C. (4th ed.), Vol. 2, p. 462 (C.A.). See *post*, p. 200.

[96] Cl. 15 (*b*) of the 1931 ed. of the R.I.B.A. form.

[97] *Vigers, Sons & Co. Ltd.* y. *Swindell* [1939] 3 All E.R. 590. In such a case the sub-contractor may have a remedy against the architect. See *post*, p. 203.

[98] See *Eccles* v. *Southern* (1861) 3 F. & F. 142.

[99] See also *ante*, p. 73.

[1] *Cf.* cll. 27, 28, 30 (5) (*c*) of the Standard form of building contract.

[2] *Crittall Manufacturing Co.* v. *L.C.C.* (1910) 75 J.P. 203.

is ordering the goods as agent of the employer.[3] In the special circumstances of
one case in which there was an option for either the contractor or the employer
to pay provisional sums to the specialists, it was held that the employer was
liable.[4]

Section 4: *Nominated Sub-Contractors and Nominated Suppliers*

It is frequently provided in the main contract that the employer may nominate
sub-contractors. It is then the normal practice for the architect to negotiate with
the sub-contractor and settle the terms of his sub-contract without prior consultation with the main contractor, merely informing him of the terms of the
proposed sub-contract at the time of nomination.[5] In such a case the architect
does not act as the agent of the employer,[6] therefore, the contractor has, apart
from any express term in the main contract, no cause of action against the
employer in respect of any delay or default on the part of the nominated sub-
contractor.[7]

1. Delay in nomination

If there is delay in making a nomination the contractor may have a claim for
payment against the employer. This may arise either as a claim for damages for
breach of an express or implied term,[8] or for reimbursement under the express
terms of the contract.[9]

2. Problems inherent in nomination

The system is an ingenious attempt to give the employer the benefit of two
opposing concepts. Theoretically it enables him to have all the advantages of
choosing his own specialist contractor and of bargaining with him for his price
and the terms of his contract and for the performance of services, such as
detailed technical design, which an architect cannot carry out, but avoids the disadvantages of a multiplicity of direct contracts. The achievement of these objects
at first sight appears incompatible.[10] The result has been, particularly in those
common form contracts where contractors were party to the drafting, provisions

---

[3] *Hampton* v. *Glamorgan County Council* [1917] A.C. 13, 22 (H.L.), overruling *Crittall's*
case and distinguishing *Hobbs* v. *Turner* (1902) 18 T.L.R. 235 (C.A.). *Young* v. *White*
(1911) 28 T.L.R. 87, which followed *Crittall's* case, will not, it is submitted, be followed;
see *Hampton's* case at p. 17. See also *Ramsden & Carr* v. *Chessum & Sons* (1913) 110 L.T.
274 (H.L.); *Bickerton, T. A. & Son Ltd.* v. *North West Regional Hospital Board* [1970] 1
W.L.R. 607, 615 (H.L.).

[4] *Hobbs* v. *Turner, supra.* See the remarks of Lord Haldane in *Hampton's* case at pp. 20,
21. See also *Milestone* v. *Yates* [1938] 2 All E.R. 439.

[5] See, *e.g.* the procedure in *Gloucestershire County Council* v. *Richardson* [1969] 1 A.C.
480 (H.L.) referred to *ante,* p. 40.

[6] *Mitchell* v. *Guildford Union (Guardians of)* (1903) 68 J.P. 84; *Leslie & Co. Ltd.* v.
Managers of Metropolitan Asylums District (1901) 68 J.P. 86 (C.A.).

[7] *Ibid.* See also *ante,* p. 38.

[8] See *ante,* p. 37.

[9] Under the Standard form of building contract there is a contractual right to payment
and, it seems, a right to damages—see clauses 3, 24 and notes thereto, *post.*

[10] See the remarks of Lord Reid, *Bickerton (T. A. & Son) Ltd.* v. *North West Metropolitan Regional Hospital Board* [1970] 1 W.L.R. 607, 611 (H.L.).

which are complex and sometimes incomplete or unsatisfactory.[11] Further, perhaps influenced to some extent by the suggestion that it seems unjust in the absence of clear language to make a contractor liable for a nominated sub-contractor over whose activities he has such little control,[12] the House of Lords has, in two important cases,[13] refused to hold a contractor liable for the default of a nominated sub-contractor or supplier even though it has left the employer without apparent remedy. The system of nomination tends to give rise to difficulties and obscurities, and problems frequently require a consideration of the terms of the main contract, of the sub-contract in question and of the surrounding circumstances, including, in particular, those relating to the actual nomination.

3. Express terms of main contract

These must be considered first. Those relating to the Standard form of building contract and the I.C.E. Conditions are discussed *post*, p. 367, and *post*, p. 522. The Government form of contract (Form GC/Works 1 (1st ed., November 1973) contains in Clause 31 (2)[14] what is, presumably, intended to be an unqualified guarantee by the contractor of the performance of nominated sub-contractors and suppliers.

Description of work in main contract

The traditional method, and that contemplated by the Standard form of building contract,[15] is to give no description of the sub-contract works, but merely to include an item such as, " P.C. Sum for steelwork to be supplied and erected by a nominated sub-contractor, £X." Therefore, at the time of entering into the contract there is an estimate of price and a description of the type of work to be ordered but not of the works themselves. Subsequently, when the nomination is made the description of the works becomes known. At this stage, by necessary implication, it is submitted, such description defines the work as between employer and contractor. But a distinction must be drawn between the description of the work and the terms of the sub-contract subject to which it is to be carried out. There is, it is submitted, ordinarily no necessity[16] to imply that the terms of the sub-contract become terms of the main contract. In *Gloucestershire County Council* v. *Richardson* (see *ante*, p. 40) the main contract was in the R.I.B.A. form (1939 ed.). The House of Lords rejected an argument that the contractor was liable to the employer for latent defects in components delivered by a nominated supplier but that his liability was limited to the same extent as the nominated supplier's liability to the contractor was limited under the express terms of the sub-contract which the contractor had been directed to enter into.

---

[11] See, *e.g.* Standard form of building contract, cll. 11, 27, 28, 30; I.C.E. Conditions, cll. 58, 59.
[12] See the *Bickerton* case.
[13] *Gloucestershire County Council* v. *Richardson* [1969] 1 A.C. 480, discussed *ante*, p. 40; *Bickerton's* case, discussed *post*, p. 182.
[14] Referred to (in an earlier version) in *Gloucestershire County Council* v. *Richardson* in the Court of Appeal [1967] 3 All E.R. 458, 473, but never tested in any reported case.
[15] See cl. 12, and " Standard Method of Measurement," cl. B 20.
[16] See *ante*, p. 35.

Description of sub-contract works written into main contract

This is sometimes done. Its advantages are that it avoids argument as to the description of the sub-contract work and it enables the contractor to know the work to be carried out at the time when he tenders instead of after the contract has been agreed. But problems of construction may arise. Thus the description of the sub-contract works may include an obligation to design. This may be express, *e.g.* to design, supply and erect a certain plant, or implied, *e.g.* to provide a heating system which will satisfy a certain defined performance specification. Are these design obligations undertaken by the contractor? Ordinarily it is thought that they are, but it is necessary to consider all the terms of the contract. Thus where the Standard form of building contract is used it is not clear beyond argument that, without amendment of the conditions of contract,[17] such a description as that given above of the sub-contract works written into the bills of quantities or specification as the case may be, necessarily imposes the obligation of design upon the contractor.

4. Implied terms of main contract

Implied terms are dealt with generally *ante,* p. 35. The contractor is ordinarily as liable for performance by his sub-contractor as by himself (see *ante,* p. 174). But the fact of and the circumstances surrounding nomination and the words of the main contract may bring about a departure, to greater or lesser degree, from the contractor's usual implied obligations. The problems are often difficult and vary according to the particular main contract and the circumstances, but an approach is suggested below.

Time

Nomination does not, it is submitted, affect the contractor's liability for delay unless the terms of the main contract[18] or the surrounding circumstances show an intention to exclude such liability.[19]

Materials

PATENT DEFECTS. There is, it is submitted, an implied term of the main contract that materials or goods supplied by nominated suppliers or used by nominated sub-contractors will not contain defects that reasonable inspection by the contractor would disclose, unless the express terms of the main contract or the surrounding circumstances show an intention to exclude such term.[20] The requisite standard of inspection will vary according to the terms of the main contract and the surrounding circumstances.

LATENT DEFECTS. There is an implied term of the main contract that the materials or goods to be supplied by nominated suppliers or used by nominated sub-contractors will be of good quality, unless the express terms of the main contract or the surrounding circumstances show an intention to exclude such

---

[17] See cl. 12, Standard form of building contract, and discussion, *post,* p. 368.
[18] *e.g.* cl. 23 (*g*), Standard form of building contract, discussed, *post,* p. 349.
[19] An analogy can be made with liability for defects in materials, see *infra.*
[20] See *Young & Marten Ltd.* v. *McManus Childs Ltd.* [1969] 1 A.C. 454 (H.L.); *Gloucestershire County Council* v. *Richardson, supra.* See also " Design," *infra.*

term.[21] This term makes the contractor liable for latent defects even where the exercise of all proper skill and care on his part at the time of delivery or fixing would not have disclosed such defects.[22] An instruction to place a sub-contract upon terms which limit the contractor's right of recourse may exclude the term.[23] Other circumstances which, in the case of the delivery of a completed component, may assist the court to find an intention to exclude the term are if the design, choice of materials, specification, quality and price were fixed without reference to the contractor and there was no right to object to the nomination or to insist on an indemnity from the supplier.[24]

Workmanship

It is, it is submitted, a question of construction in each case to determine whether the nomination of a sub-contractor excludes or reduces the liability for the workmanship of the sub-contractor which would exist if the sub-contractor had not been nominated.[25]

Fitness for purpose

Where the employer has not relied on the contractor's skill and judgment in the selection of a nominated sub-contractor or nominated supplier, or the work or materials they are to carry out or supply, a term that such work or materials will be reasonably fit for their purpose is not, it is submitted, implied, unless the express terms of the main contract or the surrounding circumstances show that the parties intended the contractor to accept such liability.[26]

The practical significance of this principle is great and frequently does not seem to be appreciated. It means that under the ordinary procedures of nomination currently in use the employer has no remedy against the contractor if, for example, tiles, bricks, windows, mechanical plant, heating or air conditioning systems or other specially designed aspects of the works, the subject of nomination, are of good quality but unfit for their purpose. Even an employer using the Government form of contract (see *supra*, p. 179) has, it seems, no remedy where the contractor has supplied what he was told to supply or the sub-contractor has built what was agreed to be built.

Design

Where there is no description of the sub-contract works in the main contract, and no express obligation to design on the part of the contractor, his liability, if any, for design by a nominated sub-contractor, must arise as an implied term of the main contract.[27] It is thought that the court would not easily find that it was

---

[21] *Ibid.*

[22] *Ibid.*

[23] *Ibid.*

[24] *Ibid.*

[25] There is no direct authority and one must proceed by way of analogy with the cases dealing with materials and fitness for purpose; see, *ante*, p. 39.

[26] *Young & Marten's* case, *supra*; *Gloucestershire County Council* case, *supra*; cases cited at notes 27 and 37, *ante*, pp. 40, 41.

[27] For implication of terms, see *ante*, p. 35. For how far the employer has a remedy against his architect when the R.I.B.A. conditions of engagement (1976 ed.) apply, see cl. 1.40, discussed, *post*, p. 212.

necessary to imply a term, in such circumstances, that the contractor was, in effect, guaranteeing any design work carried out by the nominated sub-contractor. It might more readily listen to such an argument where the contractor had accepted a nomination of a sub-contractor and the sub-contract contained an express duty to design.[28] But even here the position cannot be approached as one of general principle and must depend upon the particular circumstances.

Design of materials. In so far as the contractor is liable for the quality of materials (see *supra*) he is liable for any selection of the constituent parts of the materials and other acts by the sub-contractor or supplier which, in one sense of the term, may be considered as design.[29] Such liability is of particular significance where the materials are made-up components such as concrete columns or beams.[30]

Guarantee

A nominated sub-contractor or nominated supplier may give a guarantee or warranty to the contractor. Although every case requires consideration it is thought that the courts will ordinarily refuse to imply that the guarantee or warranty is part of the main contract. The result is that it cannot be enforced by the employer and is of no value to the employer.

5. Repudiation by nominated sub-contractor

In *Bickerton* v. *N.W. Metropolitan Regional Hospital Board*[31] the contract in use was the R.I.B.A. (1963 ed.). S Ltd. was nominated to provide the heating system. Before it had carried out any substantial part of the work S Ltd. went into voluntary liquidation and the liquidator refused to complete the sub-contract. The contractor's contentions that he was entitled to a re-nomination, to damages for delay awaiting such re-nomination and was not liable to reimburse the employer for the expenditure incurred in completing the sub-contract works over and above the original sub-contract price were upheld by the House of Lords. The contract then (as now) did not deal expressly with the points. The chief factor influencing the House seems to have been that by the terms of the main contract the contractor had neither the duty nor the right to carry out any of the work the subject-matter of the P.C. sum in the Bills and which it was stated by the contract had to be carried out by a nominated sub-contractor.[32] Although the contractor was ultimately responsible for the nominated sub-contractor such responsibility could not exist, and the contract could not be completed, unless there was a nominated sub-contractor in existence. Therefore if the first fell out another must be nominated.

---

[28] See discussion of the position under the Standard form of building contract, *post*, p. 369.

[29] See *supra*, note 26.

[30] As in the *Gloucestershire* case, *supra*, where the exact proportions of the mix of the concrete were left to the decision of the supplier and the defect in the columns was due to an excess of a chemical in the mix.

[31] [1970] 1 W.L.R. 607 (H.L.).

[32] The current version of the Standard form of building contract is unchanged, see *post*, p. 363.

While *Bickerton's* case turned upon the express words of the main contract, the principle that certain work is reserved to be carried out by a nominated sub-contractor is to be found in other forms of contract and the decision may, therefore, wholly or partly apply. The I.C.E. Conditions (see *post*, p. 518) deal elaborately with the position. The Government form of contract (Form GC/Works/1) has clauses (38 (5), 31 (2)) apparently intended to put the risk of repudiation upon the contractor.

Where the Standard form of building contract, or other contract to same effect, is used it seems that the contractor's liability in respect of bad work, or defective materials does not arise until after practical completion of the works.[33]

6. Protection of employer

As will have appeared above, the system of nomination of sub-contractors may result in the employer in some cases having no remedy in contract against the contractor in respect of defaults which may cause great loss and leave him with no remedy at all unless he can find that, fortuitously, there was an informal collateral warranty [34] given by the sub-contractor, or he can persuade the court to find that he has a remedy in negligence against the sub-contractor. The best protection against this position is to avoid nomination. The reasons advanced in favour of nomination are that it gives the employer the choice of a specialist contractor, it enables him to obtain the best price for specialist works and it enables him to have the advantage of design work which his professional advisers cannot carry out. None of these arguments are valid save the first. Even here the employer can issue a list of sub-contractors any one of whom he will approve. The rest is a matter of tendering procedure. Nomination can be avoided if specialist works are described in terms of performance specification and the main contractors invited to tender are given sufficient time to make their own inquiries among specialist sub-contractors. If there is a competition among main contractors it will reflect any competition among specialist sub-contractors. As regards the total time, there should be no difference. Instead of the architect spending months in negotiation with prospective nominated sub-contractors before sending out the invitation to tender the same period will be absorbed by the main contractor. As regards detailed design, if the specialist sub-contractors are not prepared to do this for nothing in the hope of getting the work one of them can be employed on a separate contract to carry out the design.

If there must be nominations then it would appear (although it has never been tested) that the rigorous provisions of the Government form of contract would avoid most legal problems in that they appear to have the intention of putting all the risks, save lack of fitness for purpose, upon the contractor. Presumably the pricing of government tenders (by contractors who remain solvent) reflects this acceptance of risks.

A palliative is, while retaining the nomination system, to enter into formal

---

[33] See discussion, *post*, p. 371.

[34] See *ante*, p. 174. In some cases the employer may have a remedy against his architect or other professional adviser, see Chap. 12, or, in tort, against other persons such as local authorities, see *Dutton* v. *Bognor Regis U.D.C.* [1972] 1 Q.B. 373 (C.A.); *Anns* v. *Merton London Borough Council* [1977] 2 W.L.R. 1024 (H.L.).

collateral contracts [35] between the employer and the sub-contractors or suppliers. According to the wording of the collateral contracts, and the solvency of the sub-contractors or suppliers if and when the contracts have to be enforced, this procedure is of value.

Writing the description of the sub-contract works into the main contract sometimes assists (see *ante*, p. 180) but the effect depends upon the particular words of the main contract (see discussion of the position where the Standard form of building contract is used *post*, p. 368).

A last alternative is to enter into direct contracts with specialist contractors. Various committees over the past 40 years have decried this procedure without seeming to understand the defects of the modern system of nomination. If the contractual arrangements are carefully prepared, and the employer engages an administrative contractor to co-ordinate the various direct contracts, and the administration is skilfully and conscientiously carried out this procedure is probably no worse and could well be better than the present system of nomination.

Section 5: Relationship Between Sub-Contractor and Main Contractor

The relationship between a sub-contractor and the main contractor depends upon the construction of the sub-contract.[36] The Standard form of sub-contract, commonly called "The Green Form," for use where the sub-contractor is nominated under the Standard form of building contract, is available and is printed *post*, p. 576. There is also available a Standard form of sub-contract where the sub-contractor is not nominated. The ordinary rules of construction of contracts apply,[37] and much of the law set out earlier as between contractor and employer applies, with such modifications as the context makes necessary. In a case where the contractor forfeited a sub-contract for delay the court took into account the contractor's onerous position under the main contract.[38]

Incorporation of terms

Where the parties do not sign a formal sub-contract they sometimes seek to incorporate terms of contract. This is satisfactory provided the essential terms of the sub-contract are agreed and the words relied on as having the effect of incorporation are clear. An example of a form of words which caused some argument was where an order form was expressed to be " in full accordance with the appropriate form for nominated sub-contractors (R.I.B.A. 1965 ed.)." After considering expert evidence that this would be understood in the trade to be a reference to the green form the court held that these words did incorporate the

---

[35] For forms and a commentary thereon see *post*, p. 427.

[36] For the implication of a duty to supply necessary equipment in a labour only sub-contract, see *Quinn* v. *Burch Bros. (Builders) Ltd.* [1966] 2 Q.B. 370 (C.A.). For duties as regards safety, see *Sims* v. *Foster Wheeler Ltd.* [1966] 1 W.L.R. 769 (C.A.). For the tests whether an orally engaged person is a sub-contractor or an employee, see *Ferguson* v. *John Dawson & Partners (Contractors) Ltd.* [1976] 1 W.L.R. 1213 (C.A.).

[37] See *ante*, p. 27.

[38] *Stadhard* v. *Lee* (1863) 3 B. & S. 364, referred to *ante*, pp. 75 and 121.

green form [39] including the arbitration agreement which it contains and, there being a dispute, granted a stay of proceedings. [40]

Incorporation of main contract terms

The parties sometimes seek to incorporate the main contract or some of it. This is inherently likely to cause problems having regard to the difference of subject-matter and is not to be recommended. If such incorporation is attempted each contract must be construed according to its own words, but there are some reported cases which may be of some assistance. Thus it was agreed that "the terms of payment for the work . . . shall be exactly the same as those set forth in clause 30 of the [main] . . . contract." [41] Clause 30 provided for the retention of money up to an amount exceeding the sum payable under the sub-contract, and for the repayment of such fund by certain sums at the end of various periods. It was held that effect could be given to the sub-contract clause on the basis that there should be a retention fund, and repayments in the same proportion that the sub-contract sum bore to the main contract sum. In another case [42] the main contract empowered the employer to order the contractor to remove a sub-contractor guilty of delay. The sub-contract contained a recital to the effect that the sub-contractor agreed to carry out the work in accordance with the terms of the main contract. It dealt expressly with many of the matters in the main contract but not with the power to order the removal of the sub-contractor. The main contractor upon the order of the employer removed the sub-contractor for delay, and upon the latter's action for breach of contract it was held: (i) that the sub-contract was not frustrated [43] when the employer ordered the sub-contractor's removal; (ii) that the removal clause in the main contract was not imported into the sub-contract by the recital which only meant that the sub-contractor "was to provide work of the quality and with the dispatch which was stipulated for in the head contract." [44] Therefore as the sub-contractor had not been guilty of such delay as to show an intention to repudiate the contract [45] the main contractor was in breach of the sub-contract.

In *Brightside Kilpatrick* v. *Mitchell Construction (1973) Ltd.*[46] the order referred to the conditions relating to the sub-contract "being those embodied in R.I.B.A. as above." The court, after referring to the difficulty of construing the words, held that their effect was that the sub-contractual relationship should he such as to be consistent with all those terms in the head contract that specifically dealt with matters relating to sub-contractors. This in turn led to a consideration of Clause 27 of the R.I.B.A. contract and to the ruling that it was necessary to read into the contractual relationship between the contractor and the sub-

---

[39] *Modern Building Wales Ltd.* v. *Limmer & Trinidad Co. Ltd.* [1975] 1 W.L.R. 1281 (C.A.).
[40] See *post*, p. 241 for stay of proceedings under s. 4 of the Arbitration Act 1950.
[41] *Geary, Walker & Co. Ltd.* v. *W. Lawrence & Son* (1906) H.B.C. (4th ed.), Vol. 2, p. 382 (C.A.).
[42] *Chandler Bros. Ltd.* v. *Boswell* [1936] 3 All E.R. 179 (C.A.).
[43] See *ante*, p. 101, for frustration.
[44] *Per* Greer L.J., the *Chandler Bros.* case, *supra*, at p. 185.
[45] *Ibid.*
[46] [1975] 2 Lloyd's Rep. 493 (C.A.).

contractor what would have been agreed between them in any formal contract had they entered into a contract in the green form.

Set-off

There are no special rules as to set off applicable to building contracts and, in particular, to sub-contracts. Thus if, for example, a contractor is liable to pay a nominated sub-contractor money upon the direction of the architect, but satisfies the court that there is a triable issue as to whether he has a cross-claim for delay, he is entitled to leave to defend upon an application for summary judgment.[47]

Certificate condition precedent

The terms of the sub-contract may make the certificate of the architect given under the main contract a condition precedent to payment under the sub-contract.[48]

Damages

The principles upon which damages are awarded between the parties to a sub-contract are those set out above in Chapter 9. The second rule in *Hadley* v. *Baxendale*[49] has particular application as the sub-contractor frequently knows of the losses the contractor is likely to suffer as a result of his default.[50] Thus he will be liable for the amount of liquidated damages payable as a result of his delay if he has notice of such liability in the main contract.[51] He will not be liable if he does not have notice, as the liability to pay liquidated damages is not a natural consequence of delay,[52] but he will be liable for such damages as he should, as a businessman in the trade, have contemplated as likely (or not unlikely[53]) to result from his breach,[54] and such damages may in many cases be no less than the liquidated damages payable under the main contract.[55] If the contractor has settled a claim by the employer arising out of the sub-contractor's default the amount of the settlement is admissible prima facie evidence of the amount of the loss caused by the sub-contractor, but not of his liability.[56] And if the

---

[47] *Modern Engineering (Bristol) Ltd.* v. *Gilbert-Ash (Northern) Ltd.* [1974] A.C. 689 (H.L.), overruling *Dawnays Ltd.* v. *F. G. Minter Ltd. and Trollope & Colls Ltd.* [1971] 1 W.L.R. 1205 (C.A.) and other cases referred to in the *Modern Engineering* decision. For meaning of, and some notes upon, summary judgment, see *post*, p. 278. For set-off, see *post*, p. 275.

[48] See *Dunlop & Ranken Ltd.* v. *Hendall, supra*, discussed in detail, *post*, p. 376; see also *A. Davies & Co. (Shopfitters)* v. *William Old* (1969) 113 S.J. 262.

[49] See *ante*, p. 144.

[50] See, *e.g. Hydraulic Engineering Co. Ltd.* v. *McHaffie* (1878) 4 Q.B.D. 670.

[51] *Ibid.* See *e.g.* cl. 1 of " The Green Form," *post*, p. 577.

[52] *Ibid.*; *Elbinger Actien-Gesellschaft, etc.* v. *Armstrong* (1874) L.R. 9 Q.B. 473. For a clause limiting the sub-contractor's liability, see *Prince of Wales Dock Co.* v. *Fownes Forge Co.* (1904) 90 L.T. 527 (C.A.); *cf. Saint Line Ltd.* v. *Richardsons, etc., Ltd.* [1940] 2 K.B. 99.

[53] *Koufos* v. *Czarnikow Ltd.* [1969] 1 A.C. 350, 383 (H.L.).

[54] *Victoria Laundry Ltd.* v. *Newman Ltd.* [1949] 2 K.B. 528 (C.A.) and see generally, *ante*, p. 144.

[55] Liquidated damages in main contracts are frequently modest in relation to possible losses. In any event the contractor's own losses are often heavy so that the sub-contractor's total liability is usually greater than the liquidated damages in the main contract.

[56] *Biggin & Co. Ltd.* v. *Permanite Ltd.* [1951] 2 K.B. 314 (C.A.). See also *ante*, p. 151.

contractor has to pay damages to a plaintiff arising out of the sub-contractor's breach which he would not have had to pay if there had been no breach, the contractor's right to damages from the sub-contractor for that breach is not affected by the fact that the plaintiff recovered against the contractor and another defendant; and that on an apportionment between the contractor and the other defendant under the Law Reform (Married Women and Tortfeasors) Act 1935 the contractor was held to be partly to blame.[57] But so far as the contractor relies on an express indemnity in the sub-contract he may fail completely if the loss was partly caused by himself.[58]

Inaccurate statements

The general principles are set out *ante*, p. 89.

In *Holman Construction Ltd.* v. *Delta Timber Co. Ltd.*[59] a contractor asked a sub-contractor to tender for the delivery of timber. A tender was made on September 15, 1970. In reliance on the tender price the contractor made up his contract figure and entered into a main contract on September 23, 1970. The sub-contractor then discovered an error in his tender and withdrew it before acceptance. It was held that the contractor could have no remedy in negligence against the sub-contractor on what is sometimes called the *Hedley Byrne* [60] principle.

---

[57] *Sims* v. *Foster Wheeler Ltd.* [1966] 1 W.L.R. 769 (C.A.).

[58] *A.M.F. International Ltd.* v. *Magnet Bowling Ltd.* [1968] 1 W.L.R. 1028 applying *Walters* v. *Whessoe* [1968] 2 All E.R. 816n. (C.A.), and see *ante*, p. 140.

[59] [1972] N.Z.L.R. 108—New Zealand Supreme Court.

[60] *Hedley Byrne & Co. Ltd.* v. *Heller & Partners Ltd.* [1964] A.C. 465 (H.L.).

Chapter 11

INSOLVENCY AND DEATH

THE term insolvency is used here to cover bankruptcy and the winding up of insolvent companies where the same rules in general apply as under the law of bankruptcy.[1] There are certain differences which are referred to *infra*. No exposition is given of bankruptcy or winding up in general, for which reference should be made to the standard textbooks,[2] but attention is drawn here to some matters which seem particularly relevant.

Section 1: *Insolvency of Contractor*

1. Forfeiture, lien and seizure clauses

Most contracts contain clauses purporting to give the employer certain rights upon the bankruptcy or winding up of the contractor. These clauses are not effective if they are contrary to the principles of the bankruptcy law. Thus a provision that upon the contractor's bankruptcy unfixed materials or plant should be forfeited and vest in the employer is void, for an agreement that, " upon a man's becoming bankrupt, that which was his property up to the date of the bankruptcy should go over to someone else and be taken away from his creditors, is void as being a violation of the bankrupt law." [3] For the same reason it seems that a right of lien on unfixed materials, the property of the contractor, arising on his bankruptcy is void. But if the property in the materials or plant is already vested in the employer at the time of bankruptcy a right of seizure effective on the bankruptcy is good.[4] And if a right of forfeiture or lien is enforceable upon bankruptcy and other events, and is exercised because of some other event such as delay, the forfeiture or lien is good even though the contractor has become bankrupt, provided the employer at the time of enforcing his right had no notice of any available act of bankruptcy committed before the contract was entered into.[5]

A forfeiture clause operating upon the contractor's bankruptcy and not purporting to transfer any of the contractor's property is, it seems, valid and enforceable although it may prevent the contractor's trustee from completing the contract.[6]

The principles just stated are applications of a basic rule of bankruptcy which

---

[1] Companies Act 1948, s. 317.

[2] *e.g. Williams on Bankruptcy*; *Buckley on Companies*; and the relevant sections of the Companies Acts 1948 to 1967. See also the Insolvency Act 1976.

[3] *Per* James L.J., *Re Harrison, ex p. Jay* (1880) 14 Ch.D. 19, 25 (C.A.); *Re Walker, ex p. Barter* (1884) 26 Ch.D. 510 (C.A.). The clause is void as against the trustee, but was said by Cotton L.J. in *Re Harrison* at p. 26 to be good as between the parties.

[4] *Re Walker, ex p. Barter, supra.*

[5] *Re Waugh, ex p. Dickin* (1876) 4 Ch.D. 524; Bankruptcy Act 1914, s. 45. Acts of bankruptcy have no reference to companies.

[6] *Cf. Re Walker, ex p. Barter, supra*; *New Zealand Shipping Co.* v. *Ateliers de France* [1919] 1 A.C. 1, 13 (H.L.); *Woodfall on Landlord and Tenant* (27th ed.), para. 1905, for the operation of forfeiture clauses in leases which frequently provide for re-entry in the case of the lessee's bankruptcy.

is that " . . . a man is not allowed, by stipulation with a creditor, to provide for a different distribution of his effects in the event of bankruptcy from that which the law provides." [7] The court will not give effect to such a stipulation even where it is part of arrangements and the parties " . . . had good business reasons for entering into them and did not direct their minds to the question how the arrangements might be affected by the insolvency of one or more of the parties." [8]

2. The trustee in bankruptcy and liquidator—disclaimer and completion

Upon the contractor's bankruptcy the benefit and burden of the contract passes, subject to the effect of any valid forfeiture clause, to his trustee in bankruptcy unless the contract is personal. In the latter case the trustee cannot complete, but if the contractor completes, the contract moneys are payable to, and recoverable by, the contractor's trustee.[9] In a winding up the contract remains vested in the company and the liquidator may " carry on the business of the company so far as may be necessary for the beneficial winding up thereof." [10]

Disclaimer

The trustee has a right within 12 months of his appointment [11] to disclaim any onerous property or unprofitable contract.[12] Any person interested may apply in writing to the trustee requiring him to decide whether he will disclaim or not and if he does not decide within 28 days or such further time as the court may allow he will be deemed to have adopted the contract.[13] The effect of disclaimer is to bring the contract to an end from the date of disclaimer and the employer can thereupon prove in the bankruptcy for his prospective loss.[14] The liquidator has a similar right to disclaim save that the 12-month period runs from the commencement of the winding up, and the leave of the court is required.[15]

Completion

If the trustee or liquidator adopts the contract and decides to complete, he steps into the shoes of the contractor and takes the contract and the contractor's property subject to the rights of the employer unless they contravene the bankruptcy law.[16] Thus if the trustee in completing is guilty of delay or other default

---

[7] James L.J., *Ex p. Mackay* (1873) L.R. 8 Ch. 643, 647; applied in *British Eagle Ltd.* v. *Air France* [1975] 1 W.L.R. 758 (H.L.).

[8] *British Eagle* case, *supra, per* Lord Cross at p. 780.

[9] *Emden* v. *Carte* (1881) 17 Ch.D. 768 (C.A.).

[10] Companies Act 1948, ss. 245 (1) (*b*), 303. In carrying on the business the liquidator is not ordinarily personally responsible: *Stead Hazel* v. *Cooper* [1933] 1 K.B. 840; neither is a receiver if he does not enter into any new contract, but he is liable upon any new contract: Companies Act 1948, s. 369 (2); *Re Mack Trucks (Britain) Ltd.* [1967] 1 All E.R. 977.

[11] Or within 12 months of knowledge of the contract if it does not come within his knowledge within a month of his appointment.

[12] Bankruptcy Act 1914, s. 54 (1); *Re Bastable* [1901] 2 K.B. 518 (C.A.).

[13] Bankruptcy Act 1914, s. 54 (4).

[14] *Ibid.* s. 54 (8); see further *Williams on Bankruptcy*.

[15] Companies Act 1948, s. 323.

[16] *Re Garrud, ex p. Newitt* (1881) 16 Ch.D. 522, 531 (C.A.); *Bendall* v. *McWhirter* [1952] 2 Q.B. 466, 487 (C.A.).

giving rise to a right of forfeiture, the right can be validly exercised against him.[17]

The trustee when completing becomes personally liable on the contract.[18] The liquidator is not personally liable.[19]

Where there is a provision which by its nature cannot be performed by the trustee or by the company, as where a company [20] in liquidation was under an obligation to keep the works in repair for 15 years, the court may presume that the obligation cannot be carried out and allow the employer to prove his prospective damages.[21] These damages may be set off against payments and retention money due to the contractor upon completion.[22]

The trustee may, with the permission of the committee of inspection refer any dispute to arbitration.[23] The liquidator requires the sanction of the court or of the committee of inspection.[24]

3. Arbitration clauses—bankruptcy

Where the trustee has adopted a contract which contains a term referring any differences arising thereout or in connection therewith to arbitration, such term is enforceable by or against him.[25] Even when there has been no adoption, if the contractor had become a party to an arbitration agreement [26] before the commencement of his bankruptcy the court has a discretion upon the application of the trustee with the consent of the committee of inspection, or upon the application of any other party to the agreement, to make an order that the matter be determined by arbitration.[27]

4. Stay of proceedings—winding up by the court

" When a winding-up order has been made or a provisional liquidator has been appointed, no action or proceedings shall be proceeded with or commenced against the company except by leave of the court and subject to such terms as the court may impose." [28]

5. Section 105 of the Bankruptcy Act 1914

The court has a wide power under section 105 (1) to decide any such questions, whether of law or fact, as it considers necessary for the purpose of doing complete justice or making a complete distribution of the bankrupt's property.

---

[17] *Re Keen, ex p. Collins* [1902] 1 K.B. 555 (D.C.); see also *Re Waugh, supra; Re Garrud, supra.*
[18] *Titterton* v. *Cooper* (1882) 9 Q.B.D. 473 (C.A.). He has a right to recoup himself from the assets of the estate.
[19] See *supra*, note 10.
[20] *Re Asphaltic Wood Pavement Co., ex p. Lee and Chapman* (1885) 30 Ch.D. 216 (C.A.).
[21] *Ibid.*
[22] *Ibid.*; see further *post*, p. 193, " Mutual dealings."
[23] Bankruptcy Act 1914, s. 56.
[24] Companies Act 1948, s. 245.
[25] Arbitration Act 1950, s. 3 (1). For arbitration generally, see Chap. 13.
[26] For meaning, see *post*, p. 233.
[27] Arbitration Act 1950, s. 3 (2).
[28] Companies Act 1948, s. 231. See generally ss. 218–277.

This power was exercised where, after the contractor's bankruptcy, direct payment of a provisional sum was made to a sub-contractor upon the latter giving the employer an indemnity against any claim by the contractor's trustee. There was apparently no express right of direct payment. The contractor's trustee applied to the architect for a certificate for the amount of the provisional sum and it was refused. It was held that the court had jurisdiction to order the architect to make such a certificate, and an order was made leaving the sub-contractor to prove in the bankruptcy.[29]

There is no comparable express power on a winding up and it has never been held that section 105 of the Bankruptcy Act 1914 is applied to a winding up by section 317 of the Companies Act 1948, but on these facts, the court could, it is thought, declare that the payment was not in accordance with the contract and require the employer to account to the liquidator for the sum paid.

6. Reputed ownership—bankruptcy

All the property belonging to or vested in a bankrupt at the commencement of the bankruptcy or which he afterwards acquires before his discharge vests in his trustee in bankruptcy to be distributed among his creditors.[30] In addition there may also vest in the trustee goods belonging to other persons, who will then be left to prove in the bankruptcy for the value of the goods and thus, perhaps, obtain only a small fraction of their value. Section 38 of the Bankruptcy Act 1914 provides for the vesting in the trustee of " All goods being, at the commencement of the bankruptcy, in the possession, order or disposition of the bankrupt, in his trade or business, by the consent and permission of the true owner, under such circumstances that he is the reputed owner thereof." Questions on the application of this provision may well arise in building contracts, particularly where the contractor has possession of unfixed materials the property in which has passed to the employer, but cannot arise where the contractor is a company.[31]

The position was extensively discussed in *Re Fox*,[32] where it was said, " The conditions essential to the operation of the section are that the true owner of the goods should by leaving them in the possession, order or disposition of the bankrupt put him in a position by means of them to obtain false credit." [33] The true owner must be " shown to have been guilty of some remissness in this respect." [34]

---

[29] *Re Holt, ex p. Gray* (1888) 58 L.J.Q.B. 5 (D.C.).

[30] Bankruptcy Act 1914, s. 38. For definition of property, see s. 167; for commencement of bankruptcy, see s. 37 (1) (2). In general the trustee's title relates back to the act of bankruptcy on which the receiving order was made. See also s. 45. The position is different in a winding up; there is no automatic vesting of the property and the reputed ownership doctrine does not apply; the liquidator takes the companies' property into his custody or under his control unless there is an order of the court vesting the property in the liquidator. See Part V of the Companies Act 1948 generally. and in particular, ss. 243, 244. As a matter of *fact*, not law, the liquidator may require evidence that property, apparently owned by the company, belongs to someone else.

[31] *Ibid.*

[32] *Re Fox, ex p. Oundle and Thrapston R.D.C.* [1948] Ch. 407 (D.C.).

[33] *Ibid.* at p. 414, *per* Jenkins L.J.

[34] *Ibid.*

The hypothetical inquirer. The question is whether a hypothetical inquirer *must* draw the inference that the bankrupt was the owner of the goods; if he merely might have drawn the inference then the section does not apply and the goods do not pass to the trustee.[35] This is a question of fact in each case, but where materials the property in which had passed to the employer were, by the employer's consent, allowed to remain in the contractor's yard, then the court could not " but think that any man seeing building materials stored in a builder's yard would naturally suppose that they were the builder's property and could not be expected to come to any other conclusion." [36] The goods therefore passed to the trustee. But where unfixed materials, the property of the employer, were left on the building site upon which two houses were being erected the hypothetical inquirer might think that the materials were the property of the contractor, or he might equally well consider that they had passed to the employer under a contract in the R.I.B.A. form,[37] or a similar contract.[38] As it was not a necessary inference that the bankrupt was the owner of the materials the goods did not pass under the section.[39] Some of the materials on the site which were claimed by the trustee in *Re Fox* were the property of the sub-contractor, and one cf the questions for the court was whether there was any distinction between goods which had at one time been the property of the contractor (the employer's property) and those which had never been the property of the contractor (the sub-contractor's property). It was held that no assistance could be drawn from this fact, for " the section is concerned with the apparent, not the actual, situation." [40]

The hypothetical inquirer cannot draw the inference that goods are the property of the bankrupt if he has notice of the true facts. This notice may be actual or it may be imputed to him if the true owner can prove that the bankrupt was in possession under a " notorious custom well known among traders in general." [41]

In *Re Keen* [42] materials brought on to the site were " considered to be the property " of the employer [43] although the contractor was expressly forbidden from taking the materials away without the consent of the architect. The contractor became bankrupt and the employer later, under an express term, forfeited the contract and the materials for the trustee's failure to proceed. The trustee claimed the materials on the ground that they had passed to him upon the contractor's bankruptcy by reason of section 44 of the Bankruptcy Act 1883 (the equivalent of section 38). On appeal it was held that the contractor was the true owner of the materials at the time of his bankruptcy, so that the section did not

---

[35] *Ibid.* at pp. 416, 418.

[36] *Ibid.* at p. 417.

[37] A reference to cl. 11, now cl. 14 of the Standard form of building contract.

[38] *Re Fox*, at p. 419.

[39] *Ibid.* at pp. 419, 420; distinguishing *Re Weibking, ex p. Ward* [1902] 1 K.B. 713 as a decision on the particular facts.

[40] *Ibid.* at p. 418.

[41] *Per* Chitty L.J., *Re Goetz, ex p. The Trustee* [1898] 1 Q.B. 787, 798 (C.A.); *Re Fox supra.*

[42] [1902] 1 K.B. 555 (D.C.).

[43] For the wording of vesting clauses, see *ante,* p. 125.

operate, and that the employer was entitled to seize and retain the materials under the forfeiture clause.

7. Mutual dealings

If the employer is in possession of retention money or otherwise owes money to the contractor he can, subject to the provisions of the Act, set off against it any claims he proves against the contractor for damages including unliquidated or prospective damages.[44]

8. The position of sub-contractors

A sub-contractor [45] must prove in the bankruptcy or winding up for any debts due to him from the contractor. He has no lien or charge on money due to the contractor from the employer in respect of work carried out on goods supplied,[46] but if he still retains the property in goods supplied he can retain the goods as against the trustee provided the reputed ownership provision does not apply.[47] If there is a clause in the contract providing that the employer may, if the contractor delays payment to the sub-contractor, pay the sub-contractor direct such a power is not annulled or revoked by the bankruptcy or winding up of the contractor and a sub-contractor can be paid in full in priority to the trustee.[48] " Wages " payable under a labour only sub-contract are not preferential debts under the Companies Act 1948, s. 319 (4).[49]

9. Assignees

The position of assignees of money due to the contractor is largely governed by express provisions of the Bankruptcy Act 1914.[50] Subject to the effect of these provisions, the assignee of money earned by the contractor at the date of his bankruptcy may have the right to payment of such money when it accrues due in priority to the trustee in bankruptcy,[51] and even though by completing under the original contract the trustee has perfected the title of the assignee to the

---

[44] See, *e.g. Re Asphaltic Co., ex p. Lee & Chapman* (1885) 30 Ch.D. 216 (C.A.). Reference must be made to the full wording of the Bankruptcy Act 1914, s. 31 giving such a right (applied by Companies Act 1948, s. 317, to companies) and to the cases decided thereon, in particular to the important review of such cases in *National Westminster Bank Ltd.* v. *Halesowen, etc. Ltd.* [1972] A.C. 785 (H.L.).

[45] This includes a person supplying goods under a contract with the contractor, sometimes referred to as a supplier or " nominated supplier." See, *e.g.* cl. 28, Standard form of building contract.

[46] *Pritchett, etc., Co. Ltd.* v. *Currie* [1916] 2 Ch. 515 (C.A.), doubting *Bellamy* v. *Davey* [1891] 3 Ch. 540.

[47] See, *e.g. Re Fox, ex p. the Oundle and Thrapston R.D.C.* [1948] 1 Ch. 407 (D.C.). Reputed ownership has no application in a winding up, see *ante*, p. 191.

[48] *Re Wilkinson, ex p. Fowler* [1905] 2 K.B. 713; *Re Tout and Finch* [1954] 1 W.L.R. 178. It is possible that the authority of these decisions has been affected by the approach of the House of Lords in *British Eagle Ltd.* v. *Air France* [1975] 1 W.L.R. 758 (H.L.); see *ante*, p. 189. See also *ante*, p. 176, and cl. 27 (*c*) Standard form of building contract, and see *ante*, p. 170 for equitable assignment of retention moneys.

[49] *Re C. W. & A. C. Hughes Ltd.* [1966] 1 W.L.R. 1369.

[50] See in particular, ss. 7, 37, 38, 39 (as amended), 42–45 and *Williams on Bankruptcy.*

[51] *Drew* v. *Josolyne* (1887) 18 Q.B.D. 590 (C.A.); *Re Davis & Co., ex p. Rawlings* (1888) 22 Q.B.D. 193 (C.A.); *Re Tout and Finch, supra*; *Re Warren, ex p. Wheeler* [1938] Ch. 725 (D.C.).

money.[52] In such a case the right of the assignee to payment is not defeated merely because the trustee has expended his own money in completing.[53]

An assignment of money to be earned after bankruptcy is bad as against the trustee, for it is an " established principle of bankruptcy law that it is impossible to assign the profits of a business earned after the bankruptcy." [54]

Where the money assigned is not recoverable by the assignee because of the contractor's bankruptcy he can, if he has a provable debt,[55] prove his claim in the contractor's bankruptcy.

Section 2: *Insolvency of Employer*

Upon the bankruptcy or winding up of the employer the position is governed by the ordinary law of bankruptcy and winding up. See the matters discussed *supra* in relation to the contractor, of which disclaimer is probably the most important.

If the contract requires the contractor to give credit for materials supplied or for work and labour he has, it seems, upon the employer's bankruptcy, the right to refuse further performance without payment.[56]

In a liquidation the contractor cannot claim as owner of materials incorporated in the building, but only as an unsecured creditor.[57]

Section 3: *Death*

1. The general rule

Upon the death of either the contractor or the employer the benefit and burden of existing contracts other than personal contracts pass to their respective legal, personal representatives.[58] The personal representatives must honour those obligations of the deceased that could have been enforced against him even if the contract is not beneficial to the estate,[59] although in the case of an enforceable onerous contract they ought not " to neglect any opportunity that may present itself of coming to terms with the other contracting party that may benefit the estate." [60]

2. Personal contracts

The liability of a person to perform personal obligations ceases with his death,

---

[52] *Re Toward, ex p. Moss* (1884) L.R. 14 Q.B.D. 318 (C.A.); *Re Asphaltic Wood Pavement Co., ex p. Lee and Chapman, supra*; *Drew* v. *Josolyne, supra*; *Re Tout and Finch, supra*, but see note 48 *supra*.

[53] *Re Asphaltic, supra*; *Drew* v. *Josolyne, supra*.

[54] *Per* Wynn-Parry J., *Re Tout and Finch, supra*; *Re Jones, ex p. Nichols* (1883) 22 Ch.D. 782 (C.A.); *Wilmot* v. *Alton* [1897] 1 Q.B. 17 (C.A.). The principle applies to the winding up of an insolvent company—*Re Tout and Finch, supra*.

[55] See the Bankruptcy Act 1914, s. 30, and *Williams on Bankruptcy*.

[56] *Re Edwards, ex p. Chalmers* (1873) L.R. 8 Ch. 289; *Re Sneezum, ex p. Davis* (1876) 3 Ch.D. 463, 473.

[57] *Re Yorkshire Joinery Co.* (1967) 111 S.J. 701. See also *ante*, p. 124.

[58] *Marshall* v. *Broadhurst* (1831) 1 Cr. & J. 403; *Cooper* v. *Jarman* (1866) L.R. 3 Eq. 98; *Ahmed Anguillia, etc.* v. *Estate & Trust Agencies (1927) Ltd.* [1938] A.C. 624 (P.C.).

[59] *Ahmed Anguillia, supra*, at pp. 634, 639.

[60] *Ibid.* at p. 635.

the contract having been rendered impossible.[61] Thus in a contract of master and servant the death of either party brings the contract to an end.[62] In a contract of personal confidence such as that with an architect the death of the architect brings the contract to an end,[63] although the death of the employer does not normally have this effect.[64]

The employment of the contractor may be personal in that his obligations cannot be performed by his personal representatives.[65] In such a case the position will be similar to that on the death of the architect.

---

[61] *Stubbs* v. *Holywell Ry.* (1867) L.R. 2 Ex. 311.

[62] *Farrow* v. *Wilson* (1869) L.R. 4 C.P. 744.

[63] *Stubbs* v. *Holywell, supra.*

[64] *Davison* v. *Reeves* (1892) 8 T.L.R. 391 (the employment of a civil engineer by a contractor).

[65] See *ante*, p. 167.

CHAPTER 12

ARCHITECTS, ENGINEERS AND QUANTITY SURVEYORS

Section 1: *Meaning and Use of the Term " Architect "*

1. Definition

" An ' architect ' is one who possesses, with due regard to aesthetic as well as practical considerations, adequate skill and knowledge to enable him (i) to originate, (ii) to design and plan, (iii) to arrange for and supervise the erection of such buildings or other works calling for skill and design in planning as he might in the course of his business, reasonably be asked to carry out or in respect of which he offers his services as a specialist." [1] The use of the term architect is now, in general, restricted to architects registered under statutory provisions.

2. Registration [2]

Use of title " Architect "

A person is prohibited from practising or carrying on business under any name, style or title containing the word architect [3] unless he is a person registered in the Register of Architects. [4] " ' Practising ' . . . means: Holding out for reward to act in a professional capacity in activities which form at least a material part of his business. A man is not practising who operates incidentally, occasionally, in an administrative capacity only, or in pursuit of a hobby." [5] A person if not treated as not practising by reason only that he is in the employment of another person. [6] The prohibition does not affect the use of the designation " Naval Architect," " Landscape Architect," or " Golf-Course Architect," or the validity of any building contract in customary form. [7] Further where a person on July 2, 1938, held office in the service of a local authority by virtue of which he had the control and management of the architectural work of the local authority and was a member of the Institution of Civil Engineers, or the Institution of Structural Engineers or the Chartered Surveyors Institution or the Institution of Municipal and County Engineers, there is no prohibition against the use of the word " architect " in the description of that person as the holder of such office in the service of that or any other local authority, if and so long

---

[1] This is the final sentence of the test applied by the tribunal set up under the Architects Registration Act 1938, s. 2, in deciding whether a person was an architect practising on July 29, 1938, and was cited without disapproval in *R. v. Architects' Registration Tribunal, ex p. Jaggar* [1945] 2 All E.R. 131, 134.

[2] For an account of registration and of the professional conduct of an architect, see Rimmer's *The Law relating to the Architect* (2nd ed. 1964), Chap. 7.

[3] See *Jacobowitz v. Wicks* [1956] Crim.L.R. 697 (D.C.); use of letters " Dip.Ing.Arch." not an offence.

[4] Architects Registration Act 1938, s. 1 (1). This Act together with the Architects (Registration) Act 1931 and the Architects Registration (Amendment) Act 1969, may be cited together as the Architects Registration Acts 1931 to 1969 (s. 4 of the 1969 Act).

[5] This is the other sentence of the test referred to in note 1, *supra*.

[6] Architects Registration Act 1938, s. 4 (2).

[7] *Ibid.* s. 1 (1), proviso.

as the local authority's servant or servants engaged under him for the purposes of such work is or include a person registered in the Register of Architects.[8]

A body corporate, firm or partnership may carry on business under the style or title of Architect if the business of the body corporate, firm or partnership so far as it relates to architecture is under the control and management of a superintendent who is a registered person and who does not act at the same time in a similar capacity for any other body corporate, firm or partnership; and if in every premises where such business is carried on and is not personally conducted by the superintendent such business is bona fide conducted under the direction of the superintendent by an assistant who is a registered person.[9]

Contravention of these provisions is a criminal offence, the person guilty being liable to be fined.[10] It would follow that, presumably, such a person could not sue for fees in respect of work carried out while he was contravening the Acts.[11]

The Architects' Registration Council

This is a statutory corporation one of whose duties is the maintenance of the Register of Architects.[12] It must also appoint annually a Board of Architectural Education, an Admission Committee,[13] and a Discipline Committee.[14]

The right to be registered

A person who makes an application to the Registration Council in the prescribed manner and pays the prescribed fee [15] is entitled to be registered if the Council is satisfied on a report of the Admission Committee that he is an architect member of the Royal Academy or of the Royal Scottish Academy, or that he has passed any examination in architecture which is for the time being recognised by the Council or that he possesses the prescribed qualifications.[16] Persons who were practising as architects in the United Kingdom on January 1, 1932, were entitled to be registered if they made application within two years from that date,[17] and persons who could prove to the satisfaction of the Council or, on appeal, to a tribunal of appeal, by an application made before August 1, 1940, that on July 29, 1938, they were practising as an architect in the United Kingdom or some other part of His Majesty's Dominions were also entitled to be registered.[18]

---

[8] *Ibid.* s. 1 (1) (i) (ii). " Local authority " means a local authority within the meaning of the Local Government Superannuation Act 1937: see the Architects Registration Act 1938, s. 1 (2).

[9] Architects (Registration) Act 1931, s. 17.

[10] Architects Registration Act 1938, s. 3; for certain special defences, see the proviso to s. 3. See also Rimmer, *op. cit.* p. 191.

[11] See Chap. 7 (*ante*, p. 107), " Illegality."

[12] Architects (Registration) Act 1931, s. 3 (3).

[13] *Ibid.* s. 5 (1).

[14] *Ibid.* s. 7 (2).

[15] Prescribed means prescribed by regulations made under the Act. The regulations can be obtained from the Council; Architects (Registration) Act 1931, s. 15.

[16] 1931 Act, s. 6 (1); the qualifying examinations and prescribed qualifications are set out in regulations which can be obtained from the Council, 1931 Act, s. 15.

[17] 1931 Act, s. 6 (1) (*b*).

[18] 1938 Act, s. 2. For test applied by tribunal, see *R.* v. *Architects' Registration Tribunal, ex p. Jaggar, supra.*

Removal of names from register

If a registered person is convicted of a criminal offence or if the Discipline Committee, after an inquiry, report to the Council that the person has been guilty of conduct disgraceful [19] to him in his capacity as an architect, the Council may cause his name to be removed from the register and he is then, during such period as the Council determine, disqualified for registration under the Acts. [20] Written notice served by registered post of a proposed inquiry must be served on the person concerned and he is entitled upon application to be heard by the Discipline Committee in person or by counsel or a solicitor. [21] Where the Council intends to remove the name of any person from the register for a criminal offence or professional misconduct a written notice must first be served on him by registered letter, and, on an application being made by that person in the prescribed manner within three months from the date of the service of the notice, the Council must consider any representations with regard to the matter which may be made by him to the Council either in person or by counsel or by a solicitor. [22] The Council may at any time, either of their own motion or on the application of the person concerned, cause his name to be restored to the register. [23]

For the purpose of maintaining the register the Council may at any time by notice in writing served on any registered person inquire if such person has changed his regular business address, and if no answer is received within six months from the sending of such notice, the Council must send to the said person a further notice by post as a registered letter, and if no answer is received within three months from the sending of such further notice, the Council may remove the name of such person from the register. [24]

Where the Council has removed the name of any person from the register, other than in consequence of his death, they must forthwith serve written notice by registered post on that person and where they have determined that there shall be a period of disqualification the determination must be specified in the notice. [25]

Any person aggrieved by the removal of his name from the register, or by a determination of the Council that he be disqualified for registration during any period, may, within three months from the date on which notice of the removal or determination was served on him, appeal to the High Court against the removal or determination, and on any such appeal the court may give such directions in the matter as they think proper and the order of the court is final. [26]

---

[19] For a discussion of the meaning of " disgraceful," see *Hughes* v. *Architects' Registration Council* [1957] 2 Q.B. 550 (D.C.). See also *Marten* v. *Royal College of Veterinary Surgeons' Disciplinary Committee* [1966] 1 Q.B. 1 (D.C.); *Lawther* v. *Royal College of Veterinary Surgeons* [1968] 1 W.L.R. 1441 (P.C.). For the Architects' Registration Council Code of Professional Conduct, apply to the Council.

[20] 1931 Act, s. 7 (1).

[21] *Ibid.* s. 7 (4). Service may be by recorded delivery : Recorded Delivery Service Act 1962, s. 1 and sched. 1.

[22] *Ibid.* s. 7 (5). See also note 21.

[23] *Ibid.* s. 7 (1).

[24] *Ibid.* s. 11. See also note 21.

[25] *Ibid.* s. 8. See also note 21.

[26] *Ibid.* s. 9; the appeal is to a divisional court. For procedure see R.S.C., Ords. 55, 94, r. 6. See also *Hughes* v. *Architects' Registration Council, supra.*

3. Trade Descriptions Act 1968

A false statement by a person that he was an architect was held to be an offence under this Act.[27]

Section 2: *Engineers and Others*

1. Meaning of term " engineer "

No exact definition can be given of the term " engineer." It is sometimes used to describe a person who undertakes to carry out engineering and constructional works, but such a person is a contractor in the sense in which that term is used in this book. The term " engineer " is here used to describe one who, in relation to engineering or constructional works, carries out duties analogous to those carried out by an architect in a building contract. Such engineers frequently specialise and are usually members of some professional body, but the law does not require registration or qualification before a person can practise as an engineer, or term himself " engineer."

Resident engineer

In large contracts, or overseas contracts, there is sometimes appointed, in addition to a chief engineer, a resident engineer [28] who, as his name suggests, stays continuously on the site while the works are being carried out.

2. Other persons in the position of an architect

There are many contracts in which, although the term " architect " is never used much of the law relating to architects, save registration,[29] applies. This arises where, in a contract for work and labour, there is a person placed in the position of exercising the duties of an architect. The most common example is the engineer in an engineering contract [30]; but he may be termed a surveyor,[31] a clerk of works,[32] a supervising officer,[33] or whatever else the parties choose to call him.

Section 3: *The Position of the Architect*

1. The contract with the employer

The contract may come into existence as the result of an individual agreement or a competition. The normal rules of formation of contract apply.[34] If there is a competition the rules of entry and its announcement must be studied to see whether the organisers make a definite promise to employ the successful or best

---

[27] *R.* v. *Breeze* [1973] 1 W.L.R. 994 (C.A., Crim. Div.).

[28] For the use of the term in the I.C.E. Conditions (1955 ed. and 1973 ed.), see clause 1 (1) (*d*).

[29] See *ante*, p. 196.

[30] *e.g.* the I.C.E. Conditions, *supra*.

[31] *e.g. Panamena Europea Navigacion (Compania Limitada)* v. *Frederick Leyland & Co. Ltd.* (*J. Russell & Co.*) [1947] A.C. 428 (H.L.).

[32] *e.g. Jones* v. *St. John's College* (1870) L.R. 6 Q.B. 115. Note that this is quite contrary to the usual meaning given to the term. See *post*, p. 214.

[33] *e.g. A. E. Farr Ltd.* v. *The Admiralty* [1953] 1 W.L.R. 965, a contract using Form CCC/Wks/1 General Conditions of Government Contracts.

[34] See Chap. 2.

competitor.[35] Contracts [36] need not be in writing and it is still the practice for much work to be carried out without any formal agreement, leaving the law to supply the necessary terms by implication.[37] Although there are theoretical advantages of certainty in a written contract, problems of construction can always arise; in particular the most commonly used written contract is probably the R.I.B.A. Conditions of Engagement (1976 ed., printed *post*, p. 560) and these raise certain difficult questions (discussed briefly, *post*, pp. 212–214) in relation to the extent of the architect's duties as to design and supervision.

Limited companies performing architect's duties

The professional rules relating to architects at present prohibit them from carrying on business as architects as limited companies. But providing there is no contravention of the Architects Registration Acts (see *ante*, p. 196) there is nothing to prevent limited companies performing architectural services, and many do, with or without building the works as well. Design and build contracts are discussed *ante*, p. 5. Any claim against such companies for breach of contract is governed by the ordinary rules. Often they employ architects as employees. Does such an architect owe a duty of care to the building owner for breach of which he is liable in tort? The Court of Appeal of British Columbia has said that he does not,[38] but it cannot be assumed that the English courts would take the same view.[39]

2. The architect's duty to act fairly

The architect is engaged by the employer to act as his agent for the purpose of securing the completion of the works in an economical and efficient manner. He must perform these duties properly and if he fails to do so may be liable to the employer in damages. But in performing them he must act fairly and professionally in applying the terms of the building contract. Until recently it was thought that this gave rise to a " dual capacity " as agent and as quasi-arbitrator. It has now been settled by the House of Lords that an architect acting under the ordinary building contract is the employer's agent throughout notwithstanding that in the administration of the contract he has to act in a fair and professional manner.[40]

3. The architect's authority as agent

An architect's authority is strictly limited by the terms of his employment.[41]

---

[35] *Cf. Rooke* v. *Dawson* [1895] 1 Ch. 480; *Carlill* v. *Carbolic Smokeball Co.* [1893] 1 Q.B. 256 (C.A.). The R.I.B.A. has special rules for architectural competitions.

[36] Except with corporations and made before July 19, 1960, see *ante*, p. 25, and see *Lawford* v. *Billericay R.D.C.* [1903] 1 K.B. 772 (C.A.).

[37] For general principles as to implication of terms, see *ante*, p. 35, and for the approach of the courts to the duties of architects see this section of this chapter.

[38] *Sealand of the Pacific* v. *McHaffie* (1975) 51 D.L.R. (3d) 702.

[39] See authorities cited to section " Negligent misstatement," *ante*, p. 98.

[40] *Sutcliffe* v. *Thackrah* [1974] A.C. 727 (H.L.), discussed *ante*, p. 35 and *post*, p. 216. See also *Arenson* v. *Arenson* [1975] 3 W.L.R. 815 (H.L.).

[41] *Fredk. Betts Ltd.* v. *Pickfords Ltd.* [1906] 2 Ch. 87, 95; and see cases cited, *infra*. For a statement of the law applicable in considering whether there is an agency relationship, see *Garnac Grain Co. Inc.* v. *H. M. F. Faure & Fairclough* [1968] A.C. 1130, 1137 (H.L.).

His authority as agent is a question in each case, but it is possible to consider how far in respect of certain matters which have come before the courts he has an implied authority from his position as architect.

Tenders

An architect engaged " to originate, . . . design and . . . arrange for the erection of buildings " [42] has, it is submitted, implied authority to invite tenders, and he may, in special circumstances, have implied authority to employ a quantity surveyor to prepare bills of quantities.[43] The invitation to tender may contain a specification, plans, drawings and bills of quantities. There is no implied warranty that the statements in these documents are accurate,[44] and the architect has no implied authority to warrant their accuracy.[45] If he does so and the contractor is put to extra cost as a result the employer is not, in the absence of fraud, liable for breach of warranty unless he knew of or ratified the architect's statement,[46] although the contractor may have a remedy against the architect.[47] Further, if the architect was acting within his ostensible authority in making the statements the contractor may have a remedy against the employer for innocent misrepresentation or negligent misstatement.[48]

Contracts

In the absence of some express power acceptance should be by the employer. It seems reasonably clear that an architect engaged for the purposes referred to in the last paragraph has no implied power to bind the employer by acceptance of a tender,[49] and he cannot without authority pledge the employer's credit.[50]

Supervision

It is part of the normal duties of the architect to supervise the work and there is implied such authority as is necessary to ensure that the work is carried out according to the terms of the contract.[51]

Variations

The architect has no implied authority to vary the works,[52] or to order extras,[53]

---

[42] See definition cited *ante*, p. 196.

[43] See *post-* p. 228. For the position where the R.I.B.A. conditions of engagement apply see clause 1.23, *post*, p. 562.

[44] See *ante*, p. 101.

[45] *Scrivener v. Pask* (1865) L.R. 1 C.P. 715.

[46] *Ibid.*

[47] See *infra.*

[48] See *ante*, p. 89.

[49] See *Vigers, Sons & Co. Ltd.* v. *Swindell* [1939] 3 All E.R. 590; and *cf.* estate agent cases such as *Hamer* v. *Sharp* (1874) L.R. 19 Eq. 108; *Keen* v. *Mear* (1920) 89 L.J.Ch. 513. A local authority architect may have authority as an official: *Roberts & Co. Ltd.* v. *Leicestershire C.C.* [1961] Ch. 55; *Carlton Contractors* v. *Bexley Corpn.* (1962) 106 S.J. 391.

[50] *Vigers* v. *Swindell, supra.*

[51] *R.* v. *Peto* (1826) 1 Y. & J. 37; *Kimberley* v. *Dick* (1871) L.R. 13 Eq. 1; *Brodie* v. *Cardiff Corp.* [1919] A.C. 337, 351 (H.L.).

[52] *Cooper* v. *Langdon* (1841) 9 M. & W. 60.

[53] *Ibid.*

or to order as extras works impliedly included in the work for which the contract sum is payable.[54]

An express power to order variations must be exercised within the scope of the contract.[55] Neither can the architect, without the employer's knowledge or consent, bind the employer by a promise that a condition of the contract will be waived.[56]

Arrangements with adjoining owners

An architect employed merely to superintend the construction of works has no implied power to agree with adjoining owners to vary the works in such a way as to affect the rights of his employer.[57]

Employed architect

Where the architect is an employee of the employer he has that authority which the employer holds him out as having by virtue of his position and the particular facts, so that in some cases his authority may be greater than that of an architect engaged as an independent person.[58]

4. Excess of authority by architect

Position of the employer

If the architect exceeds the authority of his employment the employer is not liable for his acts unless he is estopped from denying the authority of his architect[59] or he subsequently ratifies the architect's acts.

An example of agency by estoppel would arise if E has appointed A to be his architect with express authority to order goods on his behalf, and A has given several orders to S, a supplier. If E then withdraws his authority from A and fails to notify S, he is liable to S on any contracts which A purports to make on his behalf.[60] This principle only applies if the person dealing with the alleged agent had no notice of the lack of authority, and the burden of proof is on such person to prove the agency, or that the alleged principal is estopped from disputing it.[61]

The employer can subsequently ratify an unauthorised contract of the architect provided it was professedly made on his behalf, even though he was not named,[62] and the contract was within the powers of the employer to make,[63]

---

[54] See *ante*, p. 68.

[55] See *ante*, p. 68. For an emergency, see *ante*, p. 67 and, where the R.I.B.A. Conditions of Engagement apply, cl. 1.31, *post*, p. 562.

[56] *Sharpe* v. *San Paulo Ry.* (1873) 8 Ch.App. 597.

[57] *Fredk. Betts Ltd.* v. *Pickfords Ltd.* [1906] 2 Ch. 87.

[58] *Roberts & Co. Ltd.* v. *Leicestershire C.C., supra*; *Carlton Contractors* v. *Bexley Corporation, supra*.

[59] See *Pole* v. *Leask* (1863) 33 L.J.Ch. 155. See also *Ismail* v. *Polish Ocean Lines* [1976] 2 W.L.R. 477, 484 (C.A.).

[60] *Summers* v. *Solomon* (1857) 26 L.J.Q.B. 301; *Scarf* v. *Jardine* (1882) 7 App.Cas. 345, 349 (H.L.); *Watteau* v. *Fenwick* [1893] 1 Q.B. 346.

[61] *Pole* v. *Leask, supra*, at p. 162.

[62] *Keighley, Maxsted & Co.* v. *Durant* [1901] A.C. 240 (H.L.); see also *Warehousing & Forwarding Co., etc., Ltd.* v. *Jafferali & Sons Ltd.* [1964] A.C. 1 (P.C.).

[63] *Ashbury Ry.* v. *Riche* (1875) L.R. 7 H.L. 653 (H.L.).

and the employer was in existence at the time when the contract was made.[64] The last question may arise where a promoter of a company purports to make contracts on behalf of a company which has not yet been formed.[65]

If an agent contracts subject to ratification by his principal there is no concluded contract until ratification is obtained, and accordingly it seems that if the person who has entered into such " contract " with the agent withdraws from it before ratification it cannot be ratified so as to bring a binding contract into existence.[66]

The employer is only bound by the ratification of a contract if at the time of ratification he had full knowledge of all the material facts, or there was an intention to ratify the contract whatever it may have been.[67]

Position of the architect

The architect may be liable in an action for breach of warranty of authority at the suit of the contractor or other person who has suffered loss as a result of the architect exceeding his authority. The principle upon which the action is based is " that where a person by asserting that he has the authority of the principal induces another person to enter into any transaction which he would not have entered into but for that assertion, and the assertion turns out to be untrue, to the injury of the person to whom it is made, it must be taken that the person making it undertook that it was true and he is liable personally for the damage that has occurred." [68] The liability arises " (a) if he has been fraudulent, (b) if he has without fraud untruly represented that he had authority when he had not, and (c) also where he innocently represents that he has authority where the fact is either, (1) that he never had authority or (2) that his original authority has ceased by reason of facts of which he has not knowledge or means of knowledge." [69] Thus the architect might be liable where, without his knowledge, his authority has ceased by reason of the employer's death.[70] He cannot be liable if the other party was aware of his lack of authority,[71] so that where there is a written contract, such as the Standard form of building contract, it is difficult for a contractor to succeed in an action for breach of warranty of authority on some matter arising out of the contract, for he has notice of the extent of the architect's authority; and the same principle would apply to a sub-contractor under a sub-

---

[64] *Kelner* v. *Baxter* (1866) L.R. 2 C.P. 174.

[65] For personal liability of promoters see European Communities Act 1972, s. 9 (2) and *Halsbury's Laws* (4th ed.), Vol. 7, para. 51.

[66] *Warehousing & Forwarding Co., etc., Ltd.* v. *Jafferali & Sons Ltd., supra,* at p. 9. The point is difficult, see *Bolton Partners* v. *Lambert* (1889) 41 Ch.D. 295 (C.A.) apparently to the opposite effect, but distinguished and doubted in subsequent cases referred to in *Warehousing, etc.,* v. *Jafferali, supra.* See further *Chitty on Contracts* (24th ed.), Vol. 2, para. 2023.

[67] *Bowstead on Agency,* Chap. 2, Sect. 4, and see *Vigers* v. *Swindell* [1939] 3 All E.R. 590, the facts of which are set out at p. 177, *ante.*

[68] Per Lord Esher M.R., *Firbank's Executors* v. *Humphreys* (1886) 18 Q.B.D. 54, 60 (C.A.). See also *Randell* v. *Trimen* (1856) 18 C.B. 786; *Collen* v. *Wright* (1857) 8 E. & B. 647.

[69] Per Buckley L.J., *Yonge* v. *Toynbee* [1910] 1 K.B. 215, 227 (C.A.).

[70] *Yonge* v. *Toynbee, supra.*

[71] *Halbot* v. *Lens* [1901] 1 Ch. 344.

contract whereby he is expressed to have notice of the terms of the main contract.[72]

The starting point in considering the measure of damages in an action for breach of warranty of authority is to compare the plaintiff's position before the representation of authority with his position brought about by the representation.[73] Damages normally include the loss on any contract which cannot be enforced for the want of authority, and may include the costs of unsuccessful proceedings to enforce the contract against the alleged principal.[74] If there is doubt about an agent's authority it may be advisable, before suing the principal, to give notice to the agent that such proceedings will be at his cost if they fail for lack of authority.[75]

5. Architect's personal liability on contracts

The general rule is that an agent is not personally liable on contracts entered into on behalf of his principal, but in certain circumstances the architect may find himself liable on contracts made on behalf of his employer. This liability must be distinguished from that discussed in the last paragraph which arises not on the original contract but on an implied warranty that the architect had authority to make it.

The architect is personally liable where he contracts on behalf of the employer but without disclosing the existence of a principal.[76] In such a case the contract may be enforced either against the architect or, upon discovering his existence, against the employer.[77] And the architect may make himself personally liable if he signs a contract in his own name without excluding his liability.[78] Thus an architect was held personally liable at the suit of a sub-contractor where he had signed an order in this manner although earlier correspondence had referred to the existence of his clients and the judge considered that the main contractor was the person who ultimately ought to find the money.[79]

6. Fraudulent misrepresentation

An architect who makes a fraudulent misrepresentation,[80] intending it to be acted upon, is liable in damages for the tort of deceit at the suit of the person who acted upon the statement and suffered loss, and it is no excuse that he merely acted as the agent of his employer, for " all persons directly concerned in the commission of a fraud are to be treated as principals . . . The contract of agency

---

[72] *Cf.* the standard form of sub-contract, cl. 1.

[73] *Doyle* v. *Olby (Ironmongers) Ltd.* [1969] 2 Q.B. 158, 167 (C.A.).

[74] *Randell* v. *Trimen* (1856) 18 C.B. 786; *Simons* v. *Patchett* (1857) 7 E. & B. 568; *Collen* v. *Wright, supra.*

[75] As in *Collen* v. *Wright, supra.*

[76] See *Bowstead on Agency.*

[77] *Ibid.*

[78] See Bowstead, Art. 118; *Beigtheil & Young* v. *Stewart* (1900) 16 T.L.R. 177; *cf. Himley Brick Co. Ltd.* v. *Lamb & Sons Ltd., and Seifert* (1962) *The Guardian,* January 16— no contract where suppliers at architect's request " reserved " bricks.

[79] *Beigtheil & Young* v. *Stewart, supra, per* Bigham J.

[80] See *ante,* p. 92, for the elements of fraud.

or of service cannot impose any obligation on the agent or servant to commit or assist in the committing of a fraud." [81]

The employer is liable for his own fraudulent misrepresentations, and for any fraud committed by the architect in the execution of his authority,[82] even though it is done for the benefit of the architect,[83] and even though the employer is innocent.[84] In *Pearson* v. *Dublin Corporation* it was held that a clause providing that the contractors must not rely on any representation would not protect the employer from an action for the fraudulent misrepresentation of his engineers, but a passage from Lord Loreburn's speech suggested that it might be possible by apt and express clauses in special circumstances for an employer to guard himself from liability for the fraud of his agents.[85]

Fraud involves dishonesty, and the employer cannot be liable for his architect's misrepresentation if neither of them had a guilty mind; " you cannot add an innocent state of mind to an innocent state of mind and get as a result a dishonest state of mind." [86] So if the architect were innocently to make statements about the contract works or the site and the employer knew of facts which rendered those statements untrue but had neither authorised the architect to make the statements nor hoped, nor suspected that he would make them,[87] nor knew he had made them, then there could be no fraud.[88] There would be fraud so as to make the employer liable if he had deliberately employed the architect hoping that he would, innocently, make the false statements.[89]

A contract may be rescinded for fraud on the part of an undisclosed principal.[90]

7. Misconduct as agent

Bribes and secret commissions

Criminal liability can arise both in respect of the agents of public bodies and of private principals.[91]

Quite apart from criminal liability, any secret dealing between the architect and the contractor, or other person, is a fraud on the employer, and entitles him to dismiss the architect,[92] and recover any money paid by way of bribe or secret commission.[93] Further, the employer may rescind the contract between himself

---

[81] *Per* Lord Westbury, *Cullen* v. *Thomson's Trustees & Kerr* (1862) 4 Macq. 424, 432.

[82] *Lloyd* v. *Grace, Smith & Co.* [1912] A.C. 716 (H.L.); *Uxbridge Permanent Benefit Building Society* v. *Pickard* [1939] 2 K.B. 248 (C.A.); see also *Briess* v. *Woolley* [1954] A.C. 333 (H.L.).

[83] *Ibid.*

[84] *Pearson* v. *Dublin Corp.* [1907] A.C. 351 (H.L.); as explained in *Armstrong* v. *Strain* [1952] 1 K.B. 232, 258 (C.A.). See *ante,* p. 92, for the facts of *Pearson's* case.

[85] For the passage *in extenso,* see *ante,* p. 93. See also *Tullis* v. *Jacson* [1892] 3 Ch. 441, discussed *ante,* p. 85.

[86] *Per* Devlin J., *Armstrong* v. *Strain* [1951] 1 T.L.R. 856, 872, referred to with approval in the Court of Appeal [1952] 1 K.B. 232, 246.

[87] *Cf. Ludgater* v. *Love* (1881) 44 L.T. 694 (C.A.).

[88] *Armstrong* v. *Strain* [1952] 1 K.B. 232 (C.A.).

[89] See *supra,* note 87.

[90] *Garnac Grain Co. Inc.* v. *H. M. F. Faure & Fairclough Ltd.* [1965] 1 All E.R. 47, overruled on different points [1966] 1 Q.B. 650 (C.A.); [1968] A.C. 1130 (H.L.).

[91] For the relevant law see *Archbold* (39th ed.), paras. 3991–4002.

[92] *Temperley* v. *Blackrod Manufacturing Co. Ltd.* (1907) 71 J.P. 341.

[93] *Grant* v. *Gold Exploration, etc., Syndicate* [1900] 1 Q.B. 233.

and the contractor or other person concerned,[94] and if he has suffered loss he can recover damages from the architect and the contractor or other person.[95] Any secret dealing which might tend to prevent an agent performing his duties faithfully to his principal is presumed by the law to be corrupt.[96]

The court will not listen to evidence of a custom to pay secret commissions,[97] but an agent accused of accepting a secret commission may be able to show that his principal must have been aware that he had received a commission.[98]

It was, apparently at one time the practice in some cases for the architect to charge the contractor with the cost of taking out the quantities or measuring up deviations. Unless the architect accounts fully to the employer for any sums received, or makes it clear to the employer that he will receive such sums, giving their amount,[99] this procedure would seem to be a fraud on the employer, coming within the principles stated above.[1]

Conflict of duty and interest

An agent must not put his personal interest in conflict with his duty towards his principal.[2] Thus where a surveyor who had been engaged to supervise certain work to be carried out by a building company [3] subsequently became managing director of that company, it was said that this was a breach of duty going to the root of his contract with the employer making him liable to dismissal without, it seems, any right to payment for the work he had done, his contract being entire.[4]

The employer may waive his rights in respect of his agent's breach of duty. Thus if he engages an agent with full knowledge of circumstances which give rise to a conflict between the agent's duty and his interest, or continues to employ him without protest after discovery of such circumstances, he waives his right to rely on the breach of duty.[5]

8. The architect's duties to the employer

The architect must serve the employer faithfully as his agent.[6] Further, by holding himself out as an architect, he impliedly warrants that he possesses the requisite ability and skill.[7] Consequently if he fails to exercise such skill he is liable in damages if any loss is suffered, and may not be able to recover his fees.

---

[94] *Panama & South Pacific Telegraph Co.* v. *India Rubber, etc., Co.* (1875) L.R. 10 Ch. App. 515.

[95] *Salford Corp.* v. *Lever* [1891] 1 Q.B. 168 (C.A.).

[96] *Harrington* v. *Victoria Graving Dock Co.* (1878) 3 Q.B.D. 549; *Shipway* v. *Broadwood* [1899] 1 Q.B. 369 (C.A.).

[97] *Bulfield* v. *Fournier* (1894) 11 T.L.R. 62; affd. 11 T.L.R. 282 (C.A.).

[98] *Holden* v. *Webber* (1860) 29 Beav. 117.

[99] See *Dunne* v. *English* (1874) L.R. 18 Eq. 524.

[1] See *supra*.

[2] *Bank of Upper Canada* v. *Bradshaw* (1867) 1 L.R.P.C. 479, 489 (P.C.); *Hall* v. *Wembley Appliances Ltd.* [1947] 2 All E.R. 630 (C.A.).

[3] And thus for present purposes was in the same position as an architect.

[4] *Hall* v. *Wembley Appliances, supra,* at p. 634. For entire contracts, see *ante*, p. 50.

[5] *Hall's case, supra.*

[6] See *supra* and *Bowstead on Agency* generally.

[7] *Harmer* v. *Cornelius* (1858) 5 C.B.(N.S.) 236, 246; see also *Jones* v. *Manchester Corp.* [1952] 2 Q.B. 852, 876 (C.A.).

The degree of skill required

The architect's liability is not, in the ordinary case,[8] absolute in the sense that he is liable whenever loss results from his acts; it must be shown that he has been negligent, *i.e.* that he has failed to exercise the standard of care required, and that the negligence complained of is a matter of substance.[9] A professional man must bring to his task " a fair, reasonable and competent degree of skill," [10] and the usual way of testing this is to ask " whether other persons exercising the same profession or calling, and being men of experience and skill therein, would or would not have come to the same conclusion as the defendant." [11] If the majority of such persons would, under the circumstances, have done the same thing as the architect this normally provides a good defence,[12] for " a defendant charged with negligence can clear himself if he shows that he acted in accord with general and approved practice," [13] or, where there is no one accepted practice, if he acted in accordance with a practice accepted as proper by a responsible body of architects.[14] But if the possibility of danger emerging is reasonably apparent, then to take no precautions is negligence even though others in like circumstances have in the past not taken proper precautions.[15] Thus if the nature of the contract clearly imposes a duty on the architect, such as inspecting the work of one contractor before it is covered up by another contractor's work, then it is not sufficient that he has done what was customary if he fails to inspect.[16]

Where a building, of ordinary description, of which an architect has had abundant experience, proves a failure that is evidence of want of skill or attention.[17] But if an architect, out of the ordinary course, is employed upon a novel process, of which he does not profess experience, then its failure may be con-

---

[8] See *Greaves & Co. Ltd.* v. *Baynham Meikle* [1975] 1 W.L.R. 1905 (C.A.) discussed *infra*.

[9] *Cotton* v. *Wallis* [1955] 1 W.L.R. 1168 (C.A.); a difficult case discussed further *post*, p. 296.

[10] *Lanphier* v. *Phipos* (1831) 8 C. & P. 475, 479.

[11] *Chapman* v. *Walton* (1833) 10 Bing. 57, 63. For a discussion of whether, where a man professes a specialist skill, his client can expect a special degree of care in the exercise of that skill, see *Argyll* (*Duchess of*) v. *Beuselinck* [1972] 2 Lloyd's Rep. 172, 183; *cf. Greaves & Co. Ltd.* v. *Baynham Meikle*, discussed *infra*.

[12] *Ibid.*; see also *Slater* v. *Baker* (1767) 2 Wils. 359.

[13] *Per* Maugham L.J., *Marshall* v. *Lindsey County Council* [1935] 1 K.B. 516, 540 (C.A.), cited by Lord Porter, *Whiteford* v. *Hunter* [1950] W.N. 553 (H.L.).

[14] See *Bolam* v. *Friern Hospital Committee* [1957] 1 W.L.R. 582—a medical case; *Cotton* v. *Wallis*, *supra*; *Hasker* v. *Hall* (1958) unrep. but noted at length in the *R.I.B.A. Journal* for October 1958; *Chin Keow* v. *Govt. of Malaysia* [1967] 1 W.L.R. 813 (P.C.)—a medical case. For the meaning of " faulty design " in an insurance policy, see *Manufacturers Mutual Insurance Ltd.* v. *Queensland Government Railways* [1969] 1 Lloyd's Rep. 214 (Australia High Court—fault in operation, not in preparation).

[15] *Savory & Co.* v. *Lloyds Bank Ltd.* [1932] 2 K.B. 122, 136 (C.A.); *Markland* v. *Manchester Corp.* [1934] 1 K.B. 566; affd. [1936] A.C. 360 (H.L.). *Cf. Marfani & Co. Ltd.* v. *Midland Bank Ltd.* [1968] 1 W.L.R. 956, 972, 975, 980 (C.A.). See also *post*, p. 213.

[16] *Jameson* v. *Simon* (1899) 1 F. (Ct. of Sess.) 1211, 1222. This is a Scottish case with the work carried out by various individual contractors, and therefore requiring a higher degree of supervision than in the usual English contract where there is only one contractor. The defendant was held liable although he had followed the custom of local architects.

[17] *Per* Erle J., *Turner* v. *Garland & Christopher* (1853) H.B.C. (4th ed.), Vol. 2, p. 2. But in most cases expert evidence is necessary to prove negligence; *Worboys* v. *Acme Investment Ltd.*, *The Times*, March 26, 1969 (C.A.).

sistent with proper skill on his part.[18] He may profess experience by not refusing the job and not warning the client of his inexperience.[19]

Date of alleged negligence. What may be negligent after warning of a risk in a publication likely to be read in the profession, may not have been negligent in the light of ordinary professional knowledge at the time of the act or omission complained of.[20]

Warranty of fitness

A designer can, by contract, enter into a duty beyond that of using skill and care in his design. He may do this either expressly or, in particular circumstances it may be implied that he has warranted that he will achieve a certain result. Thus in *Greaves & Co.* v. *Baynham Meikle* [21] " package deal " contractors undertook to the employer to produce a warehouse the first floor of which would be suitable for the use of stacker trucks. They engaged the defendants, who were structural engineers, as experts to carry out the design. In all the circumstances it was held that the defendants were liable in damages for designing a floor which failed because they had impliedly warranted that the floor would be reasonably fit for the purpose for which they knew it was required. The Court of Appeal emphasised that it was not laying down any general principle as to the obligations and liabilities of professional men.

Delegation of duties

On the principles stated in Chapter 10, the contract with the architect is personal in the sense that he cannot arrange for the general performance of his duties by another person.[22] In *Moresk Cleaners Ltd.* v. *Hicks*,[23] an architect was engaged to draw up plans, specification and contracts for building an extension to the plaintiff's laundry. The work involved the design of a reinforced concrete structure on a sloping site. The architect knew of a contractor whose partners were qualified engineers, and he invited the contractor to design and build the structure. After erection it became defective because of negligent design in that, *inter alia*, purlins were not strong enough and portal frames should have been, but were not, tied together. It was contended for the architect that he was entitled to delegate certain specialist design tasks to qualified specialist sub-contractors, and, alternatively, that he was acting as agent for the employer to employ the contractor to design the structure. These arguments were rejected and it was

[18] *Ibid.* at p. 3.
[19] *Moresk Cleaners Ltd.* v. *Hicks* [1966] 2 Lloyd's Rep. 338, discussed *infra*.
[20] *Roe* v. *Minister of Health* [1954] 2 Q.B. 66 (C.A.).
[21] [1975] 1 W.L.R. 1095 (C.A.). For package deal or " design and build " contracts, see *ante*, p. 5.
[22] See *ante*, p. 167; *Hemming* v. *Hale* (1859) 7 C.B.(N.S.) 487, 498; *Stubbs* v. *Holywell Ry.* (1867) L.R. 2 Ex. 311. For analogous problems in the law of master and servant, see *Davie* v. *New Merton Board Mills* [1959] A.C. 604 (H.L.); shipowners, see *Riverstone Meat Co. Pty. Ltd.* v. *Lancashire Shipping Co. Ltd.* [1961] A.C. 807 (H.L.); estate agents, see *McCann (John) & Co.* v. *Pow* [1974] 1 W.L.R. 1643 (C.A.).
[23] [1966] 2 Lloyd's Rep. 338, Sir Walker Carter, Q.C., senior official referee. See also *Sealand of the Pacific* v. *McHaffie* (1975) 51 D.L.R. (3d) 702, British Columbia C.A.—architect negligent who relied on recommendation of suitability of materials by supplier.

said [24] ". . . if a building owner entrusts the task of designing a building to an architect he is entitled to look to that architect to see that the building is properly designed. The architect has no power whatever to delegate his duty to anybody else, certainly not to a contractor. . . . If the defendant was not able, because this form of reinforced concrete was a comparatively new form of construction, to design it himself, he had three courses open to him. One was to say: ' This is not my field.' The second was to go to the client, the building owner, and say: ' This reinforced concrete is out of my line. I would like you to employ a structural engineer to deal with this aspect of the matter.' Or he can, while retaining responsibility for the design, himself seek the advice and assistance of a structural engineer, paying for his service out of his own pocket but having at any rate the satisfaction of knowing that if he acts upon that advice and it turns out to be wrong, the person whom he employed to give the advice will owe the same duty to him as he, the architect, owes to the building owner."

The terms of employment of the architect must be considered in each case. Thus where it is known that the contract between the employer and the contractor will contain provision for measurement by a skilled quantity surveyor an architect is not, it is submitted, in the ordinary case in breach of duty to the employer if he relies on such measurements,[25] provided, it seems, that the measurements are not grossly wrong.[26] It must be a question of construction of the terms of employment of the architect in each case to determine whether the employer has expressly or impliedly agreed that the architect is not to be responsible for duties ordinarily entrusted to him.[27] One matter which the court may take into account is whether the employer has a remedy against the nominated sub-contractor or other person relied on by the architect as performing such duties,[28] for it is thought that the court would be unwilling, unless compelled to do so, to find that the employer had no remedy against anyone for bad design.

Duties in detail

The detailed duties of an architect must depend upon the application of the general principles stated above to the particular facts of the case, including special terms, if any.[29] So far as these duties are based upon what a competent experienced architect would do in the circumstances, they are a proper subject for a member of that profession rather than for a lawyer. To give some guidance there will be set out the list of duties suggested by Hudson, himself a practising architect as well as a lawyer, followed by notes on some aspects of an architect's

[24] At p. 342. See also *Hedley Byrne & Co. Ltd.* v. *Heller & Partners Ltd.* [1964] A.C. 465, 486 (H.L.).
[25] See *Clemence* v. *Clarke* (1880) H.B.C. (4th ed.), Vol. 2, pp. 54, 58 (C.A.); *R. B. Burden Ltd.* v. *Swansea Corp.* [1957] 1 W.L.R. 1167 (H.L.); both are cases on the architect's duty as certifier (see *ante*, p. 76).
[26] *Priestley* v. *Stone* (1888) H.B.C. (4th ed.), Vol. 2, pp. 134, 142 (C.A.).
[27] Consider the R.I.B.A. Conditions of Engagement, printed *post*, p. 560 and discussed *post*, p. 212. For difficulties which arise where design is carried out by nominated sub-contractors, see *ante*, p. 181.
[28] See *ante*, p. 183, " Protection of employer."
[29] *e.g.* R.I.B.A. Conditions of Engagement printed in Appendix (*post*, p. 560).

duty. It must be understood, therefore, that what follows is not an exhaustive account of the architect's duties.

Hudson's list of duties [30]

" (i) To advise and consult with the employer (not as a lawyer) as to any limitation which may exist as to the use of the land to be built on, either (*inter alia*) by restrictive covenants or by the rights of adjoining owners or the public over the land, or by statutes [31] and by-laws affecting the works to be executed.[32]

(ii) To examine the site, sub-soil and surroundings.[33]

(iii) To consult with and advise the employer as to the proposed work.

(iv) To prepare sketch plans and a specification having regard to all the conditions which exist and to submit them to the employer for approval,[34] with an estimate of the probable cost, if requested.[35]

(v) To elaborate and, if necessary, modify or amend the sketch plans as he may be instructed and prepare working drawings and a specification or specifications.

(vi) To consult with and advise the employer as to obtaining tenders, whether by invitation or by advertisement, and as to the necessity or otherwise of employing a quantity surveyor.[36] (Engineers do not so often employ a quantity surveyor.)

(vii) To supply the builder with copies of the contract drawings and specification, supply such further drawings and give such instructions as may be necessary, supervise the work, and see that the contractor performs the contract, and advise the employer if he commits any serious breach thereof.[37]

(viii) To perform his duties to his employer as defined by any contract with his employer or by the contract with the builder, and generally to act as the employer's agent in all matters connected with the work and the contract, except where otherwise prescribed by the contract with the builder, as, for instance, in cases where he has under the contract to act as arbitrator or quasi-arbitrator." [38]

NOTES

Law and practice

An architect's duties are comparable, in some aspects, to those of ecclesiastical surveyors, of whom it has been said that, they " could not be expected to supply minute and accurate knowledge of the law; but we think under the circumstances they might properly be required to know the general rules applicable to the valuation of ecclesiastical property." [39] The law of which an architect should know the general rules includes, it is submitted, all statutes and by-laws affecting

---

[30] At p. 9 of the 7th ed. The footnotes have been inserted. The last edition edited by Hudson was published in 1926.

[31] *e.g.* town and country planning legislation.

[32] For knowledge of law and practice generally, see *infra*.

[33] See *post*, p. 211.

[34] For the position where the employer does not approve, see *post*, p. 220.

[35] See *post*, pp. 211 and 220.

[36] See *post*, p. 228.

[37] For supervision, see *post*, p. 213.

[38] The term " quasi-arbitrator " is now no longer used, see *ante*, pp. 85 and 86.

[39] *Per* Jervis C.J., *Jenkins* v. *Betham* (1855) 15 C.B. 168, 189.

the building, the main principles of town and country planning law, and private rights likely to affect the works. If his working drawings, plans or directions result in a building which contravenes the by-laws or building regulations which apply to it,[40] this is, it is submitted, some but not conclusive,[41] evidence of breach of duty. It would appear to be part of his duty to keep himself informed of recent relevant changes in the law, including important decisions.[42]

Examination of site

It is part of the general duties of the architect to examine the site to ascertain whether it is suitable for the work to be carried out.[43] He must not rely on what he is told by a third person,[44] nor by a former agent of the employer.[45] It would appear also to be part of his duty to observe whether there are obvious rights of way, light [46] and other private rights which might be affected by the proposed works.

Estimates

If the architect is asked for an estimate of the cost of the proposed work he should give an honest and careful estimate.[47] He should not enter into a firm undertaking that the estimate will not be exceeded, unless he discloses this to the contractor before the building contract is entered into; failure to disclose disqualifies him from carrying out his independent duties under the contract.[48]

Plans, drawings, specifications and quantities

By analogy with the duty of a valuer it seems to be no excuse where the plans and drawings are defective that he has shown them to the employer and told him to examine them for himself.[49] If the architect's plans involve a trespass he may be personally liable to an adjoining owner for a trespass so caused.[50] The quantities are usually taken out by the quantity surveyor, but if the architect takes them out himself he is liable for loss caused by errors due to his negligence.[51] If the architect

---

[40] See *Townsends (Builders) Ltd.* v. *Cinema News, etc., Ltd.* [1959] 1 W.L.R. 119 (C.A.), discussed *ante*, p. 109 and *post*, p. 227.

[41] Even counsel may be wrong in giving legal advice but ". . . not negligent, just mistaken, as any lawyer and judge might be "—*per* Lawton J., *Cook* v. *S.* [1966] 1 W.L.R. 635, 641.

[42] See *Lee* v. *Walker* (1872) L.R. 7 C.P. 121 (patent agent). Generally for the duty to keep up to date, see *Roe* v. *Minister of Health* [1954] 2 Q.B. 66 (C.A.).

[43] *Moneypenny* v. *Hartland* (1826) 2 C. & P. 378; *Columbus Co.* v. *Clowes* [1903] 1 K.B. 244.

[44] *Columbus Co.* v. *Clowes, supra.*

[45] *Moneypenny* v. *Hartland, supra.*

[46] See, *e.g. Armitage* v. *Palmer* [1960] C.L.Y. 326 (C.A.).

[47] *Moneypenny* v. *Hartland, supra;* for the position when his estimate is substantially less than the lowest tender, see *post*, p. 220.

[48] *Kimberley* v. *Dick* (1871) L.R. 13 Eq. 1; *Kemp* v. *Rose* (1858) 1 Giff. 258; see *ante*, p. 86.

[49] *Smith* v. *Barton* (1866) 15 L.T. 294.

[50] *Monks* v. *Dillon* (1882) 12 L.R.Ir. 321.

[51] *M'Connell* v. *Kilgallen* (1878) 2 L.R.Ir. 119, 121. See also *Young* v. *Blake* (1887) H.B.C. (4th ed.), Vol. 2, p. 110; *Lansdowne* v. *Somerville* (1862) 3 F. & F. 236. For quantity surveyors, see *post*, p. 228.

were to accept from the quantity surveyor quantities which were grossly wrong he might be liable to the employer for negligence if loss resulted.[52]

Design

The usual standard of duty is that of reasonable care, but in special circumstances it may be implied that the architect has warranted that his design will be suitable for the purpose required by his client.[53] The architect cannot delegate his duty without his client's consent.[54] His duties do not end when work starts. He " is under a continuing duty to check that his design will work in practice and to correct any errors which may emerge." [55]

R.I.B.A. Conditions of Engagement. By clause 1.22 it is provided that " the architect will advise on the need for independent consultants and will be responsible for the direction and integration of their work but not for the detailed design, inspection and performance of the work entrusted to them." By clause 1.40 it is provided that " the architect may recommend that specialist sub-contractors and suppliers should design and execute any part of the work. He will be responsible for the direction and integration of their design, and for general inspection of their work in accordance with stage H of the normal services, but not for the detailed design or performance of the work entrusted to them." Clause 1.40 is important. The employer may or may not have a remedy in tort against a sub-contractor who carries out design.[56] The employer ordinarily has no claim in respect of such design against the main contractor.[57] Therefore, if no further steps were taken he would be without remedy in many cases in respect of loss caused by the detailed design carried out by specialist sub-contractors or suppliers. There are available forms of agreement between employer and sub-contractor and a warranty by a supplier which give the employer a remedy in contract against the sub-contractor or supplier in question.[58] It is thought that an architect who relied upon clause 1.40 of the Conditions of Engagement to escape liability in respect of design might be found to be in breach of duty to his client if he had failed to recommend the use of the form of agreement between sub-contractor and employer or the warranty as the case may be and had taken no other suitable step to protect his client's interests.

Inquiries as to materials

An architect may be in breach of duty to his client if he fails to make proper inquiries as to the suitability of materials and merely relies upon the recommendation of a supplier.[59]

---

[52] *Per* Lord Esher M.R., *Priestley* v. *Stone* (1888) H.B.C. (4th ed.), Vol. 2, pp. 134, 142 (C.A.); 4 T.L.R. 730.
[53] *Greaves & Co. Ltd.* v. *Baynham Meikle* [1975] 1 W.L.R. 1095 (C.A.), discussed *ante*, p. 208.
[54] *Moresk Cleaners Ltd.* v. *Hicks* [1966] 2 Lloyd's Rep. 338, discussed *ante*, p. 208.
[55] *Brickfield Properties Ltd.* v. *Newton* [1971] 1 W.L.R. 862, 873 (C.A.).
[56] See *ante*, p. 175.
[57] See discussion of position under the Standard form of building contract, *post*, p. 369.
[58] Printed and discussed, *post*, pp. 427 and 433.
[59] *Sealand of the Pacific* v. *Robert C. McHaffie* (1974) 51 D.L.R. (3d) 702, British Columbia C.A. Consider also *Clay* v. *Crump* (*A. J.*) *& Sons Ltd.* [1964] 1 Q.B. 533, 559 (C.A.).

Advising on the contract

It has been suggested that it can no longer be consistent with professional competence or duty to recommend the Standard form of building contract without drawing attention to its worst features, which are then listed.[60] It is thought that this is too wide. An architect might well give evidence saying that he was familiar with, and had considered, the various criticisms of the form but that he considered it was in the overall interests of his client to use it. He might express a view, or call evidence to the effect, that the prices in the building industry were based upon the general use of the form and his client might expect to get materially higher prices, or in boom times no tenders at all from busy contractors, if they were asked to quote upon materially different conditions of contract. Further, the court might accept the suggestion that even experienced draughtsmen when attempting to avoid one difficulty sometimes inadvertently introduce another. Nevertheless an architect should, it is submitted, always consider the form of contract to be used even if it is the Standard form. He should, it is submitted, have regard to the likelihood of a particular event occurring and the losses which might result when carrying out the project in question. Thus, for example, if interference with the works causing a suspension or delay is reasonably foreseeable as liable to occur then it may well be that he should draw this to his client's attention and suggest that appropriate amendments to the form should be considered; if nominated sub-contractors and nominated suppliers are to be engaged, he should, it is thought, ordinarily recommend the use of the form of agreement between the sub-contractor and the employer and, in some cases, of the warranty by a nominated supplier [61]; and insurance almost always requires particular consideration. Failure to take such steps may furnish some prima facie evidence of negligence.

Recommending contractors

It may, by analogy with estate agents letting houses,[62] be the duty of the architect, in recommending contractors, to make inquiries as to their solvency and capabilities.

Supervision

The architect must give such reasonable supervision to the works as enables him to give an honest certificate that the work has been properly carried out.[63] He is not required personally to measure or check every detail, but should check substantial and important matters, such as, for example, the bottoming of a cement floor, especially if failure to do so will result in the work being covered up and therefore not being capable of inspection at a later stage.[64] It may be that the standard of care required in supervision before giving his final certificate

---

[60] H.B.C. (10th ed.), p. 146. Some of the features have since been modified.
[61] See *supra*, note 58.
[62] See *Heys* v. *Tindall* (1861) 1 B. & S. 296.
[63] *Jameson* v. *Simon* (1899) 1 F. (Ct. of Sess.) 1211; *Rogers* v. *James* (1891) 8 T.L.R. 67; H.B.C. (4th ed.), Vol. 2, p. 172 (C.A.); *Cotton* v. *Wallis* [1955] 1 W.L.R. 1168 (C.A.).
[64] *Jameson* v. *Simon, supra*; see also *Florida Hotels Pty.* v. *Mayo* (1965) 113 C.L.R. 588 (Aust.); but note R.I.B.A. Conditions of Engagement, cll. 1.60 to 1.63, *post*, p. 563.

is higher where that certificate is binding on the employer [65] than where it can be appealed against to an arbitrator of the court.

It seems that, if the building contract requires the contractor to complete the works to the architect's reasonable satisfaction, the architect is, subject to clear language in the contract to the contrary, entitled to have regard to the amount of the price and is not in breach of duty to the employer for permitting a lower standard of work where the price is low than one would expect where the price is high.[66]

R.I.B.A. Conditions of Engagement. By clause 1.33 it is provided that the architect " shall make such periodic visits to the site as he considers necessary to inspect generally the progress and quality of the work and to determine in general if the work is proceeding in accordance with the contract documents." By clause 1.60 " during his on-site inspections . . . the architect shall endeavour to guard the client against defects and deficiences in the work of the contractor, but shall not be required to make exhaustive or continuous inspections to check the quality or quantity of the work." Clause 1.61 provides for the appointment of a clerk of works " where frequent or constant inspection is required " and by clause 1.62 for the appointment of a resident architect, each of whom is to be paid for by the client. It is thought that it is part of an architect's duty to advise the client when a clerk of works and a resident architect should be appointed. Clause 1.34 provides that the architect shall not " be responsible for any failure by the contractor to carry out and complete the works in accordance with the terms of the building contract between the client and the contractor." It is thought that the intention is to make it clear that the architect is not warranting compliance by the contractor with the building contract but that, read in the light of the other clauses cited, it is not intended to prevent the architect from being under a duty of care in supervising the contractor.

Clerk of works

The duties of supervision where there is a clerk of works require special consideration. His powers and duties as between himself and the contractor depend upon the terms of the building contract. In *Leicester Board of Guardians* v. *Trollope* the building contract provided that he was to " be considered and act solely as the inspector and assistant of the architect." [67] It was said that the clerk of works was appointed to protect the interests of the employer against the contractor mainly because the architect could not be there all the time,[68] and that when a clerk of works is employed the architect is responsible to see that his design is carried out, but is entitled to leave the supervision of matters of detail

---

[65] *Cf.* notes to cl. 30 (7) of the Standard form of building contract, *post*, p. 393.

[66] *Cotton* v. *Wallis*, *supra*, discussed *post*, p. 296. The Court of Appeal by a majority upheld the finding of the county court judge that an architect who passed some trifling defects but nothing " rank bad " was not guilty of negligence.

[67] The relevant clause in the building contract in *Leicester Board of Guardians* v. *Trollope* (1911) 75 J.P. 197; H.B.C. (4th ed.), Vol. 2, p. 419; the contract was described by Channell J. as of the usual character. *Cf.* cl. 10 of the Standard form of building contract and cll. 1.60, 1.61 of the R.I.B.A. Conditions of Engagement.

[68] *Per* Channell J., *Leicester Board of Guardians* v. *Trollope*, *supra*, H.B.C. at p. 423.

to the clerk of works.[69] In that case an architect was held liable in damages where dry rot resulted from a failure to carry out the design of the lower floor. A series of concrete blocks had to be made, and the architect admitted that he had not checked whether they were made properly or not. It was said by the learned judge, " if the architect had taken steps to see that the first block was all right and he had then told the clerk of works that the work in the others was to be carried out in the same way, I would have been inclined to hold that the architect had done his duty." [70] The case is of interest because it was alleged that the deviation from the design was due to the fraud of the clerk of works, and it was held that even if this were so the architect could still be liable.[71] If an architect acquiesces in the appointment and carrying out of his duties by an incompetent clerk of works, he cannot afterwards set up such incompetence as a defence.[72]

Clerk of works as agent. The terms of the building contract usually give the clerk of works very limited powers as agent and do not appoint him as agent for the architect to give instructions on his behalf.[73] Despite this, architects sometimes encourage the clerk of works to act as their agents, in particular in the giving of " site instructions." This procedure can lead to confusion and to claims for damages for delay by the contractor unless the parties carefully agree upon the status to be accorded to the clerk of works, his exact degree of authority and the procedure to be followed, such agreement to take effect by way of variation of the contract.

Clerk of works' manual.[74] The R.I.B.A. and the Institute of Clerks of Works have collaborated to produce this manual. It is specifically directed to the position and duties of the clerk of works when the Standard form of building contract is used, but it is thought that it is of general importance as stating the modern view of the clerk of works.

Liability of clerk of works to others. By analogy with the duties owed by architects the clerk of works is under a duty of care to the employer by virtue of his engagement and, it is submitted, in certain circumstances, in tort, to the contractor and, perhaps, to others, such as sub-contractors, who may be affected by his lack of care.[75]

---

[69] *Ibid.* See also *Lee* v. *Bateman (Lord), The Times,* October 31, 1893.

[70] *Leicester Board of Guardians* v. *Trollope, supra,* at p. 200 of the J.P. Report.

[71] *Ibid.* Note that one of the grounds of this part of the decision was that the fraud, if any, was perpetrated by the clerk of works in his own interest and not in the interest of his employer. Since the decision in *Lloyd* v. *Grace, Smith & Co.* [1912] A.C. 716 (H.L.), this part of the judgment cannot be supported on this ground, but it may be that it is part of the architect's duty to take reasonable care to guard against fraud by the clerk of works.

[72] *Saunders & Collard* v. *Broadstairs Local Board* (1890) H.B.C. (4th ed.), Vol. 2, p. 164 (D.C.).

[73] See his position under the Standard form of building contract, cl. 10, and for further discussion, *post,* p. 310.

[74] R.I.B.A. Publications Ltd. 1975.

[75] See *post,* p. 227, for position of architects.

Certificates

In *Sutcliffe* v. *Thackrah* [76] the official referee held than an architect had, in the circumstances, been negligent in issuing an interim certificate for an excessive sum of money. The employer paid the amount due on the certificate. The contractor went into liquidation and the employer lost the money. Reluctantly, the Court of Appeal, taking the view that it was bound by an earlier decision [77] held that the architect could not be liable for damages because he was performing his special function as a certifier. The House of Lords unanimously allowed the appeal [78] and thereby settled that an architect owes a duty of care towards his client in the performance of all his duties, including those of certifying, and is not entitled to any special exemption. They said that there was nothing incompatible between being fair and being careful. At the same time the House emphasised that proving negligence against an architect for over-certification may often be difficult. An architect, like other professional men, can be wrong without being negligent.

The architect's liability extends both to interim and final certificates and indeed to all his functions under the contract. It is accepted law that judges and arbitrators are entitled to immunity from actions for negligence in respect of the performance of their duties [79] but this does not apply to mere certifiers. It seems that there may be circumstances in which what is in effect an arbitration is not one that is within the provisions of the Arbitration Act, but for this to arise there must have been a submission either of a specific dispute or of present points of difference or of defined differences that may in the future arise and an agreement that the decision will be binding.[80] These circumstances do not apply to the role of the architect under the Standard form of building contract or ordinary forms of building contract. It is possible that they arise in respect of the engineer giving his decision under the I.C.E. Conditions of Contract.[81]

9. Breach of architect's duties to employer

Breach of the architect's duties to the employer may consist in misconduct in his position as agent or in failure to exercise the requisite skill of an architect.[82] Either breach if so serious as to go to the root of his contract with the employer [83] renders him liable to be dismissed,[84] and he may not be able to recover his fees if his work is useless or results in a loss to the employer.[85] He will be

---

[76] [1974] A.C. 727 (H.L.).

[77] *Chambers* v. *Goldthorpe* [1901] 1 K.B. 624, 638 (C.A.). For other cases which dealt with the point, see the report of *Sutcliffe* v. *Thackrah* and pp. 225–227 of the 3rd ed. of this book.

[78] Overruling *Chambers* v. *Goldthorpe* (*supra*) and distinguishing cases which appeared to follow it.

[79] *Sutcliffe* v. *Thackrah*, *supra*, pp. 726, 744, 753, 754, 757, and see *post*, p. 234.

[80] *Ibid.* p. 752.

[81] See the wording of cl. 66 of both the 1955 and 1973 editions and consider *Monmouth C.C.* v. *Costelloe & Kemple* (1965) L.G.R. 429 (C.A.).

[82] See *ante*, p. 206.

[83] See *ante*, p. 111.

[84] See *Harmer* v. *Cornelius* (1858) 5 C.B.(N.S.) 236; *Duncan* v. *Blundell* (1820) 3 Stark. 6; *Hall* v. *Wembley Appliances Ltd.* [1947] 2 All E.R. 630 (C.A.).

[85] *Moneypenny* v. *Hartland* (1826) 2 C. & P. 378; and see *post*, p. 220.

liable in an action for damages for loss suffered, the measure of damages being calculated upon the principles set out in Chapter 9, so that in some cases damages may greatly exceed the claim for fees.[86] The employer may either bring a separate action, or counterclaim to the architect's action for fees.[87] As with any breach of contract the employer must act reasonably and keep his loss resulting from the architect's breach of duty to the minimum. Thus, for example, if the plans are found to be wrong because of the architect's failure to measure the site properly the employer must normally give him the opportunity of making the plans good, and if he does not will only be awarded nominal damages.[88]

Unauthorised sub-contracting. Breach by an architect of his duty cf personal performance [89] does not of itself amount to a total failure of consideration so as to entitle the employer to the return of fees paid for work carried out; and damages, other than nominal,[90] are only recoverable to the extent that actual loss is proved.[91] If the architect is a person of fame such loss may, it is submitted, be capable of proof. Further, if the unauthorised sub-contracting is discovered, the employer can, it is submitted, require the architect to resume personal performance and failure to do so may well, it is thought, amount to a repudiation by the architect of his contract of employment with the employer rendering him liable to dismissal.

Negligent survey. Where an architect or surveyor is engaged to survey and report upon the condition of a property by an intending purchaser who acts on the report and purchases, the measure of damages arising from a negligent survey is prima facie the difference in value between the property as it was reported to be and its value in the condition it was in fact.[92] The difference may be less than the cost of amending the defects negligently omitted from the report.[93]

Liability to employer in tort. The architect may be liable in respect of statements made or acts done before entering into a contract with the employer and he may be liable after the contract in respect of matters, such as negligence causing a personal injury, which fall outside his contract of engagement.[94] The contract, when entered into, may, either expressly or by implication, deal with liabilities which have, or may have, previously arisen.[95] Can the architect be liable in tort to his client in respect of breaches of duty arising after the contract

---

[86] See *Saunders & Collard* v. *Broadstairs Local Board* (1890) H.B.C. (4th ed.), Vol. 2, p. 164. In *Florida Hotels Pty.* v. *Mayo* (1965) 113 C.L.R. 588 (Aust.) the architect was held liable for damages paid by the employer to a workman for injuries caused by the architect's negligence in supervision of the works.

[87] See, *e.g. Rogers* v. *James* (1891) H.B.C. (4th ed.), Vol. 2, p. 172 (C.A.).

[88] *Columbus Co.* v. *Clowes* [1903] 1 K.B. 244.

[89] See *ante*, p. 168.

[90] See *ante*, p. 144.

[91] *Hamlyn Construction Co. Ltd.* v. *Air Couriers* [1968] 1 Lloyd's Rep. 395, His Honour Norman Richards, Q.C., official referee.

[92] *Philips* v. *Ward* [1956] 1 W.L.R. 471 (C.A.).

[93] *Ibid.*

[94] See textbooks on tort and for pre-contract statements see *ante*, p. 98.

[95] See *Esso Petroleum Co. Ltd.* v. *Mardon* [1976] 1 Q.B. 801, 833 (C.A.).

of engagement has been entered into and which are the subject of that contract? In *Bagot* v. *Stevens Scanlan & Co. Ltd.*[96] where an architect was under contract with the employer to supervise the carrying out of building works it was held that he owed a duty to the employer only in contract and not in tort. It has since been said by Lord Denning that *Bagot* v. *Stevens Scanlan & Co. Ltd.* is in conflict with other decisions which show that " in the case of a professional man the duty to use reasonable care arises not only in contract, but is also imposed by the law apart from contract, and is therefore actionable in tort." [97] If this is correct it can affect the position as to limitation of actions (see *infra*) as to the measure of damages, because on occasion it is greater in tort than in contract,[98] and it may give rise to difficult questions as to the effect of express terms in the contract of engagement. And must a plaintiff elect which remedy he is pursuing or may he have any advantages given by the contract while avoiding the effect of, say, a shorter limitation period? [99]

Limitation of actions—contract. Actions against the architect in contract must be commenced within six years of the date of the cause of action accruing, or 12 years if his engagement is under seal.[1] Time runs, it is submitted, from the date of breach of duty and not from its discovery,[2] unless the employer can rely on the provisions of the Limitation Act 1939 relating to fraud, fraudulent concealment or mistake.[3]

When does the cause of action arise? This depends on the terms of the contract. Where an architect was engaged to prepare plans and perform the duties of supervision required by the R.I.B.A. form of building contract it was said that the correct approach was that " a claim against an architect for negligence in the design of a building raises a cause of action different from that of negligence in supervising its erection in purported compliance with that design." [4] Upon this approach where the design was carried out outside the limitation period a claim for defective design is prima facie barred notwithstanding that supervision by the architect was carried out within the period. But in such a case a claim may arise under a separate cause of action which arose after the design was originally carried out in respect of a breach of the architect's " continuing duty to check that his design will work in practice and to correct any errors which may emerge." [5] A somewhat different view has been expressed by the High Court of Southern Australia which suggested that where there is the usual engagement of an architect to design and supervise there is an entire contract "to see the business through " so that a claim for damages for defective design work carried

---

[96] [1966] 1 Q.B. 197. See also *Harbutt's " Plasticine " Ltd.* v. *Wayne Tank & Pump Co. Ltd.* [1970] 1 Q.B. 447, 464 (C.A.), *per* Lord Denning.

[97] *Esso Petroleum Co. Ltd.* v. *Mardon, supra,* p. 820.

[98] *Esso Petroleum Co. Ltd.* v. *Mardon, supra,* and see cases cited, *ante,* p. 98.

[99] For some assistance, see *Chesworth* v. *Farrar* [1967] 1 Q.B. 407, and *Salmond* (16th ed.), pp. 9–11. See also Prof. Fridman (1977) 93 L.Q.R. 422.

[1] Limitation Act 1939, s. 2. See also *ante,* p. 165.

[2] See *Cartledge* v. *Jopling (E.) & Sons Ltd.* [1963] A.C. 758, 782 (H.L.); *Bagot* v. *Stevens Scanlan & Co.* [1966] 1 Q.B. 197.

[3] s. 26. See also *ante,* p. 166.

[4] *Brickfield Properties Ltd.* v. *Newton* [1971] 1 W.L.R. 862, 869 (C.A.).

[5] See *ibid.* p. 873.

out more than six years (or 12 as the case may be) before issue of the writ would not be barred if performance of any of the architect's duties, such as those relating to supervision, was carried out within the period of six years.[6]

It is thought that when the R.I.B.A. Conditions of Engagement (*post*, p. 560) are the basis of the contract causes of action can accrue from time to time as various duties are, or should have been, performed.

Limitation of actions—tort. Subject to the provisions of the Limitation Act 1939 as to fraud, fraudulent concealment and mistake, actions must be commenced within six years of the cause of action accruing.[7] If the liability of the architect is in the tort of negligence the cause of action does not accrue until there has been both the breach of duty and damage to the plaintiff.[8] It may require careful consideration of the particular facts to decide when the damage was done.[9]

10. Duration of architect's duties

Completion of works

In the normal course, if the architect is employed to arrange for and supervise the building of a complete works, then, in the absence of express terms to the contrary, he has the right and duty to act as architect until completion of the works, and if the employer wrongfully dismisses him, the architect can recover damages for breach of contract,[10-11] or a reasonable sum for the work he has carried out.[12] In the normal course there is an implied term that the employer will not prevent the architect from earning his remuneration.[13]

Termination or resignation

The contract between architect and employer frequently provides that the engagement of the architect may be terminated at any time by either party upon giving reasonable notice.[14] In the absence of a provision for earlier termination an architect who gives up his duties without the employer's consent before the completion of a whole work which he has undertaken is in breach of contract and prima facie is not entitled to his fees.[15]

Death

The services of the architect are personal and his death therefore brings his contract with the employer to an end. The architect's personal representatives

---

[6] *Edelman* v. *Boehm* (1964) unrep., South Australia High Court.

[7] See *supra*, note 1.

[8] *Sparham-Souter* v. *Town & Country Developments (Essex) Ltd.* [1976] Q.B. 858 (C.A.); *Anns* v. *Merton London Borough Council* [1977] 2 W.L.R. 1024 (H.L.).

[9] See, *e.g.* comments of Lord Salmon, *Anns*' case, at p. 1048.

[10-11] *Thomas* v. *Hammersmith Borough Council* [1938] 3 All E.R. 203 (C.A.). *Cf. Rutledge* v. *Farnham Local Board of Health* (1861) 2 F. & F. 406. *Quaere*, whether an architect can insist on completing and charging for plans—see *White & Carter (Councils) Ltd.* v. *McGregor* [1962] A.C. 413 (H.L.) referred to *ante*, p. 114.

[12] *Elkington* v. *Wandsworth Corp.* (1924) 41 T.L.R. 76.

[13] Per Slesser L.J., *Thomas* v. *Hammersmith Borough Council, supra*.

[14] See, *e.g.* R.I.B.A. Conditions of Engagement, cl. 7.00.

[15] See *Cutter* v. *Powell* (1795) 2 Smith's *Leading Cases*, 1, and other cases discussed in Chap. 4; see also for repudiation, *ante*, p. 111.

can, subject to the effect of the Law Reform (Frustrated Contracts) Act 1943,[16] recover money earned by the architect and payable at the time of his death, and are not limited to a claim for a reasonable sum if instalments were payable under the contract.[17] The death of the employer does not normally bring the contract between himself and the architect to an end.[18]

Bankruptcy

Upon the bankruptcy of the architect the benefit and burden of his contract with the employer do not, it is submitted, vest in his trustee in bankruptcy because the architect's duties are personal; the contract therefore continues and the architect can sue for his fees although his trustee in bankruptcy can intervene in the action and claim the proceeds.[19]

11. Remuneration

Employer liable

The architect must in the absence of express terms look to the person who employed him for payment.[20]

The right to remuneration

There is a presumption that an architect is entitled to be paid something for his work for, " generally speaking, people who do work for others expect to be paid for it." [21] If there is an agreement for employment and the employer is to fix the amount, the employed person may, it seems, obtain a reasonable payment.[22] Where sketch plans and probationary drawings are sent in by way of tender, or subject to approval, and they are not accepted or approved the architect is not, it seems, entitled to be paid for such work,[23] unless there is an express agreement to pay.[24]

Estimates

Where an architect prepares plans and drawings and from them gives an estimate of the probable cost, he may not be able to recover any remuneration if the lowest tender is substantially higher than the estimate and the employer therefore refuses to carry on with the project.[25] In the absence of an express

---

[16] See *ante*, p. 106.

[17] *Stubbs* v. *Holywell Ry.* (1867) L.R. 2 Ex. 311.

[18] *Davison* v. *Reeves* (1892) 8 T.L.R. 391.

[19] *Emden* v. *Carte* (1881) 17 Ch.D. 169, 768 (C.A.); *Bailey* v. *Thurston* [1903] 1 K.B. 137 (C.A.).

[20] *Locke* v. *Morter* (1885) 2 T.L.R. 121.

[21] Per Maule J., *Moffatt* v. *Laurie* (1855) 15 C.B. 583; *Landless* v. *Wilson* (1880) 8 R. (Ct. of Sess.) 289.

[22] *Bryant* v. *Flight* (1839) 5 M. & W. 114; *Bird* v. *McGaheg* (1849) 2 C. & K. 707; *cf. Kofi Sunkersette Obu* v. *A. Strauss & Co. Ltd.* [1951] A.C. 243 (P.C.).

[23] *Moffat* v. *Dickson* (1853) 13 C.B. 543; *Moffatt* v. *Laurie*, *supra*.

[24] See *Moffat* v. *Dickson*, *supra*, at p. 575; see *ante*, p. 210; where the R.I.B.A. Conditions of Engagement apply there are express provisions for payment according to the stage of work reached.

[25] See *Nelson* v. *Spooner* (1861) 2 F. & F. 613; *Burr* v. *Ridout*, *The Times*, February 22, 1893.

term,[26] it is a question whether, in all the circumstances, it is an implied condition of the architect's employment that the works should be capable of being exercised at or reasonably near the estimated cost.[27] Where there is such an implied condition, and the tenders are too high, the employer must act reasonably to mitigate his loss so far as possible.[28] Thus it may be that the employer can substantially achieve his desired object despite reductions in the work to be done. In such a case it seems that he should give the architect the opportunity of altering his plans so as to effect the reductions.[29]

When right to remuneration arises

It is a question in each case whether the architect must complete his whole task before he has the right to call for payment of his services,[30] or whether he is entitled to be paid for what he actually does,[31] or at various stages of his work.[32]

Amount of remuneration

An express agreement may govern the amount of the remuneration. In the absence of such an agreement the architect is entitled to a reasonable sum, the amount of which is a question of fact.[33] A reasonable sum is not recoverable where an express agreement for payment exists. Thus in *Gilbert & Partners* v. *Knight* [34] a surveyor agreed to prepare drawings, arrange tenders and supervise works of alteration, the cost of which he estimated at roughly £600, for a fee of £30. The employer ordered extra work which brought the total cost to £2,283 but the surveyor did not, while the work was going on, tell the employer that he would require further fees, and was held to be bound by the existing agreement to perform services for £30 and could not recover a reasonable sum on a new implied contract.

Implication of R.I.B.A. scale. The R.I.B.A. scale, first issued in 1872, and now incorporated in the R.I.B.A. Conditions of Engagement (last revised in 1976 and printed *post*, p. 560) is based on a percentage of the cost of the works. Various attempts have been made to persuade the courts that the percentage charges from time to time set out in the scale are customary and reasonable, and should therefore, in the absence of express agreement, be payable automatically.

---

[26] See, *e.g.* the R.I.B.A. Conditions of Engagement, *post*, p. 560. It is a difficult question of construction how far, if at all, those conditions are effective in engagements to which they apply where the abandonment of the project is due to the faulty estimate of the architect.

[27] *Nelson* v. *Spooner, supra*; for the duty of the architect in giving an estimate, see *Moneypenny* v. *Hartland* (1826) 2 C. & P. 378; see also *Hunt* v. *Wimbledon Local Board* (1878) 4 C.P.D. 48.

[28] See *ante*, p. 147.

[29] See *Columbus* v. *Clowes* [1903] 1 K.B. 244, 247; for the authority of the architect to employ a quantity surveyor to reduce the quantities, see *post*, p. 229.

[30] See *Cutter* v. *Powell* (1795) 2 Smith's *Leading Cases*, 1 and generally, Chap. 4.

[31] See *Appleby* v. *Myers* (1867) L.R. 2 C.P. 651, 660.

[32] See *e.g.* R.I.B.A. Conditions of Engagement, *post*, p. 560.

[33] *Bryant* v. *Flight* (1839) 5 M. & W. 114. See also *Way* v. *Latilla* [1937] 3 All E.R. 759 (H.L.) for the position where there has been inconclusive negotiations as to fees.

[34] [1968] 2 All E.R. 248 (C.A.).

The argument has in the past been rejected [35] on the ground that the architect was only entitled to charge for work he had actually carried out.[36] However, in *Whipham* v. *Everitt*,[37] where no building contract was placed because the tenders were too high, certain eminent architects were called to prove that it was customary in such a case for the architect to be paid for the plans and preliminary work on the percentage basis set out in the R.I.B.A. scale. It was held by Kennedy J. that the architect was entitled to a reasonable remuneration, and that, although the R.I.B.A. scale was not binding in law because it was not a custom of so universal an application as to be an implied term of every contract, nevertheless it was right to take into consideration the practice adopted by the large proportion of the profession, as shown by the rules drawn up by the Council of the R.I.B.A. for the guidance of the members of the profession.

The architect may be able to prove that his particular employer was, in the circumstances of the case, well acquainted with the R.I.B.A. scale and a custom to pay according to the scale, and that he must therefore be taken to have contracted on that basis.[38] In a case concerning copyright where both parties were familiar with the R.I.B.A. Conditions of Engagement it was said that, they " may be assumed to have had regard to them," [39] and the court looked at the Conditions to assist in the determination of the issue.

Ryde's Scale. Ryde's Scale was a scale of charges for the valuation of land and the giving of expert evidence. It was based on a percentage of the total value of the land involved.[40] The courts did not readily imply a term that the scale should be paid,[41] and it was once said that the scale, " at its best, may create an unconscious bias in the mind of the witness remunerated and, at its worst, may be attended by much more serious results." [42]

Assessment of reasonable sum. A reasonable sum for the services of the architect will vary according to the circumstances of each case, and its amount is a question of fact in each case.[43] In *Brewin* v. *Chamberlain* [44] Birkett J. had to consider what

[35] *Gwyther* v. *Gaze* (1875) H.B.C. (4th ed.), Vol. 2, p. 34; *Burr* v. *Ridout, The Times,* February 22, 1893; *Farthing* v. *Tomkins* (1893) 9 T.L.R. 566; and see *Elkington* v. *Wandsworth Corp.* (1925) 41 T.L.R. 76.

[36] See Lord Coleridge, *Farthing* v. *Tomkins, supra,* objecting that where the work was abandoned, a scale charge would be unfair because it included supervision not carried out. This objection has now been met.

[37] *The Times,* March 22, 1900.

[38] *Buckland and Garrard* v. *Pawson & Co.* (1890) 6 T.L.R. 421. See also the Scottish case of *Wilkie* v. *Scottish Aviation Ltd.,* 1956 S.C. 198.

[39] *Stovin-Bradford* v. *Volpoint Properties Ltd.* [1971] 1 Ch. 1007, 1014 (C.A.). For copyright, see *post,* p. 224.

[40] See now, scales issued by the Royal Institution of Chartered Surveyors.

[41] See *Upsdell* v. *Stewart* (1793) Peake 255; *Debenham* v. *King's College, Cambridge* (1884) 1 T.L.R. 170; *Brocklebank* v. *Lancashire & Yorkshire Ry.* (1887) 3 T.L.R. 575; *Drew* v. *Josolyne* (1888) 4 T.L.R. 717; *Faraday* v. *Tamworth Union* (1916) 86 L.J.Ch. 436, 439; *cf. Att.-Gen.* v. *Drapers' Company* (1869) L.R. 9 Eq. 69.

[42] Per Younger J., *Faraday* v. *Tamworth Union* (1916) 86 L.J.Ch. 436, 439; *cf. Buckland and Garrard* v. *Pawson & Co.* (see *supra*).

[43] *Bird* v. *McGaheg* (1849) 2 C. & K. 707.

[44] Only reported by way of an excerpt from the judgment in Rimmer's *Law relating to the Architect* (2nd ed.), p. 304.

sum was payable under the then R.I.B.A. Scale of Charges for partial services. The relevant clause merely directed that the charge was on a *quantum meruit* and gave no guide as to how it was to be ascertained save that it was subject to a maximum of one-sixth of the percentage charge for complete services. The principles applied by Birkett J. are, therefore, of general application to claims on a *quantum meruit*. It was contended on behalf of the employer that the charge should be assessed on a time basis,[45] and on behalf of the architect that it should be assessed on a percentage basis. It was held that the time spent, and the fact that the maximum recoverable was expressed to be a percentage, were only two of the relevant factors, neither of which was dominant. Other relevant factors were: the work was not a straightforward ordinary task; the drawings prepared were of great merit; the architect was a person of standing [46] and had considerable experience; the amount and nature of the work which was involved, including such matters as interviews with the authorities and correspondence, as well as the work of preparing the drawings.

12. The architect's lien

Where an architect or surveyor has been employed to carry out a specific work such as preparing plans or a survey, his right to payment, in the absence of express terms, arises as soon as he has done the work satisfactorily and has given his employer a reasonable opportunity of ascertaining whether it has been properly done.[47] Thereafter he may retain the plans or survey until he is paid and does not lose his right of subsequently suing for a reasonable price because he has demanded a sum in excess of a reasonable price.[48] An employer who brings proceedings for the recovery of such documents may obtain them by paying into court the amount claimed as fees to abide the event of the action.[48a]

Discovery and inspection

The benefit of a professional man's lien may be lost if his client, in legal proceedings, can obtain inspection of the documents with the consequent right to copy them. Where accountants claimed a lien on documents, but had counterclaimed for the payment of the balance of their fees and the amount of such fees was in issue, it was held that the plaintiffs, their clients, were entitled to inspection of the documents.[49]

13. Property in plans and other documents

Property in plans

In the ordinary building contract the property in the plans and drawings prepared for and used in the works passes to the employer on payment of the remuneration provided under the contract,[50] and any custom to the effect that

---

[45] See the current R.I.B.A. Conditions of Engagement, cl. 2.10.
[46] See also *Bird* v. *McGaheg, supra.*
[47] *Hughes* v. *Lenny* (1839) 5 M. & W. 183, 191.
[48] *Ibid.*
[48a] R.S.C., Ord. 20, r. 6.
[49] *Woodworth* v. *Conroy* [1976] Q.B. 884 (C.A.).
[50] *Gibbon* v. *Pease* [1905] 1 K.B. 810, 813 (C.A.), applying *Ebdy* v. *M'Gowan* (1870) H.B.C. (4th ed.), Vol. 2, p. 9.

the architect is allowed to retain the plans is unreasonable and will not be enforced.[51]

Property in documents prepared as agent

" If an agent brings into existence certain documents while in the employment of his principal, they are the principal's documents and the principal can claim that the agent should hand them over." [52] It seems that such documents include communications with the contractor, sub-contractors, local authorities and others on behalf of the employer.[52]

Property in documents prepared as professional man

Documents prepared by the architect as a professional man for his own assistance in carrying out his expert work, are not " documents brought into existence by an agent on behalf of his principal, and therefore they cannot be said to be the property of the principal." [54] Thus the employer is not, in the absence of express agreement, entitled to demand memoranda, calculations, draft plans and other documents which the architect has prepared to assist him in carrying out his duties.[55]

14. Copyright in plans and design

Apart from any express term, " the architect owns the copyright in the plans and also in the design embodied in the owner's building. The building owner may not therefore reproduce the plans or repeat the design in a new building without the architect's express or implied consent." [56] The protection is afforded by the Copyright Act 1956 and includes literary work and artistic work.[57] Artistic work includes paintings, drawings, engravings and photographs irrespective of artistic quality, and works of architecture being either buildings or models for buildings.[58] The blueprint of a new shop-front has been held to be an original literary work, and therefore an unauthorised copying was a breach of copyright.[59]

---

[51] *Ibid.*

[52] *Per* MacKinnon L.J., *Leicester County Council* v. *Michael Faraday & Partners Ltd.* [1941] 2 K.B. 205, 216 (C.A.).

[53] *Beresford (Lady)* v. *Driver* (1852) 22 L.J.Ch. 407.

[54] See *supra*, note 52.

[55] *Leicester County Council* case, *supra*; *London School Board* v. *Northcroft* (1889) H.B.C. (4th ed.), Vol. 2, p. 147; see also *Chantrey Martin* v. *Martin* [1953] 2 Q.B. 286 (C.A.).

[56] *Per* Uthwatt J., *Meikle* v. *Maufe* [1941] 3 All E.R. 144, 152; Copyright Act 1956, Pt. 1; there is no infringement by reconstructing the original building, s. 9 (10). The law of copyright and its allied subjects is complex. Only the briefest introduction is given here. See further, *Law of Copyright* by Copinger and Skone James; *Copyright and Industrial Designs* by A. D. Russell-Clarke; Rimmer's *Law relating to the Architect* (2nd ed.), Chap. 5. For breach of confidence, see *ante*, p. 142.

[57] Copyright Act 1956, ss. 2 and 3.

[58] *Ibid.* s. 3 (1) (*a*); for registration of designs, see the Registered Designs Act 1949 and for the relation of this Act to the Copyright Act 1956, see *Dorling* v. *Honnor Marine* [1965] Ch. 1 (C.A.); *Merchant-Adventurers Ltd.* v. *M. Grew & Co. Ltd.* [1972] Ch. 242. See also Design Copyright Act 1968.

[59] *Chabot* v. *Davies* [1936] 3 All E.R. 221.

The case of *Meikle* v. *Maufe* [60] raised points of considerable interest. In 1912 H employed X, an architect, to erect certain buildings to be used as furniture showrooms. There were no express terms of engagement, and the possibility of extending the building southward was discussed. In 1935 H employed M, an architect, to extend the building southward. The extended façade consisted of repetitions, with minor alterations, of the design of the original façade, and the interior was reproduced to a substantial degree. Y, the owner of the copyright originally vested in X, brought proceedings for breach of copyright, and various arguments were raised by way of defence. It was said that there could be no separate copyright in a building as distinct from the copyright in the plans on which the building was based; alternatively, that if there was a separate copyright in the building then it was vested in the builder, and further that it was an implied term of the contract of engagement in 1912 that the employer could reproduce the design in an extension southwards. All these arguments were rejected, and it was held that the architect owned the copyright of the design embodied in the building as well as the plans.[61] Uthwatt J. referred to a feature of the architect's copyright—" apart from some express or implied bargain to the contrary, the architect is free, if so minded, to repeat the building for another owner." [62]

Implied consent to reproduce

This, it is submitted, can arise either from conduct as, *e.g.* the architect issues plans to the contractor for use in the erection of the building, or from the application of the ordinary rules as to the implication of terms [63] into the contract between architect and employer.

Partial services by architect

Two cases came before the courts dealing with the position where the architect had prepared plans but his engagement had been terminated before completion of his services and the then applicable R.I.B.A. Conditions of Engagement made no express provision for the position as to copyright. In the first [64] the architect was engaged to prepare plans suitable for obtaining full planning approval. He did this and charged the full scale fee applicable for work to that stage. Subsequently the owner sold the land and handed over the plans which were used by the builder. It was held that there was no breach of copyright because the architect had granted an implied licence to the owner. In the second case [65] the architect again agreed to carry out certain design work with a view to obtaining planning approval. Both parties were familiar with the then R.I.B.A. scale of fees. They agreed upon a fee of 100 guineas which was nominal in comparison with the full stage payment. It was held that, while it was necessarily implied

---

[60] [1941] 3 All E.R. 144. It is thought that the substance of this decision is unaffected by the Copyright Act 1956, but all cases decided before the Act must be read with and subject to the Act.

[61] See quotation *supra*. For the measure of damages, see *infra*.

[62] At p. 152.

[63] For the implication of terms, see *ante*, p. 35.

[64] *Blair* v. *Osborne & Tomkins* [1971] 2 Q.B. 78 (C.A.).

[65] *Stovin-Bradford* v. *Volpoint Properties Ltd.* [1971] 1 Ch. 1007 (C.A.).

that the architect licensed the owners to make use of the drawings for the purpose of obtaining planning permission, there was no implication, in the circumstances, of a licence to use the drawings for erecting a building. While the particular implication must depend upon the express terms and the surrounding circumstances in each case,[66] it seems, from the cases just referred to, that the implied licence, if any, will only take effect from the time of payment (or, probably, tender of payment) by the client.[67]

The current R.I.B.A. Conditions of Engagement, cll. 1.50–1.54, deal expressly with the question of copyright where there have been partial services.[68]

Remedies for breach of copyright

The remedies are damages, and, in appropriate cases, an injunction to restrain the breach[69]; but an injunction cannot be obtained if building has commenced.[70]

Measure of damages

Damages are to a certain extent at large,[71] but a sound basis from which to begin their assessment is the sum that " might fairly have been charged for a licence to use the copyright for the purpose for which it is used." [72] In *Meikle* v. *Maufe* a comparatively small sum was awarded on the grounds: (i) that the original design was such that the only market for it was its repetition in an extension of the original building; (ii) that the plaintiff, although the person in whom the copyright was now vested, was not himself personally responsible for the original design; (iii) that when the defendant, before undertaking the work, informed the plaintiff of his intention to do so, the plaintiff wrote to him wishing him well. This was not a licence, but the " defendant may well be forgiven for thinking that, *qua* Meikle,[73] Maufe had a free hand so far as the artistic necessities of the building required him to follow the original design." [74]

Flagrant infringement

In such a case the court has a special power to award punitive damages.[75]

Term of copyright

The term of the copyright is the life of the author plus 50 years after his death.[76]

---

[66] *Ibid.*

[67] The case of *Tingay* v. *Harris* [1967] 2 Q.B. 327 (C.A.) may be referred to on the point but is probably of no assistance. Lord Denning, in *Blair's* case at p. 85, *supra*, approaches the matter on the basis that the licence only takes effect on payment.

[68] *Post*, p. 562.

[69] See the Copyright Act 1956, s. 17 (1).

[70] *Ibid.* s. 17 (4).

[71] *Chabot* v. *Davies* [1936] 3 All E.R. 221; *Meikle* v. *Maufe* [1941] 3 All E.R. 144, 153. This means that the judge has a certain discretion and is not necessarily limited to pecuniary loss actually proved. For damages generally, see Chap. 9.

[72] *Per* Uthwatt J., *Meikle* v. *Maufe*, *supra*, at p. 153 referring to *Chabot* v. *Davies* where this was the basis upon which damages were assessed; *Stovin-Bradford* v. *Volpoint Properties Ltd.*, *supra*.

[73] The plaintiff.

[74] *Meikle* v. *Maufe*, *supra*, at p. 155.

[75] Copyright Act 1956, s. 17 (3); *Williams* v. *Settle* [1960] 1 W.L.R. 1072 (C.A.).

[76] Copyright Act 1956, ss. 2, 3. See Third Schedule for joint authorship, and s. 4 for the position where the author is in the employment of another under a contract of service; *Sterling Engineering Co.* v. *Patchett* [1955] A.C. 534 (H.L.).

15. Architect's duties to contractor and others [77]

Negligent misstatements

The architect can, in some circumstances, become liable to the contractor for loss caused by reliance on a negligent misstatement,[78] particularly in respect of an answer to some specific inquiry on a matter and in circumstances where the contractor reasonably relies upon the architect's judgment or skill or his ability to make careful inquiry; but the architect is not liable if he expresses his statement to be given " without responsibility " or otherwise in such a way as to show that he does not accept a duty of care towards the contractor.[79]

Negligent certificate

If the employer suffers loss the architect is liable.[80] But what if the architect under-certifies, is he liable in tort to the contractor for loss caused by diminished cash-flow? Lord Salmon has expressed the view that he is,[81] in words which lend support to an argument that he may, further, be liable in tort to others with whom he is in sufficiently close proximity such as a nominated sub-contractor whose right to payment,[82] or liability to a claim for loss,[83] is dependent upon an architect's certificate.

Negligent acts

The architect may in some circumstances be liable in tort to the contractor or others. In *Townsends Ltd.* v. *Cinema News, etc., Ltd.*,[84] there was a statutory duty upon the contractor to serve notice upon the sanitary authority before executing certain work, but there was proved to be a clear practice that the contractor " relies upon the architect to do all the work and give the notices and see that regulations are complied with." [85] The contractor relied in this case upon the architect serving the proper notice and seeing that the by-laws were complied with. The architect failed to carry out these tasks properly. There was a breach of the by-laws causing loss to the employer which he was entitled to recover from the contractor. In turn the contractor recovered this loss from the architect who, having undertaken a duty towards the contractor, was liable to him for its negligent performance although it was undertaken gratuitously. The architect's liability extends to economic loss.[86]

---

[77] This includes liability in tort generally, as to which see a standard textbook such as Salmond, Clerk & Lindsell.

[78] *Hedley Byrne & Co. Ltd.* v. *Heller & Partners Ltd.* [1964] A.C. 465, 503, 530, 539 (H.L.). *Priestley* v. *Stone* (1888) 4 T.L.R. 730; H.B.C. (4th ed.), Vol. 2, p. 134 (C.A.) so far as it is to the contrary must be considered as overruled. For a further discussion, see *ante*, p. 98.

[79] *Ibid.*

[80] See *ante*, p. 216.

[81] *Arenson* v. *Arenson* [1976] A.C. 405, 438 (H.L.).

[82] See, *e.g.* cl. 27 (*b*), Standard form of building contract.

[83] See, *e.g.* cll. 22, 27 (*d*) (ii), Standard form of building contract; cl. 8, Standard form of sub-contract.

[84] [1959] 1 W.L.R. 119 (C.A.).

[85] At p. 22 of the transcript in the Bar Library; the passage is not reported.

[86] See the *Townsend's* case; *Dutton* v. *Bognor Regis U.D.C.* [1972] 1 Q.B. 373 (C.A.); *cf.* *S.C.M. (United Kingdom) Ltd.* v. *W. J. Whittall & Son Ltd.* [1971] 1 Q.B. 337 (C.A.);

Liability for personal injuries. In certain circumstances this can arise.[87]

Fraud

The architect is liable to the contractor for a fraudulent misrepresentation upon which the contractor has relied and thereby suffered damage[88]; and he is liable to the contractor for any refusal to certify or incorrect certification which is fraudulent and collusive or corrupt.[89]

Breach of warranty of authority

This has been discussed *ante,* p. 203.

Breach of statutory duty in connection with the provision of dwellings

See *ante,* p. 44.

Section 4: *Quantity Surveyors*

1. Their duties generally

A quantity surveyor's duties in a building contract include, " taking out in detail the measurements and quantities from plans prepared by an architect, for the purpose of enabling builders to calculate the amounts for which they would execute the plans." [90] His duties may also include other works of measurement such as preparing a bill of variations. He is normally employed by the employer to whom he owes a duty of care, but he may in some cases be employed by the contractor; for example, where he is engaged by the contractor to take out quantities for the purposes of tender where quantities do not form part of the contract. The building contract may show that in regard to some matters he is intended to exercise his judgment fairly and professionally [91] as between the parties.

Qualification or registration is not required by law before a person can practise as a quantity surveyor, but there are certain well-known professional bodies to one of which most quantity surveyors belong.

2. The formation of the contract with the employer

(a) *Authority of architect to engage*

    To take out quantities for tender. Frequently the architect engages a quantity

---

Spartan Steel Ltd. v. *Martin & Co. Ltd.* [1973] 1 Q.B. 27 (C.A.). See also 92 L.Q.R. 213, and consider *Anns* v. *Merton London Borough Council* [1977] 2 W.L.R. 1024 (H.L.).

[87] See, *e.g. Clay* v. *A. J. Crump & Sons Ltd.* [1964] 1 Q.B. 533 (C.A.); *Dutton* v. *Bognor Regis U.D.C.* [1972] 1 Q.B. 373, 395, 404, 415 (C.A.); *cf. Clayton* v. *Woodman & Son (Builders) Ltd.* [1962] 1 W.L.R. 583 (C.A.).

[88] See *ante,* p. 204.

[89] *Ludbrook* v. *Barrett* (1877) 46 L.J.Q.B. 798; *Stevenson* v. *Watson* (1879) 4 C.P.D. 148; *Priestley* v. *Stone, supra*; *Re De Morgan Snell & Co.* (1892) H.B.C. (4th ed.), Vol. 2, p. 185 (C.A.).

[90] *Per* Morris J., *Taylor* v. *Hall* (1870) I.R. 4 C.L. 467, 476. Note that the reference was to a " building surveyor," but obviously referred to a " quantity surveyor " as that term is generally understood.

[91] See *Sutcliffe* v. *Thackrah* [1974] A.C. 727 (H.L.).

surveyor. His authority to do this may be express[92] or implied. It has been implied where the employer has authorised the architect to obtain tenders, and such tenders could only be obtained if quantities were prepared and issued,[93] but in modern conditions no useful rule of general application can, it is submitted, be stated. No authority to engage a quantity surveyor is implied where the employer's building plans are still inchoate and no definite order to proceed has been given to the architect.[94]

Reducing tenders. The invitation to tender may result in tenders which are higher than the amount which the employer is prepared to expend. In such a case the plans, specification and bills of quantities are sometimes altered for the purpose of obtaining smaller tenders. It is doubtful whether the architect has any implied authority to employ a quantity surveyor for this purpose.[95] There would certainly seem to be no grounds for implying such an authority where the tenders are higher than expected because of some error or miscalculation on the part of the architect.[96]

Measuring up variations. If by the building contract a quantity surveyor has to measure up variations, then, in principle, it seems that the architect has implied authority to employ a quantity surveyor. If the contract merely provides that additions or deductions shall be ascertained and valued and the contract price adjusted, with no provision for the employment of a quantity surveyor, then it seems that it is the duty of the architect to measure up,[97] and he would therefore have no implied authority to employ a quantity surveyor. In such a case the architect's right to be paid by the employer depends upon the contract, but he cannot in default of payment by the employer recover from the contractor.[98] If the contract is one of some magnitude then, even in the absence of any express provisions, the architect may by custom have authority to call in a quantity surveyor to measure up.[99] There are no modern cases reported dealing with such a custom.

(b) *Acquiescence or ratification by employer*

If the employer knows that work is being done which normally involves the

---

[92] For the position where the R.I.B.A. Conditions of Engagement apply, see cll. 1.23; 2.10.

[93] *Waghorn* v. *Wimbledon Local Board* (1877) H.B.C. (4th ed.), Vol. 2, p. 52; see also *Gwyther* v. *Gaze* (1875) H.B.C. (4th ed.), Vol. 2, p. 34; *Moon* v. *Witney Union* (1837) 3 Bing.N.C. 814.

[94] *Knox & Robb* v. *The Scottish Garden Suburb Co. Ltd.*, 1913 S.C. 872. See also *Antisell* v. *Doyle* [1899] 2 I.R. 275. For breach of warranty of authority, see *ante*, p. 204.

[95] *Evans* v. *Carte* (1881) H.B.C. (4th ed.), Vol. 2, p. 78.

[96] See *ante*, p. 220.

[97] *Cf. Beattie* v. *Gilroy* (1882) 10 R. (Ct. of Sess.) 226.

[98] *Locke* v. *Morter* (1885) 2 T.L.R. 121.

[99] *Birdseye* v. *Dover Harbour Commissioners* (1881) H.B.C. (4th ed.), Vol. 2, p. 76; *cf. Plimsaul* v. *Kilmorey (Lord)* (1884) 1 T.L.R. 48. In *Birdseye's* case the total amount of the works was about £4,000, and the custom proved applied to works over £2,000. To allow for inflation an appropriate multiplier would have to be applied to arrive at a comparable figure for today. Even so it must not be assumed that the court would, as a matter of course, find such a custom proved. For proof of custom, see *ante*, p. 29.

employment of a quantity surveyor, he may be taken to have tacitly assented to such employment, or, in any event, he may subsequently ratify it.[1]

(c) *Direct employment*

If the employer himself engages the quantity surveyor he is, of course, liable to pay him.

3. Duties to employer

The quantity surveyor, like any professional man, owes a duty to the person by whom he is employed to carry out his work with proper care. The standard of care owed may be ascertained by reference to the same principles as those governing the architect's duties.[2] In general the test is whether he has failed to take the care of an ordinarily competent quantity surveyor in the circumstances. Where the employer suffered a loss of £118 in a £12,000 contract because of an arithmetical error in the surveyor's accounts, it was held not to be negligence, for the error was due to the slip of a competent clerk who normally carried out his duties properly.[3] Negligence on the part of the surveyor makes him liable in damages upon the principles which apply in the case of architects.[4]

Lithography. The surveyor, as agent, must not make any secret commissions[5]; thus, if he employs his own lithographer for making copies of the bills of quantities, it seems that he is entitled to retain any small cash discount, but should give the employer the benefit of any commission or trade discount.[6]

Checking bills of quantities. The surveyor frequently checks bills of quantities submitted for tender for errors. It was said that if he discovers errors of substance operating against the interest of the contractor he is under a moral but not a legal duty to inform the contractor[7] and by so doing is not, it is submitted, in breach of duty to the employer. Since the decision of the House of Lords in 1964 in *Hedley Byrne & Co. Ltd.* v. *Heller*[8] it may well be that the surveyor is liable in tort to the contractor if he fails to inform him.

4. Remuneration

Amount

The amount of the remuneration may be expressly agreed; failing such agreement the surveyor is entitled to a reasonable sum for his services. The amount of such sum is a question of fact in each case, and the courts do not automatically

[1] *Evans* v. *Carte* (1881) H.B.C. (4th ed.), Vol. 2, pp. 78, 80 (D.C.). For the requisites of ratification, see *ante*, p. 202.
[2] See *ante*, p. 206.
[3] *London School Board* v. *Northcroft* (1889) H.B.C. (4th ed.), Vol. 2, p. 147.
[4] See *ante*, p. 216.
[5] See *ante*, p. 205.
[6] *London School Board* v. *Northcroft, supra.*
[7] See *Dutton* v. *Louth Corporation* (1955) 116 E.G. 128 (C.A.). See also *ante*, p. 49.
[8] See discussion, *ante*, pp. 48, 98.

award payment at the scale rates laid down by the main professional institu-
tions.[9] It has been held that a custom that a surveyor should be paid $2\frac{1}{2}$ per cent.
on the lowest tender when no tender was accepted was unreasonable, and $1\frac{1}{2}$ per
cent. was awarded.[10]

Payment by employer or contractor

In the normal modern building contract the quantity surveyor is employed by
the employer, and prima facie in such a case the employer is liable to pay him.[11]
Where the employment is by the architect on behalf of the employer it must
have been authorised or subsequently ratified or acquiesced in.[12] Certain difficulties have arisen from the fact that building contracts often used to provide that
the contractor should pay the surveyor's charges out of money to be paid to the
contractor under the building contract.[13] The position in such a case is as follows:

(i) *Contractor has received charges.* If the contractor has been paid the surveyor's
charges the surveyor can recover them from him.[14] The court may find that
instalments paid to the contractor impliedly include proportionate parts of the
surveyor's charges.[15]

(ii) *Contractor receives the equivalent to charges.* The surveyor can recover his
charges from the contractor if the latter has received something equivalent to or
in discharge of his claim for the surveyor's charges.[16] Thus the contractor was
liable where the employer was impecunious and could not pay the first instalment, whereupon the contractor accepted an assignment of the employer's
contract with the freeholder in discharge of his liabilities under the building
contract.[17] But where the contractor had not received payment of charges from
the employer, but had taken over a mortgage on the building bona fide to protect
his claim under the building contract against an insolvent employer he was held
not to be liable to the quantity surveyor.[18]

(iii) *Contractor prevents himself receiving charges.* It has been said that if the
contractor does anything which prevents himself from receiving the charges he
will be equally liable as if he had received them,[19] but it is doubtful whether this
dictum would now be followed.

---

[9] See cases cited *ante*, p. 221, upon implication of R.I.B.A. and Ryde's Scale.
[10] *Gwyther* v. *Gaze* (1875) H.B.C. (4th ed.), Vol. 2, p. 34, *per* Quain J. *Cf.* the R.I.C.S.
scales incorporated in the R.I.B.A. Conditions of Engagement, cl. 4.20.
[11] *Waghorn* v. *Wimbledon Local Board* (1877) H.B.C. (4th ed.), Vol. 2, p. 52; *Priestley* v.
Stone (1888) 4 T.L.R. 730; H.B.C. (4th ed.), Vol. 2, p. 134 (C.A.).
[12] See *ante*, p. 202.
[13] *Cf.* R.I.B.A. form (1939 ed.), cl. 10 [A]. Such a provision does not appear in the 1963
ed.—now the Standard form of building contract.
[14] *North* v. *Bassett* [1892] 1 Q.B. 333 (D.C.). The decision was given on the basis of
usage and of a contract between the contractor and the surveyor—as to such contract, see
infra. The action could, it is submitted, be based on a claim for money had and received.
[15] See the county court decision in *Payne* v. *Wheeldon* (1954) 104 L.J. 844.
[16] *Mellor* v. *Britton* (1900) 16 T.L.R. 465.
[17] *Ibid.*
[18] *Campbell* v. *Blyton* (1893) H.B.C. (4th ed.), Vol. 2, p. 234.
[19] *Campbell* v. *Blyton* (1893) H.B.C. (4th ed.), Vol. 2, pp. 234, 236. The dictum was
based on the assumption that there was a contract between the contractor and the surveyor,

(iv) *Employer fails to conclude contract.* If the employer after having employed the surveyor for the preparation of bills of quantities fails to enter into a binding contract with a contractor, the employer is liable to pay the surveyor's charges for preparing bills of quantities.[20]

(v) " *Conditional Contract.*" It has been held that where a surveyor prepares quantities for tender, and the quantities provide that his fees are to be paid by the successful tenderer, that the surveyor enters into a conditional contract in which the employer says in effect to the surveyor, " it is not intended that I shall pay you, but that the successful person shall pay you " [21] (*i.e.* the contractor who is awarded the contract). A usage or custom to this effect has been held to be reasonable and enforceable by the courts.[22] It can result in a surveyor being unable to recover from an employer who, without default, has included nothing for surveyor's fees in payments to the contractor before lawfully determining the contractor's contract.[23]

(vi) *Default of the employer.* If the employer is in default, *e.g.* if he is bound by the building contract to include the surveyor's charges in the first instalment and fails to do so, he is, it is submitted, liable to the surveyor.

Payment by architect

If the architect has personally employed the quantity surveyor for his own purposes [24] then he and not the employer is liable. He may be liable in an action for breach of warranty of authority.[25]

5. Duties to contractor and others

The position is analogous to that of the architect, as to which see *ante,* p. 227. For the discovery of errors when checking the bills, see *ante,* p. 230.

---

but ordinarily in modern conditions there is no such contract. See also *M'Connell* v. *Kilgallen* (1878) 2 L.R.Ir. 119, but this was a case where the *contractor* had employed the surveyor.

[20] *Moon* v. *Witney Union* (1837) 3 Bing.N.C. 814; *North* v. *Bassett* [1892] 1 Q.B. 333 (D.C.).

[21] *Per* Field J., *Young* v. *Smith* (1879) H.B.C. (4th ed.), Vol. 2, pp. 70, 73.

[22] *Ibid.*; see *North* v. *Bassett, supra.*

[23] *Young* v. *Smith, supra,* discussed at rather greater length in editions 1 and 2 of this book.

[24] *e.g. Plimsaul* v. *Kilmorey* (*Lord*) (1884) 1 T.L.R. 48.

[25] See *ante,* p. 203.

CHAPTER 13

ARBITRATION

THE settlement of disputes arising out of building contracts by arbitration, rather than by the ordinary courts, is common practice. The advantages of this procedure are that where the substantial questions are matters of fact a final and conclusive decision can be obtained in a manner which theoretically is quicker and cheaper [1] than the ordinary processes of law. In particular the appointment as arbitrator of a person experienced in building matters such as an architect or engineer may shorten proceedings as he will have personal knowledge, which he is entitled to use,[2] of customs and technical terms and processes. The amount of evidence required to prove a case may therefore be materially less than in proceedings in the ordinary courts where the role of the judge is to find strictly on the evidence before him.[3] The courts are no longer jealous of the jurisdiction of arbitrators and the modern tendency is, more especially in commercial arbitrations, to endeavour to uphold the awards of the skilled persons that the parties themselves have selected to decide the question at issue between them.[4]

This chapter deals with some parts of the law of arbitration which more particularly affect building contracts. For an account of arbitration procedure, the powers of the arbitrator, admissibility of evidence, costs, the arbitrator's remuneration and other general matters, reference must be made to one of the standard works such as *Russell on Arbitration.*[5]

Section 1 : *What is an Arbitration Agreement?*

1. The statutory definition

For the purposes of the Arbitration Act 1950 an arbitration agreement is a " written agreement to submit present or future differences to arbitration, whether an arbitrator is named therein or not." [6] Such an agreement is governed by the Act. The parties can agree orally to the settlement of disputes by arbitration, but oral agreements are not governed by the Act and are subject to many

---

[1] But is sometimes longer and dearer. One attraction of arbitrations is that they are private.

[2] *Mediterranean and Eastern Export Co. Ltd.* v. *Fortress Fabrics (Manchester) Ltd.* [1948] 2 All E.R. 186 (C.A.).

[3] *Metropolitan Tunnel and Public Works Ltd.* v. *London Electric Ry.* [1926] Ch. 371 (C.A.). Note that the litigant before an arbitrator must often necessarily be ignorant as to the extent of the arbitrator's technical knowledge on any particular item and how far, therefore, evidence on that item is necessary. But the arbitrator must give the parties an opportunity to deal with matters which might affect his decision—*Thomas Borthwick (Glasgow) Ltd.* v. *Faure Fairclough Ltd.* [1968] 1 Lloyd's Rep. 16; *Micklewright* v. *Mullock* (1974) 232 E.G. 237.

[4] Per Lord Goddard C.J., *Mediterranean, etc.* v. *Fortress Fabrics, supra,* at p. 189.

[5] Much of the next chapter applies to arbitrations.

[6] Arbitration Act 1950, s. 32. For foreign arbitration agreements and certain foreign awards, see the Arbitration Act 1975.

disadvantages; for example, the arbitrator's authority is revocable at any time before the award is made, and an award is not enforceable under the provisions of the Act.[7] Arbitration agreements should therefore be in writing. The distinction is sometimes made between a " submission," *i.e.* the actual submission of a particular, existing dispute or disputes, and an " agreement to refer," *i.e.* an agreement to refer future disputes, if they arise, to arbitration.[8] Both these terms are comprised in the statutory definition. Thus both the arbitration clause in the Standard form of building contract, and a written agreement to submit an existing dispute arising out of an oral contract are arbitration agreements for the purposes of the Act.

An arbitrator is not liable in damages for negligence in performing his judicial duties.[9] The House of Lords recently considered the extent of this judicial immunity. Lord Morris said, " There may be circumstances in which what is in effect an arbitration is not one that is within the provisions of the Arbitration Act. The expression quasi-arbitrator should only be used in that connection. A person will only be an arbitrator or quasi-arbitrator if there is a submission to him either of a specific dispute or of present points of difference or of defined differences that may in the future arise and if there is agreement that his decision will be binding." [10]

2.   The difference between arbitration and certification

Where clear words are used referring disputes or differences to arbitration,[11] the court will have little difficulty in deciding that there is an arbitration agreement; but sometimes a difficult question of construction arises. The intention of the parties is the test. If the intention is that differences should be settled by a judicial inquiry worked out in a judicial manner upon evidence submitted the case is one of arbitration.[12] If, on the other hand, a person is appointed " to ascertain some matter for the purpose of preventing differences from arising, not of settling them when they have arisen," [13] this is a mere valuation, or, where the matter to be ascertained is to be stated in a certificate, a certification.[14]

---

[7] See Russell (18th ed.), p. 47.

[8] *Ibid.* p. 39.

[9] *Sutcliffe* v. *Thackrah* [1974] A.C. 727, 736, 744, 753, 754, 758 (H.L.). In the subsequent decision of the House of Lords in *Arenson* v. *Arenson* [1977] A.C. 405 there are passages (pp. 430, 431, 440, 442) which indicate that the matter is not concluded, but it is submitted that the balance of judicial view supports the passage in the text, and that if there is any personal liability on the part of an arbitrator it does not arise out of mere negligence in the performance of his judicial duties but in respect of such matters as impropriety going beyond mere negligence or some deliberate non-performance of his duties causing loss to the parties.

[10] *Sutcliffe* v. *Thackrah, supra,* at p. 752; see also *Arenson* v. *Arenson, supra,* at pp. 424, 429.

[11] *e.g.* Standard form of building contract, cl. 35. See *Jowett* v. *Neath·R.D.C.* (1916) 80 J.P.J. 207; *Hobbs Padgett & Co.* v.*Kirkland* (1969) 113 S.J. 832 (C.A.), " suitable arbitration clause," valid agreement.

[12] *Per* Lord Esher M.R., *Re Carus-Wilson and Greene* (1886) 18 Q.B.D. 7, 9 (C.A.); see also Arbitration Act 1950, s. 32. See also *Sutcliffe* v. *Thackrah, supra.*

[13] *Per* Lord Esher M.R., *Re Carus-Wilson, supra.*

[14] *Ibid.*; *Northampton Gas Light Co.* v. *Parnell* (1855) 24 L.J.C.P. 60; *Scott* v. *Liverpool Corp.* (1858) 28 L.J.Ch. 230; *Wadsworth* v. *Smith* (1871) L.R. 6 Q.B. 332; *Kennedy Ltd.*

Intermediate between the case where a full judicial inquiry into a dispute is intended and that where a third person is to give a decision merely to prevent a dispute arising is the case where, " though a person is appointed to settle disputes that have arisen, still it is not intended that he shall be bound to hear evidence or arguments. In such cases it may often be difficult to say whether he is intended to be an arbitrator or to exercise some function other than that of an arbitrator. Such cases must be determined each according to its particular circumstances." [15]

Prima facie if a person is appointed to decide a matter solely by using his own eyes, knowledge and skill, he is a valuer or certifier and not an arbitrator for the purposes of the Arbitration Act.[16] But the parties may by clear words select as an arbitrator a person skilled in the trade intending that he shall decide disputes using his own skill and knowledge of the trade generally.[17] If the court finds that this was their intention the arbitrator's award will be upheld although no evidence was put before him.[18] The intention will not be presumed if the arbitrator selected is not an expert,[19] and in any event the arbitrator must hear the evidence of any party who puts in a claim to be heard.[20]

It has been held in the case of valuers that, if it appears that an arbitration was not intended, the mere fact that evidence is heard by a valuer,[21] or that an umpire is appointed,[22] or that valuers are termed arbitrators with power to appoint an umpire does not turn a valuation into an arbitration.[23]

The granting of certificates in a building contract is not an arbitration for the purposes of the Arbitration Act,[24] unless express words show that the certificates were intended to be given only after arbitration proceedings.[25] An engineer has been held not to be an arbitrator although he was termed " the exclusive judge " of a very wide range of matters arising under the contract and it was provided that his certificates should have the force and effect of awards.[26] Difficulties of terminology have arisen because a certifier must frequently act fairly and impartially in applying the terms of the contract.[27] He has sometimes

---

v. *Barrow-in-Furness Corp.* (1909) H.B.C. (4th ed.), Vol. 2, p. 411 (C.A.); *Finnegan* v. *Allen* [1943] K.B. 425, 436 (C.A.). See also *Arenson* v. *Arenson, supra.*

[15] See *supra,* note 13.

[16] *Re Dawdy* (1885) 15 Q.B.D. 426 (C.A.); *Wadsworth* v. *Smith, supra.*

[17] *Mediterranean, etc., Ltd.* v. *Fortress Fabrics Ltd.* [1948] 2 All E.R. 186 (C.A.); *Eads* v. *Williams* (1854) 24 L.J.Ch. 531. Note that it is misconduct for the arbitrator to use knowledge of special facts such as that one of the parties was bankrupt, acquired elsewhere than in the arbitration proceedings before him: *Owen* v. *Nicholl* [1948] 1 All E.R. 707 (C.A.). See also *Thomas Borthwick (Glasgow) Ltd.* v. *Faure Fairclough Ltd.* [1968] 1 Lloyd's Rep. 16, 29.

[18] *Ibid.*

[19] *Mediterranean, etc., Ltd.* v. *Fortress Fabrics, supra.*

[20] *Re Maunder* (1883) 49 L.T. 535; *cf. Johnston* v. *Cheape* (1817) 5 Dow 247.

[21] *Re Dawdy, supra.*

[22] *Re Carus-Wilson, supra; Re Hammond and Waterton* (1890) 62 L.T. 808.

[23] *Ibid.*

[24] *Sutcliffe* v. *Thackrah* [1974] A.C. 727 (H.L.).

[25] See cases cited at note 14, *supra; cf. Eaglesham* v. *McMaster* [1920] 2 K.B. 169.

[26] *Kennedy Ltd.* v. *Barrow-in-Furness Corp., supra.*

[27] See *Sutcliffe* v. *Thackrah, supra,* discussed *ante,* p. 216.

in the past been referred to as a quasi-arbitrator or even an arbitrator,[28] but these terms should no longer be used in respect of a person performing the usual duties of a certifier.[29]

Section 2: *Jurisdiction of the Arbitrator*

The extent of the arbitrator's jurisdiction depends upon the construction of the agreement in each case, but where the agreement is to refer disputes or differences which have arisen " in respect of," or " with regard to," or " under," [30] or " out of " [31] the contract, or where similar words are used, then prima facie it will give the arbitrator jurisdiction to decide all allegations of breach of the contract by either party,[32] and, it seems, allegations of tort closely connected with the allegations of breach of contract.[33]

Repudiation, frustration, and void contracts

It was at one time thought that the jurisdiction of an arbitrator under an ordinary arbitration agreement did not extend to questions of whether the contract had been repudiated,[34] but the House of Lords in *Heyman* v. *Darwins Ltd.*[35] has now finally settled that an arbitrator appointed under a clause such as that referred to in the preceding paragraph has jurisdiction to decide whether a contract has been repudiated by one of the parties. Such a clause is also effective where it is alleged that some supervening event has frustrated a contract,[36] whether partly executed or wholly executory (*i.e.* where no step has been taken towards its performance) at the date of the event.[37] But " if the dispute is whether the contract which contains the clause has ever been entered into at all, that issue cannot go to arbitration under the clause, for the party who denies that he has ever entered into the contract is thereby denying that he has ever joined in the submission. Similarly, if one party to the alleged contract is contending that it is void *ab initio* (because, for example, the making of such a contract is illegal), the arbitration clause cannot operate for on this view the clause itself also is void." [38]

Where an agreement refers disputes " arising out of " the contract the

---

[28] See *e.g. Hickman & Co.* v. *Roberts* [1913] A.C. 229 (H.L.).

[29] *Sutcliffe* v. *Thackrah, supra,* esp. at p. 752.

[30] *Heyman* v. *Darwins Ltd.* [1942] A.C. 356, 366 (H.L.); *Produce Brokers Co. Ltd.* v. *Olympia Oil & Cake Co. Ltd.* [1916] 1 A.C. 314 (H.L.); *Re Hohenzollern, etc., and City of London Corp.* (1886) 54 L.T. 596; H.B.C. (4th ed.), Vol. 2, p. 100 (C.A.).

[31] *Union of India* v. *Aaby (E. B.)'s Rederi A/S* [1975] A.C. 797 (H.L.).

[32] Authorities in notes 30, 31.

[33] *Astro Vencedor S.A.* v. *Mabanaft* [1971] 2 Q.B. 588 (C.A.).

[34] See *Hirji Mulji* v. *Cheong Yue Steamship Co.* [1926] A.C. 497 (P.C.); and for a building contract case, see Moulton L.J., *Kennedy Ltd.* v. *Barrow-in-Furness Corp.* (1909) H.B.C. (4th ed.), Vol. 2, p. 411 at p. 415 (C.A.).

[35] *Supra; Bloemen (F. J.) Pty. Ltd.* v. *Gold Coast City Council* [1973] A.C. 115, 126 (P.C.).

[36] *Heyman* v. *Darwins Ltd., supra,* at p. 366; *Gibraltar (Govt. of)* v. *Kenney* [1956] 2 Q.B. 410. For frustration generally, see *ante,* p. 101.

[37] *Kruse* v. *Questier & Co. Ltd.* [1953] 1 Q.B. 669.

[38] Per Lord Simon, *Heyman* v. *Darwins, supra,* at p. 366. See also *Toller* v. *Law Accident Ins. Soc. Ltd.* [1936] 2 All E.R. 952 (C.A.); *The Tradesman* [1962] 1 W.L.R. 61.

arbitrator has power not merely to determine whether the contract is frustrated but also to decide what is due to the parties upon such frustration.[39]

Fraud

By express words the parties can agree to submit questions of fraudulent misrepresentation to the arbitrator,[40] but in the absence of express words it seems that the ordinary clause does not extend to allegations of fraudulent misrepresentation rendering a contract voidable.[41]

Questions of law

The arbitrator under the ordinary clause can decide questions of law such as the construction of the contract, including the implication of customs,[42] but in two cases his jurisdiction has been held not to extend to rectification of the contract.[43]

Arbitrator's jurisdiction

Where the scope of the arbitration clause is in dispute, then in the absence of express language,[44] the arbitrator's decision as to the extent of his own jurisdiction is not final,[45] and an arbitrator cannot decide whether a condition precedent to his jurisdiction has been fulfilled.[46] Thus where a building contract provided that a " reference . . . shall not be opened until after the completion . . . of the works," and a reference was opened before completion it was held that the award was bad, Bankes L.J. being reported as saying " if one thing is quite plain, it is that an arbitrator cannot give himself jurisdiction by deciding in his favour some preliminary point upon which his jurisdiction depends." [47]

If the arbitrator's jurisdiction is challenged he should not refuse to act until it has been determined by some court which has power to determine it finally.

---

[39] *Gibraltar (Govt. of)* v. *Kenney, supra.* Whether the arbitrator has such a power where " under " is used has not been decided, but probably he has: see *Union of India* v. *Aaby (E. B.)'s Rederi A/S* [1975] A.C. 797 (H.L.); disapproving dicta by Lord Porter in *Heyman* v. *Darwins, supra,* at p. 398, and Sellers L.J. in *Gibraltar (Govt. of)* v. *Kenney, supra,* to the effect that " arising out of " was wider than " under."

[40] *Per* Lord Porter, *Heyman* v. *Darwins, supra,* at p. 392.

[41] *Monro* v. *Bognor U.D.C.* [1915] 3 K.B. 167; *Heyman* v. *Darwins, supra. Cf. Belcher* v. *Roedean School, etc., Ltd.* (1901) 85 L.T. 468. An allegation of fraud may in any event be a ground for refusing a stay of proceedings. See *infra.*

[42] *Produce Brokers Co.* v. *Olympia Oil and Cake Co.* [1916] 1 A.C. 314 (H.L.); *Re Hohenzollern, etc., and City of London Corp.* (1886) 54 L.T. 596; H.B.C. (4th ed.), Vol. 2, p. 100 (C.A.). This may be a ground for refusing a stay—see *infra.*

[43] *Printing Machinery Co. Ltd.* v. *Linotype & Machinery Ltd.* [1912] 1 Ch. 566; *Crane* v. *Hegeman-Harris Co. Inc.* [1939] 4 All E.R. 68. These cases leave open the question whether an arbitration clause could be so widely worded as to give the arbitrator authority to rectify the contract.

[44] *e.g. Willesford* v. *Watson* (1873) L.R. 8 Ch.App. 473. And even then, *quaere?* See cases cited in next three notes.

[45] *Pethick Bros.* v. *Metropolitan Water Board* (1911) H.B.C. (4th ed.), Vol. 2, p. 456 (C.A.); *May* v. *Mills* (1914) 30 T.L.R. 287; *Smith* v. *Martin* [1925] 1 K.B. 745; 94 L.J.K.B. 645 (C.A.).

[46] *Ibid.*

[47] *Smith* v. *Martin* (1925) 94 L.J.K.B. 645, 646 (C.A.). See also *Getreide-Import-Gesellschaft m.b.H.* v. *Contimar S.A. Cia, etc.* [1953] 1 W.L.R. 793 (C.A.); *Christopher Brown Ltd.* v. *Genossenschaft Oesterreichischer, etc.* [1954] 1 Q.B. 8.

He should inquire into the merits of the issue to satisfy himself as a preliminary matter whether he ought to get on with the arbitration or not, and if it becomes abundantly clear to him that he has no jurisdiction then he might well take the view that he should not go on with the hearing at all.[48]

Extras

Where the contract contains power to order extras, it is submitted that the ordinary arbitration clause covers disputes about such extras,[49] but the arbitration clause may expressly or by implication exclude certain matters, such as extras, from the jurisdiction of the arbitrator.[50]

Certificates

The jurisdiction of the arbitrator to decide whether a certificate was wrongly withheld, or to open up a certificate, has been considered in Chapter 6.[51]

Section 3: *The Architect as Arbitrator*

The architect to the contract has frequently in past years been appointed arbitrator under an arbitration clause in the contract. It has been said that this places him in a position " invoking and possibly involving on occasions considerable trouble." [52] It has often resulted in contractors opposing an application for a stay of proceedings,[53] or otherwise attacking the arbitration upon the grounds that the architect has an interest in the outcome of the proceedings and is therefore disqualified. The principles upon which the courts decide whether an architect is disqualified have been stated by the courts, and an account follows, but it must be read subject to the effect of section 24 of the Arbitration Act 1950, discussed below. This section makes it unlikely that parties will appoint the architect (or engineer in an engineering contract) as arbitrator. But they may; and in any event there remains a person who can be termed a quasi-arbitrator,[54] and it is submitted, that the law set out below applies to such a person.

1. Disqualification of the architect as arbitrator

(a) *Interests which have not disqualified*

Parties to an arbitration are normally entitled to an unbiased arbitrator with no interest in the result of the proceedings.[55] Where the architect of a contract

---

[48] *Per* Devlin J., the *Christopher Brown* case, *supra,* at pp. 12, 13.

[49] *Heyman* v. *Darwins Ltd.* [1942] A.C. 356 (H.L.); *Union of India* v. *Aaby's (E.B.) Rederi A/S* [1975] A.C. 797 (H.L.). *Laidlaw* v. *Hastings Pier Co.* (1874) H.B.C. (4th ed.), Vol. 2, p. 13; *cf. Pashby* v. *Birmingham Corp.* (1856) 18 C.B. 2.

[50] *Re Meadows and Kenworthy* (1896) H.B.C. (4th ed.), Vol. 2, p. 265 (C.A.), affd. (H.L.); *Taylor* v. *Western Valleys Sewerage Board* (1911) 75 J.P. 409 (C.A.); *Pashby* v. *Birmingham Corp., supra.*

[51] See *ante,* pp. 79, 87. For the Standard form of building contract see Chap. 15.

[52] *Per* Lord Shaw of Dunfermline, *Bristol Corp.* v. *Aird* [1913] A.C. 241, 252 (H.L.).

[53] See *post,* p. 241, for stay of proceedings.

[54] See passage from *Sutcliffe* v. *Thackrah* [1974] A.C. 727, 752 (H.L.) set out *ante,* p. 234.

[55] See *Russell on Arbitration* (18th ed.), p. 162.

is appointed arbitrator for that contract he is interested to the extent that he is employed and paid by one of the parties to the dispute,[56] and is probably further interested in that in his capacity as agent for the employer he may have given orders relating to the matter in dispute, or have already expressed a strong view on the subject, and may even " be judge, so to speak, in his own quarrel." [57] Such interests have been held not to be sufficient in themselves to disqualify the architect from acting as arbitrator.[58] It has been held that the contractor must show, if not that the architect is biased, that at least there is a probability that he would be biased.[59] The basis of these decisions was that " the court . . . ought to hold that nothing known at the time of the contract, nothing fairly to be expected from the position of the engineer when he becomes arbitrator, can be alleged as a ground why it should not keep the parties to their bargain." [60] It has further been held that the architect was not disqualified because he owned shares in the employer company, when the contractor knew or must be taken to have known of his ownership [61]; nor because in his capacity as agent he had previously expressed in strong language a derogatory view of a claim put forward by the contractor in the arbitration.[62]

Where, unknown to one of the parties, the other party at the time of the reference owed the arbitrator money this was held to be not sufficient reason for setting aside the award [63]; but in such a case the court will look more narrowly at the circumstances to see whether the arbitrator has acted improperly, and may find that a large advance of money by the arbitrator to one of the parties has disqualified him.[64]

Where, after the submission of a dispute between the contractor and the employer to the architect as arbitrator, the contractor's assignee commenced proceedings against the architect alleging fraudulent misrepresentation by the architect, and then applied for revocation of the submission on the ground that the architect was disqualified, his application was refused.[65]

(b) *Interests which disqualify*

In general any interest of the architect which makes him so biased, or puts

---

[56] See *Eckersley* v. *Mersey Docks and Harbour Board* [1894] 2 Q.B. 667 (C.A.).

[57] *Per* Bowen L.J., *Jackson* v. *Barry Ry.* [1893] 1 Ch. 238, 247 (C.A.).

[58] *Ibid.*; *Ives and Barker* v. *Willans* [1894] 2 Ch. 478 (C.A.); *Eckersley* v. *Mersey Docks and Harbour Board, supra.*

[59] See *supra,* note 56.

[60] *Per* Lord Moulton, *Bristol Corp.* v. *Aird, supra,* at pp. 258–259; but *cf.* Arbitration Act 1950, s. 24 (1). The engineer in the passage cited was in the position of an architect.

[61] *Ranger* v. *G.W. Ry.* (1854) 5 H.L.C. 72 (H.L.); *cf. Sellar* v. *Highland Ry.,* 1919 S.C.(H.L.) 19. In *Panamena, etc.* v. *Fredk. Leyland & Co.* [1947] A.C. 428 (H.L.) a certifier was not disqualified merely because he was president of the employer company. Note that s. 24 (1) of the Arbitration Act 1950 could not apply because the certifier was not an arbitrator for the purposes of the Act.

[62] *Scott* v. *Carluke Local Authority* (1879) 6 R. 616; *Cross* v. *Leeds Corp.* (1902) H.B.C. (4th ed.), Vol. 2, p. 339 (C.A.). These cases were referred to with approval by the House of Lords in the *Panamena* case, *supra,* as showing what conduct would not disqualify a certifier.

[63] *Morgan* v. *Morgan* (1832) 2 L.J.Ex. 56.

[64] *Malmesbury Ry.* v. *Budd* (1876) 2 Ch.D. 113.

[65] *Belcher* v. *Roedean School* (1901) 85 L.T. 468 (C.A.).

him in such a position that he cannot give substantial justice disqualifies him from acting as arbitrator.[66] Thus although vigorous actions and expressions of opinion in his capacity as agent may not disqualify him, if a matter at issue involves the architect's professional reputation and capacity the court may in its discretion refuse a stay of proceedings [67] (*i.e.* refuse to enforce the arbitration).[68] And where the dispute is such that it is difficult to see how the matter can properly be dealt with without the cross-examination of the architect in respect of duties performed as agent, it is obviously impossible to permit the architect to act as arbitrator.[69]

A secret interest [70] or fraud or collusion or failure to keep an independent attitude will disqualify [71]; so also will gifts or hospitality if they have the effect of influencing the architect's mind, but not otherwise.[72]

2. Arbitration Act 1950, s. 24

Section 24 (1) is as follows:

" Where an agreement between any parties provides that disputes which may arise in the future between them shall be referred to an arbitrator named or designated in the agreement, and after a dispute has arisen any party applies, on the ground that the arbitrator so named or designated is not or may not be impartial, for leave to revoke the authority of the arbitrator or for an injunction to restrain any other party or arbitrator from proceeding with the arbitration, it shall not be a ground for refusing the application that the said party at the time when he made the agreement knew, or ought to have known, that the arbitrator, by reason of his relation towards any other party to the agreement or of his connection with the subject referred, might not be capable of impartiality."

Where by virtue of the section the court has power to order that an arbitration agreement shall cease to have effect, or to give leave to revoke the authority of an arbitrator, it may refuse to stay proceedings brought in breach of the agreement.[73]

There have been no decisions upon the meaning of section 24 (1) [74] but its words seem to be peculiarly apt to govern the position of an architect to a building contract who is appointed arbitrator of future disputes. It should be noted that

---

[66] *Bristol Corp.* v. *Aird, supra.*
[67] See *Nuttall* v. *Manchester Corp.* (1892) H.B.C. (4th ed.), Vol. 2, p. 203 (D.C.); *cf. Re Donkin and Leeds, etc., Canal Co.* (1893) 9 T.L.R. 192; H.B.C. (4th ed.), Vol. 2, p. 239 (D.C.).
[68] See *post,* p. 241.
[69] *Freeman* v. *Chester* [1911] 1 K.B. 783, 790 (C.A.); *Bristol Corp.* v. *Aird* [1913] A.C. 241 (H.L.); *Blackwell* v. *Derby Corp.* (1909) 75 J.P. 129; H.B.C. (4th ed.), Vol. 2, p. 401 (C.A.).
[70] See *Kimberley* v. *Dick* (1871) L.R. 13 Eq. 1—firm undertaking to employer that cost should not exceed certain amount.
[71] *Hickman* v. *Roberts* [1913] A.C. 229 (H.L.); see also *ante,* p. 86.
[72] *Re Hopper* (1867) L.R. 2 Q.B. 367; *Moseley* v. *Simpson* (1873) L.R. 16 Eq. 226; *Re Maunder* (1883) 49 L.T. 535.
[73] s. 24 (3); *cf. Kruger Townswear Ltd.* v. *Northern Assurance Co. Ltd.* [1953] 1 W.L.R. 1049 (D.C.).
[74] First enacted as s. 14 (1), Arbitration Act 1934. But see the New Zealand case of *Canterbury Pipe Lines* v. *Att.-Gen.* [1961] N.Z.L.R. 785, where the engineer was the arbitrator and a stay (see *infra,* p. 241) was refused.

it only applies to arbitrators and does not apply to a certifier performing those parts of his duties where he has to act fairly in applying the terms of the contract.[75]

Section 4: *Ousting the Jurisdiction of the Court*

An arbitration agreement ousting the court's jurisdiction entirely is contrary to public policy and void.[76] Thus although the arbitrator's decision on fact is final, an agreement not to have any " recourse at all to the courts in case of error of law " is void [77]; but a provision making the arbitrator's award a condition precedent to the right to bring an action on the contract is good,[78] for the parties can agree that no action can be brought in a court of law until the amount of the liability has been settled by arbitration.[79] Further a clause providing that arbitration must commence within a limited time or not at all and that thereafter the claim will be barred is good.[80] Both these types of clauses, respectively known as *Scott and Avery* and *Atlantic Shipping* clauses, may be varied by the court in its limited discretion.[81]

Section 5: *The Right to Insist on Arbitration*

Apart from special cases where provision is made by statute,[82] the right to arbitration arises by agreement. Breach of the arbitration agreement gives rise to an action for damages, but unless costs have been thrown away it is difficult to prove any loss,[83] therefore " the appropriate remedy for breach of the agreement to arbitrate is not damages, but its enforcement." [84] This is achieved not by an order for specific performance, but by obtaining an order under section 4 of the Arbitration Act 1950 staying the legal proceedings while arbitration takes place.[85]

Section 4 of the Arbitration Act 1950

" (1) If any party to an arbitration agreement, or any person claiming through or under him, commences any legal proceedings in any court against any other party to the agreement, or any person claiming through or under him, in respect of any matter agreed to be referred, any party to those legal proceedings may at any time after appearance, and before delivering any pleadings or taking any

---

[75] See *ante*, p. 234.

[76] *Scott* v. *Avery* (1856) 5 H.L.C. 811 (H.L.); *Lee* v. *The Showmen's Guild* [1952] 2 Q.B. 329 (C.A.).

[77] *Per* Denning L.J., *Lee* v. *The Showmen's Guild, supra*, at p. 342; *Czarnikow & Co.* v. *Roth* [1922] 2 K.B. 478 (C.A.).

[78] *Scott* v. *Avery, supra.*

[79] *Czarnikow* v. *Roth, supra*, at p. 489; *Heyman* v. *Darwins Ltd.* [1942] A.C. 366, 377 (H.L.).

[80] *Atlantic Shipping and Trading Co.* v. *Louis Dreyfus & Co.* [1922] A.C. 250.

[81] Arbitration Act 1950, ss. 25 (4), 27; see *Dennehy* v. *Bellamy* [1938] 2 All E.R. 262; *Hookway & Co.* v. *Hooper & Co.* [1950] 2 All E.R. 842 (C.A.); *Liberian Shipping Corp.* v. *A. King & Sons Ltd.* [1967] 2 Q.B. 86 (C.A.); *Nea Agrex S.A.* v. *Baltic Shipping Co. Ltd.* [1976] Q.B. 933 (C.A.).

[82] This does not apply to building contracts as such.

[83] *Doleman & Sons* v. *Ossett Corp.* [1912] 3 K.B. 257, 268 (C.A.).

[84] *Per* Lord Macmillan, *Heyman* v. *Darwins Ltd.* [1942] A.C. 356, 374 (H.L.).

[85] See *supra*, note 83.

other steps in the proceedings,[86] apply to that court to stay the proceedings, and that court or a judge thereof, if satisfied that there is no sufficient reason why the matter should not be referred in accordance with the agreement, and that the applicant was, at the time when the proceedings were commenced, and still remains, ready and willing to do all things necessary to the proper conduct of the arbitration, may make an order staying the proceedings."

The court's discretion under section 4

Even where the various conditions of section 4 are complied with the court has a discretion whether or not to grant a stay and declines to fetter its discretion by laying down any fixed rules on which it will exercise it [87]; but where there is a valid arbitration agreement [88] it has a strong bias in favour of enforcing that agreement by granting a stay,[89] and the onus upon an application for a stay is upon the person opposing the stay.[90] Although these principles are applied in every case some guidance can be obtained on the manner in which the court exercises its discretion in particular cases, thus:

No dispute. If there is no dispute or difference within the meaning of the arbitration agreement a stay will not be granted.[91] And a creditor does not have to go to arbitration to collect an undisputed debt.[92]

Architect's certificate condition precedent. The contract may be such as to prevent the person applying for a stay from saying that there was a dispute because he does not have an architect's certificate in his favour. In *Brightside Kilpatrick* v. *Mitchell Construction (1973) Ltd.*[93] the point arose in relation to the Standard form of sub-contract. The Court of Appeal held that the contractor could not say that there was a dispute under clause 8 (printed *post*, p. 580) in respect of his claim for loss caused by the sub-contractor's delay because he did not have in his favour the architect's certificate required by the clause. The court distinguished, and expressed no views upon, the case of *Ramac* v. *Lesser* [94] where the Standard form of building contract was used and at the time of issue

---

[86] For meaning of " step in proceedings," see *Russell* (18th ed.), pp. 146–148, and in particular for the position where the plaintiff takes out a summons under R.S.C., Ord. 14, for leave to sign final judgment, see *Pitchers Ltd.* v. *Plaza (Queensbury) Ltd.* [1940] 1 All E.R. 151 (C.A.). For foreign arbitration agreements see the Arbitration Act 1975.

[87] *Czarnikow* v. *Roth, Schmidt & Co.* [1922] 2 K.B. 478, 488 (C.A.). See also *Kaye (P. & M.) Ltd.* v. *Hosier & Dickinson Ltd.* [1972] 1 W.L.R. 146, 152 (H.L.); *Fakes* v. *Taylor Woodrow Construction Ltd.* [1973] 1 Q.B. 436 (C.A.).

[88] Where the matter is in dispute, the court determines it on the application to stay, *Modern Buildings Ltd.* v. *Limmer and Trinidad Co. Ltd.* [1975] 1 W.L.R. 1281 (C.A.).

[89] *Bristol Corp.* v. *Aird* [1913] A.C. 241, 258–259 (H.L.); *Metropolitan Tunnel Works Ltd.* v. *London Electric Ry.* [1926] Ch. 371 (C.A.); *W. Bruce Ltd.* v. *Strong* [1951] 2 K.B. 447, 457 (C.A.).

[90] *Heyman* v. *Darwins Ltd.* [1942] A.C. 356, 388 (H.L.); *Metropolitan Tunnel* v. *London Electric Ry.*, *supra*; *Willesford* v. *Watson* (1873) L.R. 8 Ch.App. 473.

[91] *London & N.W. & G.W. Ry.* v. *Billington Ltd.* [1899] A.C. 79 (H.L.); *Nova (Jersey) Knit Ltd.* v. *Kammgarn Spinnerei GmbH* [1977] 1 W.L.R. 713 (H.L.).

[92] *London & N.W. Ry.* v. *Jones* [1915] 2 K.B. 35, 38.

[93] [1975] 2 Lloyd's Rep. 493 (C.A.).

[94] [1975] 2 Lloyd's Rep. 431.

of the writ the architect had issued a certificate for payment which deducted liquidated damages although he had not certified under clause 22.[95] It was admitted that the clause 22 certificate was a condition precedent to the right to deduct liquidated damages (see *post*, p. 342). Forbes J. held that a dispute or difference arose the minute the plaintiffs refused to accept that there was only the balance due to them on the certificate for payment and granted a stay. There is a difference of wording in the two contracts and the circumstances of the two cases are different, but it is difficult to see why the reasoning of the Court of Appeal does not apply to the absence of a clause 22 certificate.

Part of claim indisputably due. In such a case judgment may be given for that part leaving the balance, where there is a dispute, to go to arbitration.[96]

Disqualification of the arbitrator. The stay will be refused.[97]

Void or illegal contracts. A stay will not be granted to an applicant who alleges that the contract was void *ab initio*, as *e.g.* for illegality, or that he never entered into the contract at all.[98]

Point of law. Where the only point alleged is one of law such as the construction of a contract not involving the meaning of technical terms, the court may refuse a stay as the matter " will only come back to the court on a case stated " [99]; but the party opposing the stay must still discharge the burden of showing why he should not be bound by his agreement.[1] And where technical terms, or words used in a technical sense, are to be found in a contract, such as a building contract, and the parties have chosen as arbitrator an expert in the trade with personal knowledge of the meaning of the terms used, a stay will probably be granted.[2]

Fraud or misconduct of one of the parties alleged. Where a dispute involves the question whether a party has been guilty of fraud the court has an express statutory power to refuse a stay.[3] It still has a discretion and one factor it might take into account would be an express agreement to submit questions of fraud to arbitration.[4] It has been said that " where the party charged with the fraud

---

[95] If the requisite certificate is issued after the writ but before the hearing the defendant can rely upon it to resist summary judgment. See *Ramac's case*, p. 432 and R.S.C., Ord. 18, r. 9.

[96] *Ellis Mechanical Services Ltd.* v. *Wates Construction Ltd.*, *The Times*, Jan. 22, 1976 (C.A.). The position is different under the Arbitration Act 1975, *Associated Bulk Carriers Ltd.* v. *Koch Shipping Inc.*, *The Times*, August 10, 1977 (C.A.).

[97] *Bristol Corp.* v. *Aird* [1913] A.C. 241 (H.L.); Arbitration Act 1950, s. 24 (1) (3). For disqualification see *ante*, p. 238.

[98] *Heyman* v. *Darwins Ltd.* [1942] A.C. 356, 366 (H.L.). See further *ante*, p. 236.

[99] Per Lord Parker, *Bristol Corp.* v. *Aird* [1913] A.C. 241, 262 (H.L.).

[1] *Metropolitan Works Ltd.* v. *London Electric Ry.* [1926] Ch. 371 (C.A.). See also *Heyman* v. *Darwins Ltd.* [1942] A.C. 356, 369, 389 (H.L.) and *Russell on Arbitration*, Chap. 11.

[2] *Ibid.*   [3] Arbitration Act 1950, s. 24 (2) (3).

[4] See *Heyman* v. *Darwins Ltd.*, *supra*, at p. 392.

desires it . . . it is almost a matter of course to refuse the reference," [5] but there must have been raised a concrete and specific issue of fraud, and the fraud relied on must be fraud by the party opposing the stay.[6] One factor considered by the court where charges of fraud or misconduct such as negligence are made is that there is no appeal from a reference upon the finding of fact.[7] In cases of this kind an appellate court may vary the decision of the court of first instance where that court in exercising its discretion did not take into account all the relevant factors.[8]

Expiry of time for arbitration. It is not a sufficient ground for refusing a stay that the party applying for the stay relies on an *Atlantic Shipping* clause [9] to allege that the arbitration is out of time.[10]

Delay in hearing of dispute. The Standard form of sub-contract (*post*, p. 576) provides that without the consent of the parties (and subject to the important exception of certificates) arbitration cannot commence until the completion or abandonment of the main contract works.[11] It has been said that if the main contractor refuses to give his consent, thus delaying the hearing of the dispute, the courts would refuse a stay.[12]

More than two parties. In *Taunton-Collins* v. *Cromie* [13] the employer sued his architect for bad design and lack of proper supervision of his house and was met with a defence putting the blame in respect of some of the matters alleged upon the contractor. The contract between the employer and the contractor contained an arbitration clause. The refusal of an application for a stay on behalf of the contractor was upheld [14] on the grounds that if there were two proceedings before different tribunals in respect of the matters in question there would be delay, extra cost, procedural difficulties, and the danger of inconsistent findings of fact.[15] The principle of the decision applies, it is submitted, where a plaintiff

---

[5] *Per* Jessel M.R., *Russell* v. *Russell* (1880) 14 Ch.D. 471, 477.

[6] *Camilla Cotton Co.* v. *Granadex* [1973] 2 Lloyd's Rep. 10, 16 (H.L.).

[7] *Charles Osenton & Co.* v. *Johnston* [1942] A.C. 130 (H.L.); *Simplicity Products Ltd.* v. *Domestic Installations Ltd.* [1973] 1 W.L.R. 837 (C.A.)—whether there should be a reference to the official referee—but the principle applies, it is thought, to arbitrations.

[8] *Ibid.* See also as to discretion *Birkett* v. *James* [1977] 3 W.L.R. 38 (H.L.).

[9] See *ante*, p. 241.

[10] *Bruce* v. *Strong* [1951] 2 K.B. 447 (C.A.).

[11] cl. 24, *post*, p. 607.

[12] *Per* Lord Salmon, *Gilbert Ash Ltd.* v. *Modern Engineering (Bristol) Ltd.* [1974] A.C. 689, 726 (H.L.). In the unreported case of *Mitchell Construction Ltd.* v. *East Anglian Regional Hospital Board*, decided in 1966, where the 1957 revision of the R.I.B.A. form limited arbitration as to the validity of determination until after completion, save with the consent of the parties, and the employer refused consent, Goff J., after considering evidence, granted a stay.

[13] [1964] 1 W.L.R. 633 (C.A.).

[14] Distinguishing *Bruce* v. *Strong*, *supra*; applying *Halifax Overseas Freighters Ltd.* v. *Rasno Export* [1958] 2 Lloyd's Rep. 146. See also *Finer* v. *Melgrave*, *The Times*, June 4, 1959 (C.A.).

[15] These considerations may not be sufficient to get an arbitration stopped once it has been commenced, see *City Centre Properties (I.T.C. Pensions) Ltd.* v. *Matthew Hall & Co. Ltd.* [1969] 1 W.L.R. 772 (C.A.) and *post*, p. 246.

has proper [16] grounds for joining two or more defendants initially, or where one of the parties to an arbitration agreement reasonably intends to join a third party. The Arbitration Act 1950 makes no provision for proceedings between more than two parties. The parties concerned may overcome this by agreement [17] but such agreements are often difficult to make and, if made, to give effect to, as an arbitrator lacks the wide powers possessed by the court in third party proceedings.[18]

Part only of dispute referable. Where part of the dispute lies outside the jurisdiction of the arbitrator the court may, if it is convenient, sever the dispute and grant a stay only in respect of that part referable.[19] If the matter cannot be split and a substantial part of the dispute is outside the arbitration agreement a stay will be refused.[20] If only a trifling and subordinate part of the dispute is outside the agreement a stay will not be refused.[21]

Expense of reference. It has been held not to be a ground for refusing a stay that the poverty of a litigant would cause him hardship in arbitration proceedings whereas in proceedings in the High Court he could sue as a poor person.[22] But it seems that if the plaintiff can show a prima facie case that the poverty from which he suffers [23] was caused by the very breaches of contract by the defendant of which he complains, a stay will be refused.[24]

Where no stay granted

Where no stay is granted, or where proceedings are allowed to continue without an application for a stay being made, " the private tribunal of the parties, [*i.e.* arbitrator or umpire] if it has ever come into existence, is *functus officio*," [25] and an award made by the arbitrator after the commencement of the action is invalid and affords no defence to the action.[26]

Section 6: *Control by the Court*

Although the arbitrator's decision as to fact is final the court retains control over arbitration agreements and references, and a party who considers that there

---

[16] See Pearson L.J., *Taunton-Collins* v. *Cromie, supra,* at p. 637.

[17] *Ibid.*

[18] See Ord. 16, r. 4. The arbitrator's only effective power is, on the application of one of the parties, to give a notice marked " peremptory " of his intention to proceed *ex parte* if one of the parties does not attend, see *Russell on Arbitration,* p. 170.

[19] *Ives & Barker* v. *Willans* [1894] 2 Ch. 478, 489 (C.A.); *Bristol Corp.* v. *Aird* [1913] A.C. 261 (H.L.). See also *Bulk Oil Co.* v. *Trans-Asiatic Oil* [1973] 1 Lloyd's Rep. 129.

[20] *Turnock* v. *Sartoris* (1889) 43 Ch.D. 150 (C.A.).

[21] *Ives & Barker* v. *Willans, supra.*

[22] *Smith* v. *Pearl Assurance Co.* [1939] 1 All E.R. 95 (C.A.). The poor person procedure has since been replaced by legislation now consolidated in the Legal Aid Act 1974. Legal aid is still not available for arbitrations.

[23] And which may enable him to get legal aid in the High Court.

[24] *Fakes* v. *Taylor Woodrow Ltd.* [1973] Q.B. 436 (C.A.).

[25] *Doleman* v. *Ossett Corp.* [1912] 3 K.B. 257, 269 (C.A.), distinguished by the House of Lords in *Kaye (P. & M.) Ltd.* v. *Hosier & Dickinson Ltd.* [1972] 1 W.L.R. 146 holding that the court would not go behind an architect's certificate issued after service of a writ in so far as the certificate was expressed to be " conclusive evidence in any proceedings." See further, *post,* p. 395.

[26] *Ibid.*

has been some irregularity or other special reason may invoke its assistance. For a full account reference should be made to *Russell on Arbitration*; there follows an outline as a guide and to point the contrast between an arbitrator, who is subject to the control of the court, and a certifier [27] who is not, except the more limited way set out in Chapter 6.

Appointment of arbitrator

The court has power in certain circumstances to appoint an arbitrator.[28] Where arbitration is to be by someone appointed by a third person, and there is no agreement express or implied that if the third person fails to appoint, an arbitrator shall be agreed upon by the parties, the court has no power to appoint an arbitrator.[29]

Refusal to stay proceedings

See above.[30]

Revocation of arbitrator's authority

The authority of the arbitrator is, unless the agreement expresses a contrary intention, irrevocable except by leave of the court.[31] The power of the court under section 24 (1) of the Arbitration Act 1950 to revoke the authority of the arbitrator for want of impartiality has been considered above. By section 24 (2) of the Act the court may revoke the arbitrator's authority where a dispute involves the question whether any party has been guilty of fraud.[32] Subject to these provisions the court's discretion to revoke is exercised in a sparing and cautious manner,[33] and only where the court is satisfied that a substantial miscarriage of justice will take place if the application to revoke is refused.[34] The discretion is not merely the converse of that exercised under section 4 [35] (*supra*, p. 241).

Injunction to restrain arbitration proceedings

This power is analogous to that of revoking the arbitrator's authority but is rarely used.[36]

---

[27] For distinction, see *ante*, p. 234. The court's power of control over arbitrators is statutory and contained in the Arbitration Act 1950; it has no inherent power, *Exormisis S.A.* v. *Oonsoo* [1975] 1 Lloyd's Rep. 432.

[28] Arbitration Act 1950, s. 10; see also s. 25.

[29] *National Enterprises Ltd.* v. *Racal Communications Ltd.* [1975] Ch. 397 (C.A.), distinguishing *Davies Middleton & Davies Ltd.* v. *Cardiff Corporation* (1964) 62 L.G.R. 134 (C.A.) which appeared to be to contrary effect.

[30] See *ante*, p. 241.

[31] Arbitration Act 1950, s. 1.

[32] *Kruger Townswear Ltd.* v. *Northern Ass. Co.* [1953] 1 W.L.R. 1049 (D.C.).

[33] *Scott* v. *Van Sandau* (1841) 1 Q.B. 102, 110; *Belcher* v. *Roedean School* (1901) 85 L.T. 468 (C.A.); *City Centre Properties (I.T.C. Pensions) Ltd.* v. *Matthew Hall & Co. Ltd.* [1969] 1 W.L.R. 772 (C.A.).

[34] *James* v. *James* (1889) 22 Q.B.D. 669, 674 (D.C.); affd. 23 Q.B.D. 12 (C.A.).

[35] *City Centre* case *supra*. But *quaere* whether it is simply that the circumstances are usually different? In any event on the facts in the *City Centre* case a stay might have been granted.

[36] See *Russell on Arbitration*, p. 120.

Declaration as to arbitrator's jurisdiction

The court has jurisdiction to grant such a declaration.[37]

Removal of arbitrator for misconduct [38]

Where misconduct on the part of the arbitrator is alleged an application for his removal is the correct course. If the application is granted the court has power to appoint a fresh arbitrator.[39] For the meaning of " misconduct " see *Russell on Arbitration*. It seems to comprise such irregularity in the course of the proceedings, not necessarily involving moral turpitude, as will cause the court to intervene.

Removal of arbitrator for delay

An application can be made if the arbitrator fails to use all reasonable dispatch.[40]

Preservation of status quo

The court has power to order the detention, preservation or inspection of property as to which any question may arise in the reference.[41]

Statement of special case for the court

An arbitrator or umpire may, and shall if so directed by the High Court, state: (a) any question of law arising in the course of the reference; or (b) an award or any part of an award in the form of a special case for the decision of the High Court.[42] If the arbitrator refuses to state a special case an application can be made to the court for an order directing him to do so.[43] The court will not order a case where the matter is trivial, the application is an abuse of the procedure as where the point of law is too plain for serious argument, or is made as a means of delay or is otherwise not raised bona fide [44]; but it will order a case where there is substantial, clear-cut point of law, not dependent on the special expertise of the arbitrator, and its resolution is necessary for the proper determination of the case.[45] If it is thought that a point of law upon which the decision of the court is required has arisen or will arise it should be identified and the arbitrator asked to state a case before giving his award, but on the

---

[37] *Gibraltar (Govt. of)* v. *Kenney* [1956] 2 Q.B. 410; an analogous method is to proceed by way of a construction summons; see, *e.g. A. E. Farr Ltd.* v. *Ministry of Transport* [1960] 1 W.L.R. 956.

[38] Arbitration Act 1950, s. 23 (1). For a recent case discussing misconduct, see *Prodexport* v. *Man (E. D. & F.)* [1973] 1 Q.B. 389—error of law (or fact) not in itself misconduct, although illegality is. See also *GKN Centrax Gears Ltd.* v. *Matbro Ltd.* [1976] 2 Lloyd's Rep. 555 (C.A.).

[39] *Ibid.* s. 25.

[40] *Ibid.* s. 13 (3).

[41] *Ibid.* s. 12 (6) (g). For other powers of the court in relation to the reference, see this section of the act generally. Thus for an order for security for costs, see *Bilcon Ltd.* v. *Fegmay Investments Ltd.* [1966] 2 Q.B. 221.

[42] *Ibid.* s. 21 (1).

[43] See *Russell on Arbitration*, pp. 243 *et seq.*

[44] *Halfdan Greig & Co. A/S* v. *Sterling Coal* [1973] 1 Q.B. 843 (C.A.); *GKN Centrax* case, *supra,* at p. 575.

[45] *Ismail* v. *Polish Ocean Lines* [1976] Q.B. 893 (C.A.).

hearing by the court it can reformulate the questions if necessary to raise the real issues.[46]

Refusal to enforce award

An award may be enforced by leave of the High Court or a judge thereof in the same manner as a judgment or order.[47] Alternatively an action may be brought on the award.[48] The former is the ordinary procedure where there is an arbitration agreement within the meaning of the Arbitration Act 1950.[49] If there has been no application within six weeks of the award[50] to remit or set it aside[51] the award is final and binding upon the parties[52] and leave will be given to enforce the award " unless there is real ground for doubting the validity of the award."[53] An action on the award is necessary where leave is refused, and is necessary, or to be advised, in certain other circumstances.[54]

Remission or setting aside award

The court has a discretionary power, upon application being made in accordance with Rules of the Supreme Court, to remit the matters referred or any of them to the reconsideration of the arbitrator or umpire.[55] Where the arbitrator or umpire has misconducted himself or the proceedings, or an arbitration or award has been improperly procured the High Court may set the award aside.[56] A recent example of a remission of an award to an arbitrator for further argument arose when he made an unusual award as to costs without giving his reasons.[57] The award could be justified in certain circumstances; if it could not have been justified at all according to the principles governing the exercise of discretion as to costs[58] it would have been set aside as to costs.[59]

Another ground for remission or setting aside is where there is an error of law on the face of the award,[60] *i.e.* where it appears in the award or in a document incorporated into the award. Because of this (*inter alia*) arbitrators often do not give reasons.

---

[46] *Ibid.*
[47] Arbitration Act 1950, s. 26 (1). For enforcement by the County Court of awards not exceeding the County Court jurisdiction, see the Administration of Justice Act 1977, s. 17.
[48] See *Russell*, Chap. 20.
[49] For meaning, see *ante*, p. 233.
[50] See R.S.C., Ord. 75.
[51] See *infra.*
[52] Arbitration Act 1950, s. 16.
[53] *Middlemiss & Gould* v. *Hartlepool Corporation* [1972] 1 W.L.R. 1643, 1647 (C.A.). See also *Hall & Wodehouse Ltd.* v. *Panorama Ltd.* [1974] 2 Lloyd's Rep. 413 (C.A.).
[54] See *Russell*, pp. 323 *et seq.*
[55] Arbitration Act 1950, s. 22 (1). R.S.C., Ord. 75. For the court's discretion see *Universal Cargo Carriers Corp.* v. *Citati* [1957] 1 W.L.R. 979 (C.A.).
[56] *Ibid.* s. 23 (2).
[57] *Centrala Morska* v. *Cia Nacional* [1975] 2 Lloyd's Rep. 69.
[58] See *post*, p. 267.
[59] *The Erich Schroeder* [1974] 1 Lloyd's Rep. 192; see also *Pepys* v. *L.T.E.* [1975] 1 W.L.R. 234 (C.A.)—a Lands Tribunal case, but applying general principles.
[60] See discussion in *Giacomo Costa Fu Andrea* v. *British Italian Trading Co. Ltd.* [1963] 1 Q.B. 201 (C.A.); *Belsfield Court Construction Co.* v. *Pywell* [1970] 2 Q.B. 47—pleadings not to be treated as automatically incorporated; *Nippon Yusen Kaisha* v. *Acme Shipping* [1972] 1 W.L.R. 74, 79 (C.A.)—procedure not to be used instead of asking for a special case. See also *GKN Centrax Gears Ltd.* v. *Matbro Ltd.* [1976] 2 Lloyd's Rep. 555 (C.A.).

If there is material in, or incorporated into, the award upon which the court can act, it will set aside the award for an error of law which became incidentally material to the arbitrator's award.[61] " But where a question of construction is the very thing referred for arbitration, then the decision of the arbitrator upon that point cannot be set aside by the court only because the court would itself have come to a different conclusion. If it appears by the award that the arbitrator has proceeded illegally—for instance, that he has decided on evidence which in law was not admissible or on principles of construction which the law does not countenance, then there is error in law which may be a ground for setting aside the award . . ." [62]

---

[61] *Absalom (F. R.) Ltd.* v. *Great Western (London) Garden Village Society* [1933] A.C. 592, 602, 607, 615, 616 (H.L.).
[62] *Kelantan (Government of)* v. *Duff Development Co. Ltd.* [1923] A.C. 395, 409, (P.C.).

CHAPTER 14

LITIGATION

Section 1: Features of Building Contract Litigation

THERE is no special procedure laid down by the rules for building contract disputes, and any such dispute may follow its own particular course, but there are some features which so frequently recur in disputes of fact over building contracts that they are to some extent characteristic. They are as follows [1]:

(i) the contract may show that the architect's certificate is intended to be a condition precedent to the right to bring an action, or when given is binding and conclusive as between the parties unless it can be set aside [2];

(ii) the contract may contain a clause referring disputes to an arbitrator in which case if one party commences proceedings in a court of law the other can usually obtain a stay of the proceedings.[3] If there is a formal hearing by the arbitrator much of what is said in this chapter about practice and pleading will apply to the arbitration proceedings;

(iii) the technical nature of disputes often gives rise to a reference to a referee or an arbitrator [4];

(iv) the special form of pleading known as a Scott or Official Referee's Schedule is frequently used in detailed disputes of fact;

(v) costs tend to be high even when the amount in issue is small;

(vi) a set-off or counterclaim is common and may give rise to various difficult questions.

Section 2: Preparation for Trial

This section deals with some of the more important matters that must be considered in the preparation and course of a building contract dispute, and must be read with other sections of this chapter, in particular with Section 6 dealing with costs.

1. Preparation of evidence

Early and thorough preparation of evidence is advisable. When it becomes apparent that a dispute on detailed questions of fact is inevitable, each party should at an early stage select the main expert witness on whom he intends to rely. This witness should be given the pleadings, if any, and all the contractual documents or, if the contract is oral, he should be furnished with a written version, giving the fullest possible detail, of the material interviews. An inspection of the site should be arranged, the other party being notified so that he

---

[1] Unless otherwise expressly stated High Court practice and procedure and the rules of the Supreme Court are referred to, although most of the subject-matter of this chapter applies in principle to county court litigation.

[2] See generally, Chap. 6, Pt. II.

[3] See *ante*, p. 241.

[4] This and (iv), (v) and (vi) are discussed in this chapter.

may arrange his inspection at or about the same time. As a result of his observations and his study of the contract the witness should draw up a detailed
report in writing. He should endeavour to make as fair and impartial an appraisal
as possible, remembering that an over-optimistic report will probably increase
his client's costs in that reasonable offers of settlement may be refused, or too
small a sum of money may be paid into court.[5]

The details of the report must vary according to the nature of the dispute;
for example, in a claim for a reasonable sum each item must be described and
valued and a note made of the basis of the valuation, *e.g.* measurements, labour
rates, the cost of materials and percentage additions for profits and overheads.[6]
In a claim for variations where he values according to principles stated in the
contract the witness should refer to the particular clause, or sub-clause upon
which he relies.[7] In a defects case the defects must be described in detail, and
either a blank space left after each item for the insertion of the estimated cost of
rectification by a contractor, or the witness should give his own estimate making
a note of the basis of his valuation (see *supra*). The aim should be to provide a report from which both a Scott Schedule and a proof of evidence can be prepared.

The expense of preparing a report such as that suggested may be considerable.[8]
Its great advantage is that it enables the party's legal advisers to attempt a
reasonably accurate appraisal of the situation at an early stage in the proceedings,
and thus be in a position to advise whether a settlement should be sought, or a
sum of money should be paid into, or taken out of, court. Moreover, if the
matter does come to trial (this may well be one or two years (or more) after the
commencement of proceedings) the main witness is able before trial to read
through his own detailed contemporaneous account instead of having to attempt
a reconstruction of the processes of thought that led him to general conclusions
at some considerable time in the past. When in the witness-box he can refer to
his notes made at the time of his inspection to refresh his memory, and in
certain circumstances the notes can become admissible as evidence.[9]

Photographs can sometimes be very useful. They should be taken under the
supervision of the main expert witness and the photographer should keep a note
of the date and, where it may be material, the time of day and lighting conditions.
In some cases it may be desirable to make a small sketch plan showing the
relative positions of the camera and the subject-matter, and of the latter to the
whole works.

2. County court jurisdiction

The ordinary court of first instance is the High Court of Justice, but some

---

[5] See *post*, p. 270.

[6] See further, *ante*, p. 58, *Assessment of a reasonable sum.*

[7] *e.g.* Standard form of building contract, cl. 11.

[8] As to whether expert's fees properly incurred are recoverable as costs or damages,
compare *Peak Construction (Liverpool) Ltd.* v. *McKinney Foundations Ltd.* (1970) 69
L.G.R. 1, 10 (C.A.), damages, and *Bolton* v. *Mahadeva* [1972] 1 W.L.R. 1009, 1014
(C.A.), costs and see also *Manakee* v. *Brattle* [1970] 1 W.L.R. 1607.

[9] Evidence Act 1968, Pt. I; R.S.C., Ord. 38, Pt. III; for restrictions upon adducing
expert evidence, see Civil Evidence Act 1972; R.S.C., Ord. 38, Pt. IV, and *post*, p. 258.

disputes can be decided in the county courts where the costs are less and the procedure is speedier.

County courts have jurisdiction to hear and determine actions founded on contract or tort where the debt, demand, or damage claimed is not more than £2,000.[10] A plaintiff who has a cause of action for more than £2,000 may abandon the excess so as to give the court jurisdiction.[11] County courts have unlimited jurisdiction to try counterclaims unless a successful application has been made to transfer the matter to the High Court.[12]

Jurisdiction by agreement

The parties may by a written memorandum give a county court jurisdiction to hear and determine a High Court matter.[13]

Transfer from county court to High Court

Where in an action in contract or tort the plaintiff claims damages he may apply to the judge for an order transferring the action to the High Court on the ground that there is reasonable ground for supposing the amount recoverable in respect of his claim to be in excess of the amount recoverable in the county court.[14] Where in an action founded on contract the plaintiff's claim exceeds £300 the defendant may have the action transferred if he gives security approved by the registrar for such amount as the registrar may determine and the county court judge certifies that in his opinion some important question of law or fact is likely to arise.[15]

Transfer from High Court to county court

A transfer of proceedings from the High Court to a county court may take place either by agreement (see above) or as the result of a successful application by one of the parties. Such an application can be made to the High Court where the amount claimed or remaining in dispute does not exceed £2,000 and in certain other circumstances.[16]

3. Pleadings

Building contracts are contracts for work and labour, or for work and labour and materials supplied. Precedents for ordinary pleadings are not given in this book but can be found under these headings in other works.[17] The special form

---

[10] County Courts Act 1959, s. 39 (1). Administration of Justice Act 1969, s. 1; County Courts Jurisdiction Order 1977. See also *post*, p. 268, " Recovery of small sums."

[11] County Courts Act 1959, s. 41 (1); Administration of Justice Act 1969, s. 2; County Courts Jurisdiction Order 1977.

[12] *Hardwicke* v. *Gilroy* [1944] K.B. 460 (C.A.). The application is to the High Court under s. 65 of the County Courts Act 1959.

[13] County Courts Act 1959, ss. 42, 53.

[14] *Ibid.* s. 43.

[15] *Ibid.* s. 44. Administration of Justice Act 1969, s. 3; County Courts Jurisdiction Order, *ibid.*

[16] County Courts Act 1959, s. 45; Administration of Justice Act 1969; County Courts Jurisdiction Order, *ibid.* See also the current *County Court Practice.*

[17] *e.g.* Bullen & Leake and Jacob's *Precedents of Pleadings*; Smith's *County Court Pleader.*

of pleading called a Scott or Official Referee's Schedule and its relationship to the ordinary pleadings are discussed in Section 4 of this chapter.

In drafting pleadings a defendant who has a cross-claim against the plaintiff should consider which of the three different ways of pleading are open to him and whether any advantage as to costs or otherwise may be gained by the form of pleading employed.[18]

Particulars should nearly always be pressed for at an early stage, with, if possible, and in the appropriate cases, a mutual agreement for their delivery either as part of a Scott Schedule, or in a form suitable for inclusion in a Scott Schedule. Particulars may be ordered to be given in the form which is most convenient to both parties as, for example, and in a suitable case, by means of a scale plan.[19] The general rule is that a party need not, and ought not, to plead to particulars,[20] but this rule is often not followed in the Official Referee's court, particularly when an Official Referee's Schedule (see *post*, p. 262) is ordered. Damages which are not the necessary and immediate consequence of a breach must be pleaded.[21]

Causes of action

" It is a well settled rule of law that damages resulting from one and the same cause of action must be assessed and recovered once for all." [22] A contract may give rise to several causes of action,[23] but it seems that in a claim for defects in an ordinary lump sum contract the defects only give rise to one cause of action.[24] The employer therefore should make a careful examination and include all the defects in one claim. If he is sued for the balance of the contract price and is apprehensive lest all the defects have not yet manifested themselves, his correct course is to pay the claim and bring a separate action for the defects at a later stage,[25] unless the terms of the contract show that various defects or classes of defects such as latent defects are intended to give rise to separate causes of action.[26] In *Conquer* v. *Boot* [27] an employer claimed and recovered £24 damages for breach of contract to complete a bungalow " in a good and workmanlike manner." Ten months later he made an identical claim save that the words " and with proper materials " were added and recovered a further £81 damages. On appeal it was held that the second judgment must be quashed as the matter was *res judicata*, Sankey L.J. saying, " So far as the claim is

---

[18] See *post*, p. 275.

[19] *Tarbox* v. *St. Pancras Borough Council* [1952] 1 All E.R. 1306 (C.A.). See generally as to particulars, R.S.C., Ord. 18, r. 12.

[20] *Pinson* v. *Lloyds & National Provincial Foreign Bank Ltd.* [1941] 2 K.B. 72 (C.A.); *Chapple* v. *E.T.U.* [1961] 1 W.L.R. 1290.

[21] *Perestrello Ltda.* v. *United Paint Co. Ltd.* [1969] 1 W.L.R. 570 (C.A.); *Domsalla* v. *Barr* [1969] 1 W.L.R. 630 (C.A.).

[22] *Per* Bowen L.J., *Brunsden* v. *Humphrey* (1884) 14 Q.B.D. 141, 147 (C.A.).

[23] *Conquer* v. *Boot* [1928] 2 K.B. 336, 339 (D.C.).

[24] *Conquer* v. *Boot, supra*.

[25] *Davis* v. *Hedges* (1871) L.R. 6 Q.B. 687.

[26] *Cf.* Standard form of building contract, cl. 30 (7), pre-1976 revision.

[27] *Supra*.

concerned it appears to me to be identical in both places—namely, a claim for breach of contract to complete the bungalow in a good and workmanlike manner." [28]

" Two actions may be brought in respect of the same facts where those facts give rise to two distinct causes of action," [29] as, for example, where they result in injury to a man's property and injury to his person.[30]

The National House-Builders Registration Council's Scheme relating to the construction of dwellings (see *ante*, p. 44) envisages serial arbitration, so that a claimant may bring more than one arbitration in respect of alleged defects.[31]

The court will not set aside an arbitrator's award because he gives effect to a trade custom that a party may from time to time refer different disputes arising out of the same contract to arbitration.[32]

Where a contract sum is payable by instalments, each instalment may be sued for as it falls due and the contractor may in proper cases seek summary judgment under Order 14.[33] An amendment of the pleadings cannot be granted so as to include a claim for an instalment falling due after proceedings have been commenced.[34]

Interest

Under the Law Reform (Miscellaneous Provisions) Act 1934, s. 3 (1), in any proceedings tried in any court [or by arbitration [35]] for the recovery of any debt or damages the court may, if it thinks fit, order that there shall be included, in the sum for which judgment is given, interest at such rate as it thinks fit on the whole or any part of the debt or damages for the whole or any part of the period between the date when the cause of action arose and the date of judgment.

Interest cannot be awarded upon interest nor in relation to a debt upon which interest is payable as of right. [36] Therefore, it would seem, that where there is a contractual right to interest for overdue payment interest is not recoverable under the Act.

Interest can be awarded under the Act even if a claim for interest is not pleaded,[37] but where sought it is better to plead it. If pleaded there should be a

---

[28] *Ibid.* at p. 340; see also *H. E. Daniels Ltd.* v. *Carmel Exporters and Importers Ltd.* [1953] 2 Q.B. 242; County Courts Act 1959, s. 69. For the application of the doctrine to issues, see *Carl Zeiss Stiftung* v. *Rayner & Keeler Ltd. (No. 2)* [1967] 1 A.C. 853 (H.L.).

[29] *Per* Bowen L.J., *Brunsden* v. *Humphrey, supra,* at p. 146.

[30] *Brunsden* v. *Humphrey, supra*; see also *Overstone Ltd.* v. *Shipway* [1962] 1 W.L.R. 118 (C.A.)—a hire-purchase case. *Cf. Cahoon* v. *Franks* (1967) 63 D.L.R. (2d) 274 (Canadian Supreme Court).

[31] *Purser & Co. (Hillingdon)* v. *Jackson* [1977] Q.B. 166.

[32] *Brian Smith Ltd.* v. *Wheatsheaf Mills Ltd.* [1939] 2 K.B. 302.

[33] *Workman, Clark & Co. Ltd.* v. *Lloyd Brazileno* [1908] 1 K.B. 968 (C.A.).

[34] *Eshelby* v. *Federated European Bank Ltd.* [1932] 1 K.B. 423 (C.A.).

[35] *Chandris* v. *Isbrandtsen-Moller Co. Inc.* [1951] 1 K.B. 240 (C.A.). As regards personal injuries the Act has been amended by the Administration of Justice Act 1969, s. 22. Further, the Court of Appeal has set out the principles upon which interest is awarded in personal injuries actions, *Jefford* v. *Gee* [1970] 2 Q.B. 130 (C.A.).

[36] Law Reform (Miscellaneous Provisions) Act 1934, s. 3 (1), proviso; *Bushwall Properties Ltd.* v. *Vortex Properties Ltd.* [1975] 1 W.L.R. 1649, 1660.

[37] *Westminster Bank* v. *Riches* [1945] Ch. 381 (C.A.).

statement that it is claimed in accordance with statute to distinguish the claim from one which might arise under contract.[38]

Interest ordinarily awarded. Although there is a discretion there is a " well-recognised rule of practice . . . that prima facie the losing party should be ordered to pay interest at a reasonable rate running from the date when the amount or amounts due should reasonably have been paid," [39] and it is not " a proper exercise of discretion to refuse to award interest against a party who has been held liable for damages for breach of contract on the ground that the successful party has been guilty of delay, unless and until an opportunity has been given to that party to show whether indeed he has been guilty of delay. . . ." [40] An arbitrator who failed to follow the ordinary practice as to interest was held to be guilty of technical misconduct and the award was remitted to him.[41]

Rate of interest awarded. Guidance has been given as to the principles ordinarily applied by the commercial court.

It has been said that while the award of interest is discretionary, the general principle in the commercial court, unless there are special considerations affecting the particular case, has been to award interest on sums recovered so that the successful party should be compensated by the losing party for having kept him out of money to which he was entitled at or before the commencement of the proceedings.[42] The traditional practice of the court was to adopt bank rate as a broad basis. Bank rate has now been replaced by " Bank of England minimum lending rate." [43] For this and other reasons the tendency of the commercial court will increasingly be to award interest under the Act at the rate of interest allowed on money in court placed on the short term investment account.[44] It is thought that these principles will be of assistance in building cases wherever the venue,[45] although the rate allowed for money invested in court since Febuary 1, 1977 has been 10 per cent.[45a] and it is thought that in suitable cases the court or arbitrator may be satisfied that such a rate will be insufficient to compensate the successful party.

---

[38] para. 6/2/7, *Supreme Court Practice.* See further as to claims for interest, *supra,* p. 154. Income tax is deductible from interest awarded under the Act—*Riches* v. *Westminster Bank Ltd.* [1947] A.C. 390 (H.L.).

[39] *Panchaud Freres S.A.* v. *Pagnan, R. and Fratelli* [1974] 1 Lloyd's Rep. 394, 409, *per* Kerr J.; see also in the Court of Appeal pp. 411, 414.

[40] *Ibid.* p. 414 (C.A.).

[41] *Van der Zuden, P.J., Wildhandel, N.V.* v. *Tucker & Cross* [1976] 1 Lloyd's Rep. 341.

[42] *Cremer* v. *General Carriers S.A.* [1974] 1 W.L.R. 341, 355. See also *Riches* v. *Westminster Bank Ltd.* [1947] A.C. 390, 397 (H.L.); *Jefford* v. *Gee* [1970] 2 Q.B. 130 (C.A.). For the position where the successful party has received insurance money, see *Harbutt's " Plasticine "* v. *Wayne Tank & Pump Co.* [1970] 1 Q.B. 447 (C.A.); *Cousins (H.) & Co.* v. *D. & C. Carriers* [1971] 2 Q.B. 230 (C.A.).

[43] *Cremer* v. *General Carriers, supra,* at p. 356.

[44] *Ibid,* at p. 357. For the rates of interest allowed on money in court from time to time, see the *Supreme Court Practice,* para. 6/2/7C.

[45] Building cases are not " commercial actions " within R.S.C., Ord. 72, r. 1 (2), although in recent years a number of building cases involving points of law have been accepted into the commercial list.

[45a] The Supreme Court Funds (Amendment) Rules 1976 (S.I. 1976 No. 2235). Recently (September 1977) interest rates have fallen steeply.

Interest only upon judgment. No interest is recoverable upon money taken out of court [46]; neither is it recoverable upon an order giving leave to take it out,[47] and the Act gives no separate cause of action for interest.[48] Accordingly a plaintiff who wants interest must bargain for it, or pursue his claim to judgment or award.[49]

Summary judgment. Interest can be awarded under the Act.[50]

4. Discovery

In claims for a reasonable sum for cost plus profit early discovery may be very important to the employer, for the contractor's wage sheets and invoices for materials used will often show the employer the strength of the case he has to meet.[51]

5. Preliminary point

" Where there is a point of law *which, if decided in one way,* is going to be decisive of the litigation, advantage ought to be taken of the facilities afforded by the rules of court to have it disposed of at the close of pleadings or very shortly afterwards." [52] In building contract litigation the trial of preliminary issues of mixed fact and law is often ordered,[53] but careful thought is required before applying for such an order lest issues apparently separate are in reality so intermingled as to make the order inappropriate.

6. Plans, photographs, models, mechanically recorded evidence

Unless at or before the trial the court or a judge for special reasons otherwise orders or directs, no plan, photograph or model is receivable in evidence at the trial of an action unless at least 10 days before the commencement of the trial

---

[46] *Jefford* v. *Gee* [1970] 2 Q.B. 130 (C.A.). See further, as to payment-in, *post,* p. 270.

[47] *Newall* v. *Tunstall* [1970] 3 All E.R. 465; *Waite* v. *Redpath Dorman Long Ltd.* [1971] 1 Q.B. 294.

[48] *Ibid.*

[49] *Jefford* v. *Gee, supra.* For the possibility that a court might be persuaded to award interest as damages, see *ante,* p. 154.

[50] *Wallersteiner* v. *Moir (No. 2)* [1975] 1 Q.B. 373, 387 (C.A.). This decision is an example of the court awarding interest against a defaulting trustee, not under the Act, but in the exercise of its equitable jurisdiction. The rate ordered was 1 per cent. above the official bank rate or minimum lending rate in operation from time to time and with yearly rests which is usually higher than the ordinary rate; see text. For summary judgment see *post,* p. 278.

[51] For discovery and inspection of documents generally, see R.S.C., Ord. 24. For the special case of the liability of a person to discovery who has been involved in the tortious acts of others, see *Norwich Pharmacal Co.* v. *Customs and Excise* [1974] A.C. 133, (H.L.).

[52] *Per* Romer L.J., *Everett* v. *Ribbands* [1952] 2 Q.B. 198, 206 (C.A.); applied in *Carl Zeiss Stiftung* v. *Herbert Smith & Co.* [1969] 1 Ch. 93 (C.A.). See Ord. 18, r. 11; Ord. 33, rr. 3, 4 and notes thereto, in *Supreme Court Practice.* See also *Gold* v. *Patman & Fotheringham Ltd.* [1958] 2 All E.R. 497, 503 (C.A.); *Nissan* v. *Att.-Gen.* [1968] 1 Q.B. 286 (C.A.) (decisive of part only of litigation); *Goodchild* v. *Greatness Timber Co. Ltd.* [1968] 2 Q.B. 372 (C.A.); *Warner Bros. Records Inc.* v. *Rollgreen Ltd.* [1976] Q.B. 430 (C.A.).

[53] See R.S.C., Ord. 33, r. 3 and, for an example, *Gloucestershire County Council* v. *Richardson* [1969] 1 A.C. 480 (H.L.).

the parties, other than the party producing it, have been given an opportunity to inspect it and to agree to the admission thereof without further proof.[54] Mechanically recorded evidence is admissible.[55]

7. View

A view is sometimes helpful. " Where the matter for decision is one of ordinary common sense the judge of fact is entitled to form his own judgment on the real evidence of a view just as much as on the oral evidence of witnesses," [56] although " there may, of course, be cases in which the matter falls to be decided on technical evidence, and where a view will not be of any assistance, except no doubt as a substitute for a plan or photograph." [57] In any event each party must be given an opportunity of being present,[58] and it is undesirable for the judge even to have a " a mere look " without previously informing the parties.[59] It is a matter of discretion for the judge to decide whether or not to have a view.[60]

8. Various interlocutory matters

The provisions of Order 25 and a summons for directions in blank should be studied and all possible applications made on the first hearing of the summons for directions.

It may be desirable to join sub-contractors or others as third parties [61]; and in some cases, to apply for detention, preservation or inspection of property.[62]

The court can appoint a court expert,[63] or order trial with assessors,[64] or, in the Chancery Division, obtain the assistance of experts.[65] The usual practice is to refer prolonged investigations of fact [66] in building contract cases to an official referee (strictly, " a circuit judge discharging the functions of an official referee)." [67]

Delay in proceedings—the court

In recent years the courts have shown a greater willingness than in the past to strike out cases for delay. Where, as will ordinarily be the case, the limitation period has expired,[68] the order is "Draconian . . . and will not be lightly made." [69]

---

[54] Ord. 38, r. 5.
[55] *The Statue of Liberty* [1968] 2 All E.R. 195.
[56] *Buckingham* v. *Daily News Ltd.* [1956] 2 Q.B. 534, 551, *per* Denning L.J.
[57] *Ibid. per* Parker L.J. at p. 550.
[58] *Goold* v. *Evans* [1951] 2 T.L.R. 1189.
[59] *Salsbury* v. *Woodland* [1970] 1 Q.B. 324 (C.A.).
[60] R.S.C., Ord. 35, r. 8 (1); *Tito* v. *Waddell* [1975] 1 W.L.R. 1303.
[61] See R.S.C., Ord. 16 generally. For the position where there is an arbitration agreement with one but not with all the parties, see *ante*, p. 244.
[62] See R.S.C., Ord. 29, rr. 2, 3.
[63] Ord. 40; see notes in *Supreme Court Practice.*
[64] Ord. 33, r. 6.
[65] Ord. 32, r. 16.
[66] See *infra*, " References."
[67] Courts Act 1971, s. 25 (3).
[68] *Birkett* v. *James* [1977] 3 W.L.R. 38 (H.L.).
[69] *Allen* v. *Sir Alfred McAlpine & Sons Ltd.* [1968] 2 Q.B. 229, 259 (C.A.).

The court must be satisfied either (1) that the default has been intentional and contumelious ; or (2) (a) that there has been inordinate and inexcusable delay, and (b) that such delay will give rise to a substantial risk that it is not possible to have a fair trial, or is likely to cause serious prejudice to the defendants, either as between themselves and the plaintiff, or between each other, or between them and a third party.[70] The relevant principles have now been authoritatively stated by the House of Lords in *Birkett* v. *James*[70a] to the report of which reference should be made. Both the claim and counterclaim may be dismissed if plaintiff and defendant have been guilty of delay.[71] Periods of time bear a different complexion in cases of magnitude and complexity than they do in simple accident cases,[72] but building cases are not exempt from the Rules of the Supreme Court.[73]

Delay in proceedings—arbitration

An arbitrator does not have power to dismiss a claim for inordinate delay.[74] The remedy of a party prejudiced by delay is, it seems, to apply to the arbitrator to fix a date for the hearing.[75] The arbitrator should notify the parties that he will, in the absence of one of them, proceed *ex parte*, and the notice of the appointment for the hearing should be marked " peremptory." [76]

9. Restrictions on adducing expert evidence

Order 38, Pt. IV, must be referred to by any party wishing to rely on expert evidence. Very roughly, the position is that, unless all parties agree, expert evidence cannot be adduced without the leave of the court or where application has been made for a direction in respect of the evidence. The rule does not apply to evidence which is permitted to be given on affidavit, *e.g.* upon interlocutory proceedings.

10. Settlement of actions

In settling an action the parties should decide whether they wish to be able to refer the action back to the court or whether they are content to rely upon the terms of the settlement as a contract and, if necessary, bring fresh proceedings in respect of any breach of such contract.[77] Interest which might be awarded under the Law Reform (Miscellaneous Provisions) Act 1934, s. 3,[78] if the matter

---

[70] *Birkett* v. *James, supra.*

[70a] *Supra.*

[71] *Zimmer Orthopaedic* v. *Zimmer Manufacturing Co.* [1968] 1 W.L.R. 1349 (C.A.).

[72] *City General Insurance Co. Ltd.* v. *Robert Bradford & Co. Ltd.* [1970] 1 Lloyd's Rep. 520, 523 (C.A.).

[73] *Renown Investments Ltd.* v. *Shepherd Ltd., The Times,* Nov. 16, 1976 (C.A.).

[74] *Crawford* v. *Prowting (A. E. A.) Ltd.* [1973] Q.B. 1.

[75] *Ibid.*; *Russell on Arbitration* (18th ed.), p. 222.

[76] Russell, *ibid.*, cases cited at pp. 223–224.

[77] See *Green* v. *Rozen* [1955] 1 W.L.R. 741 for a most useful discussion by Mr. Justice Slade of various ways of settling actions. See also the *Encyclopaedia of Court Forms* (2nd ed.), Vol. 12, title " Compromise," where Master Jacob discusses 16 methods of settlement and provides precedents.

[78] See *ante,* p. 254.

proceeded to judgment should be taken into account. There cannot be separate proceedings to claim interest alone under the Act.[79]

Section 3: *References*

Trial

Where upon application by any party interested the court considers that, having regard to the nature of the case, it is desirable (whether on grounds of expedition, economy or convenience or otherwise) in the interest of one or more of the parties, the court may, subject to any right to a trial with a jury,[80] order that the cause or matter, or any question or issue of fact arising therein, shall be tried before an official referee,[81] with or without assessors.[82]

It was held that a claim against a supplier of tubes for central heating which affected its standing as a supply company should not be tried by an official referee because there was no appeal on the issues of fact involved.[83] But a claim against a civil engineer for £500,000 for damages for professional negligence was ordered to be tried by an official referee because there was an appeal on the issues of fact.[84]

Inquiry and report

The court may, subject as above, refer to an official referee[85] for inquiry and report any question or issue of fact arising in any cause or matter.[86]

Appeals

There is a right of appeal on a point of law or as to costs to the Court of Appeal from a decision of an official referee but no right of appeal on facts except on a decision relevant to a charge of fraud or breach of professional duty.[87]

Practice

Although an order for a reference can be made at any time and may be made at trial it is commonly made by the master at the close of pleadings upon the

---

[79] *Ibid.*

[80] See R.S.C., Ord. 33, r. 5. Trial by jury is very unusual in civil actions; see *Williams* v. *Beesley* [1973] 1 W.L.R. 1295 (H.L.).

[81] Strictly, "a circuit judge discharging the functions of an official referee": Courts Act 1971, s. 25 (3).

[82] Ord. 36, r. 1. This order provides a code relating to references and should be referred to with the notes in *Supreme Court Practice*. Only a brief outline is given here. By consent trial may be before a master or a special referee (a person not an officer of the court): Ord. 36, r. 9. The parties may by agreement refer the dispute to an arbitrator or official referee for arbitration: Arbitration Act 1950, s. 11. For county court procedure, see *The County Court Practice*. Note that the fees payable to an arbitrator or special referee are usually much higher than the court fees payable on reference to an official referee or, in the county court, to the registrar.

[83] *Simplicity Products Co.* v. *Domestic Installations Co. Ltd.* [1973] 1 W.L.R. 837 (C.A.). For appeals, see *infra.*

[84] *Scarborough R.D.C.* v. *Moore* (1968) 112 S.J. 986 (C.A.). For appeals, see *infra.*

[85] Or a special referee, Ord. 36, r. 8.

[86] Ord. 36, r. 2. Under this rule the court may act of its own motion.

[87] Ord. 58, r. 5. For the difference between fact and law, see *Instrumatic Ltd.* v. *Supabrase Ltd.* [1969] 1 W.L.R. 519 (C.A.); *Peak Construction (Liverpool) Ltd.* v. *McKinney Foundations Ltd.* (1971) 69 L.G.R. 1, 8, 13 (C.A.).

summons for directions. Alternatively it may be made upon application at any time, and in some cases it may be made upon a summons under Order 14 where leave to defend is granted, for example, where there is a claim for work done and an affidavit by the defendant alleging defective workmanship. After a reference to an official referee has been made an application for directions should be made to him within 14 days.[88] The practice thereafter particular to official referees is set out in a practice direction issued by Sir Walker Carter Q.C., Senior Official Referee, on July 8, 1968, and printed below in full.

Practice direction of July 8, 1968

" 1. At the time of the issue of the first summons before the official referee, a copy of all the pleadings, including particulars, already served should be lodged with his clerk so that they can be considered by the official referee before the hearing of the summons. Such copy of the pleadings may be collected from the clerk to the official referee for the purpose of preparing a bound copy of the pleadings for use at the trial.

2. At the hearing of the first summons before the official referee, the solicitors of the parties or their London agents should be in the position to state the nature of the claim and of the defence. Failure in this respect may result in unnecessary adjournments with attendant costs.

3. At the hearing of the first summons before him, the official referee will give the necessary directions and make the necessary orders regarding the steps in the action to be taken by the parties. It is of the utmost importance that these steps should be taken within the time-limits set by the direction or order, so that the practice of giving a fixed date for the trial may be continued.

4. Once an action has been given a fixed date for trial, no alteration will be granted except with the leave of the official referee which will be granted only in exceptional circumstances. If a fixed date for trial is vacated, the fresh date for trial may not be a fixed date.

5. Where a party intends to adduce expert evidence, he should produce to the other party his expert's statement of proposed evidence, together with any reports, plans, models, calculations, etc., relevant to it, for agreement if possible. Failing such agreement, the other party should deliver to the first party a written statement setting out particulars of the matters not agreed. Where both parties intend to adduce expert evidence, each should follow this procedure. Failure by any party to follow this procedure may result in a special order as to costs." [89]

Section 4: Contractor's " Claims "

There seems to have been an increasing tendency for contractors, or surveyors on their behalf, to spend much time in the preparation of claims for payment. The subject is dealt with briefly here because it is closely related to, and not

---

[88] Ord. 36, r. 6, and see notes in *Supreme Court Practice*. An existing summons for directions can be restored from time to time for further directions.

[89] [1968] 1 W.L.R. 1425. As to direction 5, this must be read subject to the provisions of R.S.C., Ord. 38, Pt. IV imposing restrictions on adducing expert evidence, see, *ante*, p. 258.

infrequently results in, litigation. Claims prepared without a careful consideration of the legal principles upon which they may be based are usually a waste of time of both the person who prepares them and he who is asked to read them. Further, the form in which a claim is presented can provide a discipline for the person who prepares it, so as to ensure that he obtains and sets out the necessary information and by so doing assists the person who reads it to see whether, and to what extent, if any, there are grounds for payment.

Meaning and nature of " Claims "

No exact meaning can be given to the term. It is used here to describe an application for payment by the contractor arising other than under the ordinary provisions for payment of the measured value of the work. A claim in this sense can arise either as a legally enforceable claim or *ex gratia*. The former class of claim comprises a right to payment arising under the contract, a claim for damages for breach of contract and, in the less usual case, under some other head of law as, *e.g.* in tort or for breach of copyright or for misrepresentation.

Preparing the claim

No useful general comment can be made upon the preparation of claims for *ex gratia* payments save to say that where they are made pursuant to some statement of principle accepted by the employer some of the suggestions below may be usefully adapted to bring the claim within the principles. What follows refers to the preparation of a claim enforceable in law.

The claim should be prepared from the beginning with litigation (which includes arbitration) in mind. This is for two reasons. The first is because the sanction for non-payment of a claim is that litigation will follow and if it is a good claim in law, the employer will have to pay the costs and interest as well as whatever is due on the claim. The claim document should therefore be such as to persuade him of the likelihood of this result if he does not pay. The second reason is that if litigation does follow it saves both time and money if the claims document has been so prepared that it can be used as, or form the basis for, the contractor's detailed pleading. In particular the person preparing the claim should have in mind that if there is litigation the contractor is likely to be required to set out most of the material necessary to establish his right to payment in schedule form (see next section and in particular *post*, p. 266). Therefore, whether or not the claim is originally prepared as a schedule it is usually prudent to assemble the material in such a way that it can be easily translated into a schedule. As an example, consider a claim made under clause 24 (1) (*a*) of the Standard form of building contract in respect of many cases of late delivery of drawings. It is necessary in respect of each such case to state the following: the date when the contractor specifically applied in writing for the drawing; the due time when he should have received it; the date when he actually received it; the written application for direct loss of expense; sufficient particulars of such loss or expense.

In all cases the ground in law relied on should be stated. If it is made under a clause of the contract give its number. If it is made under an implied term set

out the term and the reasons for the implication. If it is a claim for breach state
the term of which it is alleged there is a breach and set out the facts relied on as
constituting the breach. If it is a variation describe it and refer to the relevant
order. In all cases refer to relevant documents. The loss or other monetary
claim should be stated showing how it is calculated and why it was caused by
the ground relied on. Where the claim is made under various alternatives in law
say so. Thus, where the Standard form of building contract is used there may be
a claim in respect of late delivery of drawings. Such a claim is ordinarily made
for damages for breach of clause 3 (4), or alternatively under clause 24 (1) (a).

General words intended to arouse sympathy or other emotion favourable to
the contractor's interest should be used sparingly lest they excite a suspicion
that the contractor really thinks his claim is founded in mercy and not in law.
Further, it is rarely useful to say that the contractor has suffered loss due to
causes outside his control or which he did not reasonably anticipate. It is
elementary law (see Chaps. 4, 5 and 7) that such factors are not grounds in law
for extra payment unless, which is rare, they are expressly so made.[90] If put
forward as the sole ground of a claim the reader may well assume that there
are no other grounds.

Section 5: Scott or Official Referee's Schedule

The object of this document is to define and state the issues clearly by
assembling all the relevant allegations and admissions in tabular form. If carefully drawn it will prevent confusion arising from the necessity of referring to
two or three documents in respect of one matter of detail, and it will save
valuable time and energy at trial. Moreover, when it has been completed the
parties may find that over a large number of small issues there is so little difference
between them that they can be settled, thus reducing the dispute to a few
substantial matters.

Settling the headings

There is no set form prescribed by rules nor even by the textbooks, but the
forms suggested below should be found adequate for most purposes. They can,
if necessary, be adapted in accordance with common sense in order to present
the issues in any particular case as clearly and conveniently as possible. It is
often desirable for the headings of the various columns to be settled before the
official referee at the hearing of the summons for directions.[91] The parties can
then explain what the issues are and suggest the appropriate headings; in the
case of differences the official referee will determine what they should be. The
schedule either is ordered to be or is adapted from a pleading. The rules as to
pleadings (see R.S.C., Ord. 18, and ante, p. 252) apply generally, but with such
modifications as the official referee or arbitrator may order or approve.

---

[90] For an example where they are material in a claim under a contract, see I.C.E.
Conditions, cl. 12.

[91] There is no summons for directions as such in an arbitration, but a preliminary
meeting between the arbitrator and the legal representatives of the parties is usually
arranged to settle the issues and deal with the matters ordinarily dealt with on a summons
for directions.

Form 1.   Defects alleged by employer

| Serial No. | Contract Work | DEFENDANT'S | | PLAINTIFF'S | | Official Referee's Comments | |
		Com-ments	Estimate of Loss	Com-ments	Estimate of Loss		
1	2	3	4	5	6	7	8

NOTES

Generally. This form is suited to a dispute arising out of a lump sum contract where there has been substantial completion,[92] and the defendant (the employer) admits the claim subject to his counterclaim or set-off for defects and omissions.[93] The defendant will normally draft the first four columns and then submit the schedule to the plaintiff (the contractor) for his comments. Alternatively if the plaintiff has already delivered a reply dealing with the defendant's particulars the defendant can set out the relevant parts of the reply in column 5.

Column 1. Serial numbers should be inserted here for reference purposes. If items in the specification or bills of quantities are numbered the item number should appear in column 2.

Column 2. This sets out what the employer alleges should have been done to comply with the contract. If the contract is to do work according to a specification or bills of quantities it should refer to the relevant item; if the work is a variation it should state what ought to have been done as a variation, giving particulars both of the original work, and of any variation order. If the quality of the work was not specified it should set out what ought to have been done in order to complete the work in accordance with the implied terms set out *ante*, p. 38.

Column 3. This sets out the defects alleged.

Column 4. This column sets out the amount of the defendant's claim. In the ordinary lump sum contract the measure of his claim is the difference in value between the work as done and the work as it should have been done to comply with the contract, and this is usually calculated by the cost of rectification, *i.e.* the cost of making the defective work correspond with the quality and quantity of work required by the contract.[94] If all the items are calculated on this basis the headings of columns 4 and 6 can be altered to " Cost to Rectify." There may be cases in which in the circumstances the cost of rectification would be so great in comparison with the nature of the defect that a reasonable employer would be satisfied with the work as done plus damages for the difference in value

---

[92] See Chap. 4.
[93] See *Hoenig* v. *Isaacs* [1952] 2 All E.R. 176, 181 (C.A.).
[94] *Hoenig* v. *Isaacs, supra; Dakin* v. *Lee* [1916] 1 K.B. 566 (C.A.), and see *ante*, p. 148.

between the whole contract works [95] with the defects, and what their value would have been without the defects.[96] In such cases it is advisable to write " diminution of value " or otherwise to indicate that the cost of rectification is not being claimed. In cases where there is doubt about the principle by which the defendant's claim will be calculated column 4 can be subdivided and figures inserted in each subdivision, it being stated that the lesser sum is claimed as an alternative.

Column 5. In this column the plaintiff sets up his defence to the claim admitting, not admitting, or denying the allegations of defects or raising an objection in point of law.[97] He may admit that the work was carried out as alleged, but say that it was done in pursuance of a variation order, giving particulars, or he may plead an architect's final certificate, or waiver or some special defence under the contract.[98] If such a plea has been set out in full earlier in the schedule or in the body of the pleadings it is sufficient to refer back.

Column 6. Upon ordinary principles of pleading the plaintiff need insert nothing in this column.[99] But the object of the schedule is to gather together before trial in one document all the relevant contentions of the parties. In practice therefore a sum is usually inserted although there may have been a denial of liability in the previous column, the effect being of a plea in the alternative that, " if, which is denied, the plaintiff is liable, the amount recoverable is £x." The advantage of this course is that it shows to what extent the parties are at issue over damages.

Columns 7 *and* 8. These columns are inserted for the use of the official referee.

Form 2.   Defects alleged : third parties or more than one defendant

The form will be as Form 1, save that special directions may have been given by the court [1] or official referee and extra columns must usually be added for each party interested in the allegations. In some cases it may be unnecessary to insert a column for a particular party. This may arise where, for example, a main contractor has made it clear that he will at trial confine himself to the issue of a claim for an indemnity or contribution from a sub-contractor who has carried out the allegedly defective work. In such a case the main contractor and the sub-contractor may have agreed between themselves, with a view to saving costs, that only the latter should plead to the defects and call evidence about the defects.

---

[95] *i.e.* the house or whatever else was being built.

[96] There is no direct authority for this proposition but it is implicit, it is submitted, in *Dakin* v. *Lee, supra,* and *Hoenig* v. *Isaacs, supra,* it is not inconsistent with the principle of reasonable foreseeability (see *ante,* p. 145), and can be supported by reference to the principle of mitigation of damages (see *ante,* p. 147), and by analogy with the measure of damages payable by a tenant to his landlord for breach of a covenant to repair (see *Woodfall on Landlord & Tenant* (27th ed.), para. 1516).

[97] See Notes to Ord. 18, r. 13, in *Supreme Court Practice.*

[98] Such a defence may be suitable for hearing as a preliminary point. See *ante,* p. 256.

[99] See R.S.C., Ord. 18, r. 13 (4).

[1] See R.S.C., Ord. 16, r. 4 (4) for third party directions. See *ante,* p. 244 for the absence of such power on the part of an arbitrator.

Form 3.   Extras: claim by contractor

Serial No.	Extra work carried out	PLAINTIFF'S		DEFENDANT'S		Official Referee's Comments	
		Com- ments	Price	Com- ments	Price		
1	2	3	4	5	6	7	8

NOTES

Generally. This form is suitable for a dispute where the contractor is the plaintiff and has issued a claim for extras, and he will normally first draft columns 1–4.[2]

Column 2. This sets out the work alleged to have been done.

Column 3. This deals by way of reply with any special defence raised in the pleadings such as absence of an order, or of an order in writing,[3] giving where appropriate particulars of the order or other matters relied on. It may also deal by way of reply with any allegations of defective work raised by the defendant (see *infra*).

Column 5. The defendant (employer) sets up his defence in the usual way giving particulars of any defects upon which he relies by way of set-off in diminution of the price claimed. Where the plaintiff is claiming a reasonable sum and the defendant alleges that the sum claimed is excessive and unreasonable there is authority stating that the defendant need not give particulars of what he alleges to be a reasonable sum.[4] In practice it is usually desirable that the defendant should give such particulars and the official referees usually order them to be given.[5]

Counterclaim. If the defendant has a counterclaim he should, after completing columns 5 and 6, prepare a separate schedule headed "Counterclaim." This will be as in Form 1 save that the serial numbers will be continued from the schedule for the claim. The plaintiff will complete the counterclaim schedule in the usual way.

Form 4.   Reasonable sum: claim by contractor

Form 3 can be adapted by heading column 2 "Work carried out." The content of the columns will vary considerably according to the nature of the issues; thus the amount of the work may be agreed, the only dispute being

---

[2] Subject to these special notes the notes to Form 1 apply with the necessary modifications.

[3] See *ante*, p. 68 (contractual requirement of order in writing).

[4] *James* v. *Radnor C.C.* (1890) 6 T.L.R. 240 (C.A.).

[5] See para. 18/12/9, *Supreme Court Practice.*

the pricing, in which case column 2 may consist of summaries of costs, *e.g.*
" Plasterer: 24 hrs. at £1.25 per hour."

Form 5. Schedule contract: claim by contractor
NOTES

Form 3 can be adapted. In column 3 the contractor should state which part
of the contract schedule he relies on. If it provides for varying percentage
additions columns 3 and 4 can be adapted as follows:

PLAINTIFF'S			
Comments	Net Cost	Percentage Increase	Price Claimed
3	4	5	6

It will be a question in each case whether the next two columns should be
correspondingly adapted, or whether it is sufficient for the defendant to set out
his allegations as comments and the amount he considers should be paid under
" price."

In many cases the official referee may direct that a formal schedule is unnecessary and that it is sufficient if the plaintiff supplies lists of labour and
materials (see *infra*).

Cost plus profit contract

In a cost plus profit contract where the only issue is the amount claimed by
the contractor a schedule is unnecessary unless its preparation has been directed
by the official referee or the arbitrator. All that is normally required are two lists:

(i) A clear summary of the time sheets showing the time spent by each class
of labour or by each workman, showing the rates actually paid, insurances, etc.

(ii) A list of materials used upon the work setting out the actual cost; giving
references to invoices, receipts and other documents.

Form 6. Contractors " Claims "

For the meaning given to this term, see *ante*, p. 260. Form 3 can be adapted.
The nature of the claim, *i.e.* whether it is for payment under one of the terms of
the contract or for damages for breach of contract,[6] or, unusually, in respect of
some other cause of action, will suggest the nature of the headings. Examples
are given *ante*, p. 261, for a claim in respect of late issue of drawings where the
Standard form of building contract is used.[7]

Sometimes it is found convenient to use a chart as supplementary to, or, even,
as an alternative to a schedule. This may arise where, for example, there is a

---

[6] For damages, see *ante*, p. 144.
[7] For a discussion of claims for loss and expense and for damages arising under the
Standard form of building contract, see *post*, pp. 302 and 316.

contract programme in accordance with clause 14 of the I.C.E. Conditions or agreed at the commencement of the contract with the architect where the Standard form of building contract is used, and the claim is for delay in the issue of drawings or instructions. If the programme is in, or capable of being expressed in, traditional bar-chart form it will set out the periods within which various activities were to be carried out. The contractor can insert, in a colour different from that in the original bars, the periods within which the works were in fact carried out. He then inserts in respect of each activity the date when he alleges he should have received the necessary instruction or drawing referring to any relevant applications for information, and the date or dates when he in fact received them. When they were received over a long period and in various ways, *e.g.* by formal instruction, by letter, by the so-called " site instruction," by oral instruction confirmed in writing or simply by oral instruction, it may be convenient to use a code in order to achieve clarity without excessive size of the chart.

Damages. They are frequently in issue and their determination may require long and detailed inquiry. This is greatly assisted by full and detailed particulars supported by reference to the documents which will be relied on. Reference should be made to the heading " Employer's breach of contract," *ante,* p. 151, for assistance in formulating the claim. No general rule can be stated as to the use of a schedule or headings. Sometimes a schedule is not appropriate at all: in other cases it may be useful but the headings may have to be varied from time to time within the schedule according to the various heads of damage alleged. In all cases there should be a sufficient pleading of the causal connection between the quantum of damages claimed and the breach alleged, or the term of the contract under which the claim arises or other allegation of liability.

Section 6: *Costs*

Costs in building contract cases may be heavy. The pleadings are often long and involved; interlocutory applications are frequently necessary; arbitrator's fees and expenses are substantial; proceedings tend to be lengthy and expert witnesses are usually employed. In the result the total costs may exceed the sum in dispute, and it is therefore important at all stages of the proceedings to have the question of costs well in mind.[8] It is proposed here to deal with some of those matters which most frequently affect costs.[9]

1. Costs are discretionary

Costs are in the discretion of the trial judge[10]; in general, and subject to

[8] In certain circumstances the defendant may apply for security for costs, see R.S.C., Ord. 23, and in the case of companies, Companies Act, s. 447; *Sir Lindsay Parkinson & Co. Ltd.* v. *Triplan Ltd.* [1973] Q.B. 609 (C.A.); *Sloyan, T. & Sons Ltd.* v. *Brothers of Christian Instruction* [1974] 3 All E.R. 715.

[9] For a full account, see *Supreme Court Practice* in particular R.S.C., Ord. 62 and the notes thereto, and for county courts, see *The County Court Practice,* published annually. For litigants in person, see the Litigants in Person (Costs and Expenses) Act 1975.

[10] Supreme Court of Judicature (Consolidation) Act 1925, s. 50; R.S.C., Ord. 62, r. 2 (4). In the county court, see County Court Rules, Ord. 47, r. 1. An arbitrator must

certain special rules, costs follow the event,[11] that is to say, a successful party is awarded the costs unless there has been something in the nature of misconduct on his part.[12]

Two defendants. If a plaintiff reasonably joins two defendants in the alternative and succeeds against one, the court in its discretion may order the unsuccessful defendant to pay the costs of the successful defendant; the order may be for payment direct by one defendant to the other (a " Bullock " order), or for payment to the plaintiff against whom an order for payment of costs in favour of the successful defendant is made (a " Sanderson " order).[13]

Third parties. When a defendant has reasonably joined a third party, an unsuccessful plaintiff can, in the exercise of the court's discretion, be ordered to pay the third party's costs, either by adding them to the costs he has to pay to the defendant or direct to the third party.[14]

2. Recovery of small sums

A plaintiff may, if he wishes, commence proceedings in the High Court to recover sums within the jurisdiction of the county court, but he is discouraged from claiming less than £1,200 by being penalised in costs unless the claim is clearly suitable for Order 14 procedure or there are special circumstances.

Summary of section 47 of the County Courts Act 1959

Where an action founded on contract or tort is commenced in the High Court which could have been commenced in the county court and the plaintiff recovers less than £1,200, he is normally entitled only to county court costs, and if he recovers less than £350 he is normally not entitled to any costs.[14a] The section is not, however, to affect any question as to costs if it appears to the High Court, or referee or officer of the Supreme Court before whom it was tried, that there was reasonable ground for supposing the amount recoverable in respect of the plaintiff's claim to be in excess of the amount recoverable in an action commenced in the county court.[15] And the court may order costs either on the High Court scale or on a county court scale if satisfied: (a) that there was sufficient reason

---

follow the ordinary rule, see *Dineen* v. *Walpole* [1969] 1 Lloyd's Rep. 261 (C.A.) distinguishing *Perry* v. *Stopher* [1959] 1 W.L.R. 415 (C.A.). See also *The Erich Schroeder* [1974] 1 Lloyd's Rep. 192; *Centrala Morska* v. *Cia Nacional* [1975] 2 Lloyd's Rep. 69—case remitted for reconsideration by arbitrator who had not given costs to the apparently successful party.

[11] R.S.C., Ord. 62, r. 3 (2).

[12] *Ibid., Donald Campbell & Co. Ltd.* v. *Pollak* [1927] A.C. 732, 811, 812 (H.L.); R.S.C., Ord. 62, r. 7.

[13] See *Mayer* v. *Harte* [1960] 1 W.L.R. 770 and para. 62/2/39 of *Supreme Court Practice.*

[14] *Edginton* v. *Clark* [1964] 1 Q.B. 367 (C.A.); see also *L. E. Cattan Ltd.* v. *Michaelides & Co.* [1958] 1 W.L.R. 117 for observations on the position where there are " string " contracts with more than three parties involved.

[14a] s. 47 (1) as amended and set out in para. 3916, *Supreme Court Practice.* **See also** County Courts Jurisdiction Order 1977.

[15] *Ibid.*

for bringing the action in the High Court [16]; or (b) that the defendant or one of the defendants objected to the transfer of the action to a county court. Further, where the claim is for a debt or liquidated demand only for £350 or upwards, and there is no defence, fixed High Court costs may in certain circumstances be awarded.[17]

Notes to section 47

The test whether an action could have been commenced in a county court is, in money claims, whether the amount recovered on the claim, not the amount claimed on the writ, is within the jurisdiction of the county court.[18] Thus the section will operate if the defendant by way of his defence reduces the amount claimed so as to bring it within the limit. A set-off is a defence and if established may have this effect,[19] but a counterclaim is a cross-action and does not operate to reduce the claim to within the limit.[20] Thus in *Chell Engineering* v. *Unit Tool Co.*, there was a claim for £909. The defendants denied indebtedness and counterclaimed for damages for delay, work not properly done, etc., pleading in their defence and counterclaim that they " claim to set off an equal sum parcel thereof against the amount (if any) found due " to the plaintiffs and ending with a counterclaim for damages. The judge found in favour of the plaintiffs on their claim for £350 and in favour of the defendants for £300 and set off the latter figure against the claim leaving a balance in favour of the plaintiffs of £50, upon which it would appear that the plaintiffs would only recover county court costs. The Court of Appeal held that the defendants had been awarded £300 on a counterclaim and that it was not a set-off.[21] Consequently the plaintiffs had recovered £350 on a claim which could not have been commenced in the county court [22] and were therefore entitled to High Court costs on their claim, the defendants having High Court costs on their counterclaim.[23]

Section 47 does not apply to counterclaims. Therefore, where a plaintiff commences proceedings in the High Court and the defendant succeeds on a

---

[16] s. 47 (3). For an example, see *Columbus* v. *Clowes* [1903] 1 K.B. 244, referred to *ante*, p. 217.

[17] R.S.C., Ord. 62, App. 3 and see note 14a. Money due as an instalment is a debt or liquidated demand—*Workman, Clark & Co. Ltd.* v. *Lloyd Brazileno* [1908] 1 K.B. 968 (C.A.).

[18] *Lovejoy* v. *Cole* [1894] 2 Q.B. 861 (D.C.); *Solomon* v. *Mulliner* [1901] 1 K.B. 76 (C.A.).

[19] *Lovejoy* v. *Cole* [1894] 2 Q.B. 861 (D.C.); *Solomon* v. *Mulliner* [1901] 1 K.B. 76 (C.A.), disapproving *Goldhill* v. *Clarke* (1893) 68 L.T. 414. These cases were decided upon s. 57 of the County Courts Act 1888, where the wording was somewhat different. See also County Courts Act 1959, s. 39 (2).

[20] *Chell Engineering* v. *Unit Tool & Engineering Co.* [1950] 1 All E.R. 378 (C.A.).

[21] On the question whether or not the cross-claim for damages for delay, bad work, etc., could be set off, *Hanak* v. *Green* [1958] 2 Q.B. 9 (C.A.), discussed *post*, p. 276, is, it is submitted to be preferred to the *Chell Engineering* case, but the *Chell Engineering* case is authority on the effect of a counterclaim with regard to the County Courts Act 1959.

[22] The limit of county court jurisdiction then being £200.

[23] The payment into court by the defendants of £250 did not secure their costs of the claim. The current rules differ.

counterclaim for however small an amount, the defendant has his costs on the High Court scale.[24]

The distinction between a set-off and a counterclaim is discussed *post*, p. 276.

3. Notice to admit facts (Ord. 27, r. 2)

A notice to admit specific facts may save costs. If the party on whom the notice was served neglects or refuses to admit the facts he must normally pay the costs of proving the facts whatever the result of the case.[25]

4. Admissions on summons for directions

At the hearing of the summons, the master must endeavour to secure that the parties make all admissions and all agreements as to the conduct of the proceedings which ought reasonably to be made by them and may cause the order on the summons to record any admissions or agreements so made and (with a view to such special order, if any, as to costs as may be just being made at the trial) any refusal to make any admission or agreement.[26]

Judgment can be signed if there is a clear admission,[27] although an application under R.S.C., Ord. 14 for summary judgment (see *post*, p. 278) may sometimes be more appropriate.[28]

5. Payment into court

In any action for a debt or damages the defendant may at any time after appearance, upon notice to the plaintiff in a prescribed form (No. 23 in Appendix A of the Rules), pay into court a sum of money in satisfaction of the claim or, where two or more causes of action are joined in one action, a sum or sums of money in satisfaction of any or all of the causes of action.[29] The amount paid in can be increased.[30]

A plaintiff or other person made defendant to a counterclaim may pay money into court in the same manner as a defendant, the rules applying with the necessary modifications.[31]

Taking money out

Before trial. Within 21 days after receipt of the notice of payment into court or,

---

[24] *Blake* v. *Appleyard* (1878) 3 Ex.D. 195; *Amon* v. *Bobbett* (1889) 22 Q.B.D. 543 (C.A.). When faced with this prospect a payment in by the plaintiff should be considered (*infra*), or, possibly, a written offer: see p. 274.

[25] R.S.C., Ord. 62, r. 3 (5).

[26] R.S.C., Ord. 25, r. 4.

[27] *Technistudy Ltd.* v. *Kelland* [1976] 1 W.L.R. 1042 (C.A.).

[28] *Ibid.* p. 1046.

[29] R.S.C., Ord. 22, r. 1 (1) (2). See the text of the rule, and notes in *Supreme Court Practice* for the plea of tender. For the position where two or more causes of action are joined in the action, see further, Ord. 22, r. 1 (3) (4). For the right to sign judgment under Ord. 27, r. 3, on admissions, although there has been payment in under Ord. 22, r. 1, see *Lancashire Welders Ltd.* v. *Harland & Wolff Ltd.* [1950] 2 All E.R. 1096 (C.A.).

[30] R.S.C., Ord. 22, r. 1 (3). For withdrawal of the money where there is a change of circumstances, see *Peal Furniture Co. Ltd.* v. *Adrian Share (Interiors) Ltd.* [1977] 1 W.L.R. 464 (C.A.).

[31] R.S.C., Ord. 22, r. 6.

where more than one payment has been made, within 21 days after receipt of the notice of the last payment but, in any case, before the trial or hearing of the action begins, the plaintiff may, by giving notice in the prescribed form, accept the sum or sums paid in satisfaction of his claim, or of the cause or causes of action to which the sum or sums relate.[32] If the plaintiff has taken out the sum in satisfaction of his claim or if he accepts a sum or sums in respect of one or more specified causes of action and gives notice that he abandons the other cause or causes of action, he may, unless the court otherwise orders, tax and recover his costs down to the date of payment in.[33]

At trial. Where after the trial or hearing of an action has begun money is paid into court or money in court is increased, the plaintiff may accept the money within two days after receipt of the notice of payment or further payment but in any case before the judge begins to deliver judgment,[34] but is not entitled to tax his costs,[35] so that the question of costs will have to be dealt with by the judge in his discretion.

Order for payment out

If money is not taken out in accordance with the rules it can only be taken out pursuant to an order of the court made at any time before, at or after trial [36] and the court can,[37] and, it seems, normally will, upon making such order,[38] give the defendant the costs after payment in. Ordinarily, no application to take money out while the trial is proceeding should be made without the consent of the defendant.[39]

Failure to take money out—costs

The court in exercising the discretion as to costs must, to the extent, if any, appropriate in the circumstances, take into account any payment into court and its amount.[40] If the plaintiff does not take the money out and recovers more than the sum paid in, the position is unaffected by the payment in save that at the conclusion of the trial the plaintiff must ask for an order for the money to be paid out.[41] If the plaintiff recovers a sum which is no more than the amount paid in and liability was admitted so that the only issue was quantum, the defendant is the successful party after payment in and is normally granted the costs after the date of payment in, and the plaintiff the costs before payment

---

[32] R.S.C., Ord. 22, r. 3 (1). For the position where there is a counterclaim, see *post*, p. 273.

[33] R.S.C., Ord. 62, r. 10 (2).

[34] R.S.C., Ord. 22, r. 3 (2).

[35] R.S.C., Ord. 62, r. 10 (4).

[36] R.S.C., Ord. 22, r. 5.

[37] *Associated Recordings (Sales)* v. *Thomson Newspapers* (1967) 111 S.J. 376 (C.A.).

[38] *The Katcher I (No. 2)* [1968] 3 All E.R. 350, 555; *Gaskins* v. *British Aluminium Co. Ltd.* [1976] Q.B. 524, 530 (C.A.).

[39] *Gaskins* v. *British Aluminium Co. Ltd., supra,* and see *infra*, " Non-disclosure of payment into court."

[40] R.S.C., Ord. 62, r. 5 (*b*).

[41] See *supra*, n. 36.

in.[42] Where liability is denied so that there are two distinct issues, (1) liability, (2) quantum, and the plaintiff recovers no more than the amount paid in the plaintiff was, formerly, sometimes, granted the costs of the issue upon which he had succeeded, *i.e.* liability, and the costs before payment in, and the defendant, the costs incurred after payment in on the issue of damages.[43] But now the costs on both issues after payment in are normally awarded to the defendant,[44] for, " a payment into court is an offer to dispose of the action and, if accepted, prevents all further costs. A plaintiff who continues an action after payment in takes a risk and cannot normally complain if he has to pay all the costs which his acceptance of the offer would have avoided." [45]

The plaintiff is entitled to the full period of 21 days provided by the rules to consider whether or not to take the money out. If there is less than 21 days between receipt of the notice of payment in and the commencement of the trial a plaintiff who does not recover more than the amount of the payment in remains the successful party and therefore ordinarily entitled to his costs, subject always to the discretion of the court.[46]

Amendment after payment in

A payment in relates only to the causes of action pleaded at the date of payment in.[47] If the amendment does not alter the causes of action relied on but only facts or figures it seems that, while the particular circumstances must be considered,[48] the usual approach is to look at the case as pleaded at the time of payment in.[49]

Non-disclosure of payment into court

The fact that a payment of money into court has been made must not be disclosed on the pleadings, or at trial, until all questions of liability and of debt or damages have been decided.[50]

Assessment of amount

The amount of money to be paid into court should be carefully assessed.[51]

---

[42] See R.S.C., Ord. 62, r. 5 (*b*); *Findlay* v. *Railway Executive* [1950] 2 All E.R. 969 (C.A.). The matter is still within the discretion of the trial judge, but he must exercise his discretion judicially (*ibid.*). See also *Wagman* v. *Vare Motors* [1959] 1 W.L.R. 853 (C.A.).

[43] *Wagstaffe* v. *Bentley* [1902] 1 K.B. 124 (C.A.); *Fitzgerald* v. *Thomas Tilling Ltd.* (1907) 96 L.T. 718 (C.A.); *Willcox* v. *Kettell* [1937] 1 All E.R. 222. But see *British Railways Board* v. *Jenkinson* [1968] C.L.Y. 3074 (C.A.).

[44] *Hultquist* v. *Universal, etc., Engineering Co. Ltd.* [1960] 2 Q.B. 467 (C.A.). See also para. 22/5/3 in *Supreme Court Practice.*

[45] *Ibid.* at p. 481, *per* Sellers L.J.

[46] *Bowen* v. *Mills & Knight Ltd.* [1973] 1 Lloyd's Rep. 580, 590.

[47] *Tingay* v. *Harris* [1967] 2 Q.B. 327 (C.A.) decided on County Court Rules which then differed from the current Ord. 22, but the principle applies. For this purpose an amendment does not relate back to the date of issue of the writ as it does for purposes of the Limitation Act 1939.

[48] See *The Katcher I (No. 2)* [1968] 3 All E.R. 350, 356.

[49] *Cheeseman* v. *Bowaters Ltd.* [1971] 1 W.L.R. 1773, 1778 (C.A.).

[50] R.S.C., Ord. 22, r. 7. *Gaskins* v. *British Aluminium Ltd.* [1976] Q.B. 524 (C.A.). The rule does not apply where there is a defence of tender before action.

[51] See notes on preparation of evidence, *ante*, p. 250.

When a figure has been arrived at it may be thought advisable to add a substantial percentage to allow for the hazards of litigation.

Interest [52]

Nothing should be included for interest. This is because the rules make no provision for payment in of interest.[53] It follows that when a payment in is made, if the plaintiff thinks he is entitled to interest he should, in open correspondence, ask for it. If it is refused and the action proceeds he is, it is submitted, entitled to be treated as the successful party for purposes as to costs if he recovers a sum which, before taking interest into account, is the same as or more than the sum paid into court.[54] If something is offered for interest which is refused as being inadequate the position is more complex, but the following is suggested for consideration as to the exercise of the court's discretion: that the plaintiff should be treated as the successful party if the sum of the principal and interest awarded at trial exceeds such sum as offered after adjusting the amount of interest to the date of offer.

Set-off not relied on as counterclaim [55]

The defendant should first assess the amount to be paid in as if there were no set-off, then deduct the amount he expects to establish on the set-off and pay in the balance.

Counterclaim

If the defendant makes a counterclaim against the plaintiff for a debt or damages he must by his notice of payment state, if it be the case, that in making the payment he has taken into account and intends to satisfy—(a) the cause of action in respect of which he claims, or (b) where two or more causes of action are joined in the counterclaim, all those causes of action or, if not all, which of them.[56] Thus in effect he must elect whether he offers up his crossclaim in satisfaction or whether he wishes to retain it and pay in only on the claim so that if the money is taken out he can still proceed with his counterclaim.[57]

Costs. If the defendant states by his notice that he takes into account and intends to satisfy his counterclaim and the plaintiff accepts the money paid into court the " defendant shall, unless the court otherwise directs, be entitled to his costs of the counterclaim incurred to the time of receipt of the notice of acceptance

---

[52] For interest generally, see *ante*, p. 154.

[53] *Jefford* v. *Gee* [1970] 2 Q.B. 130, 149 (C.A.); *Butler* v. *Forestry Commission* (1971) 115 S.J. 9112 (C.A.). It is curious that the matter has not yet been dealt with expressly in the rules.

[54] See *Vehicle & General Insurance Company Limited* v. *Christie (H. & W.) Ltd.* [1976] 1 All E.R. 747, His Honour Judge Fay Q.C. Interest cannot be applied for after money has been taken out, see *ante*, p. 256.

[55] For distinction, see *post*, p. 276, and see Ord. 18, r. 17.

[56] R.S.C., Ord. 22, r. 2.

[57] See *Martin French* v. *Kingswood Hill Ltd.* [1961] 1 Q.B. 96 (C.A.). The *Supreme Court Practice*, para. 22/2/1, says that Ord. 22, r. 2, " removed the difficulty arising from " this case, but *quaere* whether Ord. 62, r. 3 (8), has not introduced new difficulties; see *infra*.

by the plaintiff of the money paid into court." [58] The effect of this rule requires careful consideration. Justice will often require, it is thought, that some direction be given, but it cannot, it is submitted, be given until after the money has been accepted; so a plaintiff when accepting the money does not know whether any, and if any, what, direction will be given. And it is not clear what materials the court can consider when asked to make a direction. It can look at the pleadings and, perhaps, at agreed documents. How far beyond this can it go ? Presumably it can give some consideration to the amount of the payment in but it must not, it is submitted, assume that the amount necessarily represents the merits as between the parties if the matter had been fought out.[59]

Parties liable to contribution

The provisions as to payment in only apply as between plaintiffs and defendants, but R.S.C., Order 16, r. 10, enables a third party [60] or joint tortfeasor who may be liable to another party to contribute towards any debt or damages recovered by the plaintiff, to make (without prejudice to his defence) a written offer to that party reserving the right to bring the offer to the attention of the judge. The offer must not be brought to the attention of the judge until all questions of liability and debt or damages have been decided. The judge must take the offer into account in exercising his discretion as to costs.[61]

6. Written offers

" Payment into court is not the only method a defendant may have to save costs." [62] He may make an offer. Ordinarily it should be open, for if made without prejudice it cannot be referred to [63] unless subsequently made open,[64] or to show that an agreement had been concluded.[65] In the Family Division of the High Court where there is a dispute as to property an offer can be looked at if it is expressed to be without prejudice to the matter in dispute but reserving the right to refer to the offer on the issue of costs.[66] It may well be that such an offer would be satisfactory in building disputes but the procedure suggested below is, it is thought, more in accordance with the practice.

Arbitration

The device of a sealed offer deposited with the arbitrator is sometimes used. But this conflicts with the principle of non-disclosure in court proceedings (see *supra*), and the better course is to make an open offer on terms that its existence and contents are not to be disclosed to the arbitrator until he has

---

[58] R.S.C., Ord. 62, r. 3 (8).
[59] Consider *Martin French* v. *Kingswood Hill Ltd.*, *supra*, at p. 102.
[60] See also *Bates* v. *Burchell* [1884] W.N. 108.
[61] Ord. 62, r. 5 (a).
[62] *Martin French* v. *Kingswood Hill Ltd.*, *supra*, at p. 104.
[63] *Walker* v. *Wilsher* (1889) 23 Q.B.D. 335 (C.A.); *Stotesbury* v. *Turner* [1943] K.B. 370. The privilege attaching to documents expressed to be " without prejudice " only arises if there is a dispute or negotiation, *Re Daintrey* [1893] 2 Q.B. 116, 119 (D.C.).
[64] *Blow* v. *Norfolk C.C.* [1966] 3 All E.R. 579 (C.A.).
[65] *Tomlin* v. *Standard Telephones & Cables Ltd.* [1969] 1 W.L.R. 1378 (C.A.).
[66] *Calderbank* v. *Calderbank* [1976] Fam. 93, 106 (C.A.).

determined all issues of liability and debt or damages when it will be brought to his attention on the issue of costs. The offer should state that it is intended to have the effect of a payment into court pursuant to R.S.C., Ord. 22.[67] It should say whether or not it includes interest. It should offer to pay costs up to the time of receipt of the offer (in order to preserve the analogy with payment-in), but should not make payment of the sum offered conditional upon acceptance within a particular time; the reason for this is that if an offer of payment is withdrawn before trial the judge may and, ordinarily will, it is thought, ignore it when considering costs.[68]

At the end of the hearing before the arbitrator he should, without the existence of the offer being disclosed, be asked to defer consideration of costs until he has made his award on all other issues. As he can ordinarily make only one final award this requires that his award on liability and debt or damages will be an interim award. Costs are usually of such importance and the issues relating to them are sometimes so complex that this procedure is frequently and conveniently followed even if there is no offer to be taken into account. A request to follow this procedure should not, therefore, cause the arbitrator to think that an offer has necessarily been made.

Where there is a counterclaim the offer should, by analogy with the rules, state whether it is intended to take into account and satisfy the counterclaim (see *supra*). The claimant can make an offer in respect of the counterclaim.

7. Set-off and counterclaim

Where a defendant has a cross-claim against the plaintiff he may be able to set this up in diminution of the plaintiff's claim, *i.e.* plead a set-off, or he may counterclaim [69] or he may both set off and counterclaim.[70] Where there are such cross-claims the court has a very wide discretion as to costs,[71] and will look to the substance of the matter and not be enslaved by the pleadings,[72] but it seems likely that the court's discretion will be influenced to some extent by whether the defendant succeeds upon a set off or a counterclaim which is not or cannot be set off. Thus a set-off may reduce a plaintiff's claim to a sum which does not normally carry costs,[73] while if the defendant's claim was a counterclaim a " usual order " (see below) may be made, or a sum of money which he has paid into court may be inadequate to secure him the costs of the claim.[74] And it seems that a cross-claim not the subject of a set-off will be ignored in

---

[67] See *Dineen* v. *Walpole* [1969] 1 Lloyd's Rep. 261, 263 (C.A.).

[68] See *The Toni* [1974] 1 Lloyd's Rep. 489 (C.A.).

[69] R.S.C., Ord. 15, r. 2.

[70] R.S.C., Ord. 18, r. 17.

[71] *Childs* v. *Blacker* [1954] 2 All E.R. 243, 245, *sub nom. Childs* v. *Gibson* [1954] 1 W.L.R. 809 (C.A.).

[72] *N.V. Amsterdamsche Lucifersfabrieken* v. *H. & H. Trading Agencies Ltd.* [1940] 1 All E.R. 587, 590 (C.A.).

[73] See " Recovery of small sums," *ante*, p. 268. In the county court it can affect the scale of costs, see, *e.g. Hanak* v. *Green* [1958] 2 Q.B. 9 (C.A.).

[74] See *Chell Engineering Ltd.* v. *Unit Tool, etc., Ltd.* [1950] 1 All E.R. 378 (C.A.) decided before R.S.C., Ord. 22, r. 2, was introduced (see *supra*, p. 273) but the principle applies, it is thought, where the defendant does not take into account and satisfy his counterclaim—although the relationship of the rule to set off is not altogether clear.

considering the amount of any security for costs to be ordered to be given by the plaintiff where such order is appropriate.[75]

Costs where counterclaim

Where there is a claim and counterclaim and both parties are successful it has been said that the usual and convenient order is judgment for the plaintiff for £X on the claim with costs, and for the defendant £Y on the counterclaim with costs.[76] But frequently this order should not be made,[77] for in " most of these cases it is desirable that a judge should consider whether a special order should be made as to costs because the issues are often very much interlocked, and the usual order . . . does not always give a just result," [78] and " where the counterclaim amounts to an equitable set-off,[79] it is only right that the judge should deal with the claim and cross-claim as one." [80] The nature of a special order is within the judge's discretion, but one method is " to say that the plaintiff or the defendant is to have his costs less, perhaps, some small proportion." [81]

Distinction between set-off and counterclaim

A set-off is a defence [82] and as such can be used only " as a shield, not as a sword." [83] Consequently a defendant must counterclaim if he hopes to obtain more on his cross-claim than the plaintiff will obtain on his claim, or if the defendant desires to be in a position to continue his cross-claim in the same proceedings although the plaintiff's claim is discontinued. A cross-claim by a defendant can always be made the subject of a counterclaim but not necessarily of a set-off, and it is sometimes difficult to decide whether a cross-claim can be set off. But the principles which determine whether a cross-claim can be set off were exhaustively discussed in *Hanak* v. *Green*,[84] and in short the position is that, " there may be (1) a set-off of mutual debts; (2) in certain cases a setting up of matters of complaint which, if established, reduce or even extinguish the claim, and (3) reliance upon equitable set-off and reliance upon matters of equity which formerly might have called for injunction or prohibition." [85] The

---

[75] *Sloyan, T. & Sons Ltd.* v. *Brothers of Christian Instruction* [1974] 3 All E.R. 715.

[76] *Per* Singleton L.J., the *Chell Engineering* case, *supra*, at p. 382; see also *Medway Oil and Storage Co.* v. *Continental Contractors Ltd.* [1929] A.C. 88 (H.L.).

[77] *Childs* v. *Blacker*, *supra*.

[78] *Per* Denning L.J., the *Chell Engineering* case, *supra*, at p. 383, approved by the Court of Appeal in *Childs* v. *Blacker*, *supra*; see also *Cullinane* v. *British Rema* [1953] 2 All E.R. 1257, 1271 (C.A.).

[79] See *infra*.

[80] *Per* Parker L.J., *Baylis Baxter Ltd.* v. *Sabath* [1958] 1 W.L.R. 529, 538 (C.A.).

[81] *Per* Lord Goddard C.J., *Childs* v. *Blacker*, *supra*, at p. 245. It was said (at p. 244) that the *Chell Engineering* case dealt only with the application of the County Courts Act 1934.

[82] *Hanak* v. *Green*, *supra*, at p. 26; *Modern Engineering (Bristol) Ltd.* v. *Gilbert Ash (Northern) Ltd.* [1974] A.C. 689 (H.L.). See also *Henriksens A/S* v. *Rolimpex T.H.Z.* [1974] 1 Q.B. 233 (C.A.).

[83] *Per* Cockburn C.J., *Stooke* v. *Taylor* (1880) 5 Q.B.D. 569, 575 (D.C.).

[84] *Supra*. All previous decisions should be read in the light of *Hanak* v. *Green*. See also *Hale* v. *Victoria Plumbing Co. Ltd.* [1966] 2 Q.B. 746 (C.A.).

[85] *Ibid.* at p. 23, *per* Morris L.J., in whose judgment are given examples of each class of set-off.

scope of the right of equitable set-off is wide. Thus in *Hanak* v. *Green* the employer claimed damages for failure to complete properly. Against this the contractor was held entitled to set off: a claim on a *quantum meruit* for extra work done outside the contract, loss caused by the employer's refusal to admit the contractor's workmen and damages for trespass to tools. The test whether a cross-claim can be relied on as an equitable set-off seems to be whether it is sufficiently closely connected with the claim.[86] The parties by agreement may limit their right to set-off.[87] Subject to such agreement one can apply the general principles to other kinds of cross-claim which frequently arise, thus:

Liquidated damages. Liquidated damages not exceeding the amount of the claim and arising under the same contract can, it is submitted, be set off against the claim.

Claim under other contract. It seems that an employer cannot set off a claim arising under another contract with the same contractor unless he has been given express power by the contract upon which the contractor claims,[88] or there are special circumstances sufficiently connecting the two claims.

Defects. Where there is a claim on a lump sum contract and the defendant alleges that there are defects he may either set off his loss in diminution of the claim, or he may counterclaim for damages.[89]

Admission of claim

Where the defendant by his defence admits the claim, and the action is fought solely upon his counterclaim upon which he is successful, an order frequently made[90] is that the defendant has the general costs of the action after delivery of defence, and the plaintiff has the costs of the action up to the time of such delivery, and thereafter only the costs of setting down for trial.[91] In a defects case the employer is not bound to set off the defects in diminution of the claim; he may pay or admit the claim and bring a separate cross-action.[92]

---

[86] *Ibid.* at pp. 24, 31. See also *Henriksens A/S* v. *Rolimpex (supra),* at pp. 248, 252.

[87] See *post,* p. 278, " Notes on summary judgment."

[88] *Re Asphaltic Wood Pavement Co., ex p. Lee and Chapman* (1885) 30 Ch.D. 216 (C.A.); *Ellis Mechanical Services Ltd.* v. *Wates Construction Ltd.* (1976) (C.A.), unreported on this point. *Cf. Small* v. *Middlesex Real Estates Ltd.* [1921] W.N. 245. For a clause intended to have this effect, see the clause in issue in the *Modern Engineering* case, *supra.*

[89] *Hoenig* v. *Isaacs* [1952] 2 All E.R. 176, 181 (C.A.), and see *ante,* p. 51; *Modern Engineering* case, *supra. Aries Tanker Corporation* v. *Total Transport Ltd.* [1977] 1 W.L.R. 185, 190, 194 (H.L.).

[90] The judge has a complete discretion to make a different order: *Nicholson* v. *Little* [1956] 1 W.L.R. 829 (C.A.), but note that in this case (a county court appeal) the defendant claimed to set off his counterclaim so that it was not an admission at all in the sense that judgment could have been signed on it under Ord. 27, r. 3, in High Court proceedings. See also *Proctor & Lavender* v. *G. T. Crouch* (1966) 110 S.J. 273; where claim admitted subject to set-off and counterclaim and plaintiffs awarded half costs. For application for stay of execution, see R.S.C., Ord. 47, r. 1.

[91] *N.V. Amsterdamsche, etc.* v. *H. & H., etc., Ltd.* [1940] 1 All E.R. 587 (C.A.); *Childs* v. *Blacker, supra; cf. Nicholson* v. *Little, supra.*

[92] *Davis* v. *Hedges* (1871) L.R. 6 Q.B. 687; *Moss* v. *L. & N.W. Ry.* (1874) 22 W.R. 532.

278 Litigation [Chap. 14

It seems to follow that by admitting the claim he is in no way estopped from setting up the defects on the counterclaim. If he is successful on his counterclaim for however small a sum he is normally awarded High Court costs where the claim was commenced in the High Court [93] and he is therefore in a very strong position. Where such an admission is made it becomes particularly important for the plaintiff to consider whether he ought to pay money into court in satisfaction of the counterclaim, and, if the amount remaining in dispute in respect of the counterclaim does not exceed £2,000, whether he should apply for the transfer of the counterclaim to a county court.[94]

Admission of the claim in a defects case simplifies the whole proceedings. At trial the defendant opens the case and affirmative evidence is given at once of the alleged defective work.

Section 7: Notes on Summary Judgment

It has been said that one demerit of our system of justice is that because of our rules of procedure an arbitration under a building contract of some size is likely to take years [95]; cases go to an arbitrator or an official referee and drag on and on and the cash flow is held up so that in the end there is some kind of compromise, very often not based on the merits of the case but on the economic situation of one of the parties.[96] In a proper case where one of the parties is entitled to money this result can be avoided by the application of the procedure for obtaining summary judgment under R.S.C., Ord. 14.

1. Nature of summary judgment

This is available where there is no defence to the claim, or no defence to that part of the claim for which summary judgment is sought. It enables the plaintiff in such a case, upon summons supported by affidavit, to obtain judgment summarily without having to await trial. The terms of Order 14 and the notes in the *Supreme Court Practice* must be consulted. Here there are set out some notes on matters of particular interest in building disputes.

2. Degree of proof of indebtedness required

This has been expressed in various ways. Thus the sum claimed, it has been said, must be " indisputably due," or it must be " as plain as could be " that the sum is due, or it must be established " on the evidence beyond reasonable doubt." [97] The court arrives at its decision after considering the evidence served on behalf of the plaintiff and any evidence served on behalf of the defendant. The plaintiff does not get his judgment and the defendant will have leave to defend if it is shown that " there is an issue or question in dispute which ought

---

[93] See p. 269.
[94] County Courts Act 1959, s. 45. See also *ante*, p. 252.
[95] *Per* Lawton L.J., *Ellis Mechanical Services Ltd.* v. *Wates Construction Ltd., The Times*, Jan. 22, 1976 (C.A.).
[96] *Ibid.* The court has no inherent power to order an interim payment on account of damages expected to be recovered, *Moore* v. *Assignment Courier Ltd.* [1977] 1 W.L.R. 638 (C.A.).
[97] See judgments in the *Ellis* case; see also notes in the *Supreme Court Practice*.

to be tried or that there ought for some other reason to be a trial of the claim or part . . ." (R.S.C., Ord. 14, r. 3). The court applies the same test in deciding whether leave to defend should be given as it does in deciding whether there is a dispute to be referred to arbitration and a stay of proceedings under section 4 of the Arbitration Act 1950.[98]

Evidence supporting an application for summary judgment

Every case turns upon its facts but excellent evidence is either an admission by the defendant, or his authorised agent, or a certificate by the defendant's architect or engineer, which is a special and formal kind of admission. When evidence of this nature is established the court usually requires cogent evidence from the defendant before it grants leave to defend. Thus merely to allege the existence of defects in the works, or a claim for damages for delay without in each case giving some reasonable amount of detail and of quantification is unlikely to result in leave to defend.[99] Neither, it seems, will the court accept bare allegations, without satisfactory supporting evidence, that a certificate is not in accordance with the contract or that, for example, extras, part of the subject-matter of the claim, were not sanctioned in writing.[1] It is the practice to allow additional evidence on appeal, even in the Court of Appeal.

3. Set-off

The Court of Appeal in a series of cases commencing with *Dawnays* v. *Minter* [2] developed a principle which may be summarised thus: where a contractor as against the employer, or a sub-contractor as against the contractor, was plainly entitled to money under the contract, the defendant could not ordinarily set off cross claims, which were unliquidated or not admitted, such as for delay or defects, but had to submit to summary judgment for the amount due under the contract and thereafter pursue his cross-claim by arbitration or litigation as the case may be.[3]

The House of Lords in *Modern Engineering (Bristol) Ltd.* v. *Gilbert-Ash Ltd.* [4] held that there was no such principle, that building contracts were subject to the ordinary rules as to set-off, that the right of set-off for defects and delay and otherwise was well-established in building cases and that it required clear language to exclude the right.[5] In *Mottram Consultants Ltd.* v. *Sunley (Bernard & Sons) Ltd.* [6] the House of Lords, by a majority, held that on the particular

---

[98] For stay of proceedings where there is an arbitration agreement, see *ante*, p. 241.

[99] *Dawnays Ltd.* v. *Minter, F. G. Ltd.* [1971] 1 W.L.R. 1205 (C.A.), not overruled on this point, *Modern Engineering (Bristol) Ltd.* v. *Gilbert-Ash Ltd.* [1974] A.C. 689, 713 (H.L.), *Killby & Gayford Ltd.* v. *Selincourt Ltd.* unreported save in (1973) 229 E.G. 1343 (C.A.).

[1] *Killby & Gayford* case, *supra*.

[2] *Supra*.

[3] The other cases are referred to and surveyed in the *Modern Engineering* case, *supra*.

[4] *Supra*.

[5] See also *post*, p. 376, in relation to the Standard form of building contract and Standard form of sub-contract.

[6] [1975] 2 Lloyd's Rep. 197 (H.L.).

contract in question the right of set-off had been excluded. The decision of the majority seems to have been materially affected by the special circumstance that there was a contract in being which the parties altered in such a way as to show an intention that a set-off should not be allowed. The majority expressly applied the reasoning in the *Modern Engineering* case and it is submitted that there is no clash [7] between the two decisions and that the earlier case remains the leading case upon the subject.

---

[7] *Cf.* the vigorous dissenting opinion of Lord Salmon in the *Mottram* case, in particular at p. 215 where he expressed his opinion that the effect of the majority view would be to depart from what was decided by the House in the *Modern Engineering* case. See also article by Adrian Julian (1976) *New Law Journal* 141.

CHAPTER 15

THE JCT STANDARD FORM OF BUILDING CONTRACT
(JULY 1977 REVISION)

A DOCUMENT termed " the Standard form of building contract " is issued by the Joint Contracts Tribunal. This body is neither statutory nor government sponsored. Its constituent bodies are the following: Royal Institute of British Architects; National Federation of Building Trades Employers; Royal Institution of Chartered Surveyors; Association of County Councils; Association of Metropolitan Authorities; Association of District Councils; Greater London Council; Committee of Associations of Specialist Engineering Contractors; Federation of Associations of Specialists and Sub-Contractors; Association of Consulting Engineers. The copyright is vested in the Royal Institute of British Architects and the form is printed here by kind permission. It is the document which, subject to revisions from time to time, was long known as " the R.I.B.A. contract " but its current title reflects the fact that the R.I.B.A. is now only one of the many bodies which sponsor it. The correct expression is now " JCT form." [1] It is in extensive use both privately and in contracts with local and public authorities.

History of form

In view of the complexity of rights and liabilities in building contracts it has long been thought desirable to use a standard (or common) form acceptable to the parties concerned, thus avoiding the expense and hazards of special contracts. Towards the end of the nineteenth century a standard form was apparently in fairly common use. [1a] It seems that later versions of this form were issued towards the end of the nineteenth century and successive versions have been issued in 1909, 1931, 1939 and 1963. Revision of the Standard form is made by the Joint Contracts Tribunal and this body also issues from time to time " Practice Notes."

Six variants

There are variants of the Standard form of building contract as follows:
(i) Form for use by local authorities where quantities form part of the contract.
(ii) As (i) where quantities do not form part of the contract.
(iii) Form for private use where quantities form part of the contract.
(iv) As (iii) where quantities do not form part of the contract.
(v) and (vi) As for (i) and (iii) but for use with bills of approximate quantities.

Nature of the Standard form of building contract

Each of the variants (i) to (iv) of the Standard form of building contract creates a lump sum contract, [2] that is to say, a contract to complete a whole work

---

[1] B586, Court Business, August 5, 1977.

[1a] See *Clemence* v. *Clarke* (1880) H.B.C. (4th ed.), Vol. 2, p. 54. The form is set out in H.B.C. (3rd ed.), Vol. 2, p. 630.

[2] There is also available a fixed fee form of prime cost contract which is not printed here. There is also an " Agreement for Minor Building Works "—a short version of variant (iv), *supra*. The various forms of contract and the JCT Practice Notes can be obtained from R.I.B.A. Publications Ltd., 66 Portland Place, London W1N 4AD.

for a lump sum. Thus in essence it does not differ from an oral contract to reglaze a window for £5. There is a single lump sum agreed to be payable by the employer for the completion of a defined whole work by the contractor. But whereas the law will imply many of the terms of the oral reglazing contract, the form provides in its conditions an elaborate code of private law [3] which deals with most of the problems which might arise, and the court's function is usually confined to construing those conditions, *i.e.* to discovering the intention of the parties as expressed in the conditions. The problems of the form are therefore, in the main, problems of construction. The general law becomes of importance only in so far as it assists in the solution of those problems, or fills the gaps left unfilled by the conditions of the Standard form, or overrides or controls those conditions as, for example, in matters of bankruptcy law. Where the court has once decided what the meaning is of words in the form it follows its previous decisions,[4] but great care must be taken when reading a decision on a version of the form earlier than that under consideration lest the revision makes the decision no longer applicable.

The variants (i) to (iv) of the form do not differ in principle but the description of the contract work is very exact where quantities form part of the contract [5] and may be less exact where they do not.[6] The forms for the use of local authorities do not differ in substance from the forms for private use but contain certain provisions necessary to accord with local government law and practice, *e.g.* a Fair Wages clause.

Forms (v) and (vi) are adaptations respectively of forms (i) and (iii) for use when the quantities in the bills are approximate and are all to be re-measured. Forms (v) and (vi) are not printed but Practice Note 20 is printed *post*, p. 420. (See also note 10 to clause 12.)

For a further introduction to some of the main features of the form, see note 2 to the Articles of Agreement.

Scope of chapter

The form for the use of local authorities where quantities form part of the contract is printed with a commentary. The other variants are not printed, but the commentary refers to the major differences.

Notes appear after most clauses. They are not a paraphrase of the form, but are intended as an aid to the reading of the text of the form. Footnotes forming part of the published text of the forms are indicated by asterisks, etc., and printed in italics, to distinguish them from the author's footnotes to the commentary which are indicated by numerals as before.

Reference to parts of this book dealing with the subject-matter of the various clauses may be facilitated by use of the footnotes to this chapter, the index and the table of references to the Standard form at the front of the book.

---

[3] The R.I.B.A. form has been judicially likened to a legislative code; see Denning L.J. *Amalgamated Building Contractors Ltd.* v. *Waltham Holy Cross U.D.C.* [1952] 2 All E.R. 452, 453 (C.A.).

[4] *Cf. Milestone & Sons* v. *Yates Brewery* [1938] 2 All E.R. 439, 444.

[5] Variants (i) and (iii), *supra.*

[6] Variants (ii) and (iv), *supra.*

The practice notes issued by the Joint Contracts Tribunal are not, in general, printed but Practice Notes No. 20 (approximate quantities) and No. 21 (sectional completion) are printed.

Revisions to form of contract

The current edition was issued in 1963 but amended versions have been issued from time to time. The form printed is the July 1977 revision. The basic features of the contract have not altered since the original 1963 edition was issued but, while some revisions are apparently intended only to improve the drafting, others are of considerable importance. Care must be taken when using the commentary to assist in the construction of any particular contract to see which revision the contract is. This appears on the face or back, or both, of the form. The commentary refers to the more important amendments and assistance can be obtained by reading the 1977 revision in the light of the list of amendments issued by the Joint Contracts Tribunal and printed below by kind permission.

J.C.T.'s list of amendments

List of amendments to the Standard Form of Building Contract, 1963 edition, which have been issued from time to time by the Joint Contracts Tribunal and incorporated in successive revisions up to and including the July 1975 Revision.

A. The reprint dated April 1966 incorporated the following amendments to the 1963 Edition:

    (i) Clauses 14 (2) and 30 (2A)—Offsite goods and materials. (Practice Note 10 and amendment slip published December 1965.)

    (ii) Clause 31—Redundancy Fund Contributions. (Practice Note 11 and amendment slip published April 1966.)

    (iii) Clause 25 (3)—London Government Act, 1963—Local Authority Editions (Practice Note 12.)

B. The reprint dated December 1967 incorporated the following amendments:

    (i) Clause 11 (4) (c)—Valuation—Daywork. (Amendment slip dated January 1967.)

    (ii) Clause 31—Fluctuations—revised clause. (Amendment slip dated October 1967.)

    (iii) Building Control Act, 1966—Supplemental Agreement issued March 1967 (Private Editions only.)

C. The reprint dated July 1968 incorporated the following amendments:

    (i) Clauses 19 and 20—Insurances—revised clauses. (Amendment slip No. 1 (July 1968), now incorporated.)

    (ii) Clause 14 (1)—Materials and Goods unfixed or offsite—revised clause. Consequential amendment arising out of revised clauses 19 and 20. (Amendment slip No. 1 (July 1968), now incorporated.)

    (iii) Local Authority Editions only—existing second footnote deleted and revised footnote substituted page 26 (appendix). Consequential amendment arising out of revised clauses 19 and 20. (Amendment slip No. 1 (July 1968), now incorporated in Local Authority Editions only.)

    (iv) Clause 25 (3) (b) Private Editions only and clause 25 (4) (b) Local Authority Editions only—Determination by Employer—additional wording. (Amendment slip No. 2 (July 1968), now incorporated.)

  (v) Clause 17—Assignment of Sub-Letting—additional wording. (Amendment slip No. 2 (July 1968), now incorporated.)

  (vi) Clause 27 (a)—Nominated Sub-Contractors—additional sub-clause. (Amendment slip No. 2 (July 1968), now incorporated.)

  (vii) Clause 28 (b)—Nominated Suppliers—additional sub-clause. (Amendment slip No. 2 (July 1968), now incorporated.)

  (viii) Clause 12 (1)—Contract Bills—inclusion of 5th edition. Metric, Standard Method of Measurement—With Quantities edition only.

D. The reprint dated July 1969 incorporated the following amendments which appeared on Amendment Sheet 3 (July 1969):

  (i) Clause 34—Antiquities—revised clause.

  (ii) Clause 23—Extension of time—new sub-clause (k).

  (iii) Appendix—reference to ' Period for Honouring Certificates ' deleted.

  (iv) Clause 26 (1) (a)—Determination by Contractor—consequential amendment arising out of item (iii).

  (v) Clause 30 (1), 30 (4) (b) and 30 (4) (c)—Certificates and payments—consequential amendment arising out of item (iii).

  (vi) Clause 11 (4) (c) (ii) and 11 (4) (c) (iii)—Variations, provisional and prime cost sums—amendment.

  (vii) Appendix: Footnote † Local Authority Editions Page 28 and Footnote ** Private Edition Page 26—revised.

  (viii) Clause 31—Fluctuations—revised footnotes.

  (ix) Clause 31 D (6)—Fluctuations—revised sub-clause.

E. The reprint dated July 1971 incorporated the following amendments, nos (*i*)-(*x*), of which nos (*i*)-(*ix*) appeared on Amendment Sheet 4 (July 1971):

  (i) Private Editions only—Clause 17—new sub-clause on assignment.

  (ii) Clause 19—insurance against injury to persons or property—rearrangement of sub-clauses—revised provisions if default in payment for insurance by contractor.

  (iii) Clause 23 (j)—sub-clause divided into paragraphs (i) and (ii)—footnote * revised.

  (iv) Clause 25 (2)—determination by Employer—insertion of appointment of provisional liquidator.

  (v) Local Authority Editions only—Clause 25 (3) and (4)—amendments on determination for corruption, etc. by contractor.

  (vi) Clause 31A—fluctuations—increase in labour costs—recovery of consequential increase in employers' compulsory contributions.

  (vii) Clause 31B (b) (i) and (ii)—Tax Fluctuation clause—materials—new duties or taxes.

  (viii) Clause 31A and 31B—tax, etc. fluctuations—revised basis for inclusion in tenders of increases in existing taxes, etc. or of new taxes, etc.

  (ix) Clause 35—new sub-clause (5)—proper law of contract—application of Arbitration Act 1950.

  (x) Private Editions only—Building Control Act 1966—supplemental agreement deleted.

F. The reprint dated July 1972 incorporated the following amendments. Amendment (*i*) below appeared on Amendment Sheet 5 (October 1971) and amendments (*ii*)-(*xi*) below appeared on Amendment Sheet 6 (January 1972):

  (i) Clause 13A—value added tax.

  (ii) Clause 20 [A] (2)—insurance of the works against fire etc.—in 1st line ' settlement ' changed to ' acceptance '.

  (iii) Clause 28 (a)—nominated suppliers—substitution for reference to purchase tax.

  (iv) Clause 28—nominated suppliers—new sub-clause (d)—restriction of liability.

The following Clauses were amended with regard to the provisions for Retention:—

  (v) Clause 16 (f).

  (vi) Clause 27 (a) (viii).

  (vii) Clause 27 (e).

  (viii) Clause 30 (1).

  (ix) Clause 30 (3).

  (x) Clause 30 (6).

  (xi) Appendix.

G. The reprint dated July 1973 incorporated the following amendments. Amendments (*i*)-(*viii*) concern value added tax and appeared on Amendment Sheet 7 (March 1973) and Amendments (*ix*)-(*xv*) appeared on Amendment Sheet 8 (July 1973).

  (i) Clause 4 (2)—amendment for VAT.

  (ii) Clause 13A—new revised clause referring to ' VAT Agreement '.

  (iii) Clause 26 (i) (a)—amendment for VAT.

  (iv) Clause 27 (c)—amendment for VAT.

  (v) Clause 30 (5) (c)—amendment for VAT.

  (vi) Clause 31 (D) (5)—new paragraph (d) for VAT.

  (vii) Clause 35—footnote on VAT.

  (viii) Supplemental agreement for VAT.

  (ix) Clause 11 (4) (c) and appendix—prime cost sums—amendment for daywork charges and deletion of references in appendix.

  (x) Clause 16—new side-heading.

  (xi) Clause 25 (1)—determination by Employer—drafting amendment.

  (xii) Clause 26 and appendix—determination by Contractor—amendment to sub-clauses (1) (c) (ii) and (1) (c) (iv). Appendix—new footnote * and amendments to appendix clauses.

  (xiii) Clause 30 (7)—final certificate—drafting amendment.

  (xiv) Clause 31 A (b) and 31 B (a)—fluctuations—new sub-paragraph (vi) for Reserve Pension Scheme—sub-paragraph (vi) renumbered (vii).

  (xv) Clauses 31A-D and appendix—addition of new Clause 31E—appendix—new entry for this new Clause.

H. The revision dated July 1975 incorporated the following amendments:

Amendments (i) and (ii) were issued as Amendment Sheet No. 9/1975 (Local Authorities Edition with quantities only) and Amendment Sheet No. 10/1975 (Private Edition with quantities only) dated March 1975. Amendments (iii)-(xix) were issued as Amendment Sheet No. 11/1975 dated July 1975.

  (i) Clause 31F—new clause for adjustment of Contract Sum by NEDO Formula (with quantities editions only).

  (ii) Appendix—additions for application of Formula Rules (with quantities editions only).

  (iii) Clause 4—new sub-clause—statutory undertakers.

  (iv) Clause 19 (1) (a)—clause partly revised—limit of indemnity.

  (v) Clause 19A—Excepted Risks—nuclear perils, etc.—new clause.

  (vi) Clause 20 (A) (1)—amendment re nuclear perils, etc.

  (vii) Clause 20 (A) (1)—drafting amendment line 17.

  (viii) Clause 20 (B)—Contractor to give notice of damage.

(ix) Clause 20 (C)—Contractor to give notice of damage.
  (x) Clause 20 (A) (1)—new footnote.
 (xi) Clause 20 (B)—new footnote—Private edition only.
(xii) Clause 23—new sub-clause (1)—delay by statutory undertakers.
(xiii) Clause 25 (3)—amendment to reference to Local Government Act—Local Authorities edition only.
(xiv) Clause 30 (2A)—new paragraph (i)—insurance.
 (xv) Clause 30 (5) (c)—deletion of reference to provisional sums.
(xvi) Clause 31A (c) (iii)—additional words for VAT.
(xvii) Clause 31B (b) (i) and (b) (ii)—additional words for VAT.
(xviii) Clause 31C—drafting amendment—clause references.
(xix) Appendix—addition for insurance cover—Clause 19 (1) (a).

I. The revision dated July 1976 incorporated the following amendments:

These amendments were issued as Amendment Sheets No. Q.12/1976 (with quantities edition only), No. XQ.12/1976 (without quantities edition only) and AQ.12/1976 (with approximate quantities edition only), dated July 1976.

  (i) Clause 1 (1)—Contractor's obligations redefined—definition of contract documents inserted.
 (ii) Clause 1 (2)—discrepancies in contract documents and in instructions, drawings and other documents issued by architect.
(iii) Clause 3 (3), (5) and (6)—consequential amendment deleting ' specification ' following amendment to clause 1 (1) (with quantities and with approximate quantities editions only).
(iv) Clause 4 (1)—additional sub-paragraphs (b), (c), (d) and (e)—obligations in connection with statutory requirements redefined.
 (v) Clause 11 (4) (c) (i) and (ii)—drafting amendment ' current at date of tender ' (with quantities and without quantities editions only)/clause 13 (4) (i) and (ii)—same drafting amendments (with approximate quantities only).
(vi) Clause 30 (7)—revised sub-clause in conjunction with amendment to clause 1 (1)—conclusiveness of final certificate redefined.
(vii) Appendix—Formula Rules, Rule 3—non-adjustable element reduced to 10% (with quantities and with approximate quantities Local Authorities Edition only).

J. The following notes indicate the changes from the previous revision dated July 1976.

Amendments (i) and (ii) were issued as Amendment Sheet No. 13/1976 dated November 1976. Amendment (iii) was issued as Amendment Sheet No. 14/1977 dated April 1977 (with quantities and with approximate quantities editions only).

  (i) Clause 27 (a) (i)—Sub-contractor's obligations redefined.
 (ii) Clause 30B—new clause—Finance (No. 2) Act 1975—Statutory Tax Deduction Scheme.
(iii) Clause 31F—amendment in conjunction with publication of JCT Formula Rules, Series 2 (with quantities and with approximate quantities editions only).

Criticisms of Standard form of building contract

They have been numerous, strongly expressed, judicial,[7] and otherwise.[8]

---

[7] *e.g. English Industrial Estates* v. *Wimpey (George & Co.) Ltd.* [1973] 1 Lloyd's Rep. 118, 126 (C.A.), *per* Edmund Davies, L.J., ". . . the farrago of obscurities. . . ."
[8] *e.g. Building and Civil Engineering Standard Forms,* by Duncan Wallace, Q.C. which, in addition to his own comments, contains an anthology of criticisms by others. See also

Some of these criticisms are, from time to time, dealt with by way of amendments. Others are not. The more weighty criticisms are discussed in the commentary. Before a person responsible for a building project rushes off to prepare a new and better form he should reflect upon certain matters. Thus the view has been expressed that the Standard form of building contract, together with the Standard form of sub-contract, despite their obscurities, work well enough in practice.[9] The most commonly occurring defects in the operation of the form are, it is thought, by now apparent and some steps can be taken to deal with them. At the least, those using the form should be aware of, and can price, or allow for, the risks they undertake. It has been suggested that the Government form of contract, subject only to necessary drafting modifications, is suitable for general use.[10] On certain matters it appears to be clearer than the Standard form. It requires the contractor to bear some risks which in the Standard form are borne by the employer and it deals expressly with some matters which are obscure in the Standard form. But some amendments to make it suitable for general use would be quite substantial, *e.g.* there are no provisions for insurance. The Government form has not often been before the courts[11] and there may be unrevealed obscurities. In so far as it imposes higher risks upon contractors than under the Standard form they must tender at higher rates or face the risk of insolvency which is a calamity usually not only for themselves but for the employer. So with all its faults it is thought that the extensive use of the Standard form will continue.

Section 1: Standard Form of Building Contract, 1963 Edition(July 1977 Revision)
Local Authorities Edition (With Quantities)[12]

Articles of Agreement made the ....day of
.... 19......... BETWEEN ....
(hereinafter called " the Employer ") of the one part and ....
...............of (or whose registered office is situate at)...............................
....
(hereinafter called " the Contractor ") of the other part.
WHEREAS the Employer is desirous of*...
....
(hereinafter called " the Works ") at ....
....

comments by the same learned author in his various editions of *Hudson's Building Contracts* and in other publications.

[9] *Modern Engineering (Bristol) Ltd.* v. *Gilbert-Ash (Northern) Ltd.* [1974] A.C. 689, 726 (H.L.), *per* Lord Salmon.

[10] *Further Building and Engineering Standard Forms*, by Duncan Wallace, Q.C., p. 4, referring to the CCC/Wks/1, now replaced by Form GC/Works/1, referred to *ante*, p. 179, in relation to nominated sub-contractors, but which is substantially similar.

[11] For an example see *Farr (A. E.) Ltd.* v. *The Admiralty* [1953] 1 W.L.R. 965 where, upon the then Government form (since amended), it was held that the Government was entitled negligently to knock down what the contractor had put up and then require him, at his own expense, to rebuild. It is not reported whether or not he was also held liable in liquidated damages for the delay.

[12] Copyright is vested with the Royal Institute of British Architects and the form is printed by kind permission.

* *State nature of intended Works.*

and has caused Drawings and Bills of Quantities showing and describing the work to be done to be prepared by or under the direction of.......................
....
of ....
AND WHEREAS the Contractor has supplied the Employer with a fully priced copy of the said Bills of Quantities (which copy is hereinafter referred to as " the Contract Bills ") AND WHEREAS the said Drawings numbered ....
to .... inclusive (hereinafter referred to as " the Contract Drawings ") and the Contract Bills have been signed by or on behalf of the parties hereto:
NOW IT IS HEREBY AGREED AS FOLLOWS:

1. For the consideration hereinafter mentioned the Contractor will upon and subject to the Conditions annexed hereto carry out and complete the Works shown upon the Contract Drawings and described by or referred to in the Contract Bills and in the said Conditions.

2. The Employer will pay to the Contractor the sum of ....
(£......... : ....) hereinafter referred to as " the Contract Sum " or such other sum as shall become payable hereunder at the times and in the manner specified in the said Conditions.

†*3. [A] The term " the Architect " in the said Conditions shall mean the said ....
of ....
or in the event of his death or ceasing to be the Architect for the purpose of this Contract, such other person as the Employer shall nominate for that purpose, *not being a person to whom the Contractor shall object for reasons considered to be sufficient by an arbitrator appointed in accordance with clause 35 of the said Conditions.* †† Provided always that no person subsequently appointed to be the Architect under this Contract shall be entitled to disregard or overrule any certificate or opinion or decision or approval or instruction given or expressed by the Architect for the time being.

†*3. [B] The term " the Supervising Officer " in the said Conditions shall mean the said ....
of ....
or, in the event of his death or ceasing to be the Supervising Officer for the purpose of this Contract, such other person as the Employer shall nominate for that purpose, *not being a person to whom the Contractor shall object for reasons considered to be sufficient by an arbitrator appointed in accordance with clause 35 of the said Conditions.* †† Provided always that no person consequently appointed to be the Supervising Officer under this Contract shall be entitled to disregard or overrule any certificate or opinion or decision or approval or instruction given or expressed by the Supervising Officer for the time being.

---

† *Article 3 [A] is applicable where the person concerned is entitled to the use of the name " Architect " under and in accordance with the Architects (Registration) Acts 1931 to 1938. Article 3 [B] is applicable in all other cases. Therefore complete whichever is appropriate and delete the alternative. Where Article 3 [A] is completed the expression " Supervising Officer " shall be deemed to have been deleted throughout the Conditions annexed hereto. Where Article 3 [B] is completed the expression " Architect " shall be deemed to have been deleted throughout the said Conditions.*

* *In cases where the Works are to be carried out under the direction of officials of the Local Authority, insert the names of such officials as are to perform the respective functions of the " Architect/Supervising Officer " and the " Quantity Surveyor " under this contract.*

†† *Strike out words in italics in cases where " the Architect ", " the Supervising Officer " or " the Quantity Surveyor " is an official of the Local Authority.*

*4. The term " the Quantity Surveyor " in the said Conditions shall mean

....

of ....

or, in the event of his death or ceasing to be the Quantity Surveyor for the purpose of this Contract, such other person as the Employer shall nominate for that purpose, *not being a person to whom the Contractor shall object for reasons considered to be sufficient by an arbitrator appointed in accordance with clause* 35 *of the said Conditions.*††

This page should be completed with the appropriate Attestation Clause.[12a]

COMMENTARY

(1) " *Whereas . . .* "

The passages beginning " Whereas " are recitals. For the construction of recitals, see *ante*, p. 34.

(2) " *Now it is hereby agreed . . .* "

This is the operative part of the agreement. Some of the main features of the contract appear from this part of the Articles of Agreement, thus:

(a) **It is a lump sum contract.** A lump sum contract is a contract to complete a whole work [13] for a lump sum. The " whole work " to be performed is that the contractor must " upon and subject to the Conditions . . . carry out and complete the Works shown upon the Contract Drawings and described by or referred to in the Contract Bills and in the said Conditions." From this definition important results follow as to the nature of extra work.[14] The lump sum payable is ". . . the sum of . . . (£ . . . : . . .) . . . or such other sum as shall become payable hereunder at the times and in the manner specified in the said Conditions." The manner of payment [15] is upon the architect's certificates issued in accordance with and at the times stated in clause 30. The " other sum " may arise by adjustment of the original contract sum under many clauses, *e.g.*[16] 11 (5), (6) (variations, provisional sum work); 12 (2) (correction of errors in items in contract bills); 15 (2), 15 (3) (defects); 24 (1) (certain loss and expense); 30 (5) (c) (prime cost, provisional sums); 31 in its various alternatives (fluctuations); 34 (3) (antiquities). To be distinguished from such adjustments are the express contractual rights for the employer to make certain deductions from sums certified for payment, *i.e.* under clauses 2 (1) (default); 19 (1) (c) (insurance); 20 [A] (1) (insurance); 22 (liquidated damages); 27 (c) (direct payment to nominated sub-contractor).

(b) **The right to payment.** There is a right to be paid instalments of the contract sum (adjusted as may be necessary) upon architect's certificates issued in accordance with clause 30. Therefore this is not an entire contract in the sense in which that term has been used in certain textbooks because entire completion by the contractor is not a condition precedent to the right to call for any payment; but, subject to the difficulty raised by the discovery of an irremediable breach (see note 10 to clause 15), entire completion of the works is, it is submitted,

---

[12a] The printed form has a blank page for this purpose.
[13] See *ante*, p. 50.
[15] Apart from the special case of determination—see clauses 25 and 26.
[16] For a full list, see note 12 to clause 30.
[14] See *post*, p. 290.

a condition precedent to the right to the certificate under clause 30 (4) (c) and hence to the release of the residue of the contract sum then retained in accordance with clause 30 (3).[17]

(c) **Design obligations.** The general approach of the contract is that the architect designs, and communicates his design to the contractor who must carry it out but does not accept liability for its fitness for any particular purpose. But the contractor cannot be indifferent to matters of design for the following reasons:

(i) Frequently the drawings or bills expressly impose some design responsibility on the contractor or a nominated sub-contractor.[18]

(ii) The contractor's express or implied duties as to workmanship and the supply of materials may import some element of design responsibility.[19]

(iii) The duty of conforming to Building Regulations and the like makes it necessary for the contractor to give some consideration to whether the works as designed by the architect are in accordance with statutory requirements.[20]

(iv) There are express duties under clause 1 (2) upon the contractor to bring to the architect's attention any discrepancy in or divergence between the documents there set out, and under clause 4 (1) (b) any divergence between statutory requirements and design documents. A contractor in breach of these duties may be unable to recover loss and expense under clauses 11 (6) and 24 (1) and may, in certain circumstances, be liable in damages. The duties imposed by clauses 1 (2) and 4 (1) (b) arise only if the contractor " shall find " a discrepancy or divergence. It seems clear therefore that a contractor is under no duty positively to search for errors in the architect's design or the quantity surveyor's taking-off. But considerable losses can be caused if an obvious error of design goes unnoticed until much work has been done and demolition eventually has to take place with, perhaps, consequential delay. It is an implied term of the contract that the contractor in his workmanship will use the proper skill of a contractor [21] and it is suggested that a contractor who fails to observe errors which ought to have been obvious to him as the contractor in the circumstances cannot take advantage of such failure. (See further, note 2 to clause 4.)

(d) **The nature of extra work.** The amount of work for which the contract sum is payable is defined in exact terms and not by a broad description.[22] The contractor does not undertake for the contract sum to build a house or factory but to complete certain exactly stated quantities of work. He must carry out greater quantities if required to do so by the architect (clause 2) but is in general entitled to be paid extra for them.

The contractor must bear losses from his faulty pricing unless there is a proper case for the court to grant rectification (see *ante*, p. 47), and many of the risks of unanticipated expense (see further, next subsection) fall on the contractor,

[17] Clauses 15 (4), 30 (4) (c); cf. *Hoenig* v. *Isaacs* [1952] 2 All E.R. 176, 181 (C.A.).
[18] For nominated sub-contractors, see clause 27. For the relationship between the bills and the conditions see clause 12 (1).
[19] See clause 6.
[20] See clause 4.
[21] See *ante*, p. 38.
[22] For exactly defined contracts generally, see *ante*, p. 64.

but the risk of extra cost resulting from faulty description or measurement of the works is in general borne by the employer. This appears from reference to the following:

(i) *Contract bills.* These, subject to clause 6, describe the quality and quantity of work included in the contract sum (clause 12). If more work is required to complete the works the contractor is entitled to be paid for it as a variation, provided he can bring his claim within clause 11.

(ii) *Drawings, instructions and documents.* For the contractor's duty if he finds any discrepancy or divergence see clause 1 (2). The architect must issue an instruction in regard thereto (clause 1 (2)). If an instruction involves more or less work than that set out in the contract bills it will be treated as a variation and increase or decrease the contract sum as the case may be (clauses 1 (2), 2, 11, 12) and may entitle the contractor to recover loss and expense under clause 11 (6) or 24 (1) (c).

(iii) *Architect's satisfaction.* See clause 1 (1) for where the quality and standards of materials or workmanship may have to be to the reasonable satisfaction of the architect.[23] But it is submitted that any discretion vested in the architect by these words cannot enlarge the obligations expressly undertaken by the contractor as to the description of the works [24]; see clauses 6 and 12. The result is that if the architect requires a higher quality or standard than that provided for in a description, in the bills it is a variation (see also clause 11 (2)).

(iv) *Variations.* The architect has a very wide power to give instructions requiring variations (clause 11 (1), (2)), but the contractor has, in addition to the right to payment of the cost of the variations, valuable, although carefully limited, rights to recovery of loss and expense in which he is involved by carrying out such variations (clause 11 (6)).

(e) **Unanticipated loss or expense.** The contractor may find as the works proceed that he is suffering some loss or expense which he did not anticipate at the time of entering into the contract.

The general rule is that the contractor's undertaking to complete the works requires that he must bear such loss or expense himself, but there are important exceptions to the rule which can arise under one or more of the following heads:

(i) the express provisions for adjustment of the contract sum, see note 12 to clause 30.

(ii) the express provisions for determination of the contractor's employment, see clauses 20 [c] (b) (i), 26, 32;

(iii) a claim for damages for the employer's breach of contract, see generally, *ante,* p. 151, or breach of collateral warranty, misrepresentation or negligent misstatement, see *ante,* p. 89;

(iv) a right to treat the contract as at an end by reason of the employer's repudiation, see *ante,* p. 111, or misrepresentation, see *ante,* p. 89;

---

[23] For comments on the effect of these words in an earlier version of the Standard form of building contract see *Kaye (P. & M.) Ltd.* v. *Hosier & Dickinson Ltd.* [1972] 1 W.L.R. 146, 168 (H.L.).

[24] See *Cammell Laird & Co. Ltd.* v. *The Manganese Bronze and Brass Co. Ltd.* [1934] A.C. 402, 433 (H.L.); *Minster Trust Ltd.* v. *Traps Tractors Ltd.* [1954] 1 W.L.R. 963, 974; *Cotton* v. *Wallis* [1955] 1 W.L.R. 1168 (C.A.)—but note comment on this case *post,* p. 296.

(v) frustration. Frustration is discussed *ante*, p. 102. Note in particular *Davis Contractors Ltd.* v. *Fareham U.D.C.*, *ante*, p. 103, which illustrates the difficulty of attempting to rely on frustration. Although the form of contract differed the principle applies.

For the right in certain cases to an extension of time, see clause 23, and, where there is a suspension of the works, to determine the contractor's employment, see clause 26. For fire, storm, etc., see clause 20. For the discovery of antiquities, see clause 34. For increase in prices, see clause 31, and for war, see clauses 32, 33.

(3) *" Architect and Surveyor "*

For the position of the architect generally, see *post*, p. 294, and for duties of the surveyor, see *post*, p. 314.

(a) **Architect/Supervising Officer.** The reason for the two terms appears from the footnote (see also *ante*, p. 196). No contractual difference follows from the use of the term " Supervising Officer " and all references in the commentary to " Architect " include " Supervising Officer."

(b) **Architect and surveyor same person.** The contract contemplates two persons, but sometimes one is named as performing both offices. In such a case if the person named becomes unable to perform the duties of surveyor it seems that the employer is entitled to appoint another person as surveyor.[25] Further, even if different persons are appointed as architect and surveyor respectively, the architect is, it seems, entitled to consult an independent surveyor with regard to a valuation.[26]

(c) **No architect appointed.** If the parties enter into a contract in this form agreeing that no architect shall in fact be appointed, many of the clauses are nonsense and cannot be of any effect. It is thought, however, that the court would not say that the contract was void but would enforce it as an ordinary lump sum contract ignoring those clauses which depend for their effect upon the act or decision of the architect.[27]

(4) *Without quantities version*

In general terms the difference is as follows:

(a) Bills of quantities in the with quantities version describe the work for which the contract sum is payable, and provide the basis for the measurement and valuation of variations. In the without quantities version the bills of quantities are replaced by the specification and a schedule of rates.

(b) In the without quantities version the contract drawings and specification describe the quality and quantity of the work included in the contract sum (clause 12). Any bills of quantities or other statements as to quantities of work

---

[25] *R. B. Burden Ltd.* v. *Swansea Corp.* [1957] 1 W.L.R. 1167 (H.L.).
[26] See *ibid.* in the Court of Appeal (1956) 54 L.G.R. 161, 167, 168 (C.A.). The point was not dealt with in the House of Lords and the case was decided on the 1939 form, but the principle, it is thought, applies.
[27] *Cf. Nicolene* v. *Simmonds* [1953] 1 Q.B. 543, 551, discussed *ante*, p. 19.

which may have been supplied to the contractor do not form part of the contract (clause 12 (3)). Two important results follow:

(i) *Meaning of extra work.* There is no required standard method for the preparation of contract drawings and specification. They may be in great detail so that they describe the work for which the contract sum is payable in exact terms, or they may be lacking in detail and describe the work for which the contract sum is payable in wide terms. The latter class of contract drawings and specification invites dispute as to whether work is contract work or extra work. There is greater opportunity than in the with quantities version for the employer to say of some item in dispute that, although it is not expressly shown or described in the contract documents, it is not extra work, because it is impliedly included in the work for which the original contract sum is payable (see *ante*, p. 62).

(ii) *Errors in estimates.* The contractor must bear the loss resulting from both errors in pricing and errors in the estimated quantities of work and materials (clauses 3, 10).

(c) Variations are priced by reference to the schedule of rates (clause 11).

THE CONDITIONS HEREINBEFORE REFERRED TO
Clause 1
Contractor's obligations

(1) The Contractor shall upon and subject to these Conditions carry out and complete the Works shown upon the Contract Drawings and described by or referred to in the Contract Bills and in the Articles of Agreement and these Conditions (which Drawings, Bills, Articles of Agreement and Conditions are hereinafter called " the Contract Documents ") in compliance therewith, using materials and workmanship of the quality and standards therein specified, provided that where and to the extent that approval of the quality of materials or of the standards of workmanship is a matter for the opinion of the Architect/Supervising Officer, such quality and standards shall be to the reasonable satisfaction of the Architect/Supervising Officer.

(2) If the Contractor shall find any discrepancy in or divergence between any two or more of the following documents, including a divergence between parts of any one of them or between documents of the same description, namely:

(i)   the Contract Drawings,
(ii) the Contract Bills,
(iii) any instructions issued by the Architect/Supervising Officer under these Conditions (save insofar as any such instruction requires a variation in accordance with the provisions of clause 11 (1) of these Conditions), and
(iv) any drawings or documents issued by the Architect/Supervising Officer under clause 3 (3), clause 3 (4) or clause 5 of these Conditions

he shall immediately give to the Architect/Supervising Officer a written notice specifying the discrepancy or divergence, and the Architect/Supervising Officer shall issue instructions in regard thereto.

Clause 2
Architect's/Supervising Officer's instruction

(1) The Contractor shall (subject to sub-clauses (2) and (3) of this Condition) forthwith comply with all instructions issued to him by the Architect/Supervising Officer in regard to any matter in respect of which the Architect/Supervising Officer is expressly empowered by these Conditions to issue instructions. If within seven days after receipt of a written notice from the Architect/Supervising Officer requiring compliance with an instruction the Contractor does not comply

therewith, then the Employer may employ and pay other persons to execute any work whatsoever which may be necessary to give effect to such instruction and all cost incurred in connection with such employment shall be recoverable from the Contractor by the Employer as a debt or may be deducted by him from any monies due or to become due to the Contractor under this Contract.

(2) Upon receipt of what purports to be an instruction issued to him by the Architect/Supervising Officer the Contractor may request the Architect/Supervising Officer to specify in writing the provision of these Conditions which empowers the issue of the said instruction. The Architect/Supervising Officer shall forthwith comply with any such request, and if the Contractor shall thereafter comply with the said instruction (neither party before such compliance having given to the other a written request to concur in the appointment of an arbitrator under clause 35 of these Conditions in order that it may be decided whether the provision specified by the Architect/Supervising Officer empowers the issue of the said instruction), then the issue of the same shall be deemed for all the purposes of this Contract to have been empowered by the provision of these Conditions specified by the Architect/Supervising Officer in answer to the Contractor's request.

(3) All instructions issued by the Architect/Supervising Officer shall be issued in writing. Any instruction issued orally shall be of no immediate effect, but shall be confirmed in writing by the Contractor to the Architect/Supervising Officer within seven days, and if not dissented from in writing by the Architect/Supervising Officer to the Contractor within seven days from receipt of the Contractor's confirmation shall take effect as from the expiration of the latter said seven days.

Provided always:

(a) That if the Architect/Supervising Officer within seven days of giving such an oral instruction shall himself confirm the same in writing, then the Contractor shall not be obliged to confirm as aforesaid, and the said instruction shall take effect as from the date of the Architect's/Supervising Officer's confirmation, and

(b) That if neither the Contractor nor the Architect/Supervising Officer shall confirm such an oral instruction in the manner and at the time aforesaid but the Contractor shall nevertheless comply with the same, then the Architect/Supervising Officer may confirm the same in writing at any time prior to the issue of the Final Certificate, and the said instruction shall thereupon be deemed to have taken effect on the date on which it was issued.

COMMENTARY

(1) *Generally*

These two clauses should be read together and provide for: a general description of the contractor's obligation (clause 1 (1)); discrepancies between contract drawings and contract bills and certain other discrepancies and divergences (clause 1 (2)); compliance with architect's instructions (clause 2 (1)); testing the validity of architect's instructions (clause 2 (2)); the mode of architect's instructions (clause 2 (3)).

(2) *The position of the architect*

The general position is that, " It is the function and the right of the [contractor] to carry out his own building operation as he thinks fit," [28] and the architect has

---

[28] *Clayton* v. *Woodman & Son Ltd.* [1962] 1 W.L.R. 585, 593 (C.A.), a decision in a personal injury case on the 1939 ed. of the R.I.B.A. form where, arguably, the architect's authority was wider because of his power to give " directions." See also *AMF International Ltd.* v. *Magnet Bowling Ltd.* [1968] 1 W.L.R. 1028, 1046, 1053.

no authority to tell him how to do the works,[29] although, it is submitted, the contract drawings and contract bills must be read to consider whether in respect of any particular item a right to require work to be done in a certain way is reserved to the architect.

The architect has ample power to vary the works, but, it is submitted, he has no authority to vary or to waive [30] the conditions of the contract,[31] nor to vary the whole nature of the works,[32] for example, to change works designed as a single dwelling-house into a complex block of flats. He probably cannot omit work in order to have it carried out by another contractor. He cannot nominate a sub-contractor or supplier where there is no appropriate prime cost item or provisional sum (see clauses 11 (3), 27, 28).

For variations generally, see clause 11.

The architect's duty to act fairly and professionally. The parties contract on the understanding that in all matters where the architect has to apply his professional skill he will act in a fair and unbiased manner in applying the terms of the contract. This is not limited to the issue of certificates but applies to every function where he has to form a professional opinion upon matters which will affect the amount paid to (or to be deducted from) the contractor. For a general discussion of these principles see *ante*, p. 85, " Disqualification of the certifier." See also as to the effect of disqualification under the Standard form, *post*, pp. 389, 396. Note that the duty of fairness does not have the effect, as was long thought to be the position, of making him a quasi-arbitrator, *i.e.* a person with most of the attributes of an arbitrator although not formally appointed as such. As to this, see *ante*, p. 234.

There is, it is submitted, an implied term (for implication of terms see *ante*, p. 35) that the employer will not so act as to disqualify the architect. It is thought that it is a question in each case depending upon the nature of the employer's act, and its effect whether breach of the term amounts to a repudiation of the contract (see *ante*, p. 111). Interference with or obstruction by the employer in the issue of a certificate is a ground for determination by the contractor under clause 26.

Architect's certificates. One of the features of the contract is the number of different certificates which the architect can or must issue under various clauses, thus: of practical completion (15 (1)); of completion of making good defects (15 (4)); frost damage (15 (5)); sectional completion (16); insurance monies (20 [A] (2)); liquidated damages (22); loss and expense upon determination (25 (4) (*d*)); delay by nominated sub-contractor (27 (d) (ii)); interim certificates for payment (30 (1)); (30 (4) (b)); final certificate (30 (6)).[33]

(3) *Clause* 1 (1) ". . . *such quality and standards shall be to the reasonable satisfaction of the architect* "

The current clause appeared in the 1976 revision and is materially different

---

[29] *Ibid.*

[31] *Sharpe* v. *San Paulo Ry.* (1873) L.R. 8 Ch.App. 597.

[30] For waiver, see *ante*, p. 119.

[32] *Ibid.*

[33] For a further analysis of the architect's functions, see note of counsel's speech, *Sutcliffe* v. *Thackrah* [1974] A.C. 727, 730 (H.L.).

from all earlier versions. Previously the works generally had to be to the architect's reasonable satisfaction. In a sense this continues by implication. The architect before certifying under clause 30 must be satisfied that the works accord with the contract, and he should not, it is submitted, apply unreasonable tests in deciding whether he is satisfied. But the words " to the reasonable satisfaction of the architect," as used in this form have a special meaning as applying only to that class of materials or workmanship referred to in clauses 1 (1) and 30 (7) (a) (i) where the architect's decision as expressed in his final certificate is conclusive. Whether or not the relevant quality or standards come within this class is a question of construction. As examples, consider: " finish to the architect's approval," and " to the standards of BSS No. . . ." (with no reference to the architect). The first example, it is suggested, comes within clause 30 (7) (a) (i), the second does not. Between these examples it is a question in each case but, having regard to the importance of the result, it is thought that a clear intention must be apparent to bring the materials or workmanship within the class where quality or standards are to be to the reasonable satisfaction of the architect. The contract bills are the prime documents to be read (see clauses 6 (1), 12 (1)), but they may show that it is necessary to refer to the contract drawings. Documents or evidence outside the contract documents cannot, it is submitted, be referred to unless they can be relied on in support of a collateral warranty or agreement (see *ante*, p. 100).

For a discussion of the limit of the architect's powers when considering whether he is reasonably satisfied see *ante*, p. 291. The question whether he ought reasonably to be satisfied can, it is submitted, be challenged either by contractor or employer before the arbitrator, provided the procedure required by clauses 30 (7) and 35 is followed.

Effect of price. When considering whether he is reasonably satisfied can the architect take into account the price, *i.e.* whether it is high or low in comparison with the work to be done ? The case of *Cotton* v. *Wallis*[34] seems to suggest that he can, and that in particular he can allow a certain tolerance where the price is low even though the bills of quantities expressly state, " the whole of the materials and workmanship is to be the best of their respective kinds . . ." But *Cotton* v. *Wallis* is not, it is submitted, a binding authority upon the construction of this contract in disputes between employer and contractor[35] and does not[36] affect the well-settled rule that evidence outside the document itself, *i.e.* extrinsic evidence, may not normally be adduced to add to, vary, modify or contradict the written terms of the document.[37] It is submitted that if this

---

[34] [1955] 1 W.L.R. 1168 (C.A.). See *ante*, p. 43.

[35] Because the issue was whether the architect as a professional man had been guilty of such a dereliction of duty he owed to his client, the employer, as to make him guilty of negligence. It was not, and was not argued as, a dispute between the employer and the contractor as to the construction of the building contract—the 1939 edition of the R.I.B.A. form, pre-1957 version. Presumably for this reason evidence of whether the price was high or low was, it seems, admitted without objection and in the Court of Appeal it was " conceded that . . . one could not leave out of account altogether the price at which the work was to be done " (p. 1175).

[36] *Cf. Cotton* v. *Wallis, supra*, at p. 1172.

[37] See *ante*, p. 27.

rule is applied to a dispute between the employer and the contractor where the contract bills use the words quoted above, it results in the exclusion of evidence as to the nature of the price.[38] Nevertheless, until the point comes directly before the court, *Cotton* v. *Wallis* is the only authority on the question, and it is therefore suggested that the architect in deciding whether he is reasonably satisfied as to quality or standards should first give careful consideration to the express terms of the contract, in particular to the contract bills; if, then, he is still left in doubt he may as a last resort and, always providing that it is not contrary to any express term, consider whether the contract price is high or low.

Effect of architect's inspection. The contractor cannot rely on lack of inspection by the architect, or the clerk of works if there is one, as an excuse for not complying with the contract.[39] Can he rely, and if so, to what extent, upon an architect's inspection as evidence of compliance with the contract? No general answer can be given. The facts and the contract documents must be considered in each case. An example is suggested. Say the contract bills require the finish of an item to be that of a sample to be approved by the architect and the architect approves a sample. Thereafter, it is submitted, he cannot, without requiring a variation, demand a higher standard. Neither, it is submitted, can the arbitrator substitute a different standard from that approved by the architect. The reason is that the architect's approval completes the description of the quality of the works required by the contract. Where there is no comparable bill item, approval by the architect, or even lack of comment by the architect as to work offered as finished, may be greater or less evidence of compliance with the contract, but such action or inaction by the architect does not ordinarily, it is submitted, prevent the arbitrator as a matter of law deciding whether the offered finish complied with the contract.

(4) *Clause* 1 (2) "*. . . any discrepancy in . . .*"

If the contractor fails to give notice upon finding a discrepancy or divergence and carries out work shown for example on the contract drawings but not described in the contract bills, he may lose rights to extra payment (clause 11 (4) and (5)); to loss and expense (clauses 11 (6) and 24 (1) (c)); and to extension of time (clause 23 (e)). For a discussion of how far the contractor is under a duty to look out for discrepancies or divergences see note 2 (c) to the articles of agreement. If there is no breach of such duty and the contractor only finds the discrepancy or divergence after the work has been carried out, an adjustment must be made under clauses 12 (2) and 11 (4) and he is entitled to any loss and expense in which he was involved under clause 11 (6) or 24 (1) (c) as the case may be.

(5) *Clause* 2 (1) "*. . . Contractor shall (subject to sub-clauses 2 and 3 . . .) forthwith comply with all instructions . . .*"

Failure to comply with a valid instruction is a breach of contract which can have

---

[38] *Cf.* the dissenting judgment of Denning L.J. in *Cotton* v. *Wallis.*
[39] *East Ham Corp.* v. *Bernard Sunley & Sons Ltd.* [1966] A.C. 406 (H.L.); *AMF International Ltd.* v. *Magnet Bowling Ltd.* [1968] 1 W.L.R. 1028, 1053.

serious consequences. Under this clause if the non-compliance continues for seven days after written notice from the architect the employer may employ others to give effect to the instruction and recover the cost from the contractor. If the failure to comply causes delay, there is a liability in liquidated damages (clause 22), and failure to comply with an instruction can be an element in a default giving rise to a right on the part of the employer to determine the contractor's employment (see clause 25 (1)).

The contractor is not in breach of contract if he fails to comply with what purports to be an instruction but is not authorised by the conditions. Sub-clause 2 gives him a method of challenging such a purported instruction and there can be immediate arbitration upon the dispute (clause 35 (2)). Sub-clause 2 does not, it is submitted, entitle the contractor to refuse to comply with what is subsequently determined to have been a valid instruction. It is therefore prudent for the contractor to comply with a purported instruction which he challenges, having first made the two requests referred to in clause 2 (2). If the arbitrator should subsequently find in the contractor's favour the contractor is, it is submitted, entitled to a reasonable sum for complying with the purported instruction.[40]

(6) *Clause* 2 (1) *". . . is expressly empowered by these Conditions . . ."*

The power arises under the following clauses: 1 (2) (discrepancies); 4 (1) (by-laws, etc.); 5 (errors in setting out); 6 (3) (opening up and testing); 6 (4) (removal of work or materials); 6 (5) (dismissal of employee); 11 (1) (variations); 11 (3) (prime cost and provisional sums); 15 (2), 15 (3) (defects); 20 [C] (c) (ii) (damage by fire, etc.); 21 (2) (postponement of work); 27 (nominated sub-contractors); 28 (nominated suppliers); 32 (3), 33 (1) (war); 34 (2) (antiquities).

The R.I.B.A. have issued a form of architect's instruction which provides for a reference to the relevant contract clause number.

(7) *Clause* 2 (1) *" If within seven days . . . the Contractor does not comply . . ."*

See note 5 *supra*. Instructions which are particularly appropriate are those to open up or to test (clause 6 (3)), to remove defective work, materials or goods (clause 6 (4)), or to make good defects during the defects liability period (clause 15 (3)). The power under this clause should be compared with the right of determination under clause 25 (1) (c) if " the Works are materially affected " and the service of the notice of determination is reasonable.

(8) *Clause* 2 (2) *". . . what purports to be an instruction . . ."*

See notes 5, 6 and 7, *supra*.

Orders by employer personally. The contract makes no provision for any order by the employer himself. Therefore, it is submitted, such an order amounts only to an offer by the employer to enter into a separate contract for the work in question, which offer the contractor need not accept. If the contractor

---

[40] See *Molloy* v. *Liebe* (1910) 102 L.T. 616 and cases cited *ante*, p. 69. It is useful, although not, it is thought, essential, for the contractor complying with a purported instruction, after making his requests under the sub-clause, to say that his compliance is without prejudice to his contentions.

does accept the offer it is prudent to confirm it in writing both to the employer and to the architect. The parties may then find it convenient to agree that the order shall be treated as if it is an instruction under the appropriate clause and the difficulties which otherwise may arise are avoided.

Reasonableness of instructions. Should instructions be of a reasonable number and issued at reasonable times? The court,[41] it is submitted, in the absence of words importing an absolute discretion, readily implies a duty to act reasonably, and such a duty may be implied in this contract; but in considering what is reasonable the court would have regard to all the circumstances including the terms of the contract. Thus, for example, in considering whether the architect is acting reasonably in the number and time of instructions requiring variations the court would consider the provisions for altering contract rates under clause 11 (4) and the right to loss or expense under clause 11 (6), so that a very large number of variations ordered at different times might not be unreasonable under this contract although they might be unreasonable under a different contract with less provision for financial compensation to the contractor.

(9) *Clause* 2 (3) " *Any instruction issued orally . . .*"

The first sentence of this sub-clause states the general intention, namely, that instructions should be in writing, the second sentence recognises the practice that in fact oral instructions are sometimes given. The contractor cannot ignore an oral instruction. He must confirm it within seven days and, if it is not dissented from in writing by the architect, must comply with it at the expiry of seven days from the receipt by the architect of his, the contractor's, confirmation. Meanwhile the architect himself may confirm it within seven days of giving it (proviso (a)). The contractor is at risk if before the expiry of the period of seven days from receipt by the architect of his, the contractor's, confirmation he carries out an oral instruction which has not been confirmed by the architect. He must rely upon the architect in the exercise of his discretion subsequently to confirm under proviso (b). If the architect refuses to do so, then, it is submitted, the arbitrator has jurisdiction to exercise the discretion to confirm the instruction and to make an award as if there had been such confirmation (clause 35 (3)).

Urgent oral instruction. In a case where an oral instruction is urgent and cannot await the processes of clause 2 (3), what should the contractor do? It is suggested that when receiving such oral instruction he obtains from the architect a request to comply with it forthwith and an undertaking to confirm in writing. The contractor should then as soon as possible confirm in writing to the architect the instruction, request and undertaking. If the architect contests the confirmation the contractor can go to arbitration and if he satisfies the arbitrator of the fact that there was such instruction, request and undertaking the arbitrator should, it is submitted, make an award as if the instruction had been confirmed in writing in accordance with proviso (b) to clause 2 (3).[42] For the position where there is an emergency compliance with statutory requirements see clause 4 (1) (d).

---

[41] Which includes an arbitrator determining a dispute.
[42] See *Molloy* v. *Liebe* (1910) 102 L.T. 616 and cases cited *ante*, p. 69.

Form of instructions. Instructions do not have to be in any particular form, although to avoid dispute it is good practice to use a common form such as that issued by the R.I.B.A. If this is not used the architect should use clear words. Vague words can provoke an argument as to what was intended, *e.g.* a permission or a suggestion.

Site meeting minutes. If they are kept questions sometimes arise as to their effect, if any, under clause 2. This varies according to the circumstances and any agreement between the parties. If they are prepared and sent out by the architect they will probably operate as confirmation under proviso (a) or (b). If prepared by the contractor and sent to the architect they are prima facie only the confirmation by the contractor required by clause 2 (3) and do not take effect until seven days from their receipt have elapsed or earlier confirmation in writing by the architect. For purposes of clause 2 (3) it is therefore better for site meeting minutes to be prepared and issued promptly by the architect.

Position of clerk of works. See clause 10 and notes thereto.

Work done where no instructions at all. The provisions for confirmation do not apply and prima facie the contractor can recover nothing extra although there can be two exceptions: (1) under clause 12, where the contractor, after he has carried out work in accordance with the drawings finds that the quantities exceed those in the contract bills; (2) under clause 11 (1) where the architect has a discretion to sanction in writing a variation made by the contractor otherwise than pursuant to an instruction of the architect. For the exercise of that discretion see note 2 to clause 11.

Oral instructions as defence. In *G. Bilton & Sons* v. *Mason* (1957) (unrep.) Sir Walker Carter, Q.C., official referee, held that compliance with the architect's oral instructions, unconfirmed in writing, to vary the contract, was a defence to a claim by the employer for damages for breach of contract, where the breach alleged was that the contractor had complied with those instructions and not with the original contract.[43]

Clause 3
Contract documents

(1) The Contract Drawings and the Contract Bills shall remain in the custody of the Architect/Supervising Officer or of the Quantity Surveyor so as to be available at all reasonable times for the inspection of the Employer or of the Contractor.

(2) Immediately after the execution of this Contract the Architect/Supervising Officer without charge to the Contractor shall furnish him (unless he shall have been previously furnished) with—
  (a) One copy certified on behalf of the Employer of the Articles of Agreement and of these Conditions,
  (b) two copies of the Contract Drawings, and
  (c) two copies of the unpriced Bills of Quantities, and (if requested by the Contractor) one copy of the Contract Bills.

---

[43] The contract was the 1939 R.I.B.A. form but the principle applies, it is submitted, to the current form.

(3) So soon as is possible after the execution of this Contract the Architect/ Supervising Officer without charge to the Contractor shall furnish him (unless he shall have been previously furnished) with two copies of the descriptive schedules or other like document necessary for use in carrying out the Works. Provided that nothing contained in the said descriptive schedules or other documents shall impose any obligation beyond those imposed by the Contract Documents.

(4) As and when from time to time may be necessary the Architect/Supervising Officer without charge to the Contractor shall furnish him with two copies of such drawings or details as are reasonably necessary either to explain and amplify the Contract Drawings or to enable the Contractor to carry out and complete the Works in accordance with these Conditions.

(5) The Contractor shall keep one copy of the Contract Drawings, one copy of the unpriced Bills of Quantities, one copy of the descriptive schedule or other like documents referred to in sub-clause (3) of this Condition, and one copy of the drawings and details referred to in sub-clause (4) of this Condition upon the Works so as to be available to the Architect/Supervising Officer or his representative at all reasonable times.

(6) Upon final payment under clause 30 (6) of these Conditions the Contractor shall if so requested by the Architect/Supervising Officer forthwith return to the Architect/Supervising Officer all drawings, details, descriptive schedules and other documents of a like nature which bear his name.

(7) None of the documents hereinbefore mentioned shall be used by the Contractor for any purpose other than this Contract, and neither the Employer, the Architect/Supervising Officer nor the Quantity Surveyor shall divulge or use except for the purposes of this Contract any of the prices in the Contract Bills.

(8) Any certificate to be issued by the Architect/Supervising Officer under these Conditions shall subject to clause 27 (d) hereof, be issued to the Employer, and immediately upon the issue of any certificate the Architect/Supervising Officer shall send a duplicate copy thereof to the Contractor.

COMMENTARY

(1) *Generally*

This clause deals with various matters relating to the provision, custody and use of documents connected with the carrying out of the works. In considering it reference should also be made to the Articles of Agreement, clause 1 (1) (contractor's obligations), clause 1 (2) (notification of discrepancies), clause 5 (accurately dimensioned drawings for setting out) and clauses 6 (1) and 12 (contract bills). Note that although the side note is " contract documents " they are in fact defined in clause 1.

(2) *Clause 3 (3)* ". . . *two copies of the descriptive schedules or other like documents* . . ."

By the July 1976 Revision the reference to supplying the contractor with a copy of the specification was omitted. This was a small but valuable improvement. In the with quantities version of the contract the specification was not a contract document and added nothing to the contractor's obligations. It could only cause confusion to supply him with a copy and it was better practice not to do so. The descriptive schedules and other like documents are intended to be for the assistance of the contractor and are, or should be, prepared from the contract documents, *i.e.* contract drawings and the contract bills.

(3) *Clause* 3 (4) " *As and when from time to time may be necessary. . . .*"

This clause read with clauses 23 (f) and 24 (1) (a) recognises that the contract documents are frequently not adequate to enable the works to be carried out. Necessary drawings and details should, it is submitted, be furnished at times which will enable the contractor to comply with his duties as to progress required by clause 21. Clause 3 (4) substantially displaces the term usually implied that the architect will supply necessary drawings at reasonable times (see *ante*, p. 37). For delay in the delivery of instructions and drawings there is an express contractual right under clause 24 (1) (a) to recover loss and expense subject to fulfilment of certain conditions appearing in the clause. This right is expressed to be (clause 24 (2)) without prejudice to any other rights and remedies which the contractor may possess. The contractor may therefore claim damages for breach of clause 3 (4) or of the usual implied term so far as it remains in operation. But he will ordinarily only claim damages if he cannot satisfy the requirements of a claim under clause 24 (1) (a). This is because it is usually more convenient to recover on a certificate issued under the terms of the contract than to seek damages for breach of the contract, and because a contractor who cannot prove compliance with the requirements of clause 24 (1) (a) may find it difficult to establish his loss as a matter of evidence and to defeat allegations of failure to mitigate his loss, or that the loss is too remote (see damages generally, *ante*, p. 144, and for contractor's " claims " *ante*, pp. 260 and 266 and note 9 to clause 11).

Drawings by consultants. The contract makes no provision for the position of consultants although it is common for them to design much of the structural work and specialist services. They act, it is submitted, as agents of the architect and their delay is the architect's delay (as between contractor and employer).

Drawings of nominated sub-contractors.[44] They sometimes have to supply drawings to enable the contractor to carry out work not to be performed by themselves. Delay in the provision of their drawings is discussed in note 2 to clause 27.

(4) *Clause* 3 (6) "*. . . the contractor shall if so requested . . . return . . . all drawings . . .*"

For property in the plans, the architect's lien and his copyright, see *ante*, p. 224. For remedies for breach of confidence, see *ante*, p. 142.

(5) *Private version*

By clause 3 (8) certificates are to be issued to the contractor. In practice it is a convenient course to send copies to the employer.

(6) *Without quantities version*

There are many differences to allow for the significance of the specification as a contract document. For general effect, see note 4 to articles of agreement.

---

[44] For nominated sub-contractors generally, see *ante*, p. 178.

Clause 4

Statutory obligations, notices, fees and charges

(1) (a) The Contractor shall comply with, and give all notices required by, any Act of Parliament, any instrument rule or order made under any Act of Parliament, or any regulation or byelaw of any local authority or of any statutory undertaker which has any jurisdiction with regard to the Works or with whose systems the same are or will be connected (all requirements to be so complied with being referred to in these Conditions as ' the statutory requirements ').

(b) If the Contractor shall find any divergence between the statutory requirements and all or any of the documents referred to in clause 1 (2) of these Conditions or any variation instruction issued in accordance with clause 11 (1) of these Conditions, he shall immediately give to the Architect/Supervising Officer a written notice specifying the divergence.

(c) If the Contractor gives notice under paragraph (b) of this sub-clause or if the Architect/Supervising Officer shall otherwise discover or receive notice of a divergence between the statutory requirements and all or any of the documents referred to in clause 1 (2) of these Conditions or any variation instruction issued in accordance with clause 11 (1) of these Conditions, the Architect/Supervising Officer shall within 7 days of the discovery or receipt of a notice issue instructions in relation to the divergence. If and insofar as the instructions require the Works to be varied, they shall be deemed to be Architect's/Supervising Officer's instructions issued in accordance with clause 11 (1) of these Conditions.

(d) (i) If in any emergency compliance with paragraph (a) of this sub-clause requires the Contractor to supply materials or execute work before receiving instructions under paragraph (c) of this sub-clause the Contractor shall supply such limited materials and execute such limited work as are reasonably necessary to secure immediate compliance with the statutory requirements.

(ii) The Contractor shall forthwith inform the Architect/Supervising Officer of the emergency and of the steps that he is taking under this paragraph of this Condition.

(iii) Work executed and materials supplied by the Contractor under sub-paragraph (i) of this paragraph shall be deemed to have been executed and supplied pursuant to an Architect's/Supervising Officer's instruction in accordance with clause 11 (1) of these Conditions provided that the emergency arose because of a divergence between the statutory requirements and all or any of the documents referred to in clause 1 (2) of these Conditions or any variation instruction issued in accordance with clause 11 (1) of these Conditions, and the Contractor has complied with sub-paragraph (ii).

(e) Provided that the Contractor complies with paragraph (b) of this sub-clause, the Contractor shall not be liable to the Employer under this Contract if the Works do not comply with the statutory requirements where and to the extent that such non-compliance of the Works results from the Contractor having carried out work in accordance with the documents referred to in clause 1 (2) of these Conditions or any variation instruction issued in accordance with clause 11 (1) of these Conditions.

(2) The Contractor shall pay and indemnify the Employer against liability in respect of any fees or charges (including any rates or taxes) legally demandable under any Act of Parliament, any instrument, rule or order made under any Act of Parliament, or any regulation or byelaw of any local authority or of any statutory undertaker in respect of the Works. Provided that the amount of any

such fees or charges (including any rates or taxes other than value added tax) shall be added to the Contract Sum unless they—

(a) arise in respect of work executed or materials or goods supplied by a local authority or statutory undertaker for which a prime cost sum is included in the Contract Bills or for which a prime cost sum has arisen as a result of Architect's/Supervising Officer's instructions given under clause 11 (3) of these Conditions, or

(b) are priced or stated by way of a provisional sum in the Contract Bill.

(3) None of the provisions of clause 27 (nominated sub-contractors) nor of clause 28 (nominated suppliers) of these Conditions shall apply where prime cost sums are included in the Contract Bills or arise as a result of an instruction by the Architect/Supervising Officer in regard to the expenditure of provisional sums in respect of any fees or charges for work executed or materials or goods supplied by a local authority or statutory undertaker solely in pursuance of its statutory obligations. Such fees or charges shall be dealt with under the provisions of sub-clause (2) of this Condition and any amount properly paid by the Contractor to any local authority or statutory undertaker shall be added to the amount that would otherwise be stated as due in the next interim certificate.

COMMENTARY

(1) *Clause* 4 (1) "*. . . shall comply with . . .*"

Provided he has the requisite knowledge (see *infra*) the contractor has, it is submitted, both to comply with and give statutory notices, and also, comply with relevant statutory requirements including the Building Regulations 1965, 1972, 1976 which, outside London, replace the local by-laws in force when this form was first published.[45] It seems that a flagrant breach of the building regulations or the like, apparent in the contract documents, may make the contract illegal.[46]

For the Building Control Act 1966, formerly restricting the construction of certain works of the value of more than a certain figure (in 1968, £100,000), see *ante*, p. 110.

(2) *Clause* 4 (1). *Losses caused by non-compliance with statutory requirements*

Clause 4 (1) was re-drafted in the July 1976 Revision. It now sets out a clear procedure to be followed if the contractor finds a divergence from statutory requirements. There is a useful provision for emergency compliance with statutory requirements (clause 4 (1) (d)), although in practice the same result could be arrived at under the earlier versions of the form by the architect exercising his discretionary power under clause 11 (1) to sanction an unauthorised variation.

Clause 4 (1) (e) was new in July 1976. It qualifies clause 4 (1) (a) and operates as a valuable protection to the contractor who either gives the requisite notice to the architect under clause 4 (1) (b) or who has not found any divergence between statutory requirements and the contract documents or variation instructions. For a suggestion that clause 4 (1) (e) may not protect a contractor who negligently fails to observe a divergence, see *ante*, p. 290.

In any claim for breach of contract by the employer, whether under the pre-

---

[45] See *Townsend (Builders) Ltd.* v. *Cinema News & Property Management Ltd.* [1959] 1 W.L.R. 119 (C.A.) and see *ante*, p. 109.

[46] See *ante*, p. 108.

or post-1976 versions, damages require careful consideration having regard to the contractor's right to be paid for work required to comply with statutory requirements.

Contractor's claim against architect. In certain circumstances where the contractor is liable for non-compliance with statutory requirements to the employer he may have a claim against the architect where he can prove a custom to rely upon the architect to serve the appropriate notices,[47] or otherwise can show a breach of a duty of care owed to him by the architect (see *ante*, p. 227).

(3) *Clause* 4 (1). *Unauthorised variation*

If the contractor at the request of a local official or for some other reason makes a variation in order to comply with statutory requirements without following the procedure of this clause and it is not an emergency he is in breach of contract and prima facie loses any rights under clause 11 to extra payment and loss or expense and under clause 23 to extension of time. The architect can however in his discretion sanction a variation under clause 11 (1). For the exercise of that discretion see note 2 to clause 11. In certain special circumstances where the contractor has had to pay money which the employer is legally liable to pay the contractor may be able to recover such money in a claim in restitution.[48]

(4) *Clause* 4 (1). *Major alteration necessary*

If a major and expensive alteration is necessary to comply with a statutory requirement it is possible that in certain circumstances the employer [49] might be able to say that the contract is frustrated.[50] Subject to this, if there is delay the contractor may be able to found a determination on lack of instructions (clause 26 (1) (c) (v)) or postponement of work (clause 21 (2), clause 26 (1) (c) (iv)).

(5) *Clause* 4 (2) " . . . *rates* . . . "

" Building sites themselves are not treated as rateable hereditaments while the work of building is in progress." [51] But builders' huts and sheds erected on the site may have a sufficient degree of permanency to become rateable.[52] It is a question of fact and degree in each case whether they have this degree of permanency.[53]

---

[47] See *supra*, footnote 45.

[48] See discussion in *Owen* v. *Tate* [1976] Q.B. 402 (C.A.).

[49] The contractor will not usually object to the alteration because of his rights under clauses 11, 23 and/or 24; neither will he plead frustration because of his much better position under clause 26.

[50] For frustration, see *ante*, p. 101.

[51] *L.C.C.* v. *Wilkins* [1957] A.C. 362, 380 (H.L.). For the stage when the works become rateable, see *Ravenseft Properties Ltd.* v. *Newham London Borough Council* [1976] Q.B. 464 (C.A.).

[52] *L.C.C.* v. *Wilkins, supra.*

[53] *Ibid.*; various temporary structures, used for offices, stores and a canteen, on the site for 18 months were held to be rateable.

(6) *Clause* 4 (2) *". . . taxes . . ."*

It is difficult to think of examples. It is submitted that Selective Employment Tax [54] does not come within the clause.[55]

(7) *Clause* 4 (3) *". . . work executed or materials or goods supplied by a local authority or statutory undertaker solely in pursuance of its statutory obligations "*

This sub-clause was introduced in July 1975. It deals with such matters as connections to street mains by gas boards or the provision of crossovers by local authorities. In versions before 1975 there was considerable difficulty as no clause quite fitted the situation although various devices were resorted to such as making payment as if it arose under clauses 27 or 28. For extension of time for delay by a statutory undertaker see clause 23 (1), but observe that there is no provision for payment of loss or expense under clause 24 (1).

Clause 5
Levels and setting out of the works

The Architect/Supervising Officer shall determine any levels which may be required for the execution of the Works, and shall furnish to the Contractor by way of accurately dimensioned drawings such information as shall enable the Contractor to set out the Works at ground level. Unless the Architect/Supervising Officer shall otherwise instruct, in which case the Contract Sum shall be adjusted accordingly, the Contractor shall be responsible for and shall entirely at his own cost amend any errors arising from his own inaccurate setting out.

COMMENTARY

(1) *". . . such information . . ."*

It is thought that the contractor is entitled to an indemnity from the employer in respect of claims made against him by an adjoining owner for trespass, committed by the contractor in reliance upon faulty information furnished by the architect.[56]

(2) *" Unless the Architect . . . shall otherwise instruct . . ."*

These words apparently give the architect a discretion to issue an instruction which could result in the employer becoming liable to bear the cost of amending errors due to the contractor's inaccurate setting out. It is thought that only special circumstances warrant such an instruction and it should not be given if there had been any want of proper skill on the part of the contractor. For a possible alternative construction of these words, see note 9 to clause 15.

Clause 6
Materials, goods and workmanship to conform to description, testing and inspection

(1) All materials, goods and workmanship shall so far as procurable be of the respective kinds and standards described in the Contract Bills.

---

[54] See Finance Act 1966, s. 44, Selective Employment Payments Act 1966; now no longer in force but problems may still arise.

[55] Because, *inter alia*, it is not payable " in respect of the works " but in respect of " workpeople," *cf.* clause 31A (a) (i).

[56] See *Kirby* v. *Chessum & Sons Ltd.* (1914) 79 J.P. 81 (C.A.). See also clauses 18 (2) and 19 (2) (a).

(2) The Contractor shall upon the request of the Architect/Supervising Officer furnish him with vouchers to prove that the materials and goods comply with sub-clause (1) of this Condition.

(3) The Architect/Supervising Officer may issue instructions requiring the Contractor to open up for inspection any work covered up or to arrange for or carry out any test of any materials or goods (whether or not already incorporated in the Works) or of any executed work, and the cost of such opening up or testing (together with the cost of making good in consequence thereof) shall be added to the Contract Sum unless provided for in the Contract Bills or unless the inspection or test shows that the work, materials or goods are not in accordance with this Contract.

(4) The Architect/Supervising Officer may issue instructions in regard to the removal from the site of any work, materials or goods which are not in accordance with this Contract.

(5) The Architect/Supervising Officer may (but not unreasonably or vexatiously) issue instructions requiring the dismissal from the Works of any person employed thereon.

COMMENTARY

(1) *Generally*

This clause defines the kinds and standards of materials, goods and workmanship and gives the architect certain important powers relating to the execution of the works and the contractor's employees. It should be read with clauses 1 and 2 and the notes thereto.

No standard described. To the extent that standards are not expressly described in the contract bills the ordinary implied duties of the contractor as to materials and workmanship discussed, *ante*, p. 38 apply. For a bill item which requires the architect's approval to set the standard see note 3 to clause 1 and see also clauses 1 (1), 30 (7) and 35.

Without quantities version. Substitute " specification " for " contract bills."

(2) *Clause 6 (1) ". . . so far as procurable . . ."*

It is thought that there is an implied duty upon the contractor who seeks to rely on these words to notify the architect before supplying materials, goods or workmanship not of the respective kinds and standards described in the contract bills.

(3) *Clause 6 (3) ". . . the cost of such opening up . . ."*

For the recovery of loss or expense additional to such cost and caused by the disturbance of the regular progress of the works due to an unjustified order to open up or test, see clause 24 (1) (b).

The meaning of " executed work " is, it is submitted, a question of fact and degree in each case. Thus the work may be an isolated example and no difficulty arises. But sometimes the test may be of work which is a sample of a class of executed work, *e.g.* of six out of a hundred similar piles. If two tests are bad and four good is the contractor entitled to payment for the four which were good, and of every other pile which, upon subsequent test, is found to be good, and for consequential loss under clause 24? The answer depends, it is submitted, upon whether in all the circumstances the result of the test is such as to show that the piling does not accord with the contract. This is a question of fact which, in default of agreement, must be determined by the arbitrator.

The architect does not have to order a test under clause 6 (3) before issuing instructions under clause 6 (4).

(4) *Clause 6* (4) *". . . removal from the site . . ."*

" There is no need for the architect to be given power to instruct the contractor to remedy defective work for the contractor is, by clause 1, required to complete the works to the reasonable satisfaction of the architect." [57] This was said in relation to the pre-July 1976 revision when all the works had to be carried out to the architect's reasonable satisfaction. The principle still applies. Even where materials or workmanship are not to be to the architect's reasonable satisfaction (see clause 1 (1)) the duty to complete the works requires the contractor to replace in accordance with the contract. If there is no need to, or it is impracticable to, remove defective works, *e.g.* piles, the architect, it is submitted, is not bound to do more than notify the contractor of his dissatisfaction with the defective works (see clause 1 (1)). For the architect's remedies see clauses 2 (1), 15, 25 (1) and 30 (2). There is it seems no remedy in damages for work which during the construction period before completion does not conform with the contract requirements.[58]

Nominated sub-contractors. The power to issue instructions requiring the removal of work applies, it is submitted, to the work of nominated sub-contractors. The instruction is issued to the contractor who transmits it to the nominated sub-contractor. If the latter fails to comply, the architect can reduce the sum which he would otherwise direct the contractor as due to the nominated sub-contractor under clause 27 (b) and, if necessary, advise the employer to enforce his remedy under clause 2 (1). The cost so expended can be deducted from monies which would become due to the contractor and which would otherwise be directed to be paid to the nominated sub-contractor. This course is to be preferred to that of attempting to enforce the procedure for determination of the contractor's employment under clause 25 (1) (c). That clause may not apply at all. If it does it is thought that ordinarily if the nominated sub-contractor's failure or refusal to remove defective work or materials is of such effect as to come within the provisions of clause 25 (1) (c) it is likely to be a fundamental breach (for meaning see *ante*, p. 111) of his sub-contract so that the contractor is entitled to a re-nomination (see *post*, p. 369). Therefore a purported determination of the contractor's employment where there has been no such re-nomination, at any rate after the contractor has requested such re-nomination, is not, it is submitted, valid. See further, note 4 to clause 25.

Clause 7
Royalties and patent rights

All royalties or other sums payable in respect of the supply and use in carrying

[57] *Modern Engineering (Bristol) Ltd.* v. *Gilbert-Ash (Northern) Ltd.* [1974] A.C. 689, 710 (H.L.), *per* Viscount Dilhorne.

[58] *Kaye (P. & M.) Ltd.* v. *Hosier & Dickinson Ltd.* [1972] 1 W.L.R. 146, 165 (H.L.).

out the Works as described by or referred to in the Contract Bills of any patented articles, processes or inventions shall be deemed to have been included in the Contract Sum, and the Contractor shall indemnify the Employer from and against all claims, proceedings, damages, costs and expenses which may be brought or made against the Employer or to which he may be put by reason of the Contractor infringing or being held to have infringed any patent rights in relation to any such articles, processes and inventions.

Provided that where in compliance with Architect's/Supervising Officer's instructions the Contractor shall supply and use in carrying out the Works any patented articles, processes or inventions, the Contractor shall not be liable in respect of any infringement or alleged infringement of any patent rights in relation to any such articles, processes and inventions and all royalties, damages or other monies which the Contractor may be liable to pay to the persons entitled to such patent rights shall be added to the Contract Sum.

Clause 8
Foreman-in-charge

The Contractor shall constantly keep upon the Works a competent foreman-in-charge and any instructions given to him by the Architect/Supervising Officer shall be deemed to have been issued to the Contractor.

COMMENTARY

(1) *Generally*

The foreman-in-charge is the contractor's agent to receive instructions. To avoid confusion he should be named, especially if within the contractor's organisation he is not ordinarily termed a foreman. For the procedure relating to instructions, see clauses 1 and 2 and notes thereto, particularly notes 5 to 9.

Clause 9
Access for Architect/Supervising Officer to the Works

The Architect/Supervising Officer and his representatives shall at all reasonable times have access to the Works and to the workshops or other places of the Contractor where work is being prepared for the Contract, and when work is to be so prepared in workshops or other places of a sub-contractor (whether or not a nominated sub-contractor as defined in clause 27 of these Conditions) the Contractor shall by a term in the sub-contract so far as possible secure a similar right of access to those workshops or places for the Architect/Supervising Officer and his representatives and shall do all things reasonably necessary to make such right effective.

COMMENTARY

(1) ". . . *a term in the sub-contract.* . . ."

See, *e.g.* clause 14 of the standard form of sub-contract. For the architect's duty to nominate a sub-contractor who will agree to such a term, see clause 27 (a) (ix).

Clause 10
Clerk of Works

The Employer shall be entitled to appoint a clerk of works whose duty shall be to act solely as inspector on behalf of the Employer under the directions of the

Architect/Supervising Officer and the Contractor shall afford every reasonable facility for the performance of that duty. If any directions are given to the Contractor or to his foreman upon the Works by the clerk of works the same shall be of no effect unless given in regard to a matter in respect of which the Architect/ Supervising Officer is expressly empowered by these Conditions to issue instructions and unless confirmed in writing by the Architect/Supervising Officer within two working days of their being given. If any such directions are so given and confirmed then as from the date of confirmation they shall be deemed to be Architect's/Supervising Officer's instructions.

COMMENTARY

(1) *Generally*

This clause and these notes should be read with the discussion of the position of clerks of works generally, *ante*, p. 214.

(2) "... *solely as inspector*. ..."

He is not the architect's agent to give instructions. For the effect of inspections by architect or clerk of works see note 3 to clause 1.

(3) " *If any directions*. ..."

This provision recognises that in practice a clerk of works in the course of his inspection makes comments which in ordinary language amount to " directions " and is an attempt to deal with difficulties which can then arise. When such a direction is given, *e.g.* for the removal of materials alleged to be defective, the contractor can, if he is so minded, do nothing and ignore the direction unless and until confirmed by the architect. In practice the prudent contractor in any case where he considers a direction involves a variation or extra cost for which he wishes to claim will inform the architect immediately and seek his instructions.

Sanctioning by architect of variation. The architect has power under clause 11 (1) to sanction a variation made otherwise than pursuant to an instruction of the architect. This power can, it is suggested, be used where the clerk of works in a case of urgency has directed the contractor to carry out some variation and the contractor has complied with the direction without awaiting confirmation and it was reasonable in the circumstances for the contractor to carry out the variation. If the architect refuses to exercise the power under clause 11 (1) the contractor can go to arbitration and seek an award for the cost of the variation.[59]

(4) "... *expressly empowered*. ..."

For instructions by the architect see clauses 1 and 2 and notes thereto; in particular note 6 for architect's powers to issue instructions.

Clause 11

Variations, provisional and prime cost sums

(1) The Architect/Supervising Officer may issue instructions requiring a variation and he may sanction in writing any variation made by the Contractor otherwise than pursuant to an instruction of the Architect/Supervising Officer.

[59] See clause 35 (3).

No variation required by the Architect/Supervising Officer or subsequently sanctioned by him shall vitiate this Contract.

(2) The term ' variation ' as used in these Conditions means the alteration or modification of the design, quality or quantity of the Works as shown upon the Contract Drawings and described by or referred to in the Contract Bills, and includes the addition, omission or substitution of any work, the alteration of the kind or standard of any of the materials as goods to be used in the Works, and the removal from the site of any work materials or goods executed or brought thereon by the Contractor for the purposes of the Works other than work materials or goods which are not in accordance with this Contract.

(3) The Architect/Supervising Officer shall issue instructions in regard to the expenditure of prime cost* and provisional sums included in the Contract Bills and of prime cost sums which arise as a result of instructions issued in regard to the expenditure of provisional sums.

(4) All variations required by the Architect/Supervising Officer or subsequently sanctioned by him in writing and all work executed by the Contractor for which provisional sums are included in the Contract Bills (other than work for which a tender made under clause 27 (g) of these Conditions has been accepted) shall be measured and valued by the Quantity Surveyor who shall give to the Contractor an opportunity of being present at the time of such measurement and of taking such notes and measurements as the Contractor may require. The valuation of variations and of work executed by the Contractor for which a provisional sum is included in the Contract Bills (other than work for which a tender has been accepted as aforesaid) unless otherwise agreed shall be made in accordance with the following rules:—

(a)   The prices in the Contract Bills shall determine the valuation of work of similar character executed under similar conditions as work priced therein:

(b)   The said prices, where work is not of similar character or executed under similar conditions as aforesaid, shall be the basis of prices for the same so far as may be reasonable, failing which a fair valuation thereof shall be made;

(c)   Where work cannot properly be measured and valued the Contractor shall unless otherwise agreed be allowed:

(i)   the prime cost of such work calculated in accordance with the " Definition of Prime Cost of Daywork carried out under a Building Contract " issued by the Royal Institution of Chartered Surveyors and the National Federation of Building Trades Employers and current at the date of tender as defined in clause 31D (6) (a) of these Conditions (or as defined in the Formula Rules where clause 31F of these Conditions applies) together with percentage additions to each section of the prime cost at the rates set out by the Contractor in the Contract Bills; or

(ii)   where the work is within the province of any specialist trade and the said Institution and the appropriate body representing the employers in that trade have agreed and issued a definition of prime cost of daywork, the prime cost of such work calculated in accordance with that definition and current at the date of tender as defined in clause 31D (6) (a) of these Conditions (or as defined in the Formula Rules where clause 31F of these Conditions applies) together with percentage additions on the prime cost rates set out by the Contractor in the Contract Bills.

---

* *The term " prime cost " may be indicated by the abbreviation " P.C." in any document relating to this Contract (including the Contract Bills), and wherever the abbreviation is used it shall be deemed to mean " prime cost."*

Provided that in any case vouchers specifying the time daily spent upon the work (and if required by the Architect/Supervising Officer the workmen's names) and the materials employed shall be delivered for verification to the Architect/Supervising Officer or his authorised representative not later than the end of the week following that in which the work has been executed;

(d) The prices in the Contract Bills shall determine the valuation of items omitted; provided that if omissions substantially vary the conditions under which any remaining items of work are carried out the prices for such remaining items shall be valued under rule (b) of this sub-clause.

(5) Effect shall be given to the measurement and valuation of variations under sub-clause (4) of this Condition in Interim Certificates and by adjustment of the Contract Sum; and effect shall be given to the measurement and valuation of work for which a provisional sum is included in the Contract Bills under the said sub-clause in Interim Certificates and by adjustment of the Contract Sum in accordance with clause 30 (5) (c) of these Conditions.

(6) If upon written application being made to him by the Contractor, the Architect/Supervising Officer is of the opinion that a variation or the execution by the Contractor of work for which a provisional sum is included in the Contract Bills (other than work for which a tender made under clause 27 (g) of these Conditions has been accepted) has involved the Contractor in direct loss and/or expense for which he would not be reimbursed by payment in respect of a valuation made in accordance with the rules contained in sub-clause (4) of this Condition and if the said application is made within a reasonable time of the loss or expense having been incurred, then the Architect/Supervising Officer shall either himself ascertain or shall instruct the Quantity Surveyor to ascertain the amount of such loss or expense. Any amount from time to time so ascertained shall be added to the Contract Sum, and if an Interim Certificate is issued after the date of ascertainment any such amount shall be added to the amount which would otherwise be stated as due in such Certificate.

COMMENTARY

(1) *Generally*

This clause provides for: the ordering or sanctioning of variations (clause 11 (1)); the meaning of variations (clause 11 (2)); instructions in regard to prime cost and provisional sums (clause 11 (3)); the valuation of variations and provisional sum work (clause 11 (4)); giving effect to such valuation (clause 11 (5)); claims for loss or expense additional to such valuation (clause 11 (6)). For variations generally see Chapter 5.

Without quantities version. For the general position, see note 4 to Articles of Agreement. For the schedule of rates which is the basic document for pricing variations see clause 3 (2) (b) observing that it has to be furnished by the contractor immediately after the execution of the contract. In practice it is prudent to insist upon its production before the execution of the contract so that the employer may know, before signing the contract, the rates proposed to be charged for variations. The definition of the term " variation " is not easy to follow. It seems to say that if the contractor has to do something which is not shown both on the drawings and described by or referred to in the specification then it is a variation. Compare this with clause 12 (1) which states that the quality and quantity of the work included in the contract sum is deemed to be that which is shown on the drawings *or* described in the specification. This suggests that the contractor is not entitled to extra payment if the work appears in either document, and it is submitted that this is the intention and that clause 12 (2) which provides for correction of errors in the drawings and/or specification should be read in this way.

(2) *Clause* 11 (1) *". . . he may sanction in writing any variation . . ."*
For instructions generally see clauses 1 and 2 and notes thereto, particularly note 9. These words give the architect a discretion to sanction a variation although the ordinary procedure has not been followed and even, apparently, where there was no order or request at all for the variation to be carried out.[59a] There is no express guidance as to the exercise of the discretion. It is submitted that the test is whether in all the circumstances it is reasonable to exercise it. It is suggested that matters which might be taken into account include: the interests of each of the parties; the reason why the variation was carried out; the nature and extent of the variation; the reason why the ordinary procedure was not followed. The power to sanction is not limited to claims for extra payment by the contractor but includes an unauthorised departure from the contract work not involving extra cost, and omissions which result in a reduction of the contract sum.

Arbitration. The architect's discretion is subject to arbitration (see clause 35).

" Deemed " variation under clause 12 (2).[60] This requires, it is submitted, neither an instruction nor a sanctioning in writing; but note the contractor's duty to give notice upon finding a discrepancy between the contract drawings and the contract bills (clause 1 (2) and note 4 thereto), breach of which makes him liable, it is submitted, for any loss thereby suffered, *e.g.* because it deprived the architect of the opportunity to require a variation minimising the financial result of the discrepancy.

Variation orders by the employer personally. See note 8 to clause 2.

(3) *Clause* 11 (1) *" No variation . . . shall vitiate this Contract."*
It is submitted [61] that the intention is that the ordering of substantial variations does not entitle the contractor: (i) to rescind the contract and refuse to do further work; (ii) to have the completed work measured and paid for on a *quantum meruit* [62] and not in accordance with the terms of the contract. Despite the width of the words it is submitted that there must be some limit to the nature and extent of the variations which can be ordered; see note 2 to clauses 1 and 2. For the pricing of variations not of similar character or not executed under similar conditions to work in the contract bills, see rule (b) of clause 11 (4); for

---

[59a] See *ante*, p. 66, " work done without request."
[60] See *post*, p. 319.
[61] Substantially similar words have been judicially considered but the points at issue and the wording of the contracts were such as to afford but small assistance in the construction of the Standard form of building contract. See *Dodd* v. *Churton* [1897] 1 Q.B. 562 (C.A.); *Wells* v. *Army & Navy Co-op. Society* (1902) H.B.C. (4th ed.) Vol. 2, p. 346 (C.A.) referred to *ante*, p. 161.
[62] See *ante*, p. 58.

the position where omissions substantially vary the conditions under which any remaining items of work are carried out, see rule (d) of clause 11 (4).

Reasonableness of instructions requiring variations. For a discussion see note 8 to clause 2.

Variations after practical completion. It is submitted that the architect cannot issue instructions requiring a variation after practical completion,[63] so that if thereafter the employer wishes variations to be executed they should be the subject of a separate agreement.

(4) *Clause* 11 (2) " *The term ' variation ' . . ."*

This is the definition. For the contract drawings, see the Articles of Agreement. For the contract bills, see the Articles of Agreement and clause 12. For the kind or standard of materials or goods, see clause 6 (1). For removal from the site of work materials or goods, see clause 6 (4). For the effect of authority to depart from the specified method of construction to assist the contractor, see *ante*, p. 67, and *post*, p. 368.

(5) *Clause* 11 (3) ". . . *prime cost and provisional sums* . . ."

This clause [64] should be read with clause 27 (nominated sub-contractors), clause 28 (nominated suppliers) and clause 30 (5) (c) (settlement of accounts). " Provisional sum " is used in the sense of expenditure the amount of which is unknown when the contract is entered into.[65] " Prime cost sum " is used where the amount is known or becomes known because of the nomination of a sub-contractor or supplier to do work or supply goods at a certain price.[66]

The architect has not merely a right but a duty to nominate.[67]

(6) *Clause* 11 (4) " *The valuation* . . ."

In *Burden Ltd.* v. *Swansea Corporation* [68] Lord Radcliffe said, referring to a contract not materially different from the Standard form: " Generally speaking, I regard the surveyor as the person charged with the duty of valuing the contractors' work and advising the architect as to the allowance of their claims for payment, interim and final. But I do not see anything in the contract which suggests that the architect is bound to accept the surveyor's opinions or valuations when he exercises his own function of certifying sums for payment. At that point the architect remains master in his own field." It is submitted, however, that the

---

[63] The works are complete and the procedure for measurement and valuation and preparation of priced bills of variations begins, see clauses 15 (1), 30 (5) (a) and the Appendix to the Standard form of building contract.

[64] In *Bickerton (T. A.) & Son Ltd.* v. *N.W. Metropolitan Hospital Board* [1970] 1 W.L.R. 607 (H.L.), discussed *ante*, p. 182 and *post*, p. 369 the House of Lords was considering the question of re-nomination of a sub-contractor who repudiated. Viscount Dilhorne was of the view that clause 11 (3) added nothing to clause 27 (p. 623) on the point; the rest of the House seemed to find the clause of some assistance.

[65] See further, *ante*, p. 73.

[66] See also the standard Method of Measurement (5th ed.), A7, where definitions are given for the use of quantity surveyors in preparing bills of quantities.

[67] *Bickerton's* case, *supra*.

[68] [1957] 1 W.L.R. 1167, 1173 (H.L.).

surveyor must, subject to any special agreement, value in accordance with the rules of this clause and that the architect must give proper consideration to such a valuation. For the priced bills of variation, see clause 30 (5) (a).

(7) *Clause* 11 (4) "*. . . unless otherwise agreed . . .*"

The agreement can, it is submitted, have been made, or be made, at any time. If a price is quoted by the contractor it is desirable to agree whether or not it includes any claim for loss or expense under clause 11 (6) and is otherwise inclusive of all " claims."

Authority to make agreement

There are no express words in the contract giving either the architect or the surveyor authority to make such an agreement.[69] It is thought that the architect, by implication, has such authority but that the quantity surveyor has not. If a substantial departure from the ordinary rules is contemplated it is suggested that, to avoid dispute, the agreement is made by the employer or he expressly authorises the architect or the quantity surveyor to make it.

Pricing errors in contract bills

Opportunity for errors can arise in writing down the unit price, the extension of that price and the casting up and carrying forward of the extended prices. Unless there is a case for rectification of prices [70] both parties are bound by these errors in the carrying out of the original contract work.[71] But must the errors be taken into account in pricing variations where there was no express agreement to do so [72] operating as an additional term of the contract, and the parties cannot agree upon a special method under the provisions of this clause ? Consider the following: the contractor in error inserts a unit price of £5 instead of £50 and the total of his prices (excluding preliminary items, prime cost and provisional sums) is stated to be £1,000, whereas if £50 had been written down it would have been £1,045; if five further units of the item are required by way of variation must he carry them out for 5 × £5 thus multiplying the effect of his error ?

It is thought that, having regard to clause 13, the contractor is bound by his £5 for purposes of valuation under clauses 11 (4) (a) and (b), but that the employer is not entitled to say that prices for variations shall be deemed to be reduced by 45/1045.[73] Further, it is submitted, the contractor cannot recover the difference between £5 and his intended price of £50 per unit as " loss or expense " under clause 11 (6). Conversely, if the contractor had written £50 when he intended £5 he obtains the benefit of his error.

---

[69] *Cf.* clauses 17, 31D (3).

[70] See *ante*, p. 47. See also the words of clause 13.

[71] See clause 13, but the principle would, as a matter of construction apply, it is submitted, even without clause 13. Note that the provision for correction in clause 12 (2) is for items and not prices.

[72] See *e.g.* the " Code of Procedure for Selective Tendering," published by the National Joint Consultative Committee of Architects, Surveyors and Builders. New edition is the " Code of Procedure for Single Stage Selective Tendering 1977."

[73] See discussion, *ante*, p. 71.

(8) *Clause* 11 (4) (*b*) ". . . *similar character* . . . *similar conditions* . . ."

These are alternatives. It is sufficient to establish either to obtain the valuation under this clause. Dissimilar conditions might, it is suggested, include physical site conditions such as: wet compared with dry; high compared with low; confined space compared with ample working space; winter working compared with summer working where the contract documents show that the bill prices were based on summer working. The execution of work in large, or conversely small, quantities compared with the quantities of the relevant bill item may also it is thought give rise to a right to a valuation under the clause. A contractor who is not certain whether a claim for extra payment arises under clause 11 (4) (b) or 11 (6) can make his application in the alternative.

(9) *Clause* 11 (6) ". . . *direct loss and/or expense.* . . ."

This clause, together with clause 24 (1), is the equivalent in a more restricted form to the wide but vague right to claim loss or expense which could be made under clause 1 of the 1939 Form. There is nothing exactly comparable in the I.C.E. conditions but consider clause 52. For the government contract, form GC/Works/1, see clause 9 (2) which expressly excludes claims to " expense " so that any claim must arise under clause 9 (1) (c) for a fair rate.

Preparation and presentation of claims generally. See *ante*, pp. 160 and 266.

Limitations to claim: the loss or expense must be " direct " (the meaning of which is discussed *infra*); it must not be recoverable in the valuation under clause 11 (4); there must have been an application in writing within a reasonable time of the loss having been incurred (but see note 12, *infra*). For a possible implied qualification where the loss or expense was caused by a variation due to a design defect see note 2 (*c*) to Articles of Agreement.

Delay caused by variation. This is a common (although in no way the only) ground for an application. It may also provide grounds for an extension of time under clause 23 (e), but the difference in subject-matter of the two clauses should be kept in mind. Clause 11 (6) provides grounds for extra payment; clause 23 provides relief from liability to pay liquidated damages, and circumstances can arise when only one of the clauses applies. The contractor's written communication to the architect should show clearly whether it is an application under clause 11 (6), notice under clause 23 or both.

Delay before receiving instruction requiring variation. This may give rise to a right to extension of time under clause 23 (f) and to loss and expense under clause 24 (1) (a), and should be distinguished from the delay caused by the variation. Sometimes when numerous instructions are received piecemeal over a period causing delay and loss or expense the distinction cannot be made with confidence. The contractor should then, it is suggested, state the essential facts in his communication to the architect and say that it is an application for loss or expense under clause 11 (6) and/or 24 (1) (a) and notice of causes of delay under clause 23 (e) and/or 23 (f).

Meaning of " direct loss and/or expense." There is no reported authority. The word " loss " considered by itself seems to indicate those heads of claim which would be recoverable as damages.[74] This often includes money which has been expended.[75] So presumably the words " and/or expense " are inserted to remove any doubt and do not limit the word " loss " but, if intended to have any separate effect, provide a different head of claim. The word " direct " is, it is thought, intended to govern both loss and expense. The intention is, it seems, to exclude those heads of claim which, if made as damages, would not be recoverable unless the employer, at the time of entering into the contract, knew, or must be taken to have known, as liable to result from the cause occasioning the claim.[76] Thus, it is thought that where a variation causes delay which in turn renders the contractor unable to tender for another contract where he, unknown to the employer, expected to make a specially high profit that profit, as such, is not recoverable, but, it is submitted, such loss of overheads and net profit as was reasonably foreseeable by the employer as liable to be lost because of the delay is recoverable.[77] This head of claim includes head office and supervision charges. When considering such claims it is always necessary to look carefully at what the contractor has recovered under clause 11 (4). Even the bill rate under sub-clause (a) ordinarily includes some element for overheads and profit. The daywork rate includes a substantial element although one cannot say that payment at daywork rates automatically excludes any claim under clause 11 (6). The extent to which the rate under sub-clause (b) includes overheads and profit varies according to the application of the sub-clause and the particular rate.

Extension of preliminaries. The claim for delay caused by a variation is frequently quantified in whole or in part by extending those items in the preliminary bill, or elsewhere in the contract bills, the price of which is affected by time, for the period of the delay. Sometimes there may be no delay in the completion of the works but certain items, *e.g.* for plant or equipment kept upon the site to carry out the variation, should be extended.

Uneconomic working of labour. If caused by delay, this is, it is submitted, " direct loss and/or expense." For a discussion of its assessment see *ante*, p. 153.

(10) *Clause 11 (6) ". . . the Architect . . . shall . . . ascertain the amount. . . ."*

Information necessary for ascertainment. It is, it is submitted, implied that the contractor will furnish the architect with such information as is reasonably necessary to enable him to carry out the ascertainment. (For the implication of terms see p. 35.) In so far as the ascertainment is being carried out after practical completion an express duty to provide necessary documents arises under clause 30 (5) (b).

---

[74] For damages generally, see *ante*, p. 144.

[75] See in particular *Anglia Television Ltd.* v. *Reed* [1972] 1 Q.B. 60 (C.A.). See also *Chandris* v. *Union of India* [1956] 1 W.L.R. 147, 154, 156, 159 (C.A.).

[76] See *Hadley* v. *Baxendale* (1854) 9 Ex. 341, discussed, *ante*, p. 144.

[77] For some assistance on the word " direct," see *Saint Line Ltd.* v. *Richardsons, Westgarth & Co. Ltd.* [1940] 2 K.B. 99, and for a case on a statute, *A. & B. Taxis Ltd.* v. *The Secretary of State for Air* [1922] 2 K.B. 328 (C.A.).

Provisional ascertainment. This is not expressly provided for. The architect may, soon after the carrying out of the variation, be satisfied that loss and expense has been incurred but cannot ascertain it with that degree of certainty possible at a later stage, probably when dealing with the final account. Can he make a provisional ascertainment? There may be an implied power but the position is not clear. A course he can take is to agree with the contractor to make a provisional ascertainment subject to revision up or down in or before the final certificate.

(11) *Clause* 11 (6) "... *The architect shall* ... *instruct the Quantity Surveyor to ascertain the amount* ..."

If the architect so instructs the quantity surveyor is he bound to carry the amount of such ascertainment into his certificates or can he vary it? The matter is not clear. There is an argument for saying that the architect is bound by the ascertainment so that if he disagrees with it he must nevertheless include it in his certificates and, as part of his duty to the employer, advise that it can be challenged in arbitration proceedings. But it is usually accepted that the architect can revise his own certificates other than the final certificate (see clause 30 (7)). If this is correct, and it is submitted that it is, it is difficult to see why he cannot revise the quantity surveyor's ascertainment. In practice unless the surveyor can be shown to have gone wrong in principle the amount of his ascertainment would be important evidence of the true amount.

(12) *Clause* 11 (6) "... *if an Interim Certificate is issued.* ..."

The loss and expense should not, it is submitted, be added until it has been incurred. Thus the price of items in the preliminary bill is often fixed expressly or by implication by reference to time (see note 9, *supra*). Some items, *e.g.* site huts, may be expended over the whole contract period, others, *e.g.* cranes, over part of the period. Until the relevant time has expired the contractor has not expended the item and has not been involved in further expenditure upon such item. By the same reasoning it cannot be said that notice of such loss or expense is too late if given within a reasonable time of the expiry of the relevant time. Certification of an item is evidence, but not conclusive, of expenditure (clause 30 (1), (2), (8)).

Clause 12
Contract Bills

(1) The quality and quantity of the work included in the Contract Sum shall be deemed to be that which is set out in the Contract Bills which Bills unless otherwise expressly stated in respect of any specified item or items shall be deemed to have been prepared in accordance with the principles of the Standard Method of Measurement of Building Works 5th edition Imperial, revised March 1964/5th edition Metric* by the Royal Institution of Chartered Surveyors and the National Federation of Building Trades Employers, but save as aforesaid nothing contained in the Contract Bills shall override, modify, or affect in any way whatsoever the application or interpretation of that which is contained in these Conditions.

(2) Any error in description or in quantity in or omission of items from the

* *Delete whichever edition is inapplicable.*

Contract Bills shall not vitiate this Contract but shall be corrected and deemed
to be a variation required by the Architect/Supervising Officer.

COMMENTARY

(1) *Generally*

This clause defines the work for which the contract sum is payable and provides
for the correction of errors in *items* (not prices) in the contract bills.

(2) *Clause* 12 (1) " *The quality and quantity . . .*"

On the importance of these words in defining the contractor's obligation, see
notes to the Articles of Agreement, in particular, " The nature of extra work,"
ante, p. 290. For the kinds and standards of materials, goods and workmanship
see clause 6 (1). For the without quantities version see note 4 to Articles of
Agreement and note 1 to clause 11.

(3) *Clause* 12 (1) ". . . *which Bills unless otherwise expressly stated . . .*"

Unless otherwise expressly stated in respect of any specified item or items the
contract bills are deemed to have been prepared in accordance with the principles
of the Standard Method of Measurement. If they have not been so prepared there
is an error which must be corrected and the correction is deemed to be a variation
required by the architect (clause 12 (2)). For example, if there is no provision in
the contract bills for excavation in rock but in carrying out the works it becomes
clear that such excavation is necessary then, it is submitted, the bills must be
corrected by inserting an item for excavation in rock and the contractor will be
entitled to extra payment under clause 11 (4) and, where appropriate, 11 (6) for
the Standard Method states, " excavating in rock shall be so described." [78]

(4) *Clause* 12 (1) ". . . *in respect of any specified item . . .*"

It is a question of construction of the contract bills in question whether or not
adequate notice is given of any particular departure from the principles of the
Standard Method, but it is thought that general statements at the beginning of
the bills such as, " any departure from the Standard Method of Measurement
must be accepted " are not sufficient.

For small dwellings there is a " Code for the Measurement of Building Works
in Small Dwellings," issued in 1963. The Code is based on, but modifies the
Standard Method in respect of many specified items. It is thought that a statement that the contract bills have been prepared in accordance with the code is a
sufficient compliance with clause 12 (1) which must then be read as if it referred
to the Code.

(5) *Clause* 12 (1) ". . . *the principles of the Standard Method of Measurement . . .*"

The Standard Method is a long document with detailed and exact descriptions

---

[78] See the Standard Method of Measurement, " Excavation and Earthwork," D6 (e);
see also *Bryant* v. *Birmingham H.S.F.* [1938] 1 All E.R. 503 where the contract was the
1931 edition of the R.I.B.A. form and the clause corresponding to clause 12 was somewhat
different. *Cf.* contracts where the contractor's obligation is expressed in wide terms and
quantities are not part of the contract, *e.g. McDonald* v. *Workington Corp.* (1892) H.B.C.
(4th ed.), Vol. 2, p. 228 (C.A.); *Re Nuttall & Lynton & Barnstaple Ry.* (1899) H.B.C.
(4th ed.), Vol. 2, p. 279 (C.A.); see also *ante*, p. 62.

of the methods of describing and measuring building work. It is thought that it is not the intention of these words to excuse a failure to comply with the detailed requirements of the Standard Method. Thus the introduction to the Standard Method states, " The Standard Method of Measurement provides a uniform basis for measuring building work and embodies the essentials of good practice but more detailed information than is demanded by this document should be given where necessary in order to define the precise nature and extent of the required work. The Standard Method shall apply equally to the measurement of proposed work and of executed works." Rule A (1) of the General Rules of the Standard Method states, " Bills of Quantities shall fully describe and accurately represent the works to be executed. Work which cannot be measured accurately shall be described as provisional or given in a Bill of approximate quantities."

(6) *Clause* 12 (1) "*. . . nothing contained in the Contract Bills shall override, modify, or affect in any way whatsoever . . . these Conditions.*"

These words are of the utmost importance. It may be taken now as settled that they have the effect that, contrary to the ordinary rule that written words (*i.e.* the bills) prevail over printed (*i.e.* the conditions),[79] the court (which includes the arbitrator) will not look at special provisions in the contract bills in order to ascertain the intention of the parties upon matters dealt with in the conditions such as insurance,[80] nomination of sub-contractors,[81] phasing of works,[82] possession of parts of the works.[83] In *English Industrial Estates* v. *Wimpey*,[84] where the Court of Appeal considered the point fully, Lord Denning M.R. dissented from this view and the rest of the Court were not happy about it. Nevertheless, although they looked at the bills " to follow exactly what was going on," [85] they applied the principle. This case, together with *Gleeson's* case [86] is discussed further in the notes to clause 16 (partial possession by employer).

Amendment of form. It follows that if the parties wish to have special terms on any matter dealt with in the conditions they must amend the form. It is not sufficient to say in the contract bills that the articles of agreement and conditions will be in the standard form " with certain amendments and additional clauses as set out in the bill." [87] Various methods of amendment can be used. One is to follow the procedure very common in engineering contracts of having special conditions. Another, where the amendments are contained in the preliminary bill, is expressly to write in a reference to that bill in the articles of agreement and in clause 1 which defines the contract documents.

---

[79] For the principle, see *ante*, p. 33.

[80] *Gold* v. *Patman & Fotheringham Ltd.* [1958] 1 W.L.R. 697, 701 (C.A.).

[81] *Bickerton & Son Ltd.* v. *N.W. Metropolitan Regional Hospital Board* [1969] 1 All E.R. 977, 982 (C.A.); [1970] 1 W.L.R. 607, 617 (H.L.).

[82] *Gleeson, M. J. (Contractors) Ltd.* v. *Hillingdon, London Borough of* (1970), Mocatta J. (unreported), save in (1970) 215 *Estates Gazette*, 165.

[83] *English Industrial Estates* v. *Wimpey (George) & Co. Ltd.* [1973] 1 Lloyd's Rep. 118 (C.A.).

[84] *Ibid.*

[85] *Ibid.* p. 126 ; see also p. 128.

[86] See *supra*, footnote 82.

[87] *Gleeson's* case, the words used in the bills of quantities, transcript p. 3 ; see further, *post*, p. 330.

Words defining work and not obligation. In *AMF International Ltd. v. Magnet Bowling Ltd.*[88] the contract was in the 1939 form but for present purposes the differences were immaterial. Clause 14 made the contractor liable for injury to property in words substantially the same as those of clause 18 (2) of these conditions. The bills contained certain items expressly requiring the contractor to allow for protection of all work and materials from injury by weather or any other cause, to cover and case up and protect the work whensoever needed, to provide as necessary for the diversion of storm water to requisite channels, drains etc. and that no water was to be allowed to accumulate on any part of the site. Equipment stored in a partially completed part of the works was damaged by a flood. It was held that clause 10 (the equivalent of clause 12 in these conditions) did not prevent the court from giving full effect to these items in the bills so that the contractor was liable.

(7) *Clause* 12 (2) " *Any error in description . . . of items . . . shall be corrected . . .*"
Errors can arise from discrepancies between contract drawings and contract bills,[89] from mathematical errors of measurement and from failure to prepare the contract bills in accordance with the principles of the Standard Method (see note 3, *supra*). Such failure may be due to facts unknown to the parties at the time of making the contract, *e.g.* rock or springs in the sub-soil. The result is to place upon the employer many of the risks of defects in the soil which would otherwise be upon the contractor (see *ante*, pp. 101 and 291).

Re-measurement. The correction is effected by inserting or deleting items as the case may be, or re-measuring items wrongly described, and valuing in accordance with clause 11 (4) and, where appropriate, 11 (6). If items in, or sections of, the contract bills are marked " provisional " this, it is submitted, operates as an agreement that such items or sections will be re-measured in any event. In some cases there may be a dispute as to the existence of an error, which dispute can only be resolved on a re-measurement. It is thought that there is an implied term that either party may require re-measurement of a disputed item or items but must act reasonably in the exercise of such right, so that if he unreasonably puts the other party to the expense of re-measurement he may be liable to that other party for such expense. There is always re-measurement where the form for use with approximate quantities (see *infra*) is used.

(8) *Clause* 12 (2) " *. . . shall not vitiate this Contract . . .*"
For a note on similar words in clause 11 see note 3 to clause 11.

(9) *Clause* 12 (2) " *. . . deemed to be a variation . . .*"
For the contractor's duty to report any discrepancy he finds between the contract drawings and the contract bills, see clause 1 (2) and note, " ' *deemed* ' *variation under clause* 12 (2)," *ante*, p. 313.

(10) *Bills of approximate quantities*
The Standard form of contract has, in the past, sometimes been used when

---

[88] [1968] 1 W.L.R. 1028.
[89] See clause 1 (2) and note 4 thereto.

the works have not been designed in sufficient detail to enable bills, complying with the Standard Method of Measurement, to be prepared. It is not intended for this purpose and considerable difficulties of application of various clauses, particularly those relating to variations, arise unless careful and detailed amendments are carried out. The Joint Contracts Tribunal in October 1975 issued a form for use with approximate quantities. The form is not printed but a helpful practice note, 20 was also issued and is printed *post*, p. 420.

Clause 13
Contract Sum

The Contract Sum shall not be adjusted or altered in any way whatsoever otherwise than in accordance with the express provisions of these Conditions, and subject to clause 12 (2) of these Conditions any error whether of arithmetic or not in the computation of the Contract Sum shall be deemed to have been accepted by the parties hereto.

COMMENTARY

(1) " *The Contract Sum shall not be adjusted . . .*"

Unless there is a case for the equitable remedy of rectification [90] the parties are bound by any errors incorporated into the contract sum. For the question whether pricing errors affect the pricing of variations see note 7 to clause 11. For the correction of items (not prices) in the contract bills, see clause 12 and notes thereto.

Clause 13A
Value Added Tax Supplemental Agreement

(1) (a) In this Condition and in the supplemental agreement pursuant hereto and annexed to these Conditions (hereinafter called the 'VAT Agreement') 'tax' means the value added tax introduced by the Finance Act 1972 which is under the care and management of the Commissioners of Customs and Excise (hereinafter and in the VAT Agreement called 'the Commissioners').

(b) The Employer and Contractor shall on the same date as executing the Articles of Agreement execute the VAT Agreement.

Contract Sum—Exclusive of VAT—Supplemental Agreement

(2) Any reference in these Conditions to ' Contract Sum ' shall be regarded as such Sum exclusive of any tax and recovery by the Contractor from the Employer of tax properly chargeable by the Commissioners on the Contractor under or by virtue of the Finance Act 1972 or any amendment thereof on the supply of goods and services under this Contract shall be under the provisions of this Condition and of the VAT Agreement.

(3) To the extent that after the date of tender as defined in Clause 31D (6) (a) of these Conditions the supply of goods and services to the Employer becomes exempt from the tax there shall be paid to the Contractor an amount equal to the loss of credit (input tax) on the supply to the Contractor of goods and services which contribute exclusively to the Works.

COMMENTARY

For V.A.T. see the supplemental agreement and notes thereto, *post*, p. 416.

---

[90] See *ante*, p. 47.

Clause 14
Materials and Goods unfixed or off-site

(1) Unfixed materials and goods delivered to, placed on or adjacent to the Works and intended therefor shall not be removed except for use upon the Works unless the Architect/Supervising Officer has consented in writing to such removal which consent shall not be unreasonably withheld. Where the value of any such materials or goods has in accordance with clause 30 (2) of these Conditions been included in any Interim Certificate under which the Contractor has received payment, such materials and goods shall become the property of the Employer, but subject to clause 20 [B] or clause 20 [C] of these Conditions (if applicable), the Contractor shall remain responsible for loss or damage to the same.

(2) Where the value of any materials or goods has in accordance with clause 30 (2A) of these Conditions been included in any Interim Certificate under which the Contractor has received payment, such materials and goods shall become the property of the Employer, and thereafter the Contractor shall not, except for use upon the Works, remove or cause or permit the same to be moved or removed from the premises where they are, but the Contractor shall nevertheless be responsible for any loss thereof or damage thereto and for the cost of storage, handling and insurance of the same until such time as they are delivered to and placed on or adjacent to the Works whereupon the provisions of sub-clause (1) of this clause (except the words ' where the value ' to the words ' the Employer but ') shall apply thereto.

COMMENTARY

(1) *Generally*

This clause is of a kind sometimes termed a " vesting clause " [91] and is of particular importance in the event of the contractor's default or insolvency.[92]

(2) *Clause 14 (1) " Unfixed materials . . . shall not be removed. . . ."*

A lien [93] is created in favour of the employer even before certification under clause 30 (2). The property in materials and goods passes to the employer upon payment of the architect's certificate which includes the value of such materials and goods, but both before and after certification the risk remains with the contractor. The property in materials built into the works passes to the owner of the land (see *ante*, p. 124) whether paid for or not.

(3) *Clause 14 (2) ". . . in accordance with clause 30 (2A). . . ."*

The architect is bound to certify for materials and goods properly and not prematurely delivered to or adjacent to the works (clause 30 (2)). He has a discretion whether to certify for materials and goods before delivery (clause 30 (2A)). For a discussion of the problems arising from certification of goods and materials not on the site and of the provisions of clause 30 (2A), see note 7 to clause 30.

Clause 15
Practical completion and defects liability

(1) When in the opinion of the Architect/Supervising Officer the Works are

---

[91] For ownership of materials and vesting clauses generally, see *ante*, p. 124.

[92] For default and insolvency of the contractor, see *post*, p. 353.

[93] For meaning, see *ante*, p. 126.

practically completed, he shall forthwith issue a certificate to that effect and Practical Completion of the Works shall be deemed for all the purposes of this Contract to have taken place on the day named in such certificate.

(2) Any defects, shrinkages or other faults which shall appear within the Defects Liability Period stated in the appendix to these Conditions and which are due to materials or workmanship not in accordance with this Contract or to frost occurring before Practical Completion of the Works, shall be specified by the Architect/Supervising Officer in a Schedule of Defects which he shall deliver to the Contractor not later than 14 days after the expiration of the said Defects Liability Period, and within a reasonable time after receipt of such Schedule the defects, shrinkages, and other faults therein specified shall be made good by the Contractor and (unless the Architect/Supervising Officer shall otherwise instruct, in which case the Contract Sum shall be adjusted accordingly) entirely at his own cost.

(3) Notwithstanding sub-clause (2) of this Condition the Architect/Supervising Officer may whenever he considers it necessary so to do, issue instructions requiring any defect, shrinkage or other fault which shall appear within the Defects Liability Period named in the Appendix to these Conditions and which is due to materials or workmanship not in accordance with this Contract or to frost occurring before Practical Completion of the Works, to be made good, and the Contractor shall within a reasonable time after receipt of such instructions comply with the same and (unless the Architect/Supervising Officer shall otherwise instruct, in which case the Contract Sum shall be adjusted accordingly) entirely at his own cost. Provided that no such instructions shall be issued after delivery of a Schedule of Defects or after 14 days from the expiration of the said Defects Liability Period.

(4) When in the opinion of the Architect/Supervising Officer any defects, shrinkages or other faults which he may have required to be made good under sub-clauses (2) and (3) of this Condition shall have been made good he shall issue a certificate to that effect, and completion of making good defects shall be deemed for all the purposes of this Contract to have taken place on the day named in such certificate.

(5) In no case shall the Contractor be required to make good at his own cost any damage by frost which may appear after Practical Completion of the Works, unless the Architect/Supervising Officer shall certify that such damage is due to injury which took place before Practical Completion of the Works.

COMMENTARY

(1) *Generally*

This provides for: certificate of practical completion (clause 15 (1)); a schedule of defects (clause 15 (2)); instructions requiring the making good of particular defects (clause 15 (3)); a certificate of completion of making good defects (clause 15 (4)); damage by frost (clause 15 (5)). For the contractor's obligations, see clause 1; for standards of materials, goods and workmanship, see clause 6 (1); for the way in which the provisions of this clause fit into the final account procedure, see note 11 to clause 30; for a discussion of defects clauses generally, see *ante,* p. 127.

(2) " *When . . . the Works are practically completed . . .*"

What do these words mean? No clear answer can be given. The works can be practically completed notwithstanding that there are latent defects.[94] But can

---

[94] *Jarvis (J.) & Sons Ltd.* v. *Westminster Corporation* [1969] 1 W.L.R. 1448 (H.L.).

there be practical completion when there are some, albeit minor, patent defects ? Certain statements in opinions delivered in the House of Lords seem to suggest that practical completion means completion to the architect's reasonable satisfaction in all matters other than latent defects.[95] If this is correct, it means that the architect can hold up issuing his certificate, with all the consequences as to liquidated damages, retention money, insurance and otherwise which follow, notwithstanding that there are some minor defects apparent which could easily be put right without loss to the employer during the defects liability period and which do not prevent the employer from taking possession of and using the works for their intended purpose.[96] It is thought that this cannot be the intention, and that an architect would not be in breach of duty to his client in issuing the certificate of practical completion where he is reasonably satisfied that the works accord with the contract, notwithstanding that there may be such minor defects, and always providing that he is satisfied that the retention money will cover their cost and that there is no likelihood of the employer suffering loss due to interference with his use of the works while the minor defects are remedied. Further, he should obtain a written acknowledgment of the existence of, and an undertaking to put right, the defects from the contractor (*cf.* clause 48 (1) of the I.C.E. Conditions). It is submitted that this procedure is consistent with the intention underlying the use of the word " practically " and will ordinarily protect the employer's interests adequately. But it remains that the position is obscure. Actual entry into possession by the employer is strong, but, it is submitted, not conclusive,[97] evidence of the works having reached the stage of practical completion. See note 4 to clause 21 for the relationship of practical completion to liability for liquidated damages.

(3) *Clause* 15 (1) ". . . *a certificate to that effect.* . . ."

No particular form is specified, but the certificate should be clear and definite. The R.I.B.A. makes available a common form certificate for use under this clause and, where appropriate, under clause 16.

(4) *Clause* 15 (1) ". . . *for all the purposes of this Contract.* . . ."

These are: to fix the commencement of the defects liability period (clauses 15 (2), (3); 16; and Appendix) and the period of final measurement (clause 30 (5) (a) and Appendix); to fix the dates for: the expiry of the employer's right to retain the retention percentage (clause 30 (3) (a)); the release of the first moiety of the retention (clause 30 (4) (b)); the release of the obligation to insure under clause 20 [A] where clause 20 [A] applies; the end of the contractor's liability for damage to the works (see note 4 to clause 18) arising as an incident of his obligation to complete. For the importance of the certificate in the final account procedure, see note 11 to clause 30.

---

[95] *Ibid.*, p. 646, *per* Viscount Dilhorne; *Kaye (P. & M.) Ltd.* v. *Hosier & Dickinson Ltd.* [1972] 146, 165, *per* Lord Diplock. The form did not, of course, have the 1976 revised clause 1 so that the works generally had to be to his reasonable satisfaction.

[96] Compare the view of Salmon L.J. in *Jarvis* v. *Westminster Corporation* in the Court of Appeal [1969] 1 W.L.R. 1448, 1458.

[97] The employer, in the case of the contractor's delay, may be mitigating his loss by entering premises before the stage of practical completion.

(5) *Clause* 15 (2) "... *defects, shrinkages or other faults.* ..."

Excluding frost damage, which is dealt with separately below, defects, including shrinkage or other faults, must be due to " materials or workmanship not in accordance with this contract. . . ." Thus the following are not defects: the failure of the architect's design of the works,[98] or the unsuitability for their purpose of goods or materials which are of good quality and exactly as described in the contract bills or as required by an instruction of, or sanctioned in writing by, the architect.[1]

For materials, goods or workmanship not " procurable " see clause 6 (1). For defects in the work of nominated sub-contractors and the goods of nominated suppliers see notes to clauses 27 and 28. For the limitation on the contractor's liability where final payment is made to a nominated sub-contractor see clause 27 (e). For the effect, if any, of price, see note 3 to clause 1. For the effect of inspections by the architect or the clerk of works see note 3 to clause 1.

There is no reference to " goods " in this clause although there is such reference elsewhere (clauses 6, 11, 14, 20, 27, 28, 30, 31), but it is submitted that " materials " here is used to include goods.[2]

Arbitration. Either party can, subject to clauses 30 (6), (7), challenge the architect's decision on what is or is not a defect, see clause 35.

(6) *Clause* 15 (2) "... *defects which shall appear.* ..."

Defects outstanding at the time of practical completion are not, on a literal construction of these words, within clause 15, but it is convenient to deal with them as if they are.[3] For whether the architect can issue the certificate of practical completion notwithstanding the existence of minor defects, see *supra*.

(7) *Clause* 15 (2) "... *frost.* ..."

The contractor is liable to make good at his own cost damage by frost which appears within the defects liability period but only if the architect certifies that such damage is due to injury which took place before practical completion (clause 15 (5)). It is thought that an example of the contractor's liability might arise where concrete was injured by frost when being laid, but the damage does not appear until during the defects liability period and the architect gives his certificate under clause 15 (5).

(8) *Clause* 15 (2) "... *a Schedule of Defects.* ..."

Compare clause 15 (3) which is supplementary to clause 15 (2) and which is in substantially similar terms save that it empowers the architect to issue instructions requiring the making good of defects appearing within the defects liability

---

[98] See *supra*, p. 290, for a discussion of the extent of the contractor's design responsibility.

[1] See clauses 1, 2, 6 (1), 10 and 11. See also *Young & Marten Ltd.* v. *McManus Childs Ltd.* [1969] 1 A.C. 454 (H.L.); *Gloucestershire County Council* v. *Richardson* [1969] 1 A.C. 480 (H.L.); *Cotton* v. *Wallis* [1955] 1 W.L.R. 1168 (C.A.).

[2] Under the Sale of Goods Act 1893 " goods " includes materials, but the two terms are presumably used in the clauses referred to in the text to show that both builders' materials and made-up goods come within the ambit of such clauses.

[3] A contractor who had taken the benefit of the first moiety of the retention (see clause 30 (4) (b)) would, it is submitted, be estopped from alleging that clause 15 did not apply to such defects. In any event the architect can proceed under clauses 6 (4) and 2 (1).

period " whenever he considers it necessary " but not after the delivery of a schedule under clause 15 (2). Thus, very broadly, clause 15 (3) provides for instructions for making good defects from time to time during the defects liability period and clause 15 (2) for a schedule of defects at the end of the period.

(9) *Clause* 15 (2) *". . . and (unless the Architect shall otherwise instruct. . . .) entirely at his own cost "*

The architect apparently can issue an instruction which results in the employer having to bear the cost of the contractor making good his own defect. It is thought that only special circumstances warrant such an instruction and that it should not ordinarily be issued where the defect arises out of the contractor's want of proper skill. For the comparable position under clause 5, see note 2 to that clause. The discretion of the architect in each case is, it is submitted, open to review by the arbitrator (see clause 35).

It has been suggested [4] that the words in clauses 15 (2) and 15 (3) and the similar words in clause 5 are intended to give the architect a power only to adjust the contract sum downwards where the architect, in his discretion, decides that it is unnecessary under clause 5 to amend an error of setting out, or, under this clause, to amend the defect. Even if this is thought to be a reasonable construction it is difficult to reconcile with the express words.

(10) *Clause* 15 (5) *" When . . . defects . . . shall have been made good he shall issue a certificate to that effect. . . ."*

Various matters arise, thus:

(a) **Contractual remedies for breach.** The second half of the retention is not released until this certificate is given (clause 30 (4) (c)). The issue of the final certificate, with the protection it affords the contractor (see clause 30 (7)), may be delayed by the absence of this certificate (clause 30 (6)). For breach of an instruction under clause 15 (3) or under clause 6 (4), a notice under clause 2 (1) can be given and upon non-compliance others can be employed to do the work necessary and the cost deducted from the retention. It is not clear whether the power of determination under clause 25 (1) (c) can be exercised after practical completion, but having regard to the other remedies available it is unlikely, in ordinary circumstances, to be necessary to attempt such a serious step.

(b) **Damages for breach.** Clause 15 imposes a liability and gives a right to make good defects.[5] It does not exclude a claim for damages in respect of those breaches.[6] Failure by the employer, through the architect, to make use of the provisions of the clause may it is thought put the employer at some risk as to costs and may affect damages recoverable.[7] Thus in so far as the diminution in value of the works exceeds the cost of rectification, the employer cannot recover

---

[4] I. N. Duncan Wallace, Q.C., *Building and Civil Engineering Standard Forms*, pp. 53 and 77.

[5] *Kaye (P. & M.) Ltd.* v. *Hosier & Dickinson Ltd.* [1972] 1 W.L.R. 147, 166 (H.L.), *per* Lord Diplock. See also *Hancock & Others* v. *B. W. Brazier (Anerley) Ltd.* [1966] 1 W.L.R. 1317, 1333 (C.A.).

[6] *Ibid.*

[7] For damages generally, see *ante*, p. 144.

such excess if he has not given the contractor the opportunity of remedying the defects under clause 15.[8] Conversely, if the contractor fails to comply with the architect's requirements the employer can recover the cost of remedying defects even though such cost is greater than the diminution in value as a result of the unremedied defects.[9] The employer is, it is submitted, entitled to consequential loss suffered as a result of defects appearing in the defects liability period.[10]

(c) **Effect of the certificate as to breaches.** It is some, but not conclusive, evidence of the completion of the works in accordance with the contract and of the making good of defects (clause 30 (8)). For the effect of the final certificate under clause 30, see clause 30 (7) and notes 14 and 15 to clause 30.

(d) **Irremediable breach.** The architect may include a defect in an instruction or the schedule, but then find on representation by the contractor that it cannot be remedied, or cannot be remedied except at a cost which is unreasonable in comparison with the loss to the employer and the nature of the defect. In such circumstances he has, it is thought, a discretion to issue this certificate and to make a reduction in the amount certified for payment, being the amount by which the works are reduced in value by reason of the unremedied defect[11]; but, if approval of the quality of any materials or the standards of any workmanship is a matter for his opinion, he should not issue a final certificate—see clauses 1 (1), 30 (7) and notes thereto.

(e) **Defects after certificate.** If defects appear after the certificate of completion of making good defects issued under clause 15 (4) the architect has no power to issue further instructions, but can adjust any further certificate.[12] The amount of the adjustment is, it is submitted, assessed by the cost of rectification or, where the breach is irremediable, the diminution in value of the works. It should not, it is submitted, include any deduction for consequential loss, this being the subject-matter of a claim for damages by the employer. In so far as such defects, as they appear, evidence a breach of contract by the contractor, the usual rules as to damages, including those relating to mitigation, apply so that ordinarily the employer should give the contractor an opportunity of remedying the defects if it is reasonable to do so. For circumstances in which the architect should not issue a final certificate if any defects still exist—see " irremediable breach," *supra.* If defects appear after the issue of the final certificate see clause 30, and notes thereto.

(11) *Clause 15 (4) ". . . for all the purposes of this Contract . . ."*
See clause 30 (4) (c), release of second half of retention fund; 30 (6), time of issue of final certificate and note 10, *supra.*

---

[8] See *Kaye (P. & M.) Ltd.* v. *Hosier & Dickinson Ltd., supra,* pp. 163 *et seq.* for the opinion of Lord Diplock, discussed *post,* note 15 (c) to clause 30, whence it appears that on one view his speech may be treated as dissenting, but it is thought that the better view is that, apart from the narrow point in issue, it may be treated as *obiter* of very high authority.
[9] *Ibid.*
[10] *Ibid.*
[11] See clause 30 (2), (4) (c), (6); *Cotton* v. *Wallis* [1955] 1 W.L.R. 1168 (C.A.), discussed *ante,* pp. 214, 296. See also the measure of damages for defects, *ante,* p. 264.
[12] See *supra,* footnote 8.

Clause 16
Partial possession by Employer

If at any time or times before Practical Completion of the Works the Employer with the consent of the Contractor shall take possession of any part or parts of the same (any such part being hereinafter in this clause referred to as " the relevant part ") then notwithstanding anything expressed or implied elsewhere in this Contract:—

(a) Within seven days from the date on which the Employer shall have taken possession of the relevant part the Architect/Supervising Officer shall issue a certificate stating h's estimate of the approximate total value of the said part, and for all the purposes of this Condition (but for no other) the value so stated shall be deemed to be the total value of the said part.

(b) For the purposes of sub-paragraph (ii) of paragraph (f) of this Condition and of sub-clauses (2) (3) and (5) of clause 15 of these Conditions, Practical Completion of the relevant part shall be deemed to have occurred and the Defects Liability Period in respect of the relevant part shall be deemed to have commenced on the date on which the Employer shall have taken possession thereof.

(c) When in the opinion of the Architect/Supervising Officer any defects, shrinkages or other faults in the relevant part which he may have required to be made good under sub-clause (2) or sub-clause (3) of clause 15 of these Conditions shall have been made good he shall issue a certificate to that effect.

(d) The Contractor shall reduce the value insured under clause 20 [A] of these Conditions (if applicable) by the full value of the relevant part, and the said relevant part shall as from the date of which the Employer shall have taken possession thereof be at the sole risk of the Employer as regards any of the contingencies referred to in the said clause.

(e) In lieu of any sum to be paid or allowed by the Contractor under clause 22 of these Conditions in respect of any period during which the Works may remain incomplete occurring after the date on which the Employer shall have taken possession of the relevant part there shall be paid or allowed such sum as bears the same ratio to the sum which would be paid or allowed apart from the provisions of this Condition as does the Contract Sum less the total value of the said relevant part to the Contract Sum.

(f) (i) Within fourteen days of the date on which the Employer shall have taken possession of the relevant part there shall be paid to the Contractor from the sums then retained under clause 30 (3) of these Conditions a percentage (which percentage shall be equal to half the Retention Percentage) of the total value (subject to paragraph (iii) of this sub-clause) of the relevant part.

(ii) On the expiration of the Defects Liability Period named in the Appendix to these Conditions in respect of the relevant part or on the issue of the Certificate of Completion of Making Good Defects in respect of the relevant part, whichever is the later, there shall be paid to the Contractor from the sums then retained under clause 30 (3) of these Conditions a percentage (which percentage shall be equal to half the Retention Percentage) of the total value (subject to paragraph (iii) of this sub-clause) of the relevant part.

(iii) Where the total value of the relevant part includes works in respect of which a final payment to a nominated sub-contractor has been made under the provisions of clause 27 (e) of these Conditions, the said total value for the purposes of paragraphs (i) and (ii) of this sub-clause shall be deemed to be reduced by the value of the said sub-contractor's work carried out in the relevant part.

COMMENTARY

(1) *Generally*

This provides for the situation where, before the works are completed, the employer, with the consent of the contractor, takes possession of part or parts, and for the application to each part of provisions as to practical completion defects, insurance, liquidated damages and retention fund analogous to those which apply to the whole. Clause 16 should be read with clause 15. The present clause 16 (*f*) was introduced by the July 1972 revision to accord with amendments then made to the retention provisions.

(2) " *If . . . the Employer . . . shall take possession . . .*"

The employer is not bound to take possession of a completed part of the works. But it is thought that where liquidated damages are running, an employer who, in all the circumstances, was unreasonable in refusing the contractor's offer of possession of a completed part of the work would be subject to a proportional reduction in liquidated damages for failing to mitigate his loss.

(3) *Sectional completion*

Until an amendment introduced in July 1973 the side note to this clause was " Sectional completion." This may have misled some people into thinking that it provided, in itself, for the work to be completed and handed over in sections if supplemented by provisions in the bills of quantities. In *Gleeson Ltd.* v. *Hillingdon, London Borough of*[13] it was held that this was not so. The contract was for the provision of a large number of houses to be erected in blocks. The form was, for all material purposes, the current form with one date for completion, namely 24 months after the date for possession. There were no amendments to the form. The preliminaries bill provided that the contract was to be completed and handed over in sections, some blocks after 12 months and the rest at 3 month intervals up to 24 months. The blocks due for completion after 12 months were not completed and the employer applied the liquidated damages clause which provided for liquidated damages at the rate of £5 per dwelling per week. It was held, applying clause 12 (discussed *ante*), that the employer was not entitled to make the deductions because there was no provision in the contract for the completion of the work in sections. Although this case has been the subject of some criticism (see notes to clause 12) it appears to state the law. So if a contract providing for completion of the works in sections is required the form itself must be amended.

There are two approaches to the necessary amendments. One is to state the parts of the works and the dates when they are to be completed and to use general words indicating that the contract must be read as if all necessary consequential amendments had been effected. The other approach is to carry out in detail the many amendments necessary. The Joint Contracts Tribunal, in December 1975, issued a sectional completion supplement together with practice note 21. The latter is printed *post*, p. 423.

---

[13] (1970) Mocatta J. (unreported) save in 215 *Estates Gazette* 165.

(4) *Use but not possession of part*

Where the parties intend that the employer shall have use of some parts of the works not amounting to possession it is again desirable that they should amend the form. Thus in *English Industrial Estates Corporation* v. *Wimpey (George & Co.) Ltd.*[14] provisions in the bills envisaged such use. The majority of the Court of Appeal, applying clause 12 (see *ante*) held that they could not look at the bills to construe the contract but only " in order to follow exactly what was going on." [15] There is nothing available the equivalent of the sectional completion supplement. It is suggested that the relevant items in the bills be expressly made part of the contract. Care must be taken to deal with the various consequential matters including, in particular, those relating to insurance.

(5) *Formality of transfer*

It is highly desirable that there should be some formality evidencing the transfer of possession so that the parties are in no doubt that the various consequences of taking possession have come into effect. The certificate referred to in clause 16 (a) may be insufficient for this purpose in so far as it can be given up to seven days from the date of taking possession.

Clause 17
Assignment or sub-letting

The Contractor shall not without the written consent of the Employer assign this Contract, and shall not without the written consent of the Architect/ Supervising Officer (which consent shall not be unreasonably withheld to the prejudice of the Contractor) sub-let any portion of the Works.

Provided that it shall be a condition in any sub-letting which may occur that the employment of the sub-contractor under the sub-contract shall determine immediately upon the determination (for any reason) of the Contractor's employment under this Contract.

COMMENTARY

(1) *Generally*

Breach of this clause is a ground for determination under clause 25 (1) (d). The proviso was introduced by the July 1968 amendment. See further, note 9 to clause 25.

(2) " *The Contractor shall not . . . assign this Contract . . .*"

It is arguable that " assign " is used here not in its strict sense of assigning the benefit of the contract, but in the sense of securing the vicarious performance by another contractor of the whole of the contractor's duties.[16] If this is the case no consent is required to an assignment by the contractor of moneys to be received under the contract, but probably the words are to be construed so as to restrict the contractor's rights to deal with payments received from the employer.

---

[14] [1973] 1 Lloyd's Rep. 118 (C.A.).
[15] *Ibid.* p. 126; see also p. 128.
[16] For the difference between assignment and vicarious performance, see *ante*, p. 167, and cases there cited, in particular, *Nokes* v. *Doncaster Collieries Ltd.* [1940] A.C. 1014 (H.L.).

(3) " *The Contractor shall not . . . sub-let any portion. . . .*"

The withholding of consent by the architect to sub-letting is subject to the test of reasonableness and can be challenged on arbitration, see clause 35.

(4) " *Specified* " *sub-contractors*

The contract expressly contemplates two kinds of sub-contractors. The first is chosen by the contractor and requires the architect's consent under this clause, and is commonly called an " approved " sub-contractor. The second is chosen by the architect and nominated by him in accordance with the procedure of clauses 11 (3) and 27 and subject to the provisions of clause 27 and is termed a " nominated sub-contractor." The latter is dealt with extensively in clause 27 and the notes thereto.

Sometimes the contract bills contain a provision specifying that certain work shall be carried out by a named sub-contractor but without providing any prime cost item in respect of the sub-contractor or provisional sum in respect of his work. No express provision is made for this kind of sub-contractor, commonly called a " specified " sub-contractor. In accordance with principles discussed earlier (*ante*, p. 174) the contractor will be responsible for the quality of the sub-contractor's materials and, it seems, his workmanship, but not for whether his work will be suitable for its purpose. What if such a specified sub-contractor refuses to perform his sub-contract or shows that he cannot perform it or otherwise repudiates it ? The position must depend upon the particular item in the bill but it may well be that, having regard to the principles applied in *Bickerton* (*T. A.*) & *Sons Ltd.* v. *North West Metropolitan Regional Hospital Board*[17] (see discussion, *ante*, p. 182 and *post*, p. 369) the employer must provide a new sub-contractor and that the position is otherwise similar to that of the default of a nominated sub-contractor. The reason is that, by virtue of words specifying by whom the work must be carried out, the contractor is given no right and has no duty to carry it out himself.

Further, it seems that if the employer, or his agent, the architect, knows that the specified sub-contractor will only contract upon terms which exclude or severely limit his liability the contractor may not even be responsible for latent defects in the quality of goods and materials used by the specified sub-contractor.[18]

(5) *Private form*

A 1971 amendment introduced a prohibition against the employer assigning the contract without the consent of the contractor. In this amendment there are no grounds for not construing assignment in its strict sense. The result is that the employer is in breach of contract if, without the contractor's consent, he sells the land with the benefit of the building contract. The need for the amendment is not apparent as the employer in such circumstances remains liable to the contractor after the sale (see *ante*, p. 173). Breach of this clause is not a ground for determination under clause 26.

---

[17] [1970] 1 W.L.R. 607 (H.L.).
[18] *Gloucestershire County Council* v. *Richardson* [1969] A.C. 480 (H.L.), applied; *Portsmouth Corporation* v. *Howe & Bishop Ltd.* (1974), unrep., H.H. Judge Norman Richards Q.C., transcript, p. 9. See also *ante*, p. 40.

Clause 17A

Fair Wages

(1) (a) The Contractor shall pay rates of wages and observe hours and conditions of labour not less favourable than those established for the trade or industry in the district where the work is carried out by machinery of negotiation or arbitration to which the parties are organisations of employers and trade unions representative respectively of substantial proportions of the employers and workers engaged in the trade or industry in the district.

(b) In the absence of any rates of wages, hours or conditions of labour so established the Contractor shall pay rates of wages and observe hours and conditions of labour which are not less favourable than the general level of wages, hours and conditions observed by other employers whose general circumstances in the trade or industry in which the Contractor is engaged are similar.

(2) The Contractor shall in respect of all persons employed by him (whether in carrying out this Contract or otherwise) in every factory, workshop or other place occupied or used by him for the carrying out of this Contract (including the Works) comply with the general conditions required by this Condition. The Contractor hereby warrants that to the best of his knowledge and belief he has complied with the general conditions required by this Condition for at least three months prior to the date of his tender for this Contract.

(3) In the event of any question arising as to whether the requirements of this Condition are being observed, the question shall, if not otherwise disposed of, and notwithstanding anything in clause 35 of these Conditions (the provisions of which shall not apply to this Condition), be referred through the Minister of Labour or his successor for the time being to an independent Tribunal for decision.

(4) The Contractor shall recognise the freedom of his workpeople to be members of Trade Unions.

(5) The Contractor shall at all times during the continuance of this Contract display, for the information of his workpeople, in every factory, workshop or place occupied or used by him for the carrying out of this Contract (including the Works) a copy of this Condition. Where rates of wages, hours or conditions of work have been established either by negotiation or arbitration as described in paragraph (a) of sub-clause (1) of this Condition or by any agreement commonly recognised by employers and workers in the district a copy of the award agreement or other document specifying or recording such rates hours or conditions shall also be exhibited by the Contractor or made available by him for inspection in any such place as aforesaid.

(6) The Contractor shall be responsible for the observance of this Condition by sub-contractors employed in the carrying out of this Contract, and shall if required notify the Employer of the names and addresses of all such sub-contractors.

(7) The Contractor shall keep proper wages books and time sheets showing the wages paid to and the time worked by the workpeople in his employ in and about the carrying out of this Contract, and such wages books and time sheets shall be produced whenever required for the inspection of any officer authorised by the Employer.

(8) If the Employer shall have reasonable ground for believing that the requirements of any of the preceding sub-clauses of this Condition are not being observed, he or the Architect/Supervising Officer on his behalf shall be entitled to require proof of the rates of wages paid and hours and conditions observed by the Contractor and sub-contractors in carrying out the Works.

COMMENTARY

(1) *Clause* 17A. *Fair wages*

This clause is based upon the Fair Wages Resolution of the House of Commons made in 1946. In *Racal Communications Ltd.* v. *The Pay Board* [19] the court had to consider whether payment of minimum rates established by a national agreement reached by machinery of negotiation between bodies representative respectively of the employers and employees in the industry in the district was a breach of the then anti-inflation legislation. There are useful comments upon the meaning of the words which appear in sub-clause 1 (a) and the view was expressed that the reality of the matter was that the passage of time had rendered the Fair Wages Resolution of 1946 out of date and that it was the increased bargaining power of organised labour that provided today's principal safeguard against exploitation by employers.

(2) *Clause* 17A (3) "... *notwithstanding anything in clause* 35 ... (*the provisions of which shall not apply to this Condition*). ..."

This leads to a difficulty. Failure to comply with the provisions of clause 17A is a ground for determination under clause 25 (1) (d), and a dispute whether such a determination is lawful is prima facie within the arbitration clause (see clause 35 (1)). Perhaps clause 17A (3) and clause 35 may be reconciled on the basis that the arbitrator before giving his award must await the decision under clause 17A (3) and will be bound by that decision, but has jurisdiction in respect of all other matters in dispute arising out of the alleged determination.

(3) *Private version*
There is no such clause.

Clause 18
Injury to persons and property and Employer's indemnity

(1) The Contractor shall be liable for, and shall indemnify the Employer against, any liability, loss, claim or proceedings whatsoever arising under any statute or at common law in respect of personal injury to or the death of any person whomsoever arising out of or in the course of or caused by the carrying out of the Works, unless due to any act or neglect of the Employer or of any person for whom the Employer is responsible.

(2) Except for such loss or damage as is at the risk of the Employer under clause 20 [B] or clause 20 [C] of these Conditions (if applicable) the Contractor shall be liable for, and shall indemnify the Employer against, any expense, liability, loss, claim or proceedings in respect of any injury or damage whatsoever to any property real or personal in so far as such injury or damage arises out of or in the course of or by reason of the carrying out of the Works, and provided always that the same is due to any negligence, omission or default of the Contractor, his servants or agents or of any sub-contractor his servants or agents.

COMMENTARY

(1) *Clause* 18 (1) " *The Contractor shall be liable for* . . ."

A wide liability is placed upon the contractor. If an injured person sues the employer [20] in respect of an injury within the scope of this clause the employer

---

[19] [1974] 3 All E.R. 263.
[20] For the rights of third parties to sue contractor or employer, see *ante,* p. 134. For risk and indemnity clauses generally, see *ante,* p. 138. For persons who come upon the site, see the Occupiers' Liability Act 1957.

can join the contractor as a third party to the action or bring separate proceedings against the contractor on the indemnity.

(2) *Clause* 18 (1) "*. . . unless due to any act or neglect . . .*"

This is an exception to the liability imposed by the clause. The exception does not, it is submitted, save the contractor from liability to indemnify the employer where the claim by the injured person arises out of breach of statutory duty by the employer not amounting to common law negligence.[21]

(3) *Clause* 18 (2) "*Except for such loss or damage. . . .*"

It seems that where clause 20 [B] or 20 [C] is used this exception operates to relieve the contractor from liability, even where damage resulting from one of the contingencies referred to in clause 20 [A] or 20 [B] is caused by the contractor's negligence,[22] and it is thought that the clauses would not be construed against the contractor as exclusion clauses so as to deprive them of effect in the event of destruction of the works by fire, etc., caused by such negligence.[23]

(4) *Clause* 18 (2) "*. . . the Contractor shall be liable for, and shall indemnify. . . .*"

The liability under this sub-clause in respect of damage to property appears to be less wide than in respect of injury to persons. There are the exceptions where clause 20 [B] or 20 [C] is used (see last note); further, the employer must prove that the damage to property is due to the contractor's default.

Extent of liability under sub-clause. It includes: claims by third parties whose property is damaged, claims by the employer in respect of his property and, it seems, damage to the works.

Damage to the works generally. The contractor's liability is not, it is submitted, limited to the liability which may arise under this sub-clause. He has a duty to complete the works (clause 1, Articles of Agreement; clause 1 of conditions) and, it is submitted, as an incident of that duty he must make good any damage to the works, such as that due to vandalism,[24] occurring before practical completion[25] and not caused by the employer's negligence or default,[26] nor within the risks accepted by the employer where clause 20 [B] or [C] is used. Such risks do not include theft.

---

[21] *Hosking* v. *De Havilland Ltd.* [1949] 1 All E.R. 540; *Murfin* v. *United Steel Companies Ltd.* [1957] 1 W.L.R. 104 (C.A.).

[22] See *Archdale (James) & Co. Ltd.* v. *Comservices Ltd.* [1954] 1 W.L.R. 459 (C.A.); *cf. Bucks County Council* v. *Lovell (Y. J.) & Son* [1956] J.P.L. 196. See also *Gillespie Bros. & Co. Ltd.* v. *Roy Bowles Ltd.* [1973] 1 Q.B. 400 (C.A.); *Levison* v. *Patent Steam Carpet Cleaning Co. Ltd.* [1977] 3 W.L.R. 90 (C.A.).

[23] See *Harbutt's "Plasticine" Ltd.* v. *Wayne Tank Ltd.* [1970] 1 Q.B. 447 (C.A.) discussed *ante,* pp. 114 and 141.

[24] See *Charon (Finchley)* v. *Singer Sewing Machine Co.* (1968) 112 S.J. 536.

[25] See clause 15 for the certificate of practical completion, and for the position where the employer takes possession of part of the works before the completion of the whole, see clause 16. For the special provision as to frost, see clauses 15 (1) and 15 (5).

[26] See *ante,* p. 138, and *Gold* v. *Patman & Fotheringham Ltd.* [1958] 1 W.L.R. 697, 704 (C.A.).

Damage to contractor's property. His plant, equipment and the like and unfixed materials are at his risk.

Damage to goods and materials certified for. These remain at the contractor's risk.

Frustration. For damage so extensive as to frustrate the contract see *ante*, p. 102, but consider the express provisions of clauses 20, 23 and 26.

Employer and contractor liable to third party. In *AMF International Ltd.* v. *Magnet Bowling Ltd.*[27] the plaintiff recovered damages against both the employer and the contractor for breach of duty under the Occupiers' Liability Act 1957, their liability for purposes of contribution under the Law Reform (Married Women and Tortfeasors) Act 1935 being assessed at 40 per cent. employer and 60 per cent. contractor. It was held that as the employer's liability to the plaintiff was partly the result of his negligence, through the architect, in lack of supervision of the contractor he could not recover from the contractor on the indemnity provided by this clause.[28] The facts of the case are rather special, but the principle applied is of general application and effects an important limitation in practice to the benefit conferred by the indemnity.

Clause 19
Insurance against injury to persons and property

(1) (a) Without prejudice to his liability to indemnify the Employer under clause 18 of these Conditions the Contractor shall maintain and shall cause any sub-contractor to maintain such insurances as are necessary to cover the liability of the Contractor or, as the case may be, of such sub-contractor in respect of personal injury or death arising out of or in the course of or caused by the carrying out of the Works not due to any act or neglect of the Employer or of any person for whom the Employer is responsible and in respect of injury or damage to property, real or personal, arising out of or in the course of or by reason of the carrying out of the Works and caused by any negligence, omission or default of the Contractor, his servants or agents or, as the case may be, of such sub-contractor, his servants or agents. The insurance in respect of claims for personal injury to, or the death of, any person under a contract of service or apprenticeship with the Contractor or the sub-contractor as the case may be, and arising out of and in the course of such person's employment, shall comply with the Employer's Liability (Compulsory Insurance) Act 1969 and any statutory orders made thereunder or any amendment or re-enactment thereof. For all other claims to which this sub-clause applies the insurance cover shall be the sum stated in the Appendix to these Conditions (or such greater sum as the Contractor may choose) for any one occurrence or series of occurrences arising out of one event.

(b) As and when he is reasonably required so to do by the Employer the Contractor shall produce and shall cause any sub-contractor to produce for inspection by the Employer documentary evidence that the insurances required by this sub-clause are properly maintained, but on any occasion the Employer may (but not unreasonably or vexatiously) require to have produced for his inspection the policy or policies and receipts in question.

---

[27] [1968] 1 W.L.R. 1028.

[28] Applying the *Alderslade* principle, as it was then called (see *ante*, p. 139), but, on the facts of the case and the wording of the bills, the employer recovered from the contractor the sum he had to pay the plaintiff because of the contractor's breach of contract in failing to provide for the diversion of storm water and the protection of the works as expressly required by items in the bills. For the relationship to clause 12, see *ante*, p. 321.

(c) Should the Contractor or any sub-contractor make default in insuring or in continuing or in causing to insure as provided in this sub-clause the Employer may himself insure against any risk with respect to which the default shall have occurred and may deduct a sum or sums equivalent to the amount paid or payable in respect of premiums from any monies due or to become due to the Contractor.

(2) (a) The Contractor shall maintain in the joint names of the Employer and the Contractor insurances for such amounts of indemnity as may be specified by way of provisional sum items in the Contract Bills in respect of any expense, liability, loss, claim, or proceedings which the Employer may incur or sustain by reason of damage to any property other than the Works caused by collapse, subsidence, vibration, weakening or removal of support or lowering of ground water arising out of or in the course of or by reason of the carrying out of the Works excepting damage

> (i) caused by the negligence, omission or default of the Contractor, his servants or agents or of any sub-contractor his servants or agents;
> (ii) attributable to errors or omissions in the designing of the Works;
> (iii) which can reasonably be foreseen to be inevitable having regard to the nature of the work to be executed or the manner of its execution;
> (iv) which is at the risk of the Employer under clause 20 [B] or clause 20 [C] of these Conditions (if applicable);
> (v) arising from a nuclear risk or a war risk;

(b) Any such insurance as is referred to in the immediately preceding paragraph shall be placed with insurers to be approved by the Employer, and the Contractor shall deposit with him the policy or policies and the receipts in respect of premiums paid.

(c) Should the Contractor make default in insuring or in continuing to insure as provided in this sub-clause the Employer may himself insure against any risk with respect to which the default shall have occurred and the amounts paid or payable by the Employer in respect of premiums shall not be set against the relevant provisional sum in the settlement of accounts under clause 30 (5) (c) of these Conditions.

Clause 19A
Expected risks—nuclear perils, etc.

Notwithstanding the provisions of clauses 18 (2) or 19 of these Conditions, the Contractor shall not be liable either to indemnify the Employer or to insure against any damage, loss or injury caused to the Works, the site, or any property, by the effect of ionising radiations or contamination by radioactivity from any nuclear fuel or from any nuclear waste from the combustion of nuclear fuel, radioactive toxic explosive or other hazardous properties of any explosive nuclear assembly or nuclear component thereof, pressure waves caused by aircraft or other aerial devices travelling at sonic or supersonic speeds.

(1) *Clauses* 19, 19A. *Generally*

These clauses should be read with clauses 18 (indemnities) and 20 (insurance of the works against fire, etc.). Clause 19 has been the subject of some amendment since 1963. The current 19 (1) (a) was introduced by the July 1975 amendment.

(2) *Clause* 19 (1). *Outline*

The contractor has to insure, and cause his sub-contractors to insure, against his liabilities in respect of injury and property. The wording of clause 19 (1) follows fairly closely that of clause 18. Observe that in respect of claims, other than those to which the Employers' Liability (Compulsory Insurance) Act 1969 applies, the minimum insurance cover for any one occurrence or series of occurrences is that set out in the appendix.

(3) *Clause* 19 (2). *Outline*

The origin of this clause is the case of *Gold* v. *Patman & Fotheringham Ltd.*,[29] where adjoining owners in a crowded city area suffered damage to their property from the carrying out of the contract works. There was no negligence on the part of the contractor. The employer sought to recover against the contractor on the grounds that a duty to insure, by implication, meant a duty to effect joint insurance. It was held that this was not so.

Examination of the exceptions to the duty of insurance reveals that its scope is very narrow. Further, in contrast with the obligation under clause 19 (1), the contractor's duty to insure only arises when the architect has given an instruction under clause 11 (3) for the expenditure of the insurance premium.

(4) *Private employers*

The wording of the private form does not differ but employers who are not content to act as their own insurers should consider their position. The *AMF International* case (discussed in note 4 to clause 18) shows that there may well arise claims by third parties where the employer has no claim over against the contractor. Further, the insurance under clause 19 (2) is very limited. For these reasons private employers may consider it advisable to effect their own insurance.

Clause 20
Insurance of works against fire, etc.

*[A] (1) The Contractor shall in the joint names of the Employer and Contractor insure against loss and damage by** fire, lightning, explosion, storm, tempest, flood, bursting or overflowing of water tanks, apparatus or pipes, earthquake, aircraft and other aerial devices or articles dropped therefrom, riot and civil commotion (excluding any loss or damage caused by ionising radiations or contamination by radioactivity from any nuclear fuel or from any nuclear waste from the combustion of nuclear fuel, radioactive toxic explosive or other hazardous properties of any explosive nuclear assembly or nuclear component thereof, pressure waves caused by aircraft or other aerial devices travelling at sonic or supersonic speeds) for the full value thereof (plus the percentage (if any) named in the Appendix to these Conditions to cover professional fees) all work executed and all unfixed materials and goods, delivered to, placed on or adjacent to the Works and intended therefor but excluding temporary buildings, plant, tools and equipment owned or hired by the Contractor or any sub-contractor, and shall keep such work, material and goods so insured until Practical Completion of the Works. Such insurance shall be with insurers approved by the Employer and the Contractor shall deposit with him the policy or policies and the receipts in respect of premiums paid; and should the Contractor make default in insuring or continuing to insure as aforesaid the Employer may himself insure against any risk in respect of which the default shall have occurred and deduct a sum equivalent to the amount paid by him in respect of premiums from any monies due or to become due to the Contractor. Provided always that if the Contractor shall independently of his obligations under this Contract maintain a policy of insurance which covers (*inter alia*) the said work, materials and goods against

[29] [1958] 1 W.L.R. 697 (C.A.).

the aforesaid contingencies to the full value thereof (plus the aforesaid percentage (if any)), then the maintenance by the Contractor of such policy shall, if the Employer's interest in that policy of insurance is endorsed thereon, be a discharge of the Contractor's obligation to insure in the joint names of the Employer and Contractor; if and so long as the Contractor is able to produce for inspection as and when he is reasonably required so to do by the Employer documentary evidence that the said policy is properly endorsed and maintained then the Contractor shall be discharged from his obligation to deposit a policy or policies and receipts with the Employer but on any occasion the Employer may (but not unreasonably or vexatiously) require to have produced for his inspection the policy and receipts in question.

(2) Upon acceptance of any claim under the insurances aforesaid the Contractor with due diligence shall restore work damaged replace or repair any unfixed materials or goods which have been destroyed or injured remove and dispose of any debris and proceed with the carrying out and completion of the Works. All monies received from such insurances (less only the aforesaid percentage (if any)) shall be paid to the Contractor by instalments under certificates of the Architect/Supervising Officer issued at the Period of Interim Certificates named in the Appendix to these Conditions. The Contractor shall not be entitled to any payment in respect of the restoration of work damaged, the replacement and repair of any unfixed materials or goods, and the removal and disposal of debris other than the monies received under the said insurances.

*[B]. All work executed and all unfixed materials and goods, delivered to, placed on or adjacent to the Works and intended therefor (except temporary buildings, plant, tools and equipment owned or hired by the Contractor or any sub-contractor) shall be at the sole risk of the Employer as regards loss or damage by fire, lightning, explosion, storm, tempest, flood, bursting or overflowing of water tanks, apparatus or pipes, earthquake, aircraft and other aerial devices or articles dropped therefrom riot and civil commotion (excluding any loss or damage caused by ionising radiations or contamination by radioactivity from any nuclear fuel or from any nuclear waste from the combustion of nuclear fuel, radioactive toxic explosive or other hazardous properties of any explosive nuclear assembly or nuclear component thereof, pressure waves caused by aircraft or other aerial devices travelling at sonic or supersonic speeds). If any loss or damage affecting the Works or any part thereof or any such unfixed materials or goods is occasioned by any one or more of the said contingencies then, upon discovering the said loss or damage the Contractor shall forthwith give notice in writing both to the Architect/Supervising Officer and the Employer of the extent, nature and location thereof and

(a) The occurrence of such loss or damage shall be disregarded in computing any amounts payable to the Contractor under or by virtue of this Contract.

(b) The Contractor with due diligence shall restore work damaged, replace or repair any unfixed materials or goods which have been destroyed or injured, remove, and dispose of any debris and proceed with the carrying out and completion of the Works. The restoration of work damaged, the replacement and repair of unfixed materials and goods and the removal and disposal of debris shall be deemed to be a variation required by the Architect/Supervising Officer.

*[C]. The existing structures together with the contents thereof owned by him or for which he is responsible and the Works and all unfixed materials and goods, delivered to, placed on or adjacent to the Works and intended therefor (except temporary buildings, plant, tools and equipment owned or hired by the Contractor or any sub-contractor) shall be at the sole risk of the Employer as regards loss or damage by fire, lightning, explosion, storm, tempest, flood, bursting or overflowing of water tanks, apparatus or pipes, earthquake, aircraft and other

aerial devices or articles dropped therefrom, riot and civil commotion (excluding any loss or damage caused by ionising radiations or contamination by radioactivity from any nuclear fuel or from any nuclear waste from the combustion of nuclear fuel, radioactive toxic explosive or other hazardous properties of any explosive nuclear assembly or nuclear component thereof, pressure waves caused by aircraft or other aerial devices travelling at sonic or supersonic speeds), and the Employer shall maintain adequate insurance against those risks.† If any loss or damage affecting the Works or any part thereof or any such unfixed materials or goods is occasioned by any one or more of the said contingencies then, upon discovering the said loss or damage the Contractor shall forthwith give notice in writing both to the Architect/Supervising Officer and to the Employer of the extent, nature and location thereof and

(a)   The occurrence of such loss or damage shall be disregarded in computing any amounts payable to the Contractor under or by virtue of this Contract.

(b)   (i) If it is just and equitable so to do the employment of the Contractor under this Contract may within 28 days of the occurrence of such loss or damage be determined at the option of either party by notice by registered post or recorded delivery from either party to the other. Within 7 days of receiving such a notice (but not thereafter) either party may give to the other a written request to concur in the appointment of an arbitrator under clause 35 of these Conditions in order that it may be determined whether such determination will be just and equitable.

(ii) Upon the giving or receiving by the Employer of such a notice of determination or, where a reference to arbitration is made as aforesaid, upon the arbitrator upholding the notice of determination, the provisions of sub-clause (2) (except sub-paragraph (vi) of paragraph (b)) of clause 26 of these Conditions shall apply.

(c)   If no notice of determination is served as aforesaid, or, where a reference to arbitration is made as aforesaid, if the arbitrator decided against the notice of determination, then

(i) the Contractor with due diligence shall reinstate or make good such loss or damage, and proceed with the carrying out and completion of the Works;

(ii) the Architect/Supervising Officer may issue instructions requiring the Contractor to remove and dispose of any debris; and

(iii) the reinstatement and making good of such loss or damage and (when required) the removal and disposal of debris shall be deemed to be a variation required by the Architect/Supervising Officer.

*   *Clause 20 [A] is applicable to the erection of a new building if the Contractor is required to insure against loss or damage by fire, etc.; clause 20 [B] is applicable to the erection of a new building if the Employer is to bear the risk in respect of loss or damage by fire, etc.; and clause 20 [C] is applicable to alterations of or extensions to an existing building; therefore strike out clauses [B] and [C] or clauses [A] and [C] or clauses [A] and [B] as the case may require.*
**   *In some cases it may not be possible for insurance to be taken out against certain of the risks mentioned in this clause. This matter should be arranged between the parties at the tender stage and the clause amended accordingly.*
†   *In some cases it may not be possible for the Employer to take out insurance against certain of the risks mentioned in this clause. The matter should be arranged between the parties at the tender stage and the clause amended accordingly.*

COMMENTARY

(1) *Generally*

The alternatives are described in the footnote. This clause should be read with clauses 18 and 19. Further, it is material upon extension of time under clause 23 and determination of the contractor's employment by the contractor under clause 26.

(2) *Clause* 20 *[A]* (1) " . . . *insure* . . . *for the full value thereof* . . ."

This may be more or less than the cost of the contract works. It is a continuing obligation. Therefore, in so far as the value increases, *e.g.* due to inflation, the cover should be increased.

(3) *Clause* 20 *[A]* (1) " . . . *storm, tempest, flood* . . ."

These words have been construed in relation to insurance policies.[30]

(4) *Clause* 20 *[A]* (2) " *Upon acceptance of any claim* . . . *the contractor* . . . *shall restore* . . ."

The word "acceptance" was substituted for "settlement" in the 1972 revision. The intention, presumably, was to reduce the period of delay before the contractor becomes under a duty to restore damaged work. This is of great importance because of his right to extension of time (clause 23 (c)) and to determine his own employment (clause 26 (1) (c) (ii)). The contractor's entitlement to payment is limited to the monies received under the insurances so that it is in his interest to see that the cover is adequate. Although the contractor has the right of determination for delay caused by fire, etc., the employer has no such right. In certain circumstances he might be able to claim that the contract was frustrated (see *ante,* p. 105).

(5) *Clause* 20 *[B]* " *All work* . . . *shall be at the sole risk of the employer* . . ."

In the local authorities version there is no obligation to insure but in the private version there is. The acceptance of liability by the employer includes, it is submitted, loss due to fire, etc., caused by the contractor's negligence (see notes 3 and 4 to clause 18), but damage to the works due to theft or other risks not within the named contingencies is, it is submitted, the liability of the contractor until practical completion. (See *ibid.*)

(6) *Clause* 20 *[C]* (*b*) " *If it is just and equitable so to do* . . ."

This is a special provision for determination. Delay due to fire, etc., is not one of the grounds for determination by the contractor under clause 26 when clause 20 [C] applies.

Clause 21
Possession, completion and postponement

(1) On the Date for Possession stated in the Appendix to these Conditions possession of the site shall be given to the Contractor who shall thereupon begin

---

[30] *Oddy* v. *Phoenix Assurance Company Ltd.* [1966] 1 Lloyd's Rep. 134; *S. & M. Hotels Ltd.* v. *Legal & General Assurance Society Ltd.* [1972] 1 Lloyd's Rep. 157; *Young* v. *Sun Alliance Ltd., The Times,* May 14, 1976 (C.A.).

the Works and regularly and diligently proceed with the same, and who shall complete the same on or before the Date for Completion stated in the said Appendix subject nevertheless to the provisions for extension of time contained in clauses 23 and 33 (1) (c) of these Conditions.

(2) The Architect/Supervising Officer may issue instructions in regard to the postponement of any work to be executed under the provisions of this Contract.

Clause 22
Damages for non-completion

If the Contractor fails to complete the Works by the Date for Completion stated in the Appendix to these Conditions or within any extended time fixed under clause 23 or clause 33 (1) (c) of these Conditions and the Architect/ Supervising Officer certifies in writing that in his opinion the same ought reasonably so to have been completed, then the Contractor shall pay or allow to the Employer a sum calculated at the rate stated in the said Appendix as Liquidated and Ascertained Damages for the period during which the Works shall so remain or have remained incomplete, and the Employer may deduct such sum from any monies due or to become due to the Contractor under this Contract.

Clause 23
Extension of time

Upon it becoming reasonably apparent that the progress of the Works is delayed, the Contractor shall forthwith give written notice of the cause of the delay to the Architect/Supervising Officer, and if in the opinion of the Architect/ Supervising Officer the completion of the Works is likely to be or has been delayed beyond the Date for Completion stated in the Appendix to these Conditions or beyond any extended time previously fixed under either this clause or clause 33 (1) (c) of these Conditions.

(a) by *force majeure*, or

(b) by reason of any exceptionally inclement weather, or

(c) by reason of loss or damage occasioned by any one or more of the contingencies referred to in clause 20 [A], [B] or [C] of these Conditions, or

(d) by reason of civil commotion, local combination of workmen, strike or lockout affecting any of the trades employed upon the Works or any of the trades engaged in the preparation manufacture or transportation of any of the goods or materials required for the Works, or

(e) by reason of Architect's/Supervising Officer's instructions issued under clauses 1 (2), 11 (1) or 21 (2) of these Conditions, or

(f) by reason of the Contractor not having received in due time necessary instructions, drawings, details or levels from the Architect/Supervising Officer for which he specifically applied in writing on a date which having regard to the Date for Completion stated in the Appendix to these Conditions or to any extension of time then fixed under this clause or clause 33 (1) (c) of these Conditions was neither unreasonably distant from nor unreasonably close to the date on which it was necessary for him to receive the same, or

(g) by delay on the part of nominated sub-contractors or nominated suppliers which the Contractor has taken all practicable steps to avoid or reduce, or

(h) by delay on the part of artists tradesmen or others engaged by the Employer in executing work not forming part of this Contract, or

(i) by reason of the opening up for inspection of any work covered up or of the testing of any of the work materials or goods in accordance with clause 6 (3) of these Conditions (including making good in consequence of such opening up or testing), unless the inspection or test showed that the work materials or goods were not in accordance with this Contract, or

*(j) (i) by the Contractor's inability for reasons beyond his control and which he could not reasonably have foreseen at the date of this Contract to secure such labour as is essential to the proper carrying out of the Works, or

(ii) by the Contractor's inability for reasons beyond his control and which he could not reasonably have foreseen at the date of this Contract to secure such goods and/or materials as are essential to the proper carrying out of the Works, or

(k) by reason of compliance with the provisions of clause 34 of these Conditions or with Architect's/Supervising Officer's instructions issued thereunder, or

(l) by a local authority or statutory undertaker in carrying out work in pursuance of its statutory obligations in relation to the Works, or in failing to carry out such work,

then the Architect/Supervising Officer shall so soon as he is able to estimate the length of the delay beyond the date or time aforesaid make in writing a fair and reasonable extension of time for completion of the Works. Provided always that the Contractor shall use constantly his best endeavours to prevent delay and shall do all that may reasonably be required to the satisfaction of the Architect/Supervising Officer to proceed with the Works.

* *Strike out either or both of the sub-clauses (j) (i) or (j) (ii) if not to apply.*

COMMENTARY

(1) *Generally*

Clauses 21, 22, and 23 deal with time, liquidated damages and extension of time and should be read together. For the recovery of loss or expense in certain cases where the regular progress of the works has been materially affected, see clause 24 (1); for determination for delay, see clauses 25 and 26; for time for completion generally, see *ante*, p. 59; for extension of time clauses and frustration, see *ante*, p. 104; for liquidated damages generally, see *ante*, p. 155.

Sectional completion. These clauses only deal with the position where there is intended to be one date for handing over all the site to the contractor and one date for completion of all the works. Amendment is required if this is not the case. See note 3 to clause 16 and Practice note 21 printed, *post*, p. 423.

(2) *Clause* 21 (1) ". . . *possession of the site* . . ."

Delay by the employer in giving possession of the site prima facie disentitles the employer from recovering liquidated damages,[31] but the architect can, it is submitted, give an instruction (clause 21 (2)) postponing the commencement of work (loss or expense may be recoverable under clause 24 (1) (e)), and can extend the time for completion under clause 23 (e).[32] Postponement of the commencement of the works for more than the period stated in the appendix (one month is suggested in the footnote to the form) may give the contractor a right of determination under clause 26 (1) (c) (iv). Where this may arise the architect should advise the employer who may wish to seek to make a special agreement with the contractor to avoid the possibility of such determination.

Degree of possession to be given. " The contract necessarily requires the

---

[31] *Amalgamated Building Contractors Ltd.* v. *Waltham Holy Cross U.D.C.* [1952] 2 All E.R. 452, 455 (C.A.); see *ante*, p. 160.
[32] See clause 23 and notes *infra*.

building owner to give the contractor such possession, occupation or use as is necessary to enable him to perform the contract. . . ."[33]

(3) *Clause* 21 (1) ". . . *regularly and diligently proceed with the same* . . ."

Breach of this obligation is a ground for determination under clause 25 (1) (b). There has been some judicial comment on the meaning of the words.[34]

Programme of work. This is not required by any term of the contract but in practice architects frequently request the contractor to produce a programme and the contractor provides it. Failure to comply with such a programme is not itself a breach of contract but it may be some evidence of failure to proceed regularly and diligently. The contractor may be able to rebut the inference of breach by showing, *inter alia*, that he is entitled to an extension of time under clause 23, or that delay was caused by some act of the employer not within clause 23, or that some other programme which he is following is a compliance with the duty to proceed regularly and diligently.

If the contract bills require the contractor to prepare a programme it is a breach of contract not to prepare it, but otherwise the position is as just set out. An item in the bills requiring the contractor to comply with such a programme does not, it is submitted (see clause 12 (1)), modify clause 21 (1).

(4) *Clause* 21 (1) ". . . *complete the same on or before the Date for Completion* . . ."

There is no reference to " practical " completion (see clause 15). But in *J. Jarvis & Sons Ltd.* v. *City of Westminster* [35] it was said that the obligation under clause 21 was " to complete the works in the sense in which the words ' practically completed ' and ' practical completion ' are used in clauses 15 and 16. . . ."

Date omitted. If the date for completion is not inserted liquidated damages are not payable.[36] In a proper case if the date had been clearly agreed and was accidentally omitted the court would, it is submitted, grant rectification (see *ante*, p. 47) and insert the date, but *Kemp* v. *Rose* [37] would have to be distinguished.

Unliquidated damages. If for any reason liquidated damages are not payable, unliquidated (*i.e.* ordinary) damages (see Chapter 9) are payable by the contractor for delay for which he has no excuse in law.[38]

(5) *Clause* 21 (2) ". . . *the postponement of any work* . . ."

An instruction to postpone work leading to a suspension of the works can be disastrous for the employer. The contractor may be able to determine. See clause

---

[33] *Hounslow London Borough* v. *Twickenham Garden Developments* [1971] 1 Ch. 233, 257, *per* Megarry J., the decision in which case is discussed *ante*, p. 123 and, with reference to clause 25 (determination) *post*, p. 356.

[34] *Ibid.* p. 269.

[35] (1969) unreported, p. 12 of transcript (C.A.). See also to same effect, *Hosier & Dickinson Ltd.* v. *Kaye (P. & M.) Ltd.* [1972] 1 W.L.R. 146, 165 (H.L.).

[36] *Kemp* v. *Rose* (1858) 1 Giff. 258, 266.

[37] *Ibid.*

[38] See *Peak Construction (Liverpool) Ltd.* v. *McKinney Foundations Ltd.* (1970) 69 L.G.R. 1, 11, 16 (C.A.).

26 and note 2 *supra*. In any event he will usually have a claim to loss and/or expense under clause 24 (1) (e). Can the contractor rely on such an instruction when it arose out of a defect in the works, or other matter for which he is responsible? In *Gloucestershire County Council* v. *Richardson* it was conceded in the Court of Appeal [39] that he could not. The concession was made after some forceful expressions of view during argument by the members of the Court. When the case went to the House of Lords [40] the concession was maintained but it appeared in argument that some of their Lordships were not satisfied that it was necessarily correct. The point must be regarded as open, although, having regard to the care with which the courts approach determination clauses it may well be that the concession was correct. As regards determination important amendments to clauses 26 (1) (c) (ii) and (iv) were introduced by the 1973 revision and are discussed in notes 4 and 6 to clause 26.

(6) *Clause 22 " . . . and the Architect certifies in writing . . ."*

Before the employer is entitled to liquidated damages there must in addition to delay be: (i) the architect's certificate in writing in accordance with this clause; (ii) the performance by the architect of his duties in regard to any necessary extension of time under clause 23 or 33 (1) (c). [41] For the effect of the absence of the certificate on an application to stay proceedings under section 4 of the Arbitration Act 1950, see *ante*, p. 242.

Form of certificate under clause 22 and of extension of time in writing under clause 23. None is prescribed, but in each case it must be clear that there was an intention to issue the relevant document and that it in substance expressed the architect's opinion under clause 22 or his making of an extension of time under clause 23 as the case may be. [42]

Can more than one certificate be issued? It is thought that in certain circumstances it can, *e.g.* when the completion date is past and a certificate has been issued but subsequently a cause of delay arises entitling the contractor to an extension of time. [43]

Is there a limit of time to the issue of the certificate? It is thought not. None is expressed. Time is not a matter dealt with by the final certificate (clause 30 (7)). It may be argued that upon the issue of the final certificate the architect's functions

---

[39] [1967] 3 All E.R. 458.

[40] [1969] 1 A.C. 480. The contract was the 1939 edition, where the contractor had an absolute right of determination and was not subject to the proviso of reasonableness.

[41] See *Amalgamated Building Contractors'* case, *supra*, at p. 454.

[42] *Token Construction Company Ltd.* v. *Charlton Estates Ltd.* (1973) unreported (C.A.), not overruled on this point by *Modern Engineering (Bristol) Ltd.* v. *Gilbert Ash (Northern) Ltd.* [1974] A.C. 689 (H.L.), applying a passage from *Minster Trust Ltd.* v. *Traps Tractors Ltd.* [1954] 1 W.L.R. 963, 982, *per* Devlin J. For a form of words held to be a sufficient extension in writing, see *Amalgamated Building Contractors'* case *supra*. See also *ante*, p. 83 " Form of certificate."

[43] Because if the architect cannot grant a further extension the original certificate would become bad and it follows that an architect would never be safe in issuing a certificate until after completion. This would prevent the employer recovering liquidated damages until the works are actually completed which does not appear to be the intention of the clause. See also *Amalgamated Building Contractors'* case, *supra*.

under the contract are concluded, but the contract does not say so and it is difficult to see the necessity for such implication.

Can the certificate be given for a date which differs from the extended time under clause 23 or 33 (1) (c) ? It is submitted that it cannot. The intention of the certificate appears to be to ensure that before the employer is entitled to liquidated damages the architect reviews all extensions of time. If by the certificate he could then alter his extension the result would be to give him an unlimited discretion to override his decisions under the extension of time clauses and the words are not, it is submitted, apt to achieve this effect.

(7) *Clause 22 ". . . the Contractor shall pay or allow . . ."*

The employer can either sue for the liquidated damages or deduct them from money due under the contract. In any claim by the contractor against the employer under another contract the employer can counterclaim for liquidated damages due from the contractor under this contract but cannot, it is submitted, set off the liquidated damages against the claim.[44]

Certificates should, it is submitted, be in full (see clause 30 and note 2 (a) to the Articles of Agreement) and the employer may deduct the liquidated damages from the amount so certified. The employer does not, it is submitted, waive his claim for liquidated damages by failure to deduct them from money due under the contract.[45]

(8) *Clause 22 ". . . Liquidated and Ascertained Damages . . ."*

For the significance of the sum being a genuine pre-estimate of the damage likely to be suffered by non-completion see *ante*, p. 155.

Limitation of contractor's liability. It is submitted that this clause limits the liability of the contractor for losses caused by his wrongful delay to the sum stated as liquidated damages.[46] It does not affect any claim the employer might have under the contract or at common law for the costs of having the works completed or rectified by another contractor.

Contra Proferentem ? Clauses 21 to 23 are not, it is submitted to be construed against either party as they are part of a document prepared by bodies representative of each.[46a] Neither, it is submitted are they to be construed against the contractor as exclusion clauses (see *ante*, p. 140).

(9) *Clause 23 ". . . the Contractor shall forthwith give written notice. . . ."*

It is not clear whether or not such notice is a condition precedent to the performance by the architect of his duties under this clause.[47] If the architect wrongly

---

[44] See cases cited *ante*, p. 277.

[45] *Cf. Clydebank Eng. Co.* v. *Yzquierdo y Castaneda* [1905] A.C. 6 (H.L.).

[46] *Cf. Cellulose Acetate Ltd.* v. *Widnes Foundry (1925) Ltd.* [1933] A.C. 20; *Suisse Atlantique, etc.* v. *N.V. Rotterdamsche Kolen Centrale* [1967] 1 A.C. 361, 395, 414, 421, 435 (H.L.).

[46a] *Tersons Ltd.* v. *Stevenage Development Corporation* [1963] 2 Lloyd's Rep. 333, 368 (C.A.).

[47] For the use of express language making notice a condition precedent, see clause 31D (2).

assumes that it is when it is not and in consequence refuses to perform such duties the employer loses his right to liquidated damages (note 6, *supra*). It may therefore be against the employer's interest for the architect not to consider a cause of delay of which late notice is given or of which he has knowledge despite lack of notice. In the latter case it is suggested that frequently the better course for the architect is to invite notice. When it is received, or, if not received within a reasonable time, at the expiration of such time he should consider the matter and grant any appropriate extension.[48] But in such a case or where late notice was given he should, it is submitted, take into account that the contractor was in breach of contract, and must not benefit from his breach by receiving a greater extension than he would have received had the architect upon notice at the proper time been able to avoid or reduce the delay by some instructions or reasonable requirement (see proviso to clause 23).

Application for extension of time. This is unnecessary; notice of the cause of delay is sufficient.[49] But before the architect can perform his duties under the clause he often requires information from the contractor, *e.g.* as to the exact nature of the cause of delay, and the contractor's estimate of its effect upon completion of the works. It is probably implied that the contractor shall furnish such information as the architect reasonably requires and it must be so implied when clause 31F (the " formula " fluctuations clause) applies, see clause 31F (7) (b) (ii).

Application for loss and expense. An extension of time relieves the contractor from liability for liquidated damages but gives him no right to extra payment. The causes of delay in clauses 23 (e), (f), (h) and (i) are however part of the grounds for recovery of loss or expense under clause 11 (6) or 24 (1). If the contractor wishes to claim such loss or expense he should make the appropriate applications or say that his notice of a cause of delay is also an application under clause 11 (6) or 24 (1).

(10) *Clause 23 ". . . the opinion of the Architect . . ."*

This and the decisions and the certificate of the architect under clauses 21, 22 and 23 are subject to the jurisdiction of the arbitrator which is not affected, it is submitted, by the architect's final certificate under clause 30.[50]

The architect has to decide whether the cause notified comes within the causes described in one of the sub-clauses. There is no widely expressed general ground for extension of time (*cf.* clause 44 of the I.C.E. Conditions, *post*, p. 491, and clause 28 (2) (f) of the Government conditions). In particular there is no general clause entitling the architect to extend time for delay caused by the employer. If there is such delay not falling within the sub-clauses it will invalidate the provisions for liquidated damages and the employer will have to prove general damages.[51]

---

[48] The contractor cannot, it is submitted, rely on his own breach to say that the extension is invalid.

[49] For an application for extension of time operating as a waiver to the right to object to a retrospective extension of time, see *Sattin* v. *Poole* (1901) H.B.C. (4th ed.), Vol. 2, p. 306 (D.C.), referred to in *Amalgamated Building Contractors Ltd.* v. *Waltham Holy Cross U.D.C.* [1952] 2 All E.R. 452 (C.A.).

[50] The final certificate does not deal with time.   [51] See *ante*, pp. 155 and 160.

Striking out of sub-clauses. It seems that if the parties strike out any of the sub-clauses but leave them legible the court can look at the clauses struck out in order to construe the agreement.[52] What if the parties make the struck-out clauses illegible? Is extrinsic evidence admissible to show what appears in the Standard form and to assist the construction of the agreement? The point awaits decision but it is for consideration whether the parties, by achieving illegibility, have indicated their intention that they do not wish the struck-out clauses to be referred to. The point is raised here both because there is an express invitation to strike out the part or whole of sub-clause (j) and because other sub-clauses, in particular (g), are not infrequently struck out. The matter is one of general application to any alteration or striking out of parts of the form.

(11) *Clause* 23 "... *the completion of the Works is* ... *delayed beyond the Date for Completion* ..."

Notice must be given if the progress of the works is delayed, but there is a right to an extension only if the completion of the works is likely to be or has been delayed. Thus, for example, a delay in the progress of the works at an early stage, may, by the use of the contractor's best endeavours, be reduced or eliminated.

Programme date earlier than completion date. In such a case there can be no extension of the time when the completion of the works will be delayed beyond the programme date but not beyond the date for completion. There may however, in certain circumstances, be grounds for payment under clause 11 (6) or 24 (1).

(12) *Clause* 23 (a) "... *force majeure.* ..."

This is a term of foreign law which has been introduced into English contracts.[53] A statement of its meaning in French law by Goirand, approved by McCardie J. in *Lebeaupin* v. *Crispin* [54] as applying to many English contracts, says: " *Force Majeure.* This term is used with reference to all circumstances independent of the will of man, and which it is not in his power to control. ... Thus, war, inundations and epidemics are cases of *force majeure*; it has even been decided that a strike of workmen constitutes a case of *force majeure.*"

Force majeure is wider in its meaning than the phrases " *vis major* " or " the act of God," [55] and it has been said that " Any direct legislative or administrative interference would of course come within the term: for example, an embargo." [56] But " a ' *force majeure* ' clause should be construed in each case with a close attention to the words which precede or follow it and with due regard to the nature and general terms of the contract. The effect of the clause may vary with

---

[52] See *Mottram Consultants Ltd.* v. *Bernard Sunley & Sons Ltd.* [1975] 2 Lloyd's Rep. 197, 209 (H.L.) discussed *ante,* p. 28, noting that the striking out there was by way of a variation of a contract in being.

[53] *Lebeaupin* v. *Crispin* [1920] 2 K.B. 714, 719; *cf. Hackney Borough Council* v. *Doré* [1922] 1 K.B. 431 (D.C.).

[54] *Supra,* at p. 719. See also *Hong Guan & Co. Ltd.* v. *R. Jumabhoy & Sons Ltd.* [1960] A.C. 684, 700 (P.C.).

[55] *Ibid.*; *Matsoukis* v. *Priestman* [1915] 1 K.B. 681.

[56] *Lebeaupin* v. *Crispin, supra,* at p. 719.

each instrument." [57] Thus, in the Standard form it is submitted that "*force majeure*" has a restricted meaning because matters such as war, strikes, fire and weather are expressly dealt with in the contract.

(13) *Clause* 23 (c) ". . . *the contingencies referred to in clause* 20 [A], [B] *or* [C] . . ."

See the clauses. The contingencies are very wide and their underlying causes may, in some instances, be due to acts of negligence by the contractor.

(14) *Clause* 23 (d) ". . . *civil commotion* . . ."

In the construction of insurance policies these words have been used " to indicate a stage between a riot and civil war." [58] They are also used in clause 20 which deals with insurance and it is thought that the words cited express their meaning here.

(15) *Clause* 23 (e) ". . . *Architect's instructions issued under clauses* . . . 11 (1). . . ."

This includes, it is thought, both variations sanctioned under clause 11 (1) and correction of errors in the contract bills deemed to be variations by virtue of clause 12 (2).

(16) *Clause* 23 (f) ". . . *for which he specifically applied in writing on a date* . . ."

See also clause 3 (4) for the duty of the architect to furnish drawings or details (see note 3 to clause 3). Delay in making nominations is within the clause.

For programmes see note 3, *supra*, " Programme of Work." An agreed programme will afford some assistance in determining the requisite times under this clause but it is doubtful whether the typical programme merely showing commencement of various trades or parts of the works is the specific application required by clause 23 (e).

(17) *Clause* 23 (g) ". . . *delay on the part of nominated sub-contractors or nominated suppliers* . . ."

An extension of time on this ground deprives the employer of remedy for such delay and neither the contractor nor the nominated sub-contractor or nominated supplier as the case may be has to pay the liquidated damages payable but for the extension. The apparent injustice of this and the encouragement to delay has been much criticised. One remedy is not to nominate. Another is to strike out the sub-clause. The commercial problem which arises when the invitation to tender indicates that the sub-clause will be struck out is that careful contractors sometimes either refuse to tender or demand a material increase in price. For the protection of the employer where there are nominations and the clause is retained see *post*, p. 372.

Cause of the delay. This is immaterial. It matters not whether it is due to sloth, the remedying of defects, bad luck or other cause.[59]

---

[57] *Ibid.* at p. 720. See also *Yrazu* v. *Astral Shipping Co.* (1904) 20 T.L.R. 153; *The Concadoro* [1916] 2 A.C. 199 (P.C.); *Re Podar Trading Co., etc.* [1949] 2 K.B. 277.
[58] *Levy Assicurazioni Generali* (1940) 3 All E.R. 427, 437.
[59] *Jarvis, J. & Sons Ltd.* v. *Westminster Corporation* [1970] 1 W.L.R. 637 (H.L.).

Delay after completion of sub-contract works. This is not within the clause. In *Jarvis* v. *Westminster Corporation* [60] nominated sub-contractors carried out piling. They claimed to have completed their work. The contractor accepted it as complete and so did the architect despite his suspicions as to its quality. The contractor proceeded to the next stage of the works and while carrying them out discovered substantial defects in the piling. The sub-contractors returned to the site and carried out extensive remedial work which caused a delay to the completion of the main contract works of $21\frac{1}{2}$ weeks. It was held that the contractor was not entitled to an extension of time as the sub-clause was limited in its operation to the period when the sub-contractor was carrying out the works and did not apply after purported completion of the sub-contract works accepted by the architect and the contractor.

Contractor's losses. The employer is not liable to the contractor for losses caused by nominated sub-contractors and nominated suppliers.[61]

(18) *Clause* 23 (j) "*. . . which he could not reasonably have foreseen . . .*"
These words limit the effect of the sub-clause. Thus, for example, the contractor is not, it is submitted, entitled to an extension of time in respect of shortages of labour or materials which he could by reasonable inquiry have reasonably foreseen were likely to continue or to arise.

Rate of wages. To make clause 23 (j) (i) work there must be some implication as to the rate of wages to be paid. It is suggested that it is such rate, including bonuses, as the parties must reasonably be taken to have foreseen at the time of the contract as necessary to obtain labour for the contract.

(19) *Clause* 23 "*. . . the Architect shall so soon as he is able to estimate the length of the delay . . .*"
The "length of delay" refers not to the immediate delay in the progress of the works but to the delay to the completion of the works. The architect may sometimes be unable to make his estimate until long after notification of the delay in the progress of the works, or until after completion but he will equally be unable to give his certificate under clause 22. For the position when clause 31F applies see clause 31F (7) (b) (ii) and notes thereto.

Retrospective extension. It is submitted that an extension is not invalid [62] merely because it is granted retrospectively or after completion, whether for causes within the employer's control or otherwise, and that the validity of the extension as regards the time when it was given depends upon whether, on the facts, the architect made the extension as soon as he was able to estimate the length of the delay.

(20) *Clause* 23 "*. . . a fair and reasonable extension of time . . .*"
These words acknowledge that the period of extension can rarely be arrived at

---

[60] *Ibid.*
[61] See *ante*, p. 38.
[62] If it is invalid the liquidated damages provisions cannot be enforced, see *Amalgamated Building Contractors Ltd.* v. *Waltham Holy Cross U.D.C.* [1952] 2 All E.R. 452 (C.A.), the facts of which appear *ante*, p. 162. For retrospective extension generally, see *ante*, p. 162.

by simple processes of arithmetic but has to be the result of a consideration of various factors which it is thought may include:

(i) the exact terms and application to the facts of the sub-clause in question;

(ii) the amount of the immediate delay in the progress of the works;

(iii) the effect of any causes of delay, *e.g.* inadequate supervision, which are not within clause 23;

(iv) the effect of concurrent causes of delay, whether within clause 23 or not, and whether one of them is a critical or overriding cause of delay;

(v) the extent to which the contractor has used his best endeavours to prevent delay and has done all that might reasonably be required to proceed with the works (proviso to clause 23).

(21) *Clause 23 " Provided always that the Contractor shall use constantly his best endeavours to prevent delay. . . ."*

The proviso is an important qualification to the right to an extension of time. Thus, for example, in some cases it might be the contractor's duty to re-programme the works either to prevent or to reduce delay. How far the contractor must take other steps depends upon the circumstances of each case, but it is thought that the proviso does not contemplate the expenditure of substantial sums of money.

Clause 24
Loss and expense caused by disturbance of regular progress of the Works

(1) If upon written application being made to him by the Contractor the Architect/Supervising Officer is of the opinion that the Contractor has been involved in direct loss and/or expense for which he would not be reimbursed by a payment made under any other provision in this Contract by reason of the regular progress of the Works or of any part thereof having been materially affected by:

(a) The Contractor not having received in due time necessary instructions, drawings, details or levels from the Architect/Supervising Officer for which he specifically applied in writing on a date which having regard to the Date for Completion stated in the Appendix to these Conditions or to any extension of time then fixed under clause 23 or clause 33 (1) (c) of these Conditions was neither unreasonably distant from nor unreasonably close to the date on which it was necessary for him to receive the same; or

(b) The opening up for inspection of any work covered up or the testing of any of the work materials or goods in accordance with clause 6 (3) of these Conditions (including making good in consequence of such opening up or testing), unless the inspection or test showed that the work, materials or goods were not in accordance with this Contract; or

(c) Any discrepancy in or divergence between the Contract Drawings and/or the Contract Bills; or

(d) Delay on the part of artists tradesmen or others engaged by the Employer in executing work not forming part of this Contract; or

(e) Architect's/Supervising Officer's instructions issued in regard to the postponement of any work to be executed under the provisions of this Contract;

and if the written application is made within a reasonable time of it becoming apparent that the progress of the Works or of any part thereof has been affected as aforesaid, then the Architect/Supervising Officer shall either himself ascertain or shall instruct the Quantity Surveyor to ascertain the amount of such loss and/

or expense. Any amount from time to time so ascertained shall be added to the Contract Sum, and if an Interim Certificate is issued after the date of ascertainment any such amount shall be added to the amount which would otherwise be stated as due in such Certificate.

(2) The provisions of this Condition are without prejudice to any other rights and remedies which the Contractor may possess.

COMMENTARY

(1) *Nature of clause*

During the carrying out of building contracts, particularly if they involve complex and differing operations, it is common for contractors to suffer, or to allege that they suffer, disturbance in the regular progress of the works due to causes within the employer's or the architect's control. With modern highly paid staff and costly machines such losses can be heavy. Clause 24 (1) provides a right for the contractor in certain carefully defined circumstances to obtain payment in respect of some of the more common causes of disturbance of the works by the ordinary certificate procedure of the contract. Clause 24 (2) preserves any other rights and remedies which the contractor may possess including, in particular, his right to claim damages.

Clause 24 (1) should be read with clause 11 (6) and the notes thereto. Together they are the equivalent of the wide " loss and expense " provision of clause 1 of the 1939 R.I.B.A. form but are more exact. Grounds (a) and (d) of clause 24 (1) were not in the former clause 1 but the contractor probably had broadly similar implied rights. For the claim for loss and expense arising out of the discovery of antiquities, see clause 34 (2).

The nature of contractor's " claims " generally is discussed *ante,* p. 260, together with some comments upon the preparation of a claim under this sub-clause. The pleading of a claim and the calculation of losses are discussed *ante,* p. 266, and the meaning of the expression " direct loss and/or expense " is discussed in note 9 to clause 11. For a possible implied qualification where the loss or expense was caused by a design defect, see note 2 (c) to the articles of agreement.

(2) *Clause* 24 (1) "*. . . regular progress . . .*"

See clause 21 (1) and notes thereto for the contractor's duty to proceed regularly and diligently. An agreed programme of work is, it is submitted, some, but not conclusive, evidence of what regular progress should be.

Programme date earlier than completion date. It is thought that in some circumstances loss and expense are recoverable even though the contractor can still complete within the contract period.

(3) *Clause* 24 (1) "*. . . for which he specifically applied in writing . . .*"

Sub-paragraphs (a), (b) and (d) correspond with (f), (i) and (h) of clause 23, and sub-paragraphs (c) and (e) of clause 24 (1) describe part of the causes of delay referred to in (e) of clause 23. See notes to clause 23. For a claim under clause 24 (1) (a) the contractor must show both an application at the proper time for instructions, etc., and an application at the proper time for loss and expense.

(4) *Clause 24 (1) ". . . and if the written application . . ."*

For the importance of distinguishing between this application and notice of a cause of delay under clause 23, see note 9 to clause 23.

(5) *Clause 24 (1) ". . . if an Interim Certificate is issued . . ."*

See note 12 to clause 11 for the position where preliminary items sought to be extended have not been fully expended. The architect need not certify until the appropriate item is exhausted, but the time for giving notice differs under the two clauses. Under clause 24 (1) notice must be given within a reasonable time of it becoming apparent that the progress of the works has been affected. This may be a much earlier date than that of the exhaustion of the relevant preliminary item.

(6) *Clause 24 (2) ". . . without prejudice to any other rights and remedies . . ."*

For the right of determination, see clause 26. For the right to treat the contract as at an end because of repudiation by the employer, see notes to clause 26. For a discussion of a claim for damages for breach of the relevant terms of this contract, additional to the rights to recover payment under the contract, see note 3 to clause 3.

Clause 25
Determination by Employer

(1) Without prejudice to any other rights or remedies which the Employer may possess, if the Contractor shall make default in any one or more of the following respects, that is to say:—

(a) If he without reasonable cause wholly suspends the carrying out of the Works before completion thereof, or

(b) If he fails to proceed regularly and diligently with the Works, or

(c) If he refuses or persistently neglects to comply with a written notice from the Architect/Supervising Officer requiring him to remove defective work or improper materials or goods and by such refusal or neglect the Works are materially affected, or

(d) If he fails to comply with the provisions of either clause 17 or clause 17A of these Conditions,

then the Architect/Supervising Officer may give to him a notice by registered post or recorded delivery specifying the default, and if the Contractor either shall continue such default for fourteen days after receipt of such notice or shall at any time thereafter repeat such default (whether previously repeated or not), then the Employer may within ten days after such continuance or repetition by notice by registered post or recorded delivery forthwith determine the employment of the Contractor under this Contract, provided that such notice shall not be given unreasonably or vexatiously.

(2) In the event of the Contractor becoming bankrupt or making a composition or arrangement with his creditors or having a winding up order made or (except for purposes of reconstruction) a resolution for voluntary winding up passed or a provisional liquidator receiver or manager of his business or undertaking duly appointed, or possession taken, by or on behalf of the holders of any debentures secured by a floating charge, of any property comprised in or subject to the floating charge, the employment of the Contractor under this Contract shall be forthwith automatically determined but the said employment may be reinstated and continued if the Employer and the Contractor his trustee in bankruptcy

liquidator provisional liquidator receiver or manager as the case may be shall so agree.

(3) The Employer shall be entitled to determine the employment of the Contractor under this or any other contract, if the Contractor shall have offered or given or agreed to give to any person any gift or consideration of any kind as an inducement or reward for doing or forbearing to do or for having done or forborne to do any action in relation to the obtaining or execution of this or any other contract with the Employer, or for showing or forbearing to show favour or disfavour to any person in relation to this or any other contract with the Employer, or if the like acts shall have been done by any person employed by the Contractor or acting on his behalf (whether with or without the knowledge of the Contractor), or if in relation to this or any other contract with the Employer the Contractor or any person employed by him or acting on his behalf shall have committed any offence under the Prevention of Corruption Acts 1889 to 1916, or shall have given any fee or reward the receipt of which is an offence under sub-section (2) of section 117 of the Local Government Act 1972 or any re-enactment thereof.

(4) In the event of the employment of the Contractor under this Contract being determined as aforesaid and so long as it has not been reinstated and continued, the following shall be the respective rights and duties of the Employer and Contractor:

(a) The Employer may employ and pay other persons to carry on and complete the Works and he or they may enter upon the Works and use all temporary buildings, plant, tools, equipment, goods and materials intended for, delivered to and placed on or adjacent to the Works, and may purchase all materials and goods necessary for the carrying out and completion of the Works.

(b) The Contractor shall (except where the determination occurs by reason of the bankruptcy of the Contractor or of him having a winding up order made or (except for the purposes of reconstruction) a resolution for voluntary winding up passed), if so required by the Employer or Architect/Supervising Officer within fourteen days of the date of determination, assign to the Employer without payment the benefit of any agreement for the supply of materials or goods and/or for the execution of any work for the purposes of this Contract but on the terms that a supplier or sub-contractor shall be entitled to make any reasonable objection to any further assignment thereof by the Employer. In any case the Employer may pay any supplier or sub-contractor for any materials or goods delivered or works executed for the purposes of this Contract (whether before or after the date of determination) in so far as the price thereof has not already been paid by the Contractor. The Employer's rights under this paragraph are in addition to his rights to pay nominated sub-contractors as provided in clause 27 (c) of these Conditions and payments made under this paragraph may be deducted from any sum due or to become due to the Contractor.

(c) The Contractor shall as and when required in writing by the Architect/Supervising Officer so to do (but not before) remove from the Works any temporary buildings, plant, tools, equipment, goods and materials belonging to or hired by him. If within a reasonable time after any such requirement has been made the Contractor has not complied therewith, then the Employer may (but without being responsible for any loss or damage) remove and sell any such property of the Contractor, holding the proceeds less all costs incurred to the credit of the Contractor.

(d) The Contractor shall allow or pay to the Employer in the manner hereinafter appearing the amount of any direct loss and/or damage caused to the Employer by the determination. Until after completion of the Works under paragraph (a) of this sub-clause the Employer shall not be bound by any provision of this Contract to make any further payment to the Contractor, but upon such completion and the verification within a reasonable time of the accounts therefor the Architect/Supervising Officer shall certify the amount of expenses properly incurred by the Employer and the amount of any direct loss and/or damage caused to the Employer by the determination and, if such amounts when added to the monies paid to the Contractor before the date of determination exceed the total amount which would have been payable on due completion in accordance with this Contract, the difference shall be a debt payable to the Employer by the Contractor; and if the said amounts when added to the said monies be less than the said total amount, the difference shall be a debt payable by the Employer to the Contractor.

COMMENTARY

(1) *Generally*

Outline of clause. It provides for the determination of the contractor's employment by the employer for certain specified defaults (clause 25 (1)); automatic determination upon bankruptcy, liquidation and other events symptomatic of insolvency (clause 25 (2)); determination for corruption (clause 25 (3)); the effect of determination (clause 25 (4)). The provision for determination for corruption does not appear in the private contract although at common law corruption of the employer's agent by the contractor ordinarily gives the employer the right to treat the contract as at an end (see *ante*, p. 205). There is no reference to clause 17A in clause 25 (1) (d) of the private form because there is no clause 17A.

Forfeiture clause. This is the term often loosely applied to this type of clause. Such clauses are discussed generally *ante*, p. 120. It is prudent to think carefully before using the clause because if it should turn out that the employer was not entitled to stop the contract by treating the contractor's employment as at an end the employer himself will ordinarily be guilty of repudiation and will have to pay damages or a reasonable sum to the contractor (see *ante*, p. 120).

Clause 25 (1). Events compared with repudiation

For the meaning given to the term " repudiation," see *ante*, p. 111. There may be a breach by the contractor which both falls within clause 25 (1) and is a repudiatory event. Thus an absolute refusal to carry out the works or an abandonment of the works before they are substantially completed, without any lawful excuse, is a repudiation.[63] (*Cf.* clause 25 (1) (a).) But delay, bad work and unauthorised sub-letting are not breaches any one of which is automatically a repudiation (*cf.* clause 25 (1) (b), (c) and (d),) although the effect of any such breach or more than one considered together may, in the circumstances, be fundamental so as to entitle the employer to treat the contract as at an end.[64]

If the contractor is guilty of default which is both a repudiation and also an event which gives rise to a right of determination under clause 25 (1), the employer has, it is submitted, an option. He may either follow the procedure of clause 25 (1) whereupon he will have the advantage of clause 25 (4) or he may accept the repudiation and treat the contract as at an end forthwith.

---

[63] See *ante*, pp. 115 to 116, for authorities.   [64] *Ibid.*

Immediate arbitration is available for disputes [65]

Elements of a valid determination under clause 25 (1). The following is suggested as an introduction both for an employer minded to determine and a contractor considering whether he has a defence to a purported determination. Reference to the exact words of the clause must be made. The procedure is based on two notices,[66] which have been termed " the architect's notice " and " the employer's notice." [67]

(i) *The first notice.* Is the contractor guilty of a default as specified in one of the sub-paragraphs (a) to (d)? Does the notice specify the default? Is it posted correctly?

(ii) *The second notice.* Has the contractor continued the default for a period of fourteen days after receipt of the first notice, or repeated such default at any time after the expiration of such period? Is the second notice given within ten days of such continuance or repetition? Is it posted correctly? Is it given unreasonably or vexatiously?

Interim injunction to remove contractor. It has been held that such an injunction would not be granted where the affidavits showed that the allegations of fact upon which a determination under clause 25 (1) (b) was based were hotly disputed.[68] For a discussion of this case and the reasons for a submission that it would now no longer ordinarily be followed, see *ante*, p. 123.

Drafting Amendment, 1973 Revision. The opening words of the clause down to the first comma were removed from the later part of the sub-clause and inserted at the beginning in the 1973 Revision. It is thought that this was done to achieve greater clarity and achieved no material difference in the intention of the sub-clause.

(2) *Clause* 25 (1) (a) "*. . . reasonable cause . . .*"

It is thought that such a cause is not limited to the causes of delay set out in clause 23.

(3) *Clause* 25 (1) (b) "*. . . fails to proceed regularly and diligently.*"

This ground of determination applies, it is submitted, after as well as before the completion date is past.[69] For the contractor's duty to proceed regularly and diligently, see clause 21 (1) and note 3 thereto.

(4) *Clause* 25 (1) (c) "*. . . refuses or persistently neglects . . . and . . . the works are materially affected.*"

It is arguable that this ground is only intended to be available before practical completion. For failure to comply with a notice to remedy defects under clause 15

---

[65] Under the 1939 ed. it was not. Note that in practice a long period often elapses between commencement of proceedings and the award.

[66] Save under clause 25 (1) (c) where there must additionally have been an earlier notice to remedy defects.

[67] *Hounslow, London Borough Council* v. *Twickenham Garden Developments Ltd.* [1971] 1 Ch. 233.

[68] *Ibid.*

[69] *Cf. Henshaw & Sons* v. *Rochdale Corp.* [1944] K.B. 381 (C.A.).

a less drastic remedy is available under clause 2 (1). For determination under this sub-clause there must be *three* notices.

Nominated sub-contractor. The contractor has no right or duty to carry out his work (see *ante*, p. 182) and it may be argued that this sub-clause is not intended to apply (compare clause 26 (1) (c) (ii)). But it can be said that the wording of the sub-clause is quite general and by necessary implication the contractor has power to remove the defective work of a nominated sub-contractor who refuses to do so himself. Even if this is correct there are difficulties; see note 4 to clause 6.

(5) *Clause 25 (1) ". . . determine the employment . . ."*
The contract is not determined, for on both sides important contractual rights and liabilities continue or arise.

(6) *Clause 25 (1) ". . . shall not be given unreasonably or vexatiously."*
If so given it is, it is submitted, void. This proviso imposes an important limitation on the operation of the clause. What is unreasonable or vexatious depends upon the circumstances and may give rise to a lengthy investigation in subsequent proceedings.

(7) *Clause 25 (2) ". . . automatically determined . . ."*
Upon the happening of one of the events set out in clause 25 (2) the employment of the contractor is automatically determined and no notice or other formality is required. Nevertheless as soon as the employer discovers the event, if he wishes to take advantage of the clause [70] he is well advised to tell the contractor, his trustee in bankruptcy, etc., that he relies on the clause. Otherwise if the employer takes no step and permits the contractor, his trustee in bankruptcy etc., to continue with the works it may be said that the employer has waived his rights under the clause, or, to use the words of the clause, that the employment of the contractor, his trustee in bankruptcy, etc., has been " reinstated and continued."

Right to disclaim. The trustee in bankruptcy and liquidator have a statutory right to disclaim the contract if it is unprofitable, *i.e.* to bring the contract to an end from the date of the disclaimer [71] It is doubtful how far the provisions of clause 25 (4) are good against a trustee in bankruptcy or liquidator who disclaims.
For the position as to materials upon bankruptcy or liquidation, see further, note 7 to clause 30.

(8) *Clause 25 (4) (a) ". . . use all temporary buildings, plant, tools . . ."*
This does not bind persons not party to the contract, *e.g.* owners of cranes, scaffolding and other equipment who have hired it to the contractor. See also clause 25 (4) (c).

(9) *Clause 25 (4) (b) ". . . assign to the Employer . . ."*
The words in brackets in this clause beginning " (except where . . .) ", the

---

[70] In practice it is often to the advantage of both parties for the contractor's trustee in bankruptcy, etc., to complete the contract.
[71] Bankruptcy Act 1914, s. 54 (1) (8), Companies Act 1948, s. 323 (1), and see *ante*, p. 189.

proviso to clause 17, clause 27 (a) (x) and clause 28 (b) (v) were introduced by the 1968 amendment with the intention, it is thought, of overcoming difficulties which might arise under the bankruptcy law when the sub-contractor's or supplier's employment did not determine and his sub-contract with the insolvent contractor was of value to the insolvent contractor,[72] but *quaere*, is there now anything worth assigning?

(10) *Clause 25 (4) (b) " In any case the Employer may pay any supplier or sub-contractor . . ."*

This provision gives the employer the right to pay any supplier or sub-contractor direct. Compare clause 27 (c) which gives a right of direct payment to a nominated sub-contractor in certain circumstances. The equivalent of clause 27 (c) has been approved by the courts as being good as against the contractor's trustee in bankruptcy or liquidator completing the contract.[73] This provision has not been before the courts. It is not open to the same objections as the equivalent clause in the 1939 edition of the R.I.B.A. form.[74] But in cases of determination upon insolvency under clause 25 (2) there may be difficulties and it is suggested that there should be a careful consideration of the facts before exercising this power in such a case.[75]

(11) *Clause 25 (4) (d) " Until after completion . . . the Employer shall not be bound . . . to make any further payment . . ."*

This extends, it is submitted, not only to work not certified for, but to sums certified but not due for payment under clause 30 (1) at the date of determination. It only applies if the employer proceeds to complete.

(12) *Clause 25 (4) (d) ". . . The Architect . . . shall certify the amount of expenses properly incurred by the Employer and the amount of any direct loss and/or damage caused to the Employer by the determination . . ."*

The contractor has to allow or pay these sums subject to receiving a credit for what he would have been paid if the contract had not been determined. There thus has to be taken a notional final account. Theoretically there could be something payable to the contractor. In practice this is unlikely to happen because the employer is entitled to the expenses of completing under clause 25 (4) (a) and, under the words " direct loss and/or damage " any damages which he has suffered by reason of the determination. This includes, it is submitted, damages for delay. Note the difference in wording of " direct loss and/or damage " in this clause and " direct loss and/or expense " in clauses 11, 24 and 34, the meaning of which is discussed in note 9 to clause 11. It is not thought that

[72] See statement published by the Joint Contracts Tribunal in the *Chartered Surveyor*, December 1967, p. 289. For Bankruptcy and liquidation, see *ante*, p. 188.
[73] See *Re Wilkinson, ex p. Fowler* [1905] 2 K.B. 713; *Re Tout & Finch* [1954] 1 W.L.R. 178. See also *ante*, pp. 176, 193, and reference to *British Eagle Ltd.* v. *Air France* [1975] 1 W.L.R. 758 (H.L.).
[74] See pp. 308, 309 of 2nd ed. of this book.
[75] The danger is of having to pay again to the trustee in bankruptcy or liquidator. For the objections to a clause which upon bankruptcy provides for payment to one creditor in preference to others, see *ante*, p. 188.

there is any difference in meaning between the two terms and that the word
" expense " is avoided in clause 25 in order to keep quite distinct the two
elements which go to make up the architect's certificate under clause 25 (4) (d)
as allowable or payable to the employer subject to the credit in favour of the
contractor. See also note 9 to clause 26.

Clause 26
Determination by contractor

(1) Without prejudice to any other rights and remedies which the Contractor
may possess, if

(a) The Employer does not pay to the Contractor the amount due on any
certificate (otherwise than as a result of the operation of Clause 1 (2) (b) of
the supplemental agreement annexed to these Conditions) (the VAT
Agreement) within 14 days from the issue of that certificate and continues
such default for seven days after receipt by registered post or recorded
delivery of a notice from the Contractor stating that notice of determination
under this Condition will be served if payment is not made within seven days
from receipt thereof; or

(b) The Employer interferes with or obstructs the issue of any certificate due
under this Contract; or

(c) The carrying out of the whole or substantially the whole of the uncompleted Works (other than the execution of work required under clause 15 of
these Conditions) is suspended for a continuous period of the length named in
the Appendix to these Conditions by reason of:

(i) by *force majeure*, or

(ii) loss or damage (unless caused by the negligence of the Contractor,
his servants or agents or of any sub-contractor, his servants or agents)
occasioned by any one or more of the contingencies referred to in clause
20 [A] or clause 20 [B] of these Conditions (if applicable), or

(iii) civil commotion, or

(iv) Architect's/Supervising Officer's instructions issued under clauses
1 (2), 11 (1) or 21 (2) of these Conditions unless caused by reason of
some negligence or default of the Contractor, or

(v) the Contractor not having received in due time necessary instructions, drawings, details or levels from the Architect/Supervising Officer
for which he specifically applied in writing on a date which having regard
to the Date of Completion stated in the Appendix to these Conditions or
to any extension of time then fixed under clause 23 or clause 33 (1) (c)
of these Conditions was neither unreasonably distant from nor unreasonably close to the date on which it was necessary for him to receive
the same, or

(vi) delay on the part of artists tradesmen or others engaged by the
Employer in executing work not forming part of this Contract, or

(vii) the opening up for inspection of any work covered up or of the
testing of any of the work materials or goods in accordance with clause 6
(3) of these Conditions (including making good in consequence of such
opening up or testing), unless the inspection or test showed that the
work materials or goods were not in accordance with this Contract,

then the Contractor may thereupon by notice by registered post or recorded
delivery to the Employer or Architect/Supervising Officer forthwith determine the employment of the Contractor under this Contract; provided that
such notice shall not be given unreasonably or vexatiously.

(2) Upon such determination, then without prejudice to the accrued rights or
remedies of either party or to any liability of the classes mentioned in clause 18 of

these Conditions which may accrue either before the Contractor or any sub-contractors shall have removed his temporary buildings, plant, tools, equipment, goods or materials or by reason of his or their so removing the same the respective rights and liabilities of the Contractor and the Employer shall be as follows, that is to say:—

(a) The Contractor shall with all reasonable dispatch and in such manner and with such precautions as will prevent injury, death or damage of the classes in respect of which before the date of determination he was liable to indemnify the Employer under clause 18 of these Conditions remove from the site all his temporary buildings, plant, tools, equipment, goods and materials and shall give facilities for his sub-contractors to do the same, but subject always to the provision of sub-paragraph (iv) of paragraph (b) of this sub-clause.

(b) After taking into account amounts previously paid under this Contract the Contractors shall be paid by the Employer:—

(i) The total value of work completed at the date of determination.

(ii) The total value of work begun and executed but not completed at the date of determination, the value being ascertained in accordance with clause 11 (4) of these Conditions as if such work were a variation required by the Architect/Supervising Officer.

(iii) Any sum ascertained in respect of direct loss and/or expense under clauses 11 (6), 24 and 34 (3) of these Conditions (whether ascertained before or after the date of determination).

(iv) The cost of materials or goods properly ordered for the Works for which the Contractor shall have paid or for which the Contractor is legally bound to pay, and on such payment by the Employer any materials or goods so paid for shall become the property of the Employer.

(v) The reasonable cost of removal under paragraph (a) of this sub-clause.

(vi) Any direct loss and/or damage caused to the Contractor by the determination.

Provided that in addition to all other remedies the Contractor upon such determination may take possession of and shall have a lien upon all unfixed goods and materials which may have become the property of the Employer under clause 14 of these Conditions until payment of all monies due to the Contractor from the Employer.

Commentary

(1) *Generally*

Outline of clause. It provides for the contractor, without prejudice to any other rights and remedies he may possess, determining his employment on certain specified grounds including the suspension of the work due to causes both within and beyond the control of the employer for a continuous period which the parties have stated in the appendix; by sub-clause (2) it provides for the effect of such determination in terms which, in effect, indemnify the contractor against loss. In the private version there is a further clause, 26 (1) (d), entitling the contractor to determine upon the employer's bankruptcy or other event symptomatic of insolvency.

Nature of clause. It should be compared with clause 25. It gives the contractor the right upon the happening of the events referred to and upon following the procedure set out to withdraw from the site and to have the benefit of the valuable provisions of sub-clause (2). Like clause 25 if it is relied upon it must be properly and exactly complied with for if the contractor wrongfully withdraws permanently from the site this is normally a repudiation of the contract by the contractor which not only deprives him of the benefit of sub-clause (2) but renders him liable in damages to the employer.[76]

Clause 26 events compared with repudiation. For the meaning given to the term " repudiation " see *ante*, p. 111. It will be seen that many of the events giving rise to the right to determine are not even breaches of contract at all (clause 26 (1) (c) (i), (ii), (iii)). Other events, while they may be breaches, may not necessarily be repudiatory in effect. If the event in clause 26 (1) (a) to (c) is also a repudiation the contractor has, it is submitted an option; he may determine under clause 26 or he may accept the repudiation and treat the contract as at an end and forthwith claim damages. The rights under sub-clause (2) are so extensive that it is usually in his interest to determine.

Determination of employment. The contract is not determined for on both sides important contractual rights and liabilities continue or arise.

Notice to determine " shall not be given unreasonably or vexatiously." For the importance of these words, see note 6 to clause 25.

Immediate arbitration. It is available for disputes.[77]

Procedure. It is simpler than under clause 25 (1) in that if the relevant event has occurred only one notice is required, but the event itself must be carefully studied, *e.g.* under clause 26 (1) (a) a prior written notice and a continuance thereafter of the default for seven days is required, and under clause 26 (1) (c) (v) prior written application plus the expiry of due time is necessary.

(2) *Clause* 26 (1) (a). *Failure to pay certificate*

The employer has express contractual rights to make deductions from sums certified for payment under clauses 2 (1), 19 (1) (c), 20 [A] (1), 22, 27 (c). Commenting on the version of clause 26 (1) before the 1973 Amendment it was said that the reference to " the amount due on any certificate " meant the amount due after deduction of any sums which the employer was entitled under the contract to deduct.[78] The 1973 Amendment introduced the words in brackets beginning " otherwise than . . . ." It is thought that, having regard to the special subject-matter of the amendment, these words do not affect this principle which still remains. Further, the employer has a right of set-off in respect of bad work.[79] There is no authority upon the question whether " the amount due " means the amount due after allowing for such set-off. It is arguable that it does. In any event, if the contractor knew of the employer's claim to make such set-off and it was eventually established in fact the arbitrator might well find that a contractor who had given notice in such circumstances had given it unreasonably.

---

[76] See *ante*, p. 115.
[77] See clause 35 (2), but proceedings are not always quick.
[78] *Modern Engineering (Bristol) Ltd.* v. *Gilbert Ash (Northern) Ltd.* [1974] A.C. 639 at p. 709, *per* Viscount Dilhorne.
[79] *Ibid.*

(3) *Clause* 26 (1) (b) " *The Employer interferes with or obstructs the issue of any certificate.*"

This, it is submitted, includes the employer " refusing to allow the architect to go on to the site for the purpose of giving his certificate, or directing the architect as to the amount for which he is to give his certificate or as to the decision which he should arrive at on some matter within the sphere of his independent duty." [80]

(4) *Clause* 26 (1) (c). *Suspension of the Works*

The 1973 version introduced an amendment which may have more effect than at first appears. Before the 1973 version there was express provision for inserting periods of delay in the appendix but a statement that in the event of delay by reason of loss or damage caused by the contingencies referred to in clause 20 [A] or clause 20 [B] if no period was stated it was three months and for any other reason if none was stated it was one month. The 1973 Amendment positively requires the insertion of periods. There is a footnote suggesting that they should be the same as before but such footnote has no contractual effect. It is suggested that the architect should consider carefully with his client whether the suggested periods are adequate. For example, in a project involving complex engineering a technical problem might arise making it necessary to suspend the works while it is solved. One month's suspension might be inadequate.

(5) *Clause* 26 (1) (c) (i), (iii)

For the meaning of the terms *force majeure* and " civil commotion " see notes 12 and 14 to clause 23.

(6) *Clause* 26 (1) (c) (ii) "*. . . loss or damage . . . occasioned by . . . contingencies referred to in clause* 20 [A] *or clause* 20 [B] *. . .*"

The words in brackets were introduced by the 1973 Amendment. In earlier versions if the works were damaged by fire, flood or the like the contractor could rely on the clause even though the event was due to his own negligence. Even so such negligence is a matter which can be relied on in support of an argument that the determination notice was given unreasonably. There is, in respect of the pre-1973 version, at least an argument that destruction of, and perhaps even substantial damage to, the works by the contractor's negligence is a fundamental breach entitling the employer to treat the contract as at an end and that in any event the clause cannot be relied on. (See repudiation, *ante*, p. 111 and exclusion clauses, *ante*, p. 140.)

(7) *Clause* 26 (1) (c) (iv) *Architect's instructions*

The words beginning " unless caused . . ." were introduced in the 1973 revision. See the last note. The word " default " is wide. It probably includes

---

[80] *Per* Lord Tucker, *Burden* v. *Swansea Corp.* [1957] 1 W.L.R. 1167, 1180 (H.L.), a case on the 1939 form where the clause was confined to certificates for payment, but where the principle was, it is submitted, the same. For the architect's duty to act fairly and professionally when certifying see *ante*, p. 295.

a tortious act, a breach of contract and a failure to conform with the requirements of the contract not amounting to a breach.[81]

Nominated sub-contractors and nominated suppliers. The contractor has no right or duty to carry out nominated sub-contract work himself.[82] This together with the contrast with clause 26 (1) (c) (ii) shows, it is submitted, that the negligence or default of a nominated sub-contractor is not that of the contractor. Probably, although the matter is less clear, the position is the same for a nominated supplier.

(8) *Clause* 26 (1) (c) (v) *Delay in instructions*

This ground of suspension is not subject to the qualification in clause 26 (1) (c) (iv) relating to the negligence or default of the contractor (see last note). In the appropriate case, therefore, the architect will postpone the works under clause 21 (2) and not leave a negligent or defaulting contractor the opportunity of determining under this sub-clause.

(9) *Clause* 26 (2) (b) (vi) " *Any direct loss and/or damage. . . .*"

See note 12 to clause 25. In *Wraight Ltd.* v. *P. H. & T. (Holdings) Ltd.* (1968)[83] Megaw J. rejected submissions that, because the clause could operate where there was no fault on the part of the employer, these words were to have a limited effect, different from their natural meaning. He said, " prima facie, the claimants are entitled to recover, as being direct loss and/or damage, those sums of money which they would have made if the contract had been performed, less the money which has been saved to them because of the disappearance of their contractual obligations."[84] Their entitlement included loss of profit even though they had only worked a few weeks out of a 60 week contract when work was suspended due to unanticipated difficulties.

(10) *Clause* 26 (2) *proviso*

The proviso that the contractor has a lien upon unfixed goods and materials which have become the property of the employer until payment of all monies due to him from the employer may be very important. He can prevent the employer from using, for example, specially made windows and other components until everything has been paid under the sub-clause. The proviso may not be effective against the trustee in bankruptcy or liquidator of an insolvent company (see *ante,* pp. 126 and 188).

Clause 27
Nominated sub-contractors

The following provisions of this Condition shall apply where prime cost sums are included in the Contract Bills, or arise as a result of Architect's/Supervising Officer's instructions given in regard to the expenditure of provisional sums, in

---

[81] See *Kaye* (*P. & M.*) *Ltd.* v. *Hosier & Dickinson Ltd.* [1972] 1 W.L.R. 146, 165 (H.L.) ". . . temporary disconformity . . ."

[82] *Bickerton* (*T. A.*) *& Son Ltd.* v. *N.W. Metropolitan Regional Hospital Board* [1970] 1 W.L.R. 607 (H.L.), discussed *ante,* p. 182 and *post,* p. 369.

[83] Unreported.

[84] Transcript, p. 12.

respect of persons to be nominated by the Architect/Supervising Officer to supply and fix materials or goods or to execute work.

(a) Such sums shall be deemed to include 2½ per cent. cash discount and shall be expended in favour of such persons as the Architect/Supervising Officer shall instruct, and all specialists or others who are nominated by the Architect/Supervising Officer are hereby declared to be sub-contractors employed by the Contractor and are referred to in these Conditions as ' nominated sub-contractors.' Provided that the Architect/Supervising Officer shall not nominate any person as a sub-contractor against whom the Contractor shall make reasonable objection, or (save where the Architect/Supervising Officer and Contractor shall otherwise agree) who will not enter into a sub-contract which provides (*inter alia*):—

(i) That the nominated sub-contractor shall so carry out and complete the sub-contract Works as to enable the Contractor to discharge his obligations under clause 1 (1) of these conditions so far as they relate and apply to the sub-contract Works or any portion of the same and in conformity with all the reasonable directions and requirements of the Contractor.

(ii) That the nominated sub-contractor shall observe, perform and comply with all the provisions of this Contract on the part of the Contractor to be observed, performed and complied with (other than clause 20 [A] of these Conditions, if applicable) so far as they relate and apply to the sub-contract Works or to any portion of the same.

(iii) That the nominated sub-contractor shall indemnify the Contractor against the same liabilities in respect of the sub-contract Works as those for which the Contractor is liable to indemnify the Employer under this Contract.

(iv) That the nominated sub-contractor shall indemnify the Contractor against claims in respect of any negligence, omission or default of such sub-contractor, his servants or agents or any misuse by him or them of any scaffolding or other plant, and shall insure himself against any such claims and produce the policy or policies and receipts in respect of premiums paid as and when required by either the Employer or the Contractor.

(v) That the sub-contract Works shall be completed within the period or (where they are to be completed in sections) periods therein specified, that the Contractor shall not without the written consent of the Architect/Supervising Officer grant any extension of time for the completion of the sub-contract Works or any section thereof, and that the Contractor shall inform the Architect/Supervising Officer of any representations made by the nominated sub-contractor as to the cause of any delay in the progress or completion of the sub-contract Works or of any section thereof.

(vi) That if the nominated sub-contractor shall fail to complete the sub-contract Works or (where the sub-contract Works are to be completed in sections) any section thereof within the period therein specified or within any extended time granted by the Contractor with the written consent of the Architect/Supervising Officer and the Architect/Supervising Officer certifies in writing to the Contractor that the same ought reasonably so to have been completed, the nominated sub-contractor shall pay or allow to the Contractor either a sum calculated at the rate therein agreed as liquidated and ascertained damages for the period during which the said Works or any section thereof, as the case may be, shall so remain or have remained incomplete or (where no such rate is therein agreed) a sum equivalent to any loss or damage suffered or incurred by the Contractor and caused by the failure of the nominated sub-contractor as aforesaid.

(vii) That payment in respect of any work, materials or goods comprised in the sub-contract shall be made within 14 days after receipt by the Contractor of the duplicate copy of the Architect's/Supervising Officer's certificate under clause 30 of these Conditions which states as due an amount calculated by including the total value of such work, materials or goods, and shall when due be subject to the retention by the Contractor of the sums mentioned in sub-paragraph (viii) of paragraph (a) of this Condition, and to a discount for cash of $2\frac{1}{2}$ per cent. if made within the said period of 14 days.

(viii) That the Contractor shall retain from the sum directed by the Architect/Supervising Officer as having been included in the calculation of the amount due in any certificate issued under clause 30 of these Conditions in respect of the total value of work, materials or goods executed or supplied by the nominated sub-contractor a percentage (which percentage shall be equal to the percentage currently being retained by the Employer under clause 30 of these Conditions) of such value; and that the Contractor's interest in any sums so retained (by whomsoever held) shall be fiduciary as trustee for the nominated sub-contractor (but without obligation to invest); and that the nominated sub-contractor's beneficial interest in such sums shall be subject only to the right of the Contractor to have recourse thereto from time to time for payment of any amount which he is entitled under the sub-contract to deduct from any sum due or to become due to the nominated sub-contractor; and that if and when such sums or any part thereof are released to the nominated sub-contractor they shall be paid in full less only a discount for cash of $2\frac{1}{2}$ per cent. if paid within 14 days of the date fixed for their release in the sub-contract.

(ix) That the Architect/Supervising Officer and his representatives shall have a right of access to the workshops and other places of the nominated sub-contractor as mentioned in clause 9 of these Conditions.

(x) That the employment of the nominated sub-contractor under the sub-contract shall determine immediately upon the determination (for any reason) of the Contractor's employment under this Contract.

(b) The Architect/Supervising Officer shall direct the Contractor as to the total value of the work, materials or goods executed or supplied by a nominated sub-contractor included in the calculation of the amount stated as due in any certificate issued under clause 30 of these Conditions and shall forthwith inform the nominated sub-contractor in writing of the amount of the said total value. The sum representing such total value shall be paid by the Contractor to the nominated sub-contractor within 14 days of receiving from the Architect/Supervising Officer the duplicate copy of the certificate less only (i) any retention money which the Contractor may be entitled to deduct under the terms of the sub-contract (ii) any sum to which the Contractor may be entitled in respect of delay in the completion of the sub-contract Works or any section thereof, and (iii) a discount for cash of $2\frac{1}{2}$ per cent.

(c) Before issuing any certificate under clause 30 of these Conditions the Architect/Supervising Officer may request the Contractor to furnish to him reasonable proof that all amounts included in the calculation of the amount stated as due on previous certificates in respect of the total value of work, materials or goods executed or supplied by any nominated sub-contractor have been duly discharged, and if the Contractor fails to comply with any such request the Architect/Supervising Officer shall issue a certificate to that effect and thereupon the Employer may himself pay such amounts to any nominated sub-contractor concerned (to which amounts the Employer may add an amount for any value added tax which would have been properly due to the nominated sub-contractor) and deduct the same from any sums due or to become due to the Contractor.

(d)   (i) The Contractor shall not grant to any nominated sub-contractor any extension of the period within which the sub-contract Works or (where the sub-contract Works are to be completed in sections) any section thereof is to be completed without the written consent of the Architect/Supervising Officer. Provided always that the Contractor shall inform the Architect/Supervising Officer of any representation made by the nominated sub-contractor as to the cause of any delay in the progress or completion of the sub-contract Works or of any section thereof, and that the consent of the Architect/Supervising Officer shall not be unreasonably withheld.

(ii) If any nominated sub-contractor fails to complete the sub-contract Works or (where the sub-contract Works are to be completed in sections) any section thereof within the period specified in the sub-contract or within any extended time granted by the Contractor with the written consent of the Architect/Supervising Officer, then if the same ought reasonably so to have been completed the Architect/Supervising Officer shall certify in writing accordingly; any such certificate shall be issued to the Contractor and immediately upon issue the Architect/Supervising Officer shall send a duplicate copy thereof to the nominated sub-contractor.

(e) If the Architect/Supervising Officer desires to secure final payment to any nominated sub-contractor before final payment is due to the Contractor, and if such sub-contractor has satisfactorily indemnified the Contractor against any latent defects, then the Architect/Supervising Officer may in an Interim Certificate include an amount to cover the said final payment, and thereupon the Contractor shall pay to such nominated sub-contractor the amount so certified less only a discount for cash of $2\frac{1}{2}$ per cent. Upon such final payment the Contractor shall, save for latent defects, be discharged from all liability for the work materials or goods executed or supplied by such sub-contractor under the sub-contract to which the payment relates.

(f) Neither the existence nor the exercise of the foregoing powers nor anything else contained in these Conditions shall render the Employer in any way liable to any nominated sub-contractor.

(g)   (i) Where the Contractor in the ordinary course of his business directly carries out works for which prime cost sums are included in the Contract Bills and where items of such works are set out in the Appendix to these Conditions and the Architect/Supervising Officer is prepared to receive tenders from the Contractor for such items, then the Contractor shall be permitted to tender for the same or any of them but without prejudice to the Employer's right to reject the lowest or any tender. If the Contractor's tender is accepted, he shall not sub-let the work without the consent of the Architect/Supervising Officer.

Provided that where a prime cost sum arises under Architect's/Supervising Officer's instructions issued under clause 11 (3) of these Conditions it shall be deemed for the purposes of this paragraph to have been included in the Contract Bills and the item of work to which it relates shall be deemed to have been set out in the Appendix to these Conditions.

(ii) It shall be a condition of any tender accepted under this paragraph that clause 11 of these Conditions shall apply in respect to the items of work included in the tender as if for the reference therein to the

Contract Drawings and the Contract Bills there were references to the equivalent documents included in or referred to in the tender.

COMMENTARY

(1) *Generally*

Outline of clause. This clause shouid be read with clause 11 (3) (expenditure of prime cost and provisional sums), clause 28 (nominated suppliers) and clause 30 (5) (c) (settlement of accounts). It makes an elaborate attempt to deal with some of the problems which arise when sub-contractors are nominated by the architect.[85] It provides for: the nomination of sub-contractors and the provisions of the sub-contract which should be used (clause 27 (a)); payment of nominated sub-contractors (clause 27 (b)); direct payment to nominated sub-contractors (clause 27 (c)); extension of time to nominated sub-contractors (clause 27 (d)); final payment to nominated sub-contractors (clause 27 (e)); absence of liability of employer to nominated sub-contractors (clause 27 (f)); contractor's tender for prime cost items (clause 27 (g)).

(2) *Default of nominated sub-contractor—contractor's liability*

This should be read with the general discussion *ante*, p. 174. The starting point is that, subject to the terms of the contract, the contractor is as responsible for the default of his sub-contractor as he is for his own default. This is the approach as regards sub-contractors nominated under this form.[86] But there are so many qualifications to this principle that the employer is in respect of many matters wholly without remedy unless he has entered into a collateral contract with the nominated sub-contractor (see *post*, p. 427), or can persuade the courts that he has a claim in negligence against the sub-contractor (see *ante*, p. 175), or that there is a breach of duty by his architect or other professional adviser or, in the special case, against some other person such as a local authority inspector.[87] The qualifications to the contractor's liability will be set out summarily, followed by some more detailed notes.

Summary of qualifications to contractor's liability. The contractor does not have to pay damages for delay on the part of a nominated sub-contractor. The contractor is not liable if work done or materials or plant supplied by a nominated sub-contractor is unsuitable for its purpose unless, exceptionally, the employer has relied on the contractor's skill and judgment in the selection of the relevant work, materials or plant. The contractor is not ordinarily liable for design carried out by the nominated sub-contractor. The contractor is not ordinarily liable upon a guarantee of performance given by the sub-contractor to the contractor.

The contractor is not responsible for repudiation or fundamental breach of contract on the part of the nominated sub-contractor and, on the contrary, has a claim against the employer for some part, at least, of the loss he suffers. The

---

[86] See *ante*, p. 178, for nominated sub-contractors generally.

[85] *Bickerton (T. A.) & Son Ltd.* v. *N.W. Metropolitan Regional Hospital Board* [1970] 1 W.L.R. 607, 609, 623 (H.L.).

[87] *Dutton* v. *Bognor Regis Urban District Council* [1972] 1 Q.B. 373 (C.A.); *Anns* v. *Merton London Borough* [1977] 2 W.L.R. 1024 (H.L.).

contractor's liability for defective work carried out and materials supplied by a nominated sub-contractor only arises, it seems, after practical completion save where there has been some independent breach of contract by the contractor.

Delay by nominated sub-contractor. There is a right to extension of time, see note 17 to clause 23, and see clause 27 (d). Delay by a nominated sub-contractor is not a ground upon which the contractor can rely to determine his employment under clause 26, or to claim his losses from the employer (see *ante*, p. 38). For delay causing a suspension of the works see note 7 to clause 26.

Authority by the architect to depart from the specified method of construction in order to reduce loss being caused by a nominated sub-contractor is not, it is submitted, a ground for additional payment unless it can be shown that there was an intention to issue an instruction requiring a variation under clause 11 (1).[88]

Delay relating to drawings and information. It is often necessary to obtain information from nominated sub-contractors or proposed nominated sub-contractors as to the details of their work to enable adequate instructions, details and drawings to be given to the contractor. Failure to give such information may delay the progress of the works. If the failure arises before nomination it will ordinarily give the contractor a right to an extension of time under clause 23 (f) if completion is delayed and, if he has suffered loss and expense, to payment under clause 24 (1) (a) or to claim damages for delay in the issue of drawings or instructions (see note 3 to clause 3). After nomination the position is as *supra*, " Delay by nominated sub-contractor," but the contractor may be able to show that what appears, prima facie, to be delay by the nominated sub-contractor is on the facts, delay by the architect. Thus the nominated sub-contractor may be delayed by lack of information by the architect, or the architect may have relied on the nominated sub-contractor to provide the contractor with information which he, the architect, is under a duty to provide under clause 3 (3) or an implied term (see note 3 to clause 3). In such instances the contractor can claim from the employer under clause 24 (1) (a) or for breach of contract.

Fit for purpose. It is in the nature of the system of nomination that there is no reliance upon the contractor's skill and judgment. Therefore upon principles discussed *ante*, pp. 40 and 181 the contractor is under no liability if the work, materials or plant carried out and supplied by the nominated sub-contractor are not fit for their purpose. The contractor is responsible (to the extent discussed below) if the quality of workmanship or of the materials and plant is bad. A difficult problem can arise if, as sometimes happens, the ordinary procedure for nomination is not followed. Clause 27, read with clause 12 and the Standard Method of Measurement, clause 20 [B] contemplates that there shall be no description of the obligations of the sub-contractor written into the main contract and that there will only be a statement of an allowance of a prime cost sum or provisional sums in respect of persons to be nominated to supply and fix materials or goods or execute work. But sometimes the bills of quantities set out

---

[88] See, for some assistance, *Kirk & Kirk Ltd.* v. *Croydon Corp.* [1956] J.P.L. 585; *Simplex Concrete Piles Ltd.* v. *St. Pancras Borough Council* (1958) unreported save in H.B.C. (10th ed.), p. 526; *cf. Tharsis Sulphur & Copper Co.* v. *M'Elroy & Sons* (1878) 3 App. Cas. 1040, 1053 (H.L.).

in full the technical description of the work to be performed by the nominated sub-contractor and, even, occasionally, the full terms of the sub-contract. If the contractor signs a contract with the bills in such a form what is the result ? The answer must depend upon the words used in each case. The immediate approach is to assume that the contractor has accepted every obligation written into the bills. But it is not as simple as that. The effect of clause 12 may be that his liability is restricted to such obligations as may be properly termed " the quality and quantity of the work " and that he is not liable in respect of any conditions of the sub-contract which override, modify or affect in any way whatsoever the application or interpretation of the conditions of the Standard form of building contract (see discussion of this difficult subject, note 6 to clause 12).

Design. It is becoming increasingly common to get the nominated sub-contractor to carry out important parts of the design. For the general position as to design obligations see note 2 (c) to Articles of Agreement. For the extent to which generally a main contractor is responsible for design by a nominated sub-contractor see *ante*, pp. 41 and 181 where the principles discussed apply here. Thus while every case must be considered according to the particular words in question and the circumstances the approach seems to be: if no design obligation is written either into the contract bills or into the sub-contract the contractor is not responsible for any design in fact carried out; if a design obligation is written into the sub-contract but not into the bills then, by accepting a nomination which he is entitled to refuse (see *post*, note 3), it may be that the main contractor has accepted the duty of design written into the sub-contract although the position must turn on the particular words used and the circumstances of each case; if the obligation to design is both expressly written into the sub-contract and into the bills the argument that the contractor has accepted the duty of design is even stronger but one still has to consider the effect of clause 12 discussed *supra* in relation to fitness for purpose.

Guarantee. Unless this is written into the bills, in which case the clause 12 problems (see *supra*) arise, it is submitted that the court would not imply[89] a term that the contractor accepted liability for a guarantee given to him by the sub-contractor. If this is correct the guarantee is of no value to the employer. If a guarantee is required by the employer he should obtain it direct from the nominated sub-contractor and have it expressed to be given in consideration of nomination or alternatively pay a nominal sum for it.

Repudiation. The position where a nominated sub-contractor repudiated his sub-contract was considered by the House of Lords in *Bickerton* v. *N.W. Metropolitan Regional Hospital Board*[90] the facts of which are set out *ante* p. 182 and upon which this paragraph is based. Where, in the course of, and before completion of, the sub-contract works, a nominated sub-contractor repudiates[91] his sub-contract, the employer must, unless he omits the rest of

---

[89] For implication of terms generally, see *ante*, p. 35.

[90] [1970] 1 W.L.R. 607 (H.L.).

[91] The word " repudiates " is used here in its widest sense as including both a refusal to carry out a contract and a breach of contract of such a kind as to entitle the innocent party to treat the contract as at an end. For a full discussion see *ante*, p. 111.

the sub-contract works, nominate a new sub-contractor, bearing any increased costs resulting from such re-nomination, and must pay damages to the contractor for any delay suffered by the contractor awaiting a re-nomination.[92] ". . . If a contractor wrongfully terminates the sub-contract—it may be because he thinks erroneously that the sub-contractor is in fundamental breach of the sub-contract . . . the contractor would be in breach of his contract with the employer. A new sub-contractor would have to be nominated. But the contractor would have to pay damages for his breach of contract including any loss caused to the employer by that breach." [93] It should be observed that the employer's duty of re-nomination arises upon repudiation and not liquidation as such. Neither, it is submitted, does it necessarily arise upon the occurrence of an event giving the contractor a right under an express term to determine the sub-contract or the sub-contractor's employment. Such an event may or may not amount to a repudiation by the sub-contractor.

Their Lordships in *Bickerton's* case observed that the contract made no express provision for repudiation. It still does not and there remain many problems upon which there is no direct guidance. An attempt will be made to deal with some of them.

When does the right of re-nomination arise? The first question is whether there is an effective nominated sub-contractor in existence or whether it may be said that he has dropped out. Then does the right of re-nomination arise immediately the nominated sub-contractor drops out? As to this it is thought that it cannot arise before the architect has been notified and a request made for the re-nomination. Does the architect then have a period of time in which to re-nominate? In practice the period of time necessary to negotiate and place a new sub-contract may be quite considerable. Is the employer in breach if, despite this, the architect does not re-nominate immediately? The answer is, it is submitted, that he is in breach if the absence of an effective nominated sub-contractor impedes the contractor in performance of the contract.[94] If because such performance is impeded an instruction is necessary can the contractor treat the failure to give the instruction immediately as a fundamental breach [95] entitling him to treat the contract as at an end? It is submitted that ordinarily he cannot and must rely on his remedy in damages.

Terms of new sub-contract. They must it is thought comply with the provisions of clause 27 (a) which include by sub-clauses (ii) and (v) obligations as to time. Can the contractor refuse as incompetent a nomination of a sub-contractor who is not prepared to undertake to complete by such date as will enable the contractor to complete by the contract date, subject to any existing

---

[92] See *supra*, footnote 90.

[93] *Bickerton's* case, p. 613.

[94] See *Bickerton's* case, p. 613, *per* Lord Reid in reference to the duty of the employer to make the first nomination. Further, in so far as it can be said that the contractor has made an application for an instruction, " due time " for the giving of the instruction is, it is submitted, when it is required so as to enable the contractor to proceed without interference with that progress of the works contemplated by the contract.

[95] For fundamental breach see *ante*, p. 112.

extensions of time? [96] Prima facie it would appear that he can. The architect cannot avoid the difficulty by purporting to extend time under clause 23 (g) because *ex hypothesi* there is no sub-contractor in existence during the period awaiting re-nomination, and an architect cannot extend time in respect of a cause within the employer's control save pursuant to express terms.[97-99] It may be that the problem can be met by an order postponing the execution of the works for the period of delay arising from the date of completion in the proposed new sub-contract (see clauses 21 (2), 23 (e)). The contractor would be compensated under clause 24 (1) (e). It is thought that by the application of this or by some other device, the court would welcome an opportunity of avoiding a situation where the contractor could treat the main contract as at an end. For a discussion of the position where the contractor is running behind time for reasons which do not entitle him to an extension of time see *post*, note 8.

Determination due to delay. If the contractor has specifically applied in writing for a re-nomination and the works are brought to a standstill for one month, or whatever other period is inserted in the appendix, by reason of the lack of a new nominated sub-contractor, can he serve notice of determination under clause 26? It appears that he can [1] but it is thought that the court or an arbitrator would look at the circumstances to consider whether the notice was given unreasonably and therefore was not effective.

Defects in former sub-contractor's work. It may well happen that the new sub-contractor finds much defective work carried out by the old. It is submitted that the cost of putting it right is all part of the costs of re-nomination which the employer has to bear.

More than one re-nomination. The *Bickerton* principle is not limited to one re-nomination. The employer must supply an effective nominated sub-contractor or, alternatively, omit the work.

Position after completion of sub-contract works. " Completion " is used here in the sense in which it was used in *Westminster Corporation* v. *Jarvis Ltd.*,[2] *i.e.* that the sub-contract works are apparently completed even though there may be latent defects within them. What is the position if thereafter latent defects appear and the nominated sub-contractor is asked to put them right but refuses? The position is not clear. One can approach the matter in this way: that if the main contractor is at all times entitled to a re-nomination in the event of repudiation by a nominated sub-contractor then there is little meaning in the statement, repeated in several cases, that the main contractor is responsible for the default of the nominated sub-contractor. Further, one refers to the express words of sub-

---

[96] See the position which arose in *Trollope & Colls Ltd.* v. *North West Metropolitan Regional Hospital Board* [1973] 1 W.L.R. 601 (H.L.), discussed *ante*, p. 36, but noting that the contract there was an amended version of the then current R.I.B.A. form.

[97-99] See *ante*, p. 161.

[1] See clause 26 (1) (c) (v); see also, *Gloucestershire County Council* v. *Richardson* [1969] 1 A.C. 480 (H.L.) but noting that under the form of R.I.B.A. contract there used the contractor had an absolute right of determination and was not subject to the proviso of reasonableness.

[2] [1970] 1 W.L.R. 637 (H.L.), discussed note 17 to clause 23.

clauses 15 (2) and 15 (3). These seem to impose, where the procedure under the sub-clauses is followed an absolute duty upon the contractor to make good defects, unless the architect otherwise instructs, at his own cost.[3] It seems, therefore, that, subject to the architect's discretion, if a defect appears during the defects liability period in the nominated sub-contractor's work, and the nominated sub-contractor fails or refuses to comply with the contractor's request to put it right, that the main contractor comes under a duty to put it right himself and is not entitled to a re-nomination however serious the consequences of the failure to remedy the defect. What is the position if a defect appears between completion of the sub-contract works and practical completion of the contract works [4] as a whole? One can only suggest that, following the reasoning in *Bickerton* and in the absence of an express duty such as arises under clause 15, the contractor has no right or duty to put right the defect so that if the nominated sub-contractor refuses to remedy the defect so as to repudiate his sub-contract, the contractor is entitled to a re-nomination.

It is submitted that the contractor is liable in damages in respect of defects which appear after the issue of the certificate of completion of making good defects (clauses 15 (4), 30 (3) (b)), and those which appear after the issue of the final certificate save where the final certificate is conclusive.[5]

Protection of employer. It will be apparent from the above that the provisions as to nomination may leave the employer bearing many risks which he may not expect to have to bear, especially if he has engaged an architect who relies on paragraph 1.40 of the R.I.B.A. conditions of engagement which seek to exempt the architect from liability " for the detailed design or performance " of work entrusted to specialist sub-contractors and suppliers. It is suggested that the employer should be made aware of his position and the matters discussed *ante,* p. 183 under the heading of " Protection of employer " should be considered. If it really is necessary to nominate consideration should be given to entering into collateral contracts with the nominated sub-contractor. Forms which give protection in respect of some of the matters referred to are printed and discussed *post,* p. 427.

(3) *The duty to nominate*

Where there is a prime cost sum in respect of persons to be nominated by the architect to supply and fix materials or goods or to execute work there must be a nomination or an omission of the work.[6] For the meaning of prime cost sum and provisional sum see note 5 to clause 11.

Delay in giving instructions. Delay by the architect in giving instructions

---

[3] See the remarks of Viscount Dilhorne, *Bickerton, supra,* at p. 625. See also the speech of Lord Diplock in *Kaye (P. & M.) Ltd.* v. *Hosier & Dickinson Ltd.* [1972] 1 W.L.R. 146, 163 in which he draws a clear distinction between the construction period and the defects liability period.

[4] *Cf. Westminster Corporation* v. *Jarvis, J. & Sons Ltd. (supra),* where, however, the House of Lords were dealing with an entirely different question.

[5] See clause 30 and notes thereto.

[6] See *supra,* footnote 94.

nominating a sub-contractor can give rise to claims for extension of time under
clause 23 (f), for loss or expense under clause 24 (1) (a) and probably to damages
at common law.[7] Further, it seems that such claims can arise out of delay caused
by the nomination, without the contractor's previous agreement, of a proposed
sub-contractor who is not willing to enter into the form of sub-contract by sub-
clause (a).

Nomination in respect of contractor's work. Where an item is neither a
prime cost sum nor shown as a provisional sum, the architect cannot, it seems,
nominate a sub-contractor to carry it out.[8] But if the architect wishes to make
such a nomination and the contractor does not object the parties may find it
convenient to treat it as if it were a variation under clause 11 so that loss or
expense, including any loss of profit to the contractor, becomes payable under
clause 11 (6).

Nomination of designer. The right of nomination is in respect of persons
to supply and fix materials or goods or to execute work. It does not extend to
persons who are to design and the contractor can, it is submitted, refuse
to accept a purported nomination of a person who is expressed to be required to
carry out design work. For the contractor's liability for design by a nominated
sub-contractor and the position if he accepts a nomination of a sub-contractor
who is to design or where the bills contemplate design by a nominated sub-
contractor see note 2 *supra*.

(4) *Clause* 27 (a) "... *such sums shall be deemed to include* $2\frac{1}{2}$ *per cent. cash discount* ..."

This means, it is submitted, that in the settlement of accounts (see clause 30
(5) (c)) the contractor is assumed to have had the benefit of such a cash discount
whether he has received it or not, and, if he has not received it, he cannot claim
it from the employer.

Default of employer. It is submitted that there is an exception to the
principle just stated if it is the act of the employer or the architect acting as
agent which causes the $2\frac{1}{2}$ per cent. cash discount not to be paid. In such a case
it is submitted that the contractor can recover the $2\frac{1}{2}$ per cent. from the employer
as damages or, alternatively, on an implied promise to pay arising from a variation
of the contract.[9]

Unwilling sub-contractor. If the architect directs the contractor to enter
into a sub-contract with a nominated sub-contractor who will not allow a $2\frac{1}{2}$ per
cent. cash discount the contractor should either: (1) refuse to enter into the sub-
contract (see proviso to this sub-clause) or (2) before entering into the sub-

---

[7] For the latter, see note 3 to clause 3.
[8] Because he is not empowered to do so by any clause—see clause 2 (1). Clause 11 (1)
empowers him to vary the works but not the contract conditions; see further, note 2 to
clauses 1 and 2.
[9] *Cf.* the principle that no man can rely on his own wrong; *Roberts* v. *Bury Commissioners*
(1870) L.R. 4 C.P. 755; *Amalgamated Building Contractors Ltd.* v. *Waltham Cross Holy
Cross U.D.C.* [1952] 2 All E.R. 452, 455 (C.A.).

contract, agree with the employer that the Standard form shall be varied by the employer agreeing to pay the $2\frac{1}{2}$ per cent. cash discount.

(5) *Clause* 27 (a) " *Provided that the Architect . . .*"

The parties must give effect to the proviso before the sub-contract is entered into. The provisions (i) to (x) can have no effect themselves upon any sub-contract and, ordinarily, if a contractor enters into a sub-contract containing different provisions he has no remedy against the employer. The provisions of clause 27 (1) (a) are in the main for the benefit of the contractor, but some of them, if they appear in a sub-contract assist the employer, *e.g.* (vii) and (ix).

(6) *Clause* 27 (a) ". . . *reasonable objection . . .*"

In *Bickerton & Son Ltd.* v. *N.W. Metropolitan Regional Hospital Board*[10] in the Court of Appeal it was said that the contractor cannot object merely because he considers the price is too low, and " He has no right to object to any of the details in the specification in the sub-contract, any more than to the sub-contract price. Provided the sub-contract conforms to the provisions of Condition 27, his sole right of objection is to the person nominated." [11]

No guidance is given as to the grounds which might be held to be reasonable, but it is suggested that they can include both the technical competence and financial ability of the proposed nominated sub-contractor.

(7) *Clause* 27 (a) ". . . *a sub-contract which provides . . .*"

For a sub-contract intended to accord with this proviso, see the standard form of sub-contract, printed *post*, p. 576.

(8) *Clause* 27 (a) (ii), (v), (vi) *Sub-contract period*

Reading sub-clauses (ii) and (v) together the result appears to be that the contractor can refuse to accept a nomination of a sub-contractor who is not prepared to undertake to complete the sub-contract works within such period as will enable the contractor to complete by the contract date subject to any existing extensions of time.[12] What if the contractor is apparently going to over-run the contract date or extended date for reasons which do not entitle him to an extension of time? The facts require consideration in each case but it is thought that he is not entitled to refuse a nomination on the grounds of the sub-contract period where such period will not, at the time of nomination, be a cause of a breach of the contractor's duties as to time or progress.

(9) *Clause* 27 (a) (vi) ". . . *and the Architect certifies in writing to the Contractor . . .*"

Clause 8 of the Standard form of sub-contract is such a clause as is contemplated by clause 27 (a) (vi). Clause 8 provides that the sub-contractor shall proceed with the sub-contract works with due expedition and that he shall complete the sub-contract works within the period or periods specified subject

[10] [1969] 1 All E.R. 977.
[11] Lord Justice Sachs, *ibid.* at p. 983.
[12] See " Terms of new sub-contract," *ante*, p. 370.

to extension of time. It further provides that the contractor is not entitled to claim any loss or damage under the clause unless the architect has issued a certificate stating that in his opinion the sub-contract works ought reasonably to have been completed within the specified period or periods or extensions thereof. Without such certificate the contractor cannot make out a good claim in damages for delay in completing the sub-contract works, and cannot therefore obtain a stay of the sub-contractor's proceedings for payment,[13] neither can he, it seems, set up such delay as an answer to the sub-contractor's claim for summary judgment.[14] It has been doubted whether the architect's certificate is a condition precedent to a right by the contractor to set up a cross-claim for damages for breach of the separate obligation to proceed with the sub-contract works with due expedition.[15]

See also clause 27 (d) (i) and (ii) which are the main contract provisions intended to correspond with the sub-contract. By providing that the contractor must not extend the sub-contract period without the written consent of the architect the intention is, presumably, effectively to keep the control of such extensions within the hands of the architect. What test does he apply? It is thought that the intention of clause 27 (a), particularly having regard to sub-clauses (ii) and (v) is that the grounds for extension will be those appearing in the main contract save only the sub-contractor's own default. This is the position under the Standard form of sub-contract. Presumably, therefore, the architect should approach his task of giving his certificate under clause 27 (d) (ii) in the same way as giving his certificate under clause 22. Any question whether the architect has or has not rightly or wrongly failed to certify in any respect does not arise under the sub-contract.[15a]

(10) *Clause* 27 (a) (vi) "... *as liquidated and ascertained damages* ..."

The Standard form of sub-contract does not provide for liquidated damages. There may be a substantial dispute between the contractor and the nominated sub-contractor as to the amount of the " sum equivalent to any loss or damage." It may therefore be difficult to give effect to clause 27 (d) (ii) of the main contract and clauses 8 (a), 11 (b) and 13 of the sub-contract which give the contractor the right to deduct damages for delay from sums certified for payment in favour of a nominated sub-contractor. (See also note 11, *infra*.) Note that clause 11 (e) of the standard form of sub-contract gives the sub-contractor the right to suspend work for non-payment by the contractor.

If the contractor and proposed nominated sub-contractor wish to agree upon liquidated damages the sum agreed should be a pre-estimate of the contractor's losses. This sum may be quite different from the employer's losses agreed as liquidated damages under the main contract.

For the contractor's right to an extension of time under the main contract for

---

[13] *BrightsideKilpatrick Engineering Services* v. *Mitchell Construction (1973) Ltd.* [1975] 2 Lloyd's Rep. 483 (C.A.), discussed *ante*, p. 242 in relation to arbitration.

[14] For a summary judgment see *ante*, p. 278.

[15] *Ellis Mechanical Construction Ltd.* v. *Wates Construction Ltd.*, *The Times*, Jan. 22, 1976 (C.A.), not reported on this point, but see transcript.

[15a] See footnote 13.

the delay of nominated sub-contractors and nominated suppliers, and for the contractor's losses see note 17 to clause 23.

(11) *Clause 27 (a) (vii) ". . . payment . . . shall be made within 14 days after receipt . . . of the Architect's certificate . . ."*

If this provision is a term of the sub-contract, receipt by the contractor of the architect's certificate for the value of the works, etc., is a condition precedent to the nominated sub-contractor's right to payment from the contractor.[16] A contractor who does not use the standard form of sub-contract but who wishes to make the architect's certificate a condition precedent to the sub-contractor's right to payment is advised to use clear words. Thus, in *Dunlop & Ranken Ltd.* v. *Hendall* the main contract was in the R.I.B.A. form.[17] The sub-contract said that the sub-contractor " shall observe and perform the conditions contained in the [main] contract . . . and this [sub-contract] shall be deemed to be supplemental thereto." The Divisional Court thought that this was not sufficient to make the equivalent of sub-clause (a) (vii) a term of the sub-contract. But the sub-contract went on to say: " Payment for this order is to be made . . . in accordance with the certificates and times provided in the [main] contract." This latter provision was held to have the effect of making the architect's certificate certifying the amount to be paid by the contractor to the sub-contractor a condition precedent to the sub-contractor's right to sue the contractor for payment.

Garnishee order. Judgment creditors of the sub-contractors in the *Dunlop & Ranken* case were held not entitled to a garnishee order against the main contractor in respect of money for work carried out by the sub-contractors until the architect's certificate had been given in respect of the money.[18]

Default by sub-contractor. It is in the employer's interest that the effect of clause 27 (a) (vii) should be incorporated in the sub-contract for it gives the architect a sanction against the default of the nominated sub-contractor. The architect refuses to certify in favour of the nominated sub-contractor who cannot then sue the main contractor because there is no certificate, and cannot sue the employer because there is no contract with the employer. If the nominated sub-contractor disputes the architect's view that he is in default his best course is, subject to clause 30 (7), to bring arbitration proceedings against the employer in the name of the contractor having given the contractor an indemnity as to costs. There is express provision for this course in the standard form of sub-contract.[19]

Contractor's right of set-off. Where under the terms of the sub-contract the sub-contractor has become entitled to payment there is no special rule applicable to building contracts which deprives the contractor of any right of set-off he may

---

[16] See clause 11 (b) of the Standard form of sub-contract; *Dunlop & Ranken Ltd.* v. *Hendal* [1957] 1 W.L.R. 1102, 1106 (D.C.).

[17] 1939 edition, but not, for this purpose materially different from the current form.

[18] The case was not followed in Alberta; *Sandy* v. *Yukon Construction Co. Ltd.* (1961) 26 D.L.R. 254. For garnishee orders and attachments of debts, see R.S.C., Ord. 49. For further cases on incorporation of terms of main contract see *ante*, p. 185.

[19] Clause 11 (d) and see also clause 12.

have.[20] But note in respect of claims for delay in completion of the sub-contract works the importance of the architect's certificate under clause 27 (d) (ii) of the main contract (*supra*, note 9). See further, the elaborate provisions of clauses 13 and 13 (a) of the Standard form of sub-contract introduced into the 1976 revision.

Equivalent clause in main contract. See clause 27 (b) where the architect has to direct the contractor as to the value of the nominated sub-contractor's work included in a certificate for payment issued under clause 30 of the main conditions. The contractor has to pay the sub-contractor within 14 days of receiving the duplicate copy of such certificate. For his right to be paid by the employer within 14 days of the issue of the certificate, see clause 30 (1).

(12) *Clause 27 (1) (a) (vii)* "... *a discount for cash* ..."
Any discount allowed by the nominated sub-contractor in addition to the $2\frac{1}{2}$ per cent. cash discount will go in reduction of the amount directed to be paid and is therefore for the benefit of the employer. Any reward for the contractor's services in relation to a nominated sub-contractor beyond the $2\frac{1}{2}$ per cent. must come out of his allowance for profit as shown in the contract bills (see clause 30 (5) (c)), or out of some other item in the contract bills such as that for attendance on nominated sub-contractors or out of his rates generally.

(13) *Clause 27 (a) (viii) Contractor's interest in retention to be* "... *fiduciary as trustee* ..."
This term appears in clause 11 (h) of the standard form of sub-contract. It has been held to operate as an equitable assignment to the sub-contractor, good as against the contractor's trustee in bankruptcy or liquidator, of that part of the retention money held by the employer under the main contract which is due to the sub-contractor under the sub-contract.[21] It is ineffective when the employer has properly spent the retention (see note 9 to clause 30).

(14) *Clause 27 (c) Right of direct payment to nominated sub-contractor*
Where after the contractor's insolvency the contract is continued (see clause 25 (2)), this right of direct payment is good as against the contractor's trustee in bankruptcy or liquidator (see note 10 to clause 25).

(15) *Clause 27 (c)* "... *such amounts* ..."
The right of direct payment [22] is limited to amounts previously certified but not paid, and arises only upon the issue of the architect's certificate under this sub-clause.

(16) *Clause 27 (d) (1)* " *The contractor shall not grant* ... "
If the contractor grants an extension in breach of this clause the architect can, it is thought, take the breach and its effect, if any, into account in

---

[20] *Modern Engineering (Bristol) Ltd.* v. *Gilbert-Ash (Northern) Ltd.* [1974] A.C. 689 (H.L.), overruling *Dawnays Ltd.* v. *Minter, F. G. Ltd.* [1971] 1 W.L.R. 1205 (C.A.). For further discussion see *ante*, p. 278, " Notes on summary judgment."
[21] *Re Tout & Finch* [1954] 1 W.L.R. 178. As to assignment of retention money generally see *ante*, p. 168.
[22] For direct payment clauses generally, see *ante*, p. 176.

considering any extension of time under clause 23 for the nominated sub-contractor's delay. (See generally clause 23 (g) and note 17 to clause 23.) Further, the architect is, it appears, released from the duty of granting a certificate under clause 27 (d) (ii).[23]

(17) *Clause 27 (e) Right to make early final payment to nominated sub-contractor*
This right is of particular significance where the nominated sub-contractor has completed his work at an early stage of the project. If the right is not exercised he will have to wait a disproportionate time for the release of the second part of the retention under the sub-contract. But by its release the contractor loses the security which retention affords. He is therefore entitled to be satisfactorily indemnified. What is satisfactory must depend upon the circumstances. In some cases a written indemnity by the nominated sub-contractor alone may be sufficient; in others some form of security [24] may be desirable, *e.g.* in the case of piling sub-contracts where losses due to latent defects can sometimes be enormous.

(18) *Clause 27 (f) " Neither the existence nor . . ."*
See also clause 27 (a) (". . . are hereby declared to be sub-contractors employed by the Contractor . . ."). A nominated sub-contractor is not party to the contract and clause 27 (f) negatives, it is submitted, any kind of trust on the part of the employer for the benefit of the nominated sub-contractor save as referred to in note 13 *supra.*

(19) *Clause 27 (g) " Provided that where a prime cost sum arises . . ."*
The effect is that where the architect wishes to nominate a sub-contractor to carry out the work the subject of a provisional sum [25] he must allow the contractor the benefit of clause 27 (g) (i).

Clause 28
Nominated suppliers
The following provisions of this Condition shall apply where prime cost sums are included in the Contract Bills, or arise as a result of Architect's/Supervising Officer's instructions given in regard to the expenditure of provisional sums, in respect of any materials or goods to be fixed by the Contractor.
(a) Such sums shall be deemed to include 5 per cent. cash discount and the term prime cost when included or arising as aforesaid shall be understood to mean the net cost to be defrayed as a prime cost after deducting any trade or other discount (except the said discount of 5 per cent.), and shall include any tax or duty not otherwise recoverable under this contract by whomsoever payable which is payable under or by virtue of any Act of Parliament on the import, purchase, sale, appropriation, processing, alteration, adapting for sale or use of the materials or goods to be supplied, and the cost of packing carriage and delivery. Provided that, where in the opinion of the Architect/ Supervising Officer the Contractor has incurred expense for special packing or special carriage, such special expense shall be allowed as part of the sums actually paid by the Contractor.

[23] For the importance of the certificate see note 9 *supra.*
[24] *Cf.* standard form of sub-contract, cl. 11 (g).
[25] See cl. 11 (3).

(b) Such sums shall be expended in favour of such persons as the Architect/
Supervising Officer shall instruct, and all specialists, merchants, tradesmen
or others who are nominated by the Architect/Supervising Officer to supply
materials or goods are hereby declared to be suppliers to the Contractor and
are referred to in these Conditions as ' nominated suppliers.' Provided that
the Architect/Supervising Officer shall not (save where the Architect/
Supervising Officer and the Contractor shall otherwise agree) nominate as a
supplier a person who will not enter into a contract of sale which provides
(*inter alia*):—

(i) That the materials or goods to be supplied shall be to the reasonable
satisfaction of the Architect/Supervising Officer.

(ii) That the nominated supplier shall make good by replacement or
otherwise any defects in the materials or goods supplied which appear
within such period as is therein mentioned and shall bear any expenses
reasonably incurred by the Contractor as a direct consequence of such
defects, provided that:—

(1) where the materials or goods have been used or fixed such defects
are not such that examination by the Contractor ought to have revealed
them before using or fixing;

(2) such defects are due solely to defective workmanship or materials
in the goods supplied and shall not have been caused by improper
storage by the Contractor or by misuse or by any act of neglect of
either the Contractor the Architect/Supervising Officer or the Employer or by any person or persons for whom they may be responsible.

(iii) That delivery of the materials or goods supplied shall be commenced and completed at such times as the Contractor may reasonably
direct.

(iv) That the nominated supplier shall allow the Contractor a discount
for cash of 5 per cent. if the Contractor makes payment in full within 30
days of the end of the month during which delivery is made.

(v) That the nominated supplier shall not be obliged to make any
delivery of materials or goods (except any which may have been paid for
in full less only the discount for cash) after the determination (for any
reason) of the Contractor's employment under this Contract.

(c) All payments by the Contractor for materials or goods supplied by a
nominated supplier shall be in full, and shall be paid within 30 days of the
end of the month during which delivery is made less only a discount for
cash of 5 per cent. if so paid.

(d) Where the said contract of sale between the Contractor and the nominated supplier in any way restricts, limits or excludes the liability of the
nominated supplier to the Contractor in respect of materials or goods
supplied or to be supplied, and the Architect/Supervising Officer has
specifically approved in writing the said restrictions, limitations or exclusions, the liability of the Contractor to the Employer in respect of the said
materials or goods shall be restricted, limited or excluded to the same extent.
The Contractor shall not be obliged to enter into a contract with, nor expend
prime cost sums in favour of, the nominated supplier until the Architect/
Supervising Officer has specifically approved in writing the said restrictions,
limitations or exclusions.

COMMENTARY

(1) *Generally*

Outline of clause. This clause should be read with clause 11 (3) (expenditure
of prime cost and provisional sums), clause 27 (nominated sub-contractors) and

clause 30 (5) (c) (settlement of accounts). Clause 28 deals with nominated suppliers and provides for: price (clause 28 (a)); terms of the contract which should be used with a nominated supplier (clause 28 (b)); payment to nominated suppliers (clause 28 (c)); approved limitation of contractor's liability (clause 28 (d)).

(2) *Comparison with sub-contractors nominated under clause 27*

There is a difference of subject-matter in that nominated suppliers do not fix. Further, the contract does not attempt such an elaborate regulation of matters affecting nominated suppliers as it does for nominated sub-contractors (see *infra*). But there are substantial similarities as regards the liability of the contractor for the default of the nominated supplier.

There is no right for the contractor to make reasonable objection to a nomination of a supplier; there is no equivalent to clause 27 (c) (direct payment); or 27 (d) (delay); or 27 (g) (tender by contractor); there is no equivalent to clauses 27 (b) and (e) because the contractor's duty to pay the nominated supplier is not conditional upon prior certification by the architect and a direction to pay.

(3) *Default of nominated supplier—contractor's liability*

The contractor's liability is broadly similar to that for nominated sub-contractors and reference should be made to note 2 to clause 27. This particularly applies to delay on the part of a nominated supplier, delay by a nominated supplier in giving information, fitness for purpose of the materials or goods to be supplied by the nominated supplier, design by the nominated supplier, guarantee by the nominated supplier. As to repudiation of his contract of supply by the nominated supplier there is no direct authority but reference should be made to *Bickerton* v. *N.W. Metropolitan Regional Hospital Board.*[26] It is thought that the reasoning applies to a nominated supplier so that upon repudiation there must be a re-nomination.

It seems that it is the contractor's duty to make a reasonable inspection of materials or goods delivered by a nominated supplier and that if he accepts the goods with defects which a reasonable inspection using proper skill and care would have disclosed he is liable to the employer in respect of such defects.[27]

Defects of quality. The contractor is liable, it is submitted, for defects in the quality of the materials or goods delivered even though a reasonable inspection by him at the time of delivery of the goods or materials would not have disclosed such defects of quality.[28] This submission is based upon general principles as to liability of a contractor for the supply of goods discussed *ante*, p. 39 and upon the effect of clause 28 (d). This clause, discussed further *infra*, was introduced in the July 1972 revision. It deals with the most important ground upon which the case of *Gloucestershire County Council* v. *Richardson* [29]

---

[26] [1970] 1 W.L.R. 607 (H.L.) discussed *ante*, p. 182.
[27] See *Young & Marten Ltd.* v. *McManus Childs Ltd.* [1969] 1 A.C. 454 (H.L.); *Gloucestershire County Council* v. *Richardson* [1969] 1 A.C. 480(H.L.), discussed *ante*, p. 39.
[28] *Ibid.*
[29] *Supra.*

was decided. Where the contract is in the pre-July 1972 version the position is not clear. It is thought that, on balance, the contractor is liable for latent defects in quality but there remains a doubt. This arises because the proviso to clause 28 (b) contemplates the nomination of a supplier who can impose some limitation of liability in his contract of supply.

Protection of Employer. The position is as for a nominated sub-contractor. (See note 2 to clause 27.)

(4) *The duty to nominate*
See note 3 to clause 27, reading it with the modifications made necessary by the different subject-matter.

(5) *Clause* 28 (a) " *5 per cent. cash discount* "
Reading this provision with clause 30 (5) (c), the result is the same in all ordinary cases as that in respect of the cash discount for nominated sub-contractors (see note 4 to clause 27).

(6) *Clause* 28 (b) " . . . *as* ' *nominated suppliers* ' . . ."
A nominated supplier is not party to the contract, and although there is no provision which corresponds with clause 27 (b), it is thought unlikely that a court would hold that the employer is in any way a trustee of any right which might be exercised for the benefit of the nominated supplier, *e.g.* the right of direct payment under clause 25 (4) (b).

(7) *Clause* 28 (b) " *Provided that the Architect shall not . . . nominate . . .*"
The parties must give effect to the proviso before the contract of sale is entered into. The provisions (i)–(v) can have no effect themselves upon any contract of sale. Compare clause 27 and note that under clause 28 the contractor is given no general right to make reasonable objection to the nomination.

(8) *Clause* 28 (c) " . . . *payments . . . in full . . .*"
As between himself and the employer the contractor has no right to deduct retention although payments to the contractor which include the value of materials or goods, fixed or unfixed, are subject to retention (clause 30 (3)).

(9) *Clause* 28 (d) *Exclusion clause in contract of supply*
This clause is very important. Its existence shows, it is submitted (see note 3, *supra*), that the contractor is under a liability for latent defects. Therefore he should read the offer of any proposed nominated supplier carefully. Such offers frequently contain exclusion clauses (see general discussion *ante*, p. 140).

If the contractor finds such a clause he can require the architect specifically to approve it in writing. If the architect gives such specific approval the contractor is protected by having the benefit of the exclusion clause. If the architect refuses to give the approval the contractor can reject the nomination. If the contractor does not follow this procedure and enters into a contract of supply containing an exclusion clause without referring it to the architect he is ordinarily, it is submitted (see *supra*), liable for losses suffered by the employer due to latent defects which appear even though he, the contractor, is subject to the exclusion clause as between himself and the nominated supplier.

Early nomination is desirable. If the architect is asked to approve an exclusion clause he can consult with the employer and discuss the risks involved. If the nomination was not early delay in consideration of the matter may give the contractor a claim for loss and/or expense or damages and may delay completion.

Clause 29
Artists and tradesmen

The Contractor shall permit the execution of work not forming part of this Contract by artists, tradesmen or others engaged by the Employer. Every such person shall for the purposes of clause 18 of these Conditions be deemed to be a person for whom the Employer is responsible and not to be a sub-contractor.

COMMENTARY

For delay caused by artists, tradesmen and others, see clauses 23 (h), 24 (1) (d) and 26 (1) (c) (vi).

Clause 30
Certificates and payments

(1) Interim valuations shall be made whenever the Architect/Supervising Officer considers them to be necessary for the purpose of ascertaining the amount to be stated as due in an Interim Certificate. The Architect/Supervising Officer shall from time to time as provided in this sub-clause issue Interim Certificates stating the amount due to the Contractor from the Employer, and the Contractor shall be entitled to payment therefor within 14 days from the issue of that Certificate. Before the issue of the Certificate of Practical Completion, Interim Certificates shall be issued at the Period of Interim Certificates specified in the Appendix to these Conditions. After the issue of the Certificate of Practical Completion, Interim Certificates shall be issued as and when further amounts are due to the Contractor from the Employer provided always that the Architect/Supervising Officer shall not be required to issue an Interim Certificate within one calendar month of having issued a previous Interim Certificate.

(2) The amount stated as due in an Interim Certificate shall, subject to any agreement between the parties as to stage payments, be the total value of the work properly executed and of the materials and goods delivered to or adjacent to the Works for use thereon up to and including a date not more than seven days before the date of the said certificate less any amount which may be retained by the Employer (as provided in sub-clause (3) of this Condition) and less any instalments previously paid under this Condition. Provided that such certificate shall only include the value of the said materials and goods as and from such times as they are reasonable, properly and not prematurely brought to or placed adjacent to the Works and then only if adequately protected against weather or other casualties.

(2A) The amount stated as due in an Interim Certificate may in the discretion of the Architect/Supervising Officer include the value of any materials or goods before delivery thereof to or adjacent to the Works provided that:
    (a) Such materials or goods are intended for inclusion in the Works;
    (b) Nothing remains to be done to such materials or goods to complete the same up to the point of their incorporation in the Works;

(c) Such materials or goods have been and are set apart at the premises where they have been manufactured or assembled or are stored, and have been clearly and visibly marked, individually or in sets, either by letters or figures or by reference to a pre-determined code, so as to identify:

(i) Where they are stored on premises of the Contractor, the Employer, and in any other case, the person to whose order they are held; and

(ii) Their destination as being the Works;

(d) Where such materials or goods were ordered from a supplier by the Contractor or a sub-contractor, the contract for their supply is in writing and expressly provides that the property therein shall pass unconditionally to the Contractor or the sub-contractor (as the case may be) not later than the happening of the events set out in paragraphs (b) and (c) of this sub-clause;

(e) Where such materials or goods were ordered from a supplier by a sub-contractor, the relevant sub-contract is in writing and expressly provides that on the property in such materials or goods passing to the sub-contractor the same shall immediately thereon pass to the Contractor;

(f) Where such materials or goods were manufactured or assembled by a sub-contractor, the sub-contract is in writing and expressly provides that the property in such materials or goods shall pass unconditionally to the Contractor not later than the happening of the events set out in paragraphs (b) and (c) of this sub-clause;

(g) The materials or goods are in accordance with this Contract;

(h) The Contractor furnishes to the Architect/Supervising Officer reasonable proof that the property in such materials or goods is in him and that the appropriate conditions set out in paragraphs (a) to (g) of this sub-clause have been complied with;

(i) The Contractor furnishes the Architect/Supervising Officer with reasonable proof that such materials or goods are insured against loss or damage for their full value under a policy of insurance protecting the interests of the Employer and the Contractor in respect of the contingencies referred to in clause 20 of these Conditions, during the period commencing with the transfer of property in such materials or goods to the Contractor until they are delivered to, or adjacent to, the Works.

(3) (a) In respect of any Interim Certificate issued before the issue of the Certificate of Practical Completion the Employer may, subject to paragraph (c) of this sub-clause, retain a percentage (in these Conditions called " the Retention Percentage ") of the total value of the work, materials and goods referred to in sub-clauses (2) and (2A) of this Condition.* The Retention Percentage shall be 5 per cent. unless a lower rate shall be agreed between the parties and specified in the Appendix to these Conditions as the Retention Percentage.

(b) If any Interim Certificate is issued after the issue of the Certificate of Practical Completion but before the issue of the Certificate for the residue of the amounts then so retained referred to in sub-clause (4) (c) of this Condition the Employer in respect of the said Interim Certificate may, subject to paragraph (c) of this sub-clause, retain a percentage (which percentage shall be equal to half the Retention Percentage) of the total value of the work, materials and goods referred to in sub-clauses (2) and (2A) of this Condition.

(c) The amount which the Employer may retain by virtue of paragraphs (a) and/or (b) of this sub-clause shall be reduced by the amounts of any releases of retention made to the Contractor in pursuance of clause 16 (f) and/or clause 27 (e) of these Conditions.

Where the Employer at tender stage estimates the Contract Sum to be £250,000 or over the Retention Percentage should be not more than 3 per cent.

(4) The amounts retained by virtue of sub-clause (3) of this Condition shall be subject to the following rules:—

(a) The Employer's interest in any amounts so retained shall be fiduciary as trustee for the Contractor (but without obligation to invest), and the Contractor's beneficial interest therein shall be subject only to the right of the Employer to have recourse thereto from time to time for payment of any amount which he is entitled under the provisions of this Contract to deduct from any sum due or to become due to the Contractor.

(b) On the issue of the Certificate of Practical Completion the Architect/Supervising Officer shall issue a certificate for one moiety of the total amounts then so retained and the Contractor shall be entitled to payment of the said moiety within 14 days from the issue of that certificate.

(c) On the expiration of the Defects Liability Period named in the Appendix to these Conditions, or on the issue of the Certificate of Completion of Making Good Defects, whichever is the later, the Architect/Supervising Officer shall issue a Certificate for the residue of the amounts then so retained and the Contractor shall be entitled to payment of the said residue within 14 days from the issue of that certificate.

(5) (a) The measurement and valuation of the Works shall be completed within the Period of Final Measurement and Valuation stated in the Appendix to the Conditions, and the Contractor shall be supplied with a copy of the priced Bills of Variation not later than the end of the said Period and before the issue of the Final Certificate under sub-clause (6) of this Condition.

(b) Either before or within a reasonable time after Practical Completion of the Works the Contractor shall send to the Architect/Supervising Officer all documents necessary for the purposes of the computations required by these Conditions including all documents relating to the accounts of nominated sub-contractors and nominated suppliers.

(c) In the settlement of accounts the amounts paid or payable under the appropriate contracts by the Contractor to nominated sub-contractors or nominated suppliers (including the discounts for cash mentioned in clauses 27 and 28 of these Conditions), the amounts paid or payable by virtue of clause 4 (2) of these Conditions in respect of fees or charges, the amounts paid or payable in respect of any insurances maintained in compliance with clause 19 (2) of these Conditions, the tender sum (or such other sum as is appropriate in accordance with the terms of the tender) for any work for which a tender made under clause 27 (g) of these Conditions is accepted and the value of any work executed by the Contractor for which a provisional sum is included in the Contract Bills shall be set against the relevant prime cost or provisional sum mentioned in the Contract Bills or arising under Architect's/Supervising Officer's instructions issued under clause 11 (3) of these Conditions as the case may be, and the balance, after allowing in all cases *pro rata* for the Contractor's profit at the rates shown in the Contract Bills, shall be added to or deducted from the Contract Sum. Any amount or value to be so set against the relevant prime cost or provisional sum shall be exclusive of value added tax. Provided that no deduction shall be made in respect of any damages paid or allowed to the Contractor by any sub-contractor or supplier.

(6) So soon as is practicable but before the expiration of the period the length of which is stated in the Appendix to these Conditions from the end of the Defects Liability Period also stated in the said Appendix or from completion of making good defects under clause 15 of these Conditions or from receipt by the Architect/Supervising Officer of the documents referred to in paragraph (b) of sub-clause (5) of this Condition, whichever is the latest, the Architect/Supervising Officer shall issue the Final Certificate. The Final Certificate shall state:—

(a) The sum of the amounts already paid to the Contractor under Interim Certificates and Certificates issued under sub-clauses (4) (b) and (4) (c) of this Condition, and

(b) The Contract Sum adjusted as necessary in accordance with the terms of these Conditions,

and the difference (if any) between the two sums shall be expressed in the said certificate as a balance due to the Contractor from the Employer or to the Employer from the Contractor as the case may be, and subject to any deductions authorised by these Conditions, the said balance shall as from the fourteenth day after the issue of the said certificate be a debt payable as the case may be by the Employer to the Contractor or by the Contractor to the Employer.

(7) (a) Except as provided in paragraphs (b) and (c) of this sub-clause (and save in respect of fraud), the Final Certificate shall have effect in any proceedings arising out of or in connection with this Contract (whether by arbitration under clause 35 of these Conditions or otherwise) as

(i) conclusive evidence that where the quality of materials or the standards of workmanship are to be to the reasonable satisfaction of the Architect/Supervising Officer the same are to such satisfaction, and

(ii) conclusive evidence that any necessary effect has been given to all the terms of this Contract which require an adjustment to be made of the Contract Sum save where there has been any accidental inclusion or exclusion of any work, materials, goods or figure in any computation or any arithmetical error in any computation, in which event the Final Certificate shall have effect as conclusive evidence as to all other computations.

(b) If any arbitration or other proceedings have been commenced by either party before the Final Certificate has been issued the Final Certificate shall have effect as conclusive evidence as provided in paragraph (a) of this sub-clause after either

(i) such proceedings have been concluded, whereupon the Final Certificate shall be subject to the terms of any award or judgment in or settlement of such proceedings, or

(ii) a period of twelve months during which neither party has taken any further step in such proceedings, whereupon the Final Certificate shall be subject to any terms agreed in partial settlement,

whichever shall be the earlier.

(c) If any arbitration or other proceedings have been commenced by either party within 14 days after the Final Certificate has been issued, the Final Certificate shall have effect as conclusive evidence as provided in paragraph (a) of this sub-clause save only in respect of all matters to which those proceedings relate.

(8) Save as aforesaid no certificate of the Architect/Supervising Officer shall of itself be conclusive evidence that any works materials or goods to which it relates are in accordance with this Contract.

Clause 30B

Finance (No. 2) Act 1975—Statutory Tax Deduction Scheme

Definitions

30B (1) In this Condition ' the Act ' means the Finance (No. 2) Act 1975; ' the Regulations ' means the Income Tax (Sub-Contractors in the Construction Industry) Regulations 1975 S.I. No. 1960; " ' contractor ' " means a person who is a contractor for the purposes of the Act and the Regulations; ' evidence ' means such evidence as is required by the Regulations to be produced to a ' contractor ' for the verification of a ' sub-contractor's ' tax certificate;

' statutory deduction ' means the deduction referred to in section 69 (4) of the
Act or such other deduction as may be in force at the relevant time; " ' sub-
contractor ' " means a person who is a sub-contractor for the purposes of the
Act and the Regulations; ' tax certificate ' is a certificate issuable under section 70
of the Act.

Whether Employer a ' contractor '

(2) (a) At the date of tender as defined in Clause 31D (6) (a) of these Conditions (or as defined in the Formula Rules where Clause 31F of these
Conditions applies) the Employer was a ' contractor '/was not a
' contractor '* for the purposes of the Act and the Regulations. Sub-
clauses (3) to (9) of this Condition shall not apply, if in this paragraph,
the Employer is stated not to be a ' contractor.'

(b) If in paragraph (a) hereof the words " was a ' contractor ' " are deleted
nevertheless if, at any time up to the issue and payment of the Final
Certificate, the Employer becomes such a ' contractor ', the Employer
shall so inform the Contractor and the provisions of this Condition
shall immediately thereupon become operative.

Provision of evidence—tax certificate

(3) (a) Not later than 21 days before the first payment under this Contract is
due to the Contractor or after sub-clause (2) (b) of this Condition has
become operative the Contractor shall:
either
(i) provide the Employer with the evidence that the Contractor is
entitled to be paid without the statutory deduction; or
(ii) inform the Employer in writing, and send a duplicate copy to the
Architect/Supervising Officer,† that he is not entitled to be paid without
the statutory deduction.

(b) If the Employer is not satisfied with the validity of the evidence
submitted in accordance with paragraph (a) (i) hereof, he shall within
14 days of the Contractor submitting such evidence notify the Contractor in writing that he intends to make the statutory deduction from
payments due under this Contract to the Contractor who is a ' sub-
contractor ' and give his reasons for that decision. The Employer shall
at the same time comply with sub-clause (6) (a) of this Condition.

Uncertificated Contractor obtains tax certificate

(4) (a) Where sub-clause (3) (a) (ii) applies, the Contractor shall immediately
inform the Employer if he obtains a tax certificate and thereupon sub-
clause (3) (a) (i) shall apply.

Expiry of tax certificate

(b) If the period for which the tax certificate has been issued to the
Contractor expires before the final payment is made to the Contractor
under this Contract the Contractor shall, not later than 28 days
before the date of expiry:
either
(i) provide the Employer with evidence that the Contractor from the
said date of expiry is entitled to be paid for a further period without the

---

* Strike out as applicable.
*† Delete '/Supervising Officer' where this amendment is incorporated in a Private
Edition.*

statutory deduction in which case the provisions of sub-clause (3) (b) hereof shall apply if the Employer is not satisfied with the evidence or

(ii) inform the Employer in writing that he will not be entitled to be paid without the statutory deduction after the said date of expiry.

Cancellation of tax certificate

(c) The Contractor shall immediately inform the Employer in writing if his current tax certificate is cancelled and give the date of such cancellation.

Vouchers

(5) The Employer shall, as a ' contractor ' in accordance with the Regulations, send promptly to the Inland Revenue any voucher which, in compliance with the Contractor's obligations as a ' sub-contractor ' under the Regulations, the Contractor gives to the Employer.

Statutory deduction—direct cost of materials

(6) (a) If at any time the Employer is of the opinion (whether because of the information given under sub-clause (3) (a) (ii) of this Condition or of the expiry or cancellation of the Contractor's tax certificate or otherwise) that he will be required by the Act to make a statutory deduction from any payment due to be made the Employer shall immediately so notify the Contractor in writing and require the Contractor to state not later than 7 days before each future payment becomes due (or within 10 days of such notification if that is later) the amount to be included in such payment which represents the direct cost to the Contractor and any other person of materials used or to be used in carrying out the Works.

(b) Where the Contractor complies with paragraph (a) of this sub-clause he shall indemnify the Employer against loss or expense caused to the Employer by any incorrect statement of the amount of direct cost referred to in that paragraph.

(c) Where the Contractor does not comply with paragraph (a) of this sub-clause the Employer shall be entitled to make a fair estimate of the amount of direct cost referred to in that paragraph.

Correction of errors

(7) Where any error or omission has occurred in calculating or making the statutory deduction the Employer shall correct that error or omission by repayment to, or by deduction from payments to, the Contractor as the case may be subject only to any statutory obligation on the Employer not to make such correction.

Relation to other Conditions

(8) If compliance with this Condition involves the Employer or the Contractor in not complying with any other of these Conditions, then the provisions of this Condition shall prevail.

Application of Arbitration Agreement

(9) The provisions of Clause 35 of these Conditions (arbitration) shall apply to any dispute or difference between the Employer or the Architect/Supervising

Officer† on his behalf and the Contractor as to the operation of this Condition except where the Act or the Regulations or any other Act of Parliament or statutory instrument rule or order made under an Act of Parliament provide for some other method of resolving such dispute or difference.

†*Delete '/Supervising Officer' where this amendment is incorporated in a Private Edition.*

COMMENTARY

(1) *Generally*

Outline of clause. This clause provides for: issue of interim certificates (clause 30 (1)); amount of interim certificates (clause 30 (2)); certification of materials or goods before delivery to the works (clause 30 (2A)); right to retention (clause 30 (3)); nature of retention (clause 30 (4) (a)); first moiety of retention (clause 30 (4) (b)); residue of retention (clause 30 (4) (c)); period of final measurement and valuation (clause 30 (5) (a)); documents necessary for the final account (clause 30 (5) (b)); settlement of accounts in respect of prime cost and provisional sums and the like (clause 30 (5) (c)); issue of and contents of the final certificate (clause 30 (6)); effect of final certificate (clause 30 (7)); effect of certificates (clause 30 (8)); statutory tax deduction scheme (clause 30B). Clause 30A, counter inflation, is not printed.

For the position of the architect [30] as agent and his duty to act fairly and professionally see notes to clauses 1 and 2, and for his relationship to the quantity surveyor see notes to clause 11.

(2) *Amendments*

April 1966. Clause 30 (2A) and clause 14 (2).

July 1968. Amendments to appendix and to clauses 30 (1), 30 (4) (b) and 30 (4) (c) to delete previous reference in the appendix for a period for honouring certificates so that it became in all cases 14 days.

July 1972. Clause 30 (1) amended to add current last sentence thereof. Previously it had been a matter of doubt whether the architect was empowered to issue certificates for payment, other than for release of retention, between the issue of the certificate of practical completion and the final certificate, although in practice many architects issued such certificates. Clause 30 (3) amended to provide for a fixed retention percentage. Previously there was a percentage up to a limit of retention after which the right to retention ceased. Clause 30 (6) amended to take account of the abolition of the limit of retention.

July 1973. Clause 30 (5) (c), penultimate sentence inserted to deal with value added tax.

July 1973. Two important amendments were made to clause 30 (7). In *Hosier & Dickinson* v. *Kaye Ltd.*[31] the House of Lords, on the previous wording, had

---

[30] For a discussion of the architect's certifying duties under the 1939 ed., pre-1957 revision, which is of some assistance on some of his duties under this form, see *East Ham Borough Council* v. *Bernard Sunley & Sons Ltd.* [1966] A.C. 406 (H.L.).
[31] [1972] 1 W.L.R. 146.

held that, where no effective written request to concur in the appointment of an arbitrator had been made, a final certificate was conclusive as to defects even though at the time of its issue High Court proceedings had been commenced relating to those defects. The first part of the 1973 amendment (" any proceedings . . . shall have been commenced . . .") overruled this decision. The second amendment was to give either party the right within 14 days of the issue of the final certificate to challenge it by going to arbitration. Previously only the contractor had such a right.

July 1975. This introduced a drafting amendment to clause 30 (5) (c) consequent upon the introduction of the new clause 4 relating to statutory undertakers. Further, a new, optional, price fluctuations clause was provided, clause 31F, providing for adjustment of the contract sum by the NEDO price adjustment formula for building contracts. Where this optional clause is used it effects by clause 31F (2) (a) various amendments to the contract, in particular to clause 30.

July 1976. This introduced substantial amendments to the clause dealing with the conclusive effect of the final certificate for the effect of which reference must be made to the new sub-clause (7) and to notes 14 and 15 *infra*.

July 1977. Statutory deduction scheme (clause 30B).

(3) *Clause* 30 (1) ". . . *and the Contractor shall be entitled to payment therefor* . . ."
A certificate is a condition precedent to payment under the contract.[32] But the contractor can, it is submitted, recover payment without a certificate [33] if: (i) the employer waives the requirement of a certificate; (ii) the architect is guilty of such conduct as to have become disqualified [34]; (iii) the employer prevents the issue of a certificate. " Prevents " here includes the case where the architect wrongly neglects, or deliberately, as for example, a result of a mistaken view of his powers, refuses, to issue a certificate and the employer concurs in his action and the contractor has done everything necessary for the issue of the certificate [35]; (iv) the architect ceases to act and the employer fails to appoint a new architect (see clause 3 of the Articles of Agreement); (v) if the arbitrator by his award has directed a sum of money to be paid to the contractor. A dispute whether a certificate has been improperly withheld or is not in accordance with the conditions can go to immediate arbitration.[36]

Private version. The period for payment runs from the contractor's presentation of the certificate to the employer. This applies to all certificates for payment.

---

[32] See clause 2 of the Articles of Agreement and for some assistance *Dunlop & Ranken Ltd.* v. *Hendall* [1957] 1 W.L.R. 1102, 1106 (D.C.).
[33] For a general discussion of recovery without a certificate, see *ante*, p. 78.
[34] See *ante*, p. 85.
[35] *Panamena, etc.* v. *Leyland & Co.* [1947] A.C. 428 (H.L.) discussed *ante*, p. 78, *cf.* clause 26 (1) (b).
[36] Clause 35 (2); but note that in practice a long period often elapses between commencement of proceedings and the award unless the parties agree to dispense with evidence and a formal hearing. Such an agreement may be particularly appropriate to a dispute as to valuation only.

(4) *Clause* 30 (2) ". . . *the total value* . . ."

The amount certified should take into account adjustments for variations (clause 11 (5)), loss and expense ascertained under clauses 11 (6), 24 (1) and 34 (2), and price fluctuations (clause 31D (4) or clause 31F) but, it is submitted, the certificate should not take into account sums which the employer may be entitled to deduct for, *e.g.* liquidated damages (clause 22) or direct payment to a nominated sub-contractor (clause 27 (c)).[37] In practice a convenient course for the architect who knows that the employer is entitled to make deductions from sums certified, is to include a covering letter with his certificate and copy certificate to the parties (clause 3 (8)) setting out what he considers to be the amount of the deductions.

Summary judgment on certificate. See *ante*, p. 278.

VAT. See clause 13A and the supplementary agreement *post*, p. 416.

(5) *Clause* 30 (2) ". . . *properly executed* . . ."

The architect can and should take into account whether the work is properly executed in arriving at the amount of his certificate. For the meaning and use of the term " reasonable satisfaction " and the test to be applied by the architect when certifying, see note 3 to clauses 1 and 2.

(6) *Clause* 30 (2) ". . . *delivered to or adjacent to the Works* . . ."

For the passing of property in unfixed materials which have been paid for, see clause 14 (1).

(7) *Clause* 30 (2A) " *The amount . . . may in the discretion of the Architect . . . include the value of any materials or goods before delivery* . . ."

The architect may be asked to exercise his discretion by the contractor where, *e.g.* he supplies the goods or materials from his own factory, or by a sub-contractor or supplier. In the latter cases the contractor may ask the architect *not* to exercise his discretion (see *infra*). The architect should, it is suggested, particularly have in mind whether the contractor and any sub-contractor or supplier involved are likely to remain solvent and not in default, and whether the detailed provisions (a) to (i) have been complied with. These provisions have been drafted in an attempt to deal with the problems which can arise. They should be read with clause 14 (2). Below appear notes on some of the problems.

Passing of property. The property cannot pass from the contractor to the employer (see clause 14 (2)) unless it is already vested in the contractor, hence sub-clauses (d), (e), (f) and (h).

Bankruptcy of contractor. Materials or goods upon his premises, although certified and paid for, can, where he was the reputed owner thereof, vest, upon his bankruptcy, in his trustee in bankruptcy.[38] Sub-clause (c) is particularly

[37] Other rights to make deductions arise under clauses 2 (1), 19 (1) (c) and 20 [A] (1).
[38] *Re Fox, ex p. The Oundle & Thrapston R.D.C.* [1948] Ch. 407 (D.C.). For full discussion and facts of *Re Fox*, see *ante*, pp. 191–193.

aimed at the bankruptcy of the contractor or any other person on whose premises the goods are stored.[39]

Liquidation of insolvent contractor. The doctrine of reputed ownership [40] does not apply, but the liquidator takes into his possession or control the property to which the company is or appears to be entitled [41] and there could be delay until he becomes satisfied that goods or materials apparently the property of the company are in fact the property of the employer. Again, see sub-clause (c).

Transfer of title to other persons. A person buying in good faith and without notice of the previous passing of the property to the employer, from a supplier (including in the appropriate case the contractor) who is left in possession of the goods or materials, can take a good title.[42] A bona fide purchaser from the sheriff selling under a writ of *fi. fa.* and where no claim has been made can acquire a good title.[43] In certain circumstances a landlord can distrain on the materials or goods for unpaid rent.[44] Sub-clause (c) in particular is aimed at these matters.

Defects after certification. It is submitted that nothing in the sub-clause (*e.g.* see sub-clauses (b) and (g)) affects the architect's powers under clause 6 (4) or otherwise.

The contractor's objections (see *supra*). The materials and goods remain at his risk (clause 14 (2)). If they are lost or damaged or by fraud or negligence disposed of by a sub-contractor or supplier the contractor is liable to replace them at his own cost. Further, he cannot recover any loss from the employer due to any delay which might thereby arise and he may have to pay liquidated damages for such delay. If the architect intends to certify at the request of a supplier or sub-contractor it is in the contractor's interest to see that provisions (a) to (i) are enforced and that there is adequate insurance cover.

(8) *Clause* 30 (3) (a) "*. . . may retain a percentage . . .*"
Retention may, it is submitted, be applied to the value of the work, materials and goods adjusted by the valuation of variations under clause 11 (4), and to fluctuations where clause 31F applies but not to loss or expense ascertained under clauses 11 (6), 24 (1) and 34 (2) nor to fluctuations where clause 31D applies.[45]

(9) *Clause* 30 (4) (a) "*. . . fiduciary as trustee . . .*"
The effect is, it is submitted, that upon the employer's bankruptcy or liquidation as an insolvent company, as the case may be, the trustee in bankruptcy or liquidator holds the balance, if any, of the retention, if any, as trustee for the

---

[39] Would the hypothetical inquirer (see *Re Fox*) understand the marking referred to in sub-clause (c)? It might be better to have a plain statement, *e.g.* " SOLD, the Property of . . . (the Employer) for use at . . . (the Works)."
[40] Companies Act 1948, ss. 243, 244.
[41] *Ibid.*
[42] See Sale of Goods Act 1893, s. 25 (1). Much law is involved and reference should be made to the standard textbooks, *e.g.* Chalmers; Benjamin.
[43] See Bankruptcy and Deeds of Arrangement Act 1913, s. 15.
[44] See Woodfall, *Landlord and Tenant* (27th ed.), Chap. 8.
[45] See the respective clauses referred to.

contractor, *i.e.* it must be paid over to the contractor.[46] But there must have been created and set aside an identifiable fund. In the absence of such a fund the sub-clause does not, it is submitted, achieve its purpose. The matter is of little importance in contracts with public authorities, but it is suggested that in contracts with other employers a contractor who wishes to have the benefit of this sub-clause obtains an additional term providing for the setting up of a fund in the joint names of the employer and himself (*cf.* clause 24 (d) [A] of the 1939 form), breach of which term is a ground for determination.

(10) *Clause* 30 (4) (a) ". . . *recourse . . . for payment . . . which he is entitled . . . to deduct from any sum due . . .*"

Deductions from sums certified can be made under clauses 2 (1), 19 (1) (b), 20 [A] (1), 22, 25 (4) (b), 27 (c). If the employer determines the contractor's employment under clause 25 the parties' rights become subject to clause 25 (4) (d) the effect being, it is suggested, that the trust operates (if at all—see last note) only in respect of any retention remaining [47] after taking the account under clause 25 (4) (d).

(11) *Clause* 30 (4) (b) " *On the issue of the Certificate of Practical Completion . . .*"
For this certificate see clause 15 and notes thereto.

Practical completion to final certificate. There is a detailed but clear procedure to be carried out which may be conveniently summarised as follows:

(i) Certificate of practical completion (clause 15 (1)), and at the same time,

(ii) certificate for first moiety (half) of retention (clause 30 (4) (b)),

(iii) period of final measurement and valuation runs from certificate of practical completion and if none stated is six months; contractor must be supplied with priced bills of variations within period and before the issue of the final certificate (clause 30 (5) (a), appendix),

(iv) before or within reasonable time after practical completion contractor must send all documents necessary for the final account (clause 30 (5) (b)),

(v) defects liability period begins and if none stated is six months from certificate of practical completion (clause 15 (2), appendix),

(1) schedule of defects to be delivered not later than fourteen days after expiration of defects liability period (clause 15 (2)),

(2) particular defects may be required to be made good before such schedule or such fourteen days, whichever is the earlier (clause 15 (3)),

(3) certificate of completion of making good defects when in opinion of the architect they have been made good (clause 15 (4)),

(vi) certificate for residue of retention on expiration of defects liability period or certificate of completion of making good defects whichever is the later (clause 30 (4) (c)). Note: it is contemplated that the contractor at this stage will have been paid substantially in full, and the period thereafter is for final adjustment of accounts one way or the other,

---

[46] The words of the clause are taken from clause 11 (h) of the Standard form of sub-contract. See *Re Tout & Finch* [1954] 1 W.L.R. 178, where the contractor was held to be trustee of money to be paid by the employer, *i.e.* there was a fund available.

[47] Which in most cases is unlikely.

(vii) final certificate so soon as is practicable but before the expiration of a period which, if none is stated, is three months from the latest of: end of defects liability period; completion of making good defects under clause 15; receipt of necessary documents from contractor (clause 30 (6)), (in private version there is no provision for a period other than three months),

(viii) there is a fourteen-day period from the issue of the final certificate for either party to challenge the final certificate by arbitration, or other proceedings, for the effect of which see clause 30 (7) (c) and note 14 *infra*.

(12) *Clause* 30 (6) (a), (b) ". . . *The Final Certificate shall state . . . The Contract Sum . . . adjusted as necessary . . ."*
  i.e. under clauses 4 (2), 6 (3), 7, 11 (5), 11 (6), 12 (2), 15 (2), 20 [A] (2), 20 [B], 20 [C], 24 (1), 30 (5) (c), 31 D (4) or 31 F, 32, 33, 34 (2). This is to be distinguished from the right to make deductions under clauses 2 (1), 19 (1) (b), 20 [A] (1), 22, 25 (4) (b), 27 (c).

(13) *Clause* 30 (6) (a), (b) ". . . *or to the Employer . . ."*
  See note 11 (vi), *supra*. It is a useful feature of this contract that the final certificate can be for the return of moneys overpaid in earlier certificates. The architect is not entitled to retain a sum of money merely to ensure that there is something held in favour of the employer until the final certificate. If he does so the contractor can, it is submitted, not only go to immediate arbitration but sue in a court of law, for the employer cannot rely on the absence of a certificate as an excuse for retaining the money.[48]

(14) *Clause* 30 (7) *Conclusive effect of final certificate*
  This sub-clause was substantially re-written in the 1976 Revision but it retains certain features common to the various versions of the standard form of building contract since 1957. These are: the conclusive effect of the final certificate is excluded or limited if the parties take certain steps; the final certificate is not conclusive in respect of every matter which can arise out of the contract.

  Matters upon which final certificate is conclusive. These are set out in clause 30 (7) (a) (i) and (ii). As to clause 30 (7) (a) (i) see clause 1 (1). This final certificate differs from that issued under pre-1976 versions. Its conclusive effect as regards quality of materials and standards of work applies only where those matters are to be to the reasonable satisfaction of the architect (see note 3 to clauses 1 and 2) whereas before it applied to the works generally. There are no longer exceptions as to fraudulent concealment or defects which reasonable inspection would have disclosed. The term " conclusive evidence " means, it is submitted, that no evidence may be called to contradict, or qualify in any way, the architect's decision as expressed in the certificate.[49] Upon matters with which it deals it is therefore as binding upon the parties on questions of fact as an arbitrator's award.[50] It is not, however, an award[51] and is not subject to the

[48] *Panamena, etc.* v. *Leyland* [1947] A.C. 428 and see generally, *ante*, p. 78.
[49] *Cf. Kerr* v. *John Mottram Ltd.* [1940] 1 Ch. 657, 660; *Kaye (P. & M.) Ltd.* v. *Hosier & Dickinson Ltd.* [1972] 1 W.L.R. 146, 169 (H.L.).
[50] See *Goodyear* v. *Weymouth Corp.* (1865) 35 L.J.C.P. 12, 17.
[51] *Sutcliffe* v. *Thackrah* [1974] A.C. 727 (H.L.).

provisions of the Arbitration Act 1950.[52] For architect's certificates generally, see Chapter 6.

Steps which limit conclusiveness of certificate. If no arbitration or other proceedings have been commenced by either party before the final certificate is issued (clause 30 (7) (b)) or within 14 days after its issue (clause 30 (7) (c)) the final certificate has the full conclusive effect given to it by clause 30 (7) (a).

If arbitration or other proceedings were commenced by either party before its issue then its conclusive effect is subject to clause 30 (7) (b) (i) or (ii). If clause 30 (7) (b) (i) applies the certificate remains conclusive subject only " to the terms of any award or judgment in or settlement of such proceedings." It will be most desirable that such award, judgment or settlement is a " speaking " [53] award, judgment or settlement in order to determine the remaining ambit of the conclusiveness of the final certificate.

If clause 30 (7) (b) (ii) applies then if there is " a period of twelve months during which neither party has taken any further steps in such proceedings " prima facie the full conclusive effect of the final certificate revives " subject to any terms agreed in partial settlement." The latter words are not easy to construe. Ordinarily one thinks of a matter as either being settled or not settled.[54] Perhaps what is intended is a reference to an agreement which does not dispose of the whole dispute but of some or all of the matters dealt with in the final certificate and that in respect of those matters the final certificate ceases to have any effect.

Proceedings commenced within 14 days after the final certificate. As to this see clause 30 (7) (c). The final certificate has its conclusive effect " save only in respect of all matters to which those proceedings relate." The document by which such proceedings are commenced will therefore be of the greatest importance. If sufficiently widely drawn it can deprive the final certificate of all conclusive effect. Proceedings are commenced in the High Court by the issue of a writ and by arbitration by a written request to concur in the appointment of an arbitrator (see clause 35 (1)). In the latter case it will be particularly important to define the dispute which it is requested that he should deal with in terms wide enough to prevent the claimant subsequently being hampered by the conclusive effect of the certificate. Presumably if proceedings have been commenced upon a somewhat narrow ground before the certificate is issued there is nothing to prevent either party commencing further proceedings upon wider grounds within 14 days after the issue of the final certificate.

Powers of the court. For the question of whether a court has the same powers as the arbitrator, see note 1 to clause 35.

Request to concur in the appointment of an arbitrator. A party who wishes to rely upon arbitration proceedings to defeat or attack a final certificate

---

[52] See generally, Chap. 13.

[53] *i.e.* it " speaks " on its face as to what it deals with. In the case of an award see further, *ante,* p. 247.

[54] For settlement of actions see *ante,* p. 258.

should, preferably, follow the words of clause 35 (1) closely so as to show that he is commencing arbitration proceedings and not merely threatening to commence them in the future. However, it seems that if he uses less formal language than that used by the contract but it is such as to indicate a clear requirement to refer the dispute to arbitration this will be sufficient.[55]

(15) *Attacking a final certificate*

A party wishing to attack an allegation that a certificate is conclusive evidence upon certain matters within the meaning of clause 30 (7) (a) should consider one or more of the following:

(a) Commencement of proceedings before or within 14 days after the issue of the final certificate. See note 14 *supra*.

(b) Whether the matters fall within one of the express exceptions to the conclusive effect of the certificate. Fraud is an exception. Another express exception is in clause 30 (7) (a) (ii) beginning " any accidental inclusion or exclusion. . . ."

(c) Whether the point at issue is not within the range of matters upon which the certificate is stated to be conclusive evidence. Thus, for example, it is submitted that the certificate is not conclusive evidence that the works ought reasonably to have been completed by the date in the architect's certificate, if given, under clause 22 or that the extension of time made by the architect under clause 23 is fair and reasonable.[56] Further, claims by the contractor for damages for delay in the issue of drawings and instruction (see note 3 to clause 3), and by the employer for consequential losses (see *infra*) are not within the certificate. For materials and workmanship not required to be to the architect's reasonable satisfaction, see note 14, *supra*.

In *Hosier & Dickinson Ltd.* v. *Kaye (P. & M.) Ltd.*[57] a point arose in the House of Lords which the majority of their Lordships refused to consider because it had not been argued in the courts below. Consequential loss, *i.e.* loss of profits due to defects in a warehouse, had been suffered by the employer due to defects which appeared after practical completion but before the issue of the final certificate (pre-1976 version). Two meanings were put forward as to the effect of the certificate. The first was that the whole series of building operations from beginning to end must have been deemed to have been duly carried out and completed so that any claim in respect of alleged past defects and their consequences was excluded. The second meaning was that everything which had to be done by way of building operations had now been done and all defects made good but there was no exclusion of claims in respect of alleged past defects and their consequences. Lord Diplock, contrary to the views of the majority, found it necessary to decide the point. He was in favour of the second meaning. Respectfully, it is submitted that he is correct and that an employer in such circumstances is entitled to recover consequential loss and that this remains the position under

---

[55] See *Nea Agrex S.A.* v. *Baltic Shipping* [1976] Q.B. 933 (C.A.); *cf. Surrendra Overseas Ltd.* v. *Government of Sri Lanka* [1977] 1 W.L.R. 565.

[56] Thus there is apparently a conflict. The architect's decision under clause 23 (e), (f) and (i) can be challenged, but not his corresponding decision under clauses 24 (1) (a), (b), (c) and (e).

[57] [1972] 1 W.L.R. 146 (H.L.).

the current, July 1977, revision of the form, even where the final certificate is conclusive as to the relevant materials or workmanship.[58]

(d) That there was some irregularity in the giving of the certificate so that it cannot be said to be the certificate required by the contract, in which case it will not be conclusive at all.[59] An example might be where the architect delegated his whole function of certifying.[60] It is submitted that mere delay in issuing the final certificate does not prevent it when issued from being conclusive.

(e) That the architect was disqualified at the time when he gave his certificate, in which case the certificate is of no effect.[61]

(16) *Emergency amendment No. 1/1975: additional clause* 30A

In December 1975 the JCT issued an emergency amendment together with an explanatory note none of which is printed. It was occasioned by then current counter-inflation legislation designed to prevent the recovery by an enterprise (including contractors) of remuneration (as defined) in excess of certain limits. Public sector employers were asked to provide for certificates to the effect that excess remuneration was not being paid. The amendment is for use only on firm price contracts (*i.e.* where there is no fluctuations clause or where clause 31B C, D and E apply) where it is anticipated that the contract sum will be more than £50,000 and on contracts let on a price fluctuations basis (*i.e.* where clauses 31A, C, D and E or 31F apply) where it is anticipated that the contract sum will be more than £5,000. The explanatory note states that it is not intended to incorporate the amendment in future revisions of the Standard form of building contract.

Clause 31A*
Fluctuations

The Contract Sum shall be deemed to have been calculated in the manner set out below and shall be subject to adjustment in the events specified hereunder:—

    (a) (i) The prices (including the cost of employer's liability insurance and of third party insurance) contained in the Contract Bills are based upon the rates of wages and the other emoluments and expenses (including holiday credits) which will be payable by the Contractor to or in respect of workpeople engaged upon or in connection with the Works in accordance with the rules or decisions of the National Joint Council for the Building Industry and the terms of the Building and Civil Engineering Annual and Public Holidays Agreements which will be applicable to the Works and which have been promulgated at the date of tender, or in the case of workpeople so engaged whose rates of wages and other emoluments and expenses (including holiday credits) are governed by the rules or decisions or agreements of some body other than the National Joint Council for the Building Industry, in accordance with the rules or decisions or agreements of such other body which will be applicable and which have been promulgated as aforesaid.

---

[58] The final certificate in issue was based on clause 30 (7) of the pre-1973 Revision which appears in the report and at p. 392 of the 3rd edition of this book.

[59] See generally *ante*, pp. 83, 84.

[60] *Cf. Clemence* v. *Clarke* (1880) H.B.C. (4th ed.), Vol. 2, pp. 54, 59 (C.A.); *Burden* v. *Swansea Corporation* [1957] 1 W.L.R. 1167, 1173 (H.L.), cited *ante*, p. 362.

[61] See generally, *ante*, p. 85.

(ii) If any of the said rates of wages or other emoluments and expenses (including holiday credits) are increased or decreased by reason of any alteration in the said rules, decisions or agreements promulgated after the date of tender, then the net amount of the increase or decrease in wages and other emoluments and expenses (including holiday credits) together with the net amount of any consequential increase or decrease in the cost of employer's liability insurance, of third party insurance, and of any contribution, levy or tax payable by a person in his capacity as an employer shall, as the case may be, be paid to or allowed by the Contractor.

(b) (i) The prices contained in the Contract Bills are based upon the types and rates of contribution, levy and tax payable by a person in his capacity as an employer and which at the date of tender are payable by the Contractor. A type and a rate so payable are in the next sub-paragraph referred to as a ' tender type ' and a ' tender rate '.

(ii) If any of the tender rates other than a rate of levy payable by virtue of the Industrial Training Act 1964, is increased or decreased, or if a tender type ceases to be payable, or if a new type of contribution, levy or tax which is payable by a person in his capacity as an employer becomes payable after the date of tender, then in any such case the net amount of the difference between what the Contractor actually pays or will pay in respect of workpeople whilst they are engaged upon or in connection with the Works or because of his employment of such workpeople upon or in connection with the Works, and what he would have paid had the alteration, cessation or new type of contribution levy or tax not become effective, shall, as the case may be, be paid to or allowed by the Contractor.

(iii) The prices contained in the Contract Bills are based upon the types and rates of refund of contributions, levies and taxes payable by a person in his capacity as an employer and upon the types and rates of premium receivable by a person in his capacity as an employer being in each case types and rates which at the date of tender are receivable by the Contractor. Such a type and such a rate are, in the next sub-paragraph, referred to as a ' tender type ' and a ' tender rate '.

(iv) If any of the tender rates is increased or decreased or if a tender type ceases to be payable or if a new type of refund of any contribution levy or tax payable by a person in his capacity as an employer becomes receivable or if a new type of premium receivable by a person in his capacity as an employer becomes receivable after the date of tender, then in any such case the net amount of the difference between what the Contractor actually receives or will receive in respect of workpeople whilst they are engaged upon or in connection with the Works or because of his employment of such workpeople upon or in connection with the Works, and what he would have received had the alteration, cessation or new type of refund or premium not become effective, shall, as the case may be, be allowed by or paid to the Contractor.

(v) The references in the two preceding sub-paragraphs to premiums shall be construed as meaning all payments howsoever they are described which are made under or by virtue of an Act of Parliament to a person in his capacity as an employer and which affect the cost to an employer of having persons in his employment.

(vi) Where the Contractor elects to contribute to a recognised occupational pension scheme instead of participating in the Reserve Pension Scheme established under the Social Security Act 1973 the Contractor shall for the purpose of recovery or allowance under this clause be

deemed to pay employer's contributions to the Reserve Pension Scheme.
(vii) The references in sub-paragraph (i) to (iv) and (vi) of this paragraph to contributions, levies and taxes shall be construed as meaning
all impositions payable by a person in his capacity as an employer howsoever they are described and whoever the recipient which are imposed
under or by virtue of an Act of Parliament and which affect the cost to
an employer of having persons in his employment.

   (c)   (i) The prices contained in the Contract Bills are based upon the market
prices of the materials and goods specified in the list attached thereto
which were current at the date of tender. Such prices are hereinafter
referred to as ' basic prices,' and the prices stated by the Contractor
on the said list shall be deemed to be the basic prices of the specified
materials and goods.

(ii) If after the date of tender the market price of any of the materials or
goods specified as aforesaid increases or decreases, then the net amount
of the difference between the basic price thereof and the market price
payable by the Contractor and current when the materials or goods are
bought shall, as the case may be, be paid to or allowed by the Contractor.

(iii) The references in the two preceding sub-paragraphs to ' market
prices ' shall be construed as including any duty or tax (other than any
value added tax which is treated, or is capable of being treated, as input
tax (as referred to in the Finance Act 1972) by the Contractor) by
whomsoever payable which is payable under or by virtue of any Act of
Parliament on the import, purchase, sale, appropriation, processing or
use of the materials or goods specified as aforesaid.

Clause 31B *

The Contract Sum shall be deemed to have been calculated in the manner set
out below and shall be subject to adjustment in the events specified hereunder:—

   (a)   (i) The prices contained in the Contract Bills are based upon the types
and rates of contribution, levy and tax payable by a person in his
capacity as an employer and which at the date of tender are payable by
the Contractor. A type and a rate so payable are in the next sub-
paragraph referred to as a ' tender type ' and a ' tender rate '.

(ii) If any of the tender rates other than a rate of levy payable by virtue
of the Industrial Training Act 1964, is increased or decreased, or if a
tender type ceases to be payable, or if a new type of contribution, levy
or tax which is payable by a person in his capacity as an employer
becomes payable after the date of tender, then in any such case the net
amount of the difference between what the Contractor actually pays or
will pay in respect of workpeople whilst they are engaged upon or in
connection with the Works or because of his employment of such workpeople upon or in connection with the Works, and what he would have
paid had the alteration, cessation or new type of contribution, levy or
tax not become effective, shall, as the case may be, be paid to or allowed
by the Contractor.

(iii) The prices contained in the Contract Bills are based upon the types
and rates of refund of contributions, levies and taxes payable by a person in his capacity as an employer and upon the types and rates of
premium receivable by a person in his capacity as an employer being in
each case types and rates which at the date of tender are receivable by
the Contractor. Such a type and such a rate are in the next sub-
paragraph referred to as a ' tender type ' and a ' tender rate '.

(iv) If any of the tender rates is increased or decreased or if a tender type ceases to be payable or if a new type of refund of any contribution levy or tax payable by a person in his capacity as an employer becomes receivable or if a new type of premium receivable by a person in his capacity as an employer becomes receivable after the date of tender, then in any such case the net amount of the difference between what the Contractor actually receives or will receive in respect of workpeople whilst they are engaged upon or in connection with the Works or because of his employment of such workpeople upon or in connection with the Works, and what he would have received had the alteration, cessation or new type of refund or premium not become effective, shall, as the case may be, be allowed by or paid to the Contractor.

(v) The references in the two preceding sub-paragraphs to premiums shall be construed as meaning all payments howsoever they are described which are made under or by virtue of an Act of Parliament to a person in his capacity as an employer and which affect the cost to an employer of having persons in his employment.

(vi) Where the Contractor elects to contribute to a recognised occupational pension scheme instead of participating in the Reserve Pension Scheme established under the Social Security Act 1973 the Contractor shall for the purpose of recovery or allowance under this clause be deemed to pay employer's contributions to the Reserve Pension Scheme.

(vii) The references in sub-paragraphs (i) to (iv) and (vi) of this paragraph to contributions, levies and taxes shall be construed as meaning all impositions payable by a person in his capacity as an employer howsoever they are described and whoever the recipient which are imposed under or by virtue of an Act of Parliament and which affect the cost to an employer of having persons in his employment.

(b) (i) The prices contained in the Contract Bills are based upon the types and rates of duty if any and tax if any (other than any value added tax which is treated, or is capable of being treated, as input tax (as referred to in the Finance Act 1972) by the Contractor) by whomsoever payable which at the date of tender are payable on the import, purchase, sale, appropriation, processing or use of the materials and goods specified in the list attached thereto under or by virtue of any Act of Parliament. A type and a rate so payable are in the next sub-paragraph referred to as a ' tender type ' and a ' tender rate '.

(ii) If in relation to any materials or goods specified as aforesaid a tender rate is increased or decreased, or a tender type ceases to be payable or a new type of duty or tax (other than any value added tax which is treated, or is capable of being treated, as input tax (as referred to in the Finance Act 1972) by the Contractor) becomes payable on the import, purchase, sale, appropriation, processing or use of those materials or goods, then in any such case the net amount of the difference between what the Contractor actually pays in respect of those materials or goods, and what he would have paid in respect of them had the alteration, cessation or imposition not occurred, shall, as the case may be, be paid to or allowed by the Contractor. In this sub-paragraph the expression ' a new type of duty or tax ' includes an additional duty or tax and a duty or tax imposed in regard to specified materials or goods in respect of which no duty or tax whatever was previously payable (other than any value added tax which is treated, or is capable of being treated, as input tax (as referred to in the Finance Act 1972) by the Contractor).

Clause 31C

(1) If the Contractor shall decide subject to clause 17 of these Conditions to

sublet any portion of the Works he shall incorporate in the sub-contract provisions to the like effect as the provisions of clauses 31A, 31D and 31E/clauses 31B, 31D and 31E (as applicable) which are applicable for the purposes of this Contract.

(2) If the price payable under such a sub-contract as aforesaid is decreased below or increased above the price in such sub-contract by reason of the operation of the said incorporated provisions, then the net amount of such decrease or increase shall, as the case may be, be allowed by or paid to the Contractor under this Contract.

Clause 31D

(1) The Contractor shall give a written notice to the Architect/Supervising Officer of the occurrence of any of the events referred to in such of the following provisions as are applicable for the purposes on this Contract:

    (a) Clause 31A (a) (ii);
    (b) Clause 31A (b) (ii);
    (c) Clause 31A (b) (iv);
    (d) Clause 31A (c) (ii);
    (e) Clause 31B (a) (ii);
    (f) Clause 31B (a) (iv);
    (g) Clause 31B (b) (ii);
    (h) Clause 31C (2).

(2) Any notice required to be given by the preceding sub-clause shall be given within a reasonable time after the occurrence of that to which the notice relates, and the giving of a written notice in that time shall be a condition precedent to any payment being made to the Contractor in respect of that event in question.

(3) The Quantity Surveyor and the Contractor may agree what shall be deemed for all the purposes of this Contract to be the net amount payable to or allowable by the Contractor in respect of the occurrence of any event such as is referred to in any of the provisions listed in sub-clause (1) of this Condition.

(4) Any amount which from time to time becomes payable to or allowable by the Contractor by virtue of clause 31A or clause 31B or clause 31C of these Conditions shall, as the case may be, be added to or subtracted from:

    (a) The Contract Sum; and
    (b) Any amounts payable to the Contractor and which are calculated in accordance with either sub-paragraph (i) or sub-paragraph (ii) of paragraph (b) of sub-clause 2 of clause 26 of these Conditions; and
    (c) The amount which would otherwise be stated as due in the next Interim Certificate.

Provided:

    (i) No addition to or subtraction from the amount which would otherwise be stated as due in an Interim Certificate shall be made by virtue of this sub-clause unless on or before the date as at which the total value of work, materials and goods is ascertained for the purposes of that Certificate the Contractor shall have actually paid or received the sum which is payable by or to him in consequence of the event in respect of which the payment or allowance arises.

    (ii) No addition to or subtraction from the Contract Sum made by virtue of this sub-clause shall alter in any way the amount of profit of the Contractor included in that Sum.

(5) Clause 31A, clause 31B and clause 31C shall not apply in respect of:
    (a) Work for which the Contractor is allowed daywork rates under clause 11 (4) (c) of these Conditions.

(b) Work executed or materials or goods supplied by any nominated sub-contractor or nominated supplier (fluctuations in relation to nominated sub-contractors and nominated suppliers shall be dealt with under any provision in relation thereto which may be included in the appropriate sub-contract or contract of sale), or

(c) Work executed by the Contractor for which a tender made under clause 27 (g) of these Conditions has been accepted.

(d) Changes in the rate of value added tax charged on the supply of goods or services by the Contractor to the Employer under this Contract.

(6) In clause 31A and clause 31B of these Conditions:

(a) The expression ' the date of tender ' means the date 10 days before the date fixed for the receipt of tenders by the Employer; and

(b) The expression ' materials ' and ' goods ' include timber used in form-work but do not include other consumable stores, plant and machinery.

(c) The expression ' workpeople ' means persons whose rates of wages and other emoluments (including holiday credits) are governed by the rules or decisions or agreements of the National Joint Council for the Building Industry or some other like body for trades associated with the building industry.

Clause 31E

There shall be added to the amount paid to or allowed by the Contractor under:

     (a) Clause 31A (a) (ii);
     (b) Clause 31A (b) (ii);
     (c) Clause 31A (b) (iv);
     (d) Clause 31A (c) (ii);
     (e) Clause 31B (a) (ii);
     (f) Clause 31B (a) (iv);
     (g) Clause 31B (b) (ii);

the percentage stated in the Appendix to these Conditions.

*Parts A, C, D and E should be used where the parties have agreed to allow the labour and materials cost and tax fluctuations to which Part A refers. Alternatively Part F should be used where the parties have agreed that the contract price shall be adjusted by the NEDO Formula Method under the Formula Rules.

Parts B, C, D and E should be used where neither Part A nor Part F is used.

COMMENTARY

(1) *Introduction*

The parties may, if they wish, have no price fluctuations clause at all. But the contract contemplates that ordinarily there will be such a clause. There are three variants. Clause 31F, introduced in July 1973, provides for adjustment of the contract sum by means of formulae and is annotated separately, *post*. Clause 31A is the alternative and operates upon a basis that has become traditional, *i.e.* by adjustment of rates and prices in accordance with alterations of wage rates, market values and certain statutory impositions. Clause 31B is confined to adjust-ments arising out of alterations of the statutory impositions.

(2) *Amendments to clauses 31A and 31B*

There were amendments in the 1966 version, a new clause in the 1967 version, and further amendments in the 1969, 1971, 1973 and 1975 versions.

(3) *General effect of clauses* 31A *and* 31B

They are " rise and fall " clauses. In practice, apart from one very brief period, prices have risen since the war and, at the time of writing, have been rising very fast for the past two or three years. If properly operated (*e.g.* a full and careful list of " basic prices "—see clause 31A (c) (i)) the clauses provide a considerable measure of protection for the contractor but do not provide an indemnity against increased costs. Thus they do not apply to increases in head office and administrative costs; they do not apply to consumable stores, plant and machinery other than timber used in formwork (see clause 31D (6) (b)). And it has been held that the increase in cost in operating a bonus scheme, based upon wage rates, where that scheme was voluntarily entered into by the contractors is not an increase recoverable under the clause.[62] For these reasons prudent contractors using the pre-1973 version would incorporate an adjustment to their rates to allow for non-recoverable increases. In July 1973 there was introduced clause 31E to provide expressly in the Appendix for an additional percentage.

In a price fluctuations clause in a different contract it was held that holiday payment credits do not come within the meaning of the word " wages." [63]

(4) *Generally*

Outline of clauses 31A to 31E.

(a) Clause 31A (a); alterations in wages, emoluments and expenses.

(b) Clause 31A (b) (i), (ii), (vi) and (vii); alterations in types and rates of statutory contributions, levies and taxes payable by a person in his capacity of employer.

(c) Clause 31A (b) (iii), (iv), (v), (vi) and (vii); alterations in statutory refunds or premiums receivable by a person in his capacity as an employer.

(d) Clause 31A (c); alterations in market prices in materials and goods specified in a list attached to contract bills, such alteration to include alteration of any statutory duty or tax.

(e) Clause 31B (a); alterations as in clause 31A (b).

(f) Clause 31B (b); alterations in statutory duties or taxes affecting materials or goods specified in a list attached to contract bills.

(g) Clause 31C; contractor to incorporate relevant fluctuation clause in sub-contracts with approved sub-contractors (see the standard form of sub-contract, clauses 23A, 23B, 23C, 23D).

(h) Clause 31D; contractor to give written notice of any relevant alterations; machinery for adjustment of price; clause not to apply to nominated sub-contractors, day works or work carried out under clause 27 (g).

(i) Clause 31E; percentage addition.

(5) *Clause* 31A *Contractor's delay in completion.* " *The contract sum . . . shall be subject to adjustment in the events specified hereunder . . .*"

The clause operates, it is submitted, notwithstanding that the contractor is in

---

[62] *Sindall (William) Ltd.* v. *N.W. Thames Regional Health Authority* [1977] I.C.R. 294 (H.L.).

[63] *London County Council* v. *Henry Boot & Sons Ltd.* [1959] 1 W.L.R. 1069 (H.L.).

delay in completion beyond the completion date extended as may be necessary under the contract.[64]

(6) *Clause* 31A (a) (i) ". . . *which have been promulgated at the date of tender* . . ."
The contractor cannot claim an increase in respect of a wage award which becomes effective during the course of carrying out the works but which had been promulgated at the date of tender. This was the effect of earlier versions of this clause although it was not so clearly expressed.

" Date of tender " is defined in clause 31D (6) (a) as 10 days before the date fixed for the receipt of tenders by the employer.

The reference to " workpeople " and to the machinery for fixing their wages excludes, in general, from the operation of the clause, administrative staff. Further, head office staff are usually not within the words " workpeople engaged upon or in connection with the Works."

(7) *Clause* 31D (2) *Notice condition precedent*
This is plain language, (*cf. e.g.* the requirements of clause 23 as to notice) and if the contractor does not give notice within a reasonable time after the occurrence of the event to which it relates he is not, in ordinary circumstances, entitled to payment. What is reasonable is a question of fact but it is thought that a most important consideration is whether an opportunity to check the facts has been lost.

(8) *Clause* 31D (3) *Conclusive agreement between quantity surveyor and contractor*
See further, clause 35 (3). Such an agreement cannot, it is submitted, in the absence of fraud or misrepresentation, be re-opened. It is not necessary, but it is convenient, that it should be in writing.

(9) *Clause* 31D (5) (d). *V.A.T.*
The clause does not apply because this is dealt with separately by clause 13A and the " V.A.T. agreement."

Clause 31F*
Adjustment for Contract Sum—Price Adjustment Formula for Building Contracts

   (1) (a) (i) The Contract Sum shall be adjusted in accordance with the provisions of this clause and the Formula Rules current at the Date of Tender issued for use with this clause by the Joint Contracts Tribunal for the Standard Form of Building Contract hereinafter called ' the Formula Rules '.
   (ii) Any adjustment under this clause shall be to sums exclusive of value added tax and nothing in this clause shall affect in any way the operation of clause 13A (value added tax) and the VAT Agreement executed pursuant to sub-clause (1) (b) of clause 13A of these Conditions.
   (b) The Definitions in rule 3 of the Formula Rules shall apply to this clause.
   (c) The adjustment referred to in this sub-clause shall be effected (after taking into account any Non-Adjustable Element) in all certificates for

---

[64] See *Peak Construction (Liverpool) Ltd.* v. *McKinney Foundations Ltd.* (1970) 69 L.G.R. 1 (C.A.), not a decision on the Standard form but of assistance.

payment (other than those under clause 16 (f) and clause 30 (4) of these Conditions (release of rentention)) issued under the provisions of these Conditions.

(d) If any correction of amounts of adjustment under this clause included in previous certificates is required following any operation of rule 5 of the Formula Rules such correction shall be given effect in the next certificate for payment to be issued.

*Part F is used where the parties have agreed that fluctuations should be dealt with by adjustment of the Contract Sum under the NEDO Price Adjustment Formulae for Building Contracts. Parts A–E should be deleted where Part F applies.

Amendments to clause 30 (1) (2) and (3)—interim valuations and payments and retention

(2) (a) (i) Interim valuations shall be made before the issue of each Interim Certificate and accordingly the words ' whenever the Architect/Supervising Officer considers them to be necessary ' shall be deemed to have been deleted in clause 30 (1) of these Conditions.

(ii) The words ' Before the issue of the Certificate of Practical Completion, Interim Certificates shall be issued at the Period of Interim Certificates specified in the Appendix to these Conditions. After the issue of the Certificate of Practical Completion ' in clause 30 (1) of these Conditions shall be deemed to have been deleted and the following words substituted: ' Interim Certificates shall be issued at the Period of Interim Certificates specified in the Appendix to these Conditions including any Period during which the Certificate of Practical Completion is issued. After the end of the Period of Interim Certificates in which the Certificate of Practical Completion is issued,'.

(b) The Retention Percentage referred to in clause 30 (3) (a) of these Conditions shall apply to any adjustment of the Contract Sum under this clause and accordingly

(i) the words ' together with the net total of the adjustments to the Contract Sum to be effected under clause 31F in the said Interim Certificate and effected in any previous Interim Certificates ' shall be deemed to have been added to clause 30 (2) of these Conditions after the words ' seven days before the date of the said certificate '; and

(ii) the words ' together with the net total of the adjustments ' shall be deemed to have been added to clause 30 (3) (a) of these Conditions after the words ' total value of the work, materials and goods '; and

(iii) the words ' together with the net total of the adjustments ' shall be deemed to have been added to clause 30 (3) (b) of these Conditions after the words ' total value of the work, materials and goods '.

Amendments to clause 35 (2)—arbitration before practical completion

(c) A reference to arbitration on a dispute or difference under clause 23 of these Conditions may be opened before Practical Completion or alleged Practical Completion of the Works or termination or alleged termination of the Contractor's employment under this Contract or abandonment of the Works and accordingly the words ' under clause 23 or ' shall be deemed to have been added to clause 35 (2) of these Conditions after the words ' on any dispute or difference '.

Amendments to clause 13A

(d) The words ' (or as defined in the Formula Rules where clause 31F of these Conditions applies) ' shall be deemed to have been added after the words ' of these Conditions ' in clause 13A (3) of these Conditions.

Fluctuations—articles manufactured outside the United Kingdom

(3) For any article to which rule 4 (ii) of the Formula Rules applies the Contractor shall insert in a list attached to the Contract Bills the market price of the article in sterling (that is the price delivered to the site) current at the Date of Tender. If after that Date the market price of the article inserted in the aforesaid list increases or decreases then the net amount of the difference between the cost of purchasing at the market price inserted in the aforesaid list and the market price payable by the Contractor and current when the article is bought shall, as the case may be, be paid to or allowed by the Contractor. The reference to ' market price ' in this sub-clause shall be construed as including any duty or tax (other than any value added tax which is treated, or is capable of being treated, as input tax (as defined in the Finance Act 1972) by the Contractor) by whomsoever payable under or by virtue of any Act of Parliament on the import, purchase, sale, appropriation or use of the article specified as aforesaid.

Nominated sub-contractors

(4) (a) Where the supply and fixing of any goods or the execution of any work is to be carried out by a sub-contractor nominated by the Architect/Supervising Officer the sub-contract between the Contractor and the nominated sub-contractor shall provide if required for adjustment to be made of the sub-contract sum for cost fluctuations by reference to whichever of the following has been tendered upon by the sub-contractor and approved in writing by the Architect/Supervising Officer prior to the issue of the nomination instruction:

(i) in the case of electrical installations, heating and ventilating and air conditioning installations, lift installations, structural steelwork installations and in the case of catering equipment installations, the relevant specialist formula (see the Formula Rules, rule 50, rule 54, rule 58, rule 63 and rule 69);

(ii) where none of the specialist formulae applies, the Formula in Part I of Section 2 of the Formula Rules and one or more of the Work Categories set out in Appendix A to the Formula Rules;

(iii) where neither sub-paragraph (i) nor sub-paragraph (ii) applies, some other method.

Non-nominated sub-contractors

(b) If the Contractor shall decide, subject to clause 17 of these Conditions, to sub-let any portion of the Works, he shall, unless the Contractor and sub-contractor otherwise agree, incorporate in the sub-contract provisions for formula adjustment of the sub-contract sum namely:

(i) in the case of electrical installations, heating and ventilating and air conditioning installations, lift installations, structural steelwork installations and in the case of catering equipment installations, the relevant specialist formula;

(ii) where none of the specialist formulae applies, the Formula in Part I of Section 2 of the Formula Rules and one or more of the Work Categories set out in appendix A to the Formula Rules appropriate to such sub-contract works.

Power to agree—Quantity Surveyor and Contractor

(5) The Quantity Surveyor and the Contractor may agree any alteration to the methods and procedures for ascertaining the amount of formula adjustment to be made under this clause and the amounts ascertained after the operation of such agreement shall be deemed for all the purposes of this Contract to be the amount

of formula adjustment payable to or allowable by the Contractor in respect of the provisions of this clause. Provided always:

(i) that no alteration to the methods and procedures shall be agreed as aforesaid unless it is reasonably expected that the amount of formula adjustment so ascertained will be the same or approximately the same as that ascertained in accordance with Part I or Part II of Section 2 of the Formula Rules whichever Part is stated to be appiicable in the Contract Bills; and

(ii) that any agreement under this sub-clause shall not have any effect on the determination of any adjustment payable by the Contractor to any sub-contractor to whom sub-clause (4) refers.

Position where monthly bulletins are delayed, etc.

(6) (a) If at any time prior to the issue of the Final Certificate under clause 30 (6) of these Conditions formula adjustment is not possible because of delay in, or cessation of, the publication of the Monthly Bulletins, adjustment of the Contract Sum shall be made in each Interim Certificate during such period of delay on a fair and reasonable basis.

(b) If publication of the Monthly Bulletins is recommenced at any time prior to the issue of the Final Certificate under clause 30 (6) of these Conditions the provisions of this clause and the Formula Rules shall operate for each Valuation Period as if no delay or cessation as aforesaid had occurred and the adjustment under this clause and the Formula Rules shall be substituted for any adjustment under paragraph (a) hereof.

(c) During any period of delay or cessation as aforesaid the Contractor and Employer shall operate such parts of this clause and the Formula Rules as will enable the amount of formula adjustment due to be readily calculated upon recommencement of publication of the Monthly Bulletins.

Formula adjustment—failure to complete

(7) (a) (i) If the Contractor fails to complete the Works by the Date for Completion stated in the appendix to these Conditions or within any extended time fixed under clause 23 or clause 33 (1) (c) of these Conditions formula adjustment of the Contract Sum under this clause shall (but subject to sub-clause (c) (i) hereof) be effected in all interim certificates issued after the aforesaid Date for Completion (or any extension thereof) by reference to the Index Numbers applicable to the Valuation Period in which the aforesaid Date for Completion (or any extension thereof) falls.

(ii) If for any reason the adjustment included in the amount certified in any Interim Certificate which is or has been issued after the aforesaid Date for Completion (or any extension thereof) is not in accordance with sub-paragraph (i) hereof, such adjustment shall be corrected to comply with the aforesaid sub-paragraph.

(b) Paragraph (a) of this sub-clause shall not operate unless:

(i) the printed text of clause 23 of these Conditions is unamended and sub-clause (j) (i) and sub-clause (j) (ii) of that clause have not been deleted in the Contract Conditions executed by the Employer and the Contractor; and

(ii) the Architect/Supervising Officer has, in respect of every written notification by the Contractor under clause 23 of these Conditions, made in writing such extensions of time, if any, for completion of the Works as he considers to be in accordance with that clause; provided always that the Contractor has given the Architect/Supervising Officer such information as the Architect/Supervising Officer may reasonably require for this purpose.

(c) Where the condition in paragraph (b) (i) hereof is fulfilled:

(i) any formula adjustment of the sub-contract sum in respect of any supply and fixing of goods, or the execution of any work, by a sub-contractor nominated by the Architect/Supervising Officer (hereinafter called ' the sub-contract works ') shall not be affected by any operation of paragraph (a) of this sub-clause;

(ii) subject to sub-paragraph (iii) hereof any formula adjustment of the sub-contract sum for such sub-contract works shall be effected in all Interim Certificates which include a direction within the terms of clause 27 (b) of these Conditions by reference to the Index Numbers applicable to the Valuation Period (or in the case of lift installations and structural steelwork installations the Index Numbers applicable at the relevant dates for adjustment) to which any Interim Certificate including the aforesaid direction relates;

(iii) if a nominated sub-contractor shall fail to complete the sub-contract works within the period specified in the sub-contract or within any extended time granted by the Contractor with the written consent of the Architect/Supervising Officer and the Architect/Supervising Officer has issued the certificate (with a duplicate copy to the nominated sub-contractor) referred to in clause 27 (d) (ii) of these Conditions, formula adjustment of the sub-contract sum shall (if the sub-contract so provides) be effected in all Interim Certificates which include a direction within the terms of clause 27 (b) of these Conditions and which are issued after the date in the aforesaid certificate (that is the day on which the sub-contract works ought for the purposes of clause 27 (d) (ii) of these Conditions to have been completed) issued under clause 27 (d) (ii), by reference to Index Numbers applicable to the Valuation Period in which that date falls.

(1) *Introduction*

This clause was introduced by documents issued in March 1975 incorporated in the July 1975 Revision and amended in the 1977 Revision. It follows the use by the Government of a formula method of calculating fluctuations in public sector work lasting over 12 months. The Joint Contracts Tribunal was asked to prepare contractual provisions for this form of contract for use where the parties have agreed that fluctuations in costs are to be dealt with through the formulae prepared by Government sponsored bodies and published as monthly bulletins. The JCT accordingly issued the new clause 31F together with a set of specially prepared rules, termed the " Formula Rules " which by incorporation become part of the contract where this clause is used. They further issued Practice Note 18. Those concerned with the detailed application of the clause must obtain the Formula Rules and will find it very helpful to have Practice Note 18. For reasons of space the Rules are not printed in this book, neither is the Practice Note save Appendix A to the Note which is printed *infra,* being some helpful notes on clause 31F.

Amendments to contract. They are set out in clause 31F (2). Further, consider clause 31F (7). Contrary to the position where clauses 31A to 31E are used, adjustment of the contract sum under clause 31F does not continue after the date for completion, extended as may be necessary. But this non-continuance of the application of the clause is subject to clause 23 remaining unamended with sub-clauses (j) (i) and (ii) not having been deleted and further subject to the architect, in respect of every written notification by the contractor under clause 23 having made in writing such extensions of time, if any, for completion of the works as he considers to be in accordance with that clause and always provided that the contractor has given the architect such information as he might reasonably require. So although clause 23 is not expressly amended it behoves the parties to use the utmost diligence in its operation. In particular, in days of rapid inflation, the prudent contractor is encouraged both to give prompt notice and to furnish reasonable information to the architect to enable him to perform his duties.

Power of quantity surveyor to agree variation of formula. Clause 31F (5) gives the quantity surveyor a carefully limited authority to agree such variation with the contractor.

(2) *JCT notes on clause* 31F (printed by kind permission)

Clause 31F. The clause is drafted as a further part of clause 31. Formula adjustment, being an adjustment of value and not referable in any way to actual costs incurred or saved, requires the total deletion of Parts A to E which are related to fluctuations in various actual costs. With the deletion of clause 31D containing the definition of ' Date of Tender ' it was necessary to provide an identical definition in rule 3 of the Formula Rules; see sub-clause (1) (b) which makes the Definitions in the Rules applicable to clause 31F and several of the expressions so defined occur in the clause (*e.g.* Monthly Bulletins, Valuation Period).

Sub-clause (1) (a) (i). This paragraph makes the Formula Rules part of the Contract Conditions. The ' date of tender ' is defined in rule 3 of the Formula Rules (see sub-clause (1) (b)). The Rules are dated by reference to a day and month and any subsequent editions will be similarly so dated.

Sub-clause (1) (a) (ii). This paragraph is for the avoidance of doubt and confirms that adjustment is to sums exclusive of VAT.

Sub-clause (1) (c). The text within the first parenthesis is only applicable to the Local Authorities edition and empowers the reduction of the amount of adjustment by the Non-Adjustable Element. The second parenthesis makes clear that no adjustment is to be made to sums of retention released under clauses 16 (f) or 30 (4).

Sub-clause (1) (d). This paragraph must be read in conjunction with rule 5; see paragraph 17 of the Practice Note.

Sub-clause (2) (a) and (b). This sub-clause amends clauses 30 (1), (2) and (3) by reference. No actual amendments are therefore needed. The reasons for the deemed additions and deletion are set out in this sub-clause. The amendment in sub-clause (2) (a) (ii) is to ensure that the last Interim Certificate issued at the Period of Interim Certificates will include normal formula adjustment under Part I rule 9 or Part II rule 29 even if it is issued after the Certificate of Practical Completion. Certificates issued thereafter will contain adjustments under the ' average ' basis referred to in paragraph 18 of the Practice Note.

Sub-clause (2) (c). This paragraph which amends clause 35 (2) by reference would enable any dispute as to whether the Contractor was in default over completion for the purposes of operating sub-clause (7), to be opened earlier than Practical Completion of the Works.

Sub-clause (2) (d). This paragraph is needed since clause 13A (3) only refers to the definition of ' date of tender ' in clause 31D (6) (a) and 31D will have been deleted where clause 31F is used.

Sub-clause (3). This sub-clause provides that when the Contract Bills require the Contractor to use manufactured articles specially purchased outside the U.K. (whose price is unlikely to be reflected in the Work Category Index Numbers) fluctuations for these articles shall be adjusted by reference to market price changes and not by formula.

Sub-clause (4) (a). This paragraph makes clear that clause 31F and the Formula Rules do not apply to work carried out by nominated sub-contractors whose recovery of fluctuations is to be dealt with by whatever provision has been agreed between the Architect/Supervising Officer on behalf of the Employer and the sub-contractor and included in the relevant sub-contract. Adjustments to the nominated sub-contract sum under such provision will be reflected in the main contract sum by virtue of clause 30 (5) (c). See paragraphs 11 and 12 of this Practice Note.

Sub-clause (4) (b). This provision is intended to make clear that if formula adjustment is agreed between the contractor and a non-nominated sub-contractor then the only formula adjustment appropriate for such sub-contracts is by reference to the relevant Work Category/Work Categories or to one of the specialist formulae (which are the same as those set out in Part III of the Rules).

Sub-clause (5). Attention is called to the two provisos to which any agreements between the Contractor and the Quantity Surveyor on varying the mechanics of operating formula adjustment are subject.

Sub-clause (6). As the Contractor will have tendered on the basis that the Value of Work for each Valuation Period will be adjusted under clause 31F and the Rules, provision was needed for the situation where the Index Numbers cease temporarily or for a longer period to be available. In these circumstances a fair and reasonable adjustment is made.

Sub-clause (7). The Standard Form does not provide for cessation of payment of fluctuations under clauses 31A or 31B (with 31C, D and E) during the period when the Contractor is, or is alleged to be, in default over completion. Where clause 31F is used paragraph (a) provides that adjustment during a period of default over completion is by reference to the firm Index Numbers applicable to the Valuation Period during which the Contractor became in default. Paragraph (b) makes the operation of paragraph (a) conditional on the use of clause 23 (extension of time) unamended and with the specific inclusion of clause 23 (j) (i) and (ii) (shortages of labour and materials). The reason for this condition is that with the new provisions of paragraph (a) it is necessary for full provisions in regard to extension of time to be included in the contract. Paragraph (c) deals with any similar limitation in formula adjustment of the sub-contract sum for nominated sub-contractors and provides that the application of any such limitation to the nominated sub-contract formula adjustment is quite independent of the operation of paragraph (a) on adjustments made in respect of work carried out by the main contractor; and will require appropriate provisions also in the nominated sub-contract.

Appendix: Rule 3. The Base Month will normally be the month preceding that during which tenders have to be returned. It is, however, open to the parties to specify a different Base Month and the actual month chosen must be set out in this appendix.

Clause 32*
Outbreak of hostilities

(1) If during the currency of this Contract there shall be an outbreak of hostilities (whether war is declared or not) in which the United Kingdom shall be involved on a scale involving the general mobilisation of the armed forces of the Crown, then either the Employer or the Contractor may at any time by notice by registered post or recorded delivery to the other, forthwith determine the employment of the Contractor under this Contract:
Provided that such a notice shall not be given
  (a) Before the expiration of 28 days from the date on which the order is given for general mobilisation as aforesaid, or
  (b) After Practical Completion of the Works unless the Works or any part thereof shall have sustained war damage as defined in clause 33 (4) of these Conditions.
(2) The Architect/Supervising Officer may within 14 days after a notice under this Condition shall have been given or received by the Employer issue instructions to the Contractor requiring the execution of such protective work as shall be specified therein and/or the continuation of the Works up to points of stoppage to be specified therein, and the Contractor shall comply with such instructions as if the notice of determination had not been given.
Provided that if the Contractor shall for reasons beyond his control be prevented from completing the work to which the said instructions relate within three months from the date on which the instructions were issued, he may abandon such work.
(3) Upon the expiration of 14 days from the date on which a notice of determination shall have been given or received by the Employer under this Condition or where works are required by the Architect/Supervising Officer under the preceding sub-clause upon completion or abandonment as the case may be of any such works, the provisions of sub-clause (2) (except sub-paragraph (vi) of paragraph (b)) of clause 26 of these Conditions shall apply, and the Contractor shall also be paid by the Employer, the value of any work executed pursuant to instructions given under sub-clause (2) of this clause, the value being ascertained in accordance with clause 11 (4) of these Conditions as if such work were a variation required by the Architect/Supervising Officer.

> *The parties hereto in the event of the outbreak of hostilities may at any time by agreement between them make such further or other arrangements as they may think fit to meet the circumstances.*

Clause 33
War damage

(1) In the event of the Works or any part thereof or any unfixed materials or goods intended for, delivered to and placed on or adjacent to the Works sustaining war damage then notwithstanding anything expressed or implied elsewhere in this Contract:
  (a) The occurrence of such war damage shall be disregarded in computing any amounts payable to the Contractor under or by virtue of this Contract.

(b) The Architect/Supervising Officer may issue instructions requiring the Contractor to remove and/or dispose of any debris and/or damaged work and/or to execute such protective work as shall be specified.

(c) The Contractor shall reinstate or make good such war damage and shall proceed with the carrying out and completion of the Works, and the Architect/Supervising Officer shall grant the Contractor a fair and reasonable extension of time for completion of the Works.

(d) The removal and disposal of debris or damaged work, the execution of protective works and the reinstatement and making good of such war damage shall be deemed to be a variation required by the Architect/Supervising Officer.

(2) If at any time after the occurrence of war damage as aforesaid either party serves notice of determination under clause 32 of these Conditions, the expression ' protective work ' as used in the said clause shall in such case be deemed to include any matters in respect of which the Architect/Supervising Officer can issue instructions under paragraph (b) of sub-clause (1) of this Condition and any instructions issued under the said paragraph prior to the date on which notice of determination is given or received by the Employer and which shall not then have been completely complied with shall be deemed to have been given under clause 32 (2) of these Conditions.

(3) The Employer shall be entitled to any compensation which may at any time become payable out of monies provided by Parliament in respect of war damage sustained by the Works or any part thereof or any unfixed materials or goods intended for the Works which shall at any time have become the property of the Employer.

(4) The expression ' war damage ' as used in this Condition means war damage as defined by section 2 of the War Damage Act 1943, or any amendment thereof.

COMMENTARY

Generally

For frustration, the effect of war and clauses of this kind see generally *ante*, pp. 102–107.

Clause 34
Antiquities

(1) All fossils, antiquities and other objects of interest or value which may be found on the site or in excavating the same during the progress of the Works shall become the property of the Employer, and upon discovery of such an object the Contractor shall forthwith:

(a) Use his best endeavours not to disturb the object and shall cease work if and in so far as the continuance of work would endanger the object or prevent or impede its excavation or its removal;

(b) Take all steps which may be necessary to preserve the object in the exact position and condition in which it was found; and

(c) Inform the Architect/Supervising Officer or the Clerk of Works of the discovery and precise location of the object.

(2) The Architect/Supervising Officer shall issue instructions in regard to what is to be done concerning an object reported by the Contractor under the preceding sub-clause, and (without prejudice to the generality of his power) such instructions may require the Contractor to permit the examination, excavation or removal of the object by a third party. Any such third party shall for the purposes of clause 18 of these Conditions be deemed to be a person for whom the Employer is responsible and not to be a sub-contractor.

(3) If in the opinion of the Architect/Supervising Officer compliance with the provisions of sub-clause (1) of this Condition or with an instruction issued under sub-clause (2) of this Condition has involved the contractor in direct loss and/or expense for which he would not be reimbursed by a payment made under any other provision in this Contract then the Architect/Supervising Officer shall either himself ascertain or shall instruct the Quantity Surveyor to ascertain the amount of such loss and/or expense. Any amount from time to time so ascertained shall be added to the Contract Sum, and if an Interim Certificate is issued after the date of ascertainment any such amount shall be added to the amount which would otherwise be stated as due in such a Certificate.

COMMENTARY

Clause 34 (2) "*. . . direct loss and/or expense . . .*"

For other grounds for the recovery of loss or expense see clauses 11 (6) and 24 (1) and notes thereto as to the meaning of these words.

Clause 35*
Arbitration

(1) Provided always that in case any dispute or difference shall arise between the Employer or the Architect/Supervising Officer on his behalf and the Contractor either during the progress or after the completion or abandonment of the Works, as to the construction of this Contract or as to any matter or thing of whatsoever nature arising thereunder or in connection therewith (including any matter or thing left by this Contract to the discretion of the Architect/Supervising Officer or the withholding by the Architect/Supervising Officer of any certificate to which the Contractor may claim to be entitled or the measurement and valuation mentioned in clause 30 (5) (a) of these Conditions or the rights and liabilities of the parties under clauses 25, 26, 32 or 33 of these Conditions), then such dispute or difference shall be and is hereby referred to the arbitration and final decision of a person to be agreed between the parties, or, failing agreement within 14 days after either party has given to the other a written request to concur in the appointment of an Arbitrator, a person to be appointed on the request of either party by the President or a Vice-President for the time being of the Royal Institute of British Architects.

(2) Such reference, except on article 3 or article 4 of the Articles of Agreement, or on the questions whether or not the issue of an instruction is empowered by these Conditions, whether or not a certificate has been improperly withheld or is not in accordance with these Conditions, or on any dispute or difference under clauses 32 and 33 of these Conditions, shall not be opened until after Practical Completion or alleged Practical Completion of the Works or termination or alleged termination of the Contractor's employment under this Contract or abandonment of the Works, unless with the written consent of the Employer or the Architect/Supervising Officer on his behalf and the Contractor.

(3) Subject to the provisions of clauses 2 (2), 30 (7) and 31D (3) of these Conditions the Arbitrator shall, without prejudice to the generality of his powers, have power to direct such measurements and/or valuations as may in his opinion be desirable in order to determine the rights of the parties and to ascertain and award any sum which ought to have been the subject of or included in any certificate and to open up, review and revise any certificate, opinion, decision, requirement or notice and to determine all matters in dispute which shall be submitted to him in the same manner as if no such certificate, opinion, decision, requirement or notice had been given.

(4) The award of such Arbitrator shall be final and binding on the parties.

**(5) Whatever the nationality, residence or domicile of the Employer, the Contractor, any sub-contractor or supplier or the Arbitrator, and wherever the Works, or any part thereof, are situated, the law of England shall be the proper law of this Contract and in particular (but not so as to derogate from the generality of the foregoing) the provisions of the Arbitration Act 1950 (notwithstanding anything in section 34 thereof) shall apply to any arbitration under this Contract wherever the same, or any part of it, shall be conducted.

The provisions of this Condition do not apply to any dispute that may arise between the Employer and the Contractor as referred to in Clause 3 of the Supplemental Agreement annexed to these Conditions (the VAT Agreement).

**Where the parties do not wish the proper law of the contract to be the law of England and/ or do not wish the provisions of the Arbitration Act 1950 to apply to any arbitration under the contract held under the procedural law of Scotland (or other country) appropriate amendments to this sub-clause should be made.*

COMMENTARY

(1) *Clause* 35 (1) ". . . *any dispute or difference* . . ."

This clause gives a very wide jurisdiction to the arbitrator, but note the limitations effected by clauses 2 (2), 31D (3) and in particular by clause 30 (7).

An award is not a condition precedent to the right to bring an action on the contract.

Where one party has commenced proceedings in a court of law, if the other party wishes to arbitrate he can usually insist on arbitration by taking the appropriate steps to obtain a stay of proceedings in a court of law.[65]

It is submitted that the arbitrator has jurisdiction on claims arising out of frustration,[66] and to decide whether the contract has been repudiated,[67] but not to decide an allegation that the contract was void *ab initio*, *e.g.* for illegality,[68] nor that it is voidable for fraudulent misrepresentation,[69] nor, it seems, to rectify the contract.[70]

The arbitrator cannot, it is submitted, decide finally some preliminary point upon which his jurisdiction depends.[71] A party aggrieved by the arbitrator proceeding, or, conversely, refusing to proceed, with a reference, can invoke the assistance of the court.[72]

Part of sum indisputably due. Where part of a sum for which proceedings are issued in the court is indisputably due judgment is given for that sum and the dispute as to the balance is referred to arbitration.[73]

---

[65] Arbitration Act 1950, s. 4; and see *ante*, pp. 241–245.

[66] See *Heyman* v. *Darwins Ltd.* [1942] A.C. 356, 366 (H.L.); *Govt. of Gibraltar* v. *Kenney* [1956] 2 Q.B. 410.

[67] *Heyman* v. *Darwins Ltd.*, *supra*.

[68] *Ibid.*

[69] *Ibid.*; *Monro* v. *Bognor U.D.C.* [1915] 3 K.B. 167.

[70] *Printing Machinery Co. Ltd.* v. *Linotype & Machinery Ltd.* [1912] 1 Ch. 566. But see *ante*, p. 237.

[71] See Bankes L.J., *Smith* v. *Martin* (1925) 94 L.J.K.B. 645, 646 (C.A.), referring to the 1909 edition of the R.I.B.A. form. The present arbitration clause is more widely worded (*cf. Willesford* v. *Watson* (1873) L.R. 8 Ch.App. 473), but it is thought that the principle expressed in *Smith* v. *Martin* still applies. See generally, *ante*, p. 236.

[72] See Arbitration Act 1950, Russell, *Arbitration* and for a very short account, *ante*, p. 245.

[73] *Ellis Mechanical Services Ltd.* v. *Wates Construction Ltd.*, *The Times*, January 22, 1976 (C.A.).

Employer challenging architect. The employer as well as the contractor can ask the arbitrator to open up, review and revise the architect's decisions as to certificates or upon other matters.[74]

Does the court have the arbitrator's powers? The parties sometimes expressly agree not to enforce the arbitration clause and instead to have their disputes determined by the court. Does the court have the same powers as the arbitrator? For example, a certificate is normally a condition precedent to the right to payment but the arbitrator has express power to award a sum of money notwithstanding the absence of a certificate (clause 35 (1), (3)). It is thought that ordinarily the court will hold that the parties have impliedly agreed that the court should have all the powers of the arbitrator under the clause,[75] but as the matter is not clear,[76] parties frequently enter into an express agreement to such effect for the avoidance of doubt. Where one party commences proceedings in the court and the other party alleges that there is a dispute between them but neither seeks a stay it is thought that the position is the same.[77]

(2) *Clause* 35 (1) *". . . a written request . . ."*

For service of a notice, equivalent to a writ, necessary to stop time running, see Limitation Act 1939, s. 27 (3), (4). For the degree of formality required of a request for purposes of clauses 30 (7) and 35 (3) see note 14 to clause 30, but it may be that the court would be more strict in its requirement of formality necessary to satisfy the statute than the contract.

(3) *Clause* 35 (2) *" Such reference . . . shall not be opened until after Practical Completion . . ."*

This is an important limitation, but note the exceptions. These include the issue and amount of interim certificates [78] and with a little thought a very wide range of matters can be brought within these heads.

(4) *Clause* 35 (5) *English law*

This clause was introduced by the 1971 revision after the decision of the House of Lords in *Whitworth Street Estates* v. *Miller* (*James*) & *Partners*.[79]

---

[74] *Modern Engineering (Bristol) Ltd.* v. *Gilbert-Ash (Northern) Ltd.* [1974] A.C. 689, 709 (H.L.).

[75] *East Ham Corporation* v. *Sunley (Bernard & Sons) Ltd.* [1966] A.C. 406, 447 (H.L.); *Modern Engineering (Bristol) Ltd.* v. *Gilbert-Ash (Northern) Ltd., supra* at p. 720; *Neale* v. *Richardson* [1938] 1 All E.R. 753, 758 (C.A.); see also *Robins* v. *Goddard* [1905] 1 K.B. 294 (C.A.) but note comments on this case in the *East Ham* case, *supra*.

[76] *Kaye (P. & M.) Ltd.* v. *Hosier & Dickinson Ltd.* [1972] 1 W.L.R. 146 at p. 158 where Lord Wilberforce refers to clause 35 as conferring " very wide powers upon arbitrators to open up and review certificates which a court would not have." The final certificate clause in this case and in those cited in footnote 75 differed from the current clause 30 (7).

[77] See *supra*, footnote 75.

[78] *Cf. Farr (A. E.) Ltd.* v. *Ministry of Transport* [1960] 1 W.L.R. 956—a case on the arbitration provisions of the I.C.E. conditions.

[79] [1970] A.C. 583 (H.L.).

Appendix

	Clause
Defects Liability Period [if none other stated is 6 months from the day named in the Certificate of Practical Completion of the Works].	15, 16 and 30............................
Insurance cover for any one occurrence or series of occurrences arising out of one event.	19 (1) (a) £...............
Percentage to cover Professional fees.*	20 [A] ....
Date for Possession.	21..
Date for Completion.	21..
Liquidated and Ascertained Damages.	22 at the rate of £ per............
**Period of delay: (i) by reason of loss or damage caused by any one of the contingencies referred to in clause 20 [A] or clause 20 [B] (if applicable). (ii) for any other reason.	26..
Prime cost sums for which the Contractor desires to tender.	27 (g) ....
Period of Interim Certificates [if none stated is one month].	30 (1) ....
Retention Percentage (if less than five per cent.).†	30 (3) ....
Period of Final Measurement and Valuation [if none stated is 6 months from the day named in the Certificate of Practical Completion of the Works].	30 (5) ....
Period for issue of Final Certificate [if none stated is 3 months].††	30 (6) ....
Percentage addition.	31E ....
Formula Rules	31F (1) (a) (i)
Rule 3	Base Month.....................19......
Rule 3	Non-Adjustable Element......% (not to exceed 10%)
Rules 10 and 30 (i)	***Part I/Part II of Section 2 of the Formula Rules is to apply

*Where the professional persons concerned with the Works are all employees of a Local Authority no percentage should be inserted, but care should be taken to include in the sum assured the cost to the Employer of their services.

**It is suggested that the periods should be (i) three months and (ii) one month. It is essential that periods be inserted since otherwise no period of delay would be prescribed.

***Strike out according to which method of formula adjustment (Part I—Work Category Method or Part II—Work Group Method) has been notified in the Bills of Quantities issued to tenderers.

†The Percentage will be five per cent. unless a lower rate is specified here.

††The period inserted must not exceed six months.

Supplemental Agreement (the VAT Agreement)

Supplemental Agreement made the................day of................19.........
between the Employer and the Contractor and supplemental to the Articles of
Agreement and Conditions (hereinafter called ' the Conditions ') of even date
to which this Supplemental Agreement is annexed.

WHEREAS the Employer and the Contractor have in pursuance of Clause 13A of
the Conditions agreed to enter into this Supplemental Agreement.

NOW IT IS HEREBY AGREED AS FOLLOWS

Interim Payments—Addition of VAT—Written Assessment by Contractor

1. The Employer shall pay to the Contractor in the manner hereinafter set out
any tax properly chargeable by the Commissioners on the Contractor on the
supply to the Employer of any goods and services by the Contractor under this
Contract:
    (1) The Contractor shall not later than the date for the issue of each Interim
Certificate, for the issue of each certificate releasing retention to the Contractor
and, unless the procedure set out in Sub-Clause (3) hereof shall have been completed, for the issue of the Final Certificate, give to the Employer a written
provisional assessment of the respective values (less any Retention Percentage
applicable thereto) of those supplies of goods and services for which the certificate is being issued and which will be chargeable, at the relevant time of supply
under Regulation 21 (1) (a) of the Value Added Tax (General) Regulations 1972
on the Contractor at
        (i) a zero rate of tax (Category (i)) and
        (ii) any rate or rates of tax other than zero (Category (ii)).
The Contractor shall also specify the rate or rates of tax which are chargeable on
those supplies included in Category (ii), and shall state the grounds on which he
considers such supplies are so chargeable.
    (2) (a) Upon receipt of such written provisional assessment the Employer,
        unless he has reasonable grounds for objection to that assessment, shall cal-
        culate the amount of tax due by applying the rate or rates of tax specified by
        the Contractor to the amount of the assessed value of those supplies in-
        cluded in Category (ii) of such assessment, and remit the calculated amount
        of such tax, together with the amount of the certificate issued by the
        Architect/Supervising Officer, to the Contractor within the period for
        payment of certificates set out in Clause 30 (1) of the Conditions.
        (b) If the Employer has reasonable grounds for objection to the provisional
        assessment he shall within three working days of receipt of that assessment
        so notify the Contractor in writing setting out those grounds. The Contractor
        shall within three working days of receipt of the written notification of the
        Employer reply in writing to the Employer either that he withdraws the
        assessment in which case the Employer is released from his obligation under
        paragraph (a) hereof or that he confirms the assessment. If the Contractor
        so confirms then the Contractor may treat any amount received from the
        Employer in respect of the value which the Contractor has stated to be
        chargeable on him at a rate or rates of tax other than zero as being inclusive
        of tax and issue an authenticated receipt under Sub-Clause (4) hereof.

**Written Final Statement—VAT Liability of Contractor—Recovery
from Employer**

    (3) (a) After the issue of the Certificate of Completion of Making Good
        Defects under Clause 15 (4) of the Conditions the Contractor shall as soon as

he can finally so ascertain prepare a written final statement of the respective values of all supplies of goods and services for which certificates have been or will be issued which are chargeable on the Contractor at

    (i) a zero rate (Category (i)) and

    (ii) any rate or rates of tax other than zero (Category (ii))

and shall issue such final statement to the Employer.

The Contractor shall also specify the rate or rates of tax which are chargeable on the value of those supplies included in Category (ii) and shall state the grounds on which he considers such supplies are so chargeable.

The Contractor shall also state the total amount of tax already received by the Contractor for which a receipt or receipts under sub-clause (4) hereof have been issued.

(b) The statement under paragraph (a) hereof may be issued either before or after the issue of the Final Certificate under Clause 30 (6) of the Conditions.

(c) Upon receipt of the written final statement the Employer shall, subject to Clause 3 hereof, calculate the final amount of tax due by applying the rate or rates of tax specified by the Contractor to the value of those supplies included in Category (ii) of the statement and deducting therefrom the total amount of tax already received by the Contractor specified in the statement, and shall pay the balance of such tax to the Contractor within 28 days from receipt of the statement.

(d) If the Employer finds that the total amount of tax specified in the final statement as already paid by him exceeds the amount of tax calculated under paragraph (c) of this Sub-Clause the Employer shall so notify the Contractor who shall refund such excess to the Employer within 28 days of receipt of the notification, together with a receipt under Sub-Clause (4) hereof showing a correction of the amounts for which a receipt or receipts have previously been issued by the Contractor.

Contractor to issue Receipt as Tax Invoice

(4) Upon receipt of any amount paid under certificates of the Architect/Supervising Officer and any tax properly paid under the provisions of this Clause the Contractor shall issue to the Employer a receipt of the kind referred to in Regulation 21 (2) of the Value Added Tax (General) Regulations 1972 containing the particulars required under Regulation 9 (1) of the aforesaid Regulations or any amendment thereof to be contained in a tax invoice.

Value of Supply—Liquidated Damages to be Disregarded

2. (1) If, when the Employer is obliged to make payment under Clause 1 (2) or 1 (3) of this Agreement, he is empowered under Clause 22 of the Conditions to deduct any sum calculated at the rate stated in the Appendix to those Conditions as Liquidated and Ascertained Damages from sums due or to become due to the Contractor under this Contract he shall disregard any such deduction in calculating the tax due on the value of goods and services supplied to which he is obliged to add tax under Clause 1 (2) or 1 (3) hereof.

(2) The Contractor when ascertaining the respective values of any supplies of goods and services for which certificates have been or will be issued under the Conditions in order to prepare the final statement referred to in Clause 1 (3) of this Agreement shall disregard when stating such values any deduction by the Employer of any sum calculated at the rate stated in the Appendix as Liquidated and Ascertained Damages under Clause 22 of the Conditions.

Employer's Right to Challenge Tax Claimed by Contractor

3. (1) If the Employer disagrees with the final statement issued by the Contractor under Clause 1 (3) of this Agreement he may but before any payment or refund

becomes due under Clause 1 (3) (c) or (d) hereof request the Contractor to obtain the decision of the Commissioners on the tax properly chargeable on the Contractor for all supplies of goods and services under this Contract and the Contractor shall forthwith request the Commissioners for such decision. If the Employer disagrees with such decision then, provided the Employer indemnifies and at the option of the Contractor secures the Contractor against all costs and other expenses, the Contractor shall in accordance with the instructions of the Employer make all such appeals against the decision of the Commissioners as the Employer shall request. The Contractor shall account for any costs awarded in his favour in any appeals to which this Clause applies.

(2) Where, before any appeal from the decision of the Commissioners can proceed, the full amount of the tax alleged to be chargeable on the Contractor on the supply of goods and services under the Conditions must be paid or accounted for by the Contractor, the Employer shall pay to the Contractor the full amount of tax needed to comply with any such obligation.

(3) Within 28 days of the final adjudication of an appeal (or of the date of the decision of the Commissioners if the Employer does not request the Contractor to refer such decision to appeal) the Employer or the Contractor as the case may be shall pay or refund to the other in accordance with such final adjudication any tax underpaid or overpaid as the case may be under the provisions of this Agreement and the provisions of Clause 1 (3) (d) shall apply in regard to the provision of authenticated receipts.

Discharge of Employer from Liability to Pay Tax to the Contractor

4. Upon receipt by the Contractor from the Employer or by the Employer from the Contractor as the case may be of any payment under Clause 1 (3) (c) or (d) hereof or upon final adjudication of any appeal made in accordance with the provisions of Clause 3 hereof and any resultant payment or refund under Clause 3 (3) hereof, the Employer shall be discharged from any further liability to pay tax to the Contractor in accordance with this Condition. Provided always that if after the date of discharge under this Clause the Commissioners decide to correct the tax due from the Contractor on the supply to the Employer of any goods and services by the Contractor under this Contract the amount of any such correction shall be an additional payment by the Employer to the Contractor or by the Contractor to the Employer as the case may be. The provisions of Clause 3 in regard to disagreement with any decision of the Commissioners shall apply to any decision referred to in this proviso.

Awards by Arbitrator or Court

5. If any dispute or difference is referred to an Arbitrator appointed under Clause 35 of the Conditions or to a Court then insofar as any payment awarded in such arbitration or court proceedings varies amounts certified for payment of goods or services supplied by the Contractor to the Employer under this Contract or is an amount which ought to have been so certified but was not so certified then the provisions of this Agreement shall so far as relevant and applicable apply to any such payments.

Arbitration Provisions Excluded

6. The provisions of Clause 35 of the Conditions shall not apply to any matters to be dealt with under Clause 3 of this Agreement.

Employer's Right where Receipt not Provided

7. Notwithstanding any provisions to the contrary elsewhere in the Conditions the Employer shall not be obliged to make any further payment to the Contractor

under the Conditions if the Contractor is in default in providing the receipt referred to in Clause 1 (4) hereof. Provided that this Clause shall only apply where:

> (i) The Employer can show that he requires such receipt to validate any claim for credit for tax paid or payable under this Agreement which the Employer is entitled to make to the Commissioners, and
> (ii) the Employer has paid tax in accordance with the provisional assessment of the Contractor under Clause 1 hereof unless he has sustained a reasonable objection under Clause 1 (2) hereof.

8. Where Clause 25 (4) (d) of the Conditions becomes operative there shall be added to the amount allowable or payable to the Employer in addition to the amounts certified by the Architect/Supervising Officer any additional tax that the Employer has had to pay by reason of determination under Clause 25 as compared with the tax the Employer would have paid if the determination had not occurred.

*As WITNESS the hands of the said parties

Signed by the said Employer in the
  presence of
....
....

Signed by the said Contractor in the
  presence of
....
....

  *If the contract is to be executed under seal, this clause and the words following it must be altered accordingly.

COMMENTARY

(1) *Nature of VAT*

VAT is a tax introduced by the Finance Act 1972, as amended and extended by subsequent legislation, and in orders and regulations made under the Act. It is a substantial subject and it is not appropriate to deal with it as such here. Reference should be made to the relevant statutes, orders and regulations and textbooks thereon, and assistance can be gained from the literature issued by the Commissioners of Customs and Excise who are charged with the care and management of the tax. There is also a Practice Note, No. 17, issued by the Joint Contracts Tribunal.

The tax is chargeable on the supply of all goods and services unless there is a specific provision to the contrary in the VAT legislation. Relief from VAT arises either by zero-rating or because a person supplying exempt goods or services does not have to charge his customers output tax.

(2) *Reason for supplemental agreement*

For reasons explained at length in Practice Note 17, the JCT decided that the contract sum should be exclusive of the tax. Hence there is this supplemental agreement to provide for the payment of whatever tax is payable by the employer to the contractor and for dealing with difficulties which may arise. The main body of the agreement has various references to achieve this purpose, *i.e.* in clauses 4

(2), 13A, 26 (1) (a), 27 (c), 30 (5) (c), 31D (5), 35. These references were all introduced in the July 1973 Revision, although the appropriate amendments had been
issued in March 1973. An earlier version of clause 13A appeared in the July 1972
Revision.

(3) *Zero-rating*

Reference should be made to Schedule 4 to the Finance Act 1972 as amended
and to the Commissioners' Notice No. 708, as amended. Very broadly the effect
is that the supply, in the course of the construction, alteration or demolition of
any building or of any civil engineering work and of materials or of builders'
hardware, sanitary ware or other articles of a kind ordinarily installed by builders
as fixtures in connection with the foregoing are zero-rated. But the following is
not zero-rated: services of an architect, surveyor or any person acting as consultant or in a supervisory capacity, any work of repair or maintenance, the supply
of services by a sub-contractor, the supply of services in the course of construction or alteration of any civil engineering work within the grounds or garden of a
building used or to be used wholly or mainly as a private residence (*e.g.* swimming
pools or tennis courts). The result is ordinarily that the contractor does not have to
charge the employer with tax in most cases where the work is not that of repair
or maintenance. Any sub-contractor has to charge the next contractor above him
and so up the line to the main contractor but each person in turn who has to pay
the tax can recover it, ultimately, if necessary, from the Customs and Excise.

(4) *Supplemental agreement*

If tax is chargeable it has to be charged upon each interim payment. Therefore
the agreement makes provision for a provisional assessment and a final statement
(clause 1).   Liquidated damages are disregarded (clause 2). Provision is made
in clause 3 for the employer to challenge the tax claimed by the contractor.
Effectively he has to indemnify and pay for the litigation to be carried out in the
name of the contractor. The other clauses are ancillary to the foregoing. If there
are problems about the agreement it would be prudent to consult Practice Note
17 which, it is stated, was prepared after consultation with the Commissioners of
Customs and Excise.

Section 2: JCT Practice Notes 20 and 21 [80]

Practice Note 20

Standard Form of Building Contract for use with Bills of Approximate Quantities
October 1975

**New form first issued October 1975—Private and Local Authorities
Editions—Adaptation of Standard Form WITH Quantities, 1963
edition (July 1975 revision)**

1. The Tribunal received representations that, judged by value, a considerable
volume of work, particularly in the private sector, was let on the basis of Bills of
Approximate Quantities so that the Works have to be completely re-measured.

---

[80] Printed by kind permission. Neither the appendix to Practice Note 20 nor the
supplement with Practice Note 21 is printed here.

For such work the Standard Form for use " WITH Quantities " is not suitable since it does not provide for general re-measurement of the work to be executed but only for adjustment of work executed as a result of Architect's instructions involving a variation or the expenditure of a provisional sum and the re-measurement of any sections of the quantities specifically marked as " Provisional " in the Contract Bills. The Tribunal therefore decided, with the approval of its constituent bodies, that it would be desirable to issue an adaptation of the " WITH Quantities " form to make it suitable for use where, at the tender stage, it is made clear to the Contractor that the quantities in the Bills are approximate and are all subject to re-measurement.

2. The Tribunal has now published two new forms of contract. The new forms are the " Standard Form of Building Contract with APPROXIMATE Quantities Local Authorities Edition 1975 Issue " and " Standard Form of Building Contract with APPROXIMATE Quantities Private Edition 1975 Issue." The headnote to these forms states: " This form is for use where the Works have been substantially designed but not completely detailed so that the quantities shown in the Bills are approximate and are subject to re-measurement."

3. Although the Scottish Building Contract Committee had previously advised the Tribunal that there was not sufficient demand for such a Form in Scotland for them to issue a Scottish Supplement adapting the provisions of the Form to Scottish law the Committee is, at the date of issue of this Practice Note, reviewing the matter.

4. There is appended to this Note a list of the differences between these new forms of contract and the Standard Form WITH Quantities 1963 Edition (July 1975 Revision) which should enable users to identify the changes that have been made. The following paragraphs of this Practice Note draw attention to the more important features of the two new forms and the circumstances in which the Tribunal intends them to be used. References to " the Form " are to both the Private and Local Authorities editions.

5. The recitals of the Articles of Agreement of the Form state that the Bills of Approximate Quantities describe the work to be done and are also intended to set out a reasonably accurate forecast of the quantity of the work to be done. This important qualification is intended to prevent the use of this Form for Works which have not been substantially designed with the result that the quantities do not set out a reasonably accurate forecast of the quantity of the Works. For the position where this intention is not fulfilled see both paragraph 11 below, where it refers to clause 13 (3) (b), and new paragraph (m) in clause 23 (Extension of time).

6. The new Form should NOT be used where only certain sections of the quantities in the Bills are approximate (for example: abnormal foundations; drainage; external works). In such cases the Standard Form (" WITH Quantities ") should be used and the relevant items in the Contract Bills marked " Provisional."

7. The Articles provide for the insertion of a " Tender Price " which is the total of the prices inserted by the Contractor in the Bills of Approximate Quantities. This Tender Price will always be converted as a result of the provisions on re-measurement and valuation contained in the Contract Conditions and the total payment to be made to the Contractor is referred to throughout as the " Ascertained Final Sum."

8. Clause 11, entitled " Contract Bills," deals with the priced Bills of Approximate Quantities which are defined in the Articles of Agreement as " the Contract Bills." These set out the quality and quantity of the work included in the " Tender Price." Since, however, that quantity is only approximate, the actual work carried out in accordance with the Contract Drawings and the Architect's/Supervising Officer's instructions will have to be re-measured. Such re-measurement is not a

variation under this Contract since the quantity of work to be carried out has never been fixed and cannot therefore be subject to a variation. For this reason the definition of "variation," which appears in Clause 12 (2) of the new Form, omits any reference to a change in quantity.

9. Clause 13 (1) provides for the measurement and valuation of work (except work resulting from instructions as to the expenditure of prime cost sums for nominated sub-contractors and work for which a tender from the Contractor under Clause 27 (g) is accepted).

10. Clause 13 (2) provides that the re-measurement must be carried out using the same principles of measurement as have been used for the preparation of the Contract Bills.

11. Clause 13 (3) contains the rules for the valuation of the measured work which follows those for variations in Clause 11 (4) of the Standard Form ("WITH Quantities" and "WITHOUT Quantities"). The main differences, however, are:

(*i*) *Rule* (*b*) *has been expanded to provide for the situation where, contrary to the intention expressed in the Articles (see paragraph 5 above), the approximate quantities were not a reasonably accurate forecast of the quantity of the work. In these circumstances if the rates in the Contract Bills do not provide a reasonable basis the Contractor is entitled to be paid a "fair valuation" of the actual quantity of the work he is instructed to carry out. Work whose quantity is not reasonably accurately forecast is never valued under Rule* (*a*).

(*ii*) *Rule* (*c*) *requires the Quantity Surveyor to take into account the effect of any percentage or lump sum adjustments priced in the Contract Bills in applying Rules* (*a*) *and* (*b*) *for valuation of the measured work.*

12. The adjustment of the Preliminaries section of the Contract Bills is dealt with in Clause 30 (5) (c) (vii) where it is specifically required that the Preliminaries shall be adjusted "where appropriate" to take account of the valuation of the work measured by the Quantity Surveyor under Clause 13 (3) (b). Thus where the re-measured work was not reasonably accurately forecast or is not of similar character or is not executed under similar conditions as the work in the Contract Bills, the Quantity Surveyor must take into account the pricing of any Preliminary items related thereto, and, where appropriate, make the necessary adjustment to the Preliminaries in the calculation of the Ascertained Final Sum. If the Quantity Surveyor deals with the Preliminaries element in arriving at a fair valuation under Rule 13 (3) (b) then no further adjustment would be required under Clause 30 (5) (c) (vii).

13. Clause 13 (4) deals with the valuation on a daywork basis of work which cannot properly be measured and valued by the Quantity Surveyor.

14. Clause 13 (5) provides, like Clause 11 (6) in the Standard Form ("WITH Quantities" and "WITHOUT Quantities") for the ascertainment of any direct loss and/or expense as a result of the execution by the Contractor of a variation or of work for which a provisional sum has been included in the Contract Bills; and has also been extended to apply to work by the Contractor whose quantity was not reasonably accurately forecast in the Contract Bills. The clause has not been extended to apply to work covered by approximate quantities which were a reasonably accurate forecast of the quantity of the work actually carried out since the Rules in Clause 13 (1) and (2) for valuing such work are considered to be adequate.

15. Clause 23 (m) is new and gives as ground for extension of time the execution of work the quantity of which was not reasonably accurately forecast in the Contract Bills.

16. Clause 30 (5) deals with the computation of the Ascertained Final Sum:
Paragraph (a) requires the Contractor to send all the documents and accounts necessary for the computations to the Architect or Quantity Surveyor either before or within a reasonable time after Practical Completion.

These would not include the priced Bills of Remeasurement which para-graph (b) requires the Quantity Surveyor to supply to the Contractor not later than the end of the period of Final Measurement and Valuation stated in the Appendix.

Paragraph (c) defines in detail what must be included in the Ascertained Final Sum and is drafted on the assumption that the Quantity Surveyor ' starts with a blank sheet of paper ' and builds up the Ascertained Final Sum in a logical and orderly way. The 10 sub-paragraphs of paragraph (c) should ensure that nothing which is appropriate in the Contract Bills is omitted from the Ascertained Final Sum.

The general ' sweeping-up ' provision in sub-paragraph (x) would include such items, where relevant, as:

Clauses:

5	Correction of setting out	(Deduction)
6	Testing	(Possible addition)
7	Patents/Royalties	(Possible addition)
15 (2) & (3)	Making Good Defects	(Deduction)
19 (1) (c)		
19 (2) (c)	Default in insuring	(Deduction)
20 [A]		
24	Claims for disturbance	(Addition)
25	Determination	(Deduction or addition)

17. The new Form provides only for Clause 31A—Fluctuations and Clause 31F—Formula Adjustment. Clause 31B of the Standard Form " WITH Quantities " is omitted (Clause 31B is used only with firm price contracts). Inasmuch as work under the " APPROXIMATE Quantities " Form is not such that the Contractor can himself at an early stage obtain and place firm orders for the work a ' firm price ' is inappropriate. Moreover such work would not be within the criteria for thorough pre-planning set out in the Department of the Environment Circular (Circular 85/72, Welsh Office 181/72, dated 18th September, 1972) entitled " Tender and Contract Procedures for Building and Civil Engineering Work: Firm Price Tendering." Contracts let under this Form in the public sector do not, therefore, have to be on a firm price and will thus include either Clause 31A or Clause 31F—Formula Adjustment.

18. Where Clause 31F—Formula Adjustment—is used the Formula Rules Parts I and III apply; Part II—Works Groups—is not to apply since the quantities, being approximate, would not be suitable for use in preparing the Work Group Indices which require the use of the tender weightings of the work in the various Work Categories.

Appendix

List of differences between the new Standard Form " With APPROXIMATE Quantities " and the Standard Form " WITH Quantities " 1963 Edition (July 1975 Revision) [80a]

Sectional Completion Practice Note 21 [81]

Adaptation of the Standard Form of Building Contract WITH Quantities, Private and Local Authorities editions, 1963 (July 1975 Revision)—Additional Clause incorporating modifications for use WHERE THE WORKS ARE TO BE COMPLETED BY PHASED SECTIONS.

[80a] Not printed.

[81] The supplement itself is, for reasons of space, not printed. Any who doubt the number and complexity of the detailed amendments required should obtain and consider the supplement and in particular, its " Table of Changes."

Issued: December 1975

Notes on the use of this supplement

This Supplement has been drawn up so that it may be used for adapting the Standard Form of Building Contract WITH Quantities as follows:

A. **The Recitals.** In the Contract Form to be adapted—
EITHER (1):
insert " both . . . and the division of the Works into Sections for phased completion (hereinafter referred to as ' Sections ') " as shown overleaf;
OR (2):
strike out the first page of the Form, detach this sheet from the Supplement, strike out this page, and affix the sheet to the Form so that the adaptation page overleaf may be used as the first page of the Form. If the Form is the Private edition insert as appropriate the address or registered office of the Employer and insert " his Architect " at the end of the first recital as shown on the Private edition Form itself; and delete the last five lines of the adaptation page (the second recital) which appear at the top of page 2 of the Private edition.

B. **The Articles of Agreement and Annexed Conditions.** Detach the centre pages 3–6 of this Supplement, which comprise an additional clause to be inserted in the Articles of the Form, and affix those pages between pages 2 and 3 of the Form so as to follow Clause 4 of the Articles and precede the page for signing/sealing.

C. To indicate where the Conditions have been modified by the incorporation of the additional clause and Table of Changes insert the symbol (T) (or some other indication) in the margin beside the modified clauses.

D. **Appendix.** Strike out the Appendix in the Contract Form except the reference to Formula Rules if Clause 31F is to apply and in substitution detach page 7 of this Supplement, strike out page 8 on the reverse and affix page 7 within the Contract Form.

E. The explanatory notes in Practice note 21 (set out below) are for guidance only and are not intended to form part of the modifications or of the Adapted Contract in which the modifications are incorporated.

Practice Note 21

1. The Sectional Completion Supplement first issued by the Joint Contracts Tribunal from December 1975 provides for the Standard Form of Building Contract WITH Quantities (Local Authorities and Private Editions 1963, July 1975 Revision) to be adapted so as to be suitable for use where the Works are to be completed by phased sections. The adaptation makes it necessary to substitute, or add a reference to, a " Section " or " Sections " in some of the Conditions of the Standard Form where the expression " the Works " appears. All the adapting modifications are set out seriatim in the Supplement. A Standard Form Contract so adapted is referred to in the following paragraphs as " the Adapted Contract."

Practice
2. The Adapted Contract is intended for use only where tenderers are notified that the Employer requires the Works to be carried out by phased sections of which the Employer will take possession on Practical Completion of each Section. The Adapted Contract cannot be used for contracts where the Works are not, at the tender stage, divided into Sections in the tender documents. Attention is called to Clause 16 of the Standard Form entitled " Partial possession by Employer " which is designed to meet those cases where the Adapted Contract has not been used but the Employer nevertheless, by agreement with the Con-

tractor, wishes to take possession of a part of the Works. Where the Adapted Contract is used Clause 16 is suitably modified.

3. It is essential that the tender documents (normally the Contract Drawings and the Bills of Quantities) should identify clearly the Sections which, together, comprise the whole Works. The Sections should then be serially numbered and these Section numbers inserted in the Appendix of the Adapted Contract. In this connection care should be taken, in dealing with any part of the Works which is common to all or several Sections, (such as a boiler-house serving three separate Sections, each comprising a block of flats) to put this part of the Works into a separate Section, and to ascribe to it a Section Value and Date for Possession, Date for Completion, Rate of Liquidated Damages for delay and Defects Liability Period.

The Adaptation

4. The principal modifications to the contract procedures which are connected with the Adapted Contract are the division of the Works into definite Sections in the Contract Documents, coupled with the division of the Contract Sum into corresponding Section Values and the fixing of separate completion periods for each of the Sections. The Adapted Contract remains, however, a single Contract and one final certificate is to be issued at the end; no provision is made for separate final certificates for each Section.

5. The contractual provisions of the Standard Form are, therefore, modified in the Adapted Contract as follows:

—The Contract Documents

5.1 The recitals to the Articles of Agreement declare that the Contract Drawings and Contract Bills show and describe the division of the Works into Sections. This is followed through into Clause 1 (1) of the Conditions which refers to the Appendix in which each Section is identified by reference to the Contract Drawings and Contract Bills.

5.2 The Section Value ascribed to each Section is also entered in the Appendix for the purposes of clauses 16 and 20 (A). The term " Section Value " is defined in Clause 16 (e) as " the value ascribed to the relevant Section in the Appendix," and the relevant Appendix entry indicates that each Section Value is to be the total value of the Section ascertained from the Contract Bills. The Section Values must amount in total to the Contract Sum, and should take into account the apportionment of Preliminaries and other like items priced in the Contract Bills.

5.3 Also to be entered in the Appendix separately for each Section are: Dates for Possession, Dates for Completion, Rates of Liquidated Damages for delay, and Defects Liability Periods.

—Carrying out the Works

5.4 The Contractor is to be given possession of the site for each of the Sections of the Works. He is then to begin and proceed with each Section concurrently or successively as required by and under the Contract, and to carry out each Section within the contract period stated in the Appendix (Clause 21 (1)) or as extended by the Architect* under the Contract (Clause 23). Liquidated damages for delay relate to each Section (Clause 22), and are separately calculated where there is delay in completing any Section.

—Practical Completion

5.5 On practical completion of any Section the Architect must issue a Practical

*In the Local Authorities edition the term " Supervising Officer " is substituted throughout where the official concerned is not an Architect.

Completion certificate for that Section (Clause 15). In consequence the Contractor is relieved of his corresponding duty to insure for that Section under Clause 20 (A) (if applicable), and the first moiety of Retention attributable to that Section must be released (Clause 30 (4)). A separate Defects Liability Period operates for each Section (Clause 15 (2)).

—Final Account

5.6 When all Sections have been carried out the Architect must issue a certificate to that effect (Clause 15 (6)) and the period begins to run within which the final account is to be prepared and the Final Certificate issued for the whole of the Works comprising all the Sections (Clause 30 (6)).

Insurance

6. The Architect should particularly note that when making insurance arrangements under Clause 19 (2) (a) agreement must be reached with the insurers as to whether any Section for which a Practical Completion Certificate has been issued is to be treated as continuing to be included in the Works and so not insured OR is to be treated as " property other than the Works " and so covered by the insurance.

APPROXIMATE Quantities Standard Form

7. The adaptations shown in the Sectional Completion Supplement may also be used in conjunction with the Standard Form of Building Contract for use with Bills of APPROXIMATE Quantities (first issued October 1975), but varied as necessary where there are differences between that Form and the WITH Quantities Form as indicated in the Appendix to Practice note 20.

CHAPTER 16

COLLATERAL CONTRACTS WITH
NOMINATED SUB-CONTRACTORS

THE desirability of such contracts has been discussed generally *ante*, p. 183, and in reference to the Standard form of building contract *ante*, p. 372, and is referred to in relation to the I.C.E. conditions *post*, p. 438. The courts have referred to the benefits of such contracts,[1] and their basis in law is well established.[2] This chapter deals with the R.I.B.A. form of agreement between employer and nominated sub-contractor, 1973 and the form of warranty, issued by the R.I.B.A. in 1970, to be given by a nominated supplier, and contains some notes upon the drafting of a collateral contract for use with the I.C.E. conditions.

Section 1: R.I.B.A. Form of Agreement Between Employer and Nominated Sub-Contractor, 1973

Employer/Sub-Contractor 1973 [3]

Form of Agreement between an Employer and a Sub-Contractor nominated under the Standard Form of Building Contract 1963 Edition, July 1973 revision*

* *The July 1975 revision made no change to the relevant provision.*

This Agreement is made the.................................day of

Date

...........................197......between...........................

Parties

.... (" the Employer ")

and ....

....(" the Sub-Contractor ")

WHEREAS the Sub-Contractor has been invited to tender for

Short description
of Sub-Contract   the carrying out of ....
Works

....

...........................(" the Sub-Contract Works ")
as a sub-contractor nominated or to be nominated to a
contractor under the terms of a Standard Form of Building

---

[1] *Bickerton & Son Ltd.* v. *N.W. Metropolitan Regnl. Hospital Board* [1969] 1 All E.R. 977, 982, 995 (C.A.); see also *Gloucestershire County Council* v. *Richardson* [1967] 3 All E.R. 458, 473 (C.A.).

[2] See *ante*, p. 174.

[3] This Form of Agreement is issued by the Royal Institute of British Architects, and is published for the R.I.B.A. by R.I.B.A. Publications Ltd., 66, Portland Place, London W1N 4AD. Printed here by kind permission.

Contract, 1963 edition, it being intended that the Sub-Contract Works form part of.....................................

....

to be carried out by the contractor under the terms of a contract with the Employer providing for payment by the contractor of liquidated damages for delay....................

NOW IT IS AGREED BETWEEN THE EMPLOYER AND THE SUB-CONTRACTOR that if the Sub-Contractor is at or after the date hereof nominated as sub-contractor for the Sub-Contract Works the warranties and undertakings set out as follows on pages 2 and 3 hereof shall have effect.

A   The Sub-Contractor warrants as follows:

A(1) The Sub-Contractor has exercised and will exercise all reasonable skill and care in:

(a) the design of the Sub-Contract Works insofar as the Sub-Contract Works have been or will be designed by the Sub-Contractor; and

(b) the selection of materials and goods for the Sub-Contract Works insofar as such materials and goods have been or will be selected by the Sub-Contractor; and

(c) the satisfaction of any performance specification or requirement insofar as such performance specification or requirement is included or referred to in the tender of the Sub-Contractor as part of the description of the Sub-Contract Works.

A(2) The Sub-Contractor will, save insofar as he is delayed by the events described or referred to in sub-paragraphs (i) and (ii) of Clause 8 (b) of the February 1971 revision* of the NFBTE/FASS/CASEC *Form of Sub-Contract for use where the Sub-Contractor is nominated under the* 1963 *edition of the Standard Form of Main Contract,*

(a) so supply the Architect or contractor with such information as either may reasonably require and

(b) so perform the sub-contract works

that the contractor shall not become entitled to an extension of time under Clause 23 of the main contract by the failure of the Sub-Contractor so to supply such information or by delay on the part of the Sub-Contractor provided always that no liability shall arise in respect of such delay on the part of the Sub-Contractor until he has accepted the contractor's order in respect of the Sub-Contract Works.

A(3) Nothing in the tender of the Sub-Contractor shall operate to exclude or limit his liability for breach of the warranties set out herein.

* *Footnote:* Includes a reference to any subsequent revision of the Form of Sub-Contract which repeats the same provision.

B   The Employer undertakes, without prejudice to the operation of Clause 27 (c) of the main contract apart from this undertaking, as follows:

B(1) The Architect will in accordance with Clause 27 (c) require the contractor to furnish reasonable proof that any amount included in the calculation of the amount stated as due in a certificate issued under Clause 30 of the main contract in respect of the total value of the work, materials or goods executed or supplied by the Sub-Contractor has been duly discharged for the purpose of Clause 27 (c), but so that where there is a dispute between the contractor and the Sub-Contractor the Architect may treat reasonable proof of payment by the contractor to a stakeholder pending determination of the dispute as reasonable proof of discharge to the extent of such payment to the stakeholder.

B(2) If the contractor fails to furnish such proof of discharge the Architect will issue a certificate to that effect and thereupon the Employer will deduct the amount so certified and unpaid from any sum due or to become due to the contractor under a certificate issued under Clause 30 of the main contract and the Employer will pay the amount so deducted to the Sub-Contractor.

VAT

B(3) For the purposes of this Agreement the provisions of Clause 27 (c) of the main contract enabling the Employer to pay the Sub-Contractor shall be extended in respect of any additional amount for value added tax which would have been properly due to the Sub-Contractor in the following manner:

(a) When issuing a certificate in accordance with this Agreement the Architect will issue a copy of the certificate to the Sub-Contractor.

(b) The Sub-Contractor will then in writing (of which he will send a copy to the contractor)

(i) inform the Employer of the amount of any value added tax properly chargeable on the supply to the contractor in respect of which the Employer will make a payment under paragraph B(2) of this Agreement, stating the rate of tax and any discount for cash or other amount deductible for the purposes of the tax, and

(ii) undertake

1) that no tax invoice or authenticated receipt has at that time been issued in respect of the value added tax attributable to such supply, and

2) that upon, but not before, receiving evidence (whether by copy of the next succeeding certificate issued to the contractor or otherwise) of recovery by the Employer of such additional amount (by deduction or otherwise) from the contractor, the Sub-Contractor will forthwith issue to the contractor an authenticated receipt for payment and will send, if so required, a copy of such receipt to the Employer and the Architect.

(c) Upon receipt of such notice the Employer will include the additional amount for value added tax in the amount to be deducted and paid by the Employer under paragraph B(2) of this Agreement.

Proviso   B(4) Provided always that after practical completion of the main contract works or the happening of one of the events set out in Clause 25 (2) of the main contract whichever first occurs the Employer's obligation to pay the Sub-Contractor pursuant to this Agreement will be only to pay the Sub-Contractor out of monies held by the Employer under Clause 30 of the main contract so that no payment need be made which would reduce such monies below the amount reasonably necessary

(a) to pay for the remedying of defects and omissions in any part of the main contract works, and

(b) to provide a fund for the remedying of defects and omissions in any part of the main contract works which might appear within any unexpired part of the defects liability period, and

(c) to provide a fund for the amount of the expenses and direct loss and/or damage recoverable under Clause 25 of the main contract where there has been a determination under that clause, and

(d) to pay other nominated sub-contractors under Clause 27 (c).

Signed by or on behalf   Signed by or on behalf
of the Employer   of the Sub-Contractor

COMMENTARY

(1) *History*

After certain judicial criticisms, particularly in *Westminster City Corporation* v. *Jarvis*,[4] the R.I.B.A., in 1969, issued forms of warranty to be given respectively by nominated sub-contractors and nominated suppliers in consideration of nomination. The form relating to nominated suppliers is still in use and appears in Section 2 of this chapter. The form relating to nominated sub-contractors was printed at page 525 of the third edition of this book. Subsequent to the issue of the 1969 form, relating to nominated sub-contractors, discussions took place between the R.I.B.A. and organisations representing specialist sub-contractors (F.A.S.S., C.A.S.E.C., N.F.B.T.E.) and various alterations were negotiated resulting in a form issued in 1971 substantially similar to that printed, save that it did not contain clause B3 added in 1973 to deal with VAT. Its use is strongly recommended by the R.I.B.A. in connection with nomination under the Standard form of building contract where the nominated sub-contractor is to be responsible for design, or the main contractor could become entitled to an extension of time under clause 23 (f) or 23 (g) in respect of a nominated sub-contractor's default.[5]

(2) *Recitals*

The R.I.B.A. issues a Standard form of tender for nominated sub-contractors. While the form of agreement is not dependent upon the use of this form of tender, it is helpful if it is used. It draws attention to the form of agreement and it has provision for the insertion of the rate of liquidated damages under the main contract. This fixes the sub-contractor with knowledge of the loss which the employer will suffer by reason of the sub-contractor's default resulting in the main contractor obtaining an extension of time. Therefore, ordinarily this will be the measure of damages payable by the nominated sub-contractor for his delay. If the standard form of tender is not used the amount of liquidated damages should be brought to the attention of the sub-contractor before the agreement is entered into.

(3) *Comparison with* 1969 *warranty*

The 1969 warranty to be given by nominated sub-contractors followed very closely the form of that to be given by nominated suppliers, printed *infra*.

(4) *Time of operation of agreement*

The agreement only comes into force if the sub-contractor is nominated and no liability arises in respect of his warranty relating to delay given in paragraph A (2) until the sub-contractor has accepted the contractor's order in respect of the sub-contract works.

(5) *Clause A* (1). *Design warranty*

This is the clause intended to give the employer a remedy in respect of design carried out by the sub-contractor. Note that it is not an absolute obligation but

---

[4] [1970] 1 W.L.R. 637 (H.L.).
[5] Practice note issued with the form.

is limited to all reasonable skill and care. For the importance of the distinction, see *ante*, p. 208. In the 1969 warranty, the sub-contractor was required to give an absolute warranty as to compliance with performance specifications.

(6) *Clause A* (2). *Delay by sub-contractor*

This is intended to give the employer a remedy in respect of delay by the sub-contractor which results in the main contractor obtaining an extension of time and thereby depriving the employer of liquidated damages. Note that it only comes into operation when the sub-contractor has accepted the contractor's order. The sub-contractor is entitled to extensions of time which correspond with those (other than his own delay) to be granted under the main contract.

(7) *Clause A* (3). *Limitation clauses*

Under ordinary circumstances this clause ought not to be necessary because the sub-contractor's tender is to the main contractor. However, it is inserted to deal with any argument by the sub-contractor that his exclusion or limitation clauses are part of this agreement.

(8) *Repudiation by nominated sub-contractor*

Under the Standard form of building contract the employer has to bear the losses flowing from a nominated sub-contractor's repudiation of his sub-contract (see *ante*, pp. 182 and 369). This agreement affords no express protection to the employer, although in so far as there has been a breach of duties relating to design or progress the employer may have some remedy by virtue of the warranties given in A (1) and A (2).

(9) *Clause B. Employer's undertaking to operate clause 27 (c)*

As between himself and the contractor the employer has only a discretion whether or not to operate clause 27 (c). (For the clause and its operation, see *ante*, p. 377.) As between himself and the sub-contractor he undertakes to operate it but subject to the very important proviso which protects his interests and, incidentally, those of other nominated sub-contractors. The position upon the main contractor's insolvency is not clear. The courts do not like clauses which favour one creditor or group of creditors at the expense of others (see *ante*, p. 188).

The right of direct payment under clause 27 (c) has been held to be good where, after the contractor's insolvency, the contract is continued (see *ante*, p. 377). In such circumstances it would, it appears, be prudent, if the employer is minded to make a direct payment to the sub-contractor, to make it expressly pursuant to his discretionary right under the main contract and not pursuant to his contractual duty under this agreement, in so far as this agreement has not been tested in the courts.

(10) *Performance bond and insurance*

The 1969 form of warranty provided for such bond and insurance as an optional clause. It does not appear in this form although the space after paragraph A (3) has been deliberately left so that it may be inserted if it is required.

*Section 2: Form of Warranty to be Given by Nominated Supplier in Consideration
of Nomination*

To ....
(name of Employer)

In respect of..

....

....
(description of main contract works as described in main contract) which it is
intended to have carried out by a contractor under the terms of a [Standard] form
of contract, 1963 edition, providing for payment of liquidated damages for
delay in completion at the rate of

£................................... per...
and ....

....

....
(short description of goods or materials to be supplied—hereinafter called " the
sub-contract goods "), and our tender for the sub-contract goods.

We..
(name of supplier) warrant that in consideration of your nominating us as
supplier for the sub-contract goods:

(1) We have exercised and will exercise all proper skill and care in the design of
the sub-contract goods and the selection of materials and goods therefor so far as
the sub-contract goods have been or will be designed by us and such materials
and goods have been or will be selected by us,

(2) In addition to (1) hereof the sub-contract goods shall correspond with the
description thereof including any performance specification or requirements
appearing or included in or referred to in our tender,

(3) We will save in so far as we are delayed by any cause described in Clause
23 (a) to (f) and Clause 23 (h) to (k) of the main contract as such clause applies
to us:
(a) so commence and complete delivery of the sub-contract goods and remedy
any defects therein, and
(b) supply the architect or contractor with such information as either may
reasonably require
that the contractor shall not become entitled to an extension of time under Clause
23 of the main contract by delay on our part or by our failure to supply
information as aforesaid,

* (4) We will when requested obtain a suitable performance bond in favour of
the employer for .... per cent of the anticipated sub-
contract price and/or suitable and adequate insurance cover against breach of
warranties (1) and (2) hereof in favour of the Employer from an insurance
company or finance house of repute.

(5) Nothing in our tender is intended to exclude or limit our liability for breach
of the warranties set out above.

Signed................................... Dated...
(Nominated Supplier)

---

* Delete where not required.

(1) *History*

See note (1) to the 1973 agreement, *supra*. This form of warranty has never been replaced by a document negotiated with organisations representing specialist suppliers. It is thought that it has not achieved the same degree of use as the warranty, and subsequently the agreement, relating to nominated sub-contractors.

(2) *Form of Warranty*

It is given by the supplier only. There is no undertaking by the employer, but the consideration moving from the employer is the nomination of the supplier. Therefore the warranty only takes effect upon nomination.

(3) *Liquidated damages*

This should be inserted in the form so that the supplier is fixed with notice of the amount of liquidated damages. Upon this and other comparable matters see the notes to the 1973 form making the necessary allowances for the differences.

(4) *Warranty 2. Performance specification*

There is an absolute obligation to comply with the performance specification. This is potentially a very heavy duty.

(5) *Warranty 3. Delay*

This is comparable to the warranty given under A (2) of the 1973 agreement (*supra*). The reference to clause 23 means that the supplier is entitled to extensions of time in so far as the causes therein described are applicable to him, reading them with the necessary modifications.

Section 3 : Collateral Contracts under the I.C.E. Conditions of Contract

The I.C.E. have never issued a form of warranty to be given by sub-contractors. Under the fourth and earlier editions of the Conditions the need for such a form was less apparent than under the R.I.B.A. forms. The contractor is generally responsible for delay by a nominated sub-contractor; and specialist design work is done by sub-contractors far less frequently than in building projects.

Under the 5th edition of the I.C.E. Conditions there are a number of instances in which the employer may have to bear the loss arising from the default of a nominated sub-contractor. While there may be justification for relieving the contractor of liability for a sub-contractor not of his own choosing, the sub-contractor can have no grounds for refusing to accept liability to the employer for his own default.

Notes are given below on the areas of liability which may be covered by the form of warranty, and as to factors to be considered in drafting the form. References to clauses are to the 5th edition of the I.C.E. Conditions which are printed and discussed in detail in Chapter 17, *post*.

(1) *Form of document*

This may conveniently be adapted from the R.I.B.A. form of agreement printed *supra*, Section 1. The form should provide for statement of the parties' names, for signature, and for insertion of the date of execution. For purposes of clarity, the form should be set out in separate sections containing recitals, undertakings by the sub-contractor and undertakings by the employer (if any).

(2) *Consideration by the employer*

Consideration is necessary unless the warranties are given under seal. Consideration may consist of the payment of a sum of money, which may be nominal, *e.g.* 5p; it may take the form of undertakings given by the employer, *e.g.* to make direct payments (see cl. 59C); or it may consist of the nomination. In the latter case the form must be executed before the nomination is made.

(3) *Recitals*

It is useful to set out particulars of the main contract, and of the sub-contract works in respect of which the warranties are to operate. Liquidated damages payable under the main contract should be stated in order to give the sub-contractor knowledge of the employer's loss where the contractor cannot be held responsible for delay. Where the sub-contract works form part of a section carrying its own liquidated damages, the relevant details should be set out (see cl. 47).

(4) *Operation of agreement*

The agreement may be made enforceable whether or not the prospective sub-contractor is nominated. This may be desirable where design work is undertaken. The agreement must then be supported by consideration other than the nomination, *e.g.* payment of a fee (see note (2), *supra*). Otherwise the agreement may follow the R.I.B.A. forms in which the warranties operate only upon nomination.

(5) *Time of operation*

Warranties in respect of performance of the sub-contract will normally become operative only when the sub-contractor has accepted the contractor's order (see section 1, note (4) *supra*). Where the warranty is to apply from an earlier date or is to be retrospective, this must be clearly stated, *e.g.* a design warranty, where the design work may be complete or partially complete should use words such as " has exercised and will exercise reasonable care . . ."

(6) *Design warranty*

For adequate protection, the employer requires an absolute warranty to comply with any performance specification, together with a general warranty of reasonable skill and care. Note that although the R.I.B.A. sub-contract warranty (Section 1, *supra*) refers to a performance specification the design obligation is to give reasonable skill and care only. This should not be used as a precedent if strict compliance with a specification is required (but see nominated supplier's warranty, Section 2, *supra*).

Clause 58 (3) of the Conditions envisages that the contractor may be required to accept responsibility for design in connection with a provisional sum or P.C. item. The employer may nevertheless fortify his position by obtaining a warranty direct from the sub-contractor (but see note 7, *infra*).

(7) *Limitations on the contractor's liability*

There are three important provisions in the I.C.E. Conditions which limit the employer's rights, and which may form the subject of direct warranties from the sub-contractor.

Clause 59A(6). This limits the employer's right to recover from the contractor in respect of a breach by a nominated sub-contractor.

Clause 59B(4). This may oblige the employer to reimburse the contractor in respect of loss which cannot be recovered from the sub-contractor, when the sub-contract has been terminated.

Clause 60(7). The employer may be obliged to reimburse to the contractor sums which cannot be recovered from the sub-contractor, when sums previously certified are reduced in the final certificate.

The employer may require specific indemnities in respect of any or all of these clauses, to the extent he cannot recover loss from the contractor. Alternatively the employer may require a general indemnity in respect of loss arising from acts or defaults of the sub-contractor which is not recoverable from the contractor.

The above clauses operate to protect the contractor only where he cannot recover loss. This is most likely to occur due to the sub-contractor's insolvency. The suggested indemnities are therefore likely to be worthless unless guaranteed by a performance bond or insurance.

The existence of such indemnities does not affect the liabilities of the contractor under the main contract or the sub-contractor under the sub-contract, it is submitted. In each case, the employer will not become entitled to be indemnified until the contractor has been unsuccessful in seeking to recover from the sub-contractor. The employer will, therefore, not be in a position to sue the sub-contractor directly until the contractor's action is concluded.

The employer obtains no advantage from a warranty of performance of the sub-contract, as opposed to an indemnity against loss which cannot be recovered from the contractor. Difficulties may arise if the employer seeks to enforce such a direct warranty concurrently with the main contract. This applies particularly under clause 59A(6): see also note (11) to clause 59A, Chapter 17, *post*, for a further discussion on the difficulties of this sub-clause.

(8) *Undertakings by employer*

Clause 59C of the Conditions entitles the employer to make direct payment to the sub-contractor, upon the engineer's certificate, where sums previously certified have not been paid to the sub-contractor. The engineer has a discretion whether to demand proof of payment from the contractor; and the employer has a discretion whether to make direct payments. The employer may undertake to the sub-contractor that the engineer will require proof of payment; and that the employer will, when entitled to do so, make direct payments.

Any such undertakings may be limited to ensure that the employer will have adequate retention to cover claims against the contractor after completion of the main contract works or after forfeiture or termination of the main contract.

CHAPTER 17

THE I.C.E. FORM OF CONTRACT—5TH EDITION 1973

A STANDARD form of Civil Engineering Contract, commonly known as " the I.C.E. Conditions of Contract," is issued under the sponsorship and approval of the Institution of Civil Engineers (I.C.E.), the Association of Consulting Engineers and the Federation of Civil Engineering Contractors. The form is used extensively in all types of civil engineering work, both by private employers and by local and central government departments.

History of the form

While building contracts have used standard forms since the end of the nineteenth century, the 1st edition of the I.C.E. Conditions of Contract was issued in December 1945, followed by further editions in 1950, 1951 and 1955. The latter, being the 4th edition within 10 years, stated that the issuing bodies did not contemplate a further edition for at least five years. In the event, the period between the 4th and 5th editions was 18 years, partly accounted for by one abortive attempt to agree a radical revision of the form.[1] This episode, and the existence of a generation of practising civil engineers familiar with the 4th edition, may account for the limited terms of the 5th edition drafting committee. Their task was to make alterations which were considered necessary for clarification and the removal of ambiguity, while keeping the contract recognisably the familiar I.C.E. Conditions.[2] The 5th edition, however, also contains a substantial number of additional provisions.

Whether the 5th edition as a whole is clearer and less ambiguous than the 4th is a matter for speculation. One matter which is clear is that, while much insight may be gained into the intention of the draftsmen by comparing the 4th and 5th editions, the courts will be concerned only with the intention of the parties,[3] which is expressed solely in the 5th edition. The detailed commentary on the Conditions is therefore, in general, confined to the form as it now stands. Some notes on the principal changes brought about by the 5th edition are included below.

Unlike the Standard form of building contract,[4] the I.C.E. Conditions of Contract are published in one version only, save for additional optional clauses. The 5th edition, however, states that the sponsoring bodies intend to set up a permanent joint review committee, which will produce revisions when such action seems warranted. To date, the work of the committee has been confined to producing guidance notes, and no revision to the conditions is anticipated. The following commentary applies only to the unrevised 5th edition.

---

[1] *New Civil Engineer*, Dec. 20, 1973, p. 33.
[2] *Ibid.*
[3] See *ante*, p. 27.
[4] See *ante*, Chap. 15.

Nature of the I.C.E. form

It has been suggested that the 4th edition of the I.C.E. form created a lump sum contract.[5] The 5th edition, by omitting reference to a contract sum,[6] makes it clear that the conditions create a " measure and value " or " re-measurement " contract. This means that the employer is bound to pay for the actual quantities of work done at the rates contained in the bills, but subject to many possible alterations [7] including alteration of the rates.[8]

The I.C.E. Conditions of Contract provide to an even greater extent than does the J.C.T. form, a detailed and elaborate code dealing with most incidents of the work of engineering construction. In large measure this operates by placing upon the engineer much wider powers and duties than exist in building contracts, *e.g.* powers to give instructions as to the mode of carrying out the work,[9] and duties to make adjustment of the sums payable to the contractor.[10] The engineer's powers under the 4th edition have been called " wide and arbitrary." [11] Under the 5th edition they are wider.

The drafting of the Conditions tends to be over-elaborate in less vital areas and to be vague in some areas of importance.[12] Many problems are created by the inconsistency of the drafting. This is such that it cannot be assumed that the same expression bears the same meaning throughout the contract; and conversely, a change of words does not always import a change of meaning. For a particular instance see note (1) to clause 39.

Principal changes in 5th edition

Two things which have not changed noticeably from the 4th edition are the clause numbering and the drafting. The former is achieved by a degree of rearrangement [13] and possibly by retaining clauses from the 4th edition which serve no known purpose (*e.g.* clause 18). The style of the Conditions is the same because very substantial parts of the 4th edition have been retained unaltered. A welcome drafting innovation is the extended use of the concept of reasonableness,[14] a matter with which engineers will feel instant familiarity.

The principal changes introduced in the 5th edition are set out below. Numbers in brackets thus: (14) refer to clause numbers.

Methods of construction

The engineer now has specific powers to request information to assist his decision whether to approve the contractor's methods (14); there are added provisions governing safety on the site (15 (1), 16, 19).

---

[5] Abrahamson, *Engineering Law* (2nd ed.), p. 32.
[6] See cl. 1(1) and note thereto.
[7] See cl. 52, note (7).
[8] Cll. 52(2), 56(2).
[9] Cll. 13, 14.
[10] See *supra*, note (7).
[11] *Farr* v. *M.O.T.* [1960] 1 W.L.R. 956, *per* Buckley J., at p. 964.
[12] *e.g.* the definition of " site " in cl. 1(1) (n).
[13] For a full schedule of alterations and re-numberings, see Abrahamson, *Engineering Law* (3rd ed.), pp. 284–317.
[14] For "reasonable" cost, etc., see *e.g.* cll. 7 (3), 12 (3), 13 (3); for "reasonable" notice, etc., see *e.g.* cl. 52 (4).

Insurance

There are some alterations in the placing of risk (20, 22, 24); third party insurance is to be in the contractor's sole name (23); insurance of the contractor's workmen is no longer compulsory (24) (save by statute).

Claims

Most provisions entitling the contractor to claim additional payment are now subject to a standard procedure requiring reasonable notice (52 (4)). This applies also to clause 12, under which the procedure is much simplified. Several significant new claims are introduced (7 (3), 14 (6), 31 (2), 59A (3), 59B (4)), including a general claim in respect of instructions under clause 13 (1) (13 (3)).

Time

There are detailed provisions for mandatory sectional completion, with corresponding liquidated damages (47). Interim extensions of time are provided for, but the grounds for extension remain dispersed throughout the Conditions (44).

Payment

A new system of accounting is laid down (60). Payments may include materials not on the site (54). The engineer now has express power to vary the rates for quantities differing from those stated in the bills (56 (2)), as well as for variations.

Certificates

The procedure for interim certificates and retention has been amended; there is now provision for a final account and final certificate (60). The effect of the maintenance certificate is clarified (61).

Nominated sub-contractors

This is, physically, the biggest area of amendment. The former provisions are entirely replaced by an epic code which substantially limits the contractor's responsibility, so that the employer is likely to bear loss resulting from the sub-contractor's default (59A, 59B, 60 (7)). The position is now that a nomination must be regarded as a potentially expensive luxury.

FORMS OF TENDER, AGREEMENT AND BOND

These forms are printed and bound with the Conditions of Contract. The tender and the written acceptance thereof are contract documents (cl. 1 (1) (e)). A written acceptance is not required by law but is mentioned in the tender. The appendix is to be submitted with, and forms part of the tender, and is thus incorporated into the contract. This is useful in practice because it minimises the likelihood of failure to complete the items of the appendix. Unlike the Standard form of building contract, there are no provisions to apply in default of completing the appendix.

The form of agreement, if completed (cl. 9), is a contract document (cl. 1 (1) (e)). It provides for execution under seal. The tender includes an undertaking to provide a bond in the form annexed to the conditions (see also cl. 10).

SHORT DESCRIPTION
OF WORKS:—

All Permanent and Temporary Works in connection with* ....

....

Form of Tender
(NOTE: The Appendix forms part of the Tender)

To......................

....

....

GENTLEMEN,

Having examined the Drawings, Conditions of Contract, Specification and Bill of Quantities for the construction of the above-mentioned Works (and the matters set out in the Appendix hereto), we offer to construct and complete the whole of the said Works and maintain the Permanent Works in conformity with the said Drawings, Conditions of Contract, Specification and Bill of Quantities for such sum as may be ascertained in accordance with the said Conditions of Contract.

We undertake to complete and deliver the whole of the Permanent Works comprised in the Contract within the time stated in the Appendix hereto.

If our tender is accepted we will, when required, provide two good and sufficient sureties or obtain the guarantee of a Bank or Insurance Company (to be approved in either case by you) to be jointly and severally bound with us in a sum equal to the percentage of the Tender Total as defined in the said Conditions of Contract for the due performance of the Contract under the terms of a Bond in the form annexed to the Conditions of Contract.

Unless and until a formal Agreement is prepared and executed this Tender, together with your written acceptance thereof, shall constitute a binding Contract between us.

We understand that you are not bound to accept the lowest or any tender you may receive.

† To the best of our knowledge and belief we have complied with the general conditions required by the Fair Wages Resolution for the three months immediately preceding the date of this tender.

We are, Gentlemen,

Yours faithfully,

Signature...............................

Address...............................

....

Date......................

* Complete as appropriate.
† Delete if not required.

APPENDIX

NOTE: Relevant Clause numbers are shown in brackets following the description

Amount of Bond (if any)   (10)   ....% of Tender Total

Minimum Amount of Insurance (23 (2))   £ ....

Time for Completion   (43)   Liquidated Damages for Delay (47)
                                                   Column 1
                                                   (see Clause 47 (1))

For the Whole of the Works.....(a) Weeks £...........(b) per Day/Week (c)
                                                   Column 2   Column 3
For the following Sections   (see Clause 47 (2))

   Section(d) ....   £.............   £.............

   ....——Weeks   Per Day/Week (c)   per Day/Week (c)

   Section (d) ....   £.............   £.............

   ....——Weeks   per Day/Week (c)   per Day/Week (c)

   Section (d) ....   £.............   £.............

   ....——Weeks   per Day/Week (c)   per Day/Week (c)

   Section (d) ....   £.............   £.............

   ....——Weeks   per Day/Week (c)   per Day/Week (c)

Period of Maintenance   (49 (1))   .... Weeks

Vesting of Materials not on Site   (54 (1) and 60 (1)) (e)

   1............................   4............................

   2............................   5............................

   3............................   6............................

Standard Method of Measurement adopted in preparation of Bills of Quantities

(57) (f) ....

Percentage for adjustment of P.C. Sums (59A (2) (b) and (5) (c))............%

Percentage of the Value of Goods and Mater-
   ials to be included in Interim Certificates   (60 (2) (b))   ....%

Minimum Amount of Interim Certificates   (60 (2))   £ ....

   (a) To be completed in every case (by Contractor if not already stipulated).
   (b) To be completed by Engineer in every case.
   (c) Delete which not required.
   (d) To be completed if required, with brief description.
   (e) (If used) materials to which clauses apply are to be filled in by Engineer
       prior to inviting tenders.
   (f) Insert here any amendment or modification adopted if different from
       that stated in Clause 57.

COMMENTARY

(1) *Generally*

Note the distinction drawn between construction of the works and maintenance of the permanent works.[15] The absence of reference to any specific sum emphasises that the contract is not a lump sum contract.

(2) *Guarantee*

The undertaking to provide a guarantee bond is contemplated, by clause 10 of the Conditions, as optional.

(3) " *Unless and until a formal Agreement* "

There is a practice of giving written acceptance of the tender followed by further negotiations of terms. If for any reason the formal agreement is not executed, this passage may deprive subsequently agreed terms of effect.

Form of Agreement

THIS AGREEMENT made the...............day of.................

19.... BETWEEN..

of.. in the

County of...........(hereinafter called " the Employer ") of the one part and

.... of ....

in the County of ....

...................(hereinafter called " the Contractor ") of the other part

WHEREAS the Employer is desirous that certain Works should be constructed,

viz. the Permanent and Temporary Works in connection with ....

.........................and has accepted a Tender by the Contractor for the construction and completion of such Works and maintenance of the Permanent Works

NOW THIS AGREEMENT WITNESSETH as follows:—

1. In this Agreement words and expressions shall have the same meanings as are respectively assigned to them in the Conditions of Contract hereinafter referred to.

2. The following documents shall be deemed to form and be read and construed as part of this Agreement, viz.:—
   (a) The said Tender.
   (b) The Drawings.
   (c) The Conditions of Contract.
   (d) The Specification.
   (e) The Priced Bill of Quantities.

3. In consideration of the payments to be made by the Employer to the Contractor as hereinafter mentioned the Contractor hereby covenants with the Employer to construct and complete the Works and maintain the Permanent Works in conformity in all respects with the provisions of the Contract.

---

[15] See cl. 8, note (1).

4. The Employer hereby covenants to pay to the Contractor in consideration of the construction and completion of the Works and maintenance of the Permanent Works the Contract Price at the times and in the manner prescribed by the Contract.

IN WITNESS whereof the parties hereto have caused their respective Common Seals to be hereunto affixed (or have hereunto set their respective hands and seals) the day and year first above written

The Common Seal of.....................

...................................Limited
was hereunto affixed in the presence of:—
of
SIGNED SEALED AND DELIVERED by the

said......................................

....
in the presence of:—

....

COMMENTARY

The comments in note (1) above apply equally to the agreement, save that the employer covenants to pay the " Contract Price." This is defined by clause 1 (1) (i) in terms similar to the wording of the tender.

Form of Bond

[1] Is appropriate to an individual, [2] to a Limited Company and [3] to a Firm. Strike out whichever two are inappropriate.

[4] Is appropriate where there are two individual Sureties, [5] where the Surety is a Bank or Insurance Company. Strike out whichever is inappropriate.

BY THIS BOND [1] We..........................
of... in the
County of.......... [2] We...................Limited
whose registered office is at...................... in the
County of............. [3] We.......................
and...........carrying on business in partnership under
the name or style of................................
at...in the
County of.....(hereinafter called " the Contractor ") [4] and
.... of ....
in the County of............and.....................
of...............................in the County of....
.... [5] and .... Limited
whose registered office is at.......................in the
County of................................(hereinafter
called " the [4] Sureties/Surety ") are held and firmly bound
unto .... (hereinafter
called " the Employer ") in the sum of............pounds
(£............) for the payment of which sum the Contractor and the [4] Sureties/Surety bind themselves their successors and assigns jointly and severally by these presents.
Sealed with our respective seals and dated this.........
.................day of........................ 19...
WHEREAS the Contractor by an Agreement between the Employer of the one part and the Contractor of the other part has entered into a Contract (hereinafter called

" the said Contract ") for the construction and completion of the Works and maintenance of the Permanent Works as therein mentioned in conformity with the provisions of the said Contract.

NOW THE CONDITION of the above-written Bond is such that if the Contractor shall duly perform and observe all the terms provisions and stipulations of the said Contract on the Contractor's part to be performed and observed according to the true purport intent and meaning thereof or if on default by the Contractor the Sureties/Surety shall satisfy and discharge the damages sustained by the Employer thereby up to the amount of the above-written Bond then this obligation shall be null and void but otherwise shall be and remain in full force and effect but no alteration in terms of the said Contract made by agreement between the Employer and the Contractor or in the extent or nature of the Works to be constructed completed and maintained thereunder and no allowance of time by the Employer or the Engineer under the said Contract nor any forbearance or forgiveness in or in respect of any matter or thing concerning the said Contract on the part of the Employer or the said Engineer shall in any way release the Sureties/Surety from any liability under the above-written Bond.

Signed Sealed and Delivered by the said }
   in the presence of:—   }
The Common Seal of   }
                          LIMITED   }
was hereunto affixed in the presence of:— }
(*Similar forms of Attestation Clause for the Sureties or Surety*).

COMMENTARY

The amount of the bond is intended to be the percentage of the tender total (see cl. 1 (1) (h)) named in the appendix, not exceeding 10 per cent. (cl. 10). The Conditions omit specifically to incorporate the appendix but the general incorporation is probably sufficient. The contract price may eventually bear little relation to the tender total.

CONDITIONS OF CONTRACT

Clause 1   DEFINITIONS AND INTERPRETATION

Definitions

(1) In the Contract (as hereinafter defined) the following words and expressions shall have the meanings hereby assigned to them except where the context otherwise requires:—

(a) " Employer " means...
    of ....
    and includes the Employer's personal representatives or successors;

(b) " Contractor " means the person or persons firm or company whose tender has been accepted by the Employer and includes the Contractor's personal representatives successors and permitted assigns;

(c) " Engineer " means ....
or other the Engineer appointed from time to time by the Employer and
notified in writing to the Contractor to act as Engineer for the purposes of
the Contract in place of the said ....;

(d) " Engineer's Representative " means a person being the resident engineer
or assistant of the Engineer or clerk of works appointed from time to time
by the Employer or the Engineer and notified in writing to the Contractor
by the Engineer to perform the functions set forth in Clause 2 (1);

(e) " Contract " means the Conditions of Contract Specification Drawings
Priced Bill of Quantities the Tender the written acceptance thereof and
the Contract Agreement (if completed);

(f) " Specification " means the specification referred to in the Tender and any
modification thereof or addition thereto as may from time to time be
furnished or approved in writing by the Engineer;

(g) " Drawings " means the drawings referred to in the Specification and any
modification of such drawings approved in writing by the Engineer and
such other drawings as may from time to time be furnished or approved
in writing by the Engineer;

(h) " Tender Total " means the total of the Priced Bill of Quantities at the
date of acceptance of the Contractor's Tender for the Works;

(i) " Contract Price " means the sum to be ascertained and paid in accordance
with the provisions hereinafter contained for the construction completion
and maintenance of the Works in accordance with the Contract;

(j) " Permanent Works " means the permanent works to be constructed
completed and maintained in accordance with the Contract;

(k) " Temporary Works " means all temporary works of every kind required
in or about the construction completion and maintenance of the Works;

(l) " Works " means the Permanent Works together with the Temporary
Works;

(m) " Section " means a part of the Works separately identified in the Appendix to the Form of Tender;

(n) " Site " means the lands and other places on under in or through which
the Works are to be executed and any other lands or places provided by
the Employer for the purposes of the Contract;

(o) " Constructional Plant " means all appliances or things of whatsoever
nature required in or about the construction completion and maintenance
of the Works but does not include materials or other things intended to
form or forming part of the Permanent Works.

Singular and plural

(2) Words importing the singular also include the plural and *vice-versa* where
the context requires.

Headings and Marginal Notes

(3) The headings and marginal notes in the Conditions of Contract shall not
be deemed to be part thereof or be taken into consideration in the interpretation
or construction thereof or of the Contract.

Clause references

(4) All references herein to clauses are references to clauses numbered in the
Conditions of Contract and not to those in any other document forming part
of the Contract.

Cost

(5) The word " cost " when used in the Conditions of Contract shall be

deemed to include overhead costs whether on or off the Site except where the contrary is expressly stated.

COMMENTARY

(1) *Generally*

The definitions in sub-clause (1) are not of uniform importance; (a) to (c) are matters of identification only; (d) adds nothing to clause 2; (e) (f) and (g) play an important part in the working of the contract documents; (h) and (i) are of importance in the interpretation of the contract; (j) to (o) are definitions which should be read with the appropriate clauses. Note that by sub-clause (3) headings and marginal notes are to be disregarded in construing the Conditions.

(2) *Clause 1 (1): Definitions*

(c) *Engineer.* The engineer exercises many important functions, such that much of the contract becomes inapplicable without an appointment; *e.g.* the engineer's approval is stipulated in respect of the materials, plant and labour provided (cl. 13 (2)), and of the manner or method of construction (cll. 13 (2), 14 (1)). If the parties contract with an express or implied agreement that no engineer shall be appointed, it is thought that the contract would be enforced as an entire contract subject to re-measurement, and that clauses depending for their operation on the engineer would be ignored.

(d) *Engineer's representative.* The intention of this definition is obscure. It seeks, apparently, to narrow the range of persons who may be appointed as engineer's representative. But the engineer's power to delegate under clause 2 (3) may also be exercised in respect of " any other person responsible to the engineer."

(e) *Contract.* This provision must be read with clause 5. It complements similar provisions in the tender and contract agreement. The Conditions and forms of tender and agreement are issued as one document. This also contains a " Table of Contents " and " Index to the General Conditions," neither of which form part of the contract or can be referred to in construing the contract, it is submitted.

(f), (g) *Specification, Drawings.* The reference to modified or other drawings complements the provisions of clause 7. The reference to modified or additional specification is of some importance since clause 7 curiously omits to give the engineer such power.

(h) *Tender Total.* This expression appears in clauses 10 and 60 (4). The significant omission of any formal statement of the amount of the tender total [16] in either the tender or the contract agreement, together with clauses 55 and 56, make it quite clear that the contract is a re-measurement contract. The contractor is bound to execute such quantities as are necessary to complete the works and is entitled to payment for such quantities at the rates and prices in the bill of quantities or in accordance with clause 56 (2).

---

[16] *cf.* the *Contract Price* in the 4th edition and the *Contract Sum* in the Standard form of building contract.

(i) *Contract Price.* This expression appears in the contract agreement but is not elsewhere employed or defined. The definition is not limited to " the amount which in (the engineer's) opinion is finally due under the contract." [17] It is thought that the words " sum to be ascertained " have the effect that the sum cannot become contractually due until ascertained, which may prevent time running under the Limitation Acts.

(j) (k) *Permanent works, temporary works.* These provisions appear to recognise the impossibility of a definition of universal application, despite important distinctions which are made in the Conditions between permanent and temporary works (*e.g.* in cll. 8 (2) and 20 (3)). What is left permanently in place may be more than the permanent works (*e.g.* sheet piles, coffer dams, etc.); and the obligation to maintain by no means clearly excludes the temporary works.[18] The provisions invite the definition of temporary work where included in the bills of quantities. But there must always be items incapable of precise labelling, *e.g.* ground treatment which is partly to facilitate construction (and therefore of a " temporary " nature) but which is relied upon in the design of the " permanent " parts of the works.

(l) *Works.* The basic obligation of the contract is to complete the works.[19] The term appears frequently throughout the Conditions, sometimes where a reference to the permanent works would be more apt, *e.g.* clauses 48 (1) and 49. Completion and maintenance [20] are matters which concern only the permanent works.[21] In such cases it may be necessary to invoke the qualifying words of sub-clause (1): " except where the context otherwise requires."

(m) *Section.* (See cll. 43, 44, 46, 47, 48.) This definition adds nothing since it is in effect repeated in these clauses. It is essential for any requirements as to sectional completion to be stated in the appendix (and not in some other part of the contract) otherwise these clauses cannot operate.

(n) *Site.* (See particularly cll. 11 (1), 22 (1), 32, 42, 53, 54.) This definition is surprisingly vague, considering the practical and economic consequences of the area of land available for the contractor. The words do not oblige the employer to provide " any other lands or places " so the site may consist merely in the land or space to be occupied by the works,[22] plus sufficient working areas to make the operations physically possible.[23] There is no express obligation to provide more.[24] Further, the contract documents usually require facilities to be provided for the engineer's representative and his staff within the site, which may create difficulty where the site is congested. It is suggested the site should be expressly defined by the contract documents.

---

[17] *i.e.* the amount to be stated in the final certificate under cl. 60 (3).
[18] See cl. 8, note (1) and I.C.E. standard method of measurement, cl. 9.
[19] See cl. 8 (1), tender and contract agreement.
[20] See cl. 8, note (1).
[21] Although sectional completion of a particular item of temporary work may be necessary, *e.g.* diversion of a river prior to anticipated flooding.
[22] Note that this includes temporary works by cl. 1 (1) (l).
[23] *Wells v. Army & Navy* (1902) 86 L.T. 764; H.B.C. (4th ed.), Vol. 2, p. 346. See also cl. 13 (1).
[24] See cl. 42, note (1).

(o) *Constructional Plant.* The definition clearly includes temporary works as defined in paragraph (k). But the conditions draw an equally clear and consistent distinction between temporary works and constructional plant.[25] Nothing appears to turn on this distinction, fortunately.

(3) *Clause* 1 (5)

The word " cost " appears extensively in the conditions where provision is made for contractual claims.[26] The express inclusion of overheads does not limit the generality of the term, which may further include items such as preliminaries and non-productive overtime. The term does not, however, include profit, which is specifically included in a claim under clause 12 (3).

ENGINEER'S REPRESENTATIVE

Clause 2

Functions and powers of engineer's representative

(1) The functions of the Engineer's Representative are to watch and supervise the construction completion and maintenance of the Works. He shall have no authority to relieve the Contractor of any of his duties or obligations under the Contract nor except as expressly provided hereunder to order any work involving delay or any extra payment by the Employer nor to make any variation of or in the Works.

Appointment of assistants

(2) The Engineer or the Engineer's Representative may appoint any number of persons to assist the Engineer's Representative in the exercise of his functions under sub-clause (1) of this Clause. He shall notify to the Contractor the names and functions of such persons. The said assistants shall have no power to issue any instructions to the Contractor save in so far as such instructions may be necessary to enable them to discharge their functions and to secure their acceptance of materials or workmanship as being in accordance with the Specification and Drawings and any instructions given by any of them for those purposes shall be deemed to have been given by the Engineer's Representative.

Delegation by engineer

(3) The Engineer may from time to time in writing authorise the Engineer's Representative or any other person responsible to the Engineer to act on behalf of the Engineer either generally in respect of the Contract or specifically in respect of particular Clauses of these Conditions of Contract and any act of any such person within the scope of his authority shall for the purposes of the contract constitute an act of the Engineer. Prior notice in writing of any such authorisation shall be given by the Engineer to the Contractor. Such authorisation shall continue in force until such time as the Engineer shall notify the Contractor in writing that the same is determined. Provided that such authorisation shall not be given in respect of any decision to be taken or certificate to be issued under Clauses 12 (3), 44, 48, 60 (3), 61, 63 and 66.

Reference to engineer or engineer's representative

(4) If the Contractor shall be dissatisfied by reason of any instruction of any assistant of the Engineer's Representative duly appointed under sub-clause (2)

---

[25] See cll. 8 (1), 21, 30 (2), 33, 53 (1), 60 (1) (d).
[26] See cll. 7 (3), 12 (3), 13 (3), 14 (6), 17, 27 (6), 31 (2), 36 (2), (3), 38 (2), 40 (1), 42 (1) 50, 59B (4).

of this Clause he shall be entitled to refer the matter to the Engineer's Representative who shall thereupon confirm reverse or vary such instruction. Similarly if the Contractor shall be dissatisfied by reason of any act of the Engineer's Representative or other person duly authorised by the Engineer under sub-clause (3) of this Clause he shall be entitled to refer the matter to the Engineer for his decision.

COMMENTARY

(1) *Generally*

The intention of this clause is to facilitate limited delegations of the engineer's powers, with a right to have delegated decisions reviewed.

(2) *Clause 2 (1):* " *No authority to relieve the Contractor* "

These words are unnecessary, since neither the engineer's representative nor the engineer can have such authority, unless it arises outside the contract. This may be the case where the engineer or engineer's representative is also an officer of the local authority employer.[27]

(3) *Clause 2 (1):* " *nor except as expressly provided* "

These words are unnecessary since the engineer's representative has no powers other than those delegated under, and subject to the limitations of, sub-clause (3).

(4) *Clause 2 (2):* " *necessary to enable them to discharge their function* "

Despite the reference to " their function " it is reasonably clear that this refers to the engineer's representative's function to watch and supervise, and that assistants are not to exercise any formal powers under the contract.

(5) *Clause 2 (3):* " *authorise the Engineer's Representative or any other person responsible to the Engineer* "

This provision makes the title or status of engineer's representative of little practical importance, and the definition of engineer's representative in clause 1 (1) (d) becomes irrelevant.

(6) *Clause 2 (3):* " *authorisation shall not be given . . .*"

This leaves within the sphere of delegation the important powers of certifying in respect of variations (cll. 51, 52), claims (cl. 52 (4)) and interim certificates (cl. 60 (2)).[28]

(7) *Clause 2 (4):* " *he shall be entitled to refer the matter* "

An instruction from the engineer's representative or the engineer given on appeal under this sub-clause has the same effect as an original instruction. However, a reference to the engineer under this sub-clause should be carefully distinguished from the reference of a dispute (arising from a decision of the

[27] *Carlton Contractors* v. *Bexley Corporation* (1962) 60 L.G.R. 311; *Roberts & Co. Ltd.* v. *Leicestershire County Council* [1961] Ch. 55.
[28] See also cl. 13 (1), note (5).

engineer's representative) to the engineer under clause 66, where the consequences of the engineer's decision may be more far-reaching.[29]

Assignment and Sub-letting

Clause 3
Assignment

The Contractor shall not assign the Contract or any part thereof or any benefit or interest therein or thereunder without the written consent of the Employer.

Clause 4
Sub-letting

The Contractor shall not sub-let the whole of the Works. Except where otherwise provided by the Contract the Contractor shall not sub-let any part of the Works without the written consent of the Engineer and such consent if given shall not relieve the Contractor from any liability or obligation under the Contract and he shall be responsible for the acts defaults and neglects of any sub-contractor his agents servants or workmen as fully as if they were the acts defaults or neglects of the Contractor his agents servants or workmen. Provided always that the provision of labour on a piece-work basis shall not be deemed to be a sub-letting under this Clause.

Commentary

(1) *Clauses 3 and 4*: " *without written consent* "

No implication of reasonableness as to the withholding of consent can arise, it is submitted. The Conditions provide expressly where consent may not be withheld unreasonably: see clauses 14 (4), 21, 23 (2) and 46. The withholding of consent cannot constitute a dispute and is therefore not subject to arbitration. In the event of the engineer appearing to act capriciously the contractor's only course is to appeal to the employer.

(2) *Clause 4: Remedies of the Employer*

Sub-letting " to the detriment of good workmanship or in defiance of the Engineer's instructions to the contrary " is a ground for forfeiture under clause 63 (1) (e). If the sub-letting is merely without consent the employer may, on discovering the sub-letting, require the contractor to resume performance of the work. The contractor's failure may amount to repudiation.[30]

(3) *Clause 4*: ". . . *he shall be responsible for the acts, defaults and neglects.*"

This would be the position in any event as regards performance of the contract. The words are wide enough to make the contractor liable also for the torts of the sub-contractor, contrary to the general rule.[31] For the position where the sub-contractor is nominated, see clauses 59A and 59B.

---

[29] See cl. 66, note (3).

[30] See also p. 116. If the employer or the engineer, with knowledge of the sub-letting, allows the work to proceed without protest, the breach may be waived.

[31] The general rule is subject to important exceptions which apply particularly to construction contracts; see *ante*, p. 135.

CONTRACT DOCUMENTS

Clause 5
Documents mutually explanatory

The several documents forming the Contract are to be taken as mutually explanatory of one another and in case of ambiguities or discrepancies the same shall be explained and adjusted by the Engineer who shall thereupon issue to the Contractor appropriate instructions in writing which shall be regarded as instructions issued in accordance with Clause 13.

COMMENTARY

(1) *Generally*

This clause avoids the type of problem which can arise when some documents are given precedence over others.[32] As a general rule of construction, written terms of a contract take precedence over printed terms in the event of inconsistency.[33] This provision, it is thought, may exclude the rule, so that care must be taken in drafting special terms (*e.g.* to be included in the bills) which are intended to supersede the Conditions.

The documents forming the contract consist of the conditions, specification, drawings, priced bill of quantities, the tender, the written acceptance thereof and the contract agreement (if completed): clause 1 (1) (e); together with modified or additional drawings or specification supplied by the engineer during the progress of the works: clause 1 (1) (f), (g) and clause 7 (1), (2).

(2) *" ambiguities or discrepancies "*

The ambit of these words is limited, it is submitted, to doubt or disagreement arising from the technical documents. Any wider construction would give the engineer power to vary the terms of the contract, which is quite inconsistent with the words of clause 13: " in strict accordance with the Contract," which, by necessary inference, qualify the engineer's powers under this clause.

The obligation of the engineer under this clause, it is submitted, is to issue such instructions as are necessary to secure the mutual working of the technical descriptions of the works. A failure to give such instructions constitutes a breach of contract on behalf of the employer. The same circumstances are likely to give rise to an alternative claim under clause 7 (3), but clause 5 does not require notice.

(3) *" instructions issued in accordance with clause 13 "*

This provision incorporates the right to make a claim under clause 13 (3) for additional cost or extension of time, which also confirms that an instruction under this clause may involve a variation pursuant to clause 51. Note the limitations on a claim under clause 13 (3): there is no automatic right to payment for an ambiguity or discrepancy.

Clause 6
Supply of documents

Upon acceptance of the Tender 2 copies of the drawings referred to in the

---

[32] See Standard form of building contract, cl. 12 (1) and *English Industrial Estates* v. *Wimpey & Co. Ltd.* (1973) 1 Lloyd's Rep. 118.
[33] *Robertson* v. *French* (1803) 4 East, 130; *Glynn* v. *Margetson* [1893] A.C. 351.

Specification and of the Conditions of Contract the Specification and (unpriced) Bill of Quantities shall be furnished to the Contractor free of charge. Copyright of the Drawings and Specification and of the Bill of Quantities (except the pricing thereof) shall remain in the Engineer but the Contractor may obtain or make at his own expense any further copies required by him. At the completion of the Contract the Contractor shall return to the Engineer all Drawings and the Specification whether provided by the Engineer or obtained or made by the Contractor.

Clause 7
Further drawings and instructions

(1) The Engineer shall have full power and authority to supply and shall supply to the Contractor from time to time during the progress of the Works such modified or further drawings and instructions as shall in the Engineer's opinion be necessary for the purpose of the proper and adequate construction completion and maintenance of the Works and the Contractor shall carry out and be bound by the same.

Notice by contractor

(2) The Contractor shall give adequate notice in writing to the Engineer of any further drawing or specification that the Contractor may require for the execution of the Works or otherwise under the Contract.

Delay in issue

(3) If by reason of any failure or inability of the Engineer to issue at a time reasonable in all the circumstances drawings or instructions requested by the Contractor and considered necessary by the Engineer in accordance with sub-clause (1) of this Clause the Contractor suffers delay or incurs cost then the Engineer shall take such delay into account in determining any extension of time to which the Contractor is entitled under Clause 44 and the Contractor shall subject to Clause 52(4) be paid in accordance with Clause 60 the amount of such cost as may be reasonable. If such drawings or instructions require any variation to any part of the works the same shall be deemed to have been issued pursuant to Clause 51.

One copy of documents to be kept on site

(4) One copy of the Drawings and Specification furnished to the Contractor as aforesaid shall be kept by the Contractor on the Site and the same shall at all reasonable times be available for inspection and use by the Engineer and the Engineer's Representative and by any other person authorised by the Engineer in writing.

COMMENTARY

(1) *Clause 7: Generally*

The provision of detailed working drawings and instructions is an essential part of the proper management of every civil engineering operation. The scheme of this clause is clear but the change of wording between " drawing and instruction " in sub-clauses (1) and (3) and " drawing and specification " in sub-clauses (2) and (4) detracts from its effect.[34] Sub-clauses (2) and (3) read

---

[34] It has been stated by representatives of the drafting body that " drawings(s), specification(s) and (or) instructions " should have been used in all three sub-clauses: *New Civil Engineer*, Dec. 20, 1973, p. 34.

together mean that an oral request for further instructions is sufficient to ground a claim for additional cost and extension of time, if such instructions are not delivered within a reasonable time.

The obligation of the contractor to comply with modified or further drawings is confirmed by the definitions of the contract documents.[35] The clause curiously omits to empower the engineer to supply a modified or additional specification.[36] There may be many occasions when a further or amplified specification is called for which does not entitle the contractor to the benefit of clauses 13 or 51. This lacuna is fortunately filled by clause 1 (1) (f).

(2) *Clause 7*: " *during the progress of the Works* "

This period extends only to the date of completion, it is submitted, so that there is no power under this clause to issue further drawings or instructions thereafter.[37] Further, the power to give a variation order must be exercised before completion: see clause 51 (1).

GENERAL OBLIGATIONS

Clause 8
Contractor's general responsibilities

(1) The Contractor shall subject to the provisions of the Contract construct complete and maintain the Works and provide all labour materials Constructional Plant Temporary Works transport to and from and in or about the Site and everything whether of a temporary or permanent nature required in and for such construction completion and maintenance so far as the necessity for providing the same is specified in or reasonably to be inferred from the Contract.

Contractor responsible for safety of site operations

(2) The Contractor shall take full responsibility for the adequacy stability and safety of all site operations and methods of construction provided that the Contractor shall not be responsible for the design or specification of the Permanent Works (except as may be expressly provided in the Contract) or of any Temporary Works designed by the Engineer.

COMMENTARY

(1) *Generally*

Sub-clause (1) repeats, with variation and addition, obligations set out in the tender (" construct and complete the whole of the said works and maintain the Permanent Works ") and in the form of agreement (" construct and complete the Works and maintain the Permanent Works "). This sub-clause appears to import an obligation to maintain the temporary works (see cl. 1 (1) (l)), which is supported by clauses 1 (1) (k) and 13. It is thought that the clear exclusion of maintenance of the temporary works in the tender and contract agreement (which, by clause 5, must be construed mutually with the Conditions) prevails over the loose drafting of these clauses.[38]

---

[35] See cl. 1(1) (e), (g), cl. 8 (1) and cl. 13 (1).
[36] See n. 34, *supra*.
[37] See cl. 39, note (1).
[38] See also notes to cl. 1 (1), (j) (k) (l).

(2) *Clause* 8 (1): " *so far as the necessity for providing the same is specified in . . .*"

These words add nothing to the obligation to complete the works in accordance with the contract.[39] The following words " or reasonably to be inferred from the Contract " merely express that which would otherwise apply as a matter of construction or implication.

(3) *Clause* 8 (2): " *safety of all site operations* "

The placing of contractual responsibility upon the contractor cannot affect liability under Part I of the Health and Safety at Work etc. Act 1974: see particularly ss. 4, 6 and 36. The general duty under section 4 applies to any person " who has, to any extent, control of premises " (which may include the employer) and requires that the premises and any plant, etc., should be " safe and without risks to health." [40]

(4) *Clause* 8 (2): " *methods of construction* "

This provision should be read with clauses 13 (2) (3) and 14 (1) (3) (7).

(5) *Clause* 8 (2): " *responsible for the design . . . except as may be expressly provided* "

It is reasonably plain that this refers to responsibility for the adequacy (and not mere provision) of the design. Unless otherwise expressly provided, the degree of skill required will not exceed that of an ordinary competent contractor, it is submitted. This may differ from that of an engineer. An obligation to comply with a performance specification must be expressly set out. See also clause 58 (3) as to design requirements under P.C. or provisional sum items.

(6) *Clause* 8 (2): " *Temporary Works designed by the Engineer* "

See note (6) to clause 14.

Clause 9
Contract Agreement

The Contractor shall when called upon so to do enter into and execute a Contract Agreement (to be prepared at the cost of the Employer) in the form annexed.

COMMENTARY

The form of agreement (see *ante*) adds nothing to the obligations contained in the other documents forming the contract (see cl. 1 (1) (e)), save that the form provides for execution under seal. There is nothing to prevent the parties drawing up their own form of agreement (as done by many local authorities) which may be under seal. The effect of execution under seal is to extend the period of limitation to 12 years, which will apply both to the employer's right to sue in respect of latent defects and to the contractor's right to pursue " latent " claims.[41]

---

[39] See cl. 13 (1), tender and form of agreement.
[40] See also cll. 15 and 19.
[41] For the courts' attitude towards delay and periods of limitation see *Birkett* v. *James* [1977] 3 W.L.R. 38 (H.L.); and as to arbitration see *Crawford* v. *A.E.A. Prowting* [1973] Q.B. 1.

Clause 10
Sureties

If the Tender shall contain an undertaking by the Contractor to provide when required 2 good and sufficient sureties or to obtain the guarantee of an Insurance Company or Bank to be jointly and severally bound with the Contractor in a sum not exceeding 10 per cent of the Tender Total for the due performance of the Contract under the terms of a Bond the said sureties Insurance Company or Bank and the terms of the said Bond shall be such as shall be approved by the Employer and the provision of such sureties or the obtaining of such guarantee and the cost of the Bond to be so entered into shall be at the expense in all respects of the Contractor unless the Contract otherwise provides.

COMMENTARY

(1) " *If the Tender shall contain an undertaking* "

The form of tender contains an express undertaking, which should be deleted if a bond is not required. The employer is secured against the contractor's failure to reinstate loss or damage to the works by insurance under clause 21.

(2) " *the terms of the said Bond shall be such as shall be approved by the Employer* "

The tender contains an undertaking to provide a bond in the form annexed to the Conditions. It is difficult to see any application for the power to approve the terms.

Clause 11
Inspection of site

(1) The Contractor shall be deemed to have inspected and examined the Site and its surroundings and to have satisfied himself before submitting his tender as to the nature of the ground and sub-soil (so far as is practicable and having taken into account any information in connection therewith which may have been provided by or on behalf of the Employer) the form and nature of the Site the extent and nature of the work and materials necessary for the completion of the Works the means of communication with and access to the Site the accommodation he may require and in general to have obtained for himself all necessary information (subject as above-mentioned) as to risks contingencies and all other circumstances influencing or affecting his tender.

Sufficiency of tender

(2) The Contractor shall be deemed to have satisfied himself before submitting his tender as to the correctness and sufficiency of the rates and prices stated by him in the Priced Bill of Quantities which shall (except in so far as it is otherwise provided in the Contract) cover all his obligations under the Contract.

COMMENTARY

(1) *Generally*

Sub-clause (1) deals with the important topic of responsibility for the site and sub-soil. In the absence of a warranty from the employer, or other express provision, the site and sub-soil are contractor's risks.[42] Clause 11 of the 4th edition of the Conditions (which is substantially repeated in sub-clause (1)) clearly excluded a contractual warranty. Where a limited amount of information

---

[42] *Thorn* v. *London Corporation* (1876) 1 App.Cas. 120; *Bottoms* v. *York Corporation* (1892), H.B.C. (4th ed.) Vol. 2, p. 208, and see *ante*, p. 38.

was supplied which, although not incorrect, was misleading, similar words to clause 11 of the 4th edition have been held to negative any implication of a warranty that the information supplied was complete or exhaustive.[43] Sub-clause (1) now contains additional qualifying words which relate to sub-soil information. It is necessary to consider whether these words alter the effect of the clause.

(2) *Clause* 11 (1): " *and having taken into account any information in connection therewith which may have been provided by or on behalf of the Employer* "

The effect of these words is obscure.[44] They qualify the contractor's deemed satisfaction as to " the nature of the ground and sub-soil." But there follows an unqualified deemed satisfaction as to " the form and nature of the site." The definition of " site " (cl. 1 (1) (n)) suggests there may be no practical distinction between " ground and sub-soil " and " site," unless the latter is given a re-stricted meaning.[45] The closing words of the sub-clause: " and in general to have obtained for himself all necessary information (subject as above-mentioned) as to risks . . ." suggest that the preceding words are intended to impose some qualification on the contractor's knowledge as to risks.

It is suggested that the effect of the words is that the contractor is now en-titled to take sub-soil information into account in assessing risks, and that prima facie, such information is representative of its subject-matter. Whether par-ticular information is to be taken as representative of the site must depend upon the form and nature of the information. The effect of information given will further depend upon any written terms or qualifications incorporated into it.[45a] The usual caveat that bore-hole records are given " for information only " may not defeat the effect of sub-clause (1).

(3) *Should the employer warrant the site*

If the contractor is required to bear the risk of the site, he must allow for it in his contract price. If the employer has paid for a site investigation under direct contract, it appears illogical to pay the contractor to take the risk that the information is incorrect or misleading. It is suggested that, in the interests of economy alone, the employer should warrant the site, to the extent to which he is able to obtain indemnity from the site-investigation contractor. Such a course would further give incentive to procuring the most accurate investigation rather than the cheapest, a matter on which many practising engineers have expressed strong views.

Clause 12
Adverse physical conditions and artificial obstructions

(1) If during the execution of the Works the Contractor shall encounter

---

[43] *Dillingham Construction Pty. Ltd.* v. *Downs* (1972) 2 N.S.W.L.R. 49, discussed *ante*, p. 99.

[44] *Cf.* Abrahamson, *Engineering Law* (3rd ed.), p. 52, and comments of drafting com-mittee in *New Civil Engineer*, Dec. 20, 1973, p. 39.

[45] The definitions in cl. 1 (1) are subject to the words " except where the context otherwise requires."

[45a] See also Misrepresentation Act 1967 and see *ante*, Chap. 7.

physical conditions (other than weather conditions or conditions due to weather conditions) or artificial obstructions which conditions or obstructions he considers could not reasonably have been foreseen by an experienced contractor and the Contractor is of opinion that additional cost will be incurred which would not have been incurred if the physical conditions or artificial obstructions had not been encountered he shall if he intends to make any claim for additional payment give notice to the Engineer pursuant to Clause 52 (4) and shall specify in such notice the physical conditions and/or artificial obstructions encountered and with the notice if practicable or as soon as possible thereafter give details of the anticipated effects thereof the measures he is taking or is proposing to take and the extent of the anticipated delay in or interference with the execution of the Works.

Measures to be taken

(2) Following receipt of a notice under sub-clause (1) of this Clause the Engineer may if he thinks fit *inter alia*:—

    (a) require the Contractor to provide an estimate of the cost of the measures he is taking or is proposing to take;

    (b) approve in writing such measures with or without modification;

    (c) give written instructions as to how the physical conditions or artificial obstructions are to be dealt with;

    (d) order a suspension under Clause 40 or a variation under Clause 51.

Delay and extra cost

(3) To the extent that the Engineer shall decide that the whole or some part of the said physical conditions or artificial obstructions could not reasonably have been foreseen by an experienced contractor the Engineer shall take any delay suffered by the Contractor as a result of such conditions or obstructions into account in determining any extension of time to which the Contractor is entitled under Clause 44 and the Contractor shall subject to Clause 52 (4) (notwithstanding that the Engineer may not have given any instructions or orders pursuant to sub-clause (2) of this Clause) be paid in accordance with Clause 60 such sums as represents the reasonable cost of carrying out any additional work done and additional Constructional Plant used which would not have been done or used had such conditions or obstructions or such part thereof as the case may be not been encountered together with a reasonable percentage addition thereto in respect of profit and the reasonable costs incurred by the Contractor by reason of any unavoidable delay or disruption of working suffered as a consequence of encountering the said conditions or obstructions or such part thereof.

Conditions reasonably foreseeable

(4) If the Engineer shall decide that the physical conditions or artificial obstructions could in whole or in part have been reasonably foreseen by an experienced contractor he shall so inform the Contractor in writing as soon as he shall have reached that decision but the value of any variation previously ordered by him pursuant to sub-clause (2)(d) of this Clause shall be ascertained in accordance with Clause 52 and included in the Contract Price.

COMMENTARY

(1) *Generally*

The provisions of this clause are contrary to the general law relating to contracts for the supply of labour and materials, namely that the means of carrying out the work are the contractor's responsibility. To be entitled to additional payment the contractor must bring his claim within the words of the clause. Sub-clause (1) retains the words of the 4th edition defining the circumstances in which claims may be made. The remainder of the clause has been substantially re-drafted and simplified.[46] The provision allowing interim arbitration under this clause is now included in clause 66 (2).

(2) *Clause* 12 (1): "*physical conditions*"

Prima facie these words are very wide and little guidance can be given as to matters intended to give rise to claims. The exclusion of " weather conditions " suggests that a " condition " may be transient. " Physical " appears to require some material existence. For example, the imposition of statutory controls, such as those of January 1974 (the " 3-day week "), could not rank as " physical," although they might constitute a " condition."

(3) *Clause* 12 (1): "*other than weather conditions or conditions due to weather conditions*"

This provision is apt to exclude events such as a flood following heavy rain, or ground instability attributable to ground water arising from heavy rain. It further provides some indication of what the draftsman intended by " physical condition." The words " due to weather conditions " will, it is thought, exclude any condition of which an effective cause is weather conditions.[47]

(4) *Clause* 12 (1): "*artificial obstructions*"

These words are much more specific than " physical conditions " and must be limited to non-naturally occurring events which obstruct, *i.e.* hinder or stop, some part of the contractor's operations. In their context, it is thought that the words must contemplate a physical occurrence, and not, *e.g.* statutory controls. A typical example of artificial obstruction is buried services, encountered during excavation. The obstructing event may, however, be transient.

(5) *Clause* 12 (1): "*could not reasonably have been foreseen by an experienced Contractor*"

Although expressed objectively, the intention is clearly to exclude matters which could have been foreseen by an experienced contractor having access to all sources available to the contractor.[48] This will include any ground or sub-soil information, as referred to in clause 11 (1), whether or not its accuracy is warranted. This may operate to the advantage of either party: if the information predicts the condition or obstruction encountered, the contractor cannot contend that it is not foreseeable; while if the information fails to predict the condition or obstruction, it may add weight to its non-foreseeability.

(6) *Clause* 12 (1): "*pursuant to Clause 52 (4)*"

Notice under sub-clause 52 (4) (b) is required to be given " as soon as reason-

---

[46] Although the clause remains substantially longer and more complex than the admirably concise equivalent in the International Civil Engineering Conditions (cl. 12).
[47] See MacGillivray & Parkington, *Insurance Law* (6th ed., 1975), p. 720.
[48] See *C. J. Pearce & Co.* v. *Hereford Corp.* (1968) 66 L.G.R. 647, a case on the 4th edition of the conditions.

ably possible after the happening of the events giving rise to the claim." The words of the present sub-clause require further notice (which may be given at the same time as the notice of claim), which is to give details of " the measures he is taking or is proposing to take." This provision, together with the words of clause 52 (4), makes it quite clear that a claim may be retrospective, thus reversing the effect of the equivalent clause of the 4th edition.[49]

(7) *Clause* 12 (2): " *the Engineer may if he thinks fit . . .*"

The engineer is not obliged to take any action. The contractor's right to an extension of time and to additional payment under sub-clause (3) is not dependent upon whether the engineer takes any of the measures specified.

(8) *Clause* 12 (3): " *the contractor shall be paid . . . the reasonable cost of carrying out any additional work done which would not have been done had such conditions or obstructions not been encountered* "

This formula takes no account of the contractor's estimate, or the engineer's express instruction or order, under sub-clause (2). Additional cost arising from such instruction or order would require to be considered under clauses 13 or 52; while an accepted estimate would constitute a separate contract for payment of the agreed sum.

(9) *Clause* 12 (3): " *and the reasonable costs incurred by the contractor by reason of any unavoidable delay or disruption* "

The total claims allowable under the sub-clause are equivalent to putting the contractor into the financial position he would be in had the condition or obstruction not existed, or possibly into a better position, since the contractor is allowed a profit on the cost of additional work. The claims are unlikely to be less than the damages recoverable at common law for breach of a warranty that there would be no such condition or obstruction. This fortifies the argument in favour of the employer giving an express warranty as to ground conditions: see note (3) to clause 11.

Clause 13
Work to be to satisfaction of engineer

(1) Save in so far as it is legally or physically impossible the Contractor shall construct complete and maintain the Works in strict accordance with the Contract to the satisfaction of the Engineer and shall comply with and adhere strictly to the Engineer's instructions and directions on any matter connected therewith (whether mentioned in the Contract or not). The Contractor shall take instructions and directions only from the Engineer or (subject to the limitations referred to in Clause 2) from the Engineer's Representative.

Mode and manner of construction

(2) The whole of the materials plant and labour to be provided by the Contractor under Clause 8 and the mode manner and speed of construction and

---

[49] *Monmouth County Council* v. *Costelloe & Kemple* (1964) 63 L.G.R. 429 (C.A.), reversing (1964) 63 L.G.R. 131.

maintenance of the Works are to be of a kind and conducted in a manner approved of by the Engineer.

Delay and extra cost

(3) If in pursuance of Clause 5 or sub-clause (1) of this Clause the Engineer shall issue instructions or directions which involve the Contractor in delay or disrupt his arrangements or methods of construction so as to cause him to incur cost beyond that reasonably to have been foreseen by an experienced contractor at the time of tender then the Engineer shall take such delay into account in determining any extension of time to which the Contractor is entitled under Clause 44 and the Contractor shall subject to Clause 52(4) be paid in accordance with Clause 60 the amount of such cost as may be reasonable. If such instructions or directions require any variation to any part of the Works the same shall be deemed to have been given pursuant to Clause 51.

COMMENTARY

(1) *Generally*

This clause contains a number of important general obligations and powers, and should be read with clauses 5, 8 and 14. Sub-clause (3) contains a provision for making claims, which is of potentially wide application.

(2) *Clause* 13 (1): " *legally or physically impossible* "

These words may have an important effect upon the obligations of the contractor to perform the contract. For the extent to which legal or physical impossibility may excuse non-performance, see *ante*, p. 101.

Legally impossible. This probably contemplates work which would infringe a statutory provision such as the Public Health Acts, or a private right which could be protected by injunction. But the wording leaves considerable room for doubt. If the only means of carrying out the work or some section of it would result in a prohibition notice under the Health and Safety at Work, etc., Act 1974, it is thought that the contractor would be absolved from further obligation to carry out the work or section.

Physically impossible. In the construction of the permanent works (which usually will have been planned and designed in every detail) this is likely to occur rarely, *e.g.*, where some vital material becomes totally unavailable. However, the reference to " the Works," which includes the temporary works (cl. 1 (1) (l)), gives the provision much wider scope. In most cases what is impossible is not the permanent works but the projected means of construction, *i.e.*, the provision of sufficient and adequate temporary works. Thus, where the extent or nature of the temporary works is specified in the technical documents, the clause may operate to prevent the contractor being in breach. Alternative works may constitute variations.

(3) *Clause* 13 (1): " *to the satisfaction of the Engineer* "

The preceding word " strict " indicates that these words are additional to and not in qualification of the obligation to comply with the contract. The requirement does not, it is submitted, give the engineer power to override or add to the express requirements of the contract. The engineer's power under this provision is restricted to matters which are not precisely specified or which expressly leave a degree of choice, it is submitted.

(4) *Clause* 13 (1): " *adhere strictly to the Engineer's instructions and directions on any matter connected therewith (whether mentioned in the Contract or not)* "

These words are important because they appear to give the engineer very wide powers, and because an instruction " in pursuance of " this sub-clause now entitles the contractor to make a claim under sub-clause (3). The precise meaning of the words is obscure. It is reasonably clear that they refer to an instruction or direction:

    (i) on a matter connected with the works and
    (ii) which need not be mentioned in the contract.

However the words do not mean, it is submitted, that an instruction on any matter connected with the works is an instruction under sub-clause (1). As to the ambit of instructions to be regarded as issued " in pursuance of " sub-clause (1), see note (8) *infra*.

(5) *Clause* 13 (1): " *instructions and directions only from the Engineer or . . . from the Engineer's representative* "

This provision has not been amended to take account of the re-drafting of clause 2. It means, presumably, that an instruction or direction under sub-clause (1) may not be given by an assistant of the engineer's representative or by an authorised person responsible to the engineer (see cl. 2 (2) and (3)).

(6) *Clause* 13 (2): " *materials, plant and labour . . . are to be of a kind* "

This requirement adds nothing to the more general words of sub-clause (1), and may limit the engineer's authority, under the *expressio unius* principle, in respect of the provisions of clause 8 (1) not here repeated.

(7) *Clause* 13 (2): " *approved of by the Engineer* "

The reference in sub-clause (3) to instructions or directions in pursuance of sub-clause (1) only, makes it clear that the engineer's power under this sub-clause is negative, *i.e.* in an appropriate case to signify non-approval.[50]

(8) *Clause* 13 (3): " *If in pursuance of Clause 5 or sub-clause* (1) *of this clause the Engineer shall issue instructions or directions which . . .*"

The ambit of instructions or directions which are issued " in pursuance of " sub-clause (1) is not defined. Sub-clause (1) itself gives no assistance, since practically any instruction will come within its words. References elsewhere in the conditions to instructions " pursuant to "[51] or " in accordance with "[52] clause 13, suggest that the payment provision of sub-clause (3) cannot be applied to any instruction, but only one issued specifically under this clause.[53]

---

[50] See also cl. 14, note (6).
[51] Cl. 71 (2).
[52] See cl. 5.
[53] *Cf. New Civil Engineer*, July 5, 1973, pp. 44, 46 and Nov. 1, 1973, p. 46.

Where an instruction is issued under another clause, particularly one giving a right to payment not limited as in sub-clause (3), the instruction would normally be regarded as being issued under that other clause. It is suggested that an instruction under clause 13 must be one which could not be issued under any other clause or power in the contract. The instruction requires no particular form and need not be in writing. No doubt there will be many disputes as to whether an instruction is, or should be, given under clause 13.

(9) *Clause* 13 (3): " *disrupt his arrangements or methods of construction so as to cause him to incur cost* "

There is no precise indication of what is meant by arrangements or methods. But the engineer is entitled to information as to the contractor's intention thereto, under clause 14 (1) and (3). The field within which extra cost may be recoverable is extremely wide, but the conditions under which a claim is payable are more limited than in other clauses.

(10) *Clause* 13 (3): " *beyond that reasonably to have been foreseen by an experienced Contractor* "

These words appear, curiously, to qualify " cost," despite the fact that the contractor is then entitled to payment of " such cost as may be reasonable," *i.e.*, the cost recoverable is subject to two qualifications. The intention appears to be to allow claims only where the " instruction " or the resulting " disruption " were unforeseeable, but the effect is far from clear.

Clause 14
Programme to be furnished

(1) Within 21 days after the acceptance of his Tender the Contractor shall submit to the Engineer for his approval a programme showing the order of procedure in which he proposes to carry out the Works and thereafter shall furnish such further details and information as the Engineer may reasonably require in regard thereto. The Contractor shall at the same time also provide in writing for the information of the Engineer a general description of the arrangements and methods of construction which the Contractor proposes to adopt for the carrying out of the Works.

Revision of programme

(2) Should it appear to the Engineer at any time that the actual progress of the Works does not conform with the approved programme referred to in sub-clause (1) of this Clause the Engineer shall be entitled to require the Contractor to produce a revised programme showing the modifications to the original programme necessary to ensure completion of the Works or any Section within the time for completion as defined in Clause 43 or extended time granted pursuant to Clause 44(2).

Methods of construction

(3) If requested by the Engineer the Contractor shall submit at such times and in such detail as the Engineer may reasonably require such information pertaining to the methods of construction (including Temporary Works and the use of Constructional Plant) which the Contractor proposes to adopt or use and

such calculations of stresses strains and deflections that will arise in the Permanent Works or any parts thereof during construction from the use of such methods as will enable the Engineer to decide whether if these methods are adhered to the Works can be executed in accordance with the Drawings and Specification and without detriment to the Permanent Works when completed.

Engineer's consent

(4) The Engineer shall inform the Contractor in writing within a reasonable period after receipt of the information submitted in accordance with sub-clause (3) of this Clause either:—

(a) that the Contractor's proposed methods have the consent of the Engineer; or

(b) in what respects in the opinion of the Engineer they fail to meet the requirements of the drawings or Specification or will be detrimental to the Permanent Works.

In the latter event the Contractor shall take such steps or make such changes in the said methods as may be necessary to meet the Engineer's requirements and to obtain his consent. The Contractor shall not change the methods which have received the Engineer's consent without the further consent in writing of the Engineer which shall not be unreasonably withheld.

Design criteria

(5) The Engineer shall provide to the Contractor such design criteria relevant to the Permanent Works or any Temporary Works designed by the Engineer as may be necessary to enable the Contractor to comply with sub-clauses (3) and (4) of this Clause.

Delay and extra cost

(6) If the Engineer's consent to the proposed methods of construction shall be unreasonably delayed or if the requirements of the Engineer pursuant to sub-clause (4) of this Clause or any limitations imposed by any of the design criteria supplied by the Engineer pursuant to sub-clause (5) of this Clause could not reasonably have been foreseen by an experienced contractor at the time of tender and if in consequence of any of the aforesaid the Contractor unavoidably incurs delay or cost the Engineer shall take such delay into account in determining any extension of time to which the Contractor is entitled under Clause 44 and the Contractor shall subject to Clause 52(4) be paid in accordance with Clause 60 such sum in respect of the cost incurred as the Engineer considers fair in all the circumstances.

Responsibility unaffected by approval

(7) Approval by the Engineer of the Contractor's programme in accordance with sub-clauses (1) and (2) of this Clause and the consent of the Engineer to the Contractor's proposed methods of construction in accordance with sub-clause (4) of this Clause shall not relieve the Contractor of any of his duties or responsibilities under the Contract.

COMMENTARY

(1) *Generally*

This clause covers several matters which, in the absence of express provisions, are solely within the province of the contractor. The requirements must be read in the light of the contractor's basic responsibility to complete within the time for completion, or extended time (cll. 43, 44) and to take full responsibility for the methods of construction (cl. 8 (2)).

(2) *Clause* 14 (1): "*programme showing the order of procedure*"

While the programme to be submitted may apparently consist simply of a statement of the order of activities, without indication of durations, sub-clause (2) clearly contemplates that any such programme will give durations, or at least sufficient information from which the contractor's actual progress may be judged. Further, the contractor's right to possession of the site is dependent upon what is required to comply with the programmes referred to in this clause (see cl. 42 (1)).

(3) *Clause* 14 (1): "*arrangements and methods of construction*"

These words appear in clause 13 (3), where they form the basis of the contractor's right to make a claim. The information is therefore of importance, not only in limiting the scope of possible claims under clause 13 (3) but also in giving the engineer advance notice of the possible effects of his instructions and directions under clauses 5 or 13 (1).

(4) *Clause* 14 (2): "*Should it appear . . . that the actual progress . . . does not conform*"

Failure to comply with the contractor's programme (approved or not) is not of itself a breach, and is significant only as evidence of the contractor's failure to proceed with due expedition (see cl. 41) or with due diligence (see cl. 63 (1) (d)); and possibly of a failure to achieve an adequate rate of progress (see cl. 46).

(5) *Clause* 14 (2): "*the Engineer shall be entitled to require the Contractor to produce a revised programme . . .*"

The sub-clause omits to bind the contractor to produce the revised programme. This may recognise the practical difficulties that the contractor may have applications for extensions of time which are outstanding; and further that the engineer is required under clause 44 (3) and (4) to review the grounds for extensions at or after the date or extended date for completion (whether or not extensions have been applied for) and again at the date of the certificate of completion, so that such revised programme will often be administratively impossible to produce. In these circumstances the only remedy available to the employer is forfeiture under clause 63 (1) (d). Any attempt to expedite progress by giving notice under clause 46 is subject to the same criticism as above.[54]

(6) *Clause* 14 (3) *to* (7): *Generally*

Sub-clauses (3) to (7) cover the same ground as clause 13 (2), in more specific detail. It is not clear why two separate provisions are included or why clause 13 refers to the "mode and manner" and clause 14 to the "methods" of construction (which are for practical purposes indistinguishable).

[54] See cl. 44, note (7) and cl. 46.

Sub-clause (4) contains a curious alteration of wording in that the engineer's statement of the respects in which the contractor's methods " fail to meet the requirements of the drawings or specification or will be detrimental to the permanent works " becomes in the following sentence " Engineer's requirements." This abrupt change is carried through into sub-clause (6) as " requirements of the Engineer," which form the basis of a claim to the extent they were not foreseeable. This appears to envisage requirements other than those of the drawings or specification or the avoidance of detriment to the permanent works. However, the intention of sub-clauses (3) to (7) appear to be no more than to allow the engineer to obtain sufficient information for the giving or withholding of approval or consent to the contractor's methods.[55] The contractor is responsible for his own methods (cl. 8 (2)) and for the works (cl. 20), both permanent and temporary. There is no apparent reason, therefore, why the engineer should find it necessary to make requirements outside those of the drawings or specification or the avoidance of detriment to the works, none of which can be unforeseeable.

Sub-clause (6) gives a perfectly logical right of claim in respect of unforeseeable limitations imposed by design criteria supplied under sub-clause (5). But more doubt is cast on the engineer's power to make requirements by the omission from sub-clause (6) of any claim in respect of delayed delivery of such requirements. The engineer is given no express power under this clause to make positive requirements. Any instruction given may take effect under clause 13 (1) or 51.

A further consequence of any positive instruction or requirement of the engineer as to the contractor's methods may be to relieve the contractor of responsibility for the safety of the operation or for the temporary works, under clauses 8 (2) or 20 (3). In addition, sub-clause (7) omits to deal with the effect of either the engineer's requirements or any approval given under clause 13 (2). It is suggested, therefore, that any approval of the contractor's proposed methods should be expressly given under clause 14, and that the engineer should limit his non-approval of the contractor's methods to a statement of the respects in which they fail to meet the requirements of the drawings or specification or will be detrimental to the permanent works.

Clause 15
Contractor's superintendence

(1) The Contractor shall give or provide all necessary superintendence during the execution of the Works and as long thereafter as the Engineer may consider necessary. Such superintendence shall be given by sufficient persons having adequate knowledge of the operations to be carried out (including the methods and techniques required the hazards likely to be encountered and methods of preventing accidents) as may be requisite for the satisfactory construction of the Works.

Contractor's agent

(2) The Contractor or a competent and authorised agent or representative

---

[55] See also cl. 13 (2).

approved of in writing by the Engineer (which approval may at any time be withdrawn) is to be constantly on the Works and shall give his whole time to the superintendence of the same. Such authorised agent or representative shall be in full charge of the Works and shall receive on behalf of the Contractor directions and instructions from the Engineer or (subject to the limitations of Clause 2) the Engineer's Representative. The Contractor or such authorised agent or representative shall be responsible for the safety of all operations.

Clause 16
Removal of contractor's employees

The Contractor shall employ or cause to be employed in and about the execution of the Works and in the superintendence thereof only such persons as are careful skilled and experienced in their several trades and callings and the Engineer shall be at liberty to object to and require the Contractor to remove from the Works any person employed by the Contractor in or about the execution of the Works who in the opinion of the Engineer misconducts himself or is incompetent or negligent in the performance of his duties or fails to conform with any particular provisions with regard to safety which may be set out in the Specification or persists in any conduct which is prejudicial to safety or health and such persons shall not be again employed upon the Works without the permission of the Engineer.

COMMENTARY

(1) *Generally*

The contractor's obligations in these clauses are not such as may readily give rise to damages for their breach. The employer's alternative remedy, if the breach is persistent, is forfeiture under clause 63 (1) (d).

(2) *Clause* 15 (2): " *instructions from the Engineer or the Engineer's Representative* "

As with several other clauses, this provision does not take account of the redrafting of clause 2. For an example of a provision which does take account of clause 2 as redrafted, see clause 39 (3).

(3) *Clause* 15 (2): " *the Contractor or such authorised agent or representative shall be responsible* "

It is reasonably clear that an agent or representative is intended to be provided where the contractor is other than a sole trader. However, this provision cannot bind the agent (who is not party to the contract) nor can it relieve the contractor *vis-à-vis* the employer, from responsibility under clause 8 (2). The effect of the provision is that the contractor must put his agent or representative in charge of " the safety of all operations."

Clause 17
Setting-out

The Contractor shall be responsible for the true and proper setting-out of the Works and for the correctness of the position levels dimensions and alignment of all parts of the Works and for the provision of all necessary instruments appliances and labour in connection therewith. If at any time during the progress of the Works any error shall appear or arise in the position levels dimensions or alignment of any part of the Works the Contractor on being required so to do by the Engineer shall at his own cost rectify such error to the satisfaction of the Engineer unless such error is based on incorrect data supplied in writing by the Engineer or the Engineer's Representative in which case the cost of rectifying the same shall be borne by the Employer. The checking of any setting-out or of any line or level by the Engineer or the Engineer's Representative shall not in any way relieve the Contractor of his responsibility for the correctness thereof and the Contractor shall carefully protect and preserve all bench-marks sight rails pegs and other things used in setting out the Works.

(1) *" If at any time during the progress of the Works any error shall appear "*

This period extends only to the date of completion of the works, it is submitted.[56]

(2) *" the Contractor on being required to do so by the Engineer shall rectify such error "*

This is a useful provision, apart from which the engineer's only sanction is to refuse to certify payment for the erroneous work. Equivalent provisions in respect of defective work or materials are contained in clause 39. The omission from the present clause of the power (included in cl. 39) to employ others to do the work in default of the contractor's compliance, detracts seriously from its value.

Clause 18
Boreholes and exploratory excavation

If at any time during the execution of the Works the Engineer shall require the Contractor to make boreholes or to carry out exploratory excavation such requirement shall be ordered in writing and shall be deemed to be a variation ordered under Clause 51 unless a Provisional Sum or Prime Cost Item in respect of such anticipated work shall have been included in the Bill of Quantities.

Clause 19
Safety and security

(1) The Contractor shall throughout the progress of the Works have full regard for the safety of all persons entitled to be upon the Site and shall keep the Site (so far as the same is under his control) and the Works (so far as the same are not completed or occupied by the Employer) in an orderly state appropriate to the avoidance of danger to such persons and shall *inter alia* in connection with the Works provide and maintain at his own cost all lights guards fencing warning signs and watching when and where necessary or required by the Engineer or by any competent statutory or other authority for the protection of the Works or for the safety and convenience of the public or others.

Employer's responsibilities

(2) If under Clause 31 the Employer shall carry out work on the Site with his own workmen he shall in respect of such work:—
(a) have full regard to the safety of all persons entitled to be upon the Site; and
(b) keep the Site in an orderly state appropriate to the avoidance of danger to such persons.
If under Clause 31 the Employer shall employ other contractors on the Site he shall require them to have the same regard for safety and avoidance of danger.

[56] See cl. 39, note (1).

COMMENTARY

(1) *Clause* 19: *Generally*

This clause should be read with clauses 8 (2), 15 and 22, and in the light of the law relating to occupiers of premises (see p. 136). The present clause clearly has no effect on the rights of persons suffering injury on the site. But where such injury results from breach of the contractor's obligations under this clause, the employer may have a right of indemnity in respect of his liability, against the contractor.

Clause 22 provides a much wider indemnity in respect of claims arising " out of or in consequence of the construction and maintenance of the works." The practical effect of the present clause is, therefore, to place a positive duty upon the contractor to keep the site, etc., safe. A failure to do so which is persistent or fundamental may lead to forfeiture under clause 63 (1) (d).

(2) *Clause* 19 (1): " *all persons entitled to be upon the site* "

This clearly excludes trespassers, who may give rise to a material risk [57] as regards the employer's liability. The provision does not prevent the contractor being liable to a trespasser.

(3) *Clause* 19 (1): " *the site (so far as the same is under his control)* "

It is thought that the site is under the contractor's control notwithstanding that it is occupied or under the control of a sub-contractor. This is supported by clause 42 (possession of the site) which requires the contractor to be given possession of so much of the site as is required for construction of the works, *i.e.* including sub-contracted work. It is also supported by clause 4, which renders the contractor liable for acts, defaults or neglects of a sub-contractor.

Clause 20
Care of the works

(1) The Contractor shall take full responsibility for the care of the Works from the date of the commencement thereof until 14 days after the Engineer shall have issued a Certificate of Completion for the whole of the Works pursuant to Clause 48. Provided that if the Engineer shall issue a Certificate of Completion in respect of any Section or part of the Permanent Works before he shall issue a Certificate of Completion in respect of the whole of the Works the Contractor shall cease to be responsible for the care of that Section or part of the Permanent Works 14 days after the Engineer shall have issued the Certificate of Completion in respect of that Section or part and the responsibility for the care thereof shall thereupon pass to the Employer. Provided further that the Contractor shall take full responsibility for the care of any outstanding work which he shall have undertaken to finish during the Period of Maintenance until such outstanding work is complete.

Responsibility for reinstatement

(2) In case any damage loss or injury from any cause whatsoever (save and except the Excepted Risks as defined in sub-clause (3) of this Clause) shall happen to the Works or any part thereof while the Contractor shall be responsible for the care thereof the Contractor shall at his own cost repair and make good

---

[57] See *Herrington* v. *British Railways Board* [1972] A.C. 877.

the same so that at completion the Permanent Works shall be in good order and condition and in conformity in every respect with the requirements of the Contract and the Engineer's instructions. To the extent that any such damage loss or injury arises from any of the Excepted Risks the Contractor shall if required by the Engineer repair and make good the same as aforesaid at the expense of the Employer. The Contractor shall also be liable for any damage to the Works occasioned by him in the course of any operations carried out by him for the purpose of completing any outstanding work or of complying with his obligations under Clauses 49 and 50.

Excepted risks

(3) The " Excepted Risks " are riot war invasion act of foreign enemies hostilities (whether war be declared or not) civil war rebellion revolution insurrection or military or usurped power ionising radiations or contamination by radio-activity from any nuclear fuel or from any nuclear waste from the combustion of nuclear fuel radioactive toxic explosive or other hazardous properties of any explosive nuclear assembly or nuclear component thereof pressure waves caused by aircraft or other aerial devices travelling at sonic or supersonic speeds or a cause due to use or occupation by the Employer his agents servants or other contractors (not being employed by the Contractor) of any part of the Permanent Works or to fault defect error or omission if the design of the Works (other than a design provided by the Contractor pursuant to his obligations under the Contract).

Clause 21
Insurance of works, etc.

Without limiting his obligations and responsibilities under Clause 20 the Contractor shall insure in the joint names of the Employer and the Contractor:—
  (a) the Permanent Works and the Temporary Works (including for the purposes of this Clause any unfixed materials or other things delivered to the Site for incorporation therein) to their full value;
  (b) the Constructional Plant to its full value;
against all loss or damage from whatever cause arising (other than the Excepted Risks) for which he is responsible under the terms of the Contract and in such manner that the Employer and Contractor are covered for the period stipulated in Clause 20(1) and are also covered for loss or damage arising during the Period of Maintenance from such cause occurring prior to the commencement of the Period of Maintenance and for any loss or damage occasioned by the Contractor in the course of any operation carried out by him for the purpose of complying with his obligations under Clauses 49 and 50.

Provided that without limiting his obligations and responsibilities as aforesaid nothing in this Clause contained shall render the Contractor liable to insure against the necessity for the repair or reconstruction of any work constructed with materials and workmanship not in accordance with the requirements of the Contract unless the Bill of Quantities shall provide a special item for this insurance.

Such insurances shall be effected with an insurer and in terms approved by the Employer (which approval shall not be unreasonably withheld) and the Contractor shall whenever required produce to the Employer the policy or policies of insurance and the receipts for payment of the current premiums.

COMMENTARY

(1) *Generally*

These clauses lay down responsibilities for damage to the works, by making the contractor generally liable, subject to exceptions, and provide for insurance in respect of the works. The contractor is prima facie liable for the works by virtue of the obligation to complete and deliver the works. The principal effect of clause 20 is to define the areas in which the contractor is not liable.

(2) *Clause* 20 (1): " *full responsibility for the care of the works* "

Responsibility for " care of the works " would require the employer to prove lack of care in order to make the contractor liable for loss. Such responsibility is, in effect, superseded by the wider responsibility for " the works " under sub-clause (2), which involves no question of lack of care. The effect of sub-clause (1) is to define the period during which, and the extent to which, the contractor's responsibility under sub-clause (2) is to operate.

(3) *Clause* 20 (2): " *from any cause whatsoever* "

These words exclude any implied limitation upon the contractor's responsibility. Such responsibility will include damage arising from causes outside the parties' contemplation [58] and damage arising (save for the excepted risks) from the employer's negligence.[59]

(4) *Clause* 20 (2): " *while the Contractor shall be responsible for the care thereof* "

The time during which the contractor is so responsible is defined by sub-clause (1), and excludes any section or part in respect of which a certificate of completion has been given. The " excepted risks " (sub-cl. (3)) further exclude damage due to the employer's use or occupation of part of the permanent works.

(5) *Clause* 20 (2): " *the Contractor shall at his own cost repair and make good the same* "

The words " the same " refer to " the works or any part thereof " which suffer damage loss or injury. This apparently includes any section or part in respect of which a certificate of completion is given. But it is reasonably clear that the contractor is obliged to make good only those parts for which he is responsible by virtue of the preceding words " while . . . he shall be responsible for the care thereof." [60]

(6) *Clause* 20 (3): " *fault defect error or omission in the design of the Works* "

These words are not limited to a failure by the designer to comply with accepted standards of skill, nor to a failure to take account of existing engineering knowledge. A design " fault " may exist where the only error consists in the

---

[58] *Beaumont Thomas* v. *Blue Star Line* (1939) 55 T.L.R. 852; *Chandris* v. *Isbrandtsen-Moller Co. Inc.* [1951] 1 K.B. 240.

[59] *Farr* v. *The Admiralty* [1953] 1 W.L.R. 965.

[60] The contrary is arguable, and is supported by the immediately following words: " so that at completion, the Permanent Works shall be . . . in conformity in every respect with . . . the Contract." To ensure that the contractor's obligation to make good at his own cost does not extend to sections or parts certified as complete, there should be added between " make good the same " and " so that at completion " the words: " to the extent that the Contractor is responsible for the care thereof."

inability of existing knowledge or theory, to predict the actual stresses to which a structure will be subjected.[61]

(7) *Clause* 21: " *in the joint names of the Employer and the Contractor* "

The employer, named as a principal in the policy, can bring a claim in his own name under the policy, as well as pursuing a claim against the contractor under clause 20. The employer has thus protection against the contractor's insolvency, while the insurer's right of subrogation is preserved.

(8) *Clause* 21: " *against all loss or damage . . . for which he is responsible* "

The liability to be insured is defined by clause 20 (2). Such liability excludes the " excepted risks " in respect of which the employer must make his own arrangements for insurance. One matter of particular concern is loss arising from a design defect not amounting to professional negligence (see note (6) *supra*). In such a case the employer may have no redress unless the loss is otherwise insured or the designer has given an express or implied warranty as to the design.[62]

Clause 22
Damage to persons and property

(1) The Contractor shall (except if and so far as the Contract otherwise provides) indemnify and keep indemnified the Employer against all losses and claims for injuries or damage to any person or property whatsoever (other than the Works for which insurance is required under Clause 21 but including surface or other damage to land being the Site suffered by any persons in beneficial occupation of such land) which may arise out of or in consequence of the construction and maintenance of the Works and against all claims demands proceedings damages costs charges and expenses whatsoever in respect thereof or in relation thereto. Provided always that:—

(a) the Contractor's liability to indemnify the Employer as aforesaid shall be reduced proportionately to the extent that the act or neglect of the Employer his servants or agents may have contributed to the said loss injury or damage;

(b) nothing herein contained shall be deemed to render the Contractor liable for or in respect of or to indemnify the Employer against any compensation or damages for or with respect to:—

(i) damage to crops being on the Site (save in so far as possession has not been given to the Contractor);

(ii) the use or occupation of land (which has been provided by the Employer) by the Works or any part thereof or for the purpose of constructing completing and maintaining the Works (including consequent losses of crops) or interference whether temporary or permanent with any right of way light air or water or other easement or quasi easement which are the unavoidable result of the construction of the Works in accordance with the Contract;

(iii) the right of the Employer to construct the Works or any part thereof on over under in or through any land;

(iv) damage which is the unavoidable result of the construction of the Works in accordance with the Contract;

---

[61] *Manufacturers Mutual Insurance* v. *Queensland Government Railway* (1969) 1 Lloyd's Rep. 214.

[62] See *Greaves (Contractors)* v. *Baynham Meikle* [1975] 1 W.L.R. 1095.

(v) injuries or damage to persons or property resulting from any act or neglect or breach of statutory duty done or committed by the Engineer or the Employer his agents servants or other contractors (not being employed by the Contractor) or for or in respect of any claims demands proceedings damages costs charges and expenses in respect thereof or in relation thereto.

Indemnity by employer

(2) The Employer will save harmless and indemnify the Contractor from and against all claims demands proceedings damages costs charges and expenses in respect of the matters referred to in the proviso to sub-clause (1) of this Clause. Provided always that the Employer's liability to indemnify the Contractor under paragraph (v) of proviso (b) to sub-clause (1) of this Clause shall be reduced proportionately to the extent that the act or neglect of the Contractor or his sub-contractors servants or agents may have contributed to the said injury or damage.

Clause 23

Insurance against damage to persons and property

(1) Throughout the execution of the Works the Contractor (but without limiting his obligations and responsibilities under Clause 22) shall insure against any damage loss or injury which may occur to any property or to any person by or arising out of the execution of the Works or in the carrying out of the Contract otherwise than due to the matters referred to in proviso (b) to Clause 22 (1).

Amount and terms of insurance

(2) Such insurance shall be effected with an insurer and in terms approved by the Employer (which approval shall not be unreasonably withheld) and for at least the amount stated in the Appendix to the Form of Tender. The terms shall include a provision whereby in the event of any claim in respect of which the Contractor would be entitled to receive indemnity under the policy being brought or made against the Employer the insurer will indemnify the Employer against such claims and any costs charges and expenses in respect thereof. The Contractor shall whenever required produce to the Employer the policy or policies of insurance and the receipts for payment of the current premiums.

COMMENTARY

(1) *Generally*

These clauses divide and apportion responsibility between contractor and employer, and provide for corresponding insurance of the contractor's risks, in respect of damage to persons or property other than the works (which are covered by similar provision in the two preceding clauses). Clause 22 is the main indemnity clause of the contract. Other indemnities, which may overlap with this clause, are contained in clauses 24 (injury to workmen), 26 (2) (compliance with statutes), 27 (7) (compliance with Public Utilities, etc., Act), 29 (1) (interference with traffic, etc.) and (2) (noise and disturbance), 30 (2) and (3) (claims arising from transport), 49 (5) (a) (reinstatement of highways).

(2) *Clause* 22 (1): "*other than the Works for which insurance is required under Clause* 21"

The exclusion does not apply to the constructional plant, for which insurance is also required under clause 21. The employer will have an interest in loss of or damage to plant by virtue of clause 53.[63] Clauses 21 and 23 therefore duplicate the requirement for insurance of constructional plant.

(3) *Clause* 22 (1): " *Provided always that the Contractor's liability to indemnify* . . . *shall be reduced proportionately* "

These and similar words in sub-clause (2) in respect of the employer's liability avoid, it is submitted, the possible unenforceability of the indemnity where the party entitled to indemnity has been negligent.[64]

(4) *Clause* 23

While this clause is clearly intended to complement clause 22 it should be noted that the mandatory insurance requirements do not coincide precisely with the contractor's liability under the previous clause. The requirements of clause 23 appear generally to be narrower than those of clause 22. In particular they omit reference to damage, loss or injury arising in consequence of the execution of the works.

(5) *Clause* 23 (1): " *Throughout the execution of the Works* "

The period extends from the date of commencement until the maintenance certificate, it is submitted.[65] The insurance should therefore cover this period.

(6) *Clause* 23 (2)

Under the policy envisaged by sub-clause (2) the sole principal is the contractor, who alone may enforce the policy. The insurer's obligation to indemnify the employer directly arises only when the contractor would be entitled to indemnity, *i.e.*, when the contractor is liable for the loss or damage under clause 22.

Clause 24

Accident or injury to workmen

The Employer shall not be liable for or in respect of any damages or compensation payable at law in respect or in consequence of any accident or injury to any workman or other person in the employment of the Contractor or any sub-contractor save and except to the extent that such accident or injury results from or is contributed to by any act or default of the Employer his agents or servants and the Contractor shall indemnify and keep indemnified the Employer against all such damages and compensation (save and except as aforesaid) and against all claims demands proceedings costs charges and expenses whatsoever in respect thereof or in relation thereto.

COMMENTARY

This provision should be read with clause 22. It does not affect the right of any workman or other person employed by the contractor or a sub-contractor

---

[63] Cl. 53 (9) preserves the contractor's responsibility for loss or injury to plant.
[64] *A.M.F. International Ltd.* v. *Magnet Bowling* [1968] 1 W.L.R. 1028 and see *ante*, p. 140.
[65] See cl. 39, note (1).

to bring proceedings against the employer, *e.g.* under the Occupier's Liability Act 1957. In the event of the employer being found liable, he is entitled to an indemnity from the contractor to the extent of the contractor's contributory default. Such apportionment of liability avoids the possible unenforceability of the indemnity where the employer has been negligent,[66] it is submitted.

The clause makes no requirement as to insurance. The contractor is obliged by statute [67] to insure against injury to employees arising out of the course of their employment.

Clause 25
Remedy on contractor's failure to insure

If the Contractor shall fail upon request to produce to the Employer satisfactory evidence that there is in force the insurance referred to in Clauses 21 and 23 or any other insurance which he may be required to effect under the terms of the Contract then and in any such case the Employer may effect and keep in force any such insurance and pay such premium or premiums as may be necessary for that purpose and from time to time deduct the amount so paid by the Employer as aforesaid from any monies due or which may become due to the Contractor or recover the same as a debt due from the Contractor.

COMMENTARY

" *The Employer may effect any such insurance* "

The employer has no authority to insure in the contractor's name or jointly, and must effect the necessary insurance in his own name. In the event of loss occurring for which the contractor is liable under the contract, the insurer, having paid the employer, would be entitled to sue the contractor under the right of subrogation.

Clause 26
Giving of notices and payment of fees

(1) The Contractor shall save as provided in Clause 27 give all notices and pay all fees required to be given or paid by any Act of Parliament or any Regulation or Bye-law of any local or other statutory authority in relation to the execution of the Works and by the rules and regulations of all public bodies and companies whose property or rights are or may be affected in any way by the Works. The Employer shall repay or allow to the Contractor all such sums as the Engineer shall certify to have been properly payable and paid by the Contractor in respect of such fees and also all rates and taxes paid by the Contractor in respect of the Site or any part thereof or anything constructed or erected thereon or on any part thereof or any temporary structures situate elsewhere but used exclusively for the purposes of the Works or any structures used temporarily and exclusively for the purposes of the Works.

Contractor to conform with statutes, etc.

(2) The Contractor shall ascertain and conform in all respects with the provisions of any general or local Act of Parliament and the Regulations and Bye-laws of any local or other statutory authority which may be applicable to the Works and with such rules and regulations of public bodies and companies

---

[66] *A.M.F. International* v. *Magnet Bowling, ibid.*
[67] Employer's Liability (Compulsory Insurance) Act 1969.

as aforesaid and shall keep the Employer indemnified against all penalties and liability of every kind for breach of any such Act Regulation or Bye-law. Provided always that:—

  (a) the Contractor shall not be required to indemnify the Employer against the consequences of any such breach which is the unavoidable result of complying with the Drawings Specification or instructions of the Engineer;

  (b) if the Drawings Specification or instructions of the Engineer shall at any time be found not to be in conformity with any such Act Regulation or Bye-law the Engineer shall issue such instructions including the ordering of a variation under Clause 51 as may be necessary to ensure conformity with such Act Regulation or Bye-law;

  (c) the Contractor shall not be responsible for obtaining any planning permission which may be necessary in respect of the Permanent Works or any Temporary Works specified or designed by the Engineer and the Employer hereby warrants that all the said permissions have been or will in due time be obtained.

COMMENTARY

(1) *Clause* 26 (1): " *give all notices* "

The contractor's obligation cannot be limited to notices required to be given by those carrying out the work. The principal notices to which this clause will apply are those required by the Building Regulations (or the London Building (Constructional) By-laws). While the initial application under the Regulations to obtain " approval " is termed a notice, it is thought the clear intention of this clause is to require the contractor to give only those notices which are in the nature of " notification " rather than " application." Under the London Building Acts notices are also required to be served on adjoining owners.[68] Such notices are in practice normally served by the employer or the engineer. But the contractor's obligation hereunder must include all such notices.

(2) *Clause* 26 (1): " *any temporary structures used exclusively . . . or any structure used temporarily and exclusively for the purposes of the Works* "

The reference to " temporary structures " does not appear to add anything, since " any structure " must include a temporary structure.

The ordinary meaning of " structure " is very wide, and is apt to include any building.[69] However, it is thought that use " for the purposes of the Works " (as opposed to the purposes of the contract) must be limited to use for a purpose physically connected with the construction of the works, and would therefore exclude, *e.g.* head office rates and taxes. But the position is by no means clear.

(3) *Clause* 26 (2): *Indemnity*

This should be read with the general indemnity provisions of clause 22.

(4) *Clause* 26 (2) (b): " *shall at any time be found not to be in conformity* "

The contractor's obligation, under the opening words of sub-clause (2), to " ascertain . . . the provisions of any . . . Act of Parliament " do not include an express obligation to inform the engineer of any departure from such provisions

---

[68] See London Building Acts (Amendment) Act 1939, Pt. VI.
[69] *Almond* v. *Birmingham Royal Institution for the Blind* [1968] A.C. 37, 51 (H.L.).

or to request instructions. But proviso (a), which limits the employer's indemnity, does not prevent the contractor being in breach of sub-clause (2) where work shown on the drawings, etc., does not conform to statutory requirements. The effect is therefore, it is thought, that when a disconformity is discovered the contractor is entitled only to payment for the work as varied to achieve conformity, and not to payment for any work already carried out in breach of statutory requirements; nor to the costs of demolition of the work which was in breach.

(5) *Clause* 26 (2) (c): " *the Employer hereby warrants* "

The contractor will be entitled to reimbursement of any loss or expense incurred by reason of delay due to absence of necessary planning permissions, as damages for breach of warranty. If planning permission is not obtained at all, the contractor will generally be entitled to be paid the price of the work carried out.[70]

Clause 27
Public Utilities Street Works Act 1950—Definitions
(1) For the purposes of this Clause:—
   (a) the expression " the Act " shall mean and include the Public Utilities Street Works Act 1950 and any statutory modification or re-enactment thereof for the time being in force;
   (b) all other expressions common to the Act and to this Clause shall have the same meaning as that assigned to them by the Act.

Notifications by employer to contractor
(2) The Employer shall before the commencement of the Works notify the Contractor in writing:—
   (a) whether the Works or any parts thereof (and if so which parts) are Emergency Works; and
   (b) which (if any) parts of the Works are to be carried out in Controlled Land or in a Prospectively Maintainable Highway.
If any duly authorised variation of the Works shall involve the execution thereof in a Street or in Controlled Land or in a Prospectively Maintainable Highway or are Emergency Works the Employer shall notify the Contractor in writing accordingly at the time such variation is ordered.

Service of notices by employer
(3) The Employer shall (subject to the obligations of the Contract [71] under sub-clause (4) of this Clause) serve all such notices as may from time to time whether before or during the course of or after completion of the Works be required to be served under the Act.

Notices by contractor to employer
(4) The Contractor shall in relation to any part of the Works (other than Emergency Works) and subject to the compliance by the Employer with sub-clause (2) of this Clause give not less than 21 days' notice in writing to the Employer before:—

---

[70] See *Strongman* (*1945*) v. *Sincock* [1955] 2 Q.B. 525. Work done without planning permission is not *per se* unlawful, but may be subject to enforcement proceedings.
[71] The reference should presumably be to " obligations of the Contractor."

(a) commencing any part of the Works in a Street (as defined by Sections 1 (3) and 38 (1) of the Act); or

(b) commencing any part of the Works in Controlled Land or in a Prospectively Maintainable Highway; or

(c) commencing in a Street or in Controlled Land or in a Prospectively Maintainable Highway any part of the Works which is likely to affect the apparatus of any Owning Undertaker (within the meaning of Section 26 of the Act).

Such notice shall state the date on which and the place at which the Contractor intends to commence the execution of the work referred to therein.

Failure to commence street works

(5) If the Contractor having given any such notice as is required by sub-clause (4) of this Clause shall not commence the part of the Works to which such notice relates within 2 months after the date when such notice is given such notice shall be treated an invalid and compliance with the said sub-clause (4) shall be requisite as if such notice had not been given.

Delays attributable to variations

(6) In the event of such a variation of the Works as is referred to in sub-clause (2) of this Clause being ordered by or on behalf of the Employer and resulting in delay in the execution of the Works by reason of the necessity of compliance by the Contractor with sub-clause (4) of this Clause the Engineer shall take such delay into account in determining any extension of time to which the Contractor is entitled under Clause 44 and the Contractor shall subject to Clause 52 [72] be paid in accordance with Clause 60 such additional cost as the Engineer shall consider to have been reasonably attributable to such delay.

Contractor to comply with other obligations of Act

(7) Except as otherwise provided by this Clause where in relation to the carrying out of the Works the Act imposes any requirements or obligations upon the Employer the Contractor shall subject to Clause 49 (5) comply with such requirements and obligations and shall (subject as aforesaid) indemnify the Employer against any liability which the Employer may incur in consequence of any failure to comply with the said requirements and obligations.

COMMENTARY

(1) *Generally*

The Act creates, in Part I, a statutory code (the Street Works Code) to regulate works carried out by statutory undertakers affecting highways. The purpose of the Act is to protect authorities affected by such works, and to enable certain works to be carried out on land abutting a highway (controlled land) instead of in the highway. Part II of the Act creates a code for the protection of statutory undertakers where works are carried out by an interested authority; and Part III provides for statutory undertakers' work which affects the apparatus of another undertaker (the owning undertaker).

(2) *Clause* 27 (3): " *the Employer shall . . . serve all such notices* "

The principal notices required to be served (subject to certain exceptions) by statutory undertakers are:

---

[72] This should presumably refer to cl. 52 (4); see cll. 7 (3) 13 (3), 14 (6), etc.

(i) Seven days' notice to the relevant authority prior to the start of works to which the Street Works Code relates.

(ii) Three days' notice to any other undertaker whose apparatus is likely to be affected by the proposed work.

(iii) In respect of emergency works, notices as soon as reasonably practicable, which may be given after start of the work.

Notice given under (i) above lapses if the work has not been substantially begun within two months, but by sub-clause (5) the contractor's notice to the employer will also have lapsed, so that the employer is protected against a statutory breach by the contractor.

(3) *Clause 27 (7): Indemnity*

This should be read with the general indemnity provisions of clauses 22 and 26 (2).

The requirements or obligations placed upon the employer by the Act and which, by sub-clause (7), devolve upon the contractor, may be very wide in their application. They include the obligation to make payments to undertakers who are required to move their apparatus or to authorities who elect to execute certain works to their own undertakings (see ss. 10, 12, 22).

Further obligations relating to any person executing street works are created by section 37 of the Highways Act 1971.[73]

The contractor is bound to conform with these provisions by clause 26 (2) *supra*, but subject to the exceptions provided.

Clause 28
Patent rights

(1) The Contractor shall save harmless and indemnify the Employer from and against all claims and proceedings for or on account of infringement of any patent rights design trade-mark or name or other protected rights in respect of any Constructional Plant machine work or material used for or in connection with the Works and from and against all claims demands proceedings damages costs charges and expenses whatsoever in respect thereof or in relation thereto.

Royalties

(2) Except where otherwise specified the Contractor shall pay all tonnage and other royalties rent and other payments or compensation (if any) for getting stone sand gravel clay or other materials required for the Works.

Clause 29
Interference with traffic and adjoining properties

(1) All operations necessary for the execution of the Works shall so far as compliance with the requirements of the Contract permits be carried on so as not to interfere unnecessarily or improperly with the public convenience or the access to or use or occupation of public or private roads and foot-paths or to or of properties whether in the possession of the Employer or of any other person and the Contractor shall save harmless and indemnify the Employer in respect of all claims demands proceedings damages costs charges and expenses whatsoever arising out of or in relation to any such matters.

---

[73] This amends s. 149 of the Highways Act 1959.

Noise and disturbance

(2) All work shall be carried out without unreasonable noise and disturbance. The Contractor shall indemnify the Employer from and against any liability for damages on account of noise or other disturbance created while or in carrying out the work and from and against all claims demands proceedings damages costs charges and expenses whatsoever in regard or in relation to such liability.

COMMENTARY

(1) *Clause* 29 (1)

The obligation in sub-clause (1) not to interfere with access or occupation etc., is qualified by the words " unnecessarily " and " so far as compliance with the requirements of the Contract permits." No such qualifications are expressed in the indemnity which follows. It is thought that the words " whatsoever arising out of or in relation to " mean that " any such matters " cannot be construed as subject to the qualifications above. However, this clause must be read with and subject to clause 22, particularly proviso (b) to sub-clause (1) and sub-clause (2) thereof.

(2) *Clause* 29 (2)

The comments to sub-clause (1) *supra* apply equally to sub-clause (2) save that liability in nuisance to adjoining occupiers will arise only where noise or other disturbance is unreasonable.[74] Thus, the first sentence of the sub-clause in effect qualifies any liability of the contractor under the second sentence.

Clause 30
Avoidance of damage to highways, etc.

(1) The Contractor shall use every reasonable means to prevent any of the highways or bridges communicating with or on the routes to the Site from being subjected to extraordinary traffic within the meaning of the Highways Act 1959 or in Scotland the Road Traffic Act 1930 or any statutory modification or re-enactment thereof by any traffic of the Contractor or any of his sub-contractors and in particular shall select routes and use vehicles and restrict and distribute loads so that any such extraordinary traffic as will inevitably arise from the moving of Constructional Plant and materials or manufactured or fabricated articles from and to the Site shall be limited as far as reasonably possible and so that no unnecessary damage or injury may be occasioned to such highways and bridges.

Transport of constructional plant

(2) Save insofar as the Contract otherwise provides the Contractor shall be responsible for and shall pay the cost of strengthening any bridges or altering or improving any highway communicating with the Site to facilitate the movement of Constructional Plant equipment or Temporary Works required in the execution of the Works and the Contractor shall indemnify and keep indemnified the Employer against all claims for damages to any highway or bridge communicating with the Site caused by such movement including such claims as may be made by any competent authority directly against the Employer pursuant to any Act of Parliament or other Statutory Instrument and shall negotiate and pay all claims arising solely out of such damage.

---

[74] See p. 136.

Transport of materials

(3) If notwithstanding sub-clause (1) of this Clause any damage shall occur to any bridge or highway communicating with the Site arising from the transport of materials or manufactured or fabricated articles in the execution of the Works the Contractor shall notify the Engineer as soon as he becomes aware of such damage or as soon as he receives any claim from the authority entitled to make such claim. Where under any Act of Parliament or other Statutory Instrument the haulier of such materials or manufactured or fabricated articles is required to indemnify the highway authority against damage the Employer shall not be liable for any costs charges or expenses in respect thereof or in relation thereto. In other cases the Employer shall negotiate the settlement of and pay all sums due in respect of such claim and shall indemnify the Contractor in respect thereof and in respect of all claims demands proceedings damages costs charges and expenses in relation thereto. Provided always that if and so far as any such claim or part thereof shall in the opinion of the Engineer be due to any failure on the part of the Contractor to observe and perform his obligations under sub-clause (1) of this Clause then the amount certified by the Engineer to be due to such failure shall be paid by the Contractor to the Employer or deducted from any sum due or which may become due to the Contractor.

COMMENTARY

(1) *Clause* 30 (1): " *extraordinary traffic* "

The term is not defined by the Highways Act 1959, but is subject to much case law. It means traffic "which is so exceptional in the quality or quantity of articles carried, or in the mode or time of user of the road, as substantially to alter and increase the burden imposed by ordinary traffic on the road and to cause damage and expense thereby beyond what is common." [75]

(2) *Clause* 30 (2) *and* (3)

The scheme of sub-clauses (2) and (3) is to require the contractor, subject to sub-clause (1), to accept responsibility only for damage caused by the transport of his constructional tackle *i.e.* " constructional plant, equipment and temporary works " (including temporary works designed by the engineer: see clause 8 (2)). The contractor is absolved from responsibility for damage caused by transport of materials for the permanent construction *i.e.* " materials or manufactured or fabricated articles," provided this is not due to a breach of the obligation under sub-clause (1) to use every reasonable means to prevent or limit extraordinary traffic.

The indemnities should be read with the general provisions of clause 22.

(3) *Clause* 30 (3): " *sums due in respect of such claim* "

The Highway authority is entitled to recover maintenance expenses by virtue of section 62 of the Highways Act 1959. This section further provides for agreement or determination by arbitration of a sum by way of a composition of the liability to the authority.

Clause 31
Facilities for other contractors

(1) The Contractor shall in accordance with the requirements of the Engineer

---

[75] *Hill* v. *Thomas* [1893] 2 Q.B. 333, 341.

afford all reasonable facilities for any other contractors employed by the Employer and their workmen and for the workmen of the Employer and of any other properly authorised authorities or statutory bodies who may be employed in the execution on or near the Site of any work not in the Contract or of any contract which the employer may enter into in connection with or ancillary to the Works.

Delay and extra cost

(2) If compliance with sub-clause (1) of this Clause shall involve the Contractor in delay or cost beyond that reasonably to be foreseen by an experienced contractor at the time of tender then the Engineer shall take such delay into account in determining any extension of time to which the Contractor is entitled under Clause 44 and the Contractor shall subject to Clause 52 (4) be paid in accordance with Clause 60 the amount of such cost as may be reasonable.

COMMENTARY

Clause 31 (1): " *afford all reasonable facilities* "
Sub-clause (1) should be read with clause 42.

The words are not sufficient, it is submitted, to displace the employer's obligation under clause 42 (1) to give possession of so much of the site " as may be required to enable the Contractor to proceed " in accordance with his programme and with due dispatch. If the facilities required by the engineer are such as to deprive the contractor of adequate possession of the site, he will be entitled to make a claim under clause 42 (1), in addition to any claim under clause 31 (2). A claim under the present clause is limited to delay or cost not reasonably foreseeable; while a claim under clause 42 (1) is not so limited.

Clause 32
Fossils etc.

All fossils coins articles of value or antiquity and structures or other remains or things of geological or archaeological interest discovered on the Site shall as between the Employer and the Contractor be deemed to be the absolute property of the Employer and the Contractor shall take reasonable precautions to prevent his workmen or any other persons from removing or damaging any such article or thing and shall immediately upon discovery thereof and before removal acquaint the Engineer of such discovery and carry out at the expense of the Employer the Engineer's orders as to the disposal of the same.

Clause 33
Clearance of site on completion

On the completion of the Works the Contractor shall clear away and remove from the Site all Constructional Plant surplus material rubbish and Temporary Works of every kind and leave the whole of the Site and Permanent Works clean and in a workmanlike condition to the satisfaction of the Engineer.

LABOUR

Clause 34
Rates of wages/hours and conditions of labour

(1) The Contractor shall in the execution of the Contract observe and fulfil the obligations upon contractors specified in the Fair Wages Resolution passed by the House of Commons on the 14 October 1946 of which the following is an extract:—

Extract from Fair Wages Resolution

" 1 (a) The contractor shall pay rates of wages and observe hours and conditions of labour not less favourable than those established for the trade or industry in the district where the work is carried out by machinery of negotiation or arbitration to which the parties are organisations of employers and trade unions representative respectively of substantial proportions of the employers and workers engaged in the trade or industry in the district.

"   (b) In the absence of any rates of wages, hours or conditions of labour so established the contractor shall pay rates of wages and observe hours and conditions of labour which are not less favourable than the general level of wages, hours and conditions observed by other employers whose general circumstances in the trade or industry in which the contractor is engaged are similar.

" 2   The contractor shall in respect of all persons employed by him (whether in execution of the contract or otherwise) in every factory workshop or place occupied or used by him for the execution of the contract comply with the general conditions required by this Resolution.

" 3   In the event of any question arising as to whether the requirements of this Resolution are being observed, the question shall, if not otherwise disposed of, be referred by the Minister of Labour and National Service to an independent Tribunal for decision.

" 4   The contractor shall recognise the freedom of his workpeople to be members of Trade Unions.

" 5   The contractor shall at all times during the continuance of a contract display, for the information of his workpeople, in every factory, workshop or place occupied or used by him for the execution of the contract a copy of this Resolution.

" 6   The contractor shall be responsible for the observance of this Resolution by sub-contractors employed in the execution of the contract."

Civil Engineering Construction Conciliation Board

(2) The wages hours and conditions of employment above referred to shall be those prescribed for the time being by the Civil Engineering Construction Conciliation Board for Great Britain save that the rates of wages payable to any class of labour in respect of which the said Board does not prescribe a rate shall be governed by the provisions of sub-clause (1) of this Clause.

Clause 35
Returns of labour and plant

The Contractor shall if required by the Engineer deliver to the Engineer or at his office a return in such form and at such intervals as the Engineer may prescribe showing in detail the numbers of the several classes of labour from time to time employed by the Contractor on the Site and such information respecting Constructional Plant as the Engineer may require. The Contractor shall require his sub-contractors to observe the provisions of this Clause.

WORKMANSHIP AND MATERIALS

Clause 36
Quality of materials and workmanship and tests

(1) All materials and workmanship shall be of the respective kinds described in the Contract and in accordance with the Engineer's instructions and shall be subjected from time to time to such tests as the Engineer may direct at the place of manufacture or fabrication or on the Site or such other place or places as may be specified in the Contract. The Contractor shall provide such assistance instruments machines labour and materials as are normally required for examin-

ing measuring and testing any work and the quality weight or quantity of any materials used and shall supply samples of materials before incorporation in the Works for testing as may be selected and required by the Engineer.

Cost of samples

(2) All samples shall be supplied by the Contractor at his own cost if the supply thereof is clearly intended by or provided for in the Contract but if not then at the cost of the Employer.

Cost of tests

(3) The cost of making any test shall be borne by the Contractor if such test is clearly intended by or provided for in the Contract and (in the cases only of a test under load or of a test to ascertain whether the design of any finished or partially finished work is appropriate for the purposes which it was intended to fulfil) is particularised in the Specification or Bill of Quantities in sufficient detail to enable the Contractor to have priced or allowed for the same in his Tender. If any test is ordered by the Engineer which is either:—

(a) not so intended by or provided for; or

(b) (in the cases above mentioned) is not so particularised;

then the cost of such test shall be borne by the Contractor if the test shows the workmanship or materials not to be in accordance with the provisions of the Contract or the Engineer's instructions but otherwise by the Employer.

COMMENTARY

(1) *Clause* 36 (1): " *and in accordance with the Engineer's instructions* "

These words do not give a power to issue instructions, it is submitted. Instructions must be given under express terms of the contract, *e.g.* clauses 5, 7, 13 or 51. In addition to this provision, clause 13 (2) requires the materials to be of a kind approved of by the engineer.

(2) *Clause* 36 (1): " *as are normally required for examining, measuring and testing any work and . . . materials used* "

These words must be read as descriptive and not as limiting the contractor's obligation, in view of the duty to carry out " such tests as the Engineer may direct " (sub-cl. (1) *supra*), and the references to " any " test (sub-cl. (3)).

Clause 37
Access to Site

The Engineer and any person authorised by him shall at all times have access to the Works and to the Site and to all workshops and places where work is being prepared or whence materials manufactured articles and machinery are being obtained for the Works and the Contractor shall afford every facility for and every assistance in or in obtaining the right to such access.

COMMENTARY

" *Shall at all times have access to the works and to the site and to all workshops and places* "

The effect of these words is that the contractor's possession of the works and the site are subject to the engineer's right of access (*cf.* the effect of cl. 31 (1)). The contractor is bound to afford every facility and assistance in obtaining access to his own premises and those of sub-contractors (against whom only the contractor can enforce such rights).

Clause 38
Examination of work before covering up

(1) No work shall be covered up or put out of view without the approval of the Engineer and the Contractor shall afford full opportunity for the Engineer to examine and measure any work which is about to be covered up or put out of view and to examine foundations before permanent work is placed thereon. The Contractor shall give due notice to the Engineer whenever any such work or foundations is or are ready or about to be ready for examination and the Engineer shall without unreasonable delay unless he considers it unnecessary and advises the Contractor accordingly attend for the purpose of examining and measuring such work or of examining such foundations.

Uncovering and making openings

(2) The Contractor shall uncover any part or parts of the Works or make openings in or through the same as the Engineer may from time to time direct and shall reinstate and make good such part or parts to the satisfaction of the Engineer. If any such part or parts have been covered up or put out of view after compliance with the requirements of sub-clause (1) of this Clause and are found to be executed in accordance with the Contract the cost of uncovering making openings in or through reinstating and making good the same shall be borne by the Employer but in any other case all such cost shall be borne by the Contractor.

COMMENTARY

Generally

Sub-clause (1) draws a necessary distinction between examination and measurement, and as to the latter, should be read with clause 56.

Examination. The contractor must give " due notice " when work or foundations are ready and afford full opportunity for examination. The work must not be covered up without approval. Such approval would, it is thought, be implied from the engineer's failure to examine within a reasonable time after notice. The sanction against failure by the contractor to comply with such requirements is the engineer's power under sub-clause (2) to order uncovering, etc., at the contractor's cost.

Measurement. The engineer must give reasonable notice, under clause 56 (3), when he requires to measure any part of the work. Under clause 38 (1), upon the contractor giving notice that " any such work " (*i.e.* work which is about to be covered up) is ready for examination, the engineer is bound (unless he considers it unnecessary) to attend for measurement. It is not clear whether, by virtue of the words " when he requires " in clause 56 (3), the engineer is absolved from the obligation to give notice to the contractor before measuring, under the present clause. It is thought that the sanction against the contractor's non-attendance under clause 56 (3) [76] would not be enforced without clear notice under that clause.

---

[76] The engineer's measurement is to be taken as correct.

Clause 39
Removal of improper work and materials

(1) The Engineer shall during the progress of the Works have power to order in writing:—

(a) the removal from the Site within such time or times as may be specified in the order of any materials which in the opinion of the Engineer are not in accordance with the Contract;

(b) the substitution of proper and suitable materials; and

(c) the removal and proper re-execution (notwithstanding any previous test thereof or interim payment therefor) of any work which in respect of materials or workmanship is not in the opinion of the Engineer in accordance with the Contract.

Default of contractor in compliance

(2) In case of default on the part of the Contractor in carrying out such order the Employer shall be entitled to employ and pay other persons to carry out the same and all expenses consequent thereon or incidental thereto shall be borne by the Contractor and shall be recoverable from him by the Employer or may be deducted by the Employer from any monies due or which may become due to the Contractor.

Failure to disapprove

(3) Failure of the Engineer or any person acting under him pursuant to Clause 2 to disapprove any work or materials shall not prejudice the power of the Engineer or any of them subsequently to disapprove such work or materials.

COMMENTARY

(1) *Clause* 39 (1): " *during the progress of the works* "

These words are not defined. Other references to " progress " appear in clauses 7 (1) (supply of further drawings and instructions), 14 (2) (revision of the contractor's programme), 40 (suspension of work) and 46 (expedition of the rate of progress). The conditions also use the term " during " or " throughout the execution of the works " in clauses 15 (1) (contractor's superintendence) and 23 (1) (insurance against third party claims). These different terms should presumably be construed as having consistent and different meanings.

In clause 40 it is clear that a suspension of " progress " is envisaged only before completion and not during the period of maintenance. Clauses 14 (2) and 46 also use the word " progress " in this sense. Further, clause 49 provides measures alternative to those of the present clause, which apply during the period of maintenance; and clause 60 (5) (c) contemplates that clause 39 will not be operated during the maintenance period. It is thought, therefore, that " during the progress " should be limited to the period between commencement and completion or substantial completion of the works, and the powers under this clause are exercisable accordingly.

Conversely those clauses which refer to the period of " execution of the works " do not appear to require such a limited interpretation. Insurable losses covered by clause 23 (1) may certainly occur during the period of maintenance or during making good. The word " execution " is wide enough to include construction and maintenance. It is thought, therefore, that the period of " execution " extends from the date of commencement to the date of the maintenance certificate.

One apparent flaw in the above reasoning is the use, in clause 62, of the words " during the execution of the Works or during the Period of Maintenance." However, the terms as there used do not compel a mutually exclusive construction.[77] It is thought clause 62 does not defeat the above construction, at least for the purpose of the clauses enumerated.

(2) *Clause* 39 (1): " *substitution and re-execution* "

An order for the removal of materials or work during the progress of the works must be given under an express power, since the contractor may not be in breach.[78] The powers to order substitution of proper materials and re-execution of work are absent from the equivalent provisions of the Standard form of building contract.[79]

It is suggested that the power to order substitutions, etc., should be exercised cautiously since the contractor remains under an over-riding obligation to complete the works in accordance with the contract, with a sanction of liquidated damages. An instruction for substitution, etc., may be construed as a variation. It is clear that an order for removal of materials may be given without a corresponding order for substitution. Insofar as the wording of paragraph (c) requires an order to specify removal *and* re-execution, the engineer is entitled to order " re-execution in accordance with the Contract."

(3) *Clause* 39 (2): *Remedy for default*

In addition to the power under sub-clause (2) to employ other persons, clause 63 (1) (c) gives a right of forfeiture in respect of matters falling within clause 39 (1) (a) and (c), where the contractor fails to comply with the engineer's notice for 14 days. But see notes to clause 63.

Where remedial or repair work is urgently necessary the engineer may act under clause 62 to secure the execution of such work forthwith.

(4) *Clause* 39 (3): " *Failure of the Engineer* . . . *to disapprove any work or materials shall not prejudice* "

The words imply that a positive approval would prejudice the right subsequently to disapprove the work or materials. But the contractor's obligation is to construct and complete the works in accordance with the contract. Any such implication is removed by the express words of sub-clause (1) (c).[80] The conditions provide no form of express approval of work during the progress of the works, other than by interim certificate.[81]

---

[77] There are other instances of the conditions drawing an apparent distinction between two terms, one of which includes the other; see note to cl. 1 (1) (o).

[78] See *Kaye (P. & M.) Ltd.* v. *Hosier & Dickinson* [1972] 1 W.L.R. 146 (H.L.) *per* Lord Diplock, at p. 157; a decision on the Standard form of building contract.

[79] See cl. 6 (4), *ante* p. 307.

[80] See also cll. 17, 36 (3), 38 (2) and 60 (7)

[81] *Ibid.*

Clause 40
Suspension of work

(1) The Contractor shall on the written order of the Engineer suspend the progress of the Works or any part thereof for such time or times and in such manner as the Engineer may consider necessary and shall during such suspension properly protect and secure the work so far as is necessary in the opinion of the Engineer. Subject to Clause 52 (4) the Contractor shall be paid in accordance with Clause 60 the extra cost (if any) incurred in giving effect to the Engineer's instructions under this Clause except to the extent that such suspension is:—

    (a)   otherwise provided for in the Contract; or

    (b)   necessary by reason of weather conditions or by some default on the part of the Contractor; or

    (c)   necessary for the proper execution of the work or for the safety of the Works or any part thereof inasmuch as such necessity does not arise from any act or default of the Engineer or the Employer or from any of the Excepted Risks defined in Clause 20.

The Engineer shall take any delay occasioned by a suspension ordered under this Clause (including that arising from any act or default of the Engineer or the Employer) into account in determining any extension of time to which the Contractor is entitled under Clause 44 except when such suspension is otherwise provided for in the Contract or is necessary by reason of some default on the part of the Contractor.

Suspension lasting more than three months

(2) If the progress of the Works or any part thereof is suspended on the written order of the Engineer and if permission to resume work is not given by the Engineer within a period of 3 months from the date of suspension then the Contractor may unless such suspension is otherwise provided for in the Contract or continues to be necessary by reason of some default on the part of the Contractor serve a written notice on the Engineer requiring permission within 28 days from the receipt of such notice to proceed with the Works or that part thereof in regard to which progress is suspended. If within the said 28 days the Engineer does not grant such permission the Contractor by a further written notice so served may (but is not bound to) elect to treat the suspension where it affects part only of the Works as an omission of such part under Clause 51 or where it affects the whole Works as an abandonment of the Contract by the Employer.

COMMENTARY

(1) *Generally*

The engineer is never obliged to give a suspension order, unless on the direction of the employer. In view of the extreme consequences provided by sub-clause (2), he should do so only when there is no other remedy under the contract. It is difficult to envisage instances in which the employer's interest would be served by a suspension order, save for financial circumstances or where a major change of design becomes necessary. The drafting of the present clause leaves a number of matters open to doubt.

(2) *Clause* 40 (2): "*elect to treat the suspension . . . as an abandonment of the Contract*"

The conditions make no express provision for the consequences of abandonment (*cf.* cl. 63 (1)). The contractor clearly ceases to be bound to complete the works. It is thought that the intention of sub-clause (2) is that the contractor may treat the deemed abandonment as a repudiation by the employer and be entitled to payment accordingly.[82] This view is supported by the alternative remedies provided where suspension affects part only of the works. But the position is not clear.[83]

(3) *Clause* 40 (2): " *where it affects the whole Works* "
These words define the degree of suspension which the contractor may treat as an abandonment of the contract. The apparent intention is to cover a suspension of all work upon which the contractor is engaged. But it is arguable that such a suspension does not affect the whole of the works as defined (cl. 1 (1) (j), (k), (l)) *a fortiori* since parts of the works will be already complete. The apparent intention may be achieved by construing " works " as " the uncompleted works " [84] and relying on the words of clause 1 (1): " except where the context otherwise requires ".

(4) *Clause* 40 (2): " *where it affects part only of the works as an omission of such part under clause* 51 "
This provision involves none of the difficulties referred to in notes (2) and (3) *supra*. The election to treat suspension of the progress of part of the works as an omission entitles the contractor to claim a refixing of rates under clause 52 (2), in respect of the part of the omitted work which has been done and also in respect of any other rate or price " rendered unreasonable or inapplicable " by reason of the omission.[85]

COMMENCEMENT TIME AND DELAYS
Clause 41
Commencement of Works
The Contractor shall commence the Works on or as soon as is reasonably possible after the Date for Commencement of the Works to be notified by the Engineer in writing which date shall be within a reasonable time after the date of acceptance of the Tender. Thereafter the Contractor shall proceed with the Works with due expedition and without delay in accordance with the Contract.

COMMENTARY
(1) " *The Contractor shall commence the Works* . . ."
The contractor's failure to commence may lead to two consequences. First, if the failure is " without reasonable excuse " the employer may forfeit under clause 63 (1) (b), with the financial consequences set out in clause 63 (4). Secondly, the time for completion (clause 43) runs from the date for commencement notified by the engineer, so that the contractor must make up any lost time or pay liquidated damages in lieu.

---

[82] For repudiation generally, see *ante*, p. 111.
[83] The express provisions for payment under clause 40 (1) and, in the event of abandonment by the contractor, under cl. 63 (4) tend to exclude any implied right to payment: see *Trollope & Colls* v. *North West Metropolitan Regional Hospital Board* [1973] 1 W.L.R. 601 (H.L.). The contractor's right must arise upon the construction of the clause.
[84] *Cf.* Standard form of building contract, cl. 26 (1) (c).
[85] See cl. 52, note (2).

(2) *" proceed with the Works with due expedition "*

Clause 63 (1) (c) gives a right of forfeiture where the contractor is failing to proceed with the works " with due diligence." It is thought that no significance can attach to the difference in terms, since they each apply to the obligation to proceed with the works. Failure to proceed with due expedition may therefore give the employer a right of forfeiture under clause 63 (1).

In addition to rights under the contract, the employer is entitled to terminate the contract at common law where the contractor's delay is so serious as to amount to repudiation.[86] Upon determination at common law the employer will not have the benefit of clause 63, but will retain plant and materials vested under clauses 53 and 54. An alternative less drastic remedy for delay is provided by clause 46, which permits the engineer to require the contractor to expedite progress where this is too slow to ensure timely completion.

The contractor's programme under clauses 14 (1) or 14 (2) will, it is thought, be evidence of what constitutes a proper level of expedition or diligence, against which any default of the contractor may be measured.

Clause 42
Possession of site

(1) Save in so far as the Contract may prescribe the extent of portions of the Site of which the Contractor is to be given possession from time to time and the order in which such portions shall be made available to him and subject to any requirement in the Contract as to the order in which the Works shall be executed the Employer will at the Date for Commencement of the Works notified under Clause 41 give to the Contractor possession of so much of the Site as may be required to enable the Contractor to commence and proceed with the construction of the Works in accordance with the programme referred to in Clause 14 and will from time to time as the Works proceed give to the Contractor possession of such further portions of the Site as may be required to enable the Contractor to proceed with the construction of the Works with due despatch in accordance with the said programme. If the Contractor suffers delay or incurs cost from failure on the part of the Employer to give possession in accordance with the terms of this Clause then the Engineer shall take such delay into account in determining any extension of time to which the Contractor is entitled under Clause 44 and the Contractor shall subject to Clause 52 (4) be paid in accordance with Clause 60 the amount of such cost as may be reasonable.

Wayleaves, etc.

(2) The Contractor shall bear all expenses and charges for special or temporary wayleaves required by him in connection with access to the Site. The Contractor shall also provide at his own cost any additional accommodation outside the Site required by him for the purposes of the Works.

COMMENTARY

(1) *Clause 42 (1):* " *possession of so much of the site as may be required to enable the Contractor to commence and proceed . . . in accordance with the programme referred to in Clause 14 "*

This clause is difficult to apply unless the site is expressly defined in one of the contract documents (see cl. 1 (1) (e)). Without such definition the contractor

---

[86] See *ante,* p. 111.

is entitled to sufficient access to enable him to commence and proceed in accordance with this programme.[87]

(2) *Clause* 42 (1): " *so much of the site* "

These words indicate that the area or extent of land or other places required to be given to the contractor must be determined in relation to the area or extent of the whole site as well as in relation to his programme.

These provisions should be read with clauses 31 (1) and 37. See also clause 1 note (2) (n).

Clause 43
Time for completion

The whole of the Works and any Section required to be completed within a particular time as stated in the Appendix to the Form of Tender shall be completed within the time so stated (or such extended time as may be allowed under Clause 44) calculated from the Date of Commencement of the Works notified under Clause 41.

COMMENTARY

(1) *Generally*

The employer's remedy for non-compliance is liquidated damages as provided by clause 47; or if the liquidated damages are unenforceable [88] the employer will have a right to general damages for delay.[89]

(2) " *within the time so stated* "

The contractor is entitled to complete earlier than the time stated, in which case the employer is obliged to make correspondingly earlier payments under clause 60.

Clause 44
Extension of time for completion

(1) Should any variation ordered under Clause 51 (1) or increased quantities referred to in Clause 51 (3) or any other cause of delay referred to in these Conditions or exceptional adverse weather conditions or other special circumstances of any kind whatsoever which may occur be such as fairly to entitle the Contractor to an extension of time for the completion of the Works or (where different periods for completion of different Sections are provided for in the Appendix to the Form of Tender) of the relevant Section the Contractor shall within 28 days after the cause of the delay has arisen or as soon thereafter as is reasonable in all the circumstances deliver to the Engineer full and detailed particulars of any claim to extension of time to which he may consider himself entitled in order that such claim may be investigated at the time.

Interim assessment of extension

(2) The Engineer shall upon receipt of such particulars or if he thinks fit in the absence of any such claim consider all the circumstances known to him at that time and make an assessment of the extension of time (if any) to which he

---

[87] See also *Wells* v. *Army & Navy* (1902) 86 L.T. 764, H.B.C. (4th ed.), Vol. 2, p. 346 and *Hounslow* v. *Twickenham Garden Developments* [1971] Ch. 233, 257.

[88] See cl. 44, note (3).

[89] See *ante*, Chap. 9.

considers the Contractor entitled for the completion of the Works or relevant Section and shall by notice in writing to the Contractor grant such extension of time for completion. In the event that the Contractor shall have made a claim for an extension of time but the Engineer considers the Contractor not entitled thereto the Engineer shall so inform the Contractor.

Assessment at due date for completion

(3) The Engineer shall at or as soon as possible after the due date or extended date for completion (and whether or not the Contractor shall have made any claim for an extension of time) consider all the circumstances known to him at that time and take action similar to that provided for in sub-clause (2) of this Clause. Should the Engineer consider that the Contractor is not entitled to an extension of time he shall so notify the Employer and the Contractor.

Final determination of extension

(4) The Engineer shall upon the issue of the Certificate of Completion of the Works or of the relevant Section review all the circumstances of the kind referred to in sub-clause (1) of this Clause and shall finally determine and certify to the Contractor the overall extension of time (if any) to which he considers the Contractor entitled in respect of the Works or any relevant Section. No such final review of the circumstances shall result in a decrease in any extension of time already granted by the Engineer pursuant to sub-clauses (2) or (3) of this Clause.

COMMENTARY

(1) *Generally*

The grounds upon which the contractor may be entitled to an extension of time are the following; the number in brackets is the condition giving such entitlement:

(i) variations under clause 51 (1) (44 (1));

(ii) increases in the quantities of items of work over those stated in the bill of quantities (44 (1));

(iii) failure of the engineer to issue any necessary drawings or instructions at reasonable times (7 (3));

(iv) adverse physical conditions or artificial obstructions (12 (3));

(v) instructions or directions under clauses 5 or 13 (1) (13 (3));

(vi) delay by the engineer in giving consent to the contractor's proposed methods of construction or unforeseeable requirements or design criteria supplied by the engineer (14 (6));

(vii) variations involving street works, etc., where the contractor must give notice before commencement (27 (6));

(viii) affording facilities for other contractors (31 (2));

(ix) suspension of the progress of the works (40 (1));

(x) failure to give possession of sufficient of the site (42 (1));

(xi) forfeiture of a nominated sub-contract (59B (4) (b));

(xii) exceptional adverse weather conditions (44 (1));

(xiii) other special circumstances of any kind whatsoever (44 (1)).

(2) *Clause 44 (1)*: " *such as fairly to entitle the Contractor to an extension of time for the completion of the Works* "
Sub-clauses (2) and (4) refer to extensions of time " to which he [the engineer]

considers the Contractor entitled." Sub-clause (3) refers back to (2). These provisions must be read with sub-clause (1) to define the criteria upon which the engineer must (or may) grant extensions.

The test to be applied in assessing an extension is not whether the completion of the works is likely to be delayed,[90] but whether the occurrence fairly entitles the contractor to an extension. The engineer may consider matters other than the immediate ground for extension. Under sub-clauses (2) and (3) the engineer is expressly required to consider " all the circumstances known to him at that time." Such matters or circumstances will, it is thought, include other variations which would tend to advance the date of completion, or other items of work which have decreased in quantity.

(3) *Clause* 44 (1): " *other special circumstances of any kind whatsoever* "

The words " of any kind whatsoever " prevent the operation of the *ejusdem generis* rule, so that the circumstances need not be limited to the kinds included in grounds (i) to (xii) *supra*. Whether particular facts can amount to " special circumstances " will depend upon construction of the whole contract, including any special terms or conditions. Prima facie, delay by a sub-contractor, nominated or direct, cannot rank as a special circumstance.[91]

While the provision entitles the contractor to an extension in respect of any circumstance falling within the ordinary meaning of the words, such general words will not be construed as entitling the engineer to grant an extension in respect of delay due to the employer's default.[92] Unless such default falls within an express ground for extension, such delay will invalidate the liquidated damages clause, as will the failure to grant an extension to which the contractor is entitled.[93] The engineer should therefore grant any extension, the necessity for which is a default by the employer, under any of the express grounds (i) to (xii) *supra* which apply, and should not rely upon the general ground.

(4) *Clause* 44 (1): " *or . . . of the relevant Section* "

It is similarly necessary for the engineer to be expressly entitled to grant, and in fact to grant, contractual extensions in respect of any section in order to protect the employer's right to claim liquidated damages for that section.[94]

(5) *Clause* 44 (2): " *the Contractor shall . . . deliver to the Engineer* "

The engineer is obliged to consider an interim extension under sub-clause (2) only upon receipt of particulars " or if he thinks fit." Under sub-clauses (3) and (4) the engineer is bound to make further assessments at the date or extended date for completion and at the date of actual completion.

---

[90] *Cf.* Standard form of building contract, cl. 23.

[91] See cll. 4 and 59A (4). See also *Westminster Corporation* v. *Jarvis* [1970] 1 W.L.R. 637 (H.L.), a case on the Standard form of building contract where an extension in respect of delay by a nominated sub-contractor is expressly permitted.

[92] *Peak Construction* v. *McKinney Foundations* (1970) 69 L.G.R. 1 (C.A.).

[93] *Ibid.* at p. 12.

[94] *Ibid.*

(6) *Clause* 44 (4): " *No such final review . . . shall result in a decrease in any `extension* "

The word " result " indicates that the final review is to be a proper determination of extensions to which the contractor is entitled, but is to take effect only in so far as it grants extensions additional to those granted under interim assessments.

(7) *Effect of revised extensions*

While the provisions for interim and final assessments of extensions of time recognise what must frequently occur in practice, some practical difficulties remain. If the engineer exercises his power under clause 14 (2) to require a revised programme, or under clause 46 to give notice requiring progress to be expedited, the grounds for such requirements may be subsequently removed by further extensions based upon grounds which arose before the date of the engineer's requirement or notice,[95] *e.g.* a variation already carried out.

It is thought that such requirement or notice would not be rendered invalid (and therefore *ultra vires*) by subsequent extensions. Clause 43 requires completion within the time stated " or such extended time as may be allowed under Clause 44," so that the contractor is bound, on a particular date, to complete by the time stated as extended at that date, notwithstanding that at a later date the time may be further extended.

While the above difficulty appears expressly to be recognised in clause 14 (2) by the reference to the time for completion " or extended time granted pursuant to Clause 44 (2) " (*i.e.* the interim extension), clause 46 contains no such refinement.

Clause 45
Night and Sunday work

Subject to any provision to the contrary contained in the Contract none of the Works shall be executed during the night or on Sundays without the permission in writing of the Engineer save when the work is unavoidable or absolutely necessary for the saving of life or property or for the safety of the Works in which case the Contractor shall immediately advise the Engineer or the Engineer's Representative. Provided always that this Clause shall not be applicable in the case of any work which it is customary to carry out outside normal working hours or by rotary or double shifts.

COMMENTARY

" *Without the permission in writing of the Engineer* "

The withholding of permission is not required to be reasonable, nor is any such requirement to be implied, it is submitted: see note (1) to clause 4 and the express permission contained at the end of clause 46.

Clause 46
Rate of progress

If for any reason which does not entitle the Contractor to an extension of time the rate of progress of the Works or any Section is at any time in the opinion of

---

[95] See cl. 14, note (5), and note to cl. 46.

the Engineer too slow to ensure completion by the prescribed time or extended time for completion the Engineer shall so notify the Contractor in writing and the Contractor shall thereupon take such steps as are necessary and the Engineer may approve to expedite progress so as to complete the Works or such Section by the prescribed time or extended time. The Contractor shall not be entitled to any additional payment for taking such steps. If as a result of any notice given by the Engineer under this Clause the Contractor· shall seek the Engineer's permission to do any work at night or on Sundays such permission shall not be unreasonably refused.

COMMENTARY

A notice given under this clause is likely to result in the contractor incurring expense to expedite progress. An invalid notice would entitle the contractor, it is submitted, to claim reasonable payment for the steps taken.[96] The question therefore arises whether the contractor is entitled to make such a claim, where the basis of a notice given under this clause is removed by a later review of extensions of time based upon grounds existing at the date of the notice.

It is thought that by virtue of clause 43,[97] the notice is not rendered invalid by a subsequent extension. But the contrary is arguable since, while clause 14 (2) refers expressly to the time for completion extended pursuant to clause 44 (2) (*i.e.* the interim assessment), clause 46 refers only to " any reason which does not entitle the Contractor to an extension of time," which must include extensions under sub-clauses (2) (3) and (4) of clause 44.

If the engineer fails to grant extensions as required by clause 44, it may render the employer in breach of contract. The contractor would then be entitled to recover, as damages, any additional cost of expediting the work to meet the completion date which should have been extended, whether or not notice has been given under this clause. However, it is clear that any claim based upon the engineer's failure or delay in granting extensions of time would be conditional upon the contractor having applied for extensions in accordance with clause 44 (1). Without timeous applications, any loss may be said to arise from the contractor's breach of contract.[98]

LIQUIDATED DAMAGES AND LIMITATION OF DAMAGES FOR DELAYED COMPLETION

Clause 47

Liquidated damages for whole of works

(1) (a) In the Appendix to the Form of Tender under the heading " Liquidated Damages for Delay " there is stated in column 1 the sum which represents the Employer's genuine pre-estimate (expressed as a rate per week or per day as the case may be) of the damages likely to be suffered by him in the event that the whole of the Works shall not be completed within the time prescribed by Clause 43.

Provided that in lieu of such sum there may be stated such lesser sum as represents the limit of the Contractor's liability for damages for

---

[96] *Molloy* v. *Liebe* (1910) 102 L.T. 616; and see unreported decision of H.H. Judge Sir Norman Richards, Q.C., in *Devon Contractors* v. *City of Plymouth*, Apr. 22, 1975.
[97] See cl. 44, note (7).
[98] *Roberts* v. *Bury Commissioners* (1870) L.R. 5 C.P. 310; and see *ante*, p. 78.

failure to complete the whole of the Works within the time for completion therefor or any extension thereof granted under Clause 44.

(b) If the Contractor should fail to complete the whole of the Works within the prescribed time or any extension thereof granted under Clause 44 the Contractor shall pay to the Employer for such default the sum stated in column 1 aforesaid for every week or day as the case may be which shall elapse between the date on which the prescribed time or any extension thereof expired and the date of completion of the whole of the Works. Provided that if any part of the Works not being a Section or part of a Section shall be certified as complete pursuant to Clause 48 before completion of the whole the of Works the sum stated in column 1 shall be reduced by the proportion which the value of the part completed bears to the value of the whole of the Works.

Liquidated damages for sections

(2) (a) In cases where any Section shall be required to be completed within a particular time as stated in the Appendix to the Form of Tender there shall also be stated in the said Appendix under the heading " Liquidated Damages for Delay " in column 2 the sum by which the damages stated in column 1 or the limit of the Contractor's said liability as the case may be shall be reduced upon completion of each such Section and in column 3 the sum which represents the Employer's genuine pre-estimate (expressed as aforesaid) of any specific damage likely to be suffered by him in the event that such Section shall not be completed within that time.

Provided that there may be stated in column 3 in lieu of such sum such lesser sum as represents the limit of the Contractor's liability for failure to complete the relevant Section within the relevant time.

(b) If the Contractor should fail to complete any Section within the relevant time for completion or any extension thereof granted under Clause 44 the Contractor shall pay to the Employer for such default the sum stated in column 3 aforesaid for every week or day as the case may be which shall elapse between the date on which the relevant time or any extension thereof expired and the date of completion of the relevant Section. Provided that:—

(i) if completion of a Section shall be delayed beyond the due date for completion of the whole of the Works the damages payable under sub-clauses (1) and (2) of this Clause until completion of that Section shall be the sum stated in column 1 plus in respect of that Section the sum stated in column 3 less the sum stated in column 2;

(ii) if any part of a Section shall be certified as complete pursuant to Clause 48 before completion of the whole thereof the sums stated in columns 2 and 3 in respect of that Section shall be reduced by the proportion which the value of the part bears to the value of the Section and the sum stated in column 1 shall be reduced by the same amount as the sum in column 2 is reduced; and

(iii) upon completion of any such Section the sum stated in column 1 shall be reduced by the sum stated in column 2 in respect of that Section at the date of such completion.

Damages not a penalty

(3) All sums payable by the Contractor to the Employer pursuant to this Clause shall be paid as liquidated damages for delay and not as a penalty.

Deduction of liquidated damages

(4) If the Engineer shall under Clause 44 (3) or (4) have determined and

certified any extension of time to which he considers the Contractor entitled and shall have notified the Employer and the Contractor that he is of the opinion that the Contractor is not entitled to any or any further extension of time the Employer may deduct and retain from any sum otherwise payable by the Employer to the Contractor hereunder the amount which in the event that the Engineer's said opinion should not be subsequently revised would be the amount of the liquidated damages payable by the Contractor under this Clause.

Reimbursement of liquidated damages

(5) If upon a subsequent or final review of the circumstances causing delay the Engineer shall grant an extension or further extension of time or if an arbitrator appointed under Clause 66 shall decide that the Engineer should have granted such an extension or further extension of time the Employer shall no longer be entitled to liquidated damages in respect of the period of such extension of time. Any sums in respect of such period which may have been recovered pursuant to sub-clause (3) of this Clause shall be reimbursable forthwith to the Contractor together with interest at the rate provided for in Clause 60 (6) from the date on which such liquidated damages were recovered from the Contractor.

COMMENTARY

(1) *Summary*

This clause provides for the deduction of liquidated damages upon the contractor's failure to complete, with the following refinements:

(i) provision for sums less than the " genuine pre-estimate " of damages. Such lesser sums would take effect as a limitation of liability, it is submitted. The clause would therefore not be subject to the same stringent rules of construction as where the damages are truly liquidated.[99] However, neither the clause nor the appendix provides for any indication whether the sums stated are liquidated or lesser sums;

(ii) levying of liquidated damages where different completion dates are required for any section or sections;

(iii) the reduction of liquidated damages where any " part " of the works or of any section is certified as complete under clause 48;

(iv) the deduction and reimbursement (with interest) of liquidated damages necessitated by interim, subsequent and final determination of extensions of time under clause 44.

(2) *Clause 47 (1): " genuine pre-estimate . . . of the damages "*

These are technical words which denote damages which are liquidated, as opposed to being a penalty.[1] Neither the use of such words, nor of the term " liquidated damages," nor the provisions of sub-clause (3), can prevent the court inquiring whether the payment stipulated is a penalty.[2] The words should therefore be taken as a direction for completing the appendix.

(3) *The Appendix, columns 2 and 3*

Although sub-clause (2) (a) refers simply to a section required to be

---

[99] *Peak Construction* v. *McKinney Foundations* (1970) 69 L.G.R. 1 (C.A.).
[1] *Dunlop Ltd.* v. *New Garage Co. Ltd.* [1915] A.C. 79, *per* Lord Dunedin at p. 86.
[2] *Ibid.* at p. 86.

completed " within a particular time," the following provisions are workable only if the particular time is *less* than the time for completion of the whole of the works. Thus, if one part is required to be completed later than the bulk of the work, it is the bulk which must be expressed as the section and the time for completion of the remaining part becomes the time for completion of the works.

The provision for different sums in columns 2 and 3 allow considerable flexibility. If the anticipated damages for a section in column 3 are disproportionately high, the sum in column 2 may be a lower figure, to protect the employer adequately after completion of that section. If the anticipated damages for a section are disproportionately low or the section carries a " lesser sum," the sum in column 2 will be a higher figure to ensure that the residue, after completion of that section, does not exceed a " genuine pre-estimate."

Despite the opportunity for elaborate distribution of liquidated damages between sections, the clause is equally enforceable if the total damages are merely apportioned rateably. The provisions for completion of a " part " in sub-clauses (1) (b) and (2) (b) rely on such apportionment.[3] If the sections are apportioned rateably then (subject to specifying any lesser sums in lieu of genuine pre-estimates) the sums in columns 2 and 3 will be identical.

(4) *Clause* 47 (2) (b) (i): *Section delayed beyond overall completion date*

It is difficult to understand the working of this provision. If, at the date for completion, a section remains uncompleted, the liquidated damages will increase, reversing any reductions in respect of sections already completed, to a sum which may exceed the total damages for the works. While this may be the intention of the provision, it produces difficulties. First, there appears to be no connection between the damages payable under this sub-clause and a genuine pre-estimate of damages in isolation. Secondly, there is no allowance for two or more sections being delayed beyond the overall date for completion. It may be that the reference to " the sum stated in Column 1 " is intended as subject to reduction for sections or parts already completed; but that is not what is said. The provision is likely to be unenforceable in its present form, because the damages will not be a genuine pre-estimate of loss.

(5) *Clause* 47 (4): *Engineer's notice of no further extension*

The engineer's notice under sub-clause (4) is a condition precedent to the employer's express right of deduction or set-off under this sub-clause. This emphasises that the engineer should, under clause 60, certify in full and leave the employer to deduct any such amounts.[4] However, the requirement of notice is avoided by the contractor's obligation under sub-clauses (1) (b) and (2) (b) to " pay to the Employer " the liquidated damages due. These provisions give a right of equitable set-off[5] against any sum otherwise payable, it is submitted, including sums due under certificates.[6] Even if this were not so, the only practical

---

[3] Similar provisions in the 4th ed. and in the Standard form of building contract, c. 16 (e), have not apparently been challenged in the courts.

[4] See also cl. 60, note (6).

[5] *Gilbert Ash (Northern) Ltd.* v. *Modern Engineering (Bristol) Ltd.* [1974] A.C. 689.

[6] Cl. 47 (4) appears to be aimed at avoiding the effect of *Dawnays* v. *Minter* [1971] I W.L.R. 1205, since over-ruled by *Gilbert Ash, supra.*

issue is whether the employer, without the engineer's certificate under sub-clause (4), has a right of set-off for the purposes of resisting summary judgment for sums otherwise due (since an arbitrator may override the absence of a certificate). For this purpose the courts are likely to allow a set-off where the absence of a certificate is a dispute within the arbitration clause.[7]

COMPLETION CERTIFICATE

Clause 48
Certificate of completion of works

(1) When the Contractor shall consider that the whole of the Works has been substantially completed and has satisfactorily passed any final test that may be prescribed by the Contract he may give a notice to that effect to the Engineer or to the Engineer's Representative accompanied by an undertaking to finish any outstanding work during the Period of Maintenance. Such notice and undertaking shall be in writing and shall be deemed to be a request by the Contractor for the Engineer to issue a Certificate of Completion in respect of the Works and the Engineer shall within 21 days of the date of delivery of such notice either issue to the Contractor (with a copy to the Employer) a Certificate of Completion stating the date on which in his opinion the Works were substantially completed in accordance with the Contract or else give instructions in writing to the Contractor specifying all the work which in the Engineer's opinion requires to be done by the Contractor before the issue of such certificate. If the Engineer shall give such instructions the Contractor shall be entitled to receive such Certificate of Completion within 21 days of completion to the satisfaction of the Engineer of the work specified by the said instructions.

Completion of sections and occupied parts

(2) Similarly in accordance with the procedure set out in sub-clause (1) of this Clause the Contractor may request and the Engineer shall issue a Certificate of Completion in respect of:—
(a) any Section in respect of which a separate time for completion is provided in the Appendix to the Form of Tender; and
(b) any substantial part of the Works which has been both completed to the satisfaction of the Engineer and occupied or used by the Employer.

Completion of other parts of works

(3) If the Engineer shall be of the opinion that any part of the Works shall have been substantially completed and shall have satisfactorily passed any final test that may be prescribed by the Contract he may issue a Certificate of Completion in respect of that part of the Works before completion of the whole of the Works and upon the issue of such certificate the Contractor shall be deemed to have undertaken to complete any outstanding work in that part of the Works during the Period of Maintenance.

Reinstatement of ground

(4) Provided always that a Certificate of Completion given in respect of any Section or part of the Works before completion of the whole shall not be deemed to certify completion of any ground or surfaces requiring reinstatement unless such certificate shall expressly so state.

---

[7] *Ramac* v. *Lesser* (1975) 2 Lloyd's Rep. 430; *cf. Brightside Kilpatrick* v. *Mitchell Construction* (1975) 2 Lloyd's Rep. 493; and see *ante*, p. 279.

COMMENTARY

(1) *Clause* 48 (1): " *substantially completed* "

Clauses 43 (time for completion) and 47 (liquidated damages) refer only to the works being " completed." However, reference in clause 49 (1) to the period of maintenance being calculated " from the date of completion of the Works . . . certified in accordance with Clause 48," indicates that no distinction is intended by the word " substantially." The existence of latent defects does not render a completion certificate invalid, it is submitted.[8] The employer may recover general damages for delay and loss of use of the works, while remedial works are done, after the completion certificate.[9]

(2) *Clause* 48 (1): " *undertaking to finish any outstanding work* "

This provision fortunately avoids the difficult question whether completion or substantial completion can be achieved despite the existence of patent defects.[10] " Outstanding work " may include defective work, it is submitted, provided there is an undertaking to complete in accordance with the contract.[11] The power of the engineer to specify work which in his opinion requires to be done before completion, gives a wide discretion as to the extent to which outstanding work may be allowed.

(3) *Clause* 48 (2) (3): *Completion of Parts*

Under sub-clause (2) (b) the engineer is bound to certify completion in respect of any part of the works which is (i) substantial (ii) completed and (iii) occupied or used by the employer. Under sub-clause (3) the engineer has an undefined and unlimited discretion whether to certify completion of any part, without the requirement to satisfy (ii) or (iii) *supra*.

If a part is completed and is of potential use to the employer in isolation, it is thought the engineer's unreasonable failure to certify completion and thereby to reduce liquidated damages under clause 47 (1) (b) or (2) (b) would render the liquidated damages unrecoverable to the extent of such possible reduction, as a failure to mitigate.[12]

The risk of damage to the works or any section or part passes to the employer, under clause 20, 14 days after the engineer's certificate of completion.

MAINTENANCE AND DEFECTS

Clause 49
Definition of " period of maintenance "

(1) In these Conditions the expression " Period of Maintenance " shall mean the period of maintenance named in the Appendix to the Form of Tender

---

[8] *Westminster Corporation* v. *Jarvis* [1970] 1 W.L.R. 637, a decision on the Standard form of building contract.

[9] *Kaye (P. & M.) Ltd.* v. *Hosier & Dickinson* [1972] 1 W.L.R. 146, *per* Lord Diplock at p. 165, a decision on the Standard form of building contract.

[10] See cl. 15, note (2), Standard form of building contract, *supra*, p. 324.

[11] If an item of work cannot be, or is not intended to be, completed in accordance with the contract, there must be a variation or a waiver for which the engineer must seek authority.

[12] *British Westinghouse* v. *Underground Ry. Co.* [1912] A.C. 673 at p. 689; and see Chap. 9, *ante*.

calculated from the date of completion of the Works or any Section or part thereof certified by the Engineer in accordance with Clause 48 as the case may be.

Execution of work of repair, etc.

(2) To the intent that the Works and each Section and part thereof shall at or as soon as practicable after the expiration of the relevant Period of Maintenance be delivered up to the Employer in the condition required by the Contract (fair wear and tear excepted) to the satisfaction of the Engineer the Contractor shall finish the work (if any) outstanding at the date of completion as certified under Clause 48 as soon as may be practicable after such date and shall execute all such work of repair amendment reconstruction rectification and making good of defects imperfections shrinkages or other faults as may during the Period of Maintenance or within 14 days after its expiration be required of the Contractor in writing by the Engineer as a result of an inspection made by or on behalf of the Engineer prior to its expiration.

Cost of execution of work of repair, etc.

(3) All such work shall be carried out by the Contractor at his own expense if the necessity thereof shall in the opinion of the Engineer be due to the use of materials or workmanship not in accordance with the Contract or to neglect or failure on the part of the Contractor to comply with any obligation expressed or implied on the Contractor's part under the Contract. If in the opinion of the Engineer such necessity shall be due to any other cause the value of such work shall be ascertained and paid for as if it were additional work.

Remedy on contractor's failure to carry out work required

(4) If the Contractor shall fail to do any such work as aforesaid required by the Engineer the Employer shall be entitled to carry out such work by his own work-men or by other contractors and if such work is work which the Contractor should have carried out at the Contractor's own cost shall be entitled to recover from the Contractor the cost thereof or may deduct the same from any monies due or that become due to the Contractor.

Temporary reinstatement

(5) Provided always that if in the course or for the purposes of the execution of the Works or any part thereof any highway or other road or way shall have been broken into then notwithstanding anything herein contained:—

(a) If the permanent reinstatement of such highway or other road or way is to be carried out by the appropriate Highway Authority or by some person other than the Contractor (or any sub-contractor to him) the Contractor shall at his own cost and independently of any requirement of or notice from the Engineer be responsible for the making good of any subsidence or shrinkage or other defect imperfection or fault in the temporary rein-statement of such highway or other road or way and for the execution of any necessary repair or amendment thereof from whatever cause the necessity arises until the end of the Period of Maintenance in respect of the works beneath such highway or other road or way or until the Highway Authority or other person as aforesaid shall have taken possession of the Site for the purpose of carrying out permanent reinstatement (whichever is the earlier) and shall indemnify and save harmless the Employer against and from any damage or injury to the Employer or to third parties arising out or in consequence of any neglect or failure of the Contractor to comply with the foregoing obligations or any of them and against and from all claims demands proceedings damages costs charges and expenses whatso-ever in respect thereof or in relation thereto. As from the end of such

Period of Maintenance or the taking of possession as aforesaid (whichever shall first happen) the Employer shall indemnify and save harmless the Contractor against and from any damage or injury as aforesaid arising out or in consequence of or in connection with the said permanent reinstatement or any defect imperfection or failure of or in such work of permanent reinstatement and against and from all claims demands proceedings damages costs charges and expenses whatsoever in respect thereof or in relation thereto.

(b) Where the Highway Authority or other person as aforesaid shall take possession of the Site as aforesaid in sections or lengths the responsibility of the Contractor under paragraph (a) of this sub-clause shall cease in regard to any such section or length at the time possession thereof is so taken but shall during the continuance of the said Period of Maintenance continue in regard to any length of which possession has not been so taken and the indemnities given by the Contractor and the Employer respectively under the said paragraph shall be construed and have effect accordingly.

COMMENTARY

(1) *Generally*

The clause does not affect the contractor's continuing liability for breaches of contract.[13] The obligation to put right defects is subject to notice in writing from the engineer, and will not normally be enforceable by the employer by specific performance.[14] The clause should be regarded, it is suggested, as being inserted primarily for the benefit of the contractor, to limit his liability for breach of contract [15] and to permit him to achieve completion notwithstanding that items of work are outstanding.

The contractor's obligation is to carry out repairs, etc., from whatever cause they may arise, so that the works or any section may be delivered up " in the condition required by the Contract." Under sub-clause (3) such repairs, etc., are to be at the contractor's expense if necessitated by his breach of contract, but otherwise valued as additional work, under clause 52.

(2) *Clause 49 (2) : " inspection made by or on behalf of the Engineer "*

There is nothing to prevent the Engineer from making more than one inspection, or from giving notice of defects, both during and at the end of the maintenance period.

The conditions do not expressly require the contractor to carry out repairs, etc., before the end of the period of maintenance.[16] But the power to give notice of defects during the period, and the words " To the intent that the Works . . . shall at or as soon as practicable after the expiration of the . . . Period . . . be delivered up . . . ," suggest that the contractor is obliged to carry out repairs, after receipt of notice, as soon as is practicable. This is supported by sub-clause (5) under which the contractor is responsible for repairs to the temporary reinstatement of a highway (without notice) " until the end of the period of maintenance " or earlier possession by the Highway Authority.

---

[13] See *Hancock* v. *Brazier* [1966] 2 All E.R. 901, 904 (C.A.).
[14] See *ante*, p. 164.
[15] See *Kaye* (*P. & M.*) *Ltd.* v. *Hosier & Dickinson* [1972] 1 W.L.R. 146, 166 (H.L.), a decision on the Standard form of building contract.
[16] *Cf.* Standard form of building contract, cl. 15 (3).

(3) *Clause* 49 (4): " *if the Contractor shall fail to do any such work as aforesaid* "

The contractor is bound to complete maintenance work to the approval of the engineer (clause 13 (2)) and so that the works may be delivered up " as soon as practicable after the expiration of the relevant period of maintenance " (cl. 49 (2)). If the contractor fails so to do, the employer is entitled to have the work done by others without express authority, *a fortiori* since the employer is in possession of the works. The contractor's failure does not, however, deprive him of entitlement to the balance of retention, save to the extent of the cost of the work which has not been done (cl. 60 (5) (c)).

(4) *Clause* 49 (5)

The provisions apply to temporary reinstatement of a highway, etc., where the contractor is not to carry out the permanent reinstatement. They should be read with clause 27 and the Public Utilities Street Works Act 1950 (see also note (2) *supra*). The provisions as to indemnities should be read with clause 22.

Clause 50
Contractor to search

The Contractor shall if required by the Engineer in writing carry out such searches tests or trials as may be necessary to determine the cause of any defect imperfection or fault under the directions of the Engineer. Unless such defect imperfection or fault shall be one for which the Contractor is liable under the Contract the cost of the work carried out by the Contractor as aforesaid shall be borne by the Employer. But if such defect imperfection or fault shall be one for which the Contractor is liable the cost of the work carried out as aforesaid shall be borne by the Contractor and he shall in such case repair rectify and make good such defect imperfection or fault at his own expense in accordance with Clause 49.

COMMENTARY

The exercise of the powers under this clause cannot be confined to the period of maintenance since the clause is not so limited, and the heading to clauses 49 and 50 (maintenance and defects) is not to be taken into consideration in the interpretation or construction of the Conditions (cl. 1 (3)).

ALTERATIONS ADDITIONS AND OMISSIONS
Clause 51
Ordered variations

(1) The Engineer shall order any variation to any part of the Works that may in his opinion be necessary for the completion of the Works and shall have power to order any variation that for any other reason shall in his opinion be desirable for the satisfactory completion and functioning of the Works. Such variations may include additions omissions substitutions alterations changes in quality form character kind position dimension level or line and changes in the specified sequence method or timing of construction (if any).

Ordered variations to be in writing

(2) No such variation shall be made by the Contractor without an order by the Engineer. All such orders shall be given in writing provided that if for any

reason the Engineer shall find it necessary to give any such order orally in the first instance the Contractor shall comply with such oral order. Such oral order shall be confirmed in writing by the Engineer as soon as is possible in the circumstances. If the Contractor shall confirm in writing to the Engineer any oral order by the Engineer and such confirmation shall not be contradicted in writing by the Engineer forthwith it shall be deemed to be an order in writing by the Engineer. No variation ordered or deemed to be ordered in writing in accordance with sub-clauses (1) and (2) of this Clause shall in any way vitiate or invalidate the Contract but the value (if any) of all such variations shall be taken into account in ascertaining the amount of the Contract Price.

Changes in quantities

(3) No order in writing shall be required for increase or decrease in the quantity of any work where such increase or decrease is not the result of an order given under this Clause but is the result of the quantities exceeding or being less than those stated in the Bill of Quantities.

COMMENTARY

(1) *Generally*

A clause permitting variation of the work (as opposed to the contract) is an essential feature of any construction contract. Without it the contractor is not bound to execute additional work or to make omissions or changes. Further, if extra work is carried out the contractor must seek reimbursement outside the contract. The above represents the position when variations are made without following the provisions of the variations clause, save that some matters may be brought within the clause by the award of an arbitrator under clause 66.

(2) *Clause 51 (1): " The Engineer shall order "*

These words make it necessary to consider whether the engineer can be obliged, by the contract, to order a variation. It is submitted that it is entirely in the discretion of the employer to give the engineer authority to order variations.[17] The word " shall " may assume significance when read with clause 13 (1) (which excuses performance where " legally or physically impossible "). Otherwise, the contractor remains bound to complete the works by whatever means are necessary. The effect of sub-clause (1) is to give the engineer the power and not the duty to vary the works. A general duty to vary would be void for uncertainty,[18] it is submitted.

(3) *Clause 51 (1): " necessary for completion . . . desirable for the satisfactory completion and functioning of the Works "*

These words may limit the engineer's powers to make variations. If the contractor carries out a variation order apparently outside the above limitations, without protest, he may waive his right subsequently to object.

The words appear to prevent the omission of work merely to achieve a saving. If an unauthorised omission is ordered the contractor will be entitled to damages for breach of contract, including loss of profit. In respect of other variations made outside the engineer's powers the contractor will be entitled to payment of

---

[17] But *vis-à-vis* the contractor, the employer will be bound by an act of the engineer within his express powers under the contract.
[18] See *ante*, p. 19.

an agreed or a reasonable sum. It may be that clause 13 can be relied on as giving power, to order an omission merely to achieve a saving (see clause 13 note (4)). In this event the contractor may claim additional payment under clause 13 (3).

(4) *Clause* 51 (1): "*changes in the specified sequence method or timing of construction*"

This is a new concept in variation which has no parallel in earlier editions of the form.[19] It involves three types of variation:

(i) *Specified sequence*. This presumably refers to the programme required under clause 14 (1) which must show the "order of procedure." The engineer has an implicit power of disapproval under clause 14 (1), apparently without financial consequences. The engineer should therefore ensure that the exercise of such power is not construed as a variation.

(ii) *Specified method*. The contractor is required under clause 14 (1) to provide a general description of his "arrangements and methods of construction," and further information as to the proposed "methods" under clause 14 (3). Again, the engineer is given an implicit power of disapproval under clause 13 (2) as to the "mode . . . of construction" (which is presumably the same as "method") and clause 14 (4) to (6) makes provision for a more limited right of claim arising from the engineer's requirements as to the contractor's methods. The engineer should, therefore, ensure that the power to order a variation of the contractor's "method" is not used unnecessarily.

(iii) *Specified . . . timing*. This may refer to the programme under clause 14 (1) (as to which, see the power to require a revised programme under clause 14 (2)) or the "speed of construction" under clause 13 (2). In either case the same comments apply as above. The words further appear wide enough to include the times for sectional and overall completion specified in the appendix. But it is thought this is unlikely to be the intention of the parties, since clause 52 is not appropriate to value such variations [20]; and the elaborate provisions as to time in clauses 43 to 47 make no allowance for a variation of time. The use of the words "timing of construction" rather than "of the Works" supports this view.

(5) *Clause* 51 (2): *Variation orders to be in writing*

The engineer is bound to confirm his oral order in writing. If he fails or refuses to do so, or contradicts the contractor's written confirmation, a dispute exists which may be referred, via the engineer, to arbitration under clause 66.

If a variation is carried out without an order of the engineer, the contractor's remedy lies outside the contract. If the work was ordered by the employer the contractor may bring a simple action for work done. If the engineer orders work which he contends is included in the contract, the contractor will be entitled to extra payment if he proves the work to be a variation.[21]

[19] Earlier editions of the I.C.E. Conditions included clauses equivalent to the present cl. 13 (1).
[20] There is no equivalent to the loss and/or expense claim found in cl. 11 (6) of the Standard form of building contract; but see cl. 52 (2) of the I.C.E. Conditions.
[21] *Molloy v. Liebe* (1910) 102 L.T. 616.

Clause 52
Valuation of ordered variations

(1) The value of all variations ordered by the Engineer in accordance with Clause 51 shall be ascertained by the Engineer after consultation with the Contractor in accordance with the following principles. Where work is of similar character and executed under similar conditions to work priced in the Bill of Quantities it shall be valued at such rates and prices contained therein as may be applicable. Where work is not of a similar character or is not executed under similar conditions the rates and prices in the Bill of Quantities shall be used as the basis for valuation so far as may be reasonable failing which a fair valuation shall be made. Failing agreement between the Engineer and the Contractor as to any rate or price to be applied in the valuation of any variation the Engineer shall determine the rate or price in accordance with the foregoing principles and he shall notify the Contractor accordingly.

Engineer to fix rates

(2) Provided that if the nature or amount of any variation relative to the nature or amount of the whole of the contract work or to any part thereof shall be such that in the opinion of the Engineer or the Contractor any rate or price contained in the Contract for any item of work is by reason of such variation rendered unreasonable or inapplicable either the Engineer shall give to the Contractor or the Contractor shall give to the Engineer notice before the varied work is commenced or as soon thereafter as is reasonable in all the circumstances that such rate or price should be varied and the Engineer shall fix such rate or price as in the circumstances he shall think reasonable and proper.

Daywork

(3) The Engineer may if in his opinion it is necessary or desirable order in writing that any additional or substituted work shall be executed on a daywork basis. The Contractor shall then be paid for such work under the conditions set out in the Daywork Schedule included in the Bill of Quantities and at the rates and prices affixed thereto by him in his Tender and failing the provision of a Daywork Schedule he shall be paid at the rates and prices and under the conditions contained in the " Schedules of Dayworks carried out incidental to Contract Work " issued by The Federation of Civil Engineering Contractors current at the date of the execution of the Daywork.

The Contractor shall furnish to the Engineer such receipts or other vouchers as may be necessary to prove the amounts paid and before ordering materials shall submit to the Engineer quotations for the same for his approval.

In respect of all work executed on a daywork basis the Contractor shall during the continuance of such work deliver each day to the Engineer's Representative an exact list in duplicate of the names occupation and time of all workmen employed on such work and a statement also in duplicate showing the description and quantity of all materials and plant used thereon or therefor (other than plant which is included in the percentage addition in accordance with the Schedule under which payment for daywork is made). One copy of each list and statement will if correct or when agreed be signed by the Engineer's Representative and returned to the Contractor. At the end of each month the Contractor shall deliver to the Engineer's Representative a priced statement of the labour material and plant (except as aforesaid) used and the Contractor shall not be entitled to any payment unless such lists and statements have been fully and punctually rendered. Provided always that if the Engineer shall consider that for any reason the sending of such list or statement by the Contractor in accordance with the foregoing provision was impracticable he shall nevertheless be entitled to authorise payment for such work either as daywork (on being satisfied as to the time employed and plant and materials used on such work) or at such value therefor as he shall consider fair and reasonable.

Notice of claims

(4) (a) If the Contractor intends to claim a higher rate or price than one notified to him by the Engineer pursuant to sub-clauses (1) and (2) of this Clause or Clause 56 (2) the Contractor shall within 28 days after such notification give notice in writing of his intention to the Engineer.

(b) If the Contractor intends to claim any additional payment pursuant to any Clause of these Conditions other than sub-clauses (1) and (2) of this Clause he shall give notice in writing of his intention to the Engineer as soon as reasonably possible after the happening of the events giving rise to the claim. Upon the happening of such events the Contractor shall keep such contemporary records as may reasonably be necessary to support any claim he may subsequently wish to make.

(c) Without necessarily admitting the Employer's liability the Engineer may upon receipt of a notice under this Clause instruct the Contractor to keep such contemporary records or further contemporary records as the case may be as are reasonable and may be material to the claim of which notice has been given and the Contractor shall keep such records. The Contractor shall permit the Engineer to inspect all records kept pursuant to this Clause and shall supply him with copies thereof as and when the Engineer shall so instruct.

(d) After the giving of a notice to the Engineer under this Clause the Contractor shall as soon as is reasonable in all the circumstances send to the Engineer a first interim account giving full and detailed particulars of the amount claimed to that date and of the grounds upon which the claim is based. Thereafter at such intervals as the Engineer may reasonably require the Contractor shall send to the Engineer further up to date accounts giving the accumulated total of the claim and any further grounds upon which it is based.

(e) If the Contractor fails to comply with any of the provisions of this Clause in respect of any claim which he shall seek to make then the Contractor shall be entitled to payment in respect thereof only to the extent that the Engineer has not been prevented from or substantially prejudiced by such failure in investigating the said claim.

(f) The Contractor shall be entitled to have included in any interim payment certified by the Engineer pursuant to Clause 60 such amount in respect of any claim as the Engineer may consider due to the Contractor provided that the Contractor shall have supplied sufficient particulars to enable the Engineer to determine the amount due. If such particulars are insufficient to substantiate the whole of the claim the Contractor shall be entitled to payment in respect of such part of the claim as the particulars may substantiate to the satisfaction of the Engineer.

COMMENTARY

(1) *Clause* 52 (1): " *similar character and executed under similar conditions* "

The contractor is entitled to a higher or different valuation when either of these requirements is not met. The words " similar conditions " do not, it is submitted, permit reference to the quantity of the work, since this is expressly covered by wider powers in sub-clause (2). It is thought that " conditions " must be limited to physical conditions and cannot include, *e.g.*, increases in material costs.

Sub-clause (1) contains three alternative valuations. These are: valuation at the rates and prices in the bill; valuation using such rates and prices as the basis; and a fair valuation. The third alternative applies to the extent that the second is not reasonable. Clause 52 (4) (a) requires notice where the contractor is dissatisfied with a rate or price notified by the engineer; but see note (7) *infra*.

(2) *Clause 52 (2): Engineer's power to fix rates*

The power is to fix a reasonable and proper rate to apply to the whole of an item of work, *i.e.* both the variation and the original work are to be re-rated. The power applies to " any rate or price for any item of work." It need not be the work which is varied, provided the rate or price to be fixed has been rendered unreasonable or inapplicable " by reason of " the variation. Further, this may arise from the " nature or amount " of the variation; *e.g.* if the engineer orders excavation to be carried out in heading instead of open cut, he may vary the rates for any work to be executed in the excavation.

This sub-clause forms an alternative to the " additional cost " type of claim found elsewhere and in other standard forms.[22] The contractor is not required to prove, nor the engineer to ascertain, loss arising from a variation. When the engineer is satisfied that a rate or price is rendered unreasonable or inapplicable by reason of a variation, his duty is to fix such rate or price as is " reasonable and proper," having regard to the contract work as varied, provided notice has been given.

" Additional cost " claims are in general solely for the contractor's benefit. But there is nothing so to limit the operation of sub-clause (2). The power of the engineer to reduce rates is confirmed by the provision for notice to be given by either the engineer or the contractor. The words " unreasonable or inapplicable " appear also in clause 56 (2), where there is an express power to decrease rates or prices.

It is notable that the Conditions make no further provision for the recovery of prolongation costs arising from delay caused by a variation, *e.g.* preliminaries. It is submitted that the engineer's powers under this sub-clause are wide enough to permit either an adjustment to preliminary rates or prices, or an appropriate allowance in the rates fixed. See also clause 52 (4) (a) for notice required where the contractor intends to claim a higher rate than that fixed by the engineer.

(3) *Clause 52 (2): Requirements as to notice*

Notice under sub-clause (2) must specify the rate or price to be varied. There is no express requirement to specify the rate claimed; but clause 52 (4) (a) suggests that if the engineer gives notice, he should specify the rate proposed: see note (7) *infra*.

As to the timing of notice, some assistance may be gained from the case of *Tersons* v. *Stevenage Development Corp.*,[23] where a wide interpretation was given

---

[22] See cll. 7 (3), 13 (3), etc., and *cf.* cll. 11 (6) and 24 (1) of the Standard form of building contract.

[23] (1963) 2 Lloyd's Rep. 333, a case on the 2nd edition of the Conditions, but applicable to the 4th edition. The case also dealt with general provisions as to notices in cl. 52 (4); but there is no application to the present substantially re-drafted sub-cl. (4).

to the words " as soon thereafter as is practicable," appearing in the equivalent clause of an earlier edition of the Conditions.

(4) *Clause 52 (3): Dayworks*

The engineer's power to order work to be executed on a daywork basis is subject only to the requirements that the work must be " additional or substituted work," and that the order must be " necessary or desirable." The power is very wide in practice.[24]

(5) *Clause 52 (3): " the Contractor shall not be entitled to any payment unless "*

This deprives the contractor of the right to payments where any daily list or monthly statement is incomplete or late. It must be read subject to the following proviso, which entitles the engineer to authorise payment where the sending of the list or statement, as required by the contract, was " impracticable." When the proviso applies, the engineer has a discretion to authorise payment at daywork rates or at a " fair and reasonable " valuation.

(6) *Clause 52 (4): Notice of claims*

The key to this long and difficult sub-clause is in paragraphs (f) and (e), which govern the right to interim and final payment, respectively, where the preceding requirements have not been fully complied with. The requirements of paragraphs (a) and (b) are clearly subject to (f) and (e).

(f) *interim payments.* This requires the supply of particulars, which will presumably comply concurrently with sub-clause (d), and may further constitute written notice under (b). The only practical issue will therefore be what amount the engineer " may consider due," having regard to the particulars supplied. Subject to this, the contractor is entitled to interim payments, it is submitted.

(e) *final payments.* If there has been a failure to comply with paragraphs (a) to (d), the contractor is entitled to payment " only to the extent that the engineer has not been prevented from or substantially prejudiced by such failure in investigating the claim." If the engineer excludes any part of a claim on the ground of prevention or prejudice, it is necessary to consider the rights of the contractor to submit further proof of the claim, and the effect of a reference under clause 66 to the engineer and to an arbitrator.

First, it should be noted that paragraph (e) sets no time limit on the engineer's investigation. If the contractor is able to furnish further proof, he may re-submit the claim under the relevant clause of the contract, at any time before the final certificate. Secondly, where a claim has been rejected on the ground of prevention or prejudice, the dispute to be decided under clause 66 is whether the engineer, or certifier, *has been* prevented or prejudiced. Both the arbitrator and the engineer under clause 66 are limited to such dispute, and it is irrelevant whether they are similarly prevented or prejudiced. However, they are bound, it is submitted, to consider all information made available to the engineer before

---

[24] The engineer's authority *vis-à-vis* the employer is likely to be subject to an implied requirement of reasonable economy. *Cf.* the stringent requirements of cl. 11 (4) (c) of the Standard form of building contract, that the work " cannot properly be measured and valued."

the final certificate, whether or not there has been a formal re-submission of the claim under the contract.

(7) *Clause 52 (4) (a): Claims under clause 52 (1), (2) or 56 (2)*
The effect of this provision is obscure. It suggests that the engineer may review his decision under the above clauses. But there is no express power to do so. It is thought no such power is to be implied from this provision. Further doubt is cast by the words of paragraph (b): " If the contractor intends to claim any additional payment pursuant to any clause . . . other than sub-clauses (1) and (2) of this clause."

It is suggested that the provision is intended to recognise that the engineer may wish to operate his express powers under clauses 52 (1) and (2) and 56 (2) more than once. There is nothing expressly to prevent him making an interim decision (*cf.* the power to give interim extensions of time in clause 44, a new power which recognises existing practice) followed by a further decision when more facts are known.

The requirement applies to a claim submitted as a dispute under clause 66. See note (6) as to the effect of failure to give notice.

(8) *Clause 52 (4) (b): Other claims to additional payment*
The requirements apply to any provision entitling the contractor to additional payment, whether or not expressed to be subject to clause 52 (4). The relevant provisions are set out below. Numbers in brackets indicate the clause reference.

A   Claims expressly subject to Clause 52 (4). Late issue of drawings or instructions (7 (3)); adverse physical conditions or artificial obstructions (12 (3)); instructions under clause 5 or 13 (1) (13 (3)); engineer's requirements or delayed consent to the contractor's methods (14 (6)); street works variations (27 (6)) [25]; facilities for other contractors (31 (2)); suspension of work (40 (1)); non-possession of the site (42 (1)); nominated sub-contract conditions (59A (3)); forfeiture of a nominated sub-contract (59B (4)).

B   Other claims or rights to additional payment. Setting out errors (17); reinstatement works (20 (2)); damage to persons or property (22 (2)); fees, rates, etc. (26 (1)); damage to highways (30 (3)); fossils (32); samples and tests (36 (2) and (3)); uncovering work (38 (2)); maintenance and reinstatement of highways (49 (3) and (5)); searches (50); frustration (64); war (65 (5)); tax fluctuations (69); VAT (70); metrication (71).

PROPERTY IN MATERIALS AND PLANT

Clause 53
Plant, etc.—Definitions
  (1) For the purpose of this Clause:—
    (a) the expression " Plant " shall mean any Constructional Plant Temporary Works and materials for Temporary Works but shall exclude

---

[25] Cl. 27 (6) as printed in the conditions omits reference to sub-cl. (4).

any vehicles engaged in transporting any labour plant or materials to
or from the Site;

(b) the expression " agreement for hire " shall be deemed not to include
an agreement for hire purchase.

Vesting of plant

(2) All Plant goods and materials owned by the Contractor or by any company
in which the Contractor has a controlling interest shall when on the Site be
deemed to be the property of the Employer.

Conditions of hire of plant

(3) With a view to securing in the event of a forfeiture under Clause 63 the
continued availability for the purpose of executing the Works of any hired Plant
the Contractor shall not bring on to the Site any hired Plant unless there is an
agreement for the hire thereof which contains a provision that the owner thereof
will on request in writing made by the Employer within 7 days after the date on
which any forfeiture has become effective and on the Employer undertaking to pay
all hire charges in respect thereof from such date hire such Plant to the Employer
on the same terms in all respects as the same was hired to the Contractor save
that the Employer shall be entitled to permit the use thereof by any other contractor employed by him for the purpose of completing the Works under the
terms of the said Clause 63.

Costs for purposes of clause 63

(4) In the event of the Employer entering into any agreement for the hire of
Plant pursuant to sub-clause (3) of this Clause all sums properly paid by the
Employer under the provisions of any such agreement and all expenses incurred
by him (including stamp duties) in entering into such agreement shall be deemed
for the purpose of Clause 63 to be part of the cost of completing the Works.

Notification of plant ownership

(5) The Contractor shall upon request made by the Engineer at any time in relation to any item of Plant forthwith notify to the Engineer in writing the name and
address of the owner thereof and shall in the case of hired Plant certify that the
agreement for the hire thereof contains a provision in accordance with the
requirements of sub-clause (3) of this Clause.

Irremovability of plant, etc.

(6) No Plant (except hired Plant) goods or materials or any part thereof shall be
removed from the Site without the written consent of the Engineer which consent
shall not be unreasonably withheld where the same are no longer immediately
required for the purposes of the completion of the Works but the Employer will
permit the Contractor the exclusive use of all such Plant goods and materials in
and for the completion of the Works until the occurrence of any event which gives
the Employer the right to exclude the Contractor from the Site and proceed with
the completion of the Works.

Revesting and removal of plant

(7) Upon the removal of any such Plant goods or materials as have been
deemed to have become the property of the Employer under sub-clause (2) of
this Clause with the consent as aforesaid the property therein shall be deemed to
revest in the Contractor and upon completion of the Works the property in the
remainder of such Plant goods and materials as aforesaid shall subject to Clause
63 be deemed to revest in the Contractor.

Disposal of plant

(8) If the Contractor shall fail to remove any Plant goods or materials as required pursuant to Clause 33 within such reasonable time after completion of the Works as may be allowed by the Engineer then the Employer may:—

(a)   sell any which are the property of the Contractor; and

(b)   return any not the property of the Contractor to the owner thereof at the Contractor's expense;

and after deducting from any proceeds of sale the costs charges and expenses of and in connection with such sale and of and in connection with return as aforesaid shall pay the balance (if any) to the Contractor but to the extent that the proceeds of any sale are insufficient to meet all such costs charges and expenses the excess shall be a debt due from the Contractor to the Employer and shall be deductible or recoverable by the Employer from any monies due or that may become due to the Contractor under the contract or may be recovered by the Employer from the Contractor at law.

Liability for loss or injury to plant

(9) The Employer shall not at any time be liable for the loss of or injury to any of the Plant goods or materials which have been deemed to become the property of the Employer under sub-clause (2) of this Clause save as mentioned in Clauses 20 and 65.

Incorporation of clause in sub-contracts

(10) The Contractor shall where entering into any sub-contract for the execution of any part of the Works incorporate in such sub-contract (by reference or otherwise) the provisions of this Clause in relation to Plant goods or materials brought on to the Site by the sub-contractor.

No approval by vesting

(11) The operation of this Clause shall not be deemed to imply any approval by the Engineer of the materials or other matters referred to herein nor shall it prevent the rejection of any such materials at any time by the Engineer.

COMMENTARY

(1) *Generally*

For definition of constructional plant and temporary works see clause 1 (1). The purpose of this clause is to ensure that plant and materials are available to the employer in the event of the contractor failing to perform the contract: a clause providing for vesting upon the contractor's insolvency is likely to be un-enforceable.[26] The present clause is subject to a number of difficulties as to its enforceability against the contractor; against third parties; and against a trustee in bankruptcy.

(2) *Enforceability against the Contractor*

The effectiveness of words such as " deemed to be the property of . . ." has been doubted.[27] But sub-clause (2), read with sub-clause (6) makes plain the intention that property in plant and materials is to vest in the employer when

---

[26]   *Re Harrison, ex p. Jay* (1880) 14 Ch.D. 19; *Re Walker, ex p. Barter* (1884) 26 Ch.D. 510; and see *ante*, p. 194.

[27]   *Bennett, etc. Ltd.* v. *Sugar City* [1951] A.C. 786, 814; *Re Keen, ex p. Collins* [1902] 1 K.B. 555; *cf. Re Winter, ex p. Bolland* (1878) 8 Ch.D. 225; and see *ante*, p. 133.

brought to site. It is thought the clause would be enforced against the contractor.[28]

(3) *Enforceability against Third Parties*

Sub-clause (2) cannot affect property owned by a company not party to the contract. If the contractor seeks to avoid the clause by supplying plant or material through wholly owned subsidiary companies, he will be in breach of clause 4 and, in respect of plant, the engineer may require notification of ownership under sub-clause (5).

Sub-clause (3) cannot be enforced directly against a sub-contractor, unless a direct warranty to such effect is obtained from the sub-contractor. Further difficulties may arise where the terms of hiring provide for automatic termination upon forfeiture of the main contract. The purpose of sub-clause (4) is obscure, since clause 63 in no way inhibits the hire of plant for completion of the works. It is thought that sub-clause (4) does not limit the employer's right to recover the cost of hiring plant from other suppliers.

(4) *Enforceability against the contractor's trustee in bankruptcy*

In the event of the contractor (not being a limited company) becoming bankrupt, sub-clause (2) must be read in the light of the doctrine of reputed ownership.[29] This is unlikely to affect goods or materials on site [30] but may operate to re-vest the contractor's plant in his trustee in bankruptcy.[31]

Clause 54
Vesting of goods and materials not on site

(1) The Contractor may with a view to securing payment under Clause 60 (1) (c) in respect of goods and materials listed in the Appendix to the Form of Tender before the same are delivered to the Site transfer the property in the same to the Employer before delivery to the Site provided:—
  (a) that such goods and materials have been manufactured or prepared and are substantially ready for incorporation in the Works; and
  (b) that the said goods and materials are the property of the Contractor or the contract for the supply of the same expressly provides that the property therein shall pass unconditionally to the Contractor upon the Contractor taking the action referred to in sub-clause (2) of this Clause.

Action by contractor

(2) The intention of the Contractor to transfer the property in any goods or materials to the Employer in accordance with this Clause shall be evidenced by the Contractor taking or causing the supplier of the said goods or materials to take the following action:—
  (a) provide to the Engineer documentary evidence that the property in the said goods or materials has vested in the Contractor;
  (b) suitably mark or otherwise plainly identify the said goods and materials so as to show that their destination is the Site that they are the property

---

[28] For a further discussion, see H.B.C. (10th ed.), pp. 669–673.
[29] Bankruptcy Act 1914, s. 38; and see *ante*, p. 191.
[30] *Re Fox, ex p. Oundle and Thrapston R.D.C.* [1948] Ch. 407.
[31] *Re Weibking, ex p. Ward* [1902] 1 K.B. 713.

of the Employer and (where they are not stored at the premises of the Contractor) to whose order they are held;

(c) set aside and store the said goods and materials so marked or identified to the satisfaction of the Engineer; and

(d) send to the Engineer a schedule listing and giving the value of every item of the goods and materials so set aside and stored and inviting him to inspect the same.

Vesting in employer

(3) Upon the Engineer approving in writing the said goods and materials for the purposes of this Clause the same shall vest in and become the absolute property of the Employer and thereafter shall be in the possession of the Contractor for the sole purpose of delivering them to the Employer and incorporating them in the Works and shall not be within the ownership control or disposition of the Contractor.

Provided always that:—

(a) approval by the Engineer for the purposes of this Clause or any payment certified by him in respect of goods and materials pursuant to Clause 60 shall be without prejudice to the exercise of any power of the Engineer contained in this Contract to reject any goods or materials which are not in accordance with the provisions of the Contract and upon any such rejection the property in the rejected goods or materials shall immediately revest in the Contractor;

(b) the Contractor shall be responsible for any loss or damage to such goods and materials and for the cost of storing handling and transporting the same and shall effect such additional insurance as may be necessary to cover the risk of such loss or damage from any cause.

Lien on goods or materials

(4) Neither the Contractor nor a sub-contractor nor any other person shall have a lien on any goods or materials which have vested in the Employer under sub-clause (3) of this Clause for any sum due to the Contractor sub-contractor or other person and the Contractor shall take all such steps as may reasonably be necessary to ensure that the title of the Employer and the exclusion of any such lien are brought to the notice of sub-contractors and other persons dealing with any such goods or materials.

Delivery to the employer of vested goods or materials

(5) Upon cessation of the employment of the Contractor under this contract before the completion of the Works whether as a result of the operation of Clause 63 or otherwise the Contractor shall deliver to the Employer any goods or materials the property in which has vested in the Employer by virtue of sub-clause (3) of this Clause and if he shall fail to do so the Employer may enter any premises of the Contractor or of any sub-contractor and remove such goods and materials and recover the cost of so doing from the Contractor.

Incorporation in sub-contracts

(6) The Contractor shall incorporate provisions equivalent to those provided in this Clause in every sub-contract in which provision is to be made for payment in respect of goods or materials before the same has been delivered to the Site.

COMMENTARY

(1) *Generally*

This clause should be read with clause 60 (1) (c) and (2) (b), which govern the contractor's entitlement to payment. There is no limitation on when the contractor may seek payment. But the clause applies only to goods, etc., expressly listed in the appendix. The amount to be paid is subject to the engineer's discretion [32] and to a maximum percentage stated in the appendix.[33]

Provided the engineer is satisfied that the goods comply with every requirement of this clause the engineer is bound to give his approval in writing under sub-clause (3). Such approval is, however, subject to the power subsequently to reject the goods and to withhold payment by a later certificate, under clause 60 (7).

(2) *Effect of vesting*

Compliance with the clause gives the employer title to the goods.[34] In the event of the contractor (not being a limited company) becoming bankrupt, the goods may vest in the contractor's trustee in bankruptcy [35] under the doctrine of reputed ownership.[36] The same may apply upon the bankruptcy of a sub-contractor in whose premises the goods are stored. But in this event the engineer may withdraw the payment by a subsequent interim certificate, relying on clause 60 (7). The contractor is not protected by the proviso to that sub-clause applying to nominated sub-contractors, since the goods do not constitute "work done, goods or materials supplied or services rendered," it is submitted: see proviso (a) to clause 60 (7). In cases where a nominated sub-contract may be terminated or forfeited, these provisions should be read with clause 59B.

MEASUREMENT

Clause 55
Quantities

(1) The quantities set out in the Bill of Quantities are the estimated quantities of the work but they are not to be taken as the actual and correct quantities of the Works to be executed by the Contractor in fulfilment of his obligations under the Contract.

Correction of errors

(2) Any error in description in the Bill of Quantities or omission therefrom shall not vitiate the Contract nor release the Contractor from the execution of the whole or any part of the Works according to the Drawings and Specification or from any of his obligations or liabilities under the Contract. Any such error or omission shall be corrected by the Engineer and the value of the work actually carried out shall be ascertained in accordance with Clause 52. Provided that there shall be no rectification of any errors omissions or wrong estimates in the descriptions rates and prices inserted by the Contractor in the Bill of Quantities.

Clause 56
Measurement and valuation

(1) The Engineer shall except as otherwise stated ascertain and determine by

---

[32] See cl. 60, note (5).

[33] The goods are not subject to retention: cl. 60 (2) (b).

[34] This is the position under English law. But note that the effect of the clause may not be the same under Scots law.

[35] *Re Fox, ex p. Oundle and Thrapston R.D.C.* [1948] Ch. 407.

[36] Bankruptcy Act 1914, s. 38.

admeasurement the value in accordance with the Contract of the work done in accordance with the Contract.

Increase or decrease of rate

(2) Should the actual quantities executed in respect of any item be greater or less than those stated in the Bill of Quantities and if in the opinion of the Engineer such increase or decrease of itself shall so warrant the Engineer shall after consultation with the Contractor determine an appropriate increase or decrease of any rates or prices rendered unreasonable or inapplicable in consequence thereof and shall notify the Contractor accordingly.

Attending for measurement

(3) The Engineer shall when he requires any part or parts of the work to be measured give reasonable notice to the Contractor who shall attend or send a qualified agent to assist the Engineer or the Engineer's Representative in making such measurement and shall furnish all particulars required by either of them. Should the Contractor not attend or neglect or omit to send such agent then the measurement made by the Engineer or approved by him shall be taken to be the correct measurement of the work.

Clause 57
Method of measurement

Except where any statement or general or detailed description of the work in the Bill of Quantities expressly shows to the contrary Bills of Quantities shall be deemed to have been prepared and measurements shall be made according to the procedure set forth in the " Standard Method of Measurement of Civil Engineering Quantities " issued by the Institution of Civil Engineers and reprinted in 1973 or such later or amended edition thereof as may be stated in the Appendix to the Form of Tender to have been adopted in its preparation notwithstanding any general or local custom.

COMMENTARY

(1) *Generally*

These clauses should be read together. Clause 55 (1), with clause 56 (1) and (2), make it clear that the contractor is to be paid for the actual quantities of work executed, at the billed rates or otherwise subject to clause 56 (2). Any inference to the contrary is removed by the avoidance of reference to a " Contract Sum " in the tender and the form of agreement, and to the terms " Tender Total " and " Contract Price " used therein: see clause 1 (1) (h) and (i) for definitions. The contract as drafted is a " re-measurement " and not a " lump sum " contract.

(2) *Clause 55 (2): " error in description in the Bill of Quantities or omission therefrom "*

By this provision the drawings and specification are to prevail over the bills, contrary to the general rule (cl. 5). Errors in quantity are automatically accounted for by re-measurement.

(3) *Clause 56 (2): " increase or decrease of any rates or prices "*

Although worded differently, the effect of this sub-clause is the same as clause 52 (2), save that the only consideration is the amount of variation. The power to fix rates applies not only to the items undergoing alteration in quantities but to " any rates or prices rendered unreasonable or inapplicable in consequence thereof." See clause 52 (4) (a) as to notice required of further claims; and see note (7) to clause 52.

(4) *Clause* 56 (3): *Measurements*

The provisions as to giving notice should be read with clause 38 (1), which requires the contractor to give notice before covering up any work. The engineer is then bound to attend to measure the work or advise the contractor that this is unnecessary. It is not clear whether the engineer is absolved from the obligation to give notice where the contractor has given notice under clause 38 (1). But it is thought the measurement would not be treated as binding (in the absence of the contractor or his agent) without express prior notice under this sub-clause.

The arbitrator is given no express power to re-open measurements. But there is express power to review a valuation. By virtue of the words " determine by admeasurement the value " in clause 56 (1), it is thought that a dispute as to valuation may include measurements, provided they are not to be taken as correct under sub-clause (3).

When under clause 56 (3) the engineer's measurements are to be taken as correct, the court or an arbitrator will be bound by the measurements, it is submitted.[37] However, what is to be taken as correct is limited to the physical measurements, it is submitted. Calculations made from the measurements are not binding, *e.g.* calculation of volumes of excavation and fill from linear measurements. Further, the physical measurements will not be binding to the extent there are errors on their face.

(5) *Clause* 57: *The Standard Method of Measurement*

The words " deemed to have been prepared " are, it is submitted, not sufficient to allow the Standard Method to override the clear intention of the relevant contract documents, *i.e.* the specification, drawings and bill. The Standard Method may be referred to for the purpose of resolving an ambiguity as to whether an item of work is included in the contract[38]; *e.g.* the provision of *in situ* concrete necessarily requires shuttering so that the omission to measure shuttering would not result in an extra.[39] However, the billing of precast concrete without measurement of reinforcement creates an ambiguity as to whether or not the concrete is to be reinforced. In this case, it is thought, reference to the Standard Method[40] would entitle the contractor to extra payment. Many more subtle arguments may be based on the Standard Method where the bills are not prepared precisely in accordance with its provisions.

A completely revised edition of the Standard Method was published in 1976.

---

[37] *Kaye (P. & M.) Ltd.* v. *Hosier & Dickinson* [1972] 1 W.L.R. 146 at p. 157; a case on the Standard form of building contract.

[38] See *A. E. Farr Ltd.* v. *Ministry of Transport* (1965) H.L. (unrep.) save in H.B.C. (10th ed.), p. 519.

[39] See Standard Method of Measurement of Civil Engineering Quantities (1973 reprint), cl. 52; and *cf.* 1976 ed., ss. 8 F, G, H.

[40] *Ibid.*

A new feature is the provision for "method-related" charges, which necessarily exclude valuation by admeasurement (see Section 7). Where such a form of billing is employed the Conditions require amendment.

PROVISIONAL AND PRIME COST SUMS AND
NOMINATED SUB-CONTRACTS

Clause 58
Provisional sum

(1) " Provisional Sum " means a sum included in the Contract and so designated for the execution of work or the supply of goods materials or services or for contingencies which sum may be used in whole or in part or not at all at the direction and discretion of the Engineer.

Prime cost item

(2) " Prime Cost (PC) Item " means an item in the Contract which contains (either wholly or in part) a sum referred to as Prime Cost (PC) which will be used for the execution of work or for the supply of goods materials or services for the Works.

Design requirements to be expressly stated

(3) If in connection with any Provisional Sum or Prime Cost Item the services to be provided include any matter of design or specification of any part of the Permanent Works or of any equipment or plant to be incorporated therein such requirement shall be expressly stated in the Contract and shall be included in any Nominated Sub-contract. The obligation of the Contractor in respect thereof shall be only that which has been expressly stated in accordance with this sub-clause.

Use of prime cost items

(4) In respect of every Prime Cost Item the Engineer shall have power to order the Contractor to employ a sub-contractor nominated by the Engineer for the execution of any work or the supply of any goods materials or services included therein. The Engineer shall also have power with the consent of the Contractor to order the Contractor to execute any such work or to supply any such goods materials or services in which event the Contractor shall be paid in accordance with the terms of a quotation submitted by him and accepted by the Engineer or in the absence thereof the value shall be determined in accordance with Clause 52.

Nominated sub-contractors—definitions

(5) All specialists merchants tradesmen and others nominated in the Contract for a Prime Cost Item or ordered by the Engineer to be employed by the Contractor in accordance with sub-clause (4) or sub-clause (7) of this Clause for the execution of any work or the supply of any goods materials or services are referred to in this Contract as " Nominated Sub-contractors ".

Production of vouchers, etc.

(6) The Contractor shall when required by the Engineer produce all quotations invoices vouchers sub-contract documents accounts and receipts in connection with expenditure in respect of work carried out by all Nominated Sub-contractors.

Use of provisional sums

(7) In respect of every Provisional Sum the Engineer shall have power to order either or both of the following:—

(a) work to be executed or goods materials or services to be supplied by the Contractor the value of such work executed or goods materials or services supplied being determined in accordance with Clause 52 and included in the Contract Price;

(b) work to be executed or goods materials or services to be supplied by a Nominated Sub-contractor in accordance with Clause 59A.

Clause 59A
Nominated sub-contractors—objection to nomination

(1) Subject to sub-clause (2) (c) of this Clause the Contractor shall not be under any obligation to enter into any sub-contract with any Nominated Sub-contractor against whom the Contractor may raise reasonable objection or who shall decline to enter into a sub-contract with the Contractor containing provisions:—

(a) that in respect of the work goods materials or services the subject of the sub-contract the Nominated Sub-contractor will undertake towards the Contractor such obligations and liabilities as will enable the Contractor to discharge his own obligations and liabilities towards the Employer under the terms of the Contract;

(b) that the Nominated Sub-contractor will save harmless and indemnify the Contractor against all claims demands and proceedings damages costs charges and expenses whatsoever arising out of or in connection with any failure by the Nominated Sub-contractor to perform such obligations or fulfil such liabilities;

(c) that the Nominated Sub-contractor will save harmless and indemnify the Contractor from and against any negligence by the Nominated Sub-contractor his agents workmen and servants and against any misuse by him or them of any Constructional Plant or Temporary Works provided by the Contractor for the purposes of the Contract and for all claims as aforesaid;

(d) equivalent to those contained in Clause 63.

Engineer's action upon objection

(2) If pursuant to sub-clause (1) of this Clause the Contractor shall not be obliged to enter into a sub-contract with a Nominated Sub-contractor and shall decline to do so the Engineer shall do one or more of the following:—

(a) nominate an alternative sub-contractor in which case sub-clause (1) of this Clause shall apply;

(b) by order under Clause 51 vary the Works or the work goods materials or services the subject of the Provisional Sum or Prime Cost Item including if necessary the omission of any such work goods materials or services so that they may be provided by workmen contractors or suppliers as the case may be employed by the Employer either concurrently with the Works (in which case Clause 31 shall apply) or at some other date. Provided that in respect of the omission of any Prime Cost Item there shall be included in the Contract Price a sum in respect of the Contractor's charges and profit being a percentage of the estimated value of such work goods material or services omitted at the rate provided in the Bill of Quantities or inserted in the Appendix to the Form of Tender as the case may be;

(c) subject to the Employer's consent where the Contractor declines to enter into a contract with the Nominated Sub-contractor only on the grounds of unwillingness of the Nominated Sub-contractor to contract on the basis of the provisions contained in paragraphs (a) (b) (c) or (d) of sub-clause (1) of this Clause direct the Contractor to enter into a contract with the Nominated Sub-contractor on such other terms as the Engineer shall specify in which case sub-clause (3) of this Clause shall apply;

(d) in accordance with Clause 58 arrange for the Contractor to execute such work or to supply such goods materials or services.

Direction by engineer

(3) If the Engineer shall direct the Contractor pursuant to sub-clause (2) of this Clause to enter into a sub-contract which does not contain all the provisions referred to in sub-clause (1) of this Clause:—

(a) the Contractor shall not be bound to discharge his obligations and liabilities under the Contract to the extent that the sub-contract terms so specified by the Engineer are inconsistent with the discharge of the same;

(b) in the event of the Contractor incurring loss or expense or suffering damage arising out of the refusal of the Nominated Sub-contractor to accept such provisions the Contractor shall subject to Clause 52 (4) be paid in accordance with Clause 60 the amount of such loss expense or damage as the Contractor could not reasonably avoid.

Contractor responsible for nominated sub-contracts

(4) Except as otherwise provided in this Clause and in Clause 59B the Contractor shall be as responsible for the work executed or goods materials or services supplied by a Nominated Sub-contractor employed by him as if he had himself executed such work or supplied such goods materials or services or had sub-let the same in accordance with Clause 4.

Payments

(5) For all work executed or goods materials or services supplied by Nominated Sub-contractors there shall be included in the Contract Price:—

(a) the actual price paid or due to be paid by the Contractor in accordance with the terms of the sub-contract (unless and to the extent that any such payment is the result of a default of the Contractor) net of all trade and other discounts rebates and allowances other than any discount obtainable by the Contractor for prompt payment;

(b) the sum (if any) provided in the Bill of Quantities for labours in connection therewith or if ordered pursuant to Clause 58 (7) (b) as may be determined by the Engineer;

(c) in respect of all other charges and profit a sum being a percentage of the actual price paid or due to be paid calculated (where provision has been made in the Bill of Quantities for a rate to be set against the relevant item of prime cost) at the rate inserted by the Contractor against that item or (where no such provision has been made) at the rate inserted by the Contractor in the Appendix to the Form of Tender as the percentage for adjustment of sums set against Prime Cost Items.

Breach of sub-contract

(6) In the event that the Nominated Sub-contractor shall be in breach of the sub-contract which breach causes the Contractor to be in breach of contract the Employer shall not enforce any award of any arbitrator or judgment which he may obtain against the Contractor in respect of such breach of contract except to the extent that the Contractor may have been able to recover the amount thereof from the Sub-contractor. Provided always that if the Contractor shall not comply with Clause 59B (6) the Employer may enforce any such award or judgment in full.

Clause 59B
Forfeiture of sub-contract

(1) Subject to Clause 59A (2) (c) the Contractor shall in every sub-contract with a Nominated Sub-contractor incorporate provisions equivalent to those

provided in Clause 63 and such provisions are hereinafter referred to as " the Forfeiture Clause ".

Termination of sub-contract

(2) If any event arises which in the opinion of the Contractor would entitle the Contractor to exercise his right under the Forfeiture Clause (or in the event that there shall be no Forfeiture Clause in the sub-contract his right to treat the sub-contract as repudiated by the Nominated Sub-contractor) he shall at once notify the Engineer in writing and if he desires to exercise such right by such notice seek the Employer's consent to his so doing. The Engineer shall by notice in writing to the Contractor inform him whether or not the Employer does so consent and if the Engineer does not give notice withholding consent within 7 days of receipt of the Contractor's notice the Employer shall be deemed to have consented to the exercise of the said right. If notice is given by the Contractor to the Engineer under this sub-clause and has not been withdrawn then notwithstanding that the Contractor has not sought the Employer's consent as aforesaid the Engineer may with the Employer's consent direct the Contractor to give notice to the Nominated Sub-contractor expelling the Nominated Sub-contractor from the sub-contract Works pursuant to the Forfeiture Clause or rescinding the sub-contract as the case may be. Any such notice given to the Nominated Sub-contractor is hereinafter referred to as a notice enforcing forfeiture of the sub-contract.

Engineer's action upon termination

(3) If the Contractor shall give a notice enforcing forfeiture of the sub-contract whether under and in accordance with the Forfeiture Clause in the sub-contract or in purported exercise of his right to treat the sub-contract as repudiated the Engineer shall do any one or more of the things described in paragraphs (a) (b) and (d) of Clause 59A (2).

Delay and extra cost

(4) If a notice enforcing forfeiture of the sub-contract shall have been given with the consent of the Employer or by the direction of the Engineer or if it shall have been given without the Employer's consent in circumstances which entitled the Contractor to give such a notice:—

(a) there shall be included in the Contract Price:—

    (i) the value determined in accordance with Clause 52 of any work the Contractor may have executed or goods or materials he may have provided subsequent to the forfeiture taking effect and pursuant to the Engineer's direction;

    (ii) such amount calculated in accordance with paragraph (a) of Clause 59A (5) as may be due in respect of any work goods materials or services provided by an alternative Nominated Sub-contractor together with reasonable sums for labours and for all other charges and profit as may be determined by the Engineer;

    (iii) any such amount as may be due in respect of the forfeitured sub-contract in accordance with Clause 59A (5);

(b) the Engineer shall take any delay to the completion of the Works consequent upon the forfeiture into account in determining any extension of time to which the Contractor is entitled under Clause 44 and the Contractor shall subject to Clause 52 (4) be paid in accordance with Clause 60 the amount of any additional cost which he may have necessarily and properly incurred as a result of such delay;

(c) the Employer shall subject to Clause 60 (7) be entitled to recover from the Contractor upon the certificate of the Engineer issued in accordance with Clause 60 (3):—

(i) the amount by which the total sum to be included in the Contract Price pursuant to paragraphs (a) and (b) of this sub-clause exceeds the sum which would but for the forfeiture have been included in the Contract Price in respect of work materials goods and services done supplied or performed under the forfeited sub-contract;

(ii) all such other loss expense and damage as the Employer may have suffered in consequence of the breach of the sub-contract;

all of which are hereinafter collectively called " the Employer's loss ".

Provided always that if the Contractor shall show that despite his having complied with sub-clause (6) of this Clause he has been unable to recover the whole or any part of the Employer's loss from the Sub-contractor the Employer shall allow or (if he has already recovered the same from the Contractor) shall repay to the Contractor so much of the Employer's loss as was irrecoverable from the Sub-contractor except and to the extent that the same was irrecoverable by reason of some breach of the sub-contract or other default towards the Sub-contractor by the Contractor or except to the extent that any act or default of the Contractor may have caused or contributed to any of the Employer's loss. Any such repayment by the Employer shall carry interest at the rate stipulated in Clause 60 (6) from the date of the recovery by the Employer from the Contractor of the sum repaid.

Termination without consent

(5) If notice enforcing forfeiture of the sub-contract shall have been given without the consent of the Employer and in circumstances which did not entitle the Contractor to give such a notice:—

(a) there shall be included in the Contract Price in respect of the whole of the work covered by the Nominated Sub-contract only the amount that would have been payable to the Nominated Sub-contractor on due completion of the sub-contract had it not been terminated;

(b) the Contractor shall not be entitled to any extension of time because of such termination nor to any additional expense incurred as a result of the work having been carried out and completed otherwise than by the said Sub-contractor;

(c) the Employer shall be entitled to recover from the Contractor any additional expense he may incur beyond that which he would have incurred had the sub-contract not been terminated.

Recovery of employer's loss

(6) The Contractor shall take all necessary steps and proceedings as may be required by the Employer to enforce the provisions of the sub-contract and/or all other rights and/or remedies available to him so as to recover the Employer's loss from the Sub-contractor. Except in the case where notice enforcing forfeiture of the sub-contract shall have been given without the consent of the Employer and in circumstances which did not entitle the Contractor to give such a notice the Employer shall pay to the Contractor so much of the reasonable costs and expenses of such steps and proceedings as are irrecoverable from the Sub-contractor provided that if the Contractor shall seek to recover by the same steps and proceedings any loss damage or expense additional to the Employer's loss the said irrecoverable costs and expenses shall be borne by the Contractor and the Employer in such proportions as may be fair in all the circumstances.

COMMENTARY

(1) *Generally*

These three clauses should be read together. They regulate closely the relations of the main contractor and the employer where nominated sub-contract work is to be executed. Their effect is that the employer runs a very serious risk of being obliged to bear the loss resulting from the sub-contractor's default, such that members of the drafting committee have stated that employers and engineers should consider carefully, in each case, whether nomination is in fact necessary.[41] The clauses provide expressly for many matters which, under other forms of contract, have been the subject of decisions of the courts.[42]

Clauses 59A and 59B apply only where work is to be executed by a nominated sub-contractor. This is defined by clause 58 (5) as comprising all specialist merchants, tradesmen and others who are:

(i) nominated in the contract for a P.C. item, or
(ii) nominated by the engineer for the execution of any work, etc., included in a P.C. item (sub-cl. (4)) or in a provisional sum (sub-cl. (7)).

The expression includes suppliers. There is no distinction such as that in the Standard form of building contract.[43]

Before commenting on the details of the clauses, some notes are given on the effect of nominations upon the rights of the parties in two areas.

(2) *Nominations which involve variations, delay or extra cost*

The amount or value of the work is not limited by the sum stated in the contract: clause 59A (5). Nor does any such increase or decrease in quantity require a variation order: clause 51 (3). As to the type of work which may be ordered, clause 58 (4) limits the engineer's power in respect of a P.C. item to ordering " the execution of any work . . . included therein." The more general the description of the item, therefore, the wider will be the scope of work which may be ordered. The engineer may vary the subject or content of a provisional sum or P.C. item by a variation order under clause 51 (see also cl. 59A (2) (b)). Thus, if a P.C. item specifies merely " piling," driven piles may be ordered; but if the item is for " bored piling " a variation order is required for driven piles.

May the engineer, by variation order, add a provisional sum or P.C. item not originally in the contract? The position is not clear; but reference to provisional sums and P.C. items being " in the contract " (see cl. 58 (1), (2)) suggests he cannot. It is thought that neither clause 51 (1) (see note (4) thereto) nor clause 13 (see note (3) thereto) give the engineer such power. Note that the engineer may use any provisional sum, *e.g.* one designated for " contingencies," for any nomination (clause 58 (1), (7) (b)).

The contractor is entitled to an extension of time where the engineer's order involves a variation or where the quantities of work exceed those billed in the provisional sum or P.C. item: clause 44 (1). There is no express entitlement to an extension for delay by a nominated sub-contractor and, it is submitted, no implied entitlement.

---

[41] *New Civil Engineer*, Dec. 20, 1973, p. 39.
[42] See *Gloucestershire C.C.* v. *Richardson* [1969] 1 A.C. 480; *Young & Marten* v. *McManus Childs* [1969] 1 A.C. 454; *N. W. Metropolitan Regional Hospital Board* v. *Bickerton* [1970] 1 W.L.R. 607.
[43] See cll. 27 and 28 of the Standard form of building contract.

A provisional sum may be used in whole or in part or not at all, at the discretion of the engineer: clause 58 (1). A P.C. item " will be used " (clause 58 (2)) and must therefore be omitted by variation order if not required. The contractor, on omission of a P.C. item, has no claim for payment save possibly under clause 52 (2). But if the item is omitted because of the sub-contractor's refusal to accept the required conditions (cl. 59A (1)), the contractor is entitled to his charges and profit on the estimated value of the work (cl. 59A (2) (b)).

(3) *Defects in the work of a nominated sub-contractor and remedies*

The topic is dealt with at length in clauses 59A and 59B. These clauses must be read subject to the general law governing the responsibilities of main contractors and employers.

Defective work by a nominated sub-contractor may constitute a breach of the express terms of the main contract. But an area exists in which the parties' rights will be determined by implied terms. Prima facie there will be implied terms of the main contract as to good workmanship and quality of materials and as to their fitness for purpose; but each term may be excluded by the circumstances.[44]

An implied term as to fitness for purpose will be excluded where there is no reliance upon the contractor's skill and judgment, *e.g.* where the materials are selected by the engineer.[45] Such term will be excluded to the extent that the nomination specifies the materials to be obtained. An implied term as to quality may be excluded by the contractor having imposed upon him by the nomination, terms which restrict his rights against the sub-contractor[46] or by the contractor having no right to object to the terms negotiated by the engineer.[47] For a full discussion as to implied terms where work or materials are supplied by a nominated sub-contractor, see *ante*, p. 180.

What remedies are available to the employer where a nominated sub-contractor carries out defective work? If the sub-contractor's performance amounts to a repudiation or gives a right of forfeiture, the employer has the right, subject to the contractor giving notice, to elect whether the sub-contract should be terminated. In the event that it is terminated the engineer has various powers under clause 59B (3), including a power to re-nominate. One of the most important financial aspects of the new clauses is that they are intended to secure that, upon termination of the sub-contract, the employer may recover the additional cost of re-nomination, etc., from the contractor, at least to the extent the contractor can recover from the sub-contractor (cl. 59B (4) (c); but see note (11) *infra*). This is contrary to the position under the Standard form of building contract, where the sub-contractor may often escape liability (even when solvent).[48]

Where the sub-contractor's breach does not result in termination, *e.g.* because the defects are discovered after apparent completion of the sub-contract work,

---

[44] See *ante*, p. 38.
[45] *Young & Marten* case, *supra*.
[46] *Richardson* case, *supra*, *per* Lord Upjohn and *Young & Marten* case, *supra*, *per* Lord Reid.
[47] *Ibid. per* Lords Pearce and Wilberforce.
[48] *Bickerton* case, *supra*.

the engineer's powers are less distinct. It is uncertain whether an instruction can be given under clause 39 (1), because the contractor cannot be ordered to do P.C. work himself.[49] Clause 39 (2) refers specifically to default " on the part of the contractor " as a condition precedent to the right to employ others to do the work. If the contractor gives an instruction to the sub-contractor to remedy defects, the sub-contractor's failure to comply may bring into operation the provisions of clause 59B.

If payment for the defective work has been certified to the contractor, a deduction may be made in a later interim certificate only if the contractor has not paid or is not bound to pay the sub-contractor (cl. 60 (7) (a)). Should the contractor contend that the work is not defective an *impasse* could be reached: clause 66 does not allow an immediate arbitration on such a dispute. Alternatively the engineer may refuse to certify payment for subsequent work which will need to be rectified in consequence of the defective sub-contract work (*e.g.* work built on defective piles). Such dispute may lead to an immediate arbitration.[50]

(4) *Clause* 58: *Provisional Sum and P.C. Item*

Note the distinction between the definitions: a provisional sum " may be used in whole or in part or not at all " while a P.C. item " will be used," so that its omission requires a variation order. The engineer's power to instruct the contractor to carry out a P.C. item requires consent: clause 58 (4).

(5) *Clause* 58 (3): *Design requirements " shall be expressly stated in the Contract "*

It is in the employer's interest to ensure that they are so stated. The contractor is entitled to refuse a nomination including design where there is no design requirement included in the main contract, it is submitted. The contractor's obligation as to design, is confined to " that which has been expressly stated " (clause 58 (3)). A design obligation cannot be added which was not originally in the contract, it is submitted. Whether or not the contractor is made liable, the employer may protect himself by obtaining a direct warranty of the design from the sub-contractor.[51-52]

(6) *Clause* 58 (4): " *with the consent of the Contractor to order the Contractor* "

The engineer should, on behalf of the employer seek such express consent at the time of entering into the contract, otherwise this is hardly a power.

(7) *Clause* 59A (1): " *against whom the Contractor may raise reasonable objection* "

There is no indication as to what matters may be taken into account in deciding whether the contractor's objection is reasonable. It is thought that financial issues and the competence of the sub-contractor are relevant matters. The contract no longer expressly entitles the contractor to a discount for prompt payment.[53] But the possibility of discount is recognised by clause 59A (5) (a).

---

[49] Cl. 58 (4), and note (6) *infra*; see also cl. 59A (2) (b).
[50] See cl. 66, note (8).
[51-52] See *ante*, Chap. 16.
[53] See cl. 59 (1) of 4th ed. and *cf.* cll. 27 (a) and 28 (a) Standard form of building contract.

It is thought that objection on the ground of refusal to offer any discount would, prima facie, be reasonable.

(8) *Clause* 59A (2): " *and shall decline to do so* "

If the contractor does not decline and enters into the sub-contract, he waives the right to object to the sub-contractor or to the terms, it is submitted, and loses the protection of clause 59A (3).

(9) *Clause* 59A (2) (c) *and* (3)

Sub-clause (3) applies when the sub-contractor refuses to agree to provisions set out in sub-clause (1) (a) (b) (c) or (d) *and* the contractor declines to enter into the sub-contract *and* the engineer then directs the contractor to enter into a sub-contract on " such other terms as the engineer shall specify ": clause 59A (2) (c). There is no reason why " other terms " should be limited to the most favourable terms which can be negotiated, *e.g.* where the engineer is able to secure a reduction in price for less favourable terms. There is no power for the engineer to give a direction where the contractor has declined to enter into the sub-contract on the (justified) grounds of a reasonable objection.

The intention of sub-clause (3) (a) is to exclude the contractor's liability so far as he cannot pass it on to the sub-contractor. The words " to the extent that the sub-contract terms . . . are inconsistent with the discharge of the same " would, as a matter of ordinary construction, probably be insufficient as an exclusion clause.[54] But the words are inserted to protect the contractor where his recourse against the sub-contractor has been removed by the engineer's direction. In their context it is thought the words would be construed *contra proferentem* against the employer, and that, so construed, they are sufficient to limit the contractor's liability.[55]

Sub-clause (3) (b) gives the contractor a right to claim loss, etc. (in addition to the limitation of liability under (a)), limited to that " arising out of the refusal of the Nominated Sub-contractor to accept such provisions," *i.e.* the provisions of sub-clause (1) (a), (b), (c) or (d)). This does not, apparently, cover loss arising from " such other terms " as are specified by the engineer under sub-clause (2) (c).

(10) *Clause* 59A (4): *Responsibility for Nominated Sub-contractors*

The words " except as otherwise provided " preserve the contractor's rights under clause 59A (3) and (6) and 59B (4) and (6). The contractor is liable for the quality of work and materials; but not for the suitability of the sub-contractor's work and materials.[56]

(11) *Clause* 59A (6): " *except to the extent that the Contractor may have been able to recover the amount thereof from the sub-contractor* "

This does not prevent the employer from obtaining an award or entering judgment against the contractor. But the sub-clause suffers from the objection

---

[54] See p. 31.
[55] *Richardson* case, *supra*, and see note (3) *supra*.
[56] See note (3), *supra*.

that the sub-contractor may contend, when sued, that the contractor cannot prove damage; the contractor cannot establish loss until he has recovered " the amount thereof from the Sub-contractor." [57]

On balance, it is thought the objection does not prevent the contractor recovering damages. The courts will seek to give effect to the palpable intention of the parties, which is to allow the employer's loss to be passed from the contractor to the sub-contractor. Where the sub-contractor has not been paid an argument available to the employer is that the contractor is entitled to set off the diminution in value of the work arising from the sub-contractor's breach [58] without proof of damage. The court, or arbitrator would be bound, it is thought, to treat such set-off as a " recovery." It would require clear words to make the employer's right of execution dependent on whether the sub-contractor had been paid. Further, the sub-clause operates as a limitation of liability which comes into operation upon the default of the sub-contractor. To permit the sub-contractor to rely on the clause is to allow him to take advantage of his own default.[59] The position might be clarified by express provision in the sub-contract.

A further consequence of the sub-clause is that it may, subject to the arguments above, prevent recovery of liquidated damages from the contractor where delay arises from breach by a nominated sub-contractor. The limitation on execution of an award or judgment " in respect of such breach " must, it is submitted, extend to any counterclaim by the employer, *e.g.* where the liquidated damages have been withheld and the contractor is seeking recovery.

When the sub-contractor is in breach, any action brought by the contractor is likely to include claims for his own damages. If the contractor recovers part only of the total damages awarded the sub-clause does not provide how the sum recovered is to be apportioned between employer and contractor. The drafting committee have suggested that the employer should be given priority if the matter is dealt with by amendment.[60]

(12) *Clause* 59B (1): " *Subject to Clause* 59A (2) (c) "

The effect of these words is, presumably, that if the engineer directs the contractor, under clause 59A (2) (c), to enter a sub-contract without a forfeiture clause, the contractor is not bound by clause 59B (1). In such a case the contractor is entitled to the benefit of clause 59A (3).

(13) *Clause* 59B (2) *and* (3)

Despite the elaborate provision as to notices and consents, sub-clauses (4) and (5) show that a justified forfeiture or termination notice given without consent is to be treated as equivalent to one given with consent; and further that

---

[57] See *The Albazero* [1976] 3 W.L.R. 419, which reviews the exceptions to the general rule that a party to a contract can recover as damages only such actual loss as he has himself sustained.

[58] *Mondel* v. *Steel* (1841) 8 M. & W. 858; *Hanak* v. *Green* [1958] 2 Q.B. 9; *Modern Engineering* v. *Gilbert Ash* [1974] A.C. 689.

[59] *Roberts* v. *Bury Commissioners* (1870) L.R. 5 C.P. 310; see also p. 78.

[60] *New Civil Engineer*, Dec. 20, 1973, p. 39.

the term " notice enforcing forfeiture of the sub-contract " covers also a notice given without consent, and whether or not given in circumstances which entitle the contractor to give such notice. It follows that the engineer is obliged to act under sub-clause (3) despite the notice being given without consent or being given wrongfully.

Sub-clauses (3) and (4) do not appear fully to protect the contractor against the consequences of a wrongful termination. These will include a claim for damages by the sub-contractor against the contractor, and the possibility that the sub-contractor will refuse to accept determination (see note (14)).

A further relevant matter, in view of the employer's obligation to pay additional costs under sub-clause (4), is whether the words " without consent " include a forfeiture after the engineer's express " notice withholding consent." The words cannot mean " without express consent " because sub-clause (2) provides that absence of a notice withholding consent is to be a " deemed " consent. The words could be limited to forfeiture without service of the contractor's notice under sub-clause (2) or where such notice did not request consent. But it is thought unlikely that the parties would intend to deprive the contractor of the benefit of sub-clause (4) where he has disregarded the engineer's refusal of consent, but to allow him the benefit where he has not served a notice giving the employer an opportunity to refuse consent. In the result it is submitted that " without consent " includes the express refusal of consent.

The effect of the engineer's notice withholding consent is therefore merely to put the contractor on risk that if the forfeiture is wrongful he must bear the excess cost in accordance with sub-clause (5).

(14) *Clause* 59B (4)

This omits to provide for compensation of the contractor where he is liable to the sub-contractor for wrongful termination, even where the sub-contract is terminated with the engineer's consent or direction. Sub-clause (4) (b) provides for recovery of cost, limited to that resulting from delay. The omission may be justified where the engineer has merely given consent, but where notice is given on the engineer's direction the contractor's position may be seriously prejudiced. Further, where the sub-contractor refuses to accept a wrongful notice, sub-clause (4) may be difficult to apply since the sub-contract may not have been " forfeited " (sub-cl. (4) (a) (iii)) and forfeiture cannot be said to have taken effect (sub-cl. (4) (a) (i)).

The engineer's power of direction in sub-clause (2) can arise only upon the contractor giving notice of an event which in the contractor's opinion entitles him to forfeit. To avoid the difficulties referred to in notes (13) and (14) the contractor should refrain from giving the initial notice unless there is a very clear case for forfeiture or repudiation, and the contractor wishes to forfeit.

(15) *Clause* 59B (4) (c): " *subject to clause* 60 (7) "

These words serve to restrict the employer's right to recover from the contractor on the final certificate, where the sum certified in respect of the nominated sub-contractor is reduced. Certification of " the employer's loss " may involve such a reduction.

(16) Clause 59B (4) (c): *Proviso*

The effect of this is similar to clause 59A (6). But the proviso does not involve the difficulty of proving loss, since it is expressly recognised that the employer may recover his loss from the contractor before the same is recovered from the sub-contractor. The words " except and to the extent that the same was irrecoverable by reason of some breach of the sub-contract . . . by the Contractor " emphasise that the contractor takes the risk of the forfeiture or termination being wrongful (see note (14) *supra*). The proviso produces the same difficulty as clause 59A (6), when the contractor sues for his own and the employer's loss, and recovers part only of the total damages awarded (see note (11) *supra*).

(17) Clause 59B (6): " *The Contractor shall take all necessary steps* "

The obligation is limited to taking steps, etc., to recover " the Employer's loss," as defined by clause 59B (4) (c). See also clause 60 (7).

Clause 59C
Payment to nominated sub-contractors

Before issuing any certificate under Clause 60 the Engineer shall be entitled to demand from the Contractor reasonable proof that all sums (less retentions provided for in the sub-contract) included in previous certificates in respect of the work executed or goods or materials or services supplied by Nominated Sub-contractors have been paid to the Nominated Sub-contractors or discharged by the Contractor in default whereof unless the Contractor shall:—
   (a) give details to the Engineer in writing of any reasonable cause he may have for withholding or refusing to make such payment; and
   (b) produce to the Engineer reasonable proof that he has so informed such Nominated Sub-contractor in writing;

the Employer shall be entitled to pay to such Nominated Sub-contractor direct upon the certification of the Engineer all payments (less retentions provided for in the sub-contract) which the Contractor has failed to make to such Nominated Sub-contractor and to deduct by way of set-off the amount so paid by the Employer from any sums due or which become due from the Employer to the Contractor. Provided always that where the Engineer has certified and the Employer has paid direct as aforesaid the Engineer shall in issuing any further certificate in favour of the Contractor deduct from the amount thereof the amount so paid direct as aforesaid but shall not withhold or delay the issue of the certificate itself when due to be issued under the terms of the Contract.

COMMENTARY

Generally

The right of direct payment arises only upon the certificate of the engineer, which must take into account the effect of any " reasonable cause " for withholding payment. If the contractor claims a right of set-off for breach by the sub-contractor, the engineer must, it is submitted, make an assessment of, and deduct, the reasonable amount of such set-off before certifying to the employer. The right of direct payment arises only where the sums in question have already been certified to the contractor. The provision will not, therefore, entitle the employer to make direct payments to nominated sub-contractors upon the contractor's insolvency without the engineer first certifying the sums in question to the contractor.

The right of direct payment may, it is submitted, be exercised against the contractor's trustee in bankruptcy, or liquidator, who elects to continue the contract,[61] when not forfeited under clause 63 (1).

CERTIFICATES AND PAYMENT

Clause 60
Monthly statements

(1) The Contractor shall submit to the Engineer after the end of each month a statement (in such form if any as may be prescribed in the Specification) showing:—

(a) the estimated contract value of the Permanent Works executed up to the end of that month;

(b) a list of any goods or materials delivered to the Site for but not yet incorporated in the Permanent Works and their value;

(c) a list of any goods or materials listed in the Appendix to the Form of Tender which have not yet been delivered to the Site but of which the property has vested in the Employer pursuant to Clause 54 and their value;

(d) the estimated amounts to which the Contractor considers himself entitled in connection with all other matters for which provision is made under the Contract including any Temporary Works or Constructional Plant for which separate amounts are included in the Bill of Quantities;

unless in the opinion of the Contractor such values and amounts together will not justify the issue of an interim certificate.

Amounts payable in respect of Nominated Sub-contractors are to be listed separately.

Monthly payments

(2) Within 28 days of the date of delivery to the Engineer or Engineer's Representative in accordance with sub-clause (1) of this Clause of the Contractor's monthly statement the Engineer shall certify and the Employer shall pay to the Contractor (after deducting any previous payments on account):—

(a) the amount which in the opinion of the Engineer on the basis of the monthly statement is due to the Contractor on account of sub-clause (1) (a) and (d) of this Clause less a retention as provided in sub-clause (4) of this Clause;

(b) such amounts (if any) as the Engineer may consider proper (but in no case exceeding the percentage of the value stated in the Appendix to the Form of Tender) in respect of (b) and (c) of sub-clause (1) of this Clause which amounts shall not be subject to a retention under sub-clause (4) of this Clause.

The amounts certified in respect of Nominated Sub-contracts shall be shown separately in the certificate. The Engineer shall not be bound to issue an interim certificate for a sum less than that named in the Appendix to the Form of Tender.

Final account

(3) Not later than 3 months after the date of the Maintenance Certificate the Contractor shall submit to the Engineer a statement of final account and supporting documentation showing in detail the value in accordance with the Contract of the work done in accordance with the Contract together with all further sums which the Contractor considers to be due to him under the Contract up to the

---

[61] See *Re Wilkinson, ex p. Fowler* [1905] 2 K.B. 713; *Re Tout & Finch* [1954] 1 W.L.R. 178.

date of the Maintenance Certificate. Within 3 months after receipt of this final account and of all information reasonably required for its verification the Engineer shall issue a final certificate stating the amount which in his opinion is finally due under the Contract up to the date of the Maintenance Certificate and after giving credit to the Employer for all amounts previously paid by the Employer and for all sums to which the Employer is entitled under the Contract up to the date of the Maintenance Certificate the balance if any due from the Employer to the Contractor or from the Contractor to the Employer as the case may be. Such balance shall subject to Clause 47 be paid to or by the Contractor as the case may require within 28 days of the date of the certificate.

Retention

(4) The retention to be made pursuant to sub-clause (2) (a) of this Clause shall be a sum equal to 5 per cent. of the amount due to the Contractor until a reserve shall have accumulated in the hand of the Employer up to the following limits:—
  (a)   where the Tender Total does not exceed £50,000 5 per cent of the Tender Total but not exceeding £1,500; or
  (b)   where the Tender Total exceeds £50,000 3 per cent of the Tender Total; except that the limit shall be reduced by the amount of any payment that shall have been made pursuant to sub-clause (5) of this Clause.

Payment of retention money

(5) (a)   If the Engineer shall issue a Certificate of Completion in respect of any Section or part of the Works pursuant to Clause 48 (2) or (3) there shall become due on the date of issue of such certificate and shall be paid to the Contractor within 14 days thereof a sum equal to $1\frac{1}{2}$ per cent of the amount due to the Contractor at that date in respect of such Section or part as certified for payment pursuant to sub-clause (2) of this Clause.
  (b)   One half of the retention money less any sums paid pursuant to sub-clause (5) (a) of this Clause shall be paid to the Contractor within 14 days after the date on which the Engineer shall have issued a Certificate of Completion for the whole of the Works pursuant to Clause 48 (1).
  (c)   The other half of the retention money shall be paid to the Contractor within 14 days after the expiration of the Period of Maintenance notwithstanding that at such time there may be outstanding claims by the Contractor against the Employer. Provided always that if at such time there shall remain to be executed by the Contractor any outstanding work referred to under Clause 48 or any works ordered during such period pursuant to Clauses 49 and 50 the Employer shall be entitled to withhold payment until the completion of such works of so much of the second half of retention money as shall in the opinion of the Engineer represent the cost of the works so remaining to be executed.
      Provided further that in the event of different maintenance periods having become applicable to different Sections or parts of the Works pursuant to Clause 48 the expression " expiration of the Period of Maintenance " shall for the purposes of this sub-clause be deemed to mean the expiration of the latest of such periods.

Interest on overdue payments

(6) In the event of failure by the Engineer to certify or the Employer to make payment in accordance with sub-clauses (2) (3) and (5) of this Clause the Employer shall pay to the Contractor interest upon any payment overdue thereunder at a rate per annum equivalent to $\frac{3}{4}$ per cent plus the minimum rate at which the Bank of England will lend to a discount house having access to the Discount Office of the Bank current on the date upon which such payment first becomes overdue. In the event of any variation in the said Bank Rate being announced

whilst such payment remains overdue the interest payable to the Contractor for the period that such payment remains overdue shall be correspondingly varied from the date of each such variation.

Correction and withholding of certificates

(7) The Engineer shall have power to omit from any certificate the value of any work done goods or materials supplied or services rendered with which he may for the time being be dissatisfied and for that purpose or for any other reason which to him may seem proper may by any certificate delete correct or modify any sum previously certified by him.

Provided always that:—

(a) the Engineer shall not in any interim certificate delete or reduce any sum previously certified in respect of work done goods or materials supplied or services rendered by a Nominated Sub-contractor if the Contractor shall have already paid or be bound to pay that sum to the Nominated Sub-contractor;

(b) if the Engineer in the final certificate shall delete or reduce any sum previously certified in respect of work done goods or materials supplied or services rendered by a Nominated Sub-contractor which sum shall have been already paid by the Contractor to the Nominated Sub-contractor the Employer shall reimburse to the Contractor the amount of any sum overpaid by the Contractor to the Sub-contractor in accordance with the certificates issued under sub-clause (2) of this Clause which the Contractor despite compliance with Clause 59B (6) shall be unable to recover from the Nominated Sub-contractor together with interest thereon at the rate stated in Clause 60 (6) from 28 days after the date of the final certificate issued under sub-clause (3) of this Clause until the date of such reimbursement.

Copy certificate for contractor

(8) Every certificate issued by the Engineer pursuant to this Clause shall be sent to the Employer and at the same time a copy thereof shall be sent to the Contractor.

COMMENTARY

(1) *Outline*

The clause requires the contractor to submit a monthly valuation (sub-cl. (1)) and the engineer to issue an interim certificate within 28 days (sub-cl. (2)); after issue of the maintenance certificate the engineer must give a final certificate (sub-cl. (3)). Interim payments are subject to a retention (sub-cl. (2) (a) and (4)); save that for goods and materials the amount to be certified is not subject to retention but is that which the engineer " may consider proper," not exceeding the percentages stated in the appendix (sub-cl. (2) (b)). The retention is to be paid as to half upon the certificate of completion (subject to earlier payments upon completion of any section or part), and the remainder 14 days after the expiry of the maintenance period (subject to deductions) (sub-cl. (5)). The contractor is entitled to interest on unpaid sums (sub-cl. (6)).

(2) *Clause* 60 (1) (b): " *List of any goods or materials . . . of which the property has vested* "

The list is, itself, of no effect unless the materials or goods have already complied with clause 54. This requires the engineer's approval in writing (cl. 54 (3)).

(3) *Clause* 60 (1) (d): " *all other matters for which provision is made under the Contract* "

See note (8) to clause 52 for claims which may arise under the contract. There is no power to certify any sum to the contractor where the claim is for damages for breach or where the claim arises outside the contract, *e.g.* extras ordered directly by the employer. The contractor is expressly required to submit with the monthly statement details of additions or deductions for tax fluctuations: clause 69 (6); the contractor's estimate for VAT on taxable goods and services is to be submitted separately: clause 70 (3).

(4) *Clause* 60 (2): " *the Engineer shall certify and the Employer shall pay* "

The engineer's certificate is a condition precedent to the right of payment, it is submitted.[62] This is confirmed by sub-clause (5) which makes retention payable automatically without a payment certificate. Interim payments should be certified in full, leaving the employer to deduct any sums to which he may be entitled by way of set-off.[63]

Payment may be recovered without an interim certificate if the requirement has been waived by the employer [64] or if the contractor is able to establish grounds for dispensing with the requirement of a certificate [65] or if the contractor obtains an award by arbitration. In respect of the alleged withholding by the engineer of a certificate, the parties are entitled to arbitrate forthwith despite the works being incomplete (clause 66 (2)). As to the power to modify previous interim certificates, see sub-clause (7) *infra*.

The effect of the words " the Engineer shall certify and the Employer shall pay " is that payment becomes due on the certificate 28 days after delivery of the contractor's monthly statement. In default of payment, the contractor is entitled to interest under sub-clause (6).

(5) *Clause* 60 (2) (b): " *such amounts (if any) as the Engineer may consider proper* "

These words qualify the contractor's right to payment for goods or materials not incorporated into the works, notwithstanding compliance with clause 54 in the case of goods or materials not delivered to the site. The engineer appears to have an unqualified discretion to withhold payment. This is in addition to the power, under sub-clause (7) to revise certificates in the event of deterioration or damage to goods or materials delivered to the site. In the event of the contractor (or the employer) disputing any amount certified there may be immediate arbitration (cl. 66 (2)).

(6) *Clause* 60 (3): *The final certificate*

The final certificate is of no particular evidential effect as to the sufficiency of

---

[62] *Dunlop & Ranken Ltd.* v. *Hendall* [1957] 1 W.L.R. 1102, 1106.

[63] As to the employer's right to withhold payment on an engineer's certificate see *Gilbert-Ash* v. *Modern Engineering* [1974] A.C. 689, and *cf. Kilby & Gayford* v. *Selincourt* (1973) 229 E.G. 1343 (C.A.) and *Ellis* v. *Wates* (1976) 1 C.L. 3286, *The Times*, Jan. 22, 1976 (decisions on or involving the Standard form of building contract). See also cl. 47, note (5) and cl. 66, note (1).

[64] See *ante*, p. 78.

[65] See *ante*, p. 78.

the works or their compliance with the contract. Its effect is to create a debt due to or from the contractor. It is not comparable to the effect the final certificate can have under the Standard form of building contract. The certificate in no way affects the continuing liability of the parties, it is submitted. At the date for its issue, the contractor will have received the retention money in full (sub-cl. (5)) and the maintenance certificate will have been issued, so that the final certificate is the last act of the engineer, save for any decision under clause 66. The reference to giving credit for " all sums to which the Employer is entitled under the Contract," shows that, contrary to the position with interim certificates,[66] the final certificate is to be net. But the final certificate cannot take account of a claim for damages, which is not a claim under the contract; nor can the engineer have any right to adjudicate upon such claim, save on a reference under clause 66. The words " subject to Clause 47 " indicate that liquidated damages are not to be deducted from the amount to be certified.

(7) *Clause* 60 (5): *Payment of retention*

The words " there shall become due " (para. (a)) and " shall be paid " ((b) and (c)) show that no payment certificate is required. The contractor's right to payment arises from the relevant completion or maintenance certificate.

(8) *Clause* 60 (6): *Interest on overdue payments*

This is a valuable provision which entitles the contractor to interest for any period during which he is kept out of money due. In the absence of such provision the contractor cannot recover interest if the money is paid (albeit late) before proceedings are issued.[67] Interest will not accrue, it is submitted, where the employer claims a right of contractual [68] or equitable set-off, since the payment cannot then be said to be due.

(9) *Clause* 60 (7): *Correction and withholding of certificates*

It is clear from the contract [69] that an interim certificate is not binding either as to the engineer's satisfaction or as to compliance with the contract of any work, materials or services. The words " for any other reason which to him may seem proper " do not, it is submitted, entitle the engineer to withhold payments otherwise contractually due; *e.g.* the engineer cannot withhold payment for goods brought prematurely to the site unless they have deteriorated or become damaged (but see note (5) *supra*). A perfectly proper reason may be that the goods were previously over-valued.

(10) *Clause* 60 (7): *Reduction of sums in respect of nominated sub-contractors*

In respect of interim certificates, no reduction is permitted where the contractor has paid or is " bound to pay " the sums proposed to be reduced or deleted. Where the sub-contractor is in breach, giving the contractor a right of set-off, the engineer may reduce a sum previously certified, to the extent of the

---

[66] See note (4), *supra*.
[67] The courts will not entertain a claim for interest only; see p. 254 *ante*.
[68] See, *e.g.* cll. 39 (2), 47 (4).
[69] See also cll. 17, 39 (1) and 53 (11).

set-off, it is submitted, since the contractor is not bound to pay the amount of set-off.

The engineer has an unrestricted right by the final certificate, to reduce sums previously certified. But this is subject to reimbursement of any sum overpaid on interim certificates which the contractor " despite compliance with clause 59B (6) shall be unable to recover." The duty imposed by clause 59B (6) appears to be limited to recovering " the Employer's loss," as defined by clause 59B (4) (c).[70] However, the words " shall be unable " are sufficient, it is thought, to impose a duty on the contractor to mitigate his loss as a condition of reimbursement. It is thought that the contractor may " recover " the amount overpaid by a set-off, *e.g.* against a justified claim for disruption by the sub-contractor.

(11) *Clause* 60 (7): " *The Employer shall reimburse* "

This provision is not subject to the same difficulty as clause 59A (6),[71] where the contractor seeks to recover damages. The contractor will suffer loss when the sum previously certified and paid to the sub-contractor is reduced. Clause 60 (7) (b) gives the contractor an alternative remedy if such loss cannot be recovered from the sub-contractor.

Clause 61
Maintenance certificate

(1) Upon the expiration of the Period of Maintenance or where there is more than one such period upon the expiration of the latest period and when all outstanding work referred to under Clause 48 and all work of repair amendment reconstruction rectification and making good of defects imperfections shrinkages and other faults referred to under Clauses 49 and 50 shall have been completed the Engineer shall issue to the Employer (with a copy to the Contractor) a Maintenance Certificate stating the date on which the Contractor shall have completed his obligations to construct complete and maintain the Works to the Engineer's satisfaction.

Unfulfilled obligations

(2) The issue of the Maintenance Certificate shall not be taken as relieving either the Contractor or the Employer from any liability the one towards the other arising out of or in any way connected with the performance of their respective obligations under the Contract.

COMMENTARY

The maintenance certificate signifies completion to the engineer's satisfaction. It further sets in train the procedure under clause 60 (3) for the certificate of final payment. The giving or withholding of the certificate may be challenged under clause 66. Clause 61 (2) makes clear that the certificate does not relieve either party from any liability under the contract.

The engineer is entitled to withhold the maintenance certificate if defects become apparent after the maintenance period but before completion of outstanding work or work of repair etc. since the contractor will not have completed

---

[70] See cl. 59B, note (17).
[71] See cl. 59A, note (11).

his obligations to construct complete and maintain the works. The contractor has no duty or right to carry out further remedial work if defects appear after the maintenance period: clause 49 (2). The employer's remedy is in damages unless further remedial work is done by agreement.

REMEDIES AND POWERS

Clause 62
Urgent repairs

If by reason of any accident or failure or other event occurring to in or in connection with the Works or any part thereof either during the execution of the Works or during the Period of Maintenance any remedial or other work or repair shall in the opinion of the Engineer be urgently necessary and the Contractor is unable or unwilling at once to do such work or repair the Employer may by his own or other workmen do such work or repair as the Engineer may consider necessary. If the work or repair so done by the Employer is work which in the opinion of the Engineer the Contractor was liable to do at his own expense under the Contract all costs and charges properly incurred by the Employer in so doing shall on demand be paid by the Contractor to the Employer or may be deducted by the Employer from any monies due or which may become due to the Contractor. Provided always that the Engineer shall as soon after the occurrence of any such emergency as may be reasonably practicable notify the Contractor thereof in writing.

COMMENTARY

(1) *Generally*

Where remedial or repair work is not required to be done at once, the engineer may act under clause 39 (1) before the completion certificate, or under clause 49 (2) during the period of maintenance. The power under clause 62 may be exercised at any time before the maintenance certificate.[72]

(2) " *Provided always that the Engineer shall notify the Contractor in writing* "

Written notice from the engineer is a condition upon which operation of the clause depends, it is submitted. If notice is not given the employer is not entitled to set off the cost of the work (subject to the engineer's opinion) unless he can show that the contractor was in breach of contract. This may be difficult or impossible where the repairs are required to work in course of erection.[73]

Clause 63
Forfeiture

(1) If the Contractor shall become bankrupt or have a receiving order made against him or shall present his petition in bankruptcy or shall make an arrangement with or assignment in favour of his creditors or shall agree to carry out the Contract under a committee of inspection of his creditors or (being a corporation) shall go into liquidation (other than a voluntary liquidation for the purposes of amalgamation or reconstruction) or if the Contractor shall assign the Contract without the consent in writing of the Employer first obtained or shall have an execution levied on his goods or if the Engineer shall certify in writing to the Employer that in his opinion the Contractor:—

---

[72] See also cl. 39, note (1).
[73] See *Kaye (P. & M.) Ltd.* v. *Hosier & Dickinson* [1972] 1 W.L.R. 146, *per* Lord Diplock at p. 157.

(a) has abandoned the Contract; or

(b) without reasonable excuse has failed to commence the Works in accordance with Clause 41 or has suspended the progress of the Works for 14 days after receiving from the Engineer written notice to proceed; or

(c) has failed to remove goods or materials from the Site or to pull down and replace work for 14 days after receiving from the Engineer written notice that the said goods materials or work have been condemned and rejected by the Engineer; or

(d) despite previous warning by the Engineer in writing is failing to proceed with the Works with due diligence or is otherwise persistently or fundamentally in breach of his obligations under the Contract; or

(e) has to the detriment of good workmanship or in defiance of the Engineer's instruction to the contrary sub-let any part of the Contract;

then the Employer may after giving 7 days' notice in writing to the Contractor enter upon the Site and the Works and expel the Contractor therefrom without thereby avoiding the Contract or releasing the Contractor from any of his obligations or liabilities under the Contract or affecting the rights and powers conferred on the Employer or the Engineer by the Contract and may himself complete the Works or may employ any other contractor to complete the Works and the Employer or such other contractor may use for such completion so much of the Constructional Plant Temporary Works goods and materials which may have been deemed to become the property of the Employer under Clauses 53 and 54 as he or they may think proper and the Employer may at any time sell any of the said Constructional Plant Temporary Works and unused goods and materials and apply the proceeds of sale in or towards the satisfaction of any sums due or which may become due to him from the Contractor under the Contract.

Assignment to employer

(2) By the said notice or by further notice in writing within 14 days of the date thereof the Engineer may require the Contractor to assign to the Employer and if so required the Contractor shall forthwith assign to the Employer the benefit of any agreement for the supply of any goods or materials and/or for the execution of any work for the purposes of this Contract which the Contractor may have entered into.

Valuation at date of forfeiture

(3) The Engineer shall as soon as may be practicable after any such entry and expulsion by the Employer fix and determine *ex parte* or by or after reference to the parties or after such investigation or enquiries as he may think fit to make or institute and shall certify what amount (if any) had at the time of such entry and expulsion been reasonably earned by or would reasonably accrue to the Contractor in respect of work then actually done by him under the Contract and what was the value of any unused or partially used goods and materials any Constructional Plant and any Temporary Works which have been deemed to become the property of the Employer under Clauses 53 and 54.

Payment after forfeiture

(4) If the Employer shall enter and expel the Contractor under this Clause he shall not be liable to pay to the Contractor any money on account of the Contract until the expiration of the Period of Maintenance and thereafter until the costs of completion and maintenance damages for delay in completion (if any) and all other expenses incurred by the Employer have been ascertained and the amount thereof certified by the Engineer. The Contractor shall then be entitled to receive only such sum or sums (if any) as the Engineer may certify would have been due to him upon due completion by him after deducting the said amount. But if such

amount shall exceed the sum which would have been payable to the Contractor on due completion by him then the Contractor shall upon demand pay to the Employer the amount of such excess and it shall be deemed a debt due by the Contractor to the Employer and shall be recoverable accordingly.

COMMENTARY

(1) *Forfeiture*

The term appears solely in the clause headings and is used to mean entry upon the site by the employer, and expulsion of the contractor. It does not involve discharge of either party from obligations or rights under the contract, save as provided by this clause.

(2) *Clause* 63 (1): *Grounds of forfeiture requiring the Engineer's certificate*

(a) " *has abandoned the Contract.*" Abandonment of the contract is referred to in clause 40 (2) in connection with a suspension of the progress of the works. It is thought that abandonment denotes at least a suspension coupled with an apparent intention not to continue.

(b) " *without reasonable excuse has failed to commence the works in accordance with Clause* 41 . . ." Clause 41 requires the contractor to commence " as soon as is reasonably possible " after the date notified by the engineer. The engineer must therefore be satisfied that it is reasonably possible to commence *and* that there is no reasonable excuse for delay.

(c) " *has failed to remove goods* . . ." In this case forfeiture is an extreme remedy, since the contractor may not be in breach of contract by supplying the condemned goods.[74] Clause 39 gives the engineer power to require removal of materials or work not in accordance with the contract, and contains its own remedy for default. Clause 39 (2) provides no time limit for compliance with the engineer's order under clause 39 (1) so the contractor may not be in breach of that clause at the time the right of forfeiture arises. It is uncertain whether the power under this provision is limited to a notice given under clause 39. The two provisions are differently worded. Notice under this clause must expressly state that the goods or materials are " condemned and rejected." A notice expressly requiring removal within 14 days would not be within the scope of clause 39. Such notice could, it is thought, be given under the general power of clause 13 (1).

(d) " *despite previous warning* . . . *is failing to proceed with the works with due diligence or is otherwise persistently or fundamentally in breach* . . ." The clause gives no indication as to whether the previous warning may, in the case of lack of diligence, be a notice under clause 46. It is thought the warning must state that the contractor is not proceeding with due diligence, or specify the breach complained of; and that a reasonable time must elapse before forfeiture may be enforced.

It is thought that " due diligence " is for all practical purposes the same as " due expedition."[75] The terms "persistently" and "fundamentally" are alternatives. A breach need not be serious in order to be persistent. But it is

---

[74] *Kaye (P. & M.) Ltd.* v. *Hosier & Dickinson* [1972] 1 W.L.R. 146, 157.
[75] See cl. 41, note (2).

submitted that a persistent breach must involve a continuing commission (*e.g.* repeated execution of bad work) rather than a single breach which persists (*e.g.* one item of bad work which is not rectified). " Fundamentally " indicates a breach which goes to the root of the contract.[76]

(e) " *sub-let any part of the Contract.*" See clause 4. Sub-letting in breach of contract is not ordinarily a repudiation.[77] There is no power to forfeit solely upon the ground of sub-letting without consent. There is, further, no express power to give instructions not to sub-let.

The power in respect of sub-letting " to the detriment of good workmanship " applies whether or not the engineer has given consent. But in this case there must be a causal connection between the sub-letting and the bad work, *e.g.* the sub-contractor being obviously incompetent.

It is thought that this paragraph is aimed solely at sub-letting of work which would otherwise be done by the contractor; and that it has no application where the sub-contractor is nominated.

(3) *Clause* 63 (1): " *the Employer may after giving 7 days' notice in writing to the Contractor* . . .*"*

The notice must comply precisely with the contract (see cl. 68). If it does not, the employer's entry upon the site is likely to constitute repudiation. It is not clear whether it is the notice or the employer's entry which is to operate as the act of forfeiture. Sub-clause (2) suggests the former. But the employer retains the option after giving notice, so that it must be the employer's act of entry and expulsion which binds the parties to the provisions of this clause. It is submitted, therefore, that the contractor must be in default from the date of the notice until the date of the employer's entry. If the contractor ceases to be in default before the employer's entry the right to forfeit is lost.

(4) *Clause* 63 (1): " *enter upon the Site and the Works and expel the Contractor therefrom* . . .*"*

The contract gives no guide as to what is meant by " enter " or " expel." Since the employer may be an occupier of the site,[78] entry must contemplate an act which goes beyond mere occupation. The employer must, it is suggested, take such action as will leave no doubt that the right of entry and expulsion has been exercised (which may be fortified by written notice to that effect), *e.g.* by locking up the site and turning away the contractor's employees.

(5) *Clause* 63 (1): *The right to use or sell the Contractor's plant and materials*

These provisions are subject to the enforceability of clauses 53 and 54. The power to appropriate the proceeds of sale in satisfaction of sums " which may become due " is a valuable addition to the right to recover excess cost from the contractor under sub-clause (4) since the latter cannot be recovered until the end of the maintenance period.

---

[76] See p. 111.
[77] *Thomas Feather & Co.* v. *Keighley Corp.* (1954) 52 L.G.R. 30.
[78] *A.M.F.* v. *Magnet Bowling* [1968] 1 W.L.R. 1028; *Harris* v. *Birkenhead Corp.* [1976] 1 W.L.R. 279, 287.

(6) *Clause* 63 (2): " *By the said notice . . . the Engineer may require the Contractor to assign . . .*"

This provision appears to ignore the fact that the " said notice " (presumably that giving seven days' notice before entry under sub-clause (1)) is to be given by the employer. The engineer's notice under sub-clause (1) is given to the employer. The option of giving a further notice within 14 days ignores the fact that the employer is not required to act immediately (or at all) upon expiry of the seven days' notice. He must act within seven days of its expiry if notice is to be given under sub-clause (2).

Note that the sub-clause does not apply to hiring of plant (see cl. 53). An assignment of the benefit only of a sub-contract gives the employer the right to have the goods delivered or work done without the obligation to pay, which remains with the contractor. Such an arrangement is likely to be unworkable, at least where the contractor is insolvent. A further matter which may defeat the employer's right to an assignment is where the sub-contract is subject to termination upon the contractor's insolvency or upon forfeiture of the main contract. Such difficulties may be overcome by the employer taking an assignment of the burden of the sub-contract (*i.e.* the obligation to pay), which requires the sub-contractor's assent (see *ante*, p. 167).

(7) *Clause* 63 (4): *Payment after forfeiture*

The final payment is unlikely to be in favour of the contractor, particularly where forfeiture was initiated by the contractor's default. Whatever the reason for the forfeiture there is likely to be a substantial claim for damages for delay in addition to the excess cost of completion. This sub-clause creates a serious obstacle to the employer, who cannot recover any sum until after completion. The effect may be mitigated by exercise of the right to sell the contractor's plant or unused materials under sub-clause (1). The employer is not entitled to proceed with an arbitration until completion of the works (cl. 66 (1)).

FRUSTRATION

Clause 64

Payment in event of frustration

In the event of the Contract being frustrated whether by war or by any other supervening event which may occur independently of the will of the parties the sum payable by the Employer to the Contractor in respect of the work executed shall be the same as that which would have been payable under Clause 65 (5) if the Contract had been determined by the Employer under Clause 65.

COMMENTARY

The term " frustrated " is used, it is thought, in the legal sense, meaning a supervening event which renders performance or further performance of the contract radically different to that contemplated.[79] The words " by any other supervening event which may occur independently of the will of the parties " confirm what would otherwise be the position at common law.

---

[79] See p. 102.

The provision entitling the contractor to payment of a sum calculated by reference to clause 65 (5) excludes the operation of section 1 of the Law Reform (Frustrated Contracts) Act 1943,[80] it is submitted. This section would otherwise require the contractor to prove that the employer had obtained a valuable benefit from the contract, in order to be able to recover even the value of the work done.[81]

WAR CLAUSE

Clause 65
Works to continue for 28 days on outbreak of war

(1) If during the currency of the Contract there shall be an outbreak of war (whether war is declared or not) in which Great Britain shall be engaged on a scale involving general mobilisation of the armed forces of the Crown the Contractor shall for a period of 28 days reckoned from midnight on the date that the order for general mobilisation is given continue so far as is physically possible to execute the Works in accordance with the Contract.

Effect of completion within 28 days

(2) If at any time before the expiration of the said period of 28 days the Works shall have been completed or completed so far as to be usable all provisions of the Contract shall continue to have full force and effect save that:—

(a) the Contractor shall in lieu of fulfilling his obligations under Clauses 49 and 50 be entitled at his option to allow against the sum due to him under the provisions hereof the cost (calculated at the prices ruling at the beginning of the said period of 28 days) as certified by the Engineer at the expiration of the Period of Maintenance of repair rectification and making good any work for the repair rectification or making good of which the Contractor would have been liable under the said Clauses had they continued to be applicable;

(b) the Employer shall not be entitled at the expiration of the Period of Maintenance to withhold payment under Clause 60 (5) (c) of the second half of the retention money or any part thereof except such sum as may be allowable by the Contractor under the provisions of the last preceding paragraph which sum may (without prejudice to any other mode of recovery thereof) be deducted by the Employer from such second half.

Right of employer to determine contract

(3) If the Works shall not have been completed as aforesaid the Employer shall be entitled to determine the Contract (with the exception of this Clause and Clauses 66 and 68) by giving notice in writing to the Contractor at any time after the aforesaid period of 28 days has expired and upon such notice being given the Contract shall (except as above mentioned) forthwith determine but without prejudice to the claims of either party in respect of any antecedent breach thereof.

Removal of plant on determination

(4) If the Contract shall be determined under the provisions of the last preceding sub-clause the Contractor shall with all reasonable despatch remove from the Site all his Constructional Plant and shall give facilities to his sub-contractors to remove similarly all Constructional Plant belonging to them and in the event of any failure so to do the Employer shall have the like powers as are contained

[80] By s. 2 (3).
[81] s. 1 (2), (3).

in Clause 53 (8) in regard to failure to remove Constructional Plant on completion of the Works but subject to the same condition as is contained in Clause 53 (9).

Payment on determination

(5) If the Contract shall be determined as aforesaid the Contractor shall be paid by the Employer (insofar as such amounts or items shall not have been already covered by payment on account made to the Contractor) for all work executed prior to the date of determination at the rates and prices provided in the Contract and in addition:—

    (a) the amounts payable in respect of any preliminary items so far as the work or service comprised therein has been carried out or performed and a proper proportion as certified by the Engineer of any such items the work or service comprised in which has been partially carried out or performed;

    (b) the cost of materials or goods reasonably ordered for the Works which shall have been delivered to the Contractor or of which the Contractor is legally liable to accept delivery (such materials or goods becoming the property of the Employer upon such payment being made by him);

    (c) a sum to be certified by the Engineer being the amount of any expenditure reasonably incurred by the Contractor in the expectation of completing the whole of the Works in so far as such expenditure shall not have been covered by the payments in this sub-clause before mentioned;

    (d) any additional sum payable under sub-clause (6) (b) (c) and (d) of this Clause;

    (e) the reasonable cost of removal under sub-clause (4) of this Clause.

Provisions to apply as from outbreak of war

(6) Whether the Contract shall be determined under the provisions of sub-clause (3) of this Clause or not the following provisions shall apply or be deemed to have applied as from the date of the said outbreak of war notwithstanding anything expressed in or implied by the other terms of the Contract *viz.*:—

    (a) The Contractor shall be under no liability whatsoever whether by way of indemnity or otherwise for or in respect of damage to the Works or to property (other than property of the Contractor or property hired by him for the purposes of executing the Works) whether of the Employer or of third parties or for or in respect of injury or loss of life to persons which is the consequence whether direct or indirect of war hostilities (whether war has been declared or not) invasion act of the Queen's enemies civil war rebellion revolution insurrection military or usurped power and the Employer shall indemnify the Contractor against all such liabilities and against all claims demands proceedings damages costs charges and expenses whatsoever arising thereout or in connection therewith.

    (b) If the Works shall sustain destruction or any damage by reason of any of the causes mentioned in the last preceding paragraph the Contractor shall nevertheless be entitled to payment for any part of the Works so destroyed or damaged and the Contractor shall be entitled to be paid by the Employer the cost of making good any such destruction or damage so far as may be required by the Engineer or as may be necessary for the completion of the Works on a cost basis plus such profit as the Engineer may certify to be reasonable.

    (c) In the event that the Contract includes the Contract Price Fluctuations Clause the terms of that Clause shall continue to apply but if subsequent to the outbreak of war the index figures therein referred to shall cease to be published or in the event that the contract shall not include a Price Fluctuations Clause in that form the following paragraph shall have effect:—

If under decision of the Civil Engineering Construction Conciliation Board or of any other body recognised as an appropriate body for regulating the rates of wages in any trade or industry other than the Civil Engineering Construction Industry to which Contractors undertaking works of civil engineering construction give effect by agreement or in practice or by reason of any Statute or Statutory Instrument there shall during the currency of the Contract be any increase or decrease in the wages or the rates of wages or in the allowances or rates of allowances (including allowances in respect of holidays) payable to or in respect of labour of any kind prevailing at the date of outbreak of war as then fixed by the said Board or such other body as aforesaid or by Statute or Statutory Instrument or any increase in the amount payable by the Contractor by virtue or in respect of any Scheme of State Insurance or if there shall be any increase or decrease in the cost prevailing at the date of the said outbreak of war of any materials consumable stores fuel or power (and whether for permanent or temporary works) which increase or increases decrease or decreases shall result in an increase or decrease of cost to the Contractor in carrying out the Works the net increase or decrease of cost shall form an addition or deduction as the case may be to or from the Contract Price and be paid to or allowed by the Contractor accordingly.

(d) If the cost of the Works to the Contractor shall be increased or decreased by reason of the provisions of any Statute or Statutory Instrument or other Government or Local Government Order or Regulation becoming applicable to the Works after the date of the said outbreak of war or by reason of any trade or industrial agreement entered into after such date to which the Civil Engineering Construction Conciliation Board or any other body as aforesaid is party or gives effect or by reason of any amendment of whatsoever nature of the Working Rule Agreement of the said Board or of any other body as aforesaid or by reason of any other circumstance or thing attributable to or consequent on such outbreak of war such increase or decrease of cost as certified by the Engineer shall be reimbursed by the Employer to the Contractor or allowed by the Contractor as the case may be.

(e) Damage or injury caused by the explosion whenever occurring of any mine bomb shell grenade or other projectile missile or munition of war and whether occurring before or after the cessation of hostilities shall be deemed to be the consequence of any of the events mentioned in sub-clause (6) (a) of this Clause.

COMMENTARY

This clause applies to a war involving general mobilisation. Its only practical importance is that upon the contract being frustrated under clause 64, the payment provisions of sub-clause (5) are to apply.

SETTLEMENT OF DISPUTES

Clause 66
Settlement of disputes—Arbitration

(1) If any dispute or difference of any kind whatsoever shall arise between the Employer and the Contractor in connection with or arising out of the Contract or the carrying out of the Works including any dispute as to any decision opinion instruction direction certificate or valuation of the Engineer (whether during the progress of the Works or after their completion and whether before or after the determination abandonment or breach of the Contract) it shall be referred to and settled by the Engineer who shall state his decision in writing and give notice of

the same to the Employer and the Contractor. Unless the Contract shall have been already determined or abandoned the Contractor shall in every case continue to proceed with the Works with all due diligence and he shall give effect forthwith to every such decision of the Engineer unless and until the same shall be revised by an arbitrator as hereinafter provided. Such decisions shall be final and binding upon the Contractor and the Employer unless either of them shall require that the matter be referred to arbitration as hereinafter provided. If the Engineer shall fail to give such decision for a period of 3 calendar months after being requested to do so or if either the Employer or the Contractor be dissatisfied with any such decision of the Engineer then and in any such case either the Employer or the Contractor may within 3 calendar months after receiving notice of such decision or within 3 calendar months after the expiration of the said period of 3 months (as the case may be) require that the matter shall be referred to the arbitration of a person to be agreed upon between the parties or (if the parties fail to appoint an arbitrator within one calendar month of either party serving on the other party a written notice to concur in the appointment of an arbitrator) a person to be appointed on the application of either party by the President for the time being of the Institution of Civil Engineers. If an arbitrator declines the appointment or after appointment is removed by order of a competent court or is incapable of acting or dies and the parties do not within one calendar month of the vacancy arising fill the vacancy then the President for the time being of the Institution of Civil Engineers may on the application of either party appoint an arbitrator to fill the vacancy. Any such reference to arbitration shall be deemed to be a submission to arbitration within the meaning of the Arbitration Act 1950 or the Arbitration (Scotland) Act 1894 as the case may be or any statutory re-enactment or amendment thereof for the time being in force. Any such reference to arbitration may be conducted in accordance with the Institution of Civil Engineers' Arbitration Procedure (1973) or any amendment or modification thereof being in force at the time of the appointment of the arbitrator and in cases where the President of the Institution of Civil Engineers is requested to appoint the arbitrator he may direct that the arbitration is conducted in accordance with the aforementioned Procedure or any amendment or modification thereof. Such arbitrator shall have full power to open up review and revise any decision opinion instruction direction certificate or valuation of the Engineer and neither party shall be limited in the proceedings before such arbitrator to the evidence or arguments put before the Engineer for the purpose of obtaining his decision above referred to. The award of the arbitrator shall be final and binding on the parties. Save as provided for in sub-clause (2) of this Clause no steps shall be taken in the reference to the arbitrator until after the completion or alleged completion of the Works unless with the written consent of the Employer and the Contractor. Provided always:—

  (a) that the giving of a Certificate of Completion under Clause 48 shall not be a condition precedent to the taking of any step in such reference;

  (b) that no decision given by the Engineer in accordance with the foregoing provisions shall disqualify him from being called as a witness and giving evidence before the arbitrator on any matter whatsoever relevant to the dispute or difference so referred to the arbitrator as aforesaid.

Interim arbitration

  (2) In the case of any dispute or difference as to any matter arising under Clause 12 or the withholding by the Engineer of any certificate or the withholding of any portion of the retention money under Clause 60 to which the Contractor claims to be entitled or as to the exercise of the Engineer's power to give a certificate under Clause 63 (1) the reference to the arbitrator may proceed notwithstanding that the Works shall not then be or be alleged to be complete.

Vice-President to act

(3) In any case where the President for the time being of the Institution of Civil Engineers is not able to exercise the functions conferred on him by this Clause the said functions may be exercised on his behalf by a Vice-President for the time being of the said Institution.

COMMENTARY

(1) *Generally*

This clause requires disputes to be settled by the engineer with a subsequent right to arbitrate. It is not clear in what capacity the engineer is required to act, nor to what extent his decision is a condition precedent to the right of arbitration or litigation.

It is clear that the engineer is not acting as an arbitrator under this clause. It has been suggested that he acts as certifier or quasi-arbitrator [82] but the latter term has been disapproved.[83] The engineer probably acts as no more than a certifier, having a duty to hold the scales fairly between the parties.[84]

No form is laid down for the engineer's decision under this clause; it should make quite clear that it is a decision under clause 66.[85] If it does not, the decision may be invalid; or if it is valid, any ambiguity may found an application to extend the time limit for submission to arbitration.[86]

The engineer's decision is a condition precedent to the right to arbitrate. But it is not clear if the condition applies when one party seeks to proceed in the courts instead of by arbitration. Where the only dispute is whether the employer can establish a counterclaim and set-off against a sum otherwise due, the contractor is not obliged to refer his claim to the engineer, it is submitted, since there is no dispute as to the claim. The contractor may, in such circumstances, proceed under Order 14 for summary judgment and the employer will be entitled to defend or stay the claim only to the extent there is a defence of set-off.[87]

But the position is not clear where the employer has a substantive defence, *e.g.* to a claim for extra payment, and the employer either cannot or does not stay the action. Can the employer rely on the absence of an engineer's decision as a complete defence? It is thought he cannot. To allow such reliance would be to give the engineer's reference greater effect than the arbitration provision itself. If a dispute is proceeding in the courts words such as " shall be referred to arbitration " are of no effect.[88] The engineer's reference does not have the effect of a *Scott* v. *Avery* clause [89] since the parties are in no way bound by the decision in subsequent proceedings. But as a matter of caution, disputes should be referred to the engineer when possible.

---

[82] H.B.C. (10th ed.), p. 627.
[83] *Sutcliffe* v. *Thackrah* [1974] A.C. 727, *per* Lord Reid at p. 737.
[84] *Ibid.* at p. 737. For the duties of the engineer *vis-à-vis* the employer see *ante*, p. 199.
[85] *Monmouth C.C.* v. *Costelloe & Kemple Ltd.* (1965) 63 L.G.R. 429.
[86] Arbitration Act 1950, s. 27; see note (3), *infra*.
[87] *Dawnays* v. *Minter* [1971] 1 W.L.R. 1205; *Brightside Kilpatrick* v. *Mitchell Construction* (1975) 2 Lloyd's Rep. 493; see also *Kilby & Gayford* v. *Selincourt* (1973) 229 E.G. 1343, C.A.
[88] See cl. 35 (1) Standard form of building contract.
[89] See *ante*, p. 241.

(2) *Clause 66 (1): Scope of the arbitration clause*

The clause covers " any dispute or difference . . . in connection with or arising out of the Contract or the carrying out of the works . . ." This includes a dispute as to whether the contract has been frustrated (see clause 64) or repudiated [90] but not a dispute as to whether the contract is void for illegality.[91]

The arbitrator has no jurisdiction to rectify the contract,[92] save under the limited scope of clause 5. The words " or the carrying out of the works " envisage disputes arising otherwise than out of the contract. These may include claims in tort [93]; but it is difficult to imagine such a claim which is not at least connected with the contract.

(3) *Clause 66 (1): " the decision of the Engineer . . . shall be final and binding upon the Contractor and the Employer unless either of them shall require that the matter be referred to arbitration . . . within three calendar months after receiving notice of such decision . . ."*

These words do not appear as one sentence; they are of vital importance to the parties. Where a decision is deemed to be final and binding on the parties it cannot be re-opened by the court.[94] There are no exceptions provided in the contract. Two exceptions which may arise are: first, if the decision contains a patent error or embodies a point of law it may be challenged in the courts [95]; secondly, the time for reference to arbitration may be extended under section 27 of the Arbitration Act 1950 if " undue hardship would otherwise be caused." But undue hardship means more than the loss of a potential claim, and the court will not lightly extend time.[96]

(4) *Clause 66 (1): The Institution of Civil Engineers Arbitration Procedure (1973)*

The procedure governs the steps leading up to the hearing, including the appointment of an arbitrator, preliminary meetings and orders for directions. The conduct of the hearing is covered only by " Notes for Guidance " which do not form part of the procedure. The powers of the arbitrator are in no way limited by the procedure.

(5) *Clause 66 (1): " Such arbitrator shall have full power to open up review and revise any decision opinion instruction direction certificate or valuation "*

These words emphasise that the arbitrator is in no way bound either by interim certificates, the final certificate or the maintenance certificate. An omission from

---

[90] *Heyman* v. *Darwins* [1942] A.C. 356 at p. 366; as to powers of the arbitrator generally, see *ante*, p. 236.

[91] *Ibid.*

[92] *Printing Machinery Co.* v. *Linotype* [1912] 1 Ch. 566; but see *ante*, p. 237.

[93] See *Esso Petroleum* v. *Mardon* [1976] Q.B. 801.

[94] *Kaye* (P. & M.) *Ltd.* v. *Hosier & Dickinson* [1972] 1 W.L.R. 146; a decision on the Standard form of building contract.

[95] *Collier* v. *Mason* (1858) 25 Beau. 200; *Dean* v. *Prince* [1954] Ch. 409; but see *Campbell* v. *Edwards* [1976] 1 W.L.R. 403 at p. 407; *Toepfer* v. *Continental Grain Co.* [1974] 1 Lloyd's Rep. 11 and *Kaye* (P. & M.) *Ltd.* v. *Hosier & Dickinson* [1972] 1 W.L.R. 146 at p. 157. See *ante*, p. 84.

[96] See *Liberian Shipping Corp.* " *Pegasus* " v. *A King & Sons Ltd.* [1967] 2 Q.B. 86 , and note (1), *supra*; see also *ante*, p. 241.

the arbitrator's express powers is that of directing measurements. The intention of clause 56 (2) is that measurements shall either be agreed or deemed to be correct. But where there is a dispute it is thought that the power to review valuations would allow the arbitrator to open up measurements, subject to clause 56 (3) (see note (4) to cl. 56).

(6) *Clause 66 (1):* "*no steps shall be taken in the reference . . . until after the completion or alleged completion . . .*"

A step, it is thought, includes any application to the arbitrator, *e.g.* for directions.[97] Paragraph (a) to the proviso does not add anything to the right to proceed upon " alleged completion " save to emphasise that completion of the works may be an issue in the arbitration.

(7) *Clause 66 (1), Proviso (b): The Engineer as a witness*

The engineer would be a competent (and often vital) witness, apart from this proviso, as to matters concerning the carrying out of the works. What is not clear is whether the engineer may be asked (by either party to the reference) questions as to his disputed decision under this clause.[98]

The reference to matters " relevant to the dispute or difference so referred to the arbitrator " suggests that the engineer's evidence should be confined to matters arising prior to the reference under clause 66. It is thought that evidence as to details of the engineer's decision under sub-clause (1) would be inadmissible as irrelevant to the dispute.[99]

(8) *Clause 66 (2): Interim arbitration*

The words " the withholding by the Engineer of any certificate " are not limited to certificates under clause 60. The contractor is entitled to immediate arbitration whenever there is a genuine dispute as to a certificate which, on one view of the question, ought to have been given.[1]

APPLICATION TO SCOTLAND
Clause 67
Application to Scotland

If the Works are situated in Scotland the Contract shall in all respects be construed and operate as a Scottish contract and shall be interpreted in accordance with Scots law.

COMMENTARY

(1) *Generally*

The contract is intended for use under Scots law without amendment. It

---

[97] *Cf.* s. 4 (1) of the Arbitration Act 1950.

[98] *Cf.* discussion on the position of the engineer as arbitrator, *ante*, p. 238.

[99] As a matter of policy, the admission in evidence of particulars of decisions under cl. 66 would tend to inhibit the engineer from doing other than affirming his previous decision under the contract. See also 2 Halsbury's Laws (4th ed.) para. 616, as to an arbitrator giving evidence of an award.

[1] *A. E. Farr* v. *Minister of Transport* [1960] 1 W.L.R. 956, a decision on cl. 66 of the 4th ed. of the Conditions, which does not differ materially from the present clause.

should be noted that clauses which depend for their effect upon the operation of law (as opposed to matters of construction) may not have the same effect under Scots law, *e.g.* clause 54 (vesting of goods not on site) is drafted to operate under English law.

(2) *Arbitration*

When Scots law is the proper law of the contract it does not follow that an arbitration must be conducted in accordance with Scots law.[2] References in clause 66 to the Arbitration (Scotland) Act 1894, and in the ICE Arbitration Procedure 1973 to a Scottish Arbitration, are not conclusive. The parties may choose or submit to a particular arbitration law, but prima facie the conduct and procedure of the arbitration will be governed by the law of the country in which the arbitration is held.[3]

NOTICES

Clause 68
Service of notice on contractor

(1) Any notice to be given to the Contractor under the terms of the Contract shall be served by sending the same by post to or leaving the same at the Contractor's principal place of business (or in the event of the Contractor being a Company to or at its registered office).

Service of notice on employer

(2) Any notice to be given to the Employer under the terms of the Contract shall be served by sending the same by post to or leaving the same at the Employer's last known address (or in the event of the Employer being a Company to or at its registered office).

COMMENTARY

Service of notices in accordance with the contract may be a matter of vital importance, *e.g.* where the employer is entitled to operate the forfeiture clause, but the grounds do not amount to common law repudiation by the contractor. If the employer acts without correctly served notices, he will himself have repudiated the contract.

This clause is ambiguous in its application when the employer or contractor is a company. It is not clear whether the words in brackets in each sub-clause give a choice or whether good service may be effected only at the registered office. In many cases the previous dealings between the parties will operate as a waiver of the right to object to the place of service. But in case of doubt, notice should be given to the registered office.

An important effect of the clause is that it appears to require any notice, to the contractor or to the employer, to be given in writing, otherwise it cannot be said to be " served " or " sent " or " left."

The clause does not cover notices to be given to the engineer. Clause 52 (4) lays down general requirements as to notices of claims; and there are many

---

[2] *James Miller & Partners* v. *Whitworth Street Estates* [1970] A.C. 583, H.L.
[3] *Ibid.*

other clauses which contain their own requirements, *e.g.* clause 7 (2): " adequate notice in writing " of further drawings, etc.; and clause 38 (1): " due notice " of work ready for examination, etc.

TAX MATTERS

Clause 69
Tax fluctuations

(1) The rates and prices contained in the Bill of Quantities take account of the levels and incidence at the date for return of tenders (hereinafter called " the relevant date ") of the taxes levies and contributions (including national insurance contributions but excluding income tax and any levy payable under the Industrial Training Act 1964) which are by law payable by the Contractor in respect of his workpeople and the premiums and refunds (if any) which are by law payable to the Contractor in respect of his workpeople. Any such matter is hereinafter called " a labour-tax matter."

The rates and prices contained in the Bill of Quantities do not take account of any level or incidence of the aforesaid matters where at the relevant date such level or incidence does not then have effect but although then known is to take effect at some later date. The taking effect of any such level or incidence at the later date shall for the purposes of sub-clause (2) of this Clause be treated as the occurrence of an event.

(2) If after the relevant date there shall occur any of the events specified in sub-clause (3) of this Clause and as a consequence thereof the cost to the Contractor of performing his obligations under the Contract shall be increased or decreased then subject to the provisions of sub-clause (4) of this Clause the net amount of such increase or decrease shall constitute an addition to or deduction from the sums otherwise payable to the Contractor under the Contract as the case may require.

(3) The events referred to in the preceding sub-clause are as follows:—
    (a) any change in the level of any labour-tax matter;
    (b) any change in the incidence of any labour-tax matter including the imposition of any new such matter or the abolition of any previously existing such matter.

(4) In this Clause workpeople means persons employed by the Contractor on manual labour whether skilled or unskilled but for the purpose of ascertaining what if any additions or deductions are to be paid or allowed under this Clause account shall not be taken of any labour-tax matter in relation to any workpeople of the Contractor unless at the relevant time their normal place of employment is the Site.

(5) Subject to the provisions of the Contract as to the placing of sub-contracts with Nominated Sub-contractors the Contractor may incorporate in any sub-contract made for the purpose of performing his obligations under the Contract provisions which are *mutatis mutandis* the same as the provisions of this Clause and in such event additions or deductions to be made in accordance with any such sub-contract shall also be made under the Contract as if the increase or decrease of cost to the sub-contractor had been directly incurred by the Contractor.

(6) As soon as practicable after the occurrence of any of the events specified in sub-clause (3) of this Clause the Contractor shall give the Engineer notice thereof. The Contractor shall keep such contemporary records as are necessary for the purpose of ascertaining the amount of any addition or deduction to be made in accordance with this Clause and shall permit the Engineer to inspect such records. The Contractor shall submit to the Engineer with his monthly statements full details of every addition or deduction to be made in accordance with

this Clause. All certificates for payment issued after submission of such details shall take due account of the additions or deductions to which such details relate. Provided that the Engineer may if the Contractor fails to submit full details of any deduction nevertheless take account of such deduction when issuing any certificate for payment.

COMMENTARY

(1) *Generally*

This clause provides for the adjustment of payments to the contractor in respect of fluctuations in any " labour-tax matter." The contractor is entitled to the net increase (or the employer to the net decrease) in the " cost to the Contractor of performing his obligation under the Contract " (sub-cl. (2)) in respect of manual labour whose normal place of employment is the site (sub-cl. (4)).

(2) *Clause* 69 (2): *" Cost to the Contractor "*

" Cost " is defined by clause 1 (5) to include overhead costs " except where the contrary is expressly stated." Sub-clauses (1) and (4) operate as an express statement to the contrary, it is submitted.

(3) *Clause* 69 (6): *Failure to give notice of deductions*

The engineer is empowered to make an informed guess if the contractor fails to submit full details of deductions. The contractor clearly has no ground of objection to the deduction of estimated sums since the necessity arises from his own breach.

Clause 70
Value Added Tax

(1) In this Clause " exempt supply " " invoice " " tax " " taxable person " and " taxable supply " have the same meanings as in Part I of the Finance Act 1972 (hereinafter referred to as " the Act ") including any amendment or re-enactment thereof and any reference to the Value Added Tax (General) Regulations 1972 (S.I. 1972/1147) (hereinafter referred to as the V.A.T. Regulations) shall be treated as a reference to any enactment corresponding to those regulations for the time being in force in consequence of any amendment or re-enactment of those regulations.

(2) The Contractor shall be deemed not to have allowed in his tender for the tax payable by him as a taxable person to the Commissioners of Customs and Excise being tax chargeable on any taxable supplies to the Employer which are to be made under the Contract.

(3) (a) The Contractor shall not in any statement submitted under Clause 60 include any element on account of tax in any item or claim contained in or submitted with the statement.

     (b) The Contractor shall concurrently with the submission of the statement referred to in sub-clause (3) (a) of this Clause furnish the Employer with a written estimate showing those supplies of goods and services and the values thereof included in the said statement and on which tax will be chargeable under Regulation 21 of the V.A.T. Regulations at a rate other than zero.

(4) At the same time as payment (other than payment in accordance with this sub-clause) for goods or services which were the subject of a taxable supply

provided by the Contractor as a taxable person to the Employer is made in accordance with the Contract there shall also be paid by the Employer a sum (separately identified by the Employer and in this Clause referred to as " the tax payment ") equal to the amount of tax payable by the Contractor on that supply. Within seven days of each payment the Contractor shall:—

(a) if he agrees with that tax payment or any part thereof issue to the Employer an authenticated receipt of the kind referred to in Regulation 21 (2) of the V.A.T. Regulations in respect of that payment or that part; and

(b) if he disagrees with that tax payment or any part thereof notify the Employer in writing stating the grounds of his disagreement.

(5) (a) If any dispute difference or question arises between the Employer and the Contractor in relation to any of the matters specified in Section 40 (1) of the Act then:—

(i) if the Employer so requires the Contractor shall refer the matter to the said Commissioners for their decision on it

(ii) if the Contractor refers the matter to the said Commissioners (whether or not in pursuance of sub-paragraph (i) above) and the Employer is dissatisfied with their decision on the matter the Contractor shall at the Employer's request refer the matter to a Value Added Tax Tribunal by way of appeal under Section 40 of the Act whether the Contractor is so dissatisfied or not

(iii) a sum of money equal to the amount of tax which the Contractor in making a deposit with the said Commissioners under Section 40 (3) (a) of the Acts is required so to deposit shall be paid to the Contractor; and

(iv) if the Employer requires the Contractor to refer such a matter to the Tribunal in accordance with sub-paragraph (ii) above then he shall reimburse the Contractor any costs and any expenses reasonably and properly incurred in making that reference less any costs awarded to the Contractor by the Tribunal and the decision of the Tribunal shall be binding on the Employer to the same extent as it binds the Contractor.

(b) Clause 66 shall not apply to any dispute difference or question arising under paragraph (a) of this sub-clause.

(6) (a) The Employer shall without prejudice to his rights under any other Clause hereof be entitled to recover from the Contractor:—

(i) any tax payment made to the Contractor of a sum which is in excess of the sum (if any) which in all the circumstances was due in accordance with sub-clause (4) of this Clause

(ii) in respect of any sum of money deposited by the Contractor pursuant to sub-clause (5) (a) (iii) of this Clause a sum equal to the amount repaid under Section 40 (4) of the Act together with any interest thereon which may have been determined thereunder.

(b) If the Contractor shall establish that the Commissioners have charged him in respect of a taxable supply for which he has received payment under this Clause tax greater in amount than the sum paid to him by the Employer the Employer shall subject to the provisions of sub-clause (5) of this Clause pay to the Contractor a sum equal to the difference between the tax previously paid and the tax charged to the Contractor by the Commissioners.

(7) If after the date for return of tenders the descriptions of any supplies of goods or services which at the date of tender are taxable or exempt supplies are with effect after the date for return of tenders modified or extended by or under the Act and that modification or extension shall result in the Contractor having to pay either more or less tax or greater or smaller amounts attributable to tax

and that tax or those amounts as the case may be shall be a direct expense or direct saving to the Contractor in carrying out the Works and not recoverable or allowable under the Contract or otherwise then there shall be paid to or allowed by the Contractor as appropriate a sum equivalent to that tax or amounts as the case may be.

Provided always that before that tax is included in any payment by the Employer or those amounts are included in any certificate by the Engineer as the case may be the Contractor shall supply all the information the Engineer requires to satisfy himself as to the Contractor's entitlement under this sub-clause.

(8) The Contractor shall upon demand pay to the Employer the amount of any sum due in accordance with sub-clauses (6) and (7) of this Clause and it shall be deemed a debt due by the Contractor to the Employer and shall be recoverable accordingly.

COMMENTARY

(1) *Generally*

The effect of these provisions is that sums payable under the contract are to be net of VAT. Concurrently with the submission to the engineer of any monthly or final statement under clause 60, the contractor is to submit direct to the employer a separate estimate of taxable goods and services. The employer must remit the amount of tax payable with payments under clause 60.[4] The engineer is given no duties as regards VAT on taxable supplies to be made under the contract.

(2) *Clause* 70 (7): " *direct expense . . . not recoverable . . . under the contract* "

This is equivalent to the " additional cost " claims which occur throughout the contract. It applies to variations in VAT elements not directly recoverable under the contract. The expense must be " direct " and must arise from " carrying out the works." The sub-clause is equally likely to result in claims by the employer or by the contractor, so that it should not be construed *contra proferentem*. It is thought the provision will cover site preliminaries and overheads, but not head office overheads.

METRICATION

Clause 71

Metrication

(1) If any materials described in the Contract or ordered by the Engineer are described by dimensions in the metric or imperial measure and having used his best endeavours the Contractor cannot without undue delay or additional expense or at all procure such materials in the measure specified in the Contract but can obtain such materials in the other measure to dimensions approximating to those described in the Contract or ordered by the Engineer then the Contractor shall forthwith give written notice to the Engineer of these facts stating the dimensions to which such materials are procurable in the other measure. Such notice shall where practicable be given in sufficient time to enable the Engineer to consider and if necessary give effect to any design change which may be required and to avoid delay in the performance of the Contractor's other obligations under the Contract. Any additional cost or expense incurred by the Contractor as a result of any delay arising out of the Contractor's default under this sub-clause shall be borne by the Contractor.

---

[4] For a very short note on VAT generally, see *ante*, p. 419.

(2) As soon as practicable after the receipt of any such notice under the preceding sub-clause the Engineer shall if he is satisfied that the Contractor has used his best endeavours to obtain materials to the dimensions described in the Contract or ordered by the Engineer and that they are not obtainable without undue delay or without putting the Contractor to additional expense either:—

(a) instruct the Contractor pursuant to Clause 13 to supply such materials (despite such delay or expense) in the dimensions described in the Contract or originally ordered by the Engineer; or

(b) give an order to the Contractor pursuant to Clause 51:—

    (i) to supply such materials to the dimensions stated in his said notice to be procurable instead of to the dimensions described in the Contract or originally ordered by the Engineer; or

    (ii) to make some other variation whereby the need to supply such materials to the dimensions described in the Contract or originally ordered by the Engineer will be avoided.

(3) This Clause shall apply irrespective of whether the materials in question are to be supplied in accordance with the Contract directly by the Contractor or indirectly by a Nominated Sub-contractor.

COMMENTARY

(1) *Generally*

The principal effect of this clause is to require the contractor to give notice of the relevant facts. Upon receipt of notice (provided the contractor has used his best endeavours) the engineer is then bound to give an instruction under clause 13 or 51 (variations). An instruction under clause 13 allows the contractor to claim additional cost and an extension of time under clause 13 (3). This requires proof that the instruction (under cl. 13 (1)) has involved the contractor in delay, etc. It is clearly arguable that it is the unavailability of the specified materials which have involved the contractor in delay, etc., and not the instruction. But it is thought that the intention of clause 71 is sufficiently clear to allow the contractor's claim for additional cost and an extension of time: it is implicit in the final sentence of sub-clause (1) that the contractor, not being in default, shall not bear the additional cost of delay.

(2) *Clause 71 (1): ". . . or at all procure such materials"*

The possibility of the specified materials not being available at all is not covered by sub-clause (2). In this event the contractor is absolved from the obligation to supply the materials in question by virtue of clause 13 (1), it is submitted. The engineer may then give a variation order for alternative materials, and/or treat the original materials as omitted from the works.

SPECIAL CONDITIONS

Clause 72

Special conditions

The following special conditions form part of the Conditions of Contract.

(Note: Any special conditions which it is desired to incorporate in the conditions of contract should be numbered consecutively with the foregoing conditions of contract.)

COMMENTARY

It is not essential to write special conditions into the Conditions. They will be equally effective if written into the specification or bills,[5] or otherwise expressly incorporated into the contract.

CONTRACT PRICE FLUCTUATIONS [6]

(1) The amount payable by the Employer to the Contractor upon the issue by the Engineer of an interim certificate pursuant to Clause 60 (2) or of the final certificate pursuant to Clause 60 (3) (other than amounts due under this Clause) shall be increased or decreased in accordance with the provisions of this Clause if there shall be any changes in the following Index Figures compiled by the Department of the Environment and published by Her Majesty's Stationery Office (HMSO) in the Monthly Bulletin of Construction Indices (Civil Engineering Works):—

  (a) the Index of the Cost of Labour in Civil Engineering Construction;

  (b) the Index of the Cost of Providing and Maintaining Constructional Plant and Equipment;

  (c) the Indices of Constructional Material Prices applicable to those materials listed in sub-clause (4) of this Clause.

The net total of such increases and decreases shall be given effect to in determining the Contract Price.

(2) For the purpose of this Clause:—

  (a) " Final Index Figure " shall mean any Index Figure appropriate to sub-clause (1) of this Clause not qualified in the said Bulletin as provisional;

  (b) " Base Index Figure " shall mean the appropriate Final Index Figure applicable to the date 42 days prior to the date for the return of tenders;

  (c) " Current Index Figure " shall mean the appropriate Final Index Figure to be applied in respect of any certificate issued or due to be issued by the Engineer pursuant to Clause 60 and shall be the appropriate Final Index Figure applicable to the date 42 days prior to:—

      (i) the due date (or extended date) for completion; or

     (ii) the date certified pursuant to Clause 48 of completion of the whole of the Works; or

    (iii) the last day of the period to which the certificate relates;

whichever is the earliest.

      Provided that in respect of any work the value of which is included in any such certificate and which work forms part of a Section for which the due date (or extended date) for completion or the date certified pursuant to Clause 48 of completion of such Section precedes the last day of the period to which the certificate relates the Current Index Figure shall be the Final Index Figure applicable to the date 42 days prior to whichever of these dates is the earliest.

  (d) The " Effective Value " in respect of the whole or any Section of the Works shall be the difference between:—

      (i) the amount which in the opinion of the Engineer is due to the Contractor under Clause 60 (2) (before deducting retention) or the amount due to the Contractor under Clause 60 (3) (but in each case before deducting sums previously paid on account) less any amounts for Dayworks Nominated Sub-contractors or any other items based on actual cost or current prices and any sums for increases or decreases in the Contract Price under this Clause;

and:—

     (ii) the amount calculated in accordance with (i) above and included

---

[5] See cll. 1 (1) (e) and 5.

[6] Revised March 1976.

in the last preceding interim certificate issued by the Engineer in accordance with Clause 60.

Provided that in the case of the first certificate the Effective Value shall be the amount calculated in accordance with sub-paragraph (i) above.

(3) The increase or decrease in the amounts otherwise payable under Clause 60 pursuant to sub-clause (1) of this Clause shall be calculated by multiplying the Effective Value by a Price Fluctuation Factor which shall be the net sum of the products obtained by multiplying each of the proportions given in (a) (b) and (c) of sub-clause (4) of this Clause by a fraction the numerator of which is the relevant Current Index Figure minus the relevant Base Index Figure and the denominator of which is the relevant Base Index Figure.

(4) For the purpose of calculating the Price Fluctuation Factor the proportions referred to in sub-clause (3) of this Clause shall (irrespective of the actual constituents of the work) be as follows and the total of such proportions shall amount to unity:—

(a)　o.　* in respect of labour and supervision costs subject to adjustment by reference to the Index referred to in sub-clause (1) (a) of this Clause;

(b)　o.　* in respect of costs of provision and use of all civil engineering plant road vehicles etc. which shall be subject to adjustment by reference to the Index referred to in sub-clause (1) (b) of this Clause:—

(c)　the following proportions in respect of the materials named which shall be subject to adjustment by reference to the relevant indices referred to in sub-clause (1) (c) of this Clause:—

　　o.　* in respect of Aggregates

　　o.　* in respect of Bricks and Clay Products generally

　　o.　* in respect of Cements

　　o.　* in respect of Cast Iron products

　　o.　* in respect of Coated Roadstone for road pavements and bituminous products generally

　　o.　* in respect of Fuel for plant to which the Gas Oil Index will be applied

　　o.　* in respect of Timber generally

　　o.　* in respect of Reinforcing steel and other metal sections

　　o.　* in respect of Fabricated Structural Steel;

(d)　o.10　in respect of all other costs which shall not be subject to any adjustment;

Total 1.00 .

(5) Provisional Index Figures in the Bulletin referred to in sub-clause (1) of this Clause may be used for the provisional adjustment of interim valuations but such adjustments shall be subsequently recalculated on the basis of the corresponding Final Index Figures.

(6) Clause 69—Tax Fluctuations—shall not apply except to the extent that any matter dealt with therein is not covered by the Index of the Cost of Labour in Civil Engineering Construction.

CONTRACT PRICE FLUCTUATIONS
FABRICATED STRUCTURAL STEELWORK [6]

(1) The amount payable by the Employer to the Contractor upon the issue by

---

* *To be filled in by the Employer prior to inviting tenders.*

the Engineer of an interim certificate pursuant to Clause 60 (2) or of the final certificate pursuant to Clause 60 (3) (other than amounts due under this Clause) shall be increased or decreased in accordance with the provisions of this Clause if there shall be any changes in the following Index Figures compiled by the Department of the Environment and published by Her Majesty's Stationery Office (HMSO) in the Monthly Bulletin of Construction Indices (Civil Engineering Works):—

(a) the Index of the Cost of labour in fabrication of steelwork and erection of steelwork;

(b) the Index for Structural Steel.

The net total of such increases and decreases shall be given effect to in determining the Contract Price.

(2) For the purpose of this Clause

(a) " Fabricated Structural Steelwork " shall mean those items of work listed in sub-clause (6) of this Clause and shall include any variations as may be ordered under Clause 51 involving work of a description which in the opinion of the Engineer is similar to the description of the items so listed;

(b) " Final Index Figure " shall mean any Index Figure appropriate to sub-clause (1) of this Clause not qualified in the said Bulletin as provisional;

(c) " Base Index Figure " shall mean the appropriate Final Index Figure applicable to the date 42 days prior to the date for the return of tenders;

(d) " Current Index Figure " shall mean the appropriate Final Index Figure to be applied in respect of any certificate issued or due to be issued by the Engineer pursuant to Clause 60 being such figure applicable

    (i) in respect of labour employed in fabrication—to a date 56 days prior to the last day of the period to which the certificate relates; or

    (ii) in respect of labour employed in erection—to a date 14 days prior to the last day of the period to which the certificate relates; or

    (iii) in respect of materials specifically purchased for inclusion in the Works—to the date of delivery to the fabricator's premises (of which date the Contractor shall produce such evidence relating to gross tonnages delivered as the Engineer may reasonably require); or

    (iv) in respect of materials (if any) not specifically purchased for inclusion in the Works—to the date of the last of the deliveries referred to in sub-paragraph (iii) of this paragraph

as the case may be.

Provided always that should the due date (or extended date) for completion or the date certified pursuant to Clause 48 of completion of the whole of the Works precede any of the aforesaid then such due date or extended date or certified date whichever is earliest shall be substituted for those aforesaid

Provided further that if in respect of any work which forms part of a Section and whose value is included in any certificate the due date (or extended date) for completion of that Section or the date certified pursuant to Clause 48 of completion of that Section precedes any of the dates aforesaid in sub-paragraphs (i) to (iv) above then such due date or extended date or certified date whichever is the earliest in respect of that Section shall be substituted for those aforesaid.

(e) The " Effective Value " in respect of the whole or any Section of the works shall be the difference between

(i) the amount which in the opinion of the Engineer is due to the Contractor under Clause 60 (2) (before deducting retention) or the amount due to the Contractor under Clause 60 (3) (but in each case before deducting sums previously paid on account) less any amounts for Dayworks Nominated Sub-contractors or any other items based on actual cost or current prices and any sums for increases or decreases in the Contract Price under this Clause;
and—

(ii) the amount calculated in accordance with (i) above and included in the last preceding interim certificate issued by the Engineer in accordance with Clause 60.

Provided that in the case of the first certificate the Effective Value shall be the amount calculated in accordance with sub-paragraph (i) above.

(3) The Effective Value shall be apportioned between labour and materials in the following manner that is to say

(a) labour

employed in erection—by multiplying the total tonnage erected during the period to which the certificate relates by the average cost per tonne entered at (b) of sub-clause (5) of this Clause

employed in fabrication and delivery—by deducting the summation of the values calculated in respect of materials and in respect of labour employed in erection from the Effective Value

(b) materials

by multiplying the total tonnage of steel delivered to the Site for inclusion in the Works during the period to which the certificate relates by the average price per tonne entered at (a) of sub-clause (5) of this Clause.

(4) (a) The increase or decrease in the amounts otherwise payable under Clause 60 pursuant to sub-clause (1) of this Clause shall be calculated by multiplying each portion of the Effective Value by a fraction the numerator of which is the product of 0.90 and the difference between the relevant Current Index Figure and the relevant Base Index Figure and the denominator of which is the relevant Base Index Figure.

(b) The relevant indices to be used in connection with this sub-clause are

(i) for labour and supervision in fabrication and erection—the Index referred to in sub-clause (1) (a) of this Clause

(ii) for materials—the Index referred to in sub-clause (1) (b) of this Clause.

(5) For the purpose of the apportionment in sub-clause (3) of this Clause the full average costs per tonne (inclusive of all associated labour plant power maintenance overheads and profit) to be used are

(a) Materials delivered to fabricator's premises.........£   per tonne†
(b) Erection   ......£   per tonne†
(c) Subject to sub-paragraphs (i) (ii) and (iii) of this paragraph the relevant figures to be used in connection with this sub-clause shall be in accordance with sub-clause (2) of this Clause

Provided that in respect of materials

(i) the Current Index Figure subsequent to the first established Current Index Figure pursuant to sub-clause (2) (d) (iii) of this Clause shall not be used until a tonnage of steel greater than the tonnage to which the Current Index Figure first established applies has been delivered to the Site for inclusion in the Works and

(ii) for the purpose of establishing the appropriate subsequent Current Index Figures to apply to all later deliveries of steel to the Site for

---

† To be filled in by the Contractor at time of tendering.

inclusion in the Works the provisions of sub-paragraph (i) of this paragraph shall apply *mutatis mutandis* and

(iii) the Current Index Figure referred to in sub-clause (2) (d) (iv) of this Clause shall not be used until the total tonnage of steel delivered to the Site for inclusion in the Works exceeds the total tonnage of steel specifically purchased for inclusion in the Works and delivered to the fabricator's premises.

(6) For the purposes of this Clause the expression Fabricated Structural Steelwork shall comprise those items listed hereunder

Bill No. *	Page No. *	Item No. *

(7) Provisional Index Figures in the Bulletin referred to in sub-clause (1) of this Clause may be used for the provisional adjustment of interim valuations but such adjustments shall be subsequently recalculated on the basis of the corresponding Final Index Figures.

(8) Clause 69—Tax Fluctuations—shall not apply except to the extent that any matter dealt with therein is not covered by the Index of the Cost of labour in fabrication of steelwork and erection of steelwork.

CONTRACT PRICE FLUCTUATIONS
CIVIL ENGINEERING WORK AND FABRICATED
STRUCTURAL STEELWORK [6]

(1) This Clause shall apply only to those contracts which incorporate the Contract Price Fluctuation Clause Revised March 1976 for Civil Engineering work and the Fabricated Structural Steelwork Clause Revised March 1976 (referred to in this Clause as the CE Clause and the FSS Clause respectively).

(2) For the purposes of this Clause

(a) " Civil Engineering work " shall mean all Works with the exception of Fabricated Structural Steelwork;

(b) " Fabricated Structural Steelwork " shall mean the work defined in sub-clause (2) (a) of the FSS Clause.

(3) The Effective Value (as defined in both the CE Clause and the FSS Clause) shall be sub-divided to show the amounts included in respect of the Civil Engineering work and the Fabricated Structural Steelwork. The amount in respect of the former shall then be treated as if it were the Effective Value as defined in the CE Clause and adjusted in accordance with the provisions of that Clause. The amount in respect of the latter shall then be treated as if it were the Effective Value as defined in the FSS Clause and adjusted in accordance with that Clause.

COMMENTARY

(1) *Generally*

These three clauses, unnumbered, are published separately from the Conditions of contract and forms. Their use is optional. They are referred to, respectively, as the CE clause, the FSS clause and the CE/FSS clause. The CE clause is appropriate for use alone where the works consist of civil engineering work

---

* To be filled in by the Employer prior to inviting tenders.

which is not predominantly fabricated structural steelwork. The FSS clause is appropriate for use alone where the work consists solely or predominantly of fabricated structural steelwork. Where it is desired to make special provision for price fluctuation in respect of both civil engineering work and fabricated structural steelwork, both clauses should be used together with the CE/FSS clause.

(2) *Outline of clauses*

The price fluctuation clauses operate as simple algebraic formulae, whose resolution depends upon current cost and price index figures published by H.M.S.O. and upon the net value of work carried out by the contractor. Irrespective of the actual constituents of that work in labour, plant costs and materials, the increase or decrease payable depends upon a predetermined apportionment. In the CE clause that apportionment is itself subject to adjustment by reference to the relevant published indices.

In the CE clause fluctuations are calculated on index changes from a base date 42 days prior to the date for return of tenders, to 42 days prior to the end of the period to which the certificate relates (sub-cl. (2)). In the FSS clause the relevant date for the current index figures depends upon whether the labour is engaged in fabrication or erection and upon whether materials are purchased specifically for the works or not (sub-cl. (2)). In each case fluctuations are to be allowed only up to the due date, or extended date for completion, so that if the contractor over-runs, no further increases in price will be allowable.

(3) *CE and FSS clauses, sub-clause* (1): " *The amount payable shall be increased or decreased in accordance with the provisions of this clause . . .*"

These words appear in both the CE and FSS clauses. They give the contractor a right to payment of fluctuations in addition to the sums to be certified under clause 60 (2) (interim) or 60 (3) (final certificate). The right to interim payments of fluctuations arises by virtue of these words and not by virtue of clause 60. Consequently fluctuation payments are not subject to retention, it is submitted (see cl. 60 (1) (d) and (2) (a)).

(4) *CE and FSS clauses, sub-clause* (2): " *less any amounts for Dayworks Nominated Sub-contractors or any other items based on actual cost or current prices*"

These words appear in both the CE and FSS clauses. They define the " effective value " on which fluctuations are to be calculated. It is submitted that these words exclude from the effective value, goods or materials delivered to the site or in which the property has vested in the employer, pursuant to clause 60 (1) (b) or (c), since the sums due are based on the value of the goods or materials.[7] Such goods or materials will be included in the effective value and therefore qualify for fluctuation payments, when incorporated into the works.

---

[7] See also Guidance Note (GN) 1: Contract Price Fluctuations, issued by the I.C.E. Conditions of Contract Standing Joint Committee.

R.I.B.A. CONDITIONS OF ENGAGEMENT [1]

Introduction

THESE Conditions of Engagement are for the mutual benefit of clients and architects. They determine the minimum fees for which members of the RIBA may undertake work and describe the professional services which clients may expect in return.

Members of the RIBA are governed by the Charters, Byelaws and Code of Professional Conduct of the Royal Institute, which determine their relationships with the public and their professional colleagues.

It is the duty of members to uphold and apply the Conditions of Engagement adopted by the Royal Institute. The engagement of a member shall therefore be in accordance with these Conditions and the fees and charges herein shall apply. However, higher fees and charges may be appropriate depending on the circumstances of the work, in which case these shall be agreed between architect and client when the former is engaged.

Members may not work speculatively nor compete with one another in respect of percentage fees or time charges. Where a prospective client is considering the engagement of one of a number of firms the members concerned may give guidance on the engagement of architects but shall not submit estimates of fees for competitive purposes.

Members whose engagements have been confirmed may give provisional forecasts of the cost of their professional services, quoting the appropriate percentage fees on any constructional cost limits supplied by the client and the estimated time charges on all other work. The services to be provided should always be clearly stated, and estimates of time charges should indicate the time likely to be spent by principals and staff at various salary levels.

Members working outside the United Kingdom shall apply these Conditions wherever they are recognised. Where they are not recognised, members shall apply the locally recognised scale or, in the absence of such a scale, may determine their own fees. Where members are likely to be in competition with their fellow members abroad they are advised to consult the RIBA.

Any value added tax chargeable on the services of the architect is chargeable to the client at the appropriate rate current at the time the tax is charged in accordance with Clause 1.16 of these Conditions. Clients who are taxable persons under the Finance Act 1972 will be able to recover such input tax from Customs and Excise.

The Government has agreed to apply these scales to their commissions.

---

[1] First printed May 1, 1971 (effective July 12, 1971). This edition reprinted May 1976, and printed here by kind permission of the Royal Institute of British Architects. See also amendments July 21, 1977, *post*, p. 575.

Part I:
General

1.00 This part deals with general conditions of engagement and will apply irrespective of the nature or extent of services to be provided.

1.1 Remuneration

1.10 The services normally provided by an architect in studying his client's needs, advising him, preparing, directing and co-ordinating design and inspecting work executed under a building contract are described in Part 2: Normal Services. Other services that an architect may provide are described in Part 4: Other Services.

1.11 The Normal Services for a building project are divided into a sequence of Work Stages A to H through which the architect's work progresses, augmented by services which vary widely in nature and extent with the circumstances of the project.

1.12 Fees for Work Stages C to H are generally calculated as a percentage of the total construction cost of the works, as described in Part 3 of these Conditions.

1.13 Fees for Work Stages A and B and for those other services which are likely to vary widely in nature and extent are charged additionally on a time basis, as described in Part 5 of these Conditions.

1.14 In exceptional circumstances, any service normally charged on a percentage fee may, by prior written agreement between architect and client, be charged on a time basis, in which case the architect shall notify the RIBA.

1.15 The minimum fees and charges described in these Conditions may not be sufficient in all circumstances, in which case higher fees and charges may be agreed between the client and architect when the architect is commissioned.

1.16 The amount of any value added tax on the services of the architect arising under the Finance Act 1972 (or any statutory modification or re-enactment thereof) shall be chargeable to the client in addition to the amount of the architect's fees and charges calculated in accordance with Part 3, Part 5 and Part 6 of these Conditions.

1.2 Consultants

1.20 Normal Services do not include quantity surveying, town planning, civil, structural, mechanical, electrical or heating and ventilating engineering or similar consultants' services. Where the provision of such services is within the competence of the architect's own office or where they are provided by consultants in association with the architect, fees shall be in accordance with the scales of fees of the appropriate professional bodies, but all time charges shall be in accordance with Part 5 of these Conditions.

1.21 Where the services of more than one profession are provided by a single firm or consortium, fees shall be the sum of the appropriate fees for the individual professional services rendered.

1.22 The architect will advise on the need for independent consultants and will be responsible for the direction and integration of their work but not for the detailed design, inspection and performance of the work entrusted to them.

1.23 Independent consultants and quantity surveyors should be nominated or approved by the architect in agreement with the client. They should be appointed and paid by the client.

1.3　Responsibilities

1.30 The architect must have the authority of his client before initiating any service or Work Stage.

1.31 The architect shall not make any material alteration, addition to or omission from the approved design without the knowledge and consent of the client, except if found necessary during construction for constructional reasons in which case he shall inform the client without delay.

1.32 The architect shall inform the client if he has reason to believe the total authorised expenditure or contract period are likely to be materially varied.

1.33 The architect shall advise on the selection and appointment of the contractor and shall make such periodic visits to the site as he considers necessary to inspect generally the progress and quality of the work and to determine in general if the work is proceeding in accordance with the contract documents.

1.34 The architect shall not be responsible for the contractor's operational methods, techniques, sequences or procedures, nor for safety precautions in connection with the work, nor shall he be responsible for any failure by the contractor to carry out and complete the work in accordance with the terms of the building contract between the client and the contractor.

1.4　Specialist sub-contractors and suppliers

1.40 The architect may recommend that specialist sub-contractors and suppliers should design and execute any part of the work. He will be responsible for the direction and integration of their design, and for general inspection of their work in accordance with Stage H of the Normal Services, but not for the detailed design or performance of the work entrusted to them.

1.5　Copyright

1.50 The provisions of this Section shall apply without prejudice to the architect's lien on drawings against unpaid fees.

1.51 In accordance with the provisions of the Copyright Act 1956, copyright in all drawings and in the work executed from them, except drawings and works for the Crown, will remain the property of the architect unless otherwise agreed.

1.52 Where an architect has completed Stage D or where an architect provides detail design in Stage EFG the client unless otherwise agreed shall, on payment or tender or any fees due to the architect, be entitled to reproduce the design by proceeding to execute the project, but only on the site to which the design relates.

1.53 Where an architect has not completed Stage D or where he and his client have agreed that Clause 1.52 shall not apply, the client may not reproduce the design by proceeding to execute the project without the consent of the architect and payment of any additional fee that may be agreed in exchange for the architect's consent.

1.54 The architect shall not unreasonably withhold his consent under Clause 1.53 but where his services are limited to making and negotiating town planning consents he may withhold his consent unless otherwise determined by an arbitrator appointed in accordance with Clause 7.50.

1.6 Inspection

1.60 During his on-site inspections made in accordance with Clause 1.33 the architect shall endeavour to guard the client against defects and deficiencies in the work of the contractor, but shall not be required to make exhaustive or continuous inspections to check the quality or quantity of the work.

1.61 Where frequent or constant inspection is required a clerk or clerks of works should be employed. He shall be nominated or approved by the architect and be under the architect's direction and control. He may be appointed and paid by the client or employed by the architect.

1.62 Where the need for frequent or constant on-site inspection by the architect is agreed to be necessary, a resident architect shall be appointed by the architect.

1.63 Where the architect employs a resident architect or a clerk or clerks of works he shall be reimbursed by the client in accordance with Part 5 and Part 6 of these Conditions.

1.7 Delay and changes in instructions

1.70 Extra work and expense caused in any Stage resulting from delay in receiving instructions, delays in building operations, changes in the client's instructions, phased contracts, bankruptcy or liquidation of the contractor or any other cause beyond the control of the architect, shall be additionally charged on a time basis.

1.8 Fees for hospital works

1.80 The Health Ministers have asked Hospital Boards to apply these scales to their commissions and have advised the Boards that claims from architects to renegotiate terms for long-running commissions, from some suitable future breakpoint and where a substantial amount of work remains to be done, should be considered.

1.81 For certain hospitals, defined as "hospital projects let as a single commission by a client representing a University and a Hospital Board" with a total construction cost of £7,000,000 or more, the minimum fee shall be 6 per cent. Such projects are to be assessed as a whole, irrespective of any phases into which they might be divided for building.

1.9 Signing buildings

1.90 Where the architect has been commissioned for and is completing the Normal Services he shall be entitled at his own expense, or may be required by the client at the client's expense, to sign the building by inscription or otherwise on a suitable and reasonably visible part of the permanent fabric of the building.

Part 2:
Normal Services

2.00 This part describes the services normally provided by an architect for a building project. The fees for Work Stages C to H are generally charged on a percentage basis as described in Part 3 of these Conditions. Stage C begins where the architect's brief has been determined in sufficient detail. Fees otherwise, including work in Stages A and B to determine the architect's brief, are charged additionally on a time basis as described in Part 5 of these Conditions. Initial consultations may be given free of charge.

2.1 Work Stages

2.10 Work Stages charged on a time basis:

A Inception

Receiving an initial statement of requirements, outlining possible courses of action, and advising on the need for a quantity surveyor and consultants. Determining the brief in sufficient detail for subsequent Stages to begin.

B Feasibility studies

Undertaking a preliminary technical appraisal of a project sufficient to enable the client to decide whether and in what form to proceed, and making town planning inquiries or application for outline town planning approval. Such an appraisal may include an approximation of the cost of meeting the client's requirements, a statement on the need for consultants, an outline timetable and a suggested contract procedure.

2.11 Work Stages normally charged on a **percentage** basis:

C Outline proposals

Analysing the client's requirements and where necessary instructing the quantity surveyor and consultants. Preparing, describing and illustrating outline proposals, including an approximation of the cost of meeting them. Informing the client of any major decisions which are needed and receiving any amended instructions.

D Scheme design

Preparing in collaboration with the quantity surveyor, and consultants if appropriate, a scheme design consisting of drawings, and outline specification sufficient to indicate spatial arrangements, materials and appearance. Presenting a report on the scheme, the estimated cost and timetable for the project, for the client's approval.

E F G Detailed design, production drawings, specifications and bills of quantities

Completing a detailed design, incorporating any design work done by consultants, nominated sub-contractors and suppliers. Carrying out cost checks as necessary. Obtaining quotations and other information from nominated sub-contractors and suppliers. Preparing production drawings and specification of materials and workmanship required. Supplying information necessary for the preparation of bills of quantities, if any.

H Tender action to completion

Obtaining and advising on tenders and preparing and advising on the contract and the appointment of the contractor. Supplying information to the contractor, arranging for him to take possession of the site and examining his programme. Making periodic visits to the site as described in Clause 1.33; issuing certificates and other administrative duties under the contract. Accepting the building on behalf of the client, providing scale drawings showing the main lines of drainage and obtaining drawings of other services as executed, and giving initial guidance on maintenance.

2.2 Development studies

To be charged on a time basis

2.20 Services where a client's initial statement of requirements in Stage A requires a special service (such as operational research) before consideration of the brief and development of outline proposals as described in Stage C can begin.

2.3 Development plans

To be charged on a time basis

2.30 Preparing development plans for any large building or complex of buildings which will be carried out in phases over a number of years.

2.31 Preparing a layout only, or preparing a layout for a greater area than that which is to be developed immediately.

2.4 Sites and buildings

To be charged on a time basis

2.40 Advising on the selection and suitability of sites, conducting negotiations concerned with sites or buildings, making measured surveys, taking levels and preparing plans of sites and buildings or existing buildings.

2.41 Making inspections, preparing reports or giving general advice on the condition of premises.

2.42 Work in connection with soil investigations.

2.5 Constructional research

To be charged on a time basis

2.50 Research where the development of a scheme design in Stage D involves special constructional research, including the design, construction or testing of prototype buildings or models.

2.6 Negotiations

To be charged on a time basis

2.60 Exceptional negotiations such as those arising from applications for Town Planning, Building Byelaw, Building Act or Building Regulations approvals.

2.61 Providing information, making all applications other than those covered by the Normal Services, such as those including applications for licences, negotiations in connection with party walls and grant aids.

2.62 Submission to the Royal Fine Art Commission and town planning appeals.

2.7 Special drawings

To be charged on a time basis

2.70 Preparing any special drawings, models or technical information specially for the use of the client, or for Town Planning, Byelaw and Building Regulations approvals; for negotiations with ground landlords, adjoining owners, public authorities, licensing authorities, mortgagors and others.

2.8 Furnishings and works of art

To be charged on a time basis

2.80 Advising on the selection and suitability of loose furniture, fittings and soft furnishings, on the commissioning or selection of works of art, obtaining tenders and supervising their installation.

2.9 Approvals in the normal services

2.90 Except in Scotland, Stages C to E F G of the Normal Services include the duty of making and negotiating applications for Town Planning consents, Building Byelaw, Building Act and Building Regulations approvals, as appropriate. All work in connection with these applications will not necessarily be included in any particular Stage.

2.91 In Scotland, the normal Services cover the duty of preparing drawings and technical information necessary for submission of applications for licences, Town Planning and Building (Scotland) Act approvals as appropriate. The and technical information necessary for submission of applications for licences, Town Planning and Building (Scotland) Act approvals as appropriate. The actual completion of the application and its presentation to the appropriate Court is not part of the architect's responsibility.

Part 3:
Fees for the Normal Services

3.00 This part describes how the percentage fees for the Normal Services are calculated and may be varied, and when they and other charges are due. Percentage fees are based on the total construction cost of the works and on the issue of the final certificate shall be re-calculated on the actual total construction cost.

3.1 Total construction cost

3.10 The total construction cost shall be the cost, as certified by the architect, of all works (including site works) executed under his direction, subject to the following conditions:

3.101 The total construction cost shall include the cost of all work designed or supervised by consultants which the architect is responsible for directing and co-ordinating in accordance with Clause 1.22, irrespective of whether such work is carried out under separate building contracts for which the architect may not be responsible. The architect shall be informed of the cost of any such separate contracts.

3.102 The total construction cost shall not include nominated sub-contractors' design fees for work on which consultants would otherwise have been employed. Where such fees are not known, the architect shall estimate a reduction from the total construction cost.

3.103 For the purpose of calculating the appropriate fees, the total construction cost shall include the actual or estimated cost of any work executed which is excluded from the contract but otherwise designed by the architect.

3.104 The total construction cost shall include the cost of built-in furniture and equipment. Where the cost of any special equipment is excluded from the total construction cost, the architect shall charge for work in connection with such items on a time basis.

3.105 Where appropriate the cost of old materials used in the work shall be calculated as if they were new.

3.106 Where any material, labour or carriage are supplied by a client who is not the builder, the cost shall be estimated by the architect as if they were supplied by the builder and included in the total cost.

3.107 Where the client is the builder, a statement of the ascertained gross cost of the works may be used in calculating the total construction cost of the works. In the absence of such a statement, the architect's own estimate shall be used. In both a statement of the ascertained gross cost and an architect's estimate there shall be included an allowance for the builder's profit and overheads.

3.11 The fee for any part of the work omitted on the client's instruction shall be calculated in accordance with Section 3.5 of these Conditions.

3.2 New works

3.20 Fees for new works generally are shown in Table 1 on page 570.[2]

3.3 Works to existing buildings

3.30 Higher percentages are chargeable for works to existing buildings and are shown in Table 2 on page 570.[3]

3.31 The percentage in Table 2 will not necessarily be sufficient for alterations to all buildings, especially those of historic importance, and higher fees may be appropriate.

3.32 Where extensions to existing buildings are substantially independent, fees may be as for new works, but the fee for those sections of works which marry existing buildings to the new shall be charged separately at the fee in Table 2 applicable to an independent commission of similar value.

---

[2] The reference in the original is to the page of the booklet containing the Conditions of Engagement.

[3] *Ibid.*

3.4 Repetition

3.40 Where a building is repeated for the same client fees for the superstructures excluding all work below the top of ground floor slabs may be reduced as follows:

3.401 On all except the first three of any houses of the same design.

3.402 On all except the first, i.e. the prototype, of all other building types to the same design.

3.41 Where a single building incorporates a number of identical compartments such as floors in multi-storey or complete structural bays in single-storey buildings, fees may be reduced on all identical compartments in excess of 10 provided that the building does not otherwise attract fee reductions and that it is completed in a single contract.

3.42 Reductions shall not be made for repeated individual dwelling units in multi-storey housing schemes but such schemes may qualify for fee reductions under Sub-clause 3.402 or Clause 3.41.

3.43 Reductions in accordance with Clauses 3.40 and 3.41 shall be made by waiving the fee for either Stages D and E F G of the Normal Services where a complete design can be re-used without modification other than the handing of plans, or Stage E F G where a complete design can be re-used with only minor modification.

3.44 The handing of a plan shall not constitute a modification.

3.45 The total construction cost of the works shall be taken first and the fee for normal or partial services calculated thereon. The appropriate reduction shall then be applied to the cost of the repeated superstructures or sections and the result deducted from the full fee.

3.46 Screen walls and outbuildings and garages shall be excluded from the construction cost of works on which fees are waived unless they are included in the type drawings and specifications.

3.47 The fees for work in Stage H of the Normal Services shall not be reduced for repetitive works or repeated buildings, and any additional work arising out of repetition shall be charged on a time basis.

3.5 Partial services

3.50 Where for any reason the architect provides only part of the Normal Services described in Part 2 of these Conditions he shall be entitled to commensurate remuneration, and his fees and charges shall be calculated as follows:

3.501 Where an architect completes the work described in any of Stages C to E F G he shall be entitled to the appropriate proportion of the full percentage fee for the service in accordance with Table 3.

3.502 Where an architect is commissioned to undertake only the work described in Stage H, whether in whole or part, fees shall be on a time basis.

3.503 Where an architect originally engaged to provide the Normal Services does part only of the work described in Stage H, he shall be entitled to not less than the percentage fee otherwise due to him under Clause 3.61.

3.504 Where an architect provides part only of the services described in Stages C to E F G, fees for service in any Stage which is incomplete shall be on a time basis, except by prior written agreement in accordance with Clause 3.51.

3.505 Where an architect has previously completed the work described in Stages C to E F G on a commission which has been abandoned under the terms of Part 7 of these Conditions and the commission is resumed within two years, fees for the work in Stage H shall be on a percentage basis. Where the commission is resumed after two years Sub-Clause 3.502 will apply.

3.51 Where work done by a client results in the omission of part of Stages C to H described in Part 2 of these Conditions or a sponsored constructional method is used, a commensurate reduction in fees may be made by prior written agreement, provided each such agreement specifies in sufficient detail the work to be done by the client, which would otherwise have formed part of the Stages provided by the architect, and is either made in accordance with the RIBA Memorandum on the application of this Clause or is approved by the RIBA.

3.52 All percentage fees for partial service shall be based on the architect's current estimate of the total construction cost of the work. Such estimates may be based on an accepted tender, or subject to Clause 3.53, on the lowest of unaccepted tenders.

3.53 Where partial service is provided in respect of works for which the executed cost is not known and no tender has been accepted, percentage fees shall be based either on the architect's estimated total construction cost or the most recent cost limit agreed with the client, whichever is the lower.

3.6   Mode and time of payment

3.60 On completion of each Stage of Stages C to H of the Normal Services described in Part 2 of these Conditions, the appropriate proportion of the full percentage fee calculated on the current estimated construction cost of the works, plus any other fee and out of pocket expenses which have accrued, shall be due for payment.

3.61 Notwithstanding Clause 3.60, fees in respect of Stages E F G and H shall be due for payment in instalments proportionate to the drawings and other work completed or value of the works certified from time to time.

3.62 Alternatively, the architect and client may arrange for interim payment of fees and charges during all Stages of the work, including payment during Stage H by instalment other than those related to the value of the works certified from time to time.

3.63 On the issue of the final certificate the final instalment of all fees and other charges shall then be due for payment.

Percentage fees for the Normal Services

Minimum charges are laid down in Tables 1 and 2 so that a fee shall not be less than the fee for works having a lower construction cost.

Table 1:
New works

Total construction cost	Minimum % rate	Minimum charges for work stages completed up to and including:			
		H	EFG	D	C
Up to £2,500	10.0	—	—	—	—
£2,500–£8,000	8.5	£250	£187.50	£87.50	£37.50
£8,000–£14,000	7.5	£680	£510.00	£238.00	£102.00
£14,000–£25,000	6.5	£1,050	£787.50	£367.50	£157.50
£25,000–£750,000	6.0	£1,625	£1,218.75	£568.75	£243.75
*£750,000–£1,750,000	5.75	£45,000	£33,750.00	£15,750.00	£6,750.00
*£1,750,000 and over	5.5	£100,625	£75,468.75	£35,218.75	£15,093.75

*Does not apply to certain hospitals (see Clause 1.81) or to works for which the fee is reduced for repetition as provided in Section 3.4. In those cases the minimum fee for works having a total construction cost of £25,000 and over shall be 6.0%.

Table 2:
Works to existing buildings

Total construction cost	Minimum % rate	Minimum charges for work stages completed up to and including:			
		H	EFG	D	C
Up to £2,500	13.0	—	—	—	—
£2,500–£8,000	12.5	£325	£243.75	£113.75	£48.75
£8,000–£14,000	12.0	£1,000	£750.00	£350.00	£150.00
£14,000–£25,000	11.0	£1,680	£1,260.00	£588.00	£252.00
£25,000 and over	10.0	£2,750	£2,062.50	£962.50	£412.50

Table 3:
Apportionment of fees between stages of service

On completion of each Stage of the Normal Services described in Part 2 of these Conditions, the following proportions of the cumulative fee shown in Tables 1 and 2 are payable.

Work stage	Proportion of fee	Cumulative total
C	15%	15%
D	20%	35%
E F G	40%	75%
H	25%	100%

Part 4:
Other services

4.00 This part describes other services which may be provided by the architect. Unless otherwise stated fees for these services shall be on a time basis.

4.1 Town planning

4.10 Fees for town planning work shall be in accordance with the Professional Charges of the Royal Town Planning Institute, except that all layouts shall be charged on a time basis and all time charges shall be in accordance with Part 5 of these Conditions.

4.2 Quantity surveying, valuing and surveying

4.20 Fees for preparing bills of quantities, valuing works executed where no quantity surveyor is employed, valuation of properties and other surveying work not described elsewhere in these Conditions, shall be in accordance with the Professional Charges of the Royal Institution of Chartered Surveyors.

4.3 Garden and landscape design

4.30 Fees for garden and landscape design executed under separate landscape contracts shall be in accordance with the Scale of Professional Charges of the Institute of Landscape Architects.

4.4 Building surveys and structural investigations

4.40 Preparing schedules of dilapidations and negotiating them on behalf of landlords or tenants; taking particulars on site, preparing specifications for repairs or restoration work, inspecting their execution.

4.41 Making structural investigations, the limits of which shall be clearly defined and agreed in writing, such as are necessary to ascertain whether or not there are defects in the walls, roof, floors, and drains of a building which may materially affect its life and value.

4.5 Separate trades contracts

4.50 Where there are separate contracts for each trade the fees shall be determined by prior written agreement and shall not be less than 20 per cent higher than the fee for Stages C to H of the Normal Services.

4.6 Interior design, shopfitting and furniture design

4.60 Fees may be charged on a percentage or time basis for the following work. Where percentage fees are charged, rates up to double those shown in Table 1 will normally be appropriate:

4.601 Special services, including the provision of special sketch studies, detailed advice on the selection of furniture, fittings and soft furnishings and inspection of making up such furnishings for interior design work executed under a special building contract or subcontract or a contract separate from that for other works on which the architect may be employed.

4.602 Works of a special quality, such as special shopfitting, fronts and interiors, exhibition design and similar works, including both the remodelling of existing shops and the design of new units, both independently and within the shell of an existing building, irrespective of whether the architect is employed for shopfitting design only or the work forms part of a general building contract.

4.61 Where all shopfitting drawings are provided by specialist sub-contractors the fee shall be as for the Normal Services described in Part 2 of these Conditions.

4.62 For the design of special items of furniture and fittings for limited production only, i.e. not more than 49 off, the percentage fee shall be either 15 per cent of the total production cost or calculated on a time basis.

4.63 Payment for the design of mass-produced items of furniture may be by royalty, or by time charges and sale of copyright. Fees for the design of prototypes shall be either on a time basis or an advance on royalties.

4.7 Building systems and components

4.70 For the development of building systems, percentage fees on the total production cost may be agreed specially. Otherwise, fees shall be either on a time basis or an advance on royalties.

4.71 Payment for the design of mass-produced building components may be by royalty or by time charges and sale of copyright. Fees for development work in connection with the design of prototypes shall be either on a time basis or an advance on royalties.

4.72 Where an architect recommends to an independent client the use of a building system or components on which he is receiving royalties, the client shall be so informed. The total construction cost shall not be reduced but the architect may reduce his fees to the extent of the royalties received.

4.8 Litigation and arbitration

4.80 For qualifying to give evidence, settling proofs, conferences with solicitors and counsel, attendance in court or at arbitrations or town planning inquiries, or before other tribunals, for services in connection with litigation, and for arbitration, fees shall be on a time basis.

4.81 Time charges shall be in accordance with Part 5 except that the time rate for arbitrators shall be not less than £6 per hour with a minimum fee of £30.

4.9 Consultancy

4.90 For acting as consultant architect, fees shall be on a time basis.

4.91 Where an architect is retained to provide consultancy or other services on a regular or intermittent basis, annual retention fees may be charged, and where appropriate may be merged with subsequent percentage fees or time charges.

Part 5:
Time charges

5.00 Time charges are based on a minimum hourly rate for principals and other operational staff. In assessing the rate at which time should be charged, all relevant factors should be considered, including the complexity of the work, the qualifications, experience and responsibility of the architect, and the character of any negotiations.

5.1 Hourly rates

5.10 The minimum hourly rate for principals shall be £5 per hour.

5.11 The minimum hourly rate for architectural and other operational staff, including resident architects and clerks of works not appointed and paid direct by the client, shall be 15 pence per hour for each £100 of gross annual salary, which shall include bonus payments and the employer's share of other overheads such as national insurance and occupational pensions schemes.

5.2 Travelling time

5.20 Where work is being charged on a time basis, travelling time shall be charged in accordance with Section 5.1 of these Conditions.

5.21 Where work charged on a percentage fee is at such a distance that an exceptional amount of time is spent in travelling, additional charges may be made by prior written agreement.

Part 6:
Out of pocket expenses

6.00 In addition to the fees under any other part of these Conditions, the architect shall be reimbursed for all reasonable out of pocket expenses actually and properly incurred in connection with the commission. Such expenses include the following:

6.1 Drawings and documents

6.10 Printing, reproduction or purchase costs of all documents, drawings, maps, models, photographs, and other records, including all those used in communication between architect, client, quantity surveyor, consultants and contractors, and for enquiries to contractors, sub-contractors and suppliers, notwithstanding any obligation on the part of the architect to supply such documents to those concerned, except that contractors and suppliers will pay for any prints additional to those to which they are entitled under the contract.

6.2 Hotel and travelling expenses

6.20 Hotel and travelling expenses, including mileage allowances for cars at recognised rates, and other similiar disbursements.

6.3 Disbursements

6.30 All payments made on behalf of the client, including expenses incurred in advertising for tenders, clerks of works, and other resident site staff, including the time and expenses of interviews and reasonable expenses for interviewees.

6.31 Fees and other charges for specialist professional advice, including legal advice, which have been incurred by the architect with the specific authority of the client.

6.32 Postage and telephone charges incurred by the architect may be charged by prior written agreement.

6.4 Compounding of expenses

6.40 By prior written agreement, expenses may be estimated or standardised in whole or part, or compounded for an increase in the percentage fee.

Part 7:
Termination, abandoned commissions and interpretation

7.00 An engagement entered into between the architect and the client may be terminated at any time by either party on the expiry of reasonable notice, when the architect shall be entitled to remuneration in accordance with Section 3.5.

7.1 Abandoned commissions

7.10 Where the construction of works is cancelled or postponed on the client's instructions, or the architect is instructed to stop work indefinitely at any time, the commission may be deemed to be abandoned and fees for partial service shall be due.

7.11 Notwithstanding Clause 7.10, if instructions necessary for the architect to continue work are not received from the client six months after such instructions were requested, the commission shall be deemed to have been abandoned.

7.12 Where a commission is abandoned or any part of the works is omitted at any time before completion, fees for partial service in respect of the whole or part of the works shall be charged for all service provided with due authority.

7.2 Resumed commissions

7.20 If a commission which has been abandoned is resumed without substantial alterations within six months, any fees paid under Section 7.1 shall rank solely as payments on account toward the total fees payable on the execution of the works and calculated on their total construction cost.

7.21 Where a commission which has been abandoned is resumed at any time with substantial alteration or is resumed after six months, any fees paid under Section 7.1 above shall be regarded as final payment for the service originally rendered. The resumed commission shall then be deemed separate, and fees charged in accordance with Section 3.5 of these Conditions.

7.22 All additional work arising out of a commission which is resumed in accordance with Clause 7.20 shall be charged on a time basis.

7.3 Interpretation

7.30 Any question arising out of these Conditions may be referred in writing by architect or client to the RIBA for advice provided always that any difference or dispute between them is determined in accordance with either Clause 7.40 or 7.50.

7.4 Disputes

7.40 Any difference or dispute on the application of these Conditions to fees charged by a member of the RIBA may by agreement between the parties be referred to the RIBA for an opinion, provided always that such opinion is sought on a joint statement of undisputed facts and the parties undertake to accept it as final.

7.5 Arbitration

7.50 Where any difference or dispute arising out of these Conditions cannot be resolved in accordance with Clause 7.40, it shall be referred to the arbitration of a person to be agreed between the parties, or, failing agreement within 14 days after either party has given to the other a written request to concur in the appointment of an arbitrator, a person to be nominated at the request of either party by the President of the Institute of Arbitrators, provided that in a difference or dispute arising out of the provisions of Section 1.5 the arbitrator shall, unless otherwise agreed, be a chartered architect.

Amendments, July 21, 1977

The RIBA Council, at its meeting on 21 July 1977, agreed amendments to the Conditions of Engagement to withdraw all mention of a minimum figure for principal's hourly rates where work is charged on a time basis. The amendments are as follows:

Clause 5.00, line 1: *delete* " a minimum hourly rate " and *substitute* " hourly rates."

Clause 5.10: *delete* " minimum "; delete " £5 per hour " and *substitute* " agreed."

Clause 4.81: *delete* " except that the time rate for arbitrators shall be not less than £6 per hour with a minimum fee of £30."

These amendments are published in the October reprint of the Conditions of Engagement and are effective from October 1, 1977. It will now be necessary for members to inform their clients at the time they are commissioned the hourly rate they will charge for services for which a time charge is appropriate. The hourly rate for staff is not affected.

FOR USE WHERE THE SUB-CONTRACTOR IS NOMINATED UNDER THE STANDARD FORM OF BUILDING CONTRACT ISSUED BY THE JOINT CONTRACTS TRIBUNAL
(1963 edition as revised) [1]

This Sub-Contract is made the................................day of..............................., 19...................

BETWEEN ....

....

of or whose registered office is situate at ....

....

(hereinafter called " the Contractor ") of the one part and ....

....

of or whose registered office is situate at ....

....

(hereinafter called " the Sub-Contractor ") of the other part;

SUPPLEMENTAL to an Agreement (hereinafter referred to as " the Main Contract ") made the...day of..............................., 19...................

between ....

....

(hereinafter called " the Employer ") of the one part and the Contractor of the other part;

WHEREAS the Contractor desires to have executed the Works of which particulars are set out in Part I of the Appendix to this Sub-Contract (hereinafter referred to as " the Sub-Contract Works ") and which form part of the Works (hereinafter referred to as " the Main Contract Works ") comprised in and to be executed in accordance with the Main Contract, and any authorised variations of the Sub-Contract Works;

AND WHEREAS the Sub-Contractor has had reasonable opportunity of inspecting the Main Contract or a copy thereof except the detailed prices of the Contractor included in schedules and bills of quantities;

---

[1] Issued under the sanction of and approved by the National Federation of Building Trades Employers and the Federation of Associations of Specialists and Sub-Contractors Approved by the Committee of Associations of Specialist Engineering Contractors. This issue was revised in June 1976 and is printed here by kind permission.

*AND WHEREAS the Sub-Contractor is exempt from the operation of the provisions of Sections 29–31 of the Finance Act, 1971.

Now it is hereby Agreed and Declared as follows:

Notice of the Main Contract to the Sub-Contractor

1 The Sub-Contractor shall be deemed to have notice of all the provisions of the Main Contract except the detailed prices of the Contractor included in schedules and bills of quantities.

Execution of the Sub-Contract Works

2 The Sub-Contractor shall execute and complete the Sub-Contract Works subject to and in accordance with this Sub-Contract in all respects to the reasonable satisfaction of the Contractor and of the Architect for the time being under the Main Contract (hereinafter called the "Architect" which expression where the context of the Main Contract so requires shall mean and include the Supervising Officer) and in conformity with all the reasonable directions and requirements of the Contractor including all reasonable rules of the Contractor (so far as they may apply) for the time being regulating the due carrying out of the Main Contract Works.

Sub-contractor's liability under incorporated provisions of the main contract

3 The Sub-Contractor shall:

(a) Observe, perform and comply with all the provisions of the Main Contract on the part of the Contractor to be observed, performed and complied with so far as they relate and apply to the Sub-Contract Works (or any portion of the same) and are not repugnant to or inconsistent with the express provisions of this Sub-Contract as if all the same were severally set out herein; and

(b) Indemnify and save harmless the Contractor against and from:

(i) any breach, non-observance or non-observance or non-performance by the Sub-Contractor, his servants or agents of the said provisions of the Main Contract or any of them; and

(ii) any act or omission of the Sub-Contractor, his servants or agents which involves the Contractor in any liability to the Employer under the Main Contract; and

(iii) any claim, damage, loss or expense due to or resulting from any negligence or breach of duty on the part of the Sub-Contractor, his servants or agents (including any wrongful user by him or them of the scaffolding referred to in Clause 18 of this Sub-Contract or other property belonging to or provided by the Contractor); and

(iv) any loss or damage resulting from any claim under any statute in force for the time being by an employee of the Sub-Contractor in

---

*This recital should be deleted if it is **not** to be a condition that the sub-contractor should be exempt under sections 29–31 of the Finance Act 1971.

respect of personal injury arising out of or in the course of his employment.

PROVIDED that nothing in this Sub-Contract contained shall impose any liability on the Sub-Contractor in respect of any negligence or breach of duty on the part of the Employer the Contractor, his other sub-contractors or their respective servants or agents nor create any privity of contract between the Sub-Contractor and the Employer or any other sub-contractor.

Insurance against injury to persons and property

4 The Contractor and Sub-Contractor respectively shall, so far as is reasonably practicable, effect and keep in force during all material times policies of insurance with such insurance company or other insurers and of such an amount as shall be approved by the other against their respective liabilities under any statute in force for the time being in respect of injuries to persons and at Common Law in respect of injuries to persons or property arising out of and in the course of the execution of the Main Contract Works and the Sub-Contract Works and/or arising out of and in the course of the employment of any workmen employed by them respectively or caused thereby or due thereto respectively; subject as regards loss or damage by fire, explosion, bursting or overflowing of water tanks, apparatus or pipes, storm, tempest, lightning, flood, earthquake, aircraft or anything dropped therefrom, aerial objects, riot and civil commotion to the provisions of Clause 5 of this Sub-Contract.

Loss or damage by fire, etc., to sub-contract works and materials and goods properly on site—contractor's responsibility

5* (a) The Sub-Contract Works and the materials and goods of the Sub-Contractor properly on site for incorporation in the Sub-Contract Works shall as regards loss or damage by fire, explosion, storm, tempest, flood, bursting or overflowing of water tanks, apparatus or pipes (whether such fire, explosion, bursting or overflowing, or loss or damage from storm, tempest or flood be caused by the negligence of the Sub-Contractor or those for whose actions the Sub-Contractor is responsible or otherwise) lightning, earthquake, aircraft or anything dropped therefrom, aerial objects, riot and civil commotion be at the sole risk of the Contractor.

Loss of or damage to sub-contract materials or goods during the progress of the sub-contract works—contractor's and sub-contractor's responsibility

(b)** Subject to sub-clause (a) hereof:

(i) The Sub-Contractor shall be responsible for loss or damage to any

*Under the Main Contract the risks referred to in sub-clause (a) may be taken by the Employer and the Sub-Contractor should so ascertain. Where the risks are so taken and the Employer is a Local Authority it may be under no obligation to insure and the Contractor will then have no obligation to the Sub-Contractor to insure against them or to procure their insurance by the Employer. See also sub-clause (d).

**The Sub-Contractor should consider whether the risks he assumes under his sub-clause *e.g.* theft or vandalism should be covered by his taking out insurance.

materials or goods on site for the Sub-Contractor's use until such materials and goods have been fully, finally and properly incorporated into the Works except for any loss or damage due to any negligence, omission or default of the Contractor, his servants or agents, or any other sub-contractor of the Contractor engaged upon the Works or any part thereof, his servants or agents or of the Employer or any person for whom the Employer is responsible.

(ii) Where materials or goods have been fully, finally and properly incorporated into the Works before completion of the Sub-Contract Works the Contractor will be responsible for loss or damage to such materials and goods except for any loss or damage caused thereto by the Sub-Contractor, his servants or agents.

(iii) Upon completion of the Sub-Contract Works, the Contractor will be responsible for loss or damage to the Sub-Contract Works properly completed and handed over except for any loss or damage caused thereto by the Sub-Contractor, his servants or agents.

Loss or damage by fire, etc., to sub-contract works and materials and goods properly on site

(c) In the event of any loss or damage by any one or more of the risks referred to in sub-clause (a) hereof being caused to the Sub-Contract Works (including any of the materials or goods of the Sub-Contractor properly on site for incorporation in the Sub-Contract Works) the Contractor to the extent of such loss or damage shall pay to the Sub-Contractor the full value of the same, such value to be calculated in accordance with Clause 10 hereof together with the cost of any professional fees or services.

Insurance for fire, etc., risks

(d) The Contractor shall for the benefit of himself and the Sub-Contractor at all material times insure for the full value thereof together with the cost of any professional fees or services the Sub-Contract Works (including materials and goods properly on site for incorporation in the Sub-Contract Works) or procure their insurance by the Employer† and keep them or have them kept insured against loss or damage by any of the risks referred to in sub-clause (a) hereof.

(e) The Sub-Contractor shall observe and comply with the conditions contained in the policy of insurance of the Contractor or of the Employer, as the case may be, against loss or damage by any of the risks referred to in sub-clause (a) hereof.

(f) Nothing in sub-clause (b) to (e) hereof shall in any way modify the Sub-Contractor's obligations in regard to defects in the Sub-Contract Works as set out in Clause 9 hereof.

---

†See footnote on page 578. [The reference in the original is to the page of the printed form.]

Policies of insurance

6 The Contractor and Sub-Contractor shall each respectively as and when reasonably required to do so by the other produce for his inspection documentary evidence that the insurances referred to in Clauses 4 and 5 of this Sub-Contract are properly maintained, but on any occasion the Contractor or Sub-Contractor may (but not unreasonably or vexatiously) require to have produced for his inspection the policy or policies and receipts in question.

Variations, etc.

7 (1) In the event of the Contractor issuing in writing to the Sub-Contractor a copy of any instructions of the Architect (whether written, or if oral, subsequently confirmed in writing either by the Architect or the Contractor and, in the latter case, not dissented from by the Architect within seven days) in relation to the Sub-Contract Works (whether in regard to variations or in regard to any other matter in respect of which the Architect is expressly empowered by the conditions of the Main Contract to issue instructions), then the Sub-Contractor shall forthwith comply with and carry out the same in all respects accordingly. Save as aforesaid no variation of the Sub-Contract Works shall be made or allowed by the Sub-Contractor.

The expression " variation " shall have the same meaning assigned to it as in the Main Contract.

(2) Upon receipt of what purports to be an instruction of the Architect issued in writing by the Contractor to the Sub-Contractor, the Sub-Contractor may require the Contractor to request the Architect to specify in writing the provision of the Main Contract which empowers the issue of the said instruction. The Contractor shall forthwith comply with any such requirement and deliver to the Sub-Contractor a copy of the Architect's answer to the Contractor's request.

If the Sub-Contractor shall thereafter comply with the said instruction, then the issue of the same shall be deemed for all purposes of this Sub-Contract to have been empowered by the provision of the Main Contract specified by the Architect in answer to the Contractor's request.

PROVIDED always that if before such compliance the Sub-Contractor shall have made a written request to the Contractor to request the Employer to concur in the appointment of an arbitrator under the Main Contract in order that it may be decided whether the provision specified by the Architect empowers the issue of the said instruction, then, subject to the Sub-Contractor giving the Contractor such indemnity and security as the Contractor may reasonably require, the Contractor shall allow the Sub-Contractor to use the Contractor's name and if necessary will join with the Sub-Contractor in arbitration proceedings by the Sub-Contractor to decide the matter as aforesaid.

Completion

8 (a)   The Sub-Contractor shall commence the Sub-Contract Works within an agreed time or, if none is agreed, then within a reasonable time after the receipt by him of an order in writing under this Sub-Contract

from the Contractor to that effect and shall proceed with the same with due expedition.

The Sub-Contractor shall complete the Sub-Contract Works and each section thereof within the period specified in Part II of the Appendix to this Sub-Contract or within such extended period or periods as may be granted pursuant to the provisions hereinafter contained.

If the Sub-Contractor fails to complete the Sub-Contract Works or any section thereof within the period specified or any extended period or periods as hereinafter provided, he shall pay or allow to the Contractor a sum equivalent to any loss or damage suffered or incurred by the Contractor and caused by the failure of the Sub-Contractor as aforesaid. The Contractor shall at the earliest opportunity give reasonable notice to the Sub-Contractor that loss or damage as aforesaid is being or has been suffered or incurred.

PROVIDED that the Contractor shall not be entitled to claim any loss or damage under this clause unless the Architect shall have issued to the Contractor (with a duplicate copy to the Sub-Contractor) a certificate in writing stating that in his opinion the Sub-Contract Works or the relevant section thereof ought reasonably to have been completed within the specified period or within any extended period or periods as the case may be.

(b) Upon it becoming reasonably apparent that the progress of the Sub-Contract Works is delayed, the Sub-Contractor shall forthwith give written notice of the cause of the delay in the progress or completion of the Sub-Contract Works or any section thereof to the Contractor, who shall inform the Architect thereof and of any representations made to him by the Sub-Contractor as to such cause as aforesaid.

If on receipt of such information and representations as aforesaid the Architect is of the opinion that the completion of the Sub-Contract Works is likely to be or has been delayed beyond the periods or period stated in Part II of the Appendix hereto or beyond any extended periods previously fixed under this Clause,

(i) by reason of any of the matters specified in Clause 7 (1) of this Sub-Contract or by any act or omission of the Contractor, his sub-contractors his or their respective servants or agents; or

(ii) for any reason (except delay on the part of the Sub-Contractor) for which the Contractor could obtain an extension of time for completion under the Main Contract

then the Contractor shall, but not without the written consent of the Architect, grant a fair and reasonable extension of the said period or periods for completion of the Sub-Contract Works or each section thereof (as the case may require) and such extended period or periods shall be the period or periods for completion of the same respectively and this clause shall be read and construed accordingly.

PROVIDED always that if the Sub-Contractor shall feel aggrieved by a failure of the Architect to give his written consent to the Contractor granting an extension of the said period or periods for completion of the Sub-Contract Works, then, subject to the Sub-Contractor giving to the Contractor such indemnity and security as the Contractor may reasonably require, the Contractor shall allow the Sub-Contractor to use the Contractor's name and if necessary will join with the Sub-Contractor as plaintiff in any arbitration proceedings by the Sub-Contractor in respect of the said complaint of the Sub-Contractor.

Loss and expense—disturbance of regular progress—benefit of main contract claims

(c) (i) The Contractor shall subject to Clause 12 of this Sub-Contract enforce and make available to the Sub-Contractor the benefit of any right under the Main Contract to claim for loss and expense caused by disturbance to regular progress of the Main Contract Works and the Sub-Contractor shall comply with all requirements reasonably necessary to enable the Contractor to obtain the aforesaid benefit.

Regular progress of sub-contract works—sub-contractor's claims

(ii) If the regular progress of the Sub-Contract Works is materially affected by any act, omission or default of the Contractor, his servants or agents, or any sub-contractor employed on the Works the Sub-Contractor shall as soon as such material effect becomes apparent give written notice thereof to the Contractor and the agreed amount of any direct loss or expense thereby caused to the Sub-Contractor shall be added to the Sub-Contract Sum and regarded as a debt due to the Sub-Contractor.

Regular progress of the main contract works—contractor's claims

(iii) If the regular progress of the Main Contract Works (including any part thereof which is sub-contracted) is materially affected by any act, omission or default of the Sub-Contractor, his servants or agents, or any sub-contractor employed by him on the Sub-Contract Works, the Contractor shall as soon as such material effect becomes apparent give written notice thereof to the Sub-Contractor and the agreed amount of any direct loss or expense thereby caused to the Contractor (whether suffered or incurred by the Contractor or by sub-contractors employed by the Contractor on the Main Contract Works from whom claims under similar provisions in the relevant sub-contracts have been agreed by the Contractor, sub-contractor and the Sub-Contractor named in these conditions) shall be regarded as a debt due to the Contractor and deducted from the Sub-Contract Sum.

Defects, shrinkages, etc.

9 (a) All defects, shrinkages or other faults in the Sub-Contract Works which the Contractor (whether at his own cost or not) shall be liable

Standard Form of Sub-Contract

to make good under the Main Contract shall be made good by the Sub-Contractor within a reasonable time after the receipt by him from the Contractor of the Architect's written instructions or a copy thereof relating to the same.

PROVIDED that where the Contractor is liable to make good such defects, shrinkages or other faults but not at his own cost, then the Contractor shall secure a similar benefit to the Sub-Contractor and shall account to the Sub-Contractor for any money actually received by him in respect of the same.

(b)   If the Contractor (whether by himself or any other sub-contractor) shall execute any work (whether permanent or temporary) to the Main Contract Works or to any part of the same required by the Architect or rendered necessary by reason of defects, shrinkages or other faults in the Sub-Contract Works due to materials or workmanship not being in accordance with this Sub-Contract or to frost occurring before the completion of the Sub-Contract Works, then the Sub-Contractor shall pay to the Contractor the cost of the execution of such work.

PROVIDED that if the Contractor shall pay or allow to the Employer the value of or other agreed sum (not exceeding such cost as aforesaid) in respect of such work instead and in satisfaction of executing the same, then the Sub-Contractor shall pay to the Contractor such value or other agreed sum as aforesaid.

PROVIDED also that the Sub-Contractor shall not be required to pay the cost of any works to make good damage by frost which may appear after completion unless the Architect shall decide that such damage is due to injury which occurred before completion which the Contractor before commencing the same has not given reasonable notice to the Sub-Contractor.

(c)   If the Sub-Contractor shall execute any work to or in connection with the Sub-Contract Works (whether permanent or temporary) required by the Architect or rendered necessary by reason of any defects, shrinkages or other faults in the Main Contract Works due to materials or workmanship not being in accordance with the Main Contract or to frost occurring before the completion of the Main Contract Works, then the Contractor shall pay to the Sub-Contractor the cost of the execution of such work.

PROVIDED that if instead of the Sub-Contractor actually executing such work and in satisfaction of the same the Contractor shall pay or allow to the Employer the value of or other agreed sum (not exceeding such cost as aforesaid) in respect of such work, then the Contractor shall indemnify the Sub-Contractor against any claim, damage or loss in respect of failure to execute such work.

PROVIDED also that the Contractor shall not be required to pay the cost of any work to make good damage by frost which may appear after completion unless the Architect shall decide that such damage was due to injury which took place before completion or of which the Sub-Contractor before commencing the same has not given reasonable notice to the Contractor.

Sub-contract sum—valuation of variations

10 (a) The price of the Sub-Contract Works (hereinafter referred to as the "Sub-Contract Sum") shall be the sum named in or determined by the provisions of Part III of the Appendix to this Sub-Contract or such other sum as shall become payable by reason of any authorised variations, fluctuations or amount ascertained under Clause 8 (c) hereof.

(b) The valuation of all variations authorised under Clause 7 hereof and all work executed by the Sub-Contractor in accordance with the instructions of the Architect as to the expenditure of Provisional Sums shall be determined by the Quantity Surveyor for the time being under the Main Contract (or if none the Architect) and where such valuation shall require any measurement or re-measurement of the Sub-Contract Work the Contractor shall give to the Sub-Contractor an opportunity of being present at the time of such measurement and of taking such notes and measurements as the Sub-Contractor may require. Such valuation (unless otherwise agreed by the Contractor and Sub-Contractor and approved by the Architect) shall be in accordance with the following rules:

(1) The price in bills of quantities or other document forming part of the Sub-Contractor's tender shall determine the valuation of work of similar character executed under similar conditions as work priced therein;

(2) The said prices, where work is not of a similar character or executed under similar conditions as aforesaid, shall be the basis of prices for the same so far as may be reasonable, failing which a fair valuation shall be made;

(3) Where work cannot be properly measured and valued the Sub-Contractor shall be allowed:

(A) The prime cost of such work calculated in accordance with the "Definition of Prime Cost of Daywork" last before issue by the Royal Institution of Chartered Surveyors and the National Federation of Building Trades Employers together with percentage additions to each section of the prime cost at the rates set out by the Sub-Contractor in his tender and recorded in Part IIIA of the Appendix to this Sub-Contract;

(B) Where an appropriate body representing the Employers in a specialist trade and the aforesaid Royal Institution have agreed and issued a Definition of Prime Cost for that specialist trade, the prime cost of such work calculated in accordance with that Definition as last before issued together with percentage additions to each section of the prime cost at the rates set out by the Sub-Contractor in his tender and recorded in Part IIIA of the Appendix to this Sub-Contract; PROVIDED that in any case vouchers specifying the time daily spent upon the work (and if requested by both the Contractor and the Architect the workmen's names) and the materials employed shall be delivered for verification to the Contractor for transmission to the Architect or his authorised representative not later than the end of the week following that in which the work has been executed.

(4) The valuation of items omitted shall be in accordance with the prices referred to in rule (1) hereof; provided that if omissions substantially vary the conditions under which any remaining items are carried out the prices for such remaining items shall be valued under rule (2) hereof.

Save that where the Sub-Contractor has with the agreement of the Contractor and Architect annexed to this sub-contract a schedule of prices for measured work and/or a schedule of daywork prices, such prices shall be allowed to the Sub-Contractor in determining the value of authorised variations in substitution for any prices which would otherwise be applicable under the aforementioned rules.

(c) Any amount ascertained under the provisions of sub-clause (b) hereof shall be added to or deducted from the Sub-Contract Sum and if a certificate under Clause 11 (a) is issued after the date of ascertainment the Contractor shall apply to the Architect for any such amount to be added to or deducted from the amount which would otherwise be included in respect of the Sub-Contract Works in such certificate.

(d) If upon written application being made to him by the Contractor on behalf of the Sub-Contractor the Architect for the time being under the Main Contract is of the opinion that a variation or the execution of work by the Sub-Contractor in accordance with Architects' instructions as to the expenditure of provisional sums has involved the Sub-Contractor in direct loss and/or expense for which he would not be reimbursed by payment in respect of a valuation made in accordance with sub-clause (b) of this Clause and if the said application is made within a reasonable time of the loss or expense having been incurred then the Contractor shall request the Architect either himself to ascertain or to instruct the Quantity Surveyor to ascertain the amount of such loss or expense. Any amount from time to time so ascertained shall be added to the Sub-Contract Sum and if a certificate under Clause 11 (a) hereof is issued after the date of ascertainment the Contractor shall apply to the Architect for any such amount to be added to the amount which would otherwise be included in respect of the Sub-Contract Works in such certificate.

Value Added Tax

10A (1) In this Clause "tax" means the value added tax introduced by the

Finance Act, 1972, which is under the care and management of the Commissioners of Customs and Excise (hereinafter called "the Commissioners").

(2) Any reference in this Sub-Contract to "Sub-Contract Sum" shall be regarded as such Sum exclusive of any tax and recovery by the Sub-Contractor from the Contractor of tax properly chargeable by the Commissioners on the Sub-Contractor under or by virtue of the Finance Act, 1972, or any amendment thereof on the supply of goods and services under this Sub-Contract shall be under the provisions of this Clause.

(3) Supplies of goods and services under this Sub-Contract are supplies under a contract providing for periodical payment for such supplies within the meaning of Regulation 21 (1) of the Value Added Tax (General) Regulations, 1972, or any amendment thereof.

(4) (a) (i) The Contractor shall pay to the Sub-Contractor in the manner hereinafter set out any tax chargeable by the Commissioners on the Sub-Contractor on the supply to the Contractor of any goods and services by the Sub-Contractor under this Sub-Contract.

   (ii) The Sub-Contractor shall, not later than seven days before the date when payment is due to the Sub-Contractor under Clause 11 (b) hereof, give to the Contractor a written provisional assessment of the respective values (less the Retention Money and cash discount referred to in Clause 11 (b) (i) and (ii) hereof) of those supplies of goods and services for which payment is due as aforesaid and which will be chargeable at the relevant time of supply under the aforesaid Regulation 21 (1) on the Sub-Contractor at a zero rate of tax (category one) and any rate or rates of tax other than zero (category two). The Sub-Contractor shall also specify the rate or rates of tax which are chargeable on those supplies included in category two and shall state the grounds on which he considers such supplies are so chargeable.

(b) The Contractor shall in relation to any amount payable in accordance with the provisions of Clause 11 hereof calculate at the rate applicable to the sub-contract supply, in accordance with the written provisional assessment notified by the Sub-Contractor under paragraph (a) (ii) of this sub-clause, the tax properly chargeable on such supply and remit such tax to the Sub-Contractor within the period prescribed by Clause 11 hereof for the payment of the amount in relation to which the tax was calculated.

(c) Upon receipt of the amounts referred to in Clause 11 hereof and paragraph (b) of this sub-clause the Sub-Contractor shall immediately issue to the Contractor a receipt as referred to in Regulation 21 as aforesaid containing the particulars required under Regulation 9 (1) of the Value Added Tax (General) Regulations, 1972, or any amendment thereof.

(d) If the Sub-Contractor disallows any cash discount claimed by the Contractor under Clause 11 (b) (ii) hereof and the Contractor pays the amount of such discount to the Sub-Contractor the provisions of paragraph (b) of this sub-clause shall not apply to such payment.

(e) If, for any reason the amount paid under paragraph (b) of this sub-clause is not the amount of tax properly chargeable on the Sub-Contractor by the Commissioners, the Sub-Contractor shall notify the Contractor and the Contractor shall forthwith make any adjustment that may be necessary.

(5) Notwithstanding any provisions to the contrary elsewhere in this Sub-Contract, the Contractor, if he has not received from the Sub-Contractor, within twenty-one days of a payment to which this Clause refers, any receipt or receipts due under sub-clause (4) (c) hereof, may so notify the Sub-Contractor in writing and shall be entitled in that notice to state that he will withhold further payments to the Sub-Contractor unless the receipt or receipts outstanding as specified in the aforesaid written notice have been received before the next payment becomes due. If any payment to the Sub-Contractor is withheld in accordance with this sub-clause, such payment shall be released to the Sub-Contractor immediately upon receipt by the Contractor of the outstanding receipt or receipts as specified in the aforesaid written notice. This sub-clause does not entitle the Contractor to withhold any payment due to the Sub-Contractor on any ground other than that specified in this sub-clause.

Value Added Tax—Special arrangement—VAT (General) Regulations 1972, Regulations 8 (3) and 21

*10B (1) In this Clause "tax" means the value added tax introduced by the Finance Act, 1972, which is under the care and management of the Commissioners of Customs and Excise (hereinafter called "the Commissioners").

(2) Any reference in this Sub-Contract to "Sub-Contract Sum" shall be regarded as such sum exclusive of any tax and recovery by the Sub-Contractor from the Contractor of tax properly chargeable by the Commissioners on the Sub-Contract under or by virtue of the Finance Act, 1972, or any amendment thereof on the supply of goods and services under this Sub-Contract shall be under the provisions of this Clause.

(3) Supplies of goods and services under this Sub-Contract are supplies under a contract providing for periodical payment for such supplies within the meaning of Regulation 21 (1) of the Value Added Tax (General) Regulations 1972 or any amendment thereof.

(4) (a) The Contractor shall pay to the Sub-Contractor in the manner hereinafter set out any tax chargeable by the Commissioners on the Sub-Contractor on the supply to the Contractor of any goods and services by the Sub-Contractor under this Sub-Contract.

---

*This Clause can only be used where the Contractor under the Value Added Tax (General) Regulations 1972, Reg. 8 (3) has been allowed to prepare the tax documents in substitution for an authenticated receipt issued by the Sub-Contractor under Regulation 21 (2) of the above Regulations and the Sub-Contractor has consented to the use of this method. Where this Clause applies, Clause 10A must be deleted and its deletion initialled by the parties.

(b) The Sub-Contractor shall, not later than seven days before the date when payment is due to the Sub-Contractor under Clause 11 (b) hereof, give to the Contractor a written provisional assessment of the respective values (less the Retention Money and cash discount referred to in Clause 11 (b) (i) and (ii) hereof) of those supplies of goods and services for which payment is due as aforesaid and which will be chargeable at the relevant time of supply under the aforesaid Regulation 21 (1) on the Sub-Contractor at a zero-rate of tax (category one) and any rate or rates of tax other than zero (category two). The Sub-Contractor shall also specify the rate or rates of tax which are chargeable on those supplies included in category two and shall state the grounds on which he considers such supplies are so chargeable.

(c) The Contractor shall in relation to any amount payable in accordance with the provisions of Clause 11 hereof calculate at the rate applicable to the sub-contract supply in accordance with the written provisional assessment notified by the Sub-Contractor under paragraph (b) of that sub-clause, the tax properly chargeable on such supply and remit such tax to the Sub-Contractor within the period prescribed by Clause 11 hereof for the payment of the amount in respect of which the tax was calculated.

(d) The contractor shall together with the payment under Clause 11 hereof and the payment of tax under this sub-clause issue to the Sub-Contractor a document approved by the Commissioners under Regulation 8 (3) of the Value Added Tax (General) Regulations 1972, and shall not insert in this document any date or other writing which purports to represent for any purposes whatsoever the time of supply in respect of which the Sub-Contractor becomes liable for tax (output tax) on the relevant supplies of goods and services to the Contractor. Without prejudice to the above obligation of the Contractor in relation to the time of supply by the Sub-Contractor, the Contractor shall insert in this document the date of despatch of the document to the Sub-Contractor. Provided always that the payment including tax referred to in the document (or reconciled with the actual payment received by the Sub-Contractor in accordance with sub-clause (4) (e) of this Clause) has been received, the Sub-Contractor shall insert on this document in a space left thereon for this purpose by the Contractor the date of receipt of the document by the Sub-Contractor. If such payment has not been received the Sub-Contractor shall immediately reject the document and explain to the Contractor the reasons for such rejection.

(e) If deductions are made so that the payment received is different from that stated in the document issued by the Contractor to the Sub-Contractor, the Contractor shall issue with that document a reconciliation statement.

(5) (a) If at any time the Commissioners withdraw the approval referred to in sub-clause (4) (d) hereof the Contractor shall immediately so inform

the Sub-Contractor in writing whereupon Clause 10A shall be deemed to be incorporated in this Sub-Contract in respect of payment and tax thereon for any supplies of goods and services remaining to be supplied and/or paid for under this Sub-Contract.

(b) If the Sub-Contractor, during the currency of this Sub-Contract withdraws his consent to the procedure referred to in sub-clause (4) (d) hereof, he shall not be in breach of this Sub-Contract; and immediately on his notifying the Commissioners and the Contractor of such withdrawal of consent, Clause 10A shall be deemed to be incorporated in this Sub-Contract in respect of payment and tax thereon for any supplies of goods and services remaining to be supplied and/or paid for under this Sub-Contract.

(6) Subject to sub-clause (5) hereof, the Sub-Contractor shall at no time in respect of supplies of goods and services under this Sub-Contract issue a document which is or purports to be an authenticated receipt within the meaning of Regulation 21 (2) of the Value Added Tax (General) Regulations 1972.

(7) It is hereby agreed and declared that in issuing any documents referred to in this Clause the Contractor is not acting as agent for the Sub-Contractor.

Contractor to apply for certificates of payment

11 (a) The Contractor shall subject to and in accordance with the Main Contract from time to time make application (of which prior thereto the Contractor shall give to the Sub-Contractor at least seven days' notice unless otherwise agreed between the Contractor and Sub-Contractor) to the Architect for Certificates of Payment and for the inclusion therein of the amount which at the date thereof fairly represents the total value of the Sub-Contract Works and of any variations authorised and executed, and of fluctuations or amounts ascertained under Clause 8 (c) (i) hereof and of the value of materials and goods delivered upon the site for use in the Sub-Contract Works, together with any amounts properly payable under the Main Contract in respect of off-site materials and goods for use in the Sub-Contract Works, provided always that the Sub-Contractor shall observe any relevant conditions set out in the Main Contract which have to be fulfilled before the Architect is empowered to include the value of any off-site materials or goods in interim certificates; and provided that (other than in the case of the said off-site materials and goods) the application shall only include the value of any materials and goods as and from such time as they are reasonable, properly and not prematurely brought upon the site and then only if adequately stored and/or protected against weather and other casualties. The Contractor shall also embody in or annex to the said application any representations of the Sub-Contractor in regard to such value.

Interim payments to the sub-contractor

(b) Within fourteen days of the receipt by the Contractor of any certificate or duplicate copy thereof from the Architect the Contractor shall notify and pay to the Sub-Contractor the total value certified therein in respect of the Sub-Contract Works and in respect of any authorised variations thereof and in respect of any fluctuations or amounts ascertained under Clause 8 (c) hereof less:—

(i) Retention Money, that is to say the proportion attributable to the Sub-Contract Works of the amount retained by the Employer in accordance with the Main Contract and at the rate specified in Part IV of the Appendix to this Sub-Contract;

(ii) A cash discount of $2\frac{1}{2}\%$, if payment is made within fourteen days, on the difference between the said total value and the said Retention Money; and

(iii) The amounts previously paid.

Retention money

(c) The Retention Money referred to above shall be dealt with in the following manner: on the issue by the Architect to the Contractor of any certificate or duplicate copy thereof which includes in accordance with the Main Contract the amount or any part thereof retained by the Employer under the Main Contract the Contractor shall pay to the Sub-Contractor such part of the Retention Money as is included in the Certificate or duplicate copy thereof (with interest if any), less a cash discount of $2\frac{1}{2}\%$, if payment is made within fourteen days of the receipt by the Contractor of that certificate or duplicate copy thereof.

Dispute as to certificate

(d) If the Sub-Contractor shall feel aggrieved by the amount certified by the Architect or by his failure to certify, then, subject to the Sub-Contractor giving to the Contractor such indemnity and security as the Contractor shall reasonably require, the Contractor shall allow the Sub-Contractor to use the Contractor's name and if necessary will join with the Sub-Contractor as claimant in any arbitration proceedings by the Sub-Contractor in respect of the said matters complained of by the Sub-Contractor.

Right of sub-contractor to suspend execution of sub-contract works

(e) Without prejudice to any other rights or remedies which the Sub-Contractor may possess, if the Contractor shall fail to make any payment to the Sub-Contractor as hereinbefore provided and such failure shall continue for seven days after the Sub-Contractor shall have given the Contractor written notice of the same, then the Sub-Contractor may suspend the further execution of the Sub-Contract Works until such payment shall be made and such period of suspension as aforesaid shall be deemed to be an extension of and be added to the period or periods for completion (as the case may be) as provided in Part II of the Appendix to this Sub-Contract or to any extended period or periods previously authorised under Clause 8 (b) hereof and shall not be deemed a delay for which the Sub-Contractor is liable under this Sub-Contract.

Special interim payment

(f) If after the issue of the last Interim Certificate but before the issue of the Final Certificate in accordance with the Main Contract the Sub-Contractor shall have completed any work upon the Sub-Contract Works, he may request the Contractor in writing to make application to the Architect for certificates certifying the value of the work executed upon the Sub-Contract Works, and the Contractor shall make such application.

The provisions of this clause shall apply to such certificates as if they were certificates of payment expressly provided for in the Main Contract.

Final payment to the sub-contractor

(g) If before the issue of a Final Certificate to the Contractor under the Main Contract the Architect desires to secure final payment to the Sub-Contractor on completion of the Sub-Contract Works and in accordance with and subject to the provisions of the Main Contract relating to prime cost sums issues a certificate to the Contractor including an amount to cover such final payment, then the Contractor shall pay to the Sub-Contractor the amount so certified by the Architect as aforesaid: but such payment shall only be made if the Sub-Contractor indemnifies and secures the Contractor to the reasonable satisfaction of the Contractor against all latent defects in the Sub-Contract Works and if by such final payment the Contractor will be discharged under the Main Contract from all liabilities in respect of the Sub-Contract Works except for any latent defects.

(h) The Contractor's interest in the Retention Money, whether it is included in the amounts retained by the Employer under the Main Contract and held by him or whether it is held by the Contractor, is fiduciary as trustee for the Sub-Contractor (without obligation to invest) and if the Contractor attempts or purports to mortgage or otherwise charge such interest or his interest in the whole of the amount retained as aforesaid (otherwise than by floating charge if the Contractor is a limited company), the Contractor shall thereupon immediately set aside and become a trustee for the Sub-Contractor of a sum equivalent to the Retention Money and shall pay the same to the Sub-Contractor on demand; provided that upon payment of the same to the Sub-Contractor the amount due to the Sub-Contractor upon final payment under this Sub-Contract shall be reduced accordingly by the amount so paid.

Sub-contractor's claim to rights and benefits under the main contract

12 The Contractor will so far as he lawfully can at the request and cost of the

Sub-Contractor obtain for him any right or benefits of the Main Contract so far as the same are applicable to the Sub-Contract Works but not further or otherwise.

Contractor's right to set-off—agreed amounts—amounts awarded in arbitration or litigation

13A (1) The Contractor shall be entitled to deduct from any money (including any retention money) otherwise due under this Sub-Contract any amount agreed by the Sub-Contractor as due to the Contractor, or finally awarded in arbitration or litigation in favour of the Contractor, and which arises out of or under this Sub-Contract.

Contractor's right to set-off—amounts not agreed

(2) The Contractor shall be entitled to set-off against any money (including any retention money) otherwise due under this Sub-Contract the amount of any claim for loss and/or expense which has actually been incurred by the Contractor by reason of any breach of, or failure to observe the provisions of, this Sub-Contract by the Sub-Contractor, Provided,

(a) that no set-off relating to any delay shall be made unless, in accordance with the proviso to Clause 8 (a) of this Sub-Contract, the certificate of the Architect referred to therein has been issued to the Contractor with a duplicate copy to the Sub-Contractor;

(b) the amount of such set-off has been quantified in detail and with reasonable accuracy by the Contractor; and

(c) the Contractor has given to the Sub-Contractor notice in writing specifying his intention to set-off the amount quantified in accordance with proviso (b) of this sub-clause and the grounds on which such set-off is claimed to be made. Such notice shall be given not less than seventeen days before the money from which the amount is to be set-off becomes due and payable to the Sub-Contractor; provided that such written notice shall not be binding insofar as the Contractor may amend it in preparing his pleadings for any Arbitration pursuant to the notice of arbitration referred to in Clause 13B (1) (a) (i) of this Sub-Contract.

(3) Any amount set-off under the provisions of sub-clause (2) hereof is without prejudice to the rights of the Contractor or Sub-Contractor in any subsequent negotiations, arbitration proceedings or litigation to seek to vary the amount claimed and set-off by the Contractor under sub-clause (2) hereof.

(4) The rights of the parties to this Sub-Contract in respect of set-off are fully set out in these Conditions and no other rights whatsoever shall be implied as terms of this Sub-Contract relating to set-off.

Contractor's claims not agreed by the sub-contractor—appointment of adjudicator

13B (1) (a) If the Sub-Contractor, at the date of the written notice of the Contractor issued under Clause 13A (2) (c) of this Sub-Contract, does not agree the amount (or any part thereof) specified in that notice which the Contractor intends to set-off, the Sub-Contractor may, within 14 days of receipt by him of such notice, send to the Contractor by registered post or recorded delivery a written statement setting out the reasons for such inability to agree and particulars of any counterclaim against the Contractor arising out of this Sub-Contract to which the Sub-Contractor considers he is entitled, provided always that he shall have quantified such counterclaim in detail and with reasonable accuracy (which statement and counter-claim, if any, shall not however be binding insofar as the Sub-Contractor may amend it in preparing his pleadings for the Arbitration pursuant to the notice of arbitration referred to in sub-paragraph (i) of this paragraph) and shall at the same time

(i) give notice of arbitration to the Contractor; and

(ii) request action by an Adjudicator in accordance with the right given in paragraph (b) hereof (and immediately inform the Contractor of such request) and send by registered post or recorded delivery a copy of the aforesaid statement and the written notice of the Contractor to which that statement relates and the aforesaid counterclaim (if any).

(b) Subject to the provisions of this Clause and of Clauses 11 (b) and 13A of this Sub-Contract the Sub-Contractor shall be entitled to request

.... *

(hereinafter called "the Adjudicator") to act as the Adjudicator to decide those matters referable to the Adjudicator under the provisions of this Clause. In the event of the above-named being unable or unwilling to act as the Adjudicator a person appointed by the above-named shall be the Adjudicator in his place. Provided that no Adjudicator shall be appointed who has any interest in this Sub-Contract or the main contract of which this Sub-Contract is part or in other contracts or sub-contracts in which the Contractor or the Sub-Contractor is engaged.

Contractor's written statement

(2) Upon receipt of the aforesaid statement the Contractor may within fourteen days from the date of such receipt send to the Adjudicator by registered post or recorded delivery a written statement with a copy to the Sub-Contractor setting out brief particulars of his defence to any counterclaim by the Sub-Contractor.

Decision of adjudicator

(3) (a) Within seven days of receipt of any written statement by the Contractor under sub-clause (2) hereof or on the expiry of the time limit to the Contractor referred to in sub-clause (2) hereof whichever is the earlier

*Name and address to be inserted by the parties.

the Adjudicator, without requiring any further statements than those submitted to him under sub-clause (1) and where relevant sub-clause (2) hereof (save only such further written statements as may appear to the Adjudicator to be necessary to clarify or explain any ambiguity in the written statements of either the Contractor or the Sub-Contractor) and without hearing the Contractor or Sub-Contractor in person, shall, subject to paragraph (b) hereof, in his absolute discretion and without giving reasons decide in respect of the amount notified by the Contractor under Clause 13A (2) (b) of this Sub-Contract:

(i) whether the whole or any part of such amount shall be retained by the Contractor; or

(ii) whether the whole or any part of such amount shall be deposited, for security pending arbitration, by the Contractor with the Trustee-Stakeholder referred to in sub-clause (5) hereof; or

(iii) whether the whole or any part of such amount shall be paid by the Contractor to the Sub-Contractor.

The Adjudicator's decision shall be binding upon the Contractor and the Sub-Contractor until the matters upon which he has given his decision have been settled by agreement or determined by an arbitrator or the courts.

(b) The Adjudicator shall reach such decision under paragraph (a) hereof as he considers in all the circumstances of the dispute to be fair, reasonable and necessary, and such decision shall deal with the whole amount set-off by the Contractor under Clause 13A (2) of this Sub-Contract.

(c) The Adjudicator shall immediately notify in writing the Contractor and the Sub-Contractor of his decision under paragraph (a) hereof.

Implementation of adjudicator's decision

(4) (a) Where any decision of the Adjudicator notified under sub-clause (3) (c) hereof requires the Contractor to deposit an amount with the Trustee-Stakeholder referred to in sub-clause (5) hereof, the Contractor shall thereupon pay such amount to the payee named in Part XIII of the Appendix to this Sub-Contract to hold upon the terms hereinafter expressed provided that the Contractor shall not be obliged to pay a sum greater than the amount due from the Contractor under Clause 11 (b) of this Sub-Contract in respect of which the Contractor has exercised the right of set-off referred to in Clause 13A (2) of this Sub-Contract.

(b) Where any decision of the Adjudicator notified under sub-clause (3) (c) hereof requires the Contractor to pay an amount to the Sub-Contractor, such amount shall be paid by the Contractor immediately upon receipt of the decision of the Adjudicator but subject to the same proviso as set out in paragraph (a) hereof.

Trustee-stakeholder

(5) The payee named in Part XIII of the Appendix to this Sub-Contract (in

this Sub-Contract called " the Trustee-Stakeholder ") shall hold any sum received under the provisions of sub-clauses (3) and (4) hereof in trust for the Contractor and Sub-Contractor until such time as:

the Arbitrator appointed pursuant to the notice of arbitration given by the Sub-Contractor under sub-clause (1) (a) (i) hereof; or

the Contractor and Sub-Contractor in a joint letter signed by each of them or on their behalf

shall otherwise direct and shall, in either of the above cases, dispose of the said sums as may be directed by the Arbitrator, or, failing any direction by the Arbitrator, as the Contractor and Sub-Contractor shall jointly determine. The Trustee-Stakeholder shall deposit the sum received in a Deposit Account in the name of the Trustee-Stakeholder but shall add the interest to the sum deposited. The Trustee-Stakeholder shall be entitled to deduct his reasonable and proper charges from the sum deposited (including any interest added thereto). The Sub-Contractor shall notify the Trustee-Stakeholder of the name and address of the Adjudicator and Arbitrator referred to above.

Adjudicator's decision—power of arbitrator

(6) The Arbitrator appointed pursuant to such Notice of Arbitration may in his absolute discretion at any time before his final award on the application of either party vary or cancel the decision of the Adjudicator given under sub-clause (3) hereof if it appears just and reasonable to him so to do.

Further sums—set-off and counterclaims

(7) Any action taken by the Contractor under Clause 13A (2) of this Sub-Contract and by the Sub-Contractor in respect of any counterclaim under Sub-Clause (1) (a) hereof is without prejudice to similar action by the Contractor or Sub-Contractor as the case may be if and when further sums become due to the Sub-Contractor.

Adjudicator's fee

(8) The Fee of the Adjudicator shall be paid by the Sub-Contractor within 28 days of the date of the decision of the Adjudicator given under sub-clause (3) hereof but the Arbitrator appointed pursuant to the notice of arbitration under sub-clause (1) (a) (i) hereof shall in his final award settle the responsibility of the Contractor or Sub-Contractor or both for payment of the fee or any part thereof and where relevant for the charges of the Trustee-Stakeholder or any part thereof.

Right of access of contractor and architect

14 The Contractor and the Architect and all persons duly authorised by them or either of them shall at all reasonable time have access to any work which is being prepared for or will be utilised in the Sub-Contract Works, unless the Architect shall certify in writing that the Sub-Contractor has reasonable grounds for refusing such access.

Sub-letting of sub-contract works

15 The Sub-Contractor shall not assign this Sub-Contract nor sub-let the Sub-Contract Works or any portion of the same without the written consent of both the Contractor and the Architect; provided that the consent of the Contractor shall not be unreasonably withheld, and that in case of any difference of opinion between the Contractor and the Architect the opinion of the Architect shall prevail.

Provision of water, etc., for sub-contract works

16 (a) If and so far as it is so provided in the Main Contract (but not otherwise) the Contractor shall supply at his own cost all necessary water, lighting, watching and attendance for the purposes of the Sub-Contract Works. Subject as aforesaid the Sub-Contractor shall make all necessary provision in regard to the said matters and each of them.

Temporary workshops etc.

(b) Save as otherwise provided in the Main Contract the Sub-Contractor at his own expense shall provide and erect all necessary workshops, sheds or other buildings for his employees and workmen at such places on the site as the Contractor shall appoint and the Contractor agrees to give all reasonable facilities to the Sub-Contractor for such erection.

Sub-contractor's user of scaffolding of contractor

17 The Sub-Contractor, his employees and workmen in common with all other persons having the like right shall for the purposes of the Sub-Contract Works (but not further or otherwise) be entitled to use any scaffolding belonging to or provided by the Contractor, while it remains so erected upon the site. PROVIDED that such user as aforesaid shall be on the express condition that no warranty or other liability on the part of the Contractor or of his other sub-contractors shall be created or implied in regard to the fitness, condition or suitability of the said scaffolding.

Contractor and sub-contractor not to make wrongful use of or interfere with the property of the other

18 The Contractor and the Sub-Contractor respectively their respective servants or agents shall not wrongfully use or interfere with the plant, ways, scaffolding, temporary works, appliances, or other property respectively belonging to or provided by the other of them or be guilty of any breach or infringement of any Act of Parliament or bye-law, regulation, order or rule made under the same or by any local or other public or competent authority; provided that nothing herein contained shall prejudice or limit the rights of the Contractor or of the Sub-Contractor in the carrying out of their respective statutory duties or contractual duties under this Sub-Contract or under the Main Contract.

Plant, tools, etc.—responsibility of sub-contractor

19 (a) Without prejudice to the obligations under Clause 5 hereof, the plant, tools, equipment or other property belonging to or provided by the Sub-Contractor, his servants or agents and in any case any materials which are not properly on site for incorporation in the Sub-Contract Works shall be at the sole risk of the Sub-Contractor, and any loss or damage to the same or caused by the same shall except for any loss or damage due to any negligence omission or default of the Contractor, his servants or agents be the sole liability of the Sub-Contractor who shall indemnify the Contractor against any loss, claim or proceedings in respect thereof.

(b) Any insurance against any loss or damage to or caused by the aforesaid plant, tools, equipment or other property shall be the sole concern of the Sub-Contractor.

Determination of this sub-contract by the contractor

20 (a) Without prejudice to any other rights or remedies which the Contractor may possess if the Sub-Contractor shall make default in any of the following respects viz.:

(i) if without reasonable cause he wholly suspends the carrying out of the Sub-Contract Works before completion thereof; or

(ii) if he fails to proceed regularly and diligently with the Sub-Contract Works; or

(iii) if he refuses or persistently neglects after notice in writing from the Contractor to remove defective work or improper material;

then, if such default shall continue for ten days after a notice by registered post or recorded delivery specifying the default has been given to him by the Contractor, the Contractor may thereupon by notice by registered post or recorded delivery determine the employment of the Sub-Contractor under this Sub-Contract; provided that notice in pursuance of this clause shall not be given unreasonably or vexatiously and shall be void if the Contractor is at the time of the notice in breach of this Sub-Contract; or

(b) If the Sub-Contractor commits an act of bankruptcy or makes or enters into any deed of arrangement or composition with his creditors or being a company suffers liquidation, whether compulsory or voluntary, except liquidation for purposes of reconstruction, the appointment of a provisional liquidator, or suffers or allows any execution whether legal or equitable, to be levied on his property or obtained against him* *or if at any time he ceases to be exempt from the provisions of Sections 29–31 of the Finance Act, 1971, or any statutory modification or re-enactment thereof for the time being in force*, then the Contractor may without prejudice to any other rights or remedies by written notice forthwith determine the employment of the Sub-Contractor under this Sub-Contract.

(c) In case of the employment of the Sub-Contractor under this Sub-

---

*The words in italics should be deleted if it is *not* to be a condition that the Sub-Contractor should be exempt under sections 29–31 of the Finance Act 1971.

Contract being determined under (a) or (b) of this Clause, then the Sub-Contractor shall be deemed to be in breach of this Sub-Contract and the Contractor shall only be liable for the value of any work actually and properly executed and not paid for at the date of such determination, such value to be calculated in accordance with Clause 10 of this Sub-Contract, for the value of any unfixed materials and goods delivered upon the site for use in the Sub-Contract Works the property in which has passed to the Employer under the terms of the Main Contract and for no other sum or sums whatsoever; and the Contractor shall have the right to recover, or to deduct from or set off against any such amount, the amount of damage suffered and/or of loss and expense incurred by him by reason of the determination of the employment of the Sub-Contractor under this Sub-Contract.

Determination of the main contract

21 If for any reason the Contractor's employment under the Main Contract is determined (whether by the Contractor or by the Employer and whether due to any default of the Contractor or otherwise), then the employment of the Sub-Contractor under this Sub-Contract shall thereupon also determine and the Sub-Contractor shall be entitled to be paid:—

  (i) The value of the Sub-Contract Works completed at the date of such determination, such value to be calculated according to Clause 10 of this Sub-Contract.

  (ii) The value of work begun and executed but not completed at the date of such determination, such value to be calculated according to Clause 10 of this Sub-Contract.

  (iii) The value of any unfixed materials and goods delivered upon the site for use in the Sub-Contract Works, the property in which has passed to the Employer under the terms of the Main Contract.

  (iv) The cost of materials or goods properly ordered for the Sub-Contract Works for which the Sub-Contractor shall have paid or of which he is legally bound to accept delivery. On such payment by the Contractor any materials or goods so paid for shall become the property of the Contractor.

  (v) Any reasonable cost of removal from the site of his temporary buildings, plant, machinery, appliances, goods and materials.

Wages and conditions

22 (a) During the continuance of this Sub-Contract the wages and conditions of employment of the employees and workmen of the Contractor and Sub-Contractor engaged on the Main Contract Works and the Sub-Contract Works respectively shall be such as may from time to time be prescribed by the competent authority in the industry or trade to which such employees or workmen belong and the Contractor and Sub-Contractor hereby mutually agree respectively to pay and observe the same accordingly.

PROVIDED that in default of such wages and/or conditions of employ-ment being prescribed as aforesaid the Contractor and Sub-Contractor as a condition of this Sub-Contract shall pay such wages and/or observe such conditions of employment as may be generally from time to time prevailing in industries or trades of a similar or comparable nature in the districts in which their respective employees and work-men are engaged.

(b) If either party shall commit a breach of this clause, then the other party shall be entitled (without prejudice to any other right or remedy) to be indemnified by the party so in breach against any loss or damage accruing from or arising out of or connected with such breach.

Fluctuations

*23A The Sub-Contract Sum shall be deemed to have been calculated in the manner set out below and shall be subject to adjustment in the events specified hereunder:

(a) (i) The Sub-Contract Sum (including the cost of employers' liability insurance and of third party insurance) is based upon the rates of wages and the other emoluments and expenses (including holiday credits) which will be payable by the Sub-Contractor to or in respect of workpeople engaged upon or in connection with the Sub-Contract Works in accordance with any rules, decisions or agreements of the recognised wage fixing body of the trade concerned applicable to the Sub-Contract Works and promulgated at the date of tender.

(ii) If any of the said rates of wages or other emoluments and expenses (including holiday credits) are increased or decreased by reason of any alteration in the said rules, decisions or agreements promulgated after the date of tender, then the net amount of the increase or decrease in wages and other emoluments and expenses (including holiday credits) together with the net amount of any consequential increase or decrease in the cost of employers' liability insurance, of third party insurance and of any contribution, levy, or tax payable by a person in his capacity as an employer shall, as the case may be, be paid to or allowed by the Sub-Contractor.

(b) (i) The Sub-Contract Sum is based upon the types and rates of contribu-tion, levy and tax payable by a person in his capacity as an employer and which at the date of tender are payable by the Sub-Contractor. A type and a rate so payable are in the next sub-paragraph referred to as a " tender-type " and a " tender rate ".

(ii) If any of the tender rates other than a rate of levy payable by virtue of the Industrial Training Act 1964, is increased or decreased, or if a tender type ceases to be payable, or if a new type of contribution, levy or tax which is payable by a person in his capacity as an employer becomes payable after the date of tender, then in any such case the net

---

*Parts A, C, D and E should be used where the parties have agreed to allow full fluctuations. Parts B, C, D and E should be used wherever Part A is not applicable.

amount of the difference between what the Sub-Contractor actually pays or will pay in respect of work-people whilst they are engaged upon or in connection with the Sub-Contract Works or because of his employment of such work people upon or in connection with the Sub-Contract Works, and what he would have paid had the alteration, cessation, or new type of contribution, levy or tax not become effective, shall, as the case may be, be paid to or allowed by the Sub-Contractor.

(iii) The Sub-Contract Sum is based upon the types and rates of refund of contributions, levies and taxes payable by a person in his capacity as an employer and upon the types and rates of premium receivable by a person in his capacity as an employer being in each case types and rates which at the date of tender are receivable by the Sub-Contractor. Such a type and such a rate are in the next sub-paragraph referred to as a " tender type " and a " tender rate ".

(iv) If any of the tender rates is increased or decreased or if a tender type ceases to be payable or if a new type of refund of any contribution, levy or tax payable by a person in his capacity as an employer becomes receivable or if a new type of premium receivable by a person in his capacity as an employer becomes receivable after the date of tender, then in any such case the net amount of the difference between what the Sub-Contractor actually receives or will receive in respect of workpeople whilst they are engaged upon or in connection with the Sub-Contract Works or because of his employment of such workpeople upon or in connection with the Sub-Contract Works, and what he would have received had the alteration, cessation or new type of refund or premium not become effective, shall, as the case may be, be allowed by or paid to the Sub-Contractor.

(v) The references in the two preceding sub-paragraphs to premiums shall be construed as meaning all payments howsoever they are described which are made under or by virtue of an Act of Parliament to a person in his capacity as an employer and which affect the cost to an employer of having persons in his employment.

(vi) Where the Sub-Contractor elects to contribute to a recognised alternative pension scheme instead of participating in the Reserve Pension Scheme established under the Social Security Act, 1973, the Sub-Contractor shall for the purpose of recovery or allowance under this Clause be deemed to pay Employers' contributions to the Reserve Pension Scheme.

(vii) The references in sub-paragraphs (i) to (iv) and (vi) of this paragraph to contributions, levies and taxes shall be construed as meaning all impositions payable by a person in his capacity as an employer howsoever they are described and whoever the recipient which are imposed under or by virtue of an Act of Parliament and which affect the cost to an employer of having persons in his employment.

(c) (i) The Sub-Contract Sum is based upon the market prices of the materials and goods specified in Part V of the Appendix to this Sub-Contract which were current at the date of tender. Such prices are hereinafter referred to as ' basic prices ', and the prices stated in the said Part V of the Appendix shall be deemed to be the basic prices of the specified materials and goods.

(ii) If after the date of tender the market price of any of the materials or goods specified as aforesaid increases or decreases, then the net amount of the difference between the basic price thereof and the market price payable by the Sub-Contractor and current when the materials or goods are bought shall, as the case may be, be paid to or allowed by the Sub-Contractor.

(iii) The references in the two preceding sub-paragraphs to "market prices " shall be construed as including any duty or tax by whomsoever payable which is payable under or by virtue of any Act of Parliament on the import, purchase, sale, appropriation, processing or use of the materials or goods specified as aforesaid.

*23B The Sub-Contract sum shall be deemed to have been calculated in the manner set out below and shall be subject to adjustment in the events specified hereunder:—

(a) (i) The Sub-Contract Sum is based upon the type and rates of contribution, levy and tax payable by a person in his capacity as an employer and which at the date of tender are payable by the Sub-Contractor. A type and a rate so payable are in the next sub-paragraph referred to as a " tender type " and a " tender rate ".

(ii) If any of the tender rates, other than a rate of levy payable by virtue of the Industrial Training Act 1964, is increased or decreased, or, if a tender type ceases to be payable, or if a new type of contribution, levy or tax which is payable by a person in his capacity as an employer becomes payable after the date of tender, then in any such case the net amount of the difference between what the Sub-Contractor actually pays or will pay in respect of workpeople whilst they are engaged upon or in connection with the Sub-Contract Works or because of his employment of such workpeople upon or in connection with the Sub-Contract Works, and what he would have paid had the alteration, cessation or new type of contribution, levy, or tax not become effective, shall, as the case may be, be paid to or allowed by the Sub-Contractor.

(iii) The Sub-Contract Sum is based upon the types and rates of refund of contributions, levies and taxes payable by a person in his capacity as an employer and upon the types and rates of premium receivable by a person in his capacity as an employer being in each case types and rates which at the date of tender are receivable by the Sub-Contractor. Such a type and such a rate are in the next sub-paragraph referred to as a " tender type " and a " tender rate ".

---

*Parts, A, C, D and E should be used where the parties have agreed to allow full fluctuations. Parts B, C, D and E should be used whenever Part A is not applicable.

(iv) If any of the tender rates is increased or decreased or if a tender type ceases to be payable or if a new type of refund of any contribution, levy or tax payable by a person in his capacity as an employer becomes receivable or if a new type of premium receivable by a person in his capacity as an employer becomes receivable, then in any such case the net amount of the difference between what the Sub-Contractor actually receives or will receive in respect of workpeople whilst they are engaged upon or in connection with the Sub-Contract Works or because of his employment of such workpeople upon or in connection with the Sub-Contract Works, and what he would have received had the alteration, cessation or new type of refund or premium not become effective, shall as the case may be, be allowed by or paid to the Sub-Contractor.

(v) The references in the two preceding sub-paragraphs to premiums shall be construed as meaning all payments howsoever they are described which are made under or by virtue of an Act of Parliament to a person in his capacity as an employer and which affect the cost to an employer of having persons in his employment.

(vi) Where the Sub-Contractor elects to contribute to a recognised alternative pensions scheme instead of participating in the Reserve Pension Scheme established under the Social Security Act, 1973, the Sub-Contractor shall for the purpose of recovery or allowance under this Clause be deemed to pay Employers' contributions to the Reserve Pension Scheme.

(vii) The references in sub-paragraphs (i) to (iv) and (vi) of this paragraph to contributions, levies and taxes shall be construed as meaning all impositions payable by a person in his capacity as an employer howsoever they are described and whoever the recipient which are imposed under or by virtue of an Act of Parliament and which affect the cost to an employer of having persons in his employment.

(b) (i) The Sub-Contract Sum is based upon the types and rates of duty if any and tax if any by whomsoever payable which are payable at the date of tender on the import, purchase, sale, appropriation, processing or use of the materials and goods specified in Part VI of the Appendix to this Sub-Contract under or by virtue of any Act of Parliament. A type and a rate so payable are in the next sub-paragraph referred to as a " tender type " and a " tender rate ".

(ii) If in relation to any materials or goods specified as aforesaid a tender rate is increased or decreased, or a tender type ceases to be payable or a new type of duty or tax becomes payable on the import, purchase, sale, appropriation, processing or use of those materials or goods, then in any such case the net amount of the difference between what the Sub-Contractor actually pays in respect of those materials or goods and what he would have paid in respect of them had the alteration, cessation or imposition not occurred, shall, as the case may be, be paid to or allowed by the Sub-Contractor. In this sub-paragraph the expression " a new type of duty or tax " includes an additional duty or tax and a duty or tax imposed in regard to specified materials or goods in respect of which no duty or tax whatever was previously payable.

23C (1) If the Sub-Contractor shall decide subject to Clause 15 of this Sub-Contract to sub-let any portion of the Sub-Contract Works he shall incorporate in the sub-contract provisions to the like effect as the provisions of Clauses 23A, 23D and 23E/Clauses 23B, 23D and 23E whichever may be applicable for the purposes of this Sub-Contract.

(2) If the price payable under such a sub-contract as aforesaid is decreased below or increased above the price in such sub-contract by reason of the operation of the said incorporated provisions, then the net amount of such decrease or increase shall, as the case may be, be allowed by or paid to the Sub-Contractor under this Sub-Contract.

23D (1) The Sub-Contractor shall give a written notice to the Contractor of the occurrence of any of the events referred to in such of the following provisions as are applicable for the purposes of this Sub-Contract:—

(a)   Clause 23A (a) (ii);
(b)   Clause 23A (b) (ii)
(c)   Clause 23A (b) (iv);
(d)   Clause 23A (c) (ii);
(e)   Clause 23B (a) (ii);
(f)   Clause 23B (a) (iv);
(g)   Clause 23B (b) (ii);
(h)   Clause 23C (2).

(2) Any notice required to be given by the preceding sub-clause shall be given within a reasonable time after the occurrence of that to which the notice relates, and the giving of a written notice in that time shall be a condition precedent to any payment being made to the Sub-Contractor in respect of the event in question.

(3) The Contractor on behalf of and in consultation with the Sub-Contractor may agree with the Quantity Surveyor or, if there is no Quantity Surveyor appointed, with the Architect what shall be deemed for all the purposes of this Sub-Contract to be the net amount payable to or allowable by the Sub-Contractor in respect of the occurrence of any event such as is referred to in any of the provisions listed in sub-clause (1) of this Clause.

(4) Any amount which from time to time becomes payable to or allowance by the Sub-Contractor by virtue of Clause 23A or Clause 23B or Clause 23C of this Sub-Contract shall, as the case may be, be added to or subtracted from:—

(a)   The Sub-Contract Sum; and
(b)   Any amounts payable to the Sub-Contractor and which are calculated in accordance with Clause 21 of this Sub-Contract; and
(c)   The amount which would otherwise be stated as due in the next interim payment.

PROVIDED:—

(i)   No addition to or subtraction from the amount which would otherwise be stated as due in an interim payment shall be made by virtue of this sub-clause unless on or before the date as at which the total value of work, materials and goods is ascertained for the purposes of that payment the Sub-Contractor shall have actually paid or received the sum which is payable by or to him in consequence of the event in respect of which the payment or allowance arises.

(ii) No addition to or subtraction from the Sub-Contract Sum made by virtue of this sub-clause shall alter in any way the amount of profit of the Sub-Contractor included in that Sum.

(5) Clause 23A, Clause 23B and Clause 23C shall not apply in respect of:

    (a) work for which the Sub-Contractor is allowed daywork rates under Clause 10 of this Sub-Contract.

    (b) changes in the rate of value added tax charged on the supply of goods and services by the Sub-Contractor to the Contractor under this Sub-Contract.

(6) In Clause 23A and Clause 23B of this Sub-Contract;

    (a) the expression "the date of tender" means the date stated in Part VII of the Appendix to this Sub-Contract;

    (b) the expressions "materials" and "goods" include timber used in formwork but do not include other consumable stores, plant and machinery;

    (c) the expression "workpeople" means persons whose rates of wages and other emoluments (including holiday credits) are governed by the Rules or decisions or agreements of the National Joint Council for the Building Industry or some other like body for trades associated with the building industry.

23E There shall be added to the amount paid to or allowed by the Sub-Contractor under:

    (a) Clause 23A (a) (ii);
    (b) Clause 23A (b) (ii);
    (c) Clause 23A (b) (iv);
    (d) Clause 23A (c) (ii);
    (e) Clause 23B (a) (ii);
    (f) Clause 23B (a) (iv);
    (g) Clause 23B (b) (ii);

the percentage stated in Part VIII of the Appendix to this Sub-Contract.

Adjustment of sub-contract sum—NEDO price adjustment formulae for building contracts

***23F** (1) The Contractor shall ensure that the Sub-Contract Sum shall be adjusted in accordance with the following provisions of this Clause and the "Nominated Sub-Contract Formula Rules" identified at Part IX of the Appendix to this Sub-Contract; that is to say

---

*Part F is to be used where the Sub-Contractor and the Employer through his professional agents have at the tender stage agreed that fluctuations should be dealt with by adjustment of the Sub-Contract Sum under the NEDO Price Adjustment Formulae. Parts A–E should be deleted where Part F applies.

(a) Where the sub-contract is for the supply and fixing of materials or goods or the execution of works to which one or more of the Work Categories referred to in Part I of the Nominated Sub-Contract Formula Rules applies, adjustment shall be under the formulae (but subject to Rule 9) in that Part of the Rules.

(b) Where the sub-contract is for the supply and fixing of materials or goods or the execution of works to which one of the formulae set out in Part II of the Nominated Sub-Contract Formula Rules applies adjustment shall be under the relevant formulae in that Part of the Rules.

(2) (i) Any adjustment under this Clause shall be to sums exclusive of value added tax and nothing in this Clause shall affect in any way the operation of Clause 10A (Value Added Tax).

(ii) The Definitions in Rule 1 of the Nominated Sub-Contract Formula Rules shall apply to this Clause.

(iii) Where Clause 31F of the Standard Form of Building Contract does not apply to the Main Contract but this clause applies to adjustment of the Sub-Contract Sum, valuations shall be made for the purpose of calculating formula adjustment due under this Clause.

(3) (i) The Contractor shall ensure that the adjustment referred to in this Clause shall be effected in all Certificates of Payments to which Clause 11 of this Sub-Contract applies and the word "fluctuations" in Clause 11 shall be deemed to have been deleted and the words "any adjustment under Clause 23F" inserted.

(ii) The Sub-Contractor shall be entitled, through the Contractor, to make to the Architect any representations in regard to such adjustment which he considers necessary including representations on the value of the work to which formula adjustment is to be made and the concluding sentence of Clause 11 (a) shall have effect subject to this paragraph. The Contractor shall forthwith pass such representations to the Architect.

(iii) Where any Non-Adjustable Element (as defined in the Formula Rules to which Clause 31F of the Standard Form of Building Contract refers) applies to formula adjustment under the Main Contract, the amount of the Non-Adjustable Element recorded in Part X of the Appendix to this sub-contract shall apply to the amount of adjustment to the Sub-Contract Sum.

(iv) Where Clause 31 of the Standard Form of Building Contract does not apply to the Main Contract but this Clause applies to adjustment of the Sub-Contract Sum and it has been agreed that a Non-Adjustable Element shall apply the amount of the Non-Adjustable Element (defined as aforesaid) shall be inserted in Part X of the Appendix to this Sub-Contract provided that this paragraph shall only apply where the Main Contract is let on the terms of the Local Authorities Edition of the Standard Form of Building Contract.

(v) The deductions referred to in paragraphs (i) to (iii) inclusive in Clause 11 (b) of this Sub-Contract shall apply to any amount of adjustment included in the total value certified in respect of the Sub-Contract Works in any certificate of the Architect to which that Clause refers.

Fluctuations—articles manufactured outside the United Kingdom

(4) For any article to which Rule (3) (ii) of the Nominated Sub-Contract Formula Rules applies the Sub-Contractor shall insert in Part XI of the Appendix to this Sub-Contract the market price of the article in sterling (that is the price delivered to the site) current at the Date of Tender. If after that Date the market price of the article inserted in the aforesaid Part increases or decreases then the net amount of the difference between the cost of purchasing at the market price inserted in the aforesaid Part and the market price payable by the Sub-Contractor and current when the article is bought shall, as the case may be, be paid to or allowed by the Sub-Contractor. The reference to "market price" in this sub-clause shall be construed as including any duty or tax (other than any value added tax which is treated, or is capable of being treated, as input tax [as defined in the Finance Act 1972] by the Sub-Contractor) by whomsoever payable under or by virtue of any Act of Parliament on the import purchase sale appropriation or use of the articles specified as aforesaid.

Power to agree—quantity surveyor, contractor and sub-contractor

(5) The Contractor on behalf of and with the consent of the Sub-Contractor may agree with the Quantity Surveyor any alteration to the methods and pro-cedures for ascertaining the amount of formula adjustment to be made under the formulae to which this Clause refers and the amounts ascertained in accordance with and from the effective date of such agreement shall be deemed for all the purposes of this Sub-Contract to be the amount of formula adjustment payable to or allowable by the Sub-Contractor in respect of the provisions of this Clause. Provided always that no alteration to the methods and procedures shall be agreed as aforesaid unless it is reasonably expected that the amount of formula adjustment will be the same or approximately the same as that ascertained in accordance with Part I or Part II of the Nominated Sub-Contract Formula Rules whichever Part is stated in the sub-contract documents to be applicable.

Position where monthly bulletins are delayed

(6) (a) If at any time prior to the final payment of the Sub-Contractor under Clause 11 of this Sub-Contract formula adjustment is not possible because of delay in, or cessation of, the publication of the Monthly Bulletins adjust-ment of the Sub-Contract Sum shall be made in each Interim Payment during such period of delay on a fair and reasonable basis.

(b) If publication of the Monthly Bulletins is recommenced at any time prior to the final payment of the Sub-Contractor under Clause 11 of this Sub-Contract the provisions of this Clause and the Nominated Sub-Contract Formula Rules shall operate for each Valuation Period as if no delay or cessation as aforesaid had occurred and the adjustment under this Clause and the aforesaid Rules shall be substituted for any adjustment under paragraph (a) hereof.

(c) During any period of delay or cessation as aforesaid the Employer the Contractor and the Sub-Contractor shall operate such parts of this Clause and the aforesaid Rules as will enable the amount of formula adjustment to be readily calculated upon recommencement of publication of the Monthly Bulletins.

Formula adjustment—failure to complete

(7) (a) If the Sub-Contractor fails to complete the sub-contract Works within the period specified in Part II of the Appendix to this Sub-Contract or within any extended time granted under Clause 8 (b) of this Sub-Contract but not otherwise formula adjustment of the Sub-Contract Sum shall be affected in all payments in favour of the Sub-Contractor directed in Interim Certificates issued after the expiry of the aforesaid period (or any extension thereof) by reference to the Index Numbers applicable to the Valuation Period in which the aforesaid date of expiry (or any extension thereof) falls.

(b) Paragraph (a) of this sub-clause shall not operate unless the Architect has, in respect of every written notification under Clause 8 (b) of this Sub-Contract, expressed to the Contractor his consent or otherwise to any extension of the period for completion referred to in Part II of the Appendix to this Sub-Contract and has sent to the Sub-Contractor the duplicate copy of the certificate referred to in Clause 27 (d) (ii) of the Standard Form of Building Contract.

(c) For the purposes of the operation of paragraph (a) of this sub-clause but not further or otherwise the reference in Clause 8 (b) (ii) of this Sub-Contract shall be deemed to be a reference to the printed text of Clause 23 of the Standard Form of Building Contract current at the date of this Sub-Contract, unamended and including sub-clauses (j) (i) and (j) (ii) whether or not the Main Contract includes Clause 23 unamended and whether or not it includes the aforesaid sub-clauses.

Amendment to Clause 24—arbitration before practical completion of main contract

(8) For the purposes of this clause a reference to arbitration on a dispute or difference under Clause 8 of the Sub-contract Conditions may be opened before Practical Completion or alleged Practical Completion of the Main Contract Works, or termination or alleged termination of the Sub-contractors employment under this Sub-Contract, or abandonment of the Main Contract Works, and accordingly the words " or on any dispute or difference under Clause 8 of the Sub-contract Conditions " shall be deemed to have been added at the end of the first Proviso of Clause 24 of the Sub-Contract conditions.

Arbitration

24 In the event of any dispute or difference between the Contractor and the

Sub-Contractor whether arising during the execution or after the completion or abandonment of the Sub-Contract Works or after the determination of the employment of the Sub-Contractor under this Sub-Contract (whether by breach or in any other manner), in regard to any matter or thing of whatsoever nature arising out of this Sub-Contract or in connection therewith, then either party shall give to the other notice in writing of such dispute or difference and such dispute or difference shall be and is hereby referred to the arbitration of such person as the parties hereto may agree to appoint as Arbitrator or failing such agreement as may be appointed on the request of either party by the President for the time being of the Royal Institution of Chartered Surveyors provided that if the dispute or difference between the Contractor and Sub-Contractor is substantially the same as a matter which is a dispute or difference between the Contractor and Employer under the Main Contract the Contractor and Sub-Contractor hereby agree that such dispute or difference shall be referred to an Arbitrator appointed or to be appointed pursuant to the terms of the Main Contract who shall have power to make such directions and all necessary awards in the same way as if the procedure in the High Court as to joining one or more defendants or joining co-defendants or third parties was available to the parties and to him and in any case the Award of such Arbitrator shall be final and binding upon the parties.

PROVIDED that such Arbitrator shall not without the written consent of the Architect or the Contractor and in any such case of the Sub-Contractor enter on the arbitration until after the completion or abandonment of the Main Contract Works, except to arbitrate upon the question whether or not a certificate has been improperly withheld or is not in accordance with the terms of the Main Contract, or in respect of matters, for which a notice of arbitration under Clause 13B (1) (a) (i) of this Sub-Contract has been given, or on any dispute or difference under Clause 8 of the Sub-Contract conditions.

AND PROVIDED further that in any such arbitration as is provided for in this clause any decision of the Architect which is final and binding on the Contractor under the Main Contract shall also be and be deemed to be final and binding between and upon the Contractor and the Sub-Contractor.

†AND PROVIDED that whatever the nationality, residence or domicile of the Employer, the Contractor, the Sub-Contractor, any sub-contractor or supplier of the Arbitrator, and wherever the Main Contract or Sub-Contract Works, or any part thereof, are situated, the law of England shall be the proper law of this Sub-Contract and in particular (but not so as to derogate from the generality of the foregoing) the provisions of the Arbitration Act, 1950 (notwithstanding anything in Section 34 thereof) shall apply to any arbitration under this Sub-Contract wherever the same, or any part of it, shall be conducted.

---

†Where the parties do not wish the proper law of the Sub-Contract to be the law of England and/or do not wish the provisions of the Arbitration Act, 1950, to apply to any arbitration under the sub-contract held under Scottish procedural law appropriate amendments to this paragraph should be made. It is essential, in any event, that in regard to these matters there should be no conflict between the provisions of Clause 24 and the corresponding provisions contained in the arbitration clause (if any) under the main contract.

*(A) IN WITNESS WHEREOF the parties hereto have hereunto set their hands the day and year first above written.

*(B) IN WITNESS WHEREOF *the Common Seals of the parties hereto have hereunto been affixed the day and year above written.*

---

*Where the Sub-Contract is executed under hand alternative (B) above should be struck out and the parties should sign in the spaces provided in alternative (A) on page 614. Where the Sub-Contract is executed under seal alternative (A) above should be struck out and the Common Seals of the parties should be affixed in the spaces provided in alternative (B) on page 614. A 50p stamp must be impressed within thirty days of the date of the Sub-Contract. [The page references in the original are to the printed form.]

Appendix

Part I

First Recital   Particulars of the Works (being a part of the Works comprised in the Main Contract in this Sub-Contract referred to as " the Sub-Contract Works ", viz.:

....

....

....

....

Part II

Description of work	Completion period or periods of the Sub-Contract Works or respective sections thereof

Clause 8

....	....
....	....
....	....
....	....
....	....

Part III

Clause 10   Value of the Sub-Contract Works

....

....

....

....£

Part IIIA

Clause 10(b)(3)   Percentage Additions on Prime Cost
(A)
    (i) Labour...%
    (ii) Materials ....%
    (iii) Plant ....%

Clause 10(b)(3)   Percentage Additions on Prime Cost
(B)
    (i) Labour...%
    (ii) Materials ....%
    (iii) Plant ....%

PART IV

Clause 11 Retention Percentage ....

(**Note.**—*The percentage of payments to be retained by the Contractor should be stated, this should not exceed the Retention Percentage under the Main Contract.*)

PART V

Clause 23A(c)(i) List of Basic Prices

....

....

....

....

....

....

....

....

....

....

PART VI

Clause 23B(b)(i) List of Materials and Goods

....

....

....

....

....

....

....

PART VII

Clause 23D(6) Date of Tender

The date of tender is the................day of................19..........

PART VIII

Clause 23E Percentage Addition on Fluctuations ....%

PART IX

Clause 23F(1)   The Nominated Sub-Contract Formula Rules are those dated
....**197** ....

PART X

Clause 23F(3)   Non-Adjustable Element ....%
(iii) and (3)(iv)   (Not to exceed 15%)

(**Note.**—*Only applicable where the Main Contract is let on the Standard Form of Building Contract, Local Authorities Edition, WITH Quantities.*)

PART XI

Clause 23F(4)   List of Market Prices

....

....

PART XII

Nominated Sub-Contract Formula Rules

Rule 1   Part I only: Balance of Adjustable Work; any measured work not allocated to a Work Category

....

Base month   ....

Date of Tender ....

Rule 11(b)   Part I only: the Work Categories applicable to the Sub-Contract Works

....

....

....

Rule 15   Method of dealing with " Fix-only " work

....

....

....

Rule 26   Part II only: Weightings of Labour and materials—Electrical Installations *or* Heating, Ventilating and Air Conditioning Installations

Labour ....%

Materials...%

Rule 40   Part II only: Lift Installations—Element of Sub-Contract Sum subject to formula adjustment

....

....

....

Rule 45   Part II only: Structural Steelwork Installations:

(i) Average price per tonne of steel delivered to fabricator's works £ ....

(ii) Average price per tonne for erection of steelwork £...............

PART XIII

Clause 13B(4)(a)   Trustee-Stakeholder ....

....

....

(A) **Signed** by the above-named Contractor in the presence of

....

....

Signed by the above-named Sub-Contractor in the presence of

....

....

OR,*

(B) **The Common Seal** *of the above-named Contractor was hereunto affixed in the presence of*

....

....

L.S.

The Common Seal *of the above-named Sub-Contractor was hereunto affixed in the presence of*

....

....

L.S.

---

* *See* Footnote *on page* 609.

NOTES

(These notes do not form part of the Sub-Contract)

A   The reprint dated June, 1968, incorporated the following amendment to the 1963 Edition:—

Amendment 1—Revised Clause 23—Fluctuations.

B   The reprint dated January, 1970, incorporated the following amendments to the 1968 revision as Amendment 2:—

(i) Clause 4—Insurance against injury to persons or property—extension of list of accepted risks.

(ii) Clause 5—Damage by Fire—extensions of list of accepted risks.

(iii) Clause 6—Policies of Insurance—documentary evidence of maintaining insurance.

(iv) Clause 10 and Appendix Part III—Sub-Contract Sum—Valuation of Variations—" Definition of Prime Cost of Daywork "—New Part IIIA to Appendix III for percentage additions of Prime Cost.

(v) Clause 23A & B Fluctuations—revised footnotes.

(vi) Appendix Part IV—Reduction of percentage of certified value retained.

C   The reprint dated February, 1971, incorporated the following amendments to the 1970 revision as Amendment 3:—

(i) Clauses 5 and 19:—Amended to clarify the position as to responsibility for sub-contract goods and materials prior to and after incorporation in the sub-contract works.

(ii) Clause 8 (c)—Amended to clarify the position as to claim by main contractor or sub-contractor for loss or expenses caused by disturbance of regular progress.

(iii) Clause 11 (a):—Amended to permit payment for off-site goods and materials.

(iv) Clause 24—Arbitration—Amendment to allow joinder of parties where dispute similar to dispute under main contract.

D   The changes from the reprint dated February, 1971, which are incorporated in this revision, and which appear in Amendment Sheet 4, issued in August, 1971, are indicated as follows:—

(i) Clause 5 (b)—Responsibility—materials and goods on site—amendments to clarify position after fixing.

(ii) Clause 20 (b)—Determination by contractor—appointment of provisional liquidator.

(iii) Clause 23A—Fluctuations—increase in labour costs—recovery of consequential increase in employers' compulsory contributions.

(iv) Clause 23B (b) (i) and (ii)—Tax Fluctuations Clause—materials—new duties or taxes.

(v) Clause 23A and 23B—Tax, etc., fluctuations—revised basis for inclusion in tenders of increases in existing taxes, etc., or of new taxes, etc.

(vi) Clause 25D—Definition of " Workpeople ".

(vii) Clause 24—Arbitration—new paragraph—proper law of contract—application of Arbitration Act, 1950.

E   The changes from the reprint dated October, 1971, which are incorporated in this revision and which appear on amendment sheets 5 and 6, issued in November, 1971 and April, 1972, respectively, are as follows:—

(i) Recitals and Clause 20 (b)—Amended to require sub-contractor's exemption under Sections 29–31, Finance Act, 1971.

(ii) Clause 10A—New clause and footnote regarding VAT.

(iii) Clause 11 (b) (i) and Appendix IV—revised retention provisions—" retention percentage ".

F The changes from the reprint dated May, 1972, which are incorporated in this revision and which appear in Amendment Sheets 7 and 8 issued in February 1973 and July 1973 respectively are indicated as follows:—

  (i) Revised Clause 10A—Value Added Tax.
  (ii) Clause 23D (5)—Fluctuations—divided into paragraphs (*a*) and (*b*).
      (*a*) reproducing former sub-clause and
      (*b*) a new provision on Value Added Tax.
  (iii) Clause 11 (*e*) and Clause 20 (*a*)—drafting improvements.
  (iv) Clause 23A (*b*) and Clause 23B (*a*)—Fluctuations—New Sub-clause (vi)—contributions by Sub-Contractor to alternative pension scheme under Social Security Act, 1973.
  (v) Clause 23E and Appendix Part VIII—new provisions—percentage addition on fluctuations.

G The changes from the reprint dated July, 1973, which are incorporated in this revision and which appear in Amendment Sheet 9 issued in May 1975, are indicated as follows:—

  (i) Clause 23A (*a*) (i) and (ii)—textual amendments.
  (ii) Clause 23F new clause and appendices for the adjustment of Sub-Contract Sum by NEDO Formula.

H The changes from the reprint dated July, 1975, which are incorporated in this revision and which appear in Amendment Sheet 10 and in Amendment Sheet 11 issued in November 1975 and February 1976 respectively are indicated as follows:—

  (i) Clause 10A (Value Added Tax)—additional provisions in Sub-Clause (4)—written provisional assessment of values at zero and positive rates.
  (ii) New Clause 10B (Value Added Tax—Special Arrangement)—additional provisions in Sub-Clause (4) now inserted on written provisional assessment of values at zero and positive rates.
  (iii) Clause 23C (1)—references to Clauses 23D and 23E included.
  (iv) Replacement of Clause 13 by Clause 13A—Contractor's right to set-off —and Clause 13B—Contractor's claims not agreed by Sub-Contractor and appointment of adjudicator.

UNFAIR CONTRACT TERMS ACT 1977

(C.50)

An Act to impose further limits on the extent to which under the law of England and Wales and Northern Ireland civil liability for breach of contract, or for negligence or other breach of duty, can be avoided by means of contract terms and otherwise, and under the law of Scotland civil liability can be avoided by means of contracts terms.

[26th October 1977]

PART I [1]

AMENDMENT OF LAW FOR ENGLAND AND WALES AND NORTHERN IRELAND

Introductory

Scope of Part I

1.—(1) For the purposes of this Part of this Act, " negligence " means the breach—

(a) of any obligation, arising from the express or implied terms of a contract, to take reasonable care or exercise reasonable skill in the performance of the contract;

(b) of any common law duty to take reasonable care or exercise reasonable skill (but not any stricter duty);

(c) of the common duty of care imposed by the Occupiers' Liability Act 1957 or the Occupiers' Liability Act (Northern Ireland) 1957.

(2) This Part of this Act is subject to Part III; and in relation to contracts, the operation of sections 2 to 4 and 7 is subject to the exceptions made by Schedule I.

(3) In the case of both contract and tort, sections 2 to 7 apply (except where the contrary is stated in section 6(4)) only to business liability, that is liability for breach of obligations or duties arising—

(a) from things done or to be done by a person in the course of a business (whether his own business or another's); or

(b) from the occupation of premises used for business purposes of the occupier;

and references to liability are to be read accordingly.

(4) In relation to any breach of duty or obligation, it is immaterial for any purpose of this Part of this Act whether the breach was inadvertent or intentional, or whether liability for it arises directly or vicariously.

---

[1] Part II dealing with Scotland is not printed.

Avoidance of liability for negligence, breach of contract, etc.

Negligence liability

2.—(1) A person cannot by reference to any contract term or to a notice given to persons generally or to particular persons exclude or restrict his liability for death or personal injury resulting from negligence.

(2) In the case of other loss or damage, a person cannot so exclude or restrict his liability for negligence except in so far as the term or notice satisfies the requirement of reasonableness.

(3) Where a contract term or notice purports to exclude or restrict liability for negligence a person's agreement to or awareness of it is not of itself to be taken as indicating his voluntary acceptance of any risk.

Liability arising in contract

3.—(1) This section applies as between contracting parties where one of them deals as consumer or on the other's written standard terms of business.

(2) As against that party, the other cannot by reference to any contract term—

(a) when himself in breach of contract, exclude or restrict any liability of his in respect of the breach; or

(b) claim to be entitled—

(i) to render a contractual performance substantially different from that which was reasonably expected of him, or

(ii) in respect of the whole or any part of his contractual obligation, to render no performance at all,

except in so far as (in any of the cases mentioned above in this subsection) the contract term satisfies the requirement of reasonableness.

Unreasonable indemnity clauses

4.—(1) A person dealing as consumer cannot by reference to any contract term be made to indemnify another person (whether a party to the contract or not) in respect of liability that may be incurred by the other for negligence or breach of contract, except in so far as the contract term satisfies the requirement of reasonableness.

(2) This section applies whether the liability in question—

(a) is directly that of the person to be indemnified or is incurred by him vicariously;

(b) is to the person dealing as consumer or to someone else.

Liability arising from sale or supply of goods

" Guarantee " of consumer goods

5.—(1) In the case of goods of a type ordinarily supplied for private use or consumption, where loss or damage—

(a) arises from the goods proving defective while in consumer use; and

(b) results from the negligence of a person concerned in the manufacture or distribution of the goods,

liability for the loss or damage cannot be excluded or restricted by reference to any contract term or notice contained in or operating by reference to a guarantee of the goods.

(2) For these purposes—

(*a*) goods are to be regarded as " in consumer use " when a person is using them, or has them in his possession for use, otherwise than exclusively for the purposes of a business; and

(*b*) anything in writing is a guarantee if it contains or purports to contain some promise or assurance (however worded or presented) that defects will be made good by complete or partial replacement, or by repair, monetary compensation or otherwise.

(3) This section does not apply as between the parties to a contract under or in pursuance of which possession or ownership of the goods passed.

Sale and hire-purchase

6.—(1) Liability for breach of the obligations arising from—

(*a*) section 12 of the Sale of Goods Act 1893 (seller's implied undertakings as to title, etc.);

(*b*) Section 8 of the Supply of Goods (Implied Terms) Act 1973 (the corresponding thing in relation to hire-purchase),

cannot be excluded or restricted by reference to any contract term.

(2) As against a person dealing as consumer, liability for breach of the obligations arising from—

(*a*) section 13, 14 or 15 of the 1893 Act (seller's implied undertakings as to conformity of goods with description or sample, or as to their quality or fitness for a particular purpose);

(*b*) section 9, 10 or 11 of the 1973 Act (the corresponding things in relation to hire-purchase),

cannot be excluded or restricted by reference to any contract term.

(3) As against a person dealing otherwise than as consumer, the liability specified in subsection (2) above can be excluded or restricted by reference to a contract term, but only in so far as the term satisfies the requirement of reasonableness.

(4) The liabilities referred to in this section are not only the business liabilities defined by section 1(3), but include those arising under any contract of sale of goods or hire-purchase agreement.

Miscellaneous contracts under which goods pass

7.—(1) Where the possession or ownership of goods passes under or in pursuance of a contract not governed by the law of sale of goods or hire-purchase, subsections (2) to (4) below apply as regards the effect (if any) to be given to contract terms excluding or restricting liability for breach of obligation arising by implication of law from the nature of the contract.

(2) As against a person dealing as consumer, liability in respect of the goods' correspondence with description or sample, or their quality or fitness for any particular purpose, cannot be excluded or restricted by reference to any such term.

(3) As against a person dealing otherwise than as consumer, that liability can be excluded or restricted by reference to such a term, but only in so far as the term satisfies the requirement of reasonableness.

(4) Liability in respect of—

    (*a*) the right to transfer ownership of the goods, or give possession; or

    (*b*) the assurance of quiet possession to a person taking goods in pursuance of the contract,

cannot be excluded or restricted by reference to any such term except in so far as the term satisfies the requirement of reasonableness.

(5) This section does not apply in the case of goods passing on a redemption of trading stamps within the Trading Stamps Act 1964 or the Trading Stamps Act (Northern Ireland) 1965.

Other provisions about contracts

Misrepresentation

8.—(1) In the Misrepresentation Act 1967, the following is substituted for section 3—

" Avoidance of provision excluding liability for misrepresentation

3. If a contract contains a term which would exclude or restrict—

    (*a*) any liability to which a party to a contract may be subject by reason of any misrepresentation made by him before the contract was made; or

    (*b*) any remedy available to another party to the contract by reason of such a misrepresentation,

that term shall be of no effect except in so far as it satisfies the requirement of reasonableness as stated in section 11(1) of the Unfair Contract Terms Act 1977; and it is for those claiming that the term satisfies that requirement to show that it does.".

(2) The same section is substituted for section 3 of the Misrepresentation Act (Northern Ireland) 1967.

Effect of breach

9.—(1) Where for reliance upon it a contract term has to satisfy the requirement of reasonableness, it may be found to do so and be given effect accordingly notwithstanding that the contract has been terminated either by breach or by a party electing to treat it as repudiated.

(2) Where on a breach the contract is nevertheless affirmed by a party entitled to treat it as repudiated, this does not of itself exclude the requirement of reasonableness in relation to any contract term.

Evasion by means of secondary contract

10.—A person is not bound by any contract term prejudicing or taking away rights of his which arise under, or in connection with the performance of, another contract, so far as those rights extend to the enforcement of another's liability which this Part of this Act prevents that other from excluding or restricting.

Explanatory provisions

The " reasonableness " test

11.—(1) In relation to a contract term, the requirement of reasonableness for the purposes of this Part of this Act, section 3 of the Misrepresentation Act 1967 and section 3 of the Misrepresentation Act (Northern Ireland) 1967 is that the term shall have been a fair and reasonable one to be included having regard to the circumstances which were, or ought reasonably to have been, known to or in the contemplation of the parties when the contract was made.

(2) In determining for the purposes of section 6 or 7 above whether a contract term satisfies the requirement of reasonableness, regard shall be had in particular to the matters specified in Schedule 2 to this Act; but this subsection does not prevent the court or arbitrator from holding, in accordance with any rule of law, that a term which purports to exclude or restrict any relevant liability is not a term of the contract.

(3) In relation to a notice (not being a notice having contractual effect), the requirement of reasonableness under this Act is that it should be fair and reasonable to allow reliance on it, having regard to all the circumstances obtaining when the liability arose or (but for the notice) would have arisen.

(4) Where by reference to a contract term or notice a person seeks to restrict liability to a specified sum of money, and the question arises (under this or any other Act) whether the term or notice satisfies the requirement of reasonableness, regard shall be had in particular (but without prejudice to subsection (2) above in the case of contract terms) to—

(a) the resources which he could expect to be available to him for the purpose of meeting the liability should it arise; and

(b) how far it was open to him to cover himself by insurance.

(5) It is for those claiming that a contract term or notice satisfies the requirement of reasonableness to show that it does.

" Dealing as consumer "

12.—(1) A party to a contract " deals as consumer " in relation to another party if—

(a) he neither makes the contract in the course of a business nor holds himself out as doing so; and

(b) the other party does make the contract in the course of a business; and

(c) in the case of a contract governed by the law of sale of goods or hire-purchase, or by section 7 of this Act, the goods passing under or in pursuance of the contract are of a type ordinarily supplied for private use or consumption.

(2) But on a sale by auction or by competitive tender the buyer is not in any circumstances to be regarded as dealing as consumer.

(3) Subject to this, it is for those claiming that a party does not deal as consumer to show that he does not.

Varieties of exemption clause

13.—(1) To the extent that this Part of this Act prevents the exclusion or restriction of any liability it also prevents—

(a) making the liability or its enforcement subject to restrictive or onerous conditions;

(b) excluding or restricting any right or remedy in respect of the liability, or subjecting a person to any prejudice in consequence of his pursuing any such right or remedy;

(c) excluding or restricting rules of evidence or procedure; and (to that extent) sections 2 and 5 to 7 also prevent excluding or restricting liability by reference to terms and notices which exclude or restrict the relevant obligation or duty

(2) But an agreement in writing to submit present or future differences to arbitration is not to be treated under this Part of this Act as excluding or restricting any liability.

Interpretation of Part I

14.—In this Part of this Act—

"business" includes a profession and the activities of any government department or local or public authority;

"goods" has the same meaning as in the Sale of Goods Act 1893;

"hire-purchase agreement" has the same meaning as in the Consumer Credit Act 1974;

"negligence" has the meaning given by section 1(1);

"notice" includes an announcement, whether or not in writing, and any other communication or pretended communication; and

"personal injury" includes any disease and any impairment of physical or mental condition.

. . . .

PART III

PROVISIONS APPLYING TO WHOLE OF UNITED KINGDOM

Miscellaneous

International supply contracts

26.—(1) The limits imposed by this Act on the extent to which a person may exclude or restrict liability by reference to a contract term do not apply to liability arising under such a contract as is described in subsection (3) below.

(2) The terms of such a contract are not subject to any requirement of reasonableness under section 3 or 4: and nothing in Part II of this Act shall require the incorporation of the terms of such a contract to be fair and reasonable for them to have effect.

(3) Subject to subsection (4), that description of contract is one whose characteristics are the following—

(*a*) either it is a contract of sale of goods or it is one under or in pursuance of which the possession or ownership of goods passes; and

(*b*) it is made by parties whose places of business (or, if they have none, habitual residences) are in the territories of different States (the Channel Islands and the Isle of Man being treated for this purpose as different States from the United Kingdom).

(4) A contract falls within subsection (3) above only if either—

(*a*) the goods in question are, at the time of the conclusion of the contract, in the course of carriage, or will be carried, from the territory of one State to the territory of another; or

(*b*) the acts constituting the offer and acceptance have been done in the territories of different States; or

(*c*) the contract provides for the goods to be delivered to the territory of a State other than that within whose territory those acts were done.

Choice of law clauses

27.—(1) Where the proper law of a contract is the law of any part of the United Kingdom only by choice of the parties (and apart from that choice would be the law of some country outside the United Kingdom) sections 2 to 7 and 16 to 22 of this Act do not operate as part of the proper law.

(2) This Act has effect notwithstanding any contract term which applies or purports to apply the law of some country outside the United Kingdom, where (either or both)—

(*a*) the term appears to the court, or arbitrator or arbiter to have been imposed wholly or mainly for the purpose of enabling the party imposing it to evade the operation of this Act or

(*b*) in the making of the contract one of the parties dealt as consumer, and he was then habitually resident in the United Kingdom, and the essential steps necessary for the making of the contract were taken there, whether by him or by others on his behalf.

(3) In the application of subsection (2) above to Scotland, for paragraph (*b*) there shall be substituted—

" (*b*) the contract is a consumer contract as defined in Part II of this Act, and the consumer at the date when the contract was made was habitually resident in the United Kingdom, and the essential steps necessary for the making of the contract were taken there, whether by him or by others on his behalf.".

28. Temporary provision for sea carriage of passengers [2]

Saving for other relevant legislation

29.—(1) Nothing in this Act removes or restricts the effect of, or prevents reliance upon, any contractual provision which—

(*a*) is authorised or required by the express terms or necessary implication of an enactment; or

---

[2] Not printed.

(*b*) being made with a view to compliance with an international agreement to which the United Kingdom is a party, does not operate more restrictively than is contemplated by the agreement.

(2) A contract is to be taken—

(*a*) for the purposes of Part I of this Act, as satisfying the requirement of reasonableness; and

(*b*) for those of Part II, to have been fair and reasonable to incorporate,

if it is incorporated or approved by, or incorporated pursuant to a decision or ruling of, a competent authority acting in the exercise of any statutory jurisdiction or function and is not a term in a contract to which the competent authority is itself a party.

(3) In this section—

" competent authority " means any court, arbitrator or arbiter, government department or public authority;

" enactment " means any legislation (including subordinate legislation) of the United Kingdom or Northern Ireland and any instrument having effect by virtue of such legislation; and

" statutory " means conferred by an enactment.

Obligations under Consumer Protection Acts

30.—(1) In section 3 of the Consumer Protection Act 1961 (Provisions against marketing goods which do not comply with safety requirements), after subsection (1) there is inserted—

" (1A) Any term of an agreement which purports to exclude or restrict, or has the effect of excluding or restricting, any obligation imposed by or by virtue of that section, or any liability for breach of such an obligation, shall be void.".

(2) The same amendment is made in section 3 of the Consumer Protection Act (Northern Ireland) 1965.

General

Commencement; amendments; repeals

31.—(1) This Act comes into force on 1st February 1978.

(2) Nothing in this Act applies to contracts made before the date on which it comes into force; but subject to this, it applies to liability for any loss or damage which is suffered on or after that date.

(3) The enactments specified in Schedule 3 to this Act are amended as there shown.

(4) The enactments specified in Schedule 4 to this Act are repealed to the extent specified in column 3 of that Schedule.

Citation and extent

32.—(1) This Act may be cited as the Unfair Contract Terms Act 1977.

(2) Part I of this Act extends to England and Wales and to Northern Ireland; but it does not extend to Scotland.

(3) Part II of this Act extends to Scotland only.

(4) This Part of this Act extends to the whole of the United Kingdom.

SCHEDULES

SCHEDULE 1

Scope of sections 2 to 4 and 7

1. Sections 2 to 4 of this Act do not extend to—
   (a) any contract of insurance (including a contract to pay an annuity on human life);
   (b) any contract so far as it relates to the creation or transfer of an interest in land, or to the termination of such an interest, whether by extinction, merger, surrender, forfeiture or otherwise;
   (c) any contract so far as it relates to the creation or transfer of a right or interest in any patent, trade mark, copyright, registered design, technical or commercial information or other intellectual property, or relates to the termination of any such right or interest;
   (d) any contract so far as it relates—

      (i) to the formation or dissolution of a company (which means any body corporate or unincorporated association and includes a partnership), or
      (ii) to its constitution or the rights or obligations of its corporators or members;
   (e) any contract so far as it relates to the creation or transfer of securities or of any right or interest in securities.

2. Section 2(1) extends to—
   (a) any contract of marine salvage or towage;
   (b) any charterparty of a ship or hovercraft; and
   (c) any contract for the carriage of goods by ship or hovercraft;
but subject to this sections 2 to 4 and 7 do not extend to any such contract except in favour of a person dealing as consumer.

3. Where goods are carried by ship or hovercraft in pursuance of a contract which either—
   (a) specifies that as the means of carriage over part of the journey to be covered, or
   (b) makes no provision as to the means of carriage and does not exclude that means,
then section 2(2), 3 and 4 do not, except in favour of a person dealing as consumer, extend to the contract as it operates for and in relation to the carriage of the goods by that means.

4. Section 2(1) and (2) do not extend to a contract of employment, except in favour of the employee.

5. Section 2(1) does not affect the validity of any discharge and indemnity given by a person, on or in connection with an award to him of compensation for pneumoconiosis attributable to employment in the coal industry, in respect of any further claim arising from his contracting that disease.

SCHEDULE 2

Sections 11(2)
and 24(2).

" Guidelines " for Application of Reasonableness Test

The matters to which regard is to be had in particular for the purposes of sections 6(3), 7(3) and (4), 20 and 21 are any of the following which appear to be relevant—

(a) the strength of the bargaining positions of the parties relative to each other, taking into account (among other things) alternative means by which the customer's requirements could have been met;

(b) whether the customer received an inducement to agree to the term, or in accepting it had an opportunity of entering into a similar contract with other persons, but without having to accept a similar term;

(c) whether the customer knew or ought reasonably to have known of the existence and extent of the term (having regard, among other things, to any custom of the trade and any previous course of dealing between the parties);

(d) where the term excludes or restricts any relevant liability if some condition is not complied with, whether it was reasonable at the time of the contract to expect that compliance with that condition would be practicable;

(e) whether the goods were manufactured, processed or adapted to the special order of the customer.

SCHEDULE 3

Section 31(3).

Amendment of Enactments

In the Sale of Goods Act 1893—

(a) in section 55(1), for the words " the following provisions of this section " substitute " the provisions of the Unfair Contract Terms Act 1977 ";

(b) in section 62(1), in the definition of " business ", for " local authority or statutory undertaker " substitute " or local or public authority ".

In the Supply of Goods (Implied Terms) Act 1973 (as originally enacted and as substituted by the Consumer Credit Act 1974)—

(a) in section 14(1) for the words from " conditional sale " to the end substitute " a conditional sale agreement where the buyer deals as consumer within Part I of the Unfair Contract Terms Act 1977 or, in Scotland, the agreement is a consumer contract within Part II of that Act ";

(b) in section 15(1), in the definition of " business ", for " local authority or statutory undertaker " substitute " or local or public authority ".

SCHEDULE 4

REPEALS

Chapter	Short title	Extent of repeal
56 & 57 Vict. c. 71	Sale of Goods Act 1893.	In section 55, subsections (3) to (11). Section 55A. Section 61(6). In section 62(1) the definition of " contract for the international sale of goods."
1962 c. 46.	Transport Act 1962	Section 43(7).
1967 c. 45.	Uniform Laws on International Sales Act 1967.	In section 1(4), the words " 55 and 55A ".
1972 c. 33.	Carriage by Railway Act 1972.	In section 1(1), the words from " and shall have " onwards.
1973 c. 13.	Supply of Goods (Implied) Terms) Act 1973.	Section 5(1). Section 6. In section 7(1), the words from " contract for the international sale of goods " onwards. In section 12, subsections (2) to (9). Section 13. In section 15(1), the definition or " consumer sale "

The repeals in sections 12 and 15 of the Supply of Goods (Implied Terms) Act 1973 shall have effect in relation to those sections as originally enacted and as substituted by the Consumer Credit Act 1974.

INDEX

The *Standard Form of Building Contract, 1977 revision* and notes thereto in Chapter 15 are indexed separately at pp. 650–659.

The *I.C.E. Form of Contract, 1973 edition* and notes thereto in Chapter 17 are indexed separately at pp. 638–645.

629